P9-CKX-464

The Next Generation CDMA Technologies

The Next Generation CDMA Technologies

Hsiao-Hwa Chen
National Cheng Kung University, Taiwan

John Wiley & Sons, Ltd

Email (for orders and customer service enquiries): cs-books@wiley.co.uk
Visit our Home Page on www.wileyeurope.com or www.wiley.com

Other Wiley Editorial Offices

John Wiley & Sons Inc., 111 River Street, Hoboken, NJ 07030, USA

Jossey-Bass, 989 Market Street, San Francisco, CA 94103-1741, USA

Wiley-VCH Verlag GmbH, Boschstr. 12, D-69469 Weinheim, Germany

John Wiley & Sons Australia Ltd, 42 McDougall Street, Milton, Queensland 4064, Australia

John Wiley & Sons (Asia) Pte Ltd, 2 Clementi Loop #02-01, Jin Xing Distripark, Singapore 129809

John Wiley & Sons Canada Ltd, 6045 Freemont Blvd, Mississauga, Ontario, L5R 4J3, Canada

Wiley also publishes its books in a variety of electronic formats. Some content that appears
in print may not be available in electronic books.

Anniversary Logo Design: Richard J. Pacifico

Library of Congress Cataloging-in-Publication Data

Chen, Hsiao-Hwa.
The Next generation CDMA technologies / Hsiao-Hwa Chen.
p. cm.
ISBN 978-0-470-02294-8 (cloth)
1. Code division multiple access. I. Title
TK5103.45.C53 2007
621.3845—dc22

2007010094

British Library Cataloguing in Publication Data

A catalogue record for this book is available from the British Library

ISBN 978-0-470-02294-8 (HB)

Typeset in 9/11pt Times by Laserwords Private Limited, Chennai, India.
Printed and bound in Great Britain by Antony Rowe Ltd, Chippenham, Wiltshire.
This book is printed on acid-free paper responsibly manufactured from sustainable forestry
in which at least two trees are planted for each one used for paper production.

Contents

Preface

This book addresses the issues on the development of next generation CDMA technologies and contains a lot of information on the subject from both the open literature and my own research activities in the last fifteen years.

When I initially agreed with the publisher to write a book on the next generation CDMA technologies in 2003, CDMA technology was just at its climax of popularity: everybody was talking about CDMA, and its applications could be found then in various wireless and wired communication systems, virtually everywhere. It seemed to me at that time that CDMA technology would stay in its leading position for a long time. However, recently CDMA technology has faced a serious challenge from other multiple access technologies, in particular from orthogonal frequency division multiple access (OFDMA) technology, and many people have turned away from CDMA to OFDMA and even to some other multiple access technologies.

There are many reasons why CDMA technology has become less popular than it was a few years ago. One of the most plausible reasons is that, as quoted from some people's opinion, the concept of CDMA technology was developed more than ten years ago and it is suited well only to slow-speed and continuous-time signal transmissions, which are relevant to voice-centric services, as carried by most second generation mobile cellular systems, such as IS-95, etc. Now, we are talking about high-speed burst-type traffic (such as Beyond 3G (B3G) wireless applications) in wireless channels, and thus CDMA technology is obviously not suitable. For almost the same reason, the OFDMA technology came onto the stage and aims at replacing CDMA as the prime multiple access technology for B3G wireless applications. Yes, everything seems to be perfectly right.

However, behind the above explanation on why CDMA technology can not continue taking the lead we have sensed some unrevealed truth, which might also be the cause that has made CDMA technology lag behind. Let us take a look at mobile cellular communication technologies, which have gone through 2G and 3G since the first commercial CDMA cellular system was launched more than ten years ago. In Taiwan, as well as many other regions or countries, we have actually entered the 3.5G era with High Speed Downlink Packet Access (HSDPA) being put in place by several mobile service providers. On the other hand, CDMA technology has remained static (with almost the same core technologies being used in 2G and most 3G systems) and we have not seen any substantial technological advancement related to CDMA so far. Therefore, it is natural and understandable that people have turned to some other better multiple access technologies to replace CDMA, if the CDMA technology itself is not advancing as fast as expected. Why has CDMA stayed in the same place for such a long time? Is it because CDMA technology itself has fully grown up to be mature enough, such that it does not have any room for improvement? It seems that neither of the above questions can be answered with 'yes', as shown by the facts to be revealed in this book. Then, what is the real cause slowing down the evolution of CDMA technology? We would like to leave this question open here and will try to explain it in the introduction part of the book.

I have to confess that it has been a painstaking process to write this book in the last three years. In fact, all materials included in this book have already been there for some time. As a matter of fact, we have generated much more information on next generation CDMA technology than what can be accommodated in this book due to the page budget limitation. The problem is that I had to translate most of them from various technical reports and documentations written in Chinese into English page by page, which is a very time consuming process and it took me a long time to finish it. Therefore, while very much enjoying the rich culture in a Chinese community, such as what we have here in Taiwan, sometimes I also feel very sorry to work in such an environment where Chinese has to be an instructing or working language, especially when I write my books and thus I have to dig out a lot of information from all those archival materials written in the Chinese langauge. Much time has to be spent due simply to the language problem, instead of the technical problems, sadly to say.

Nevertheless, I am really fascinated by the research works on the next generation CDMA technologies as I have obtained a great amount of interesting data and results. It will be shown in this book that CDMA technology will have a great opportunity to stay as a leading multiple access technology for different communication systems (wireless and wired) if we can continue working hard to make it happen. It will be shown through many examples and results given in this book that it is definitely possible to make CDMA systems interference-free (instead of being always interference-limited), which is one of the most important characteristic features for next generation CDMA technology.

This is a research oriented book with in total ten chapters, which contain a lot of state-of-the-art research results on next generation CDMA technology. However, this book can also serve as a supplementary teaching material for any communications-related courses taught for senior undergraduate students or postgraduate students, who major in electrical and computer engineering, computer science, or telecommunication systems. If it is used as a teaching material for senior undergraduate students, the best effect will be achieved if the students have already taken some prerequisites, such as 'Signals and Systems', 'Digital Communications', and 'Spread Spectrum Communications', etc. A good background knowledge of engineering mathematics of the students will also be desirable for them to follow the more advanced part of the materials presented in this book. In addition, this book can also be successfully used as the main teaching material for professional training courses, which may cover as long as a full semester or term.

I am very grateful to my family for their consistent support throughout this book project. In particular, I would like to thank my dear wife, Tsuiping, for her patience and compassion during the holidays and weekends I spent working on this book. I would like also to thank my daughter, Cindy, and my son, Peter, for their understanding rendered to me for not being able to play with them at weekends and holidays.

Many people have helped me during the manuscript preparation of this book. Especially, I would like to thank my students, Jin-Xian Lin, Shin-Wei Chu, Yu Hsin Lin, Chien Yao Chao, Yu Ching Yeh, Guan-Ting Chen, Tsung-Chi Tsai, Hsiang Yi Shih, Cheng Lung Wu, Yao Lin Tsao, I-Lin Sung, Yen-Han Huang, Jen-Ting Liu, Hui-Chin Kuo, Yi-Chang Wu, and Hung-Lun Chen, for helping me in various ways to collect the data and references, etc. Some of the works included in this book partly result from their theses research works.

Hsiao-Hwa Chen
Tainan, Taiwan

About the Author

Hsiao-Hwa Chen was the founding Director of the Institute of Communications Engineering, National Sun Yat-Sen University, Taiwan. He received BSc and MSc degrees with the highest honor from Zhejiang University, China, and a PhD from the University of Oulu, Finland, in 1982, 1985, and 1990, respectively, all in electrical engineering. He worked with the Academy of Finland as a Research Associate from 1991 to 1993 and the National University of Singapore as a Lecturer and then a Senior Lecturer from 1992 to 1997. He joined the Department of Electrical Engineering, National Chung Hsing University, Taiwan, as an Associate Professor in 1997 and was promoted to a full-Professor in 2000. In 2001 he joined the National Sun Yat-Sen University, Taiwan, as a founding Director of the Institute of Communications Engineering of the University.

Under his strong leadership the institute was ranked in the second position in the country in terms of SCI journal publications and National Science Council funding per faculty member in 2004. In particular, National Sun Yat-Sen University was ranked first in the world in terms of the number of SCI journal publications in wireless LANs research papers during 2004 to mid-2005, according to a Research Report (www.onr.navy.mil/sci_tech/special/354/technowatch/textmine.asp) released by the Office of Navel Research, USA.

He was a visiting Professor to the Department of Electrical Engineering, University of Kaiserslautern, Germany, in 1999, the Institute of Applied Physics, Tsukuba University, Japan, in 2000, the Institute of Experimental Mathematics, University of Essen, Germany in 2002 (under DFG Fellowship), the Chinese University of Hong Kong in 2004, and the City University of Hong Kong in 2007, respectively.

His current research interests include wireless networking, MIMO systems, next generation CDMA technologies, information security, and Beyond 3G wireless communications.

He is a recipient of numerous Research and Teaching Awards from the National Science Council, the Ministry of Education, and other professional groups in Taiwan. He has authored or co-authored over 200 technical papers in major international journals and conferences, five books, and several book chapters in the areas of communications, including *Next Generation Wireless Systems and Networks* (John Wiley & Sons, Ltd, 2006, 512 pages). He has been an active volunteer for IEEE various technical activities for over 15 years. Currently, he is serving as the Chair of IEEE Communications Society Radio Communications Committee. He served or is serving as symposium chair/co-chair of many major IEEE conferences, including IEEE VTC 2003 Fall, IEEE ICC 2004, IEEE Globecom 2004, IEEE ICC 2005, IEEE Globecom 2005, IEEE ICC 2006, IEEE Globecom 2006, IEEE ICC 2007, and IEEE WCNC 2007. He served or is serving as Editorial Board Member or/and Guest Editor of *IEEE Communications Letters*, *IEEE Communications Magazine*, *IEEE Wireless Communications Magazine*, *IEEE JSAC*, *IEEE Network Magazine*, *IEEE Transactions on Wireless Communications*, and *IEEE Vehicular Technology Magazine*. He is serving as the Chief Editor (Asia and Pacific) for Wiley's *Wireless Communications and Mobile Computing (WCMC) Journal* and Wiley's *International Journal of Communication Systems*. His original work in CDMA wireless networks, digital communications and radar systems has resulted in five US patents, two Finnish patents, three Taiwanese patents, and two Chinese patents, some of which have been licensed to industry for commercial applications. He is also an adjunct Professor of Zhejiang University, China, and Shanghai Jiao Tung University, China.

1

Introduction

The world's first cellular network (i.e., Advanced Mobile Phone System, AMPS) was put into service in the early 1980s, and it was built based on analog radio transmission technologies. Within few years of launching the services, the cellular network began to hit a capacity ceiling as millions of new subscribers signed up for mobile voice services, demanding more and more airtime. Dropped calls and network busy signals became commonplace in many areas covered by mobile cellular networks.

To accommodate more traffic within a limited amount of radio spectrum, the industry developed a new set of digital wireless technologies called time division multiple access (TDMA). DAMPS (Digital AMPS) and GSM (Global System for Mobile) then came onto the stage. DAMPS and GSM use a time-sharing protocol to provide three to four times more capacity than the analog systems (for instance, AMPS systems). But just as DAMPS was being standardized in North America, an even better solution was found, and that is CDMA technology.

The most important milestone in the application of CDMA technologies is the time when Qualcomm successfully developed the first CDMA-based civilian mobile cellular communication standard in the 1990s, which is commonly called IS-95. In fact, the first CDMA network was commercially launched in 1995, and provided roughly ten times more capacity than analog networks, more than TDMA-based DAMPS or GSM. Since then, CDMA-based mobile cellular has become the fastest growing of all wireless technologies, with over 100 million subscribers worldwide today. In addition to supporting more traffic, CDMA-based mobile cellular systems bring many other benefits to carriers and consumers, including better voice quality, broader coverage, lower average power emission, stronger security, and smoother/easier evolutionary upgrading of the networks.

Since then, it has been successfully demonstrated in theory as well as in practice that a CDMA system based on the direct sequence (DS) spreading technique can in fact offer a higher bandwidth efficiency than its predecessors, such as the frequency division multiple access (FDMA) and TDMA techniques, in addition to many other extremely useful technical features, such as low probability of interception, privacy, good protection against multipath interference, attractive overlay operation with existing radio systems, etc., as to be discussed in Chapter 2. Today, DS-based CDMA technology has become one of the prime multiple access radio technologies for many wireless networks and mobile cellular standards, such as cdma2000, W-CDMA, and TD-SCDMA. CDMA technology reached its climax at the beginning of this century. As a direct beneficiary of the great success of CDMA technology, Qualcomm has enjoyed a huge amount of licensing incomes from the applications of the technology even from many other companies in the same industry.

Then, it has been commonly known that the use of CDMA technology has become a very expensive business exercise, and it is to a company's best interest not to use any CDMA-related

The Next Generation CDMA Technologies Hsiao-Hwa Chen

technologies such that the company could effectively reduce the cost of the development process of any wireless communication products. Under such circumstances, the technological evolution of CDMA itself has been affected and most companies in the industry do not want to touch CDMA any more. The investment from the telecommunication industry in CDMA-related technologies has substantially shrunk, especially after the 3G mobile cellular standardization process came to an end. Instead, they would very much like to find some other competing technology which can offer equally good performance for wireless applications. Orthogonal frequency division multiplex (OFDM) or orthogonal frequency division multiple access (OFDMA) technology came to the stage partly because of this reason.

Since the first release of the IS-95A standard in 1995,[1] more than ten years have passed, during which mobile cellular standards have gone through at least two generations, from 2G to 3G, both of which have been widely deployed throughout the world. As a matter of fact, 3.5G technologies have also been put into service in many countries in the world. For instance, several mobile cellular service operators in Taiwan have started to provide their subscribers with 3.5G technology based on the High-Speed Downlink Packet Access (HSDPA) technique,[2] which was developed by 3GPP, to offer high-speed data access for mobile users, especially those who often need to use their notebooks or laptops on the move.

In contrast to the fact that mobile cellular has advanced to its 3.5G technology, it is very sad to see that CDMA technology itself has stayed virtually at the same place, or in its first generation based on the same core techniques, such as direct-sequence spreading, application of unitary spreading codes (which work on a one-code-per-user basis), closed-loop and open-loop power-control, etc., with a strictly interference-limited performance. The sluggishness in CDMA technological evolution has given us a lesson, which teaches us how to create the best environment possible for a technology to continue its evolution without being stopped by unnecessary barriers on its evolutionary path. Technically speaking, we all know that CDMA technology is a powerful and promising technology, which should be paid enough attention for its further advancement. Economically speaking, however, due to the problems with transfer of intellectual property rights (IPR) and associated huge licensing fees, many people have turned away from CDMA to search for some other cheaper and better replacement technologies (such as OFDMA, etc.) for next generation wireless applications. Politically speaking, technology is only technology, which always has its pros and cons, but the most important concern for a company or a country/region is that home-grown technologies/standards should never rely heavily on others' IPRs. Under this rationale, the technological evolution of CDMA has been effectively handicapped, without being given an opportunity to evolve into its next generation.

---

[1]Interim Standard 95 (IS-95) is the first CDMA-based digital cellular standard pioneered by Qualcomm. The brand name for IS-95 now is cdmaOne. IS-95 is also known as TIA-EIA-95. cdmaOne's technical history is reflective of both its birth as a Qualcomm internal project, and the world of then-unproven competing digital cellular standards under which it was developed. The term IS-95 generically applies to the earlier set of protocol revisions, namely P_REV's one through five. P_REV=1 was developed under an ANSI standards process with documentation reference J-STD-008. J-STD-008, published in 1995, was only defined for the then-new North American PCS band (Band Class 1, 1900 MHz). The term IS-95 properly refers to P_REV=1, developed under the Telecommunications Industry Association (TIA) standards process, for the North American cellular band (Band Class 0, 800 MHz) within roughly the same time frame. IS-95 offered interoperation (including handoff) with the analog cellular network. For digital operation, IS-95 and J-STD-008 have most technical details in common. The immature style and structure of both documents are highly reflective of the 'standardizing' of Qualcomm's internal project.

[2]HSDPA is a mobile telephony protocol, a 3.5G technology, which provides a smooth evolutionary path for UMTS-based 3G networks allowing for higher data transfer speeds. Current HSDPA deployments support 1.8 MBit/s or 3.6 MBit/s in downlink. Further steps to 7.2 MBit/s and beyond are planned for the future. As an evolution of the W-CDMA standard, HSDPA achieves the increase in the data transfer speeds by defining a new W-CDMA channel: a high-speed downlink shared channel (HS-DSCH) that operates in a different way from all existing W-CDMA channels and is used for downlink communications to the mobile.

Also, in such a circumstance (in which all people try to avoid using CDMA as much as possible in order not to be liable for license fee charges), people have turned to other replacement air-link architecture to develop their own Beyond 3G wireless systems. This has been reflected in most Beyond 3G wireless applications developed recently. One of the most important standardization efforts in this respect should be long-term evolution (LTE) and evolved UTRAN (E-UTRAN) technology proposed by 3GPP [1]. Very likely, this proposed 4G standard will use single-carrier FDMA for its uplink channel technique and OFDMA for its downlink channel air-link scheme, without using CDMA technology. The reasons for its reluctance to use CDMA technology are very complex, but one of them for sure is just to avoid possible IPR conflicts with the company which owns most CDMA IPRs.

Now, the question is, at least from the technical point of view, whether or not CDMA-based technology is inferior to OFDMA. The answer may not be obvious. It is noted that, although many wireless products on the market (mostly developed for WLANs and digital broadcasting applications) have been using OFDM or its related techniques, the OFDMA and OFDM technologies have not been fully tested and widely deployed in a relative large system/network such as mobile cellular applications. Therefore, the robustness of OFDMA and OFDM technologies for their applications in a mobile cellular communication system to cover large areas is still an unclarified concern to many people, especially its operation under severe weather conditions.

I would like to share the experience of using OFDM-based DVB-T services at my home in Taiwan. In fact, the DVB-T standard is an European standard developed for digital television broadcasting services and it can be effectively viewed as an analogy to the downlink channel transmission in a mobile cellular system, although there are still some differences between the two. Nevertheless, the DVB-T standard basically uses 4096 point IFFT/FFT as a major signal multiplex scheme to encode baseband television signals into frames before sending them into channels via amplitude modulation, which is quite similar to the technique used in downlink channels of a cellular system. Of course, the OFDMA signaling format used in downlink channels in a mobile cellular system may adopt much more signal protective schemes against channel impairment factors, such as multipath and Doppler effects.

Since I installed a set-top box for DVB-T services at my home, I have been enjoying free high-quality digital TV channels from the service providers, but only under good weather conditions. As the signal reception quality in the DVB-T is much more susceptible to weather conditions than a traditional analog TV tuner, I have to retain my old analog TV set in case no signal is available from my DVB-T set-top box, especially in the summer seasons when we usually have a lot of thunder storms with very heavy rain in Taiwan. The susceptibility to severe weather conditions of the DVB-T set-top box has much to do with the amplitude modulation (AM) used in all OFDM- or OFDMA-based air-link technologies. It is a well-known fact that AM is extremely sensitive to noise and interference because it carries information on its carrier's amplitude. In addition, there is no processing gain available in those OFDM- or OFDMA-based schemes and thus it is impossible to gain any extra protection from spectrum expansion. On the other hand, a CDMA technology can offer numerous operational advantages which OFDM- or OFDMA-based schemes lack.

Obviously, the main objective of this book is not to compare the operational advantages of CDMA and OFDMA technologies. Instead, this book wants to convey a clear and strong signal that CDMA is not a legacy technology. It is not true that CDMA has inherent problems impossible to be overcome by itself and thus has to be replaced by some other emerging technology like OFDMA. In fact, CDMA is still a viable and strong candidate for wide application in Beyond 3G wireless systems. CDMA technology should not be considered as a technology owned by only very few companies and others should be afraid of using it due to the IPR issues. The IPRs should be used to encourage more research initiatives and free competition, instead of building up high barriers to slow down technological evolution of the technology.

The motivation for writing this book is to encourage more initiatives to push CDMA technology to its second and third generations, just like mobile cellular technologies. Since its concept was first implemented in the IS-95 standard, CDMA technology unfortunately has basically stayed at the same place. The identical core CDMA technologies have been repeatedly used in 2G and 3G mobile cellular systems. We would like to call it 'the first generation CDMA technology,' which should be innovated and evolved into next generation. What, then, is the next generation CDMA technology (which is the focus of this book)?

I have been working on CDMA technology since my PhD research carried out in the Telecommunications Laboratory, University of Oulu, Finland, in 1988,[3] which was the time when CDMA technology was just being brought forward for discussions on its possible applications in commercial mobile cellular systems. The first generation CDMA technology can be characterized by the following key techniques:

- Unitary spreading codes/sequences, which work on a one-code-per-user basis and have been used by all currently existing CDMA-based mobile cellular systems, such as IS-95, cdma2000, W-CDMA, and TD-SCDMA. Those codes/sequences include Gold codes, Kasami codes, m-sequences, Walsh-Hadamard sequences, and orthogonal variable spreading factor (OVSF) codes.

- Direct sequence (DS) spreading modulation, which is used to spread the bandwidth of the original data information into wideband signal by covering a complete spreading code/sequence onto a bit duration.[4]

- Precision power control technique, in which both open-loop and closed-loop power control will be used to adjust mobile transmission power level such that all signals from different mobiles will reach roughly the same level viewed at a base station receiver. Power control is a must for all current CDMA systems to operate successfully due to the near-far effect in a CDMA system based on traditional unitary codes.

- RAKE receiver, which has been used in all traditional CDMA systems to overcome multipath-induced inter-symbol interference (ISI) or simply multipath interference (MI). A RAKE receiver consists of several 'fingers,' each of which is made up of a correlator or code-matched filter to capture a particular multipath return. All captured multipath returns will then be coherently or non-coherently combined to form a strengthened decision variable. Therefore, RAKE receiver is one of the most important components in first generation CDMA technology.

- Multi-user detection (MUD) schemes, which are useful to detect multi-user signals through signal decorrelation processes carried out in a CDMA receiver. The commonly used MUD schemes include decorrelating detection (DD), minimum mean squared error (MMSE) detector, parallel interference cancelation (PIC) detector, and serial interference cancelation (SIC) detector.

- Multi-carrier parallel transmission, which consists of a serial-to-parallel converter, followed by a multi-carrier modulator. A multi-carrier modem can split up a wideband data signal stream into several narrowband sub-streams, each of which carries part of the original data stream and occupies a much narrower bandwidth than the original signal. In multi-carrier transmissions, each of the sub-streams is less likely to suffer frequency selective fading than the original wideband data stream. Even if a sub-stream falls into a fading null, the errors can be recovered by using some proper interleaving and error-correcting coding schemes.

With the help of all aforementioned techniques, a communication system based on the first generation CDMA technology can offer bandwidth efficiency and detection efficiency better than the one based

---

[3]Therefore, it has been widely believed that the initial concept of CDMA cellular was conceived also in November 1988.

[4]In this book, we consider only short-code spreading modulations, in which one spreading code will cover a complete bit duration. We do not consider the long-code scrambling operation, in which a very long spreading sequence is used to cover many bits.

on FDMA and TDMA technologies. However, the performance of a communication system based on the first generation CDMA technology can only offer a strictly interference-limited capacity, meaning that the capacity of a mobile cellular system based on the IS-95 standard, for example, can only support a number of users far less than the processing gain of the spreading codes used by the system.

Many problems of a communication system based on the first generation CDMA technology in fact stem from the unitary spreading codes/sequences. Those unitary codes include many famous user-separation codes, such as Gold codes, Kasami codes, m-sequences, Walsh-Hadamard codes, and OVSF codes, all of which work on a one-code-per-user basis. They were proposed a relatively long time ago by researchers working in information theory. The problem is that people working in information theory then might not have had sufficient knowledge on wireless channels, in which many impairing factors exist, such as external interferences, multipath propagation, Doppler effect, etc. All of those spreading codes used in the first generation CDMA systems, such as IS-95, cdma2000, W-CDMA, and TD-SCDMA, were proposed much earlier than the time when the CDMA cellular concept was conceived. The most serious problem with these unitary spreading codes is that their correlation properties are far from ideal. Here, what we mean in terms of correlation properties stands for the auto-correlation function of a code and the cross-correlation function between any two codes in the same code family or set. In other words, the orthogonality of all those codes is bad in general, and some of them are not orthogonal at all when they are used in asynchronous transmission channels, such as uplink channels in a mobile cellular system. Unfortunately, both 2G and 3G mobile cellular systems based on the first generation CDMA technology have used these unitary codes for CDMA purposes. In this sense, their strictly interference-limited performance is inevitable.

To develop the next generation CDMA technologies, much innovation is required in spreading code design approaches. We have been working hard for years to search for new approaches to generate innovative spreading codes/sequences. Many interesting results have been obtained and will be presented in this book. Those results include many promising spreading codes/sequences, which possess much better correlation properties than all existing unitary codes. Those proposed codes include super complementary codes, generalized pair-wise complementary codes, column-wise complementary codes, optical complementary codes, etc. Among them, the super and column-wise complementary codes are perfectly orthogonal codes in the sense that they offer zero cross-correlation functions between any two codes for any relative shift in both synchronous and asynchronous transmission channels. With this desirable property, a CDMA system can achieve multiple access interference (MAI) free operation for both uplink and downlink transmissions. In addition, the super and column-wise complementary codes can offer an ideal auto-correlation property such that their auto-correlation functions will be zero for all relative shifts except zero shift in both synchronous and asynchronous transmission modes. The ideal auto-correlation in the super and column-wise complementary codes ensures multipath interference-free operation in both uplink and downlink channels. The joint effect of ideal cross-correlation functions and ideal auto-correlation functions makes a CDMA system using them virtually interference-free, making our dream come true: to make a truly noise-limited CDMA system!

On the other hand, the generalized pair-wise complementary code is not a perfectly orthogonal spreading code. However, each user is only assigned a pair of codes for CDMA and thus the CDMA transceiver can be made much simpler, being able to be implemented using a single carrier modem with the help of two orthogonal carriers, i.e., $\sin \omega_c t$ and $\cos \omega_c t$. In addition, the correlation properties of the generalized pair-wise complementary codes are much better than those of all traditional unitary spreading codes, helping to effectively improve the overall performance of a CDMA system using generalized pair-wise complementary codes.

The optical complementary codes were developed by us for their applications in next generation optical CDMA systems. The design approach of the optical complementary codes is based on the ideal auto-correlation function (with the auto-correlation functions for any relative shifts being zero except for the zero shift) and minimized cross-correlation function ($\lambda_c = 1$). In this way, an optical

CDMA system using the resultant optical complementary codes offers a performance much better than all existing optical CDMA systems.

In particular, Chapter 6 of this book will introduce a unique joint code and system design approach, which is called the real environment adaptation linearization (REAL) approach and is used to design spreading codes/sequences by taking into account almost all real operational conditions in a wireless communication system, such as multipath propagation, random signs in continuous bit stream, bursty traffic, etc. Thus, the obtained spreading codes/sequences can inherently address almost all of those impairing factors without using other external auxiliary sub-systems to overcome those impairments. This revolutionary approach also gives us two important conclusions. First, it proves that an interference-free CDMA is possible if and only if orthogonal complementary codes are used. Second, the maximal number of users supportable in such an interference-free CDMA is equal to the flock size of the orthogonal complementary codes. This is the first time that the existence of an interference-free CDMA and its close relationship with the orthogonal complementary codes has been shown in the literature. The conclusions made from the REAL approach are significant and have laid the foundation for development of the next generation CDMA technologies.

This book will also introduce a type of very interesting orthogonal complementary code in Chapter 6, called column-wise complementary codes because they can be constructed based on their column-wise correlation properties. Based on this particular code-construction process, we are allowed to view their unique characteristics from their orthogonality formulation process, such that we can find many interesting applications for them. More specifically, it is seen from the column-wise complementary codes that some of them can establish their orthogonality based on either time-domain correlation or frequency-domain correlation or both. For example, if the orthogonality of a column-wise complementary code set is established purely on the frequency-domain correlation, we can apply the code set in those applications where time-selective fading is a serious problem, such as in vehicle-to-vehicle (V2V) communications, high-speed railway communications, etc. In this way, CDMA systems based on column-wise codes can offer robust performance even under a very large Doppler spread. On the other hand, if the orthogonality of a column-wise complementary code set is based mainly on the time-domain correlation, the codes can be used in those applications where frequency-selective fading is a big problem. Therefore, the next generation CDMA technologies can be tailor-made for different applications by carefully choosing the appropriate column-wise complementary spreading code sets.

In addition to the spreading codes/sequences, spreading modulation is another important issue which should be addressed sufficiently in development of next generation CDMA technologies. Almost all communication systems based on the first generation CDMA technology use DS spreading modulation to spread the original data signal bandwidth and implant signatures for different users.[5] The DS spreading modulation scheme offers a relatively low spectral efficiency and rigid bandwidth spreading mechanism, such that it is very difficult to support quality-of-service (QoS) sensitive variable rate transmissions. We will introduce a parameter, called spreading efficiency (SE), in particular to measure the bandwidth efficiency of a spreading modulation scheme. The SE is measured by the number of bits of information carried by each chip. Therefore, the SE for a traditional DS spreading modulation scheme is merely $\frac{1}{N}$ bits per chip, if the spreading code length is N and every bit is spread by a complete spreading code with N chips. Obviously, there is much room left for improvement.

In Chapter 7 of this book, an innovative spreading modulation scheme, namely offset stacking (OS) spreading modulation, will be proposed. The OS spreading modulation should work jointly with orthogonal complementary codes and it offers a very high spreading efficiency, which can be up to one bit per chip, thus being N times higher than that of the traditional DS spreading modulation scheme. In addition, OS spreading offers a unique flexibility in adjustment of the data transmission rate through the change of relative offset chips between two consecutive bits. Therefore, OS spreading is in particular suitable for high-speed multimedia signal transmissions with variable QoS requirements.

---

[5]In this book, we will concentrate on DS spreading modulation and will not discuss other traditional spreading modulation schemes, such as frequency hopping (FH) and time hopping (TH).

It is also noted that OS spreading modulation is a general spreading modulation scheme, and DS spreading is only a special case of OS spreading with its relative offset chips being equal to the code length N. In this sense, study of OS spreading is theoretically important as it can help us to understand much better (in a much wider scope) how to optimize the performance of a spreading modulation scheme in terms of bandwidth efficiency and transmission flexibility.

Recently, space-time (S-T) coding techniques have been widely applied to various wireless communication systems. The MIMO technology based on S-T coding schemes will certainly play an extremely important role in future wireless communication systems, because it can help to increase the data transmission rate and improve the signal detection efficiency without consumption of scarce bandwidth and time resources. Several important S-T coding schemes have been proposed in the literature, such as space-time block coding (STBC), space-time trellis coding (STTC), and space-time differential coding (STDC), all of which have been found useful applications in many wireless systems. In Chapter 8, we will introduce a new S-T coding scheme, called space-time complementary coding (STCC), which operates based on the application of orthogonal complementary codes. A CDMA communication system based on the STCC scheme can enjoy both interference-free operation and full spatial diversity gain, such that the overall performance can be substantially improved. In fact, in Chapter 8, we will introduce two different types of STCC CC-CDMA systems, one being STCC DS/CC-CDMA and the other being STCC OS/CC-CDMA using generalized pair-wise complementary codes. We will also compare their performance with traditional S-T coded CDMA systems, such as STBC CDMA.

Another important ingredient of next generation CDMA technologies will be multi-dimensional spreading techniques, which will also be discussed in Chapter 5. As a matter of fact, all CDMA schemes proposed in this book are based largely on various complementary codes. Each user in a complementary-code-based CDMA system is assigned a flock of M element codes as its signature code, and all M element codes should be sent via different channels (either frequencies or time slots) in parallel (via frequency division multiplex) or in serial (via time division multiplex) to a receiver. If we use M sub-carriers to send M element codes, two-dimensional spreading takes place at a transmitter. Now, we have two dimensions to spread original data information, one being in the time domain through N different chips in each element code, and the other being in the frequency domain via M different sub-carriers. The vast number of different orthogonal complementary codes generated in this book allows us to choose many different orthogonal complementary code sets with different combinations of their element code length N and their flock size M, to form different $N \times M$ rectangular-shaped code matrices according to different applications. In some applications, we should reduce the spreading dimension in the frequency domain M (to avoid frequency-selective fading), while in others we should shorten the spreading dimension in the time domain N (to avoid time-selective fading), while still keeping their product $N \times M$ unchanged for a fixed processing gain. Therefore, two-dimensional spreading offers us a much greater degree of freedom in design of a wireless system based on next generation CDMA technology for a particular application. Furthermore, we can also add the third dimension (or the space domain) to the two-dimensional spreading schemes, forming three-dimensional spreading techniques, which can be another important enabling technology to implement next generation CDMA technologies.

M-ary CDMA, which is covered in Chapter 9, may be yet another important part of next generation CDMA technologies. Very much different from a normal DS-CDMA system, an M-ary CDMA system uses multiple spreading codes at each user. Data information will be directly encoded in the patterns of sent codes from a particular user. For instance, if each user is assigned H codes, then there will be $3^H - 1$ patterns of sent codes and each pattern can be used to represent one of 2^m different symbols if H and m are selected as long as they satisfy the relation $3^H - 1 \geq 2^m$. Obviously, if so there must be many ways to choose 2^m from $3^H - 1$ patterns of sent codes such that each symbol consists of m bits. This gives us a nice optimization problem to maximize the mean Euclidean distances among 2^m different constellation points, which are selected from a collection of $3^H - 1$ constellation points. An M-ary CDMA system can offer much higher bandwidth efficiency than a traditional DS-CDMA system as it can carry m bits of information in every bit duration, while a conventional DS-CDMA system can carry strictly one bit only.

In Chapter 10, we will also discuss the issues on next generation optical CDMA systems based on optical complementary codes, which were proposed in our research recently. In contrast to a wireless system, an optical communication system is very different and carries many unique properties. In particular, an optical system will send binary data through '0' and '1', instead of '-1' and '+1' as is the case for a wireless system. Therefore, this peculiarity should also be reflected in the spreading code design process for an optical CDMA system. We will propose a new spreading code, namely optical complementary code (OCC), for its applications in next generation optical CDMA (OCDMA) systems. We will also carry out performance comparison among various OCDMA systems with different optical spreading codes, to verify the superiority of the proposed optical complementary codes.

At the end of this book, a few appendices are given to explain the derivations or proofs of several important equations or relations used in the book. In addition, we have listed commonly used complete complementary codes and super complementary codes in Appendices E and F, respectively.

Before the end of the introduction of this book, I would like also to give some information about several special issues or feature topics on the related research topics in some IEEE journals or magazines, for which I was/am the Guest Editor.

I, together with two other guest editors, edited a feature topic on 'Multiple Access Technologies for B3G Wireless Communications' [2] in *IEEE Communications Magazine*, which was published in February 2005. This feature topic covers various important issues on which type of multiple access technologies should be used in Beyond 3G wireless communications.

I was the lead Guest Editor of the special issue on 'The Next Generation CDMA Technologies' [3] of *IEEE Journal on Selected Areas in Communications*, which was published in the first Quarter of 2006. This special issue is the first of its kind and the call - for - papers received a great response from the community, clearly reflected in the number of submissions received for the issue. In this issue many interesting ideas have been brought forward, and in fact we received too many papers, making the acceptance ratio for this issue very low.

Another special issue for which I was the lead Guest Editor has just been published by *IEEE Vehicular Technology Magazine*, entitled 'Evolution of Air-Interface Technologies for 4G Wireless Communications' [4].

Very recently, I have proofread the guest editorial for a special issue on 'Evolution toward 4G Wireless Networking' [5] for *IEEE Network Magazine*, which appeared in January 2007. In this issue some up-layer issues on the 4G wireless networks have been covered.

Currently, I am in the process of reviews for all submissions for a special issue on 'Next Generation CDMA vs. OFDMA for 4G Wireless Applications' [6], for *IEEE Wireless Communications Magazine*, which will be published in June 2007. This issue will be another important special issue covering the topics explicitly on two contending multiple access technologies, CDMA and OFDMA. Therefore, I believe that it will offer many informative discussions and comparisons on the two major air-link technologies.

I am deeply fascinated by the research on the next generation CDMA technologies, and a great amount of research data has been obtained, which is very encouraging. Due to the limited page budget allowed in this book, I could not put all of them inside. Hopefully, I can publish them in another book or in the revision of this one. I hope that readers will find this book informative and useful. All are warmly welcome for any comments or suggestions on this book. Please feel free to contact me via email at `hshwchen@ieee.org`. Thank you very much!

2

Basics of CDMA Communications

As its title implies, this book will be focused on the issues of the next generation CDMA technologies. Therefore, in this chapter we would like to start with discussions on code division multiple access (CDMA) technology, which was developed based on spread spectrum (SS) techniques and has become one of the most important multiple access technologies in contemporary mobile cellular communication systems, as reflected by the fact that almost all 3G standards have been engineered based on CDMA technologies.

It is to be noted that the SS techniques, as discussed in many published books [7–24], provide only a means to extend the bandwidth of transmitting signals to obtain some other extra operational advantages, which may not be possible if using only a bandwidth comparable to that spanned by the original information signals. However, by spreading the spectrum of the original data signals, we can also obtain other benefits, such as allowing more users to share the same communication medium simultaneously in time and frequency without introducing considerable interference with each other. Therefore, we should be very clear about the fact that SS techniques form the basis for the later CDMA technology, whose success was simply impossible without many pioneer research works on SS techniques [7–10].

The following are the major attractive features of CDMA technologies:

- multiple access capability
- protection against multipath interference
- privacy, interference rejection
- anti-jamming capability
- low probability of interception
- possibility to overlay with existing radio systems
- low transmit power emission, which is important to reduce health risks.

Obviously, due to its unique features CDMA has emerged as one of the most important multiple access technologies for the second and third generations (2G-3G) wireless communication systems, exemplified by its wide applications in many important mobile cellular standards, such as IS-95 [25–52], cdma2000 [53–82], UMTS-UTRA [83–88], WCDMA [89], and TD-SCDMA [90–97],

The Next Generation CDMA Technologies Hsiao-Hwa Chen

which were proposed by TIA/EIA of the USA (IS–95 and cdma2000), ETSI of Europe (UMTS-UTRA), ARIB of Japan (WCDMA), and CATT of China (TD-SCDMA), respectively. It is noted that a new type of CDMA technology has also been introduced in another China-born standard, called TD-LAS [98–104]. It is possible that CDMA will continue to be a primary air-link architecture for Beyond 3G (B3G) wireless communications, although some other multiple access technologies have also gained great attention in the community recently, such as orthogonal frequency division multiple access (OFDMA) and even some other renovated versions of TDMA.

As its name suggests, CDMA, in contrast to its predecessors (frequency division multiple access, FDMA, and time division multiple access, TDMA), is a multiple access technology that divides users based on orthogonality or quasi-orthogonality of their signature codes or simply CDMA codes. There are three primarily different types of CDMA technologies that have been extensively investigated in the past two decades: direct sequence (DS) CDMA, frequency hopping (FH) CDMA and time hopping (TH) CDMA. Each user in a DS-CDMA system should use a code to spread its information bit stream directly by multiplication or modulo-2 addition operation, which is also the simplest and most popular CDMA scheme among the three. FH-CDMA uses a multi-tone oscillator to generate multiple discrete carrier frequencies and each user in the system chooses a particular frequency-hopping pattern among those carriers that are governed by a specific sequence, which should be orthogonal or quasi-orthogonal to the others. Depending on the hopping rate relative to the data rate, FH-CDMA can also be further classified into two sub-categories: slow-hopping and fast-hopping FH-CDMA techniques. The majority of currently available FH-CDMA systems use a slow-hopping scheme. One typical example of the application of slow FH technique is the GSM system [105–129], which is a 2G digital technology based on TDMA air-link architecture. The third type of CDMA, TH-CDMA, is found to be much less widely used than the previous two due mainly to its implementation difficulties and the hardware cost associated with its transmitter, which should provide an extremely high dynamic range and very high switching speed. As mentioned in the existing literature, the ultra-wideband (UWB) technique can also be viewed as a type of TH-CDMA system.

In addition, there are also many different types of hybrid CDMA schemes, which can be formed by various combinations of DS, FH, and TH, together with multi-carrier (MC) and multi-tone (MT) techniques, as shown in Figure 2.1, where the family tree of various forms of CDMA technologies is depicted. It is stressed that this book will mainly be concerned with DS-based CDMA systems and their evolution issues. However, the conclusions drawn here may also be found equally relevant to other CDMA schemes.

One of the most important characteristics of a CDMA system is that it allows all users to send their information at the same frequency band and same time duration simultaneously but using different

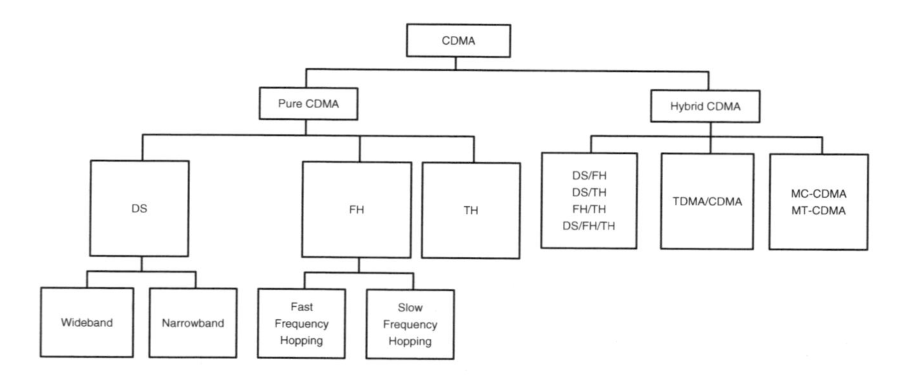

Figure 2.1 Family tree of various CDMA technologies.

codes. Therefore, it is obvious that the orthogonality or quasi-orthogonality among the codes or sequences plays an extremely important role. In fact, we should define the two important roles of the codes or sequences used in a CDMA system: one is to act as signature codes (to accomplish code division multiple access) and the other is to spread the data bits (to spread signal bandwidth to achieve a certain processing gain). It should be emphasized that the roles of the former and the latter are not necessarily given to the same code in a particular CDMA air-link architecture. For instance in the uplink channels of IS-95A/B [25–52], the signature codes are long m-sequences and the spreading codes are 64-ary Walsh-Hadamard functions. On the other hand, the downlink of the IS-95A/B standard uses 64-ary Walsh-Hadamard sequences as both the spreading codes and signature codes, due to its synchronous transmissions in the downlink channels.

Then, it comes to the question of why CDMA has become the most popular air-interface technology for the current 2G and 3G, and possibly also for B3G, wireless communications. The main reasons are summarized below. First, so far CDMA is still one of the technologies (the another candidate is OFDMA, which can also be used to effectively mitigate the intersymbol-interference (ISI) problem caused by multipath propagation) that can mitigate multipath interference (MI) using a relatively cost-effective way, which otherwise would have to be tackled by using other much more complicated sub-systems in FDMA and TDMA systems. Second, on the average current CDMA technology can offer a far better capacity than its counterparts, such as FDMA and TDMA, to meet the increasing demand for mobile cellular applications in the world. Third, the overall bandwidth efficiency of a CDMA system is much higher than that with conventional multiple access technologies, thus giving an operator a much bigger incentive to adopt it due to the extremely high price of spectra. Finally, the relatively low-peak emission power level of a CDMA transceiver offers a unique capability for CDMA-based systems to overlay existing radio services currently in operation without introducing noticeable interference with each other.

However, we have to admit that current CDMA systems are still far from perfect. It is a well-known fact that a CDMA system is always considered as an interference-limited system due mainly to the existence of multiple access interference (MAI) and MI, which are the two major causes of the limitation of capacity and performance in any CDMA-based systems currently in operation, including all mature 2G and 3G architectures. The following questions may come to the mind of anyone who has learned the basic knowledge of CDMA: (1) Do CDMA systems always deserve only interference-limited performance? (2) Why does a CDMA system have to work with so many complicated auxiliary sub-systems, such as closed-loop and open-loop power control, RAKE receiver, rate-matching algorithms, uplink synchronization control, multi-user detection, etc., to just name a few as examples? (3) Can we get rid of all of these complicated sub-systems to make a simple and yet well-performing CDMA?

Many people may think it is only a dream that never comes true to make an interference-free CDMA, but I would like to offer some different views in this book by addressing the issues related to the evolution of CDMA technologies from currently available 2G and 3G systems to the next generation CDMA technologies for the future. Here I will also present some of my thoughts on engineering a new CDMA architecture with a greatly enhanced capability to mitigate MAI and MI, a critical issue associated with next generation CDMA technology.

Several assumptions should be made to facilitate the discussions given in this book. First, we should limit our discussions to DS-CDMA systems only and we will not address the issues related to other CDMA schemes, such as FH-CDMA or TH-CDMA. Second, in such a DS-CDMA system of interest to us, data signal spreading will be performed using short codes (with the chip width being T_c), whose length is exactly the same as the duration of one entire data bit (T_b), or the processing gain (PG) of such a DS-CDMA system is equal to $N = T_b = T_c$. In other words, we will not deal with the situation where long spreading codes, whose length is longer than the width of one data bit, may be used to spread the data bit stream. Third, we will consider a wireless system with full-duplex operation, which consists of mobile terminals and a base station (BS). The transmission link from mobiles to a BS is referred to as the uplink, and the transmission link in the reverse direction is

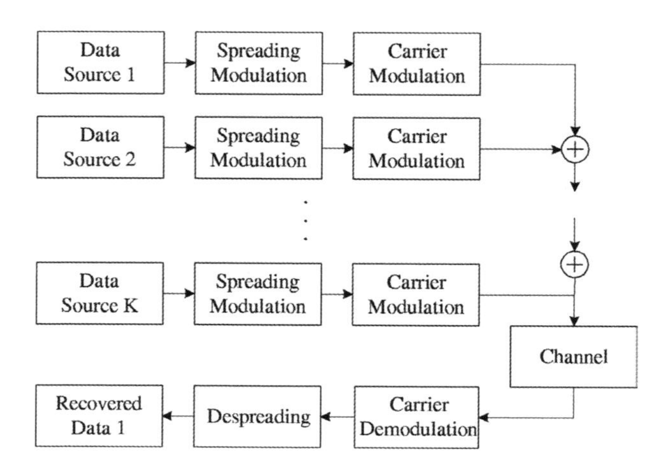

Figure 2.2 A generic K-user DS-CDMA system model, where only the uplink channels of the system are considered.

called the downlink. The block diagram of a generic DS-CDMA system considered in this chapter (as well as in this book) is shown in Figure 2.2, where we are interested in a DS-CDMA system with K users, each of which is assigned one unique code for CDMA purposes, and the signal of concern is data source 1, unless explained by additional words.

2.1 CDMA CODES AND THEIR PROPERTIES

The properties of CDMA codes will play a critical role in a CDMA system. As mentioned earlier, the fundamental principle of CDMA communications demands the use of different codes to separate different users. Therefore, a CDMA system differs from traditional FDMA and TDMA systems in terms of its unique way of separating users or channels in a communication system or network. While FDMA and TDMA systems require 'different frequency bands' and 'different time slots' to offer good separation among different channels, a CDMA system relies on the 'orthogonal properties' of CDMA codes to separate different users or channels. Therefore, the properties of CDMA codes will largely govern the performance of a CDMA system. If we choose the wrong codes, the CDMA system will never operate satisfactorily as we expect.

2.1.1 CDMA CODES

The discussions on CDMA codes will be carried out in much greater detail in later chapters, and we only give a brief introduction here.

Clearly, the CDMA codes, whose characteristics will determine the advantages and limitations of a CDMA system, play an essential role in CDMA system architecture. For instance, the use of orthogonal variable spreading factor (OVSF) codes in the UMTS-UTRA [83–88], WCDMA [89], and TD-SCDMA [90–97] standards requires that a dedicated rate-matching algorithm be used in the transceivers involved whenever the user data transmission rate changes to match a specific spreading factor or the system wants to admit as many users as possible in a cell. In addition, the rate change in UMTS-UTRA and WCDMA can be made only in multiples of two, meaning that continuous rate change is impossible. This requirement is a direct consequence of the OVSF code generation tree structure, where the codes in the upper layers bear a lower spreading factor, whereas those in the lower layers offer a higher spreading factor. Therefore, occupancy of a node in the upper layers virtually blocks all nodes in the lower layers, meaning that fewer users can be accommodated in a cell. The rate-matching algorithms indeed consume a great amount of hardware and software resources and affect the overall performance, such as increased computation load and processing latency. Therefore, we have a strong reason to question whether it is a wise choice to use OVSF codes in UMTS-UTRA, WCDMA, and TD-SCDMA systems. Therefore, the selection or design of CDMA codes is extremely important and should be exercised very carefully at a very early stage of CDMA system design; otherwise the shortcomings of the system architecture due to the use of unsuitable CDMA codes will carry on for ever with the standard, and there is no way to correct shortcomings due to the wrong codes.

2.1.2 PROPERTIES OF CDMA CODES

There are many different ways to characterize CDMA codes, but nothing can be more intuitive and effective than the auto-correlation function (ACF) and cross-correlation function (CCF), which are discussed in more detail below.

Auto-Correlation Function

The ACF is defined as the result of chip-wise convolution, correlation or matched-filtering operation between two time-shifted versions of the same code, which can be further classified into two sub-categories: periodic ACF and aperiodic ACF, depending on the same and different signs of two consecutive bits, respectively, as illustrated in the two left-hand branches in Figure 2.3. In a practical CDMA system, usually the periodic and aperiodic ACFs appear equally likely due to the fact that the binary data of '+1' and '−1' always appear with equal probability in binary bit streams.

The in-phase ACF or ACF peak, which is often equal to the length or PG value (N) of the codes, will affect detection efficiency of the desirable signal in a CDMA receiver, where a correlator or matched filter is used. On the other hand, the out-of-phase ACFs of a CDMA code will be harmless if no multipath effect is present and perfect synchronization is achieved. However, they will contribute to MAI and may seriously impair system performance under the influence of multiple propagation effect.

Cross-Correlation Function

The cross-correlation function (CCF) is defined as the result of a chip-wise convolution operation between two different spreading codes in a family of codes. For a similar reason to that mentioned earlier for the ACF, there are also two different types of CCF, i.e. periodic and aperiodic CCF. The former is mainly found in synchronous transmission channels, such as downlink channels in a wireless

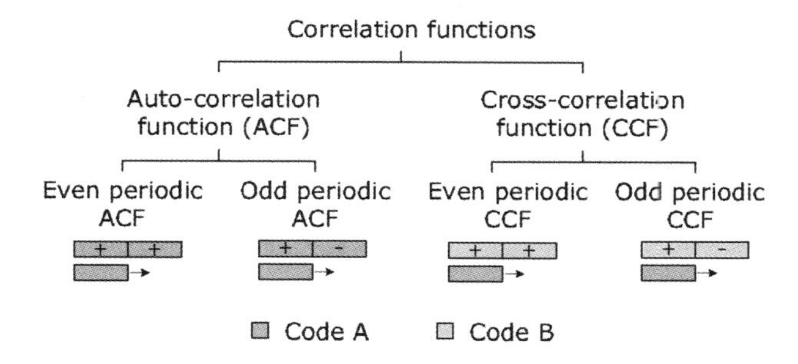

Figure 2.3 Classification of correlation functions of CDMA codes.

system, and the latter can appear in either synchronous (if MI is present) or asynchronous channels. In contrast to out-of-phase ACF, which will contribute to MAI only under multipath channels, the CCF always contributes to MAI, no matter whether or not multipath propagation is present. On the other hand, the out-of-phase ACF will become harmful if and only if a multipath channel is present; otherwise it will never yield MAI at a correlator receiver. Obviously, MAI is one of the most serious threats to jeopardize detection efficiency of a CDMA receiver using either a correlator or RAKE and thus has to be kept below a sufficiently low level to ensure satisfactory performance.

Tables 2.1 and 2.2 list all correlation functions of a CDMA code and their merit behavior in a CDMA system.

Spectral Properties of CDMA Codes

The ACF and CCF determine the time domain characteristics of a CDMA code family or set.[1] The ACF governs the performance of a CDMA system against the ISI caused by multipath propagation, while the CCF determines the capability of a CDMA system to mitigate MAI.

Table 2.1 Auto-correlation functions (ACFs) of CDMA codes and their merit behavior in a CDMA system.

ACF	IPEP[1]-ACF	OPEP[2]-ACF	IPOP[3]-ACF	OPOP[4]-ACF
Cause	Correlator	MPC[5]	Correlator	MPC
Frequency	Once a bit	High	Once a bit	High
Behavior	Wanted signal	MI	Wanted signal	MI
ITD[6]	Enhance	Impair	Enhance	Impair

[1]IPEP: In-phase even periodic.
[2]OPEP: Out-of-phase even periodic.
[3]IPOP: In-phase odd periodic.
[4]OPOP: Out-of-phase odd periodic.
[5]MPC: Multi-path channel.
[6]ITD: Impact to detection.

Table 2.2 Cross-correlation functions (CCFs) of CDMA codes and their merit behavior in a CDMA system.

CCF	IPEP[1]-CCF	OPEP[2]-CCF	IPOP[3]-CCF	OPOP[4]-CCF
Cause	Syn. channel	Asyn. channel	Syn. channel	Asyn. channel
Frequency	Once a bit	High	Once a bit	High
Behavior	MAI	MAI	MAI	MAI
ITD[5]	Impair	Impair	Impair	Impair

[1]IPEP: In-phase even periodic.
[2]OPEP: Out-of-phase even periodic.
[3]IPOP: In-phase odd aperiodic.
[4]OPOP: Out-of-phase odd aperiodic.
[5]ITD: Impact to detection.

[1]In this book, we will use either 'code family' or 'code set' to denote a collection of codes which can be used together in the same cell or sector for user separation purposes in a CDMA system. Therefore, they have the same meaning.

Another important parameter to determine the suitability of a CDMA code family is its spectral properties in the frequency domain. Obviously, the most direct way to influence the spectral occupance of CDMA signals is the chip width of a CDMA code, or T_c. The bandwidth occupancy of a CDMA signal will roughly equal the reciprocal of the chip width or $1/T_c$. As a matter of fact, the chip width of a CDMA system is not determined by the code length only. Instead, it should be determined by taking into account both peak data transmission rate $R = 1/T$ and the length of spreading codes N, assuming that each data bit should be spread by a complete code, such that the short code assumption will be applied here. Therefore, the bandwidth occupancy of a CDMA signal will be determined by $1/T_c = N/T$, where N is often called the processing gain of the CDMA system of interest.

It is noted that all the above discussions on the spectral properties of CDMA codes are based on the assumption that all chips use square waveform. Usually, this assumption does not hold well in most CDMA systems, which always use some kind of pulse shaping waveforms, such as raised-cosine waveform and so on, to shape the chip waveform to make it more spectral efficient. In fact, pulse shaping techniques have been used in all different kinds of digital communication systems, not only spread spectrum systems, to enhance their bandwidth efficiency. It is noted that pulse shaping techniques are very effective to improve the bandwidth efficiency by only using more spectral efficient pulses to shape the waveform of the chips. The improvement in the bandwidth efficiency virtually does not sacrifice any other features of the system, and thus they have found extremely wide applications in all communication systems.

We have been working in the research on spectral efficient pulse-shaping waveforms design for many years and some interesting works have been published, as shown in [130]. Pulse waveform shaping is a traditional research subject, but there is a lot of scope for futher exploration. The results from work in this area can directly help to improve the overall system bandwidth-efficiency without compromising any other system requirements of a CDMA system. Therefore, there is a huge benefit we can exploit if more bandwidth- efficient chip waveforms can be used in a CDMA system. We have also listed several other publications in this area for more information at the end of this book [131–140].

Complexity of CDMA Code Generation

Another property for a CDMA code family is its generation complexity with the help of some logic circuitry.

In fact, the generation complexity for a CDMA code family has twofold meanings. First, the complexity can be defined in terms of the computational complexity of the code design process to search for suitable codes. In this sense, the complexity level can vary from code to code, depending very much on their performance requirements, such as their ACF and CCF requirements and their processing gains, etc.

For instance, an m-sequence can be easily generated by using a shiftregister with the help of a simple feedback logic according to a particular primitive polynomial. Therefore, the m-sequence could be considered as a CDMA code set that can be generated with the least hardware complexity.

As the concept of primitive polynomials can be very useful to understand some issues related to CDMA code generation, we need to explain it a bit more here. A primitive polynomial is defined in modern algebra as a polynomial that generates all elements of an extension field from a base field. Primitive polynomials are also called 'irreducible polynomials'. For any prime or prime power q and any positive integer n, there exists a primitive polynomial of degree n over Galois field GF(q). There are

$$a_q(n) = \frac{\phi(q^n - 1)}{n} \tag{2.1}$$

primitive polynomials over GF(n), where $\phi(n)$ is the totient function.[2]

[2]The totient function $\phi(n)$, also called Euler's totient function, is defined as the number of positive integers less than n that are relatively prime to (or, do not contain any factor in common with) n, where 1 is counted as being relatively prime to all numbers.

A polynomial of degree n over the finite field GF(2) (i.e. with coefficients either 0 or 1) is primitive if it has polynomial order $2^n - 1$. For example, the polynomial $x^2 + x + 1$ has order 3 since

$$\frac{x+1}{x^2+x+1} = \frac{x+1}{x^2+x+1} \pmod 2 \tag{2.2}$$

$$\frac{x^2+1}{x^2+x+1} = 1 + \frac{x+1}{x^2+x+1} \pmod 2 \tag{2.3}$$

$$\frac{x^3+1}{x^2+x+1} = x+1 \pmod 2 \tag{2.4}$$

Plugging $q = 2$ into Equation (2.1), we have the numbers of primitive polynomials over GF(2) calculated as

$$a_2(n) = \frac{\phi(q^2-1)}{n} \tag{2.5}$$

Table 2.3 gives the primitive polynomials (mod 2) of orders 1 through 5.

For more readings on primitive polynomials, readers may find references at the end of this book [141–149].

On the other hand, we may need some more complex methods to generate the CDMA codes we want. Taking orthogonal complementary code as an example, we will show in the later chapters of this book that the generation of orthogonal complementary codes involves a much more complex process than what we have discussed earlier on the generation of m-sequences. Some special approach is needed to find orthogonal complementary codes, which can offer truly orthogonal properties unavailable in all other unitary CDMA code sets.[3]

On the other hand, although the generation process complexity of CDMA codes is the problem we need to consider, we should not overestimate its importance in deciding which CDMA codes should be used in a CDMA system, in particular under the context of implementation of next generation CDMA technologies, which is the main topic of this book. It is noted that in most communication systems spreading codes or sequences can be generated in an offline way and they will be saved in a look-up table formed by a random access memory (RAM) chip, which can be called up whenever needed. Therefore, the spreading codes do not need to be generated on a real-time basis for most CDMA application scenarios. In this sense, the orthogonal properties of the codes become much more important than their generation complexity.

Table 2.3 The primitive polynomials (mod 2) of orders 1 through 5.

n	Primitive polynomials
1	$1+x$
2	$1+x+x^2$
3	$1+x+x^3$, $1+x^2+x^3$
4	$1+x+x^4$, $1+x^3+x^4$
	$1+x^2+x^5$, $1+x+x^2+x^3+x^5$
5	$1+x^3+x^5$, $1+x+x^3+x^4+x^5$
	$1+x^2+x^3+x^4+x^5$, $1+x+x^2+x^4+x^5$

[3] A unitary CDMA code set is defined as one which works on a one-code-per-user basis in a CDMA system. Its counterpart is a complementary code set, which works on a one-flock-per-user basis.

2.2 DIRECT SEQUENCE CDMA TECHNIQUES

Basically, there are three prime CDMA technologies: direct-sequence (DS) CDMA, frequency-hopping (FH) CDMA, and time-hopping (TH) CDMA technologies. In this section, we will concentrate the discussions the DS-CDMA. Just as mentioned before, a particular CDMA technology is always developed based on its corresponding SS techniques. For example, DS-CDMA technology works based largely on the DS-SS technique. Therefore, it is always a useful approach for us to begin with the DS-SS technique, which then can be intuitively extended to the DS-CDMA technology.

Obviously, the simplest method to spread the spectrum of a data-modulated signal is to modulate the signal a second time using a wideband spreading signal, which always takes some form of sequence, i.e. a pseudo-random sequence or PN sequence for short. This second modulation usually takes some form of digital phase modulation, although analog amplitude or phase modulation is conceptually possible. This spread spectrum (SS) scheme is called direct sequence (DS) spread spectrum (or, more exactly, directly carrier-modulated, code sequence modulation), and is the best known and most widely used spread spectrum technique. This is because of its relative simplicity from the viewpoint that it does not require a high-speed, fast-settling frequency synthesizer. Nowadays, DS modulation has been used for many commercial communication systems (almost all 3G mobile cellular systems use DS-CDMA as their prime multiple access air-link architecture) and measurement instruments. It is reasonable to expect that DS modulation will continue to be a familiar form of spreading modulation scheme in the years to come due to its unique and desirable features. Even now, commercial applications of DS-SS systems are being explored, finding many applications in various wireless systems and even optical systems. Characteristic of DS spreading modulation is just exactly that modulation of a carrier by a code sequence. In the general case, the format may be amplitude modulation (AM), frequency modulation (FM), or any other amplitude- or angle-modulation form. Very often, however, it uses binary phase-shift keying (BPSK), because it can be implemented by very low cost hardware: only two balanced multiplication units are required, followed by a low-pass filter (and a decision device if it is at a receiver side). The basic form of DS signal is that produced by a simple and biphase-modulated (BPSK) carrier. Details about the BPSK DS-SS system will be introduced later.

The selection of spreading signals is of great importance in a DS-SS system to have certain properties that facilitate demodulation of the transmitted data signal by the intended receiver, and that make demodulation by an unintended receiver as hard as possible. These same properties will also make it possible for the intended receiver to discriminate between the intended signal and interference, which usually appears quite differently from that used for spreading the signal at the transmitter. If the bandwidth of the spreading signal is much larger than the original data signal bandwidth, the SS transmitting signal bandwidth will be dominated by the spreading signal and is nearly independent of the original data signal. Each element of the spreading sequences or codes is usually called a chip, whose width will determined the bandwidth of the signal after spreading modulation.

Before discussing any DS-SS communication systems, we have to introduce the most important characteristic parameter, namely processing gain (PG), which is defined as a function of the RF bandwidth of the DS signal transmitted, compared with the bandwidth of its original data information before carrier modulation. The PG is exhibited as a signal-to-interference improvement resulting from the RF-to-information bandwidth tradeoff. It will also govern its capability to mitigate many other undesirable factors appearing in communication medium and signal detection processes, such as anti-jamming property, etc. The usual assumption is that the RF bandwidth is assumed to be equal to the main lobe of the DS spectrum, which is always a $\frac{\sin x}{x}$ function. In many practical applications, the ratio between the chip rate and original data information rate can also be used as the PG. Therefore, for a DS-SS system having a 10 Mcps chip rate and a 1 kbps information rate the PG will be

$(10^7)/(10^3) = 10^4$ or about 40 dB. A more strict definition of the PG is given as

$$PG_{DS} = \frac{\text{RF bandwidth of DS/SS signal}}{\text{Baseband bandwidth of user data signal}} \cong \frac{\text{Chip rate of DS/SS signal}}{\text{User data rate}} \quad (2.6)$$

Then comes a question, why not increase the PG to a level that becomes very large to improve the performance of a DS-SS system? This question can best be answered by addressing the limitations that exist with respect to expanding the bandwidth ratio to an arbitrarily large value so that PG may be increased indefinitely. Obviously, two parameters are available to adjust PG. The first is the RF bandwidth, which depends on the chip rate used. For instance, if we have an RF (null-to-null) bandwidth 100 MHz wide, the chip rate should be at least 50 Mcps. On this basis, how wide should we make the system RF bandwidth and how much benefit can we obtain from the increase of the chip rate? To double the RF bandwidth defined by the chip rate, we can only increase PG by 3 dB. However, the cost is in the system complexity. With double chip rate, the sampling rate at a digital receiver has to be doubled at least. This will substantially increase the signal processing load at a DSP chip or CPU. It is to be noted that the increase in the computation load is not linear with the increase in the sampling rate. In other words, doubling the chip rate will probably result in trebling, quadrupling or an even higher computation load in a DSP chip. This imposes a great challenge to implement real-time based communication applications, such as multimedia services. With the decrease in chip duration (or increase in chip rate) the smallest interval to make a decision at a receiver is also reduced, leaving the result that the hardware and software have to catch up the data rate to make a sensible decision for each received bit on the basis of chips. We should admit that the channel characteristics never change with the increase of chip rate. With each chip received at a receiver, all necessary algorithms, such as channel estimation, decision feedbacks, equalization, etc., have to be carried out and finished in time before the ending of the chip of concern. It is still a great challenge to implement a full digital receiver at a chip rate of 10 Gcps using state-of-the-art microelectronics technology. Thus, it is not a wise approach to increase the PG by using a higher chip rate. On the other hand, we can easily understand that it is not a sensible way to increase the PG by reducing the user data rate either.

The most commonly used DS spreading techniques are discussed below.

BPSK Direct-Sequence Spread Spectrum

The simplest form of DS-SS employs binary phase-shift keying (BPSK) as the spreading modulation. It has to be noted that here we are talking about two modulations, i.e. the spreading modulation and carrier modulation. The former denotes the modulation of data information with a predetermined spreading code or sequence to result in bandwidth spreading, and the latter stands for modulating the baseband signal with a high frequency radio carrier, only shifting the spectrum of the original baseband signal to a certain RF frequency without yielding any bandwidth spreading. Therefore, for a BPSK DS-SS system we imply that the spreading modulation must be done using a BPSK modem. However, it is not certain whether the carrier modulation in a BPSK DS-SS system also employs a BPSK modem. As a matter of fact, a BPSK DS-SS system can also use any other modems, such as BPSK, QPSK, MSK, etc., for its carrier modulation.

Yet another important point we have to mention here is that the order of the spreading modulation and carrier modulation is irreversible in most cases, and usually the spreading modulation happens before the carrier modulation at the transmitter. In other words, the data signal should first be modulated by a spreading signal, and then the spread signal will be further modulated by a radio frequency (RF) carrier before being fed into the antenna for transmission. However, if both spreading modulation and carrier modulation use BPSK modems, the order of the two becomes interchangeable.

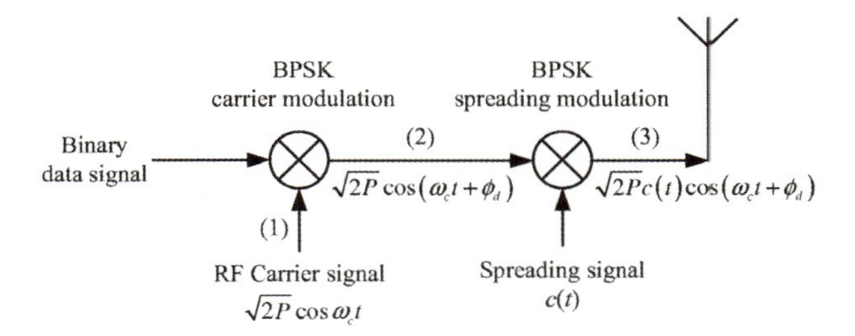

Figure 2.4 Illustration of a BPSK DS-SS transmitter.

Ideal BPSK modulation yields instantaneous phase shifts of the carrier by 0 and 180 degrees according to the signs of the binary data signal as a modulating signal. It can be mathematically expressed by a multiplication of the carrier by a function $c(i)$ that takes on the values ± 1. Let us consider a constant-envelope data-modulated carrier with power P and carrier radian frequency ω_c, defined by

$$f_d(t) = \sqrt{2P}\cos[\omega_c t + \phi_d(t)] \tag{2.7}$$

where $\phi_d(t)$ stands for the data-modulated phase, which should take two different values, either 0 or 180 degrees depending on the sign (either $+1$ or -1) of the binary data information, and the term $\sqrt{2P}$ is to give an average power P.

This signal occupies a bandwidth typically between one-half and twice the data rate prior to DS spreading modulation, depending on the details of the data modulation and the pulse shapes used in shaping the original data bits. The BPSK spreading is accomplished by simply multiplying $f_d(t)$ by a time domain signal $c(i)$ that is also called the spreading signal or spreading sequence, as illustrated in Figure 2.4.

The transmitted signal after spreading modulation becomes

$$f_s(t) = \sqrt{2P}c(t)\cos[\omega_c t + \phi_d(t)] \tag{2.8}$$

whose bandwidth is basically determined by the spectral span of the spreading signal $c(t)$, which usually is a wideband spreading sequence. It is to be noted that the process of multiplication of $c(t)$ with $f_d(t)$ will not alter the power of $f_d(t)$, but only extend the bandwidth of $f_d(t)$. This is what an SS signal means. Then, we recall the scaling property of the Fourier transform, which tells us that the extension of spectral span of a signal will equivalently make its time domain waveform shrink. From the power conservation law, the expansion in the bandwidth span of a signal in the frequency domain will reduce its peak amplitude if the total power remains the same. This effect makes an SS signal appear as a wideband noise-like interference to an unintended receiver. It is obvious that a conventional (non-spread-spectrum) receiver would not be useful for detecting the wideband noise-like signal here because it is well below the level of the real noise observed at the receiver.

The signal given in Equation (2.8) is transmitted into an additive white Gaussian noise (AWGN) channel with a transmission delay τ_d. The signal is received but contaminated by interference and channel AWGN noise. Demodulation is accomplished in part by demodulating or remodulating with the spreading code locally generated and appropriately delayed, $c(t - \tilde{\tau_d})$, as shown in Figure 2.5. This demodulation or correlation of the received signal with the delayed spreading waveform is called the despreading process and is an important process in any SS system. The signal after the despreading module in Figure 2.5 will become

$$r_1(t) = \sqrt{2P}c(t-\tau_d)c(t-\tilde{\tau_d})\cos[\omega_c t + \phi_d(t-\tau_d) + \theta] \tag{2.9}$$

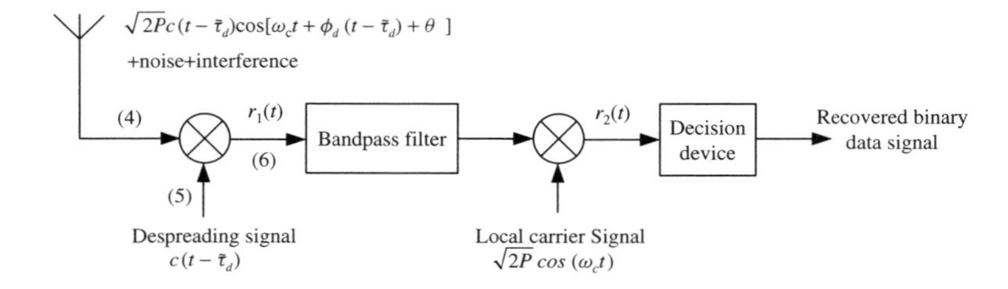

Figure 2.5 Illustration of a BPSK DS-SS receiver.

where $\tilde{\tau_d}$ is the estimated delay at the receiver, τ_d is the propagation delay that the transmitted signal experienced, and θ is the phase delay caused by the propagation delay.

If the estimated delay at the receiver is exactly the same as the real delay, or $\tilde{\tau_d} = \tau_d$, Equation (2.9) will yield

$$\sqrt{2P}\cos\left[\omega_c t + \phi_d(t - \tau_d) + \theta\right] \tag{2.10}$$

as $c(t - \tau_d)c(t - \tilde{\tau_d}) = 1$ if $\tilde{\tau_d} = \tau_d$. This despread signal has been restored into a narrowband signal, which is very similar to the original transmitted phase modulated data signal with only some difference in a delay τ_d and an extra phase θ caused by the propagation delay from the transmitter to the receiver. This despreading process plays a crucial role here to transform the received wideband signal into its original narrowband data signal.

On the other hand, if the receiver uses the wrong spreading signal or spreading sequence, say $c'(t - \tilde{\tau_d})$, to despread the received wideband signal $\sqrt{2P}c(t - \tau_d)\cos\left[\omega_c t + \phi_d(t - \tau_d) + \theta\right]$, it will never accomplish the despreading process to restore the narrowband signal correctly, because $c(t - \tau_d)c'(t - \tilde{\tau_d})$ will be another wideband sequence no matter whether $\tilde{\tau_d} = \tau_d$ or not, and thus the signal

$$\sqrt{2P}c(t - \tau_d)c'(t - \tilde{\tau_d})\cos\left[\omega_c t + \phi_d(t - \tau_d) + \theta\right] \tag{2.11}$$

will remain a wideband modulated signal. Therefore, the spreading signal $c(t)$ is usually also called the signature sequence or signature code as it behave like a key to decode or despread the received signal for recovering the original sent narrowband data signal.

There are six different time domain waveforms observed at the transmitter and the receiver, as shown in (1) to (6) in Figure 2.6. We can also allocate the corresponding observation points from Figures 2.4 and 2.5 accordingly, assuming that the binary data information in this case (shown in Figure 2.6) is a constant value of all +1 for illustration simplicity. Therefore, we can see how a BPSK DS-SS communication transceiver works step by step from the time domain perspective.

The block diagrams shown in Figures 2.4 and 2.5 illustrate a typical DS-SS communications transceiver structure. This shows that a DS-SS system can be viewed as a conventional AM or FM communication link with only an extra part added to implement spreading modulation and demodulation functionalities.[4] The baseband information is digitized and first added to the spreading sequence. For the discussion given in this section, however, we assume that the RF carrier has already been data modulated before spreading modulation, because this can simplify discussion of the modulation-demodulation process in a BPSK DS-SS system. After being amplified, a received signal is multiplied by a reference sequence generated at the receiver locally and, given that the transmitter's sequence and receiver's sequence are synchronous and the same, the carrier inversion

[4]In real applications the carrier modulation usually does not happen before spreading modulation.

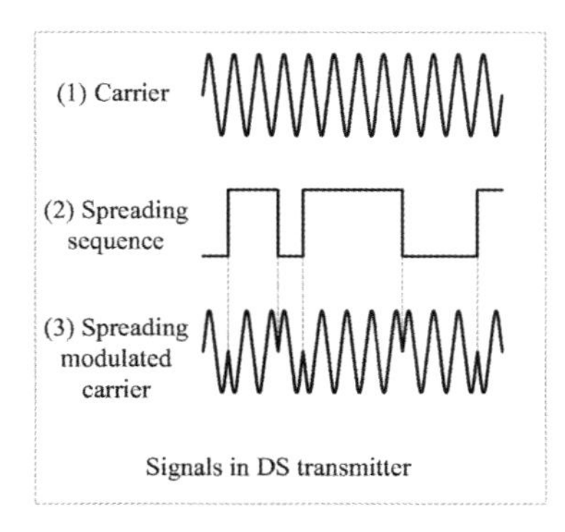

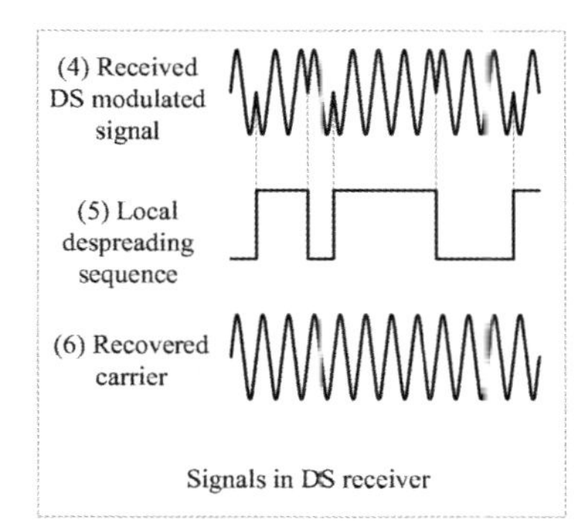

Figure 2.6 Conceptual illustration of time domain signal waveforms for a BPSK DS-SS transceiver. The waveforms shown in this graph correspond to the observation points (1) to (3) in Figure 2.4 and the points (4) to (6) in Figure 2.5.

phases (as shown in (3) and (4) in Figure 2.6) will be removed successfully and the original carrier waveform will be restored. This narrowband restored carrier can then pass through a bandpass filter designed to pass only the original data-modulated carrier.

All unwanted received signals are also treated in the same process at the receiver as the desirable signal, multiplying the received DS signal with a locally generated reference sequence. Any incoming signal not synchronous with the receiver's local reference sequence (a wideband signal) is spread to a bandwidth equal to the bandwidth of the received signal, because an unsynchronized input signal is mapped into a bandwidth at least as wide as the receiver's reference, such that the bandpass filter can reject most of the power of those undesired signals. This is the mechanism by which the process gain is realized in a DS-SS system; that is, the receiver transforms synchronous input signals from the sequence-modulated bandwidth (wideband) to the data-modulated bandwidth (narrowband). At the same time non-synchronous input signals are spread at least over the spreading-sequence-modulated bandwidth. The data-modulated bandwidth specifies the bandwidth of a bandpass filter followed by a decision device, and this bandpass filter in turn effectively controls the amount of power from an unsynchronized or unwanted signal which reaches the data demodulator. Therefore, we can see from the discussion here that the multiplication-and-filtering process before data detection at the receiver provides the desired signal with an advantage or process gain. In fact, the RF bandwidth in a DS-SS system, as discussed earlier, directly affects many capabilities of the system, such as how effectively it can reject external jamming, etc. For instance, if a maximal 10 MHz bandwidth is available, the PG possible is also limited by that 10 MHz. Several practical approaches are available in choosing the proper bandwidth in an anti-interception application; the main interest is to minimize the power transmitted in terms of watts per Hertz. When a maximum PG for interference rejection is needed, the bandwidth again should be made large enough. If either frequency allocation or the propagation medium does not permit the use of a wide RF bandwidth, some restraint must be applied. A prime consideration in SS systems (and in particular DS systems) is the bandwidth of the system with respect to the interference generated by other systems (which may not necessarily be SS systems) operating in the same or adjacent channels.

A conceptual spectral diagram of this type of DS-SS signal format is shown in Figure 2.7, where we only show the envelope of the power spectral density (PSD) function of a BPSK modulated DS-SS signal for illustration clarity. The main lobe bandwidth (null-to-null) of the signal shown is usually equal to twice the clock rate of the code sequence used as a spreading modulation signal. Each of the side lobes has a null-to-null bandwidth that is equal to the clock rate; that is, if the code

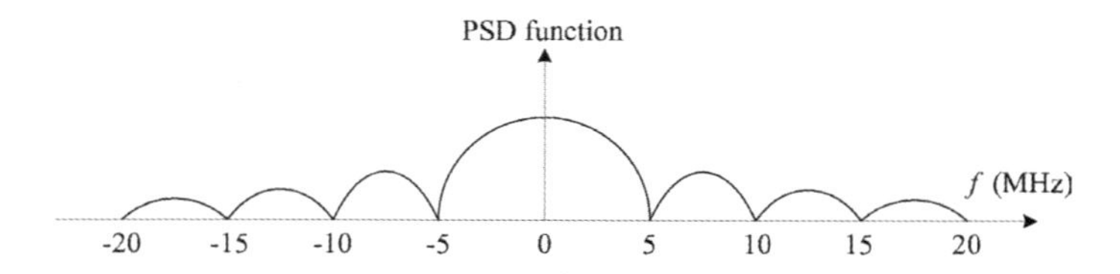

Figure 2.7 Conceptual illustration of power spectral density function for BPSK DS-SS signal. The chip rate for this system is 5 Mcps and the null-to-null bandwidth of this DS-SS system is 10 MHz.

sequence being used as a modulating waveform has a 5 Mcps operating rate,[5] the main lobe of the null-to-null bandwidth will be 10 MHz and each side lobe will be 5 MHz wide. This is exactly the case in Figure 2.7. On the other hand, in the time domain the BPSK-modulated DS-SS carrier looks like the signal shown in Figure 2.6, where the carrier is sent with 0 phase when the code sequence is a '+1', and a 180 degree phase shift when the code sequence is a '−1'.

To illustrate how a DS-SS system works in the frequency domain, we would like also to look at the issues from the perspective of the power spectral density (PSD) function as follows. Assume that the input data information stream, as shown in Figure 2.4, is a random sequence with a transmission rate of $\frac{1}{T}$ bits per second (bps), or

$$\mathrm{d}(t) = \sum_{k=-\infty}^{\infty} d_k p_T(t - kT) \tag{2.12}$$

where $d_k = \pm 1$, the bit duration is T, and $p_T(t)$ is the bit pulse waveform function. Thus, its PSD function $\varphi_d(f)$ can be written into

$$\varphi_d(f) = T\left(\frac{\sin fT}{fT}\right)^2 \tag{2.13}$$

whose shape is illustrated in Figure 2.8(a). Thus, it is seen from the figure that its bandwidth is just equal to $\frac{1}{T}$ Hz. Assume that the spreading sequence is also a random sequence and its chip rate is $\frac{1}{T_c}$. Therefore, its PSD function $\varphi_c(f)$ can be expressed by

$$\varphi_c(f) = T_c\left(\frac{\sin fT_c}{fT_c}\right)^2 \tag{2.14}$$

which forms exactly the same expression as Equation (2.13) except for the interchange of bit duration T and chip width T_c. The PSD function for the spreading sequence $\varphi_c(f)$ has also been drawn together with the PSD function of data sequence $\varphi_d(f)$ for easy comparison in Figure 2.8(a), where $T = 4T_c$ is assumed for illustration clarity. Obviously, the bandwidth of the spreading sequence is equal to $\frac{1}{T_c}$ Hz.

Now, let us consider the spreading modulation process as a simple multiplication between the data signal and spreading sequence, resulting in a PSD function expressed by

$$\varphi_{cd}(f) = T_c\left(\frac{\sin fT_c}{fT_c}\right)^2 \tag{2.15}$$

which takes exactly the same expression as Equation (2.14) and occupies the same bandwidth as that of $\varphi_c(f)$. In this way, the spreading modulation has extended the signal bandwidth to $\frac{T}{T_c} = N$

[5]Here, 'Mcps' stands for mega chips per second.

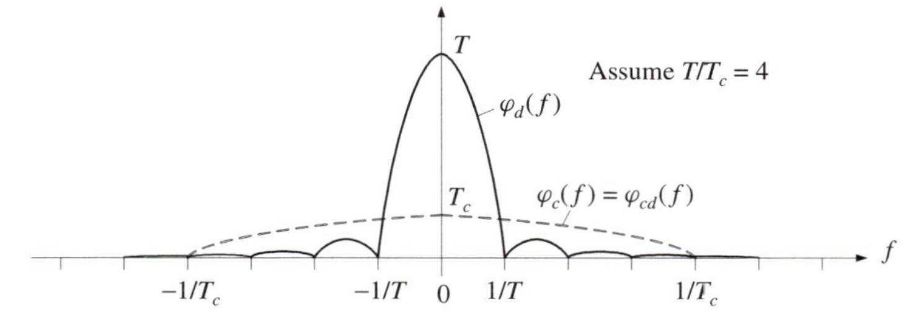

(a) PSD functions for spreading sequence and data signal

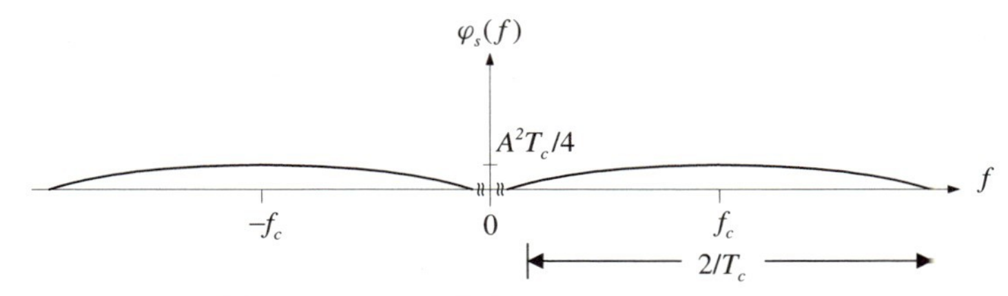

(b) PSD functions for BPSK spreading modulated signal

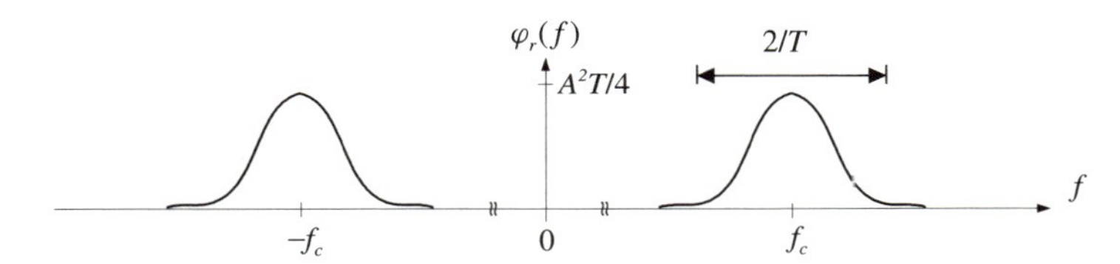

(c) The PSD for the carrier signal after despreading ($r_1(t)$ as shown in DS-SS receiver)

Figure 2.8 The PSD functions for (a) original data and spreading sequence, (b) BPSK spreading modulated signal, and (c) the carrier signal after despreading, where it is assumed that $T = 4T_c$ for illustration clarity.

(N is assumed to be 4 in Figure 2.8) times. N is just equal to the PG of this DS-SS system and it is usually a fairly large number. The carrier modulation after the spreading modulation will only shift the spectrum $\varphi_{cd}(f)$ to the center frequency f_c, but never changes the physical appearance of $\varphi_{cd}(f)$, as shown in Figure 2.8(b).

The PSD function of transmitted signal from the transmitter antenna becomes

$$\varphi_s(f) = \frac{PT_c}{2}\left\{\left[\frac{\sin(f - f_c)T_c}{(f - f_c)T_c}\right]^2 + \left[\frac{\sin(f + f_c)T_c}{(f + f_c)T_c}\right]^2\right\} \tag{2.16}$$

which is a bandpass signal, and its bandwidth is $\frac{2}{T_c}$ Hz, as shown in Figure 2.8(b). It is observed from the figure that the amplitude of $\varphi_s(f)$ is reduced by $\frac{2T}{PT_c}$ compared with $\varphi_d(f)$, whereas the width of $\varphi_s(f)$ increases by $N = \frac{T}{T_c}$ (which is just the PG value) times compared with $\varphi_d(f)$.

At the DS-SS receiver, the PSD function of the received signal has the same PSD function of transmitted signal, with only a delay and some extra phase caused also by propagation delay, as shown in Figure 2.5. The delay will never change the shape of the PSD function. It is easy to show that the PSD function of the signal after the despreading process, or signal $r_1(t)$ as indicated in Figure 2.5, can be written as

$$\varphi_r(f) = \frac{PT}{2}\left\{\left[\frac{\sin(f - f_c)T}{(f - f_c)T}\right]^2 + \left[\frac{\sin(f + f_c)T}{(f + f_c)T}\right]^2\right\} \tag{2.17}$$

which has been plotted in Figure 2.8(c). It is to be noted that Equation (2.17) has exactly the same expression as $\varphi_s(f)$, as written in Equation (2.16), except for the interchange of T_c and T. It is not surprising to us as the despreading process at the receiver will restore the original data signal bandwidth, such that most of its power can pass easily through the bandpass filter, as shown in Figure 2.5. It is seen from the figure that, similar to the $\varphi_d(f)$, the PSD function $\varphi_r(f)$ spans also a narrowband spectrum with its bandwidth being $\frac{2}{T}$, which is just double that for signal $d(t)$. The spectrum $\varphi_r(f)$ will be restored into a narrowband baseband PSD function after the carrier demodulation, which just shifts its center frequency from f_c back to zero.

QPSK Direct-Sequence Spread Spectrum

It is a well-known fact that the use of a quadrature modulation scheme can effectively improve the bandwidth efficiency of a digital modem without sacrificing the power efficiency. The two quadrature carriers, i.e. $\sin(\omega_c t)$ and $\cos(\omega_c t)$, are perfectly orthogonal with each other due to the simple fact that

$$\int_{-\infty}^{\infty} \sin(\omega_c t)\cos(\omega_c t)\mathrm{d}t = \int_{0}^{2\pi} \sin(\omega_c t)\cos(\omega_c t)\mathrm{d}t = 0 \tag{2.18}$$

It is sad to acknowledge that we can not find any more carriers other than them, i.e. $\sin(\omega_c t)$ and $\cos(\omega_c t)$, which possess such an ideal orthogonality. For instance, in an orthogonal frequency division multiplexing (OFDM) system we can use many sub-carriers to send data information in parallel. However, those sub-carriers are not orthogonal in a strict sense as each sub-carrier is always overlapped by half with its two neighboring sub-carriers, and this half-overlapping in the same signal space will introduce serious interference in many circumstances, such as in the cases of being under the influence of the multipath effect and the Doppler effect. However, the two quadrature carriers, $\sin(\omega_c t)$ and $\cos(\omega_c t)$, can work in a much more robust way against many channel impairments with their perfect orthogonality unaffected due to the property that their orthogonality is not established in the same signal plane. Instead, their orthogonality is based on a two-dimensional space, i.e. in-phase and quadrature spaces, which are vertical with each other,[6] as shown in Figure 2.9. Therefore, it is seen that the carriers, $\sin(\omega_c t)$ and $\cos(\omega_c t)$, can always keep their orthogonality even under many undesirable operational conditions due to the fact that they go in different signal spaces, which are already perfectly orthogonal with each other. The use of QPSK modulation in a digital modem can double the bandwidth efficiency with its power efficiency being kept unchanged.

The same idea can be applied to a DS-SS system to improve its bandwidth efficiency if compared with a BPSK DS-SS system. However, it has to be noted that the use of in-phase and quadrature channels in a QPSK DS-SS system should take into account the issues of assignment problems of spreading codes or sequences, i.e. should we assign two different codes to in-phase and quadrature channels, or use the same code for the two channels? Therefore, a QPSK DS-SS system should not be considered equivalently to a normal QPSK digital modulation system.

To illustrate the issue more clearly, let us consider a generic QPSK DS-SS transmitter and a receiver, as shown in Figures 2.10 and 2.11, respectively. $d(t)$ is the input information data stream defined in Equation (2.12) with its duration being T, $c_1(t)$ and $c_2(t)$ are two spreading

[6] We say the two signal spaces are vertical with each other to imply that they have $\pi/2$ phase difference.

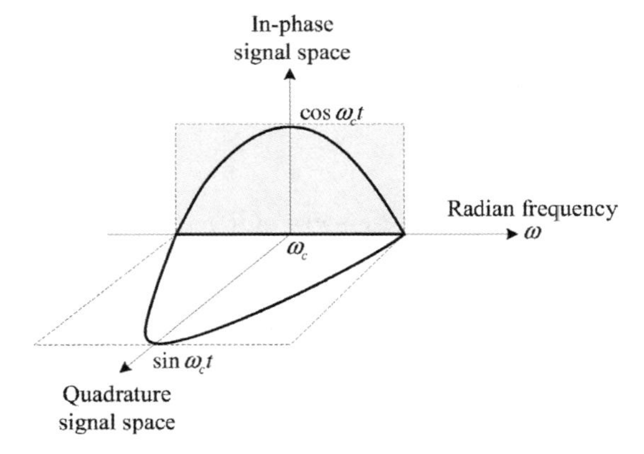

Figure 2.9 Orthogonality of $\sin(\omega_c t)$ and $\cos(\omega_c t)$ carriers in the in-phase and quadrature signal spaces in QPSK digital modulation.

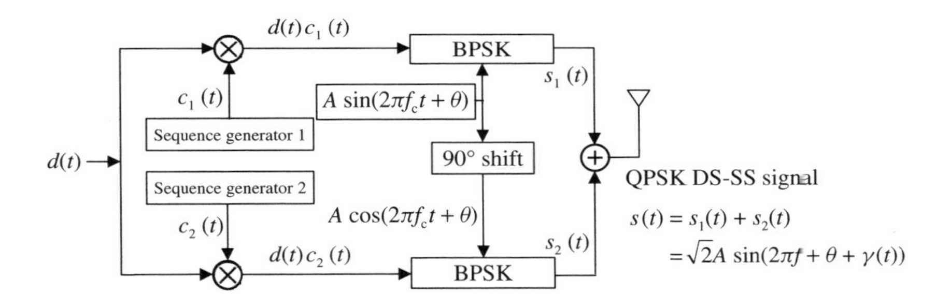

Figure 2.10 A generic QPSK DS-SS transmitter.

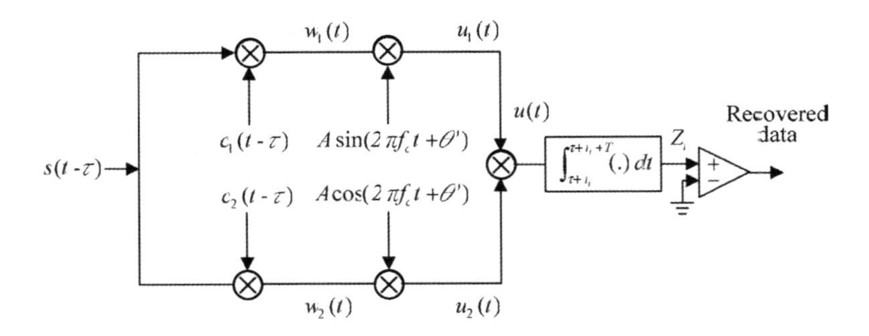

Figure 2.11 A generic QPSK DS-SS receiver.

sequences generated in the transmitter for I and Q channel spreading modulations, $A\sin(2\pi f_c t + \theta)$ and $A\cos(2\pi f_c t + \theta)$ are in-phase and quadrature carriers for QPSK modulation, where the average power of the carrier is $P = \frac{A^2}{2}$, and θ is the initial phase of the carriers. From Figure 2.10, the QPSK DS-SS signal can be expressed as

$$\begin{aligned} s(t) &= s_1(t) + s_2(t) \\ &= Ad(t)c_1(t)\sin(2\pi f_c t + \theta) + Ad(t)c_2(t)\cos(2\pi f_c t + \theta) \\ &= \sqrt{2}A\sin(2\pi f_c t + \theta + \gamma(t)) \end{aligned} \tag{2.19}$$

where the phase-modulated component can be written into

$$\begin{aligned} \gamma(t) &= \arctan\frac{c_2(t)d(t)}{c_1(t)d(t)} \\ &= \begin{cases} \frac{\pi}{4}, & \text{if } c_1(t)d(t) = +1,\ c_2(t)d(t) = +1 \\ \frac{3\pi}{4}, & \text{if } c_1(t)d(t) = -1,\ c_2(t)d(t) = +1 \\ \frac{5\pi}{4}, & \text{if } c_1(t)d(t) = -1,\ c_2(t)d(t) = -1 \\ \frac{7\pi}{4}, & \text{if } c_1(t)d(t) = +1,\ c_2(t)d(t) = -1 \end{cases} \end{aligned} \tag{2.20}$$

It is seen from Equation (2.20) that $s(t)$ will yield four different phases: $\theta + \frac{\pi}{4}$, $\theta + \frac{3\pi}{4}$, $\theta + \frac{5\pi}{4}$, and $\theta + \frac{7\pi}{4}$, according to different combinations of $d(t)c_1(t)$ and $d(t)c_2(t)$. Figure 2.12 illustrates the signal waveforms at different points of a QPSK DS-SS transceiver. It is noted that two different spreading sequences $c_1(t)$ and $c_2(t)$ have been used here to plot Figure 2.12. Of course, there are other alternatives for the assignments of two spreading sequences in both I and Q channels, resulting in very different overall performance, implementation complexity, and other characteristic features of the QPSK DS-SS system of concern. For instance, we can also choose to use the same spreading sequence for spreading modulations in both I and Q channels. In doing so, we will have the advantage that fewer sequences will be needed for each user in order to support more users in the same spread spectrum multiple access (SSMA) network.[7] The use of the same spreading sequence in both I and Q channel spreading modulations is also allowed since the in-phase and quadrature channels employ two orthogonal carriers, $\sin(2\pi f_c t + \theta)$ and $\cos(2\pi f_c t + \theta)$, which already ensure a good isolation between the two channels. However, extra protection will be given if in-phase and quadrature channels use two different spreading sequences, in case some cross-talk between the I and Q channels exists due to non-ideal operational effects, such as frequency or phase estimation inaccuracy or jitter in the local oscillator of a QPSK DS-SS receiver, etc. The price paid to have such extra protection is that the number of spreading sequences needed for the whole SSMA network will be doubled, and in many cases the family size of appropriate spreading sequences suitable for such a multiple access application is always limited.

The receiver for this generic QPSK DS-SS system is shown in Figure 2.11, where it is assumed that the receiver knows the exact propagation delay from the transmitter to receiver or τ and will generate two different spreading sequences $c_1(t - \tau)$ and $c_2(t - \tau)$ accordingly. In this case, the receiver should carry out despreading before carrier demodulation, corresponding to the order of spreading modulation and carrier modulation carried out in the transmitter. As a QPSK modem can be viewed as two BPSK modems working in parallel, we can understand that the order of spreading modulation and carrier modulation can be interchanged in both the transmitter and the receiver at the same time. It means in this case that we can also first perform carrier modulation before spreading

[7] It is to be noted that the two acronyms SSMA and CDMA are usually interchangeable in some cases. The former emphasizes the wideband nature of SS techniques, whereas the latter the user division mechanism by codes.

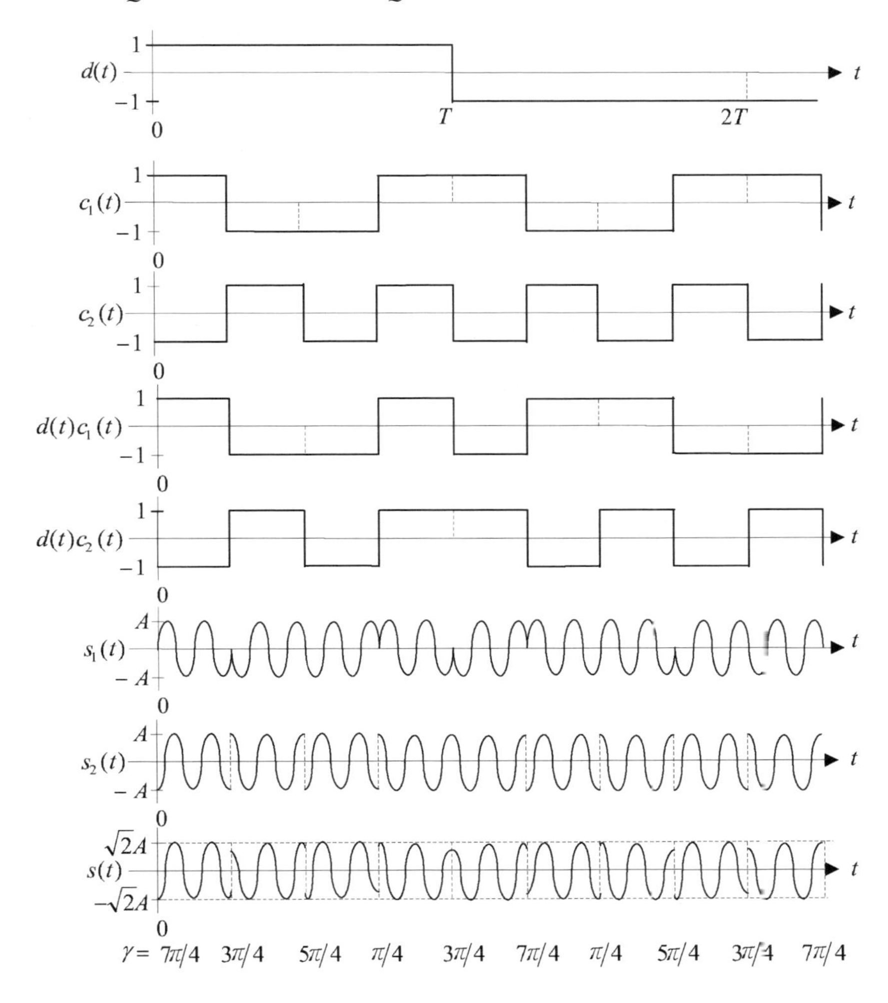

Figure 2.12 Signal waveforms in a generic QPSK DS-SS transceiver.

modulation at the transmitter, and thus first have carrier demodulation before the despreading operation at the receiver.

We also assume that the receiver knows the initial phases θ' of the in-phase and quadrature carriers in the received signal $s(t-\tau)$ such that it can regenerate the local carrier references $A\sin(2\pi f_c t+\theta')$ and $A\cos(2\pi f_c t+\theta')$ that are in-phase with the received signal, resulting in a coherent QPSK DS-SS signal reception. It is to be noted that the difference between the initial phases θ of the in-phase and quadrature carriers at the transmitter and the initial phases θ' at the receiver is due to the signal propagation delay through the channel.

After the despreading and carrier demodulation process, the signals from the I and Q channels will be combined and undergo processing in an integration unit, which functions also like a lowpass filter to remove higher frequency harmonics generated in the carrier demodulation process. The integration will take place within the duration of the ith data bit of interest from $\tau+t_i$ to $\tau+t_i+T$, where t_i

is the starting time of the ith bit and τ is the propagation delay. The output from the integrator will form a decision variable Z_i.

In the following illustration of the basic operation of a DS-SS receiver we will only consider a simple line-of-sight propagation path and will not take into account other channel impairing factors, such as multipath effect, Doppler effect, etc., for illustration simplicity. Therefore, the received signal can be written as

$$\begin{aligned} s(t-\tau) &= Ad(t-\tau)c_1(t-\tau)\sin\left(2\pi f_c t+\theta'\right) \\ &\quad + Ad(t-\tau)c_2(t-\tau)\cos\left(2\pi f_c t+\theta'\right) \end{aligned} \tag{2.21}$$

where the initial phase can also be expressed by $\theta' = \theta - 2\pi f_c \tau$, and τ is the propagation delay in the transmission path from the transmitter to the receiver. The signals in the I and Q channels after despreading and carrier demodulation become

$$\begin{aligned} u_1(t) &= Ad(t-\tau)\sin^2\left(2\pi f_c t+\theta'\right) \\ &\quad + Ad(t-\tau)c_1(t-\tau)c_2(t-\tau)\sin\left(2\pi f_c t+\theta'\right)\cos\left(2\pi f_c t+\theta'\right) \\ &= \frac{A}{2}d(t-\tau)\left[1-\cos(4\pi f_c t+2\theta')\right] \\ &\quad + \frac{A}{2}d(t-\tau)c_1(t-\tau)c_2(t-\tau)\sin(4\pi f_c t+2\theta') \end{aligned} \tag{2.22}$$

and

$$\begin{aligned} u_2(t) &= Ad(t-\tau)\cos^2\left(2\pi f_c t+\theta'\right) \\ &\quad + Ad(t-\tau)c_1(t-\tau)c_2(t-\tau)\sin\left(2\pi f_c t+\theta'\right)\cos\left(2\pi f_c t+\theta'\right) \\ &= \frac{A}{2}d(t-\tau)\left[1+\cos(4\pi f_c t+2\theta')\right] \\ &\quad + \frac{A}{2}d(t-\tau)c_1(t-\tau)c_2(t-\tau)\sin(4\pi f_c t+2\theta') \end{aligned} \tag{2.23}$$

respectively. The summation of the signals from the I and Q channels will become

$$\begin{aligned} u(t) &= u_1(t)+u_2(t) \\ &= Ad(t-\tau) + Ad(t-\tau)c_1(t-\tau)c_2(t-\tau)\sin(4\pi f_c t+2\theta') \end{aligned} \tag{2.24}$$

Obviously, after the lowpass filtering, the second term in Equation (2.24) will vanish and only the term $Ad(t-\tau)$ reflecting the data information remains, yielding the decision variable $Z_i = AT$ if $+1$ is sent or $Z_i = -AT$ if -1 is sent. Therefore, the strength of the decision variable generated from a QPSK DS-SS receiver is just equal to twice that generated from a single BPSK DS-SS receiver if they work under the same conditions (i.e. most importantly, their data transmission rates should be kept the same), implying a 3 dB gain in the signal-to-noise ratio (SNR). It is to be noted that this gain in the SNR does not pay any extra price in bandwidth efficiency as both I and Q channels occupy the same bandwidth, which is exactly the same as the bandwidth for a BPSK DS-SS system. This is really wonderful and it can happen only using the two unique orthogonal carriers, $\sin(2\pi f_c t)$ and $\cos(2\pi f_c t)$. It is pity to admit that we could not find any more such ideally orthogonal carriers than these two.

The two spreading sequences $c_1(t)$ and $c_2(t)$ applied to the I and Q channels can also be two different ones or those split up from one sequence $c(t)$, as shown in Figure 2.13, where the chip

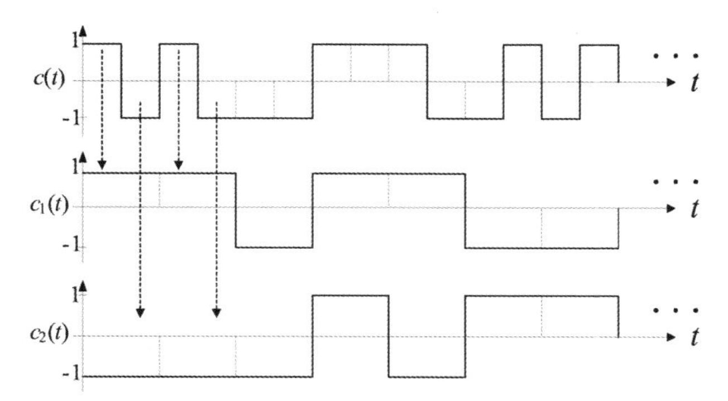

Figure 2.13 Split-up of one sequence into two spreading sequences for their use in a QPSK DS-SS system.

duration of $c(t)$ is half of that for either $c_1(t)$ or $c_2(t)$, and thus the length of either $c_1(t)$ or $c_2(t)$ is only half that of $c(t)$.

Basically, the bit error rate (BER) performance of a QPSK DS-SS system is the same as that of a BPSK DS-SS system. In fact, either the I or the Q channel can be viewed effectively as a single BPSK DS-SS system and each of them possesses the same BER as a normal BPSK DS-SS system. Therefore, two BPSK systems (in the QPSK DS-SS system of concern) working together will yield the same BER as a single BPSK system.

However, the bandwidth efficiency of a QPSK DS-SS system is double that of a single BPSK DS-SS system, and it can be explained as follows. Assume that T_c is the chip width for both $c_1(t)$ and $c_2(t)$. Thus, $s_1(t)$ and $s_2(t)$ will have the same bandwidth, equal to $\frac{2}{T_c}$. This QPSK DS-SS system has a data transmission rate of $\frac{1}{T}$ and processing gain of $PG = \frac{T}{T_c}$. The bandwidth of this QPSK DS-SS system is determined by the chip width of $c_1(t)$ and $c_2(t)$.

It is to be noted that the I and Q channels in the transmitter shown in Figure 2.10 send the same information bit stream with its data rate of $\frac{1}{T}$. However, we can also use the I and Q channels to deliver different data information to increase the transmission rate to $\frac{2}{T}$, if the bit duration in the I and Q channels is kept unchanged. In this case, the receiver structure should be modified to detect the data information in the I and Q channels separately. The modified block diagrams for a QPSK DS-SS system with double data rate are shown in Figures 2.14 and 2.15.

The following factors will affect the performance of a QPSK DS-SS system: transmission rate or bandwidth, processing gain, and SNR (or transmission power). In order to compare the performance of two DS-SS systems, such as a BPSK system and a QPSK DS-SS system, we have to concentrate on one particular parameter, with the other two fixed, to make objective comparisons. For instance, if we want to compare the BPSK and QPSK DS-SS systems shown in Figures 2.4 and 2.14, we should first fix the data rate and processing gain (thus the bandwidth), allowing us to make a fair comparison on SNR values for the two schemes. Since the same data rate is concerned, the bit duration in either the I or Q channel in Figure 2.14 will be twice as wide as that in Figure 2.4. Also, due to the fact that the same PG value is assumed for the two schemes, we can readily conclude that the bandwidth efficiency (defined by bit/s/Hz) for the two schemes should be the same. However, the power efficiency of the QPSK DS-SS system shown in Figure 2.14 is double that of the BPSK DS-SS system shown in Figure 2.4, due to its wider bit duration and thus high signal power available for the detection of each bit in the QPSK scheme, implying a higher SNR value.

The DS-SS systems using other modulation schemes, such as MSK and QAM, can also be studied using similar methodology to that illustrated in this subsection.

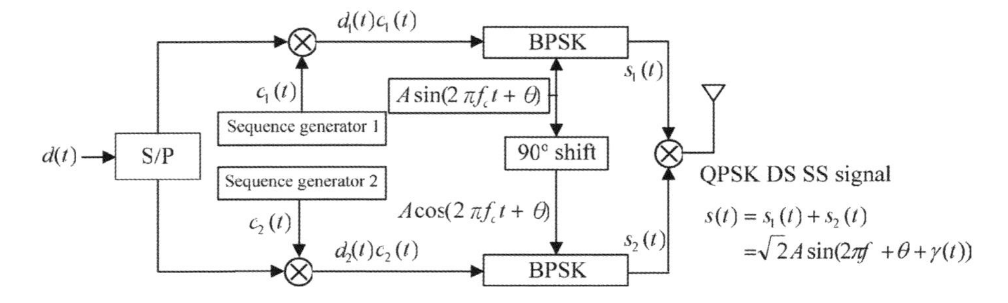

Figure 2.14 An alternative structure of QPSK DS-SS transmitter with double transmission rate.

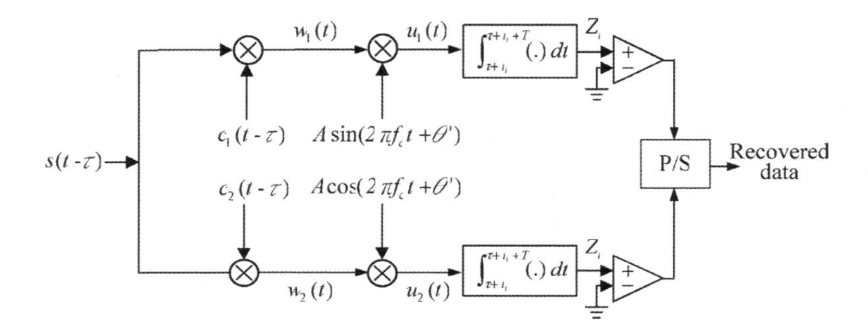

Figure 2.15 An alternative structure of QPSK DS-SS receiver with double transmission rate.

It is to be noted that one of the most successful applications for QPSK DS-SS techniques is the GPS system, which was launched initially by the US for positioning applications in military operations. Nowadays, GPS has found worldwide applications in various practical systems, most of which are civilian applications and services.

Application of DS-CDMA Techniques

After some extensive discussions on DS-SS techniques in the previous subsections, we are ready to extend the concept of DS-SS to formulate a DS-CDMA system.

A generic K-user DS-CDMA system model is shown in Figure 2.2, where only the uplink channels of the system are considered to make the illustration as general as possible, as the downlink transmissions work as a synchronous channel, being only a special case of the asynchronous channels (such as the uplink channels). To use a better example to show the typical configuration of a DS-CDMA system, we consider a CDMA-based centralized system, as shown in Figure 2.16, where two DS-CDMA systems are illustrated, one being a satellite system and the other the mobile cellular system. It is noted that the satellite in a satellite DS-CDMA system works just like a base station in a terrestrial mobile cellular system, functioning like a signal relaying station to convey the communications to and from different mobiles to the outside networks. An example of a satellite land mobile system is the Iridium system proposed initially by a consortium led by motorola. This

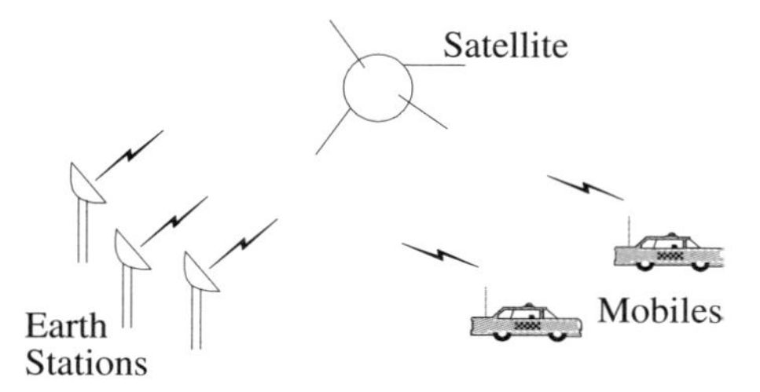

(a) DS-CDMA-based satellite system

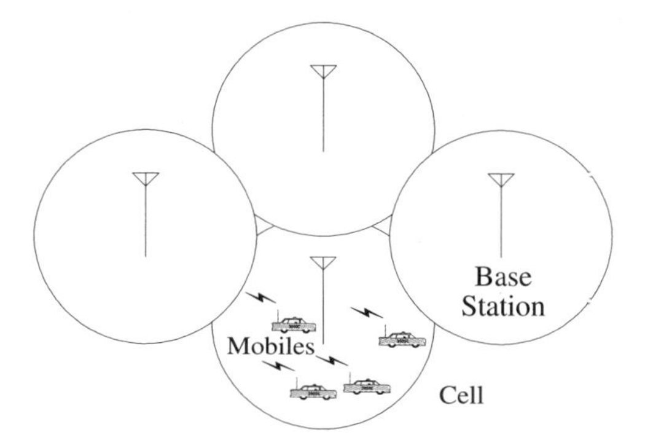

(b) DS-CDMA-based mobile cellular system

Figure 2.16 Example of communication systems using DS-CDMA technology. (a) Satellite DS-CDMA system. (b) Mobile cellular DS-CDMA system.

system had bad luck and was not successful due to many factors. Many people admitted that the failure of the Iridium system was not because of its technology, but the timing that it launched its services. Most people still have a strong belief that satellite-based systems have many important and inreplaceable applications because their operations will not be limited by the geographical locations of users. Also, their services can play an extremely important role in disaster relief operations and military operations, where the fixed terrestrial mobile cellular infrastructure will not be available.

On the other hand, an example of a DS-CDMA-based mobile cellular system is the IS-95A/B system developed by Qualcomm Inc. [25–52], and many other 3G mobile cellular standards, such as UMTS-UTRA [83–88], W-CDMA [89], cdma2000 [53–82], and TD-SCDMA [90–97].

In addition to satellite and mobile cellular systems, DS-CDMA technologies have also been widely applied to many other wireless applications, such as wireless regional area networks (WRANs) (IEEE 802.22 standard), wireless metropolitan area networks (WMANs) (IEEE 802.16 standard), wireless local area networks (WLANs) (IEEE 802.11a/b/g standards), wireless personal area networks (WPANs) (IEEE 802.15 standard) , and vehicle-to-vehicle communications (IEEE 802.11p standard).

The use of DS-CDMA can effectively enhance overall bandwidth efficiency compared with traditional multiple access schemes, such as FDMA and TDMA. Spectrum is extremely expensive; it has to be purchased from various governmental licensing authorities at auction, and sometimes those auctions have involved billions of US dollars (or equivalent monetary value in other currencies). It represents a considerable investment by a service carrier. Therefore, the bandwidth efficiency of a communication technology will be a primary concern for any network operator. The right selection of a suitable multiple access scheme to provide multi-user services is of ultimate importance. DS-CDMA-based mobile cellular carries more calls than TDMA-based technologies. Generally speaking, CDMA[8] will carry between two and three times as many calls simultaneously as TDMA in the same amount of bandwidth. This conclusion was obtained from comparison between the IS-95 and DAMPS systems, both of which were the 2G technologies widely used in North America. This is due to something known as the 'frequency reuse factor', which can be equal to one for a CDMA system. On the other hand, both FDMA and TDMA schemes will use a frequency reuse factor much larger than one, resulting in relatively low efficiency in using limited bandwidth.

The another major advantage of CDMA is its capability for dynamic allocation of bandwidth. To understand this, it is important to realize that in this context in CDMA, 'bandwidth' refers to the ability of any user to get data from one end to the other. It does not refer to the amount of spectrum used by the user, because in CDMA every terminal uses the entire spectrum of its carrier whenever it is transmitting or receiving. On the other hand, TDMA works by taking a channel with a fixed bandwidth and dividing it into several time slots. Any given mobile terminal is then given the ability to use one or more of the slots on an ongoing basis, if it is in a call. For instance, if the channel is 200 kHz wide with eight slots, and the user is allocated one of them, then the user has effective bandwidth of $200/8 = 25$ kHz. This bandwidth is allocated to that user while the call proceeds, no matter whether the user is actually using it or not. In other words, when you are in a call with a TDMA system and being silent because you are listening to the other person speak, your phone still uses that full bandwidth to transmit silence. CDMA is more efficient about that kind of thing. In both TDMA and CDMA, the outgoing voice traffic is digitized and compressed. But the CDMA codec can realize when the particular packet is noticeably simpler (e.g. silence, or a sustained tone with little change in modulation) and will compress the packet in a much better way. Thus the packet may involve fewer bits, and the phone will take less time to transmit it, saving the average bandwidth that each terminal will consume.

And that is where this odd idea of what 'bandwidth' means in CDMA comes in. For in a very real sense, bandwidth in CDMA equates to received power at the cell. CDMA systems constantly adjust power to make sure as little is used as necessary, and compensate for this by using coding gain through the use of forward error correction and other approaches, such as processing gain. The chip rate is constant, and if more actual data is carried by the constant chip rate, then there will be less coding gain. Therefore, it is necessary to use more power instead.

Conceptually, a given cell sector can tolerate a certain amount of total received power before it becomes difficult to decipher all the channels being received. If one mobile terminal uses more of that power allocation, there is less available for the others. But this is an advantage, not a disadvantage, for it can be stated in a different way: if one phone uses less of that power allocation, there is more available for the others. This is the right way to look at it, because this is going on constantly. In a TDMA system, suppose that a mobile terminal needs more or less than the 25 kHz slot. 'Less' is a non-issue because there is no way to get smaller. 'More' would require that an additional slot be allocated to the mobile terminal, which would require a protocol-level exchange: the mobile terminal says to the cell site (or base station) 'I need more bandwidth', the cell finds some other mobile terminal on that same channel and tells it to move, clearing an additional slot, then sends a message back to the mobile terminal telling it 'OK, you can use this slot in addition'. This might take quite a while, and by the time it is complete the need may have passed.

On the other hand, CDMA actually does this dynamically and on the fly. When the CDMA mobile terminal realizes that it does not need to transmit a full digital packet, it will use a 'half-rate' packet,

[8]For description simplicity, we use 'CDMA' to denote 'DS-CDMA' in the discussions following; otherwise we use FD-CDMA or TH-CDMA to represent a CDMA technology different from DS-CDMA.

or 'quarter-rate', or 'eighth-rate', and will transmit for less time. Packet transmissions happen fifty times per second in current CDMA systems (IS-95 standard), but a mobile terminal with a half-rate packet to send will pseudo-randomly send half the symbols during the 20 ms packet. Received power at the cell is an instantaneously measured quantity. If two mobile terminals are transmitting at half rate but at different times, the cell is actually only receiving power from one mobile terminal at a time. Effective bandwidth in CDMA is thus actually being dynamically allocated at all times. And when you are listening and silent, the mobile terminal drops to eighth rate and uses virtually no bandwidth at all. This is very nice for voice traffic and is an additional reason why CDMA is more efficient in use of spectrum, but where it will become particularly valuable is when data transmission becomes a significant use. That is because common data traffic is very bursty, even more than voice traffic. Consider how you use a browser, for instance: you click a URL link and in a short interval your computer downloads many kilobytes of data. You then sit and read what was downloaded, and there is virtually no data traffic going on. In a CDMA system, it would be very easy to allocate a considerable proportion of the bandwidth of a sector to a single mobile terminal for that interval. Nothing special needs to be done except to allocate that mobile terminal a considerable proportion of the power, for which it could do without requesting permission from the cell. High spectrum efficiency and dynamic allocation of bandwidth are the principal reasons why all 3G mobile cellular systems have moved to CDMA. It is noted that the second generation of ETSI system, GSM, is based on TDMA, but its third generation is using a CDMA air-interface, or UMTS-UTRA.

The wide application of DS-CDMA technology in 2G and 3G wireless communications can also be attributed to its relatively low transmit power emission compared with its counterparts, like FDMA and TDMA schemes. The relatively low average transmit power emission required in a DS-CDMA system is a direct consequence of Shannon's theory, which states that the power can be traded with bandwidth to keep a specific data transmission rate or capacity. A DS-CDMA system uses a much widebandwidth than that required by a narrowband TDMA or FDMA system. Therefore, it is natural that a DS-CDMA system needs less power to service users with the same peak data rate as that for a TDMA or FDMA system.

It is known that the average transmit power level in a DS-CDMA mobile terminal is only about one third of that required at a TDMA terminal, under the condition that both operate at the same peak data rate transmission mode. The FDMA terminal requires the highest transmit power level of the three. The lower average transmit power emission is one of the most attractive features of DS-CDMA mobile cellular systems. Many people are extremely concerned about the fact that frequent use of mobile phones in their daily life may cause some diseases, which otherwise are not easily induced, such as brain tumors, etc. The health concern over the abuse of mobile cellular phones is always a very controversial debate topic. Many agencies (mostly under the requests of mobile phone service providers) have claimed that their research has shown there is no clear evidence to prove a direct link between those diseases and the frequent use of mobile cell phones. They concluded that the use of mobile phones is safe and people do not need to worry about it at all. However, still many other independent research groups have found that RF power emission does cause many human cell malfunctions or induces their genetic mutations, which otherwise are not easily trigged.

From the health point of view, the application of DS-CDMA technology is far more attractive than any other air-link technology, such as FDMA and TDMA. Even orthogonal frequency division multiple access (OFDMA) techniques[9] are not comparable to DS-CDMA technology in terms of their average transmit power emission level. The reason is that an OFDMA transceiver uses a simple AM modem for its RF signal transmission, and the AM modem is notorious for its very low power efficiency, thus the average transmit power emission in an OFDMA transceiver must be much higher

[9]It should be noted that there is a big difference between the orthogonal frequency division multiplex (OFDM) technique and the orthogonal frequency division multiple access (OFDMA) technique. The former is only a multiplex technique which is used to multiplex different signal streams for the same user, while the latter is a multiple access technique which can be used to separate different users in a wireless system.

than that of a DS-CDMA transceiver, although the OFDMA technique has many other attractive features, which are different from those of the DS-CDMA technique.

I have done a lot of research on the comparison between DS-CDMA and OFDMA technologies. It is indeed a very interesting and timely research topic, which will have a profound impact on the evolution of future wireless technologies. I have been asked to organize a special issue entitled 'Next Generation CDMA vs. OFDMA for 4G Wireless Applications', which will be published in the June 2007 issue of IEEE Wireless Communications Magazine [6]. For other special issues I have organized on multiple access technologies, readers may refer to [2–4, 150].

2.3 FREQUENCY HOPPING CDMA TECHNIQUES

After having discussed the issues on DS-CDMA techniques, we are going to briefly introduce frequency-hopping (FH) CDMA. Compared to the DS-CDMA technique, the FH-CDMA technique is a relatively less widely used CDMA scheme in real applications. The reason for its less wide acceptance is owing to several factors. First, the FH technique requires a very accurate reference clock in the whole wireless system which uses the FH-CDMA technique for user separation. This accurate network-wide reference clock is very costly to implement using currently available digital technology. Maybe in the future the situation will be different with the advancement in micro-electronics technologies. Second, the hardware to implement an FH-CDMA is still much too complex (as further explained in the text following) compared to DS-CDMA under the same maximum data transmission rate constraint. Therefore, system designers still prefer DS-CDMA to FH-CDMA for most commercial wireless applications.

Similarly, before going in to detail on FH-CDMA technology, we would like to review the major working principles of the FH-SS technique, which is the basis for FH-CDMA technologies.

In addition to the DS-SS technique, another method to spread the spectrum of a data-modulated carrier is to switch the carrier frequency from one to another periodically. Usually, each carrier frequency is selected from a set of frequencies, which are spaced approximately the width of the data modulation bandwidth apart. The spreading code in an FH system does not directly modulate the data-modulated carrier but is instead used to control the appearance sequence of carrier frequencies, because the transmitted signal appears as a data-modulated carrier which is hopping from one frequency to another. In this sense, this type of spread spectrum is called frequency-hopping spread spectrum (FH-SS). In the receiver side, the frequency hopping is removed by mixing (down-converting) with a local oscillator signal that is hopping synchronously (which is the most difficult part of the FH system implementation) with the received signal. The difference between DS-SS and FH-SS signals in terms of time/frequency occupancy is shown in Figure 2.17.

Based on its functions and behaviors, the FH-SS technique is more accurately termed as code-controlled multi-frequency-FSK modulation. It works very similarly to a conventional frequency shift keying (FSK) modulation scheme, except that the set of frequencies is very large. On the other hand, a normal FSK modem often uses only two frequencies. For instance, f_1 is sent to denote a mark or '+1', f_2 is to signify a space or '−1'. In the FH schemes, there will be thousands of frequencies available. A number of a few hundreds to thousands is normal in a real FH system, which makes discrete frequency selections randomly on the basis of a predetermined sequence in combination with the data information conveyed. The number of frequencies and the rate of hopping from frequency to frequency in an FH-SS system are determined by operational requirements for a particular communication application.

The basic structure of an FH-SS system can be described as follows. Usually, an FH system must have a sequence generator and a frequency synthesizer, which is capable of generating the corresponding frequencies according to the sequence generator. As mentioned earlier, it is a difficult part of developing an FH-SS system to design a fast-settling frequency synthesizer with a sufficiently large number of carrier frequencies. Theoretically speaking, the output instantaneous frequency the

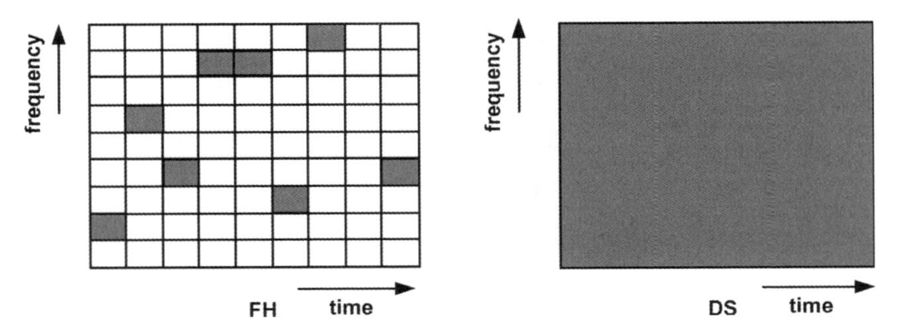

Figure 2.17 Time and frequency occupancy of FH and DS signals.

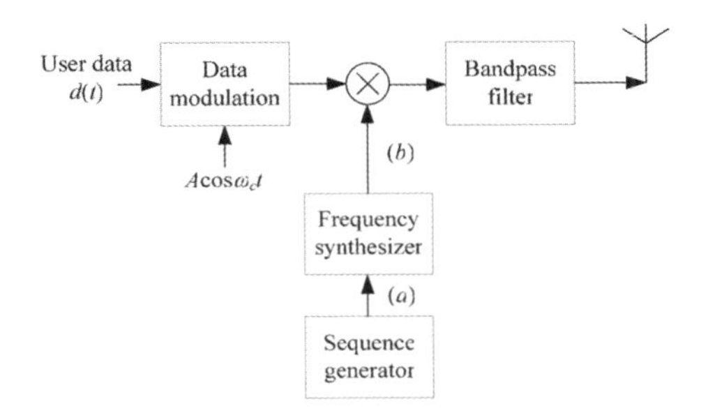

Figure 2.18 Block diagram of an FH-SS transmitter.

frequency synthesizer generates must be a single frequency. This is one of the reasons that make an FH system very difficult and costly to implement. In particular, the frequency synthesizer in a fast-hopping FH system has to work to switch from one frequency to another in a very fast and stable way, especially when the data rate is very high. However, a practical system may produce an output spectrum, which can be a composite of the desired frequency, sidebands generated by hopping, as well as some other spurious frequencies generated as by-products.

Figure 2.18 shows a conceptual block diagram of an FH-SS transmitter. The receiver of the FH-SS system is given in Figure 2.19. The waveforms generated by this simple FH-SS system (in both transmitter and receiver) are shown in Figure 2.20, where it is assumed that the data information is kept at the same level (here all bits are +1 constantly) for illustration simplification.

The FH-SS transmitter shown in Figure 2.18 consists of the following basic blocks, a data modulator, a mixer (denoted simply by a multiplier in the figure), an FH pattern sequence generator, a frequency synthesizer, a bandpass filter, and an antenna. The data modulator will perform the digital modulation between the user data $d(t)$ and a carrier $A \cos \omega_c t$, where A is its amplitude. The frequency synthesizer will work according to the hopping sequences generated by the sequence generator. Usually, the sequence generator can produce a great number of different patterns, each of which will be used by the frequency synthesizer to generate a particular carrier, which will be multiplied with the data modulated signal in the mixer to produce an up-converted transmitting signal from the antenna. Therefore, the carrier frequency of the transmission signal is under the control of the sequence generator, which can also control the FH rate from one frequency to another. The hopping

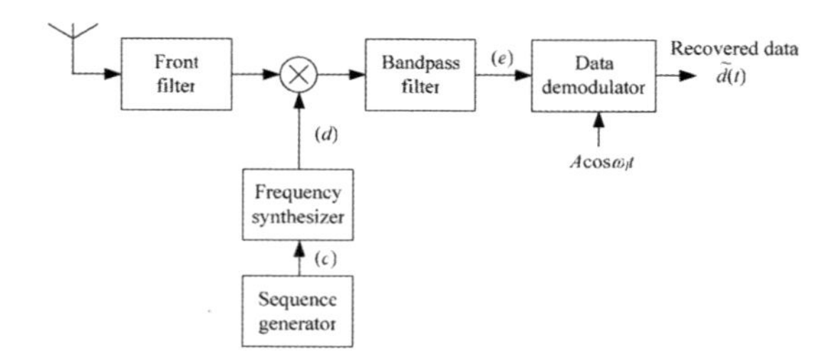

Figure 2.19 Block diagram of an FH-SS receiver.

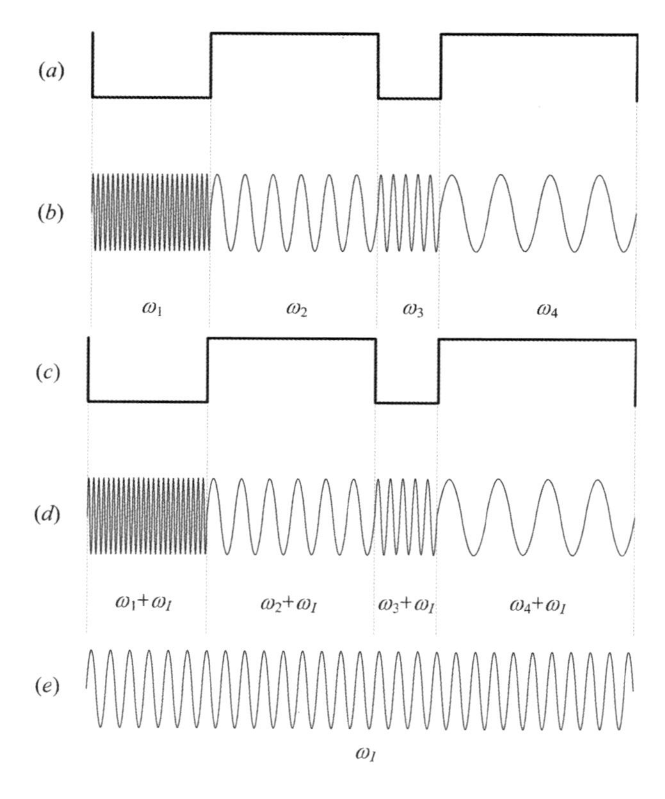

Figure 2.20 Waveforms generated in different points of an FH-SS transmitter (as shown in Figure 2.18) and receiver (as shown in Figure 2.19). (a) The output sequence generated in hopping pattern sequence generator, (b) the output signal from the frequency synthesizer, (c) sequence generated in the local sequence generator at the receiver, (d) the local carrier waveforms generated by the local frequency synthesizer, and (e) the carrier output from the mixer of the receiver.

rate is a very important parameter in an FH-SS system, which will determine if it is a fast-hopping or a slow-hopping FH system.

At the FH-SS receiver, as shown in Figure 2.19, the received signal should first go through a front-end filter, which will be used to reject the image of the carrier frequency produced in the mixer. For the same purpose, the sequence generator will produce a replica of the sequence used by the transmitter and will yield an FH pattern, which should be exactly the same as that used in the transmitter, in the output of the frequency synthesizer. The locally generated FH pattern will be mixed with the received signal to produce a narrowband data-modulated signal with a fixed carrier frequency, which should be equal to the intermediate frequency (IF) ω_I. The output IF signal will be demodulated by a data demodulator to recover the transmitted data information or $\tilde{d}(t)$.

Ideally, the spectrum generated from an FH system should be perfectly rectangular, with spectral lines distributed evenly in every predetermined frequency channel. The transmitter should also be designed to send the same amount of power in each frequency. Otherwise, the detection efficiency on different frequencies will be different, causing decision errors at a receiver.

As shown in Figure 2.19, the received frequency-hopping signal is mixed with a locally generated replica, which is offset by a fixed amount (which is equal to a carrier frequency suitable for the reception process at the receiver, ω_I) such that the output from the mixer in the receiver will produce a constant difference frequency or ω_I if transmitter and receiver code sequences are synchronous.

Similar to the case in a DS-SS system discussed in Section 2.2, any signal that is not a replica of the local reference is spread by multiplication with the local reference, and is never restored into its original narrowband waveform.The Bandwidth of an undesired signal after multiplication with the local reference is approximately equal to the bandwidth before despreading. For instant, an external sinusoidal signal received at an FH receiver will be converted into a signal that will change in the same way as the local reference (an FH carrier), and thus it will never pass the bandpass filter, which is tuned to a fixed carrier frequency or intermediate frequency, say ω_I. On the other hand, if a desirable signal appears at the input side of the receiver, the output signal from the FH despreading unit will be a narrowband signal modulated by a fix carrier ω_I, which will undergo a data demodulation process to recover the originally sent data information or $\tilde{d}(t)$, as shown in Figure 2.19.

Processing Gain of an FH System

The IF mixer and the bandpass filter in the transmitter are effective to reject undesired signal power that lies outside its bandwidth defined by the useful data signal bandwidth. Because this IF bandwidth is only a fraction of the bandwidth of the local FH carrier reference, it can be seen that almost all the undesired signal's power is rejected, whereas a desirable signal is enhanced by being correlated with the local FH carrier reference. In Section 2.2, it was illustrated that a DS-SS system operates identically from the viewpoint of undesired signal rejection and restoration of the desired signal. From this general point of view, DS and FH systems are similar. However, they are different in the details of their operation. Similar to the PG defined in the DS-SS system, we should also define the processing gain for the FH-SS system, and this value will also play an extremely important role in determining the overall performance of an FH system.

The PG of the FH systems should be defined for two different cases. One case is that all generated carrier frequencies are contiguous, and the other is non-contiguous. The non-contiguous carrier frequencies are common for applications when it is hard to find enough spectrum allocation for FH-SS communications. On the other hand, contiguous carrier frequencies are the ideal situation, which can simplify the calculation of the PG. By saying the contiguous FH spectrum, we mean the all carrier frequencies generated by the synthesizer are evenly spaced in the frequency domain.

The PG of an FH system with a contiguous spectrum can be calculated in the same way as a DS-SS system. That is,

$$PG_{FH} = \frac{BW_{RF}}{BW_{\text{information}}} \tag{2.25}$$

where BW_{RF} is the bandwidth spanned by all carrier frequencies generated by the synthesizer collectively, and $BW_{\text{information}}$ is the bandwidth given by the original baseband information signal, which is determined by the data signal $d(t)$.

On the other hand, if the carrier frequencies generated by the synthesizer are not contiguous, an objective measure of PG can be

$$PG_{FH} = \text{The total number of available frequencies} \tag{2.26}$$

which can also be used to calculate the PG value for an FH system with contiguous carrier channels. For example, an FH system containing 1000 frequencies can have 30 dB available PG. The approximation has been used to simplify the calculation for PG, as given in Equation (2.25), because all guide bands in between two carrier channels are neglected in the formula. If the guide bands in the IF bandwidth can not be omitted, Equation (2.26) should be used instead.

Hopping Rate of an FH System

The FH rate and number of carrier frequencies are determined by the sequence generator. The minimum frequency switching rate of an FH-SS system is determined by the following system parameters:

1. The bandwidth of information to be sent and its importance.
2. The amount of redundancy needed.
3. The environment where the FH system will work in terms of severity of the interferences.

Information in an FH system can be sent in any way available to the other systems. Usually, however, some form of digital signal is preferred, no matter whether the information is a digitized analog signal or data. Assume that some digital rate is prescribed and that FH has been chosen as the SS technique. Then, the question is how to determine the FH rate or the chip rate. It is to be noted that, different from a DS system, there is no chip in an FH system. The term chip is used here just for imitation.

An FH system should possess a sufficiently large number of frequencies on demand. The actual number required depends on the system operational environment. For example, two thousand frequencies will provide satisfactory operation when interference and noise are evenly distributed at every available frequency channel. For equal distribution of interference or jammers in every channel, the interference power required to block communications has to approach two thousands times that of the desired signal power. In other words, the achievable jamming margin is about 33 dB in this case. Unless some sort of redundancy that allows for bit decisions based on more than one frequency is needed, a single narrowband interferer will cause an error rate of 0.5×10^{-3}, which is generally satisfactory for normal digital data transmission. For a simple FH system without any form of transmitted data redundancy (one hop exists in each bit duration), the expected error probability can be approximated by $\frac{J}{N}$, where J is equal to the number of continuous wave (CW) interferers whose power is greater than or equal to signal power, and N is the number of frequencies available to the FH system. Error rate probability for an FH system, in which we assume that binary FSK modulation is used (here two frequencies denote binary symbols or $f_1 = +1$ and $f_0 = -1$), can be approximated by the following expression

$$P_e = \sum_{n=r}^{N_c} \binom{N_c}{r} p^n q^{N_c - n} \tag{2.27}$$

where p is the error probability of a single jamming trial, which is J/N, J denotes the number of jamming carriers, N is the number of channels available to the FH system, q is the probability of no error for a single jamming trial (we always have $q = 1 - p$), N_c is the number of hops in each bit,[10] and r is the number of wrong chip decisions necessary to cause a bit error. A chip decision error is defined as the situation that interference power in a '+1' channel exceeds the power in an intended '−1' channel (or vice versa) by some amount ε that is sufficient to cause a decision error.

An FH system is called a slow frequency hopping spread spectrum system if only one or less than one hop happens in each bit or symbol duration. Otherwise, if more than one hopping exists, a fast frequency hopping system results.

A fast-hopping FH system can offer much better performance than a slow-hopping FH system at the price of system implementation complexity. For instance, the implementation cost for a fast-hopping FH system with three hops per bit will be at least three times more complex than that of a slow-hopping FH system with one hop per bit. The synthesizer should work at least three times faster than that of a one-hop-per-bit system. The amount of data a transceiver should process would also been at least three times more than that of a one-hop-per-bit system.

The performance of a three-hop-per-bit FH system can be evaluated by the following method. Assume that the decision rule for this system is on the basis of two out of three decision rule. That is, if at least two frequencies are correct, the decision will be made in favor of the symbol or bit representing the FH patterns containing the two correct frequencies. Thus, a single jamming trial will cause not more than

$$\binom{3}{2} p^2 q^{3-2} = 3p^2 q \tag{2.28}$$

errors, where $\binom{n}{m}$ denotes the number of combinations of taking n from m items, p and q represent the probability of error caused by a single jamming trial and the probability of no error caused by a single jamming trial, respectively, and $p = 1 - q$. Consider an FH system with a total of two thousand carrier frequencies, p will be $\frac{1}{2000}$ and $q = 1 - p = 0.9995$. Thus, the error probability will become $3\left(\frac{1}{3000}\right)^2\left(1 - \frac{1}{3000}\right) \cong 7.5 \times 10^{-7}$, which is much better than 0.5×10^{-3} given by the previous one-hop-per-bit FH system.

Application of FH-CDMA: Bluetooth Technology

After having talked about the FH-SS technique, we are ready to go further to discuss the FH-CDMA technology. We will mainly address the issues on the applications of FH-CDMA.

As mentioned earlier, FH-CDMA is much less widely used than DS-CDMA due to its implementation complexity under the same maximum data transmission rate or capacity requirements. Therefore, the application of FH-CDMA is only found in some wireless systems which require a relatively small coverage area and thus relatively low transmit power. A typical example of a wireless system using FH-CDMA is Bluetooth technology [151–162].

Bluetooth is a recently conceived communication standard that allows wireless connectivity between Bluetooth-enabled computing devices. Bluetooth allows devices to communicate via short-range radio links, removing the restrictions of wires, cables, and line-of-sight requirements. The desirable features of Bluetooth include robustness, low complexity, low power, and low cost. The Bluetooth specification is primarily based on many aspects of the IEEE 802.11 standard [151]. The range of Bluetooth is about 10 meters. A range of 100 meters can also be attained by increasing the transmitter power. Bluetooth allows multiple devices to interact simultaneously in wireless networks called piconets, with each piconet consisting of two to eight devices. Each piconet has a single master device and one or more slave devices. A device may belong to more than one piconet, either

[10] The value of N_c will determine fast FH or slow FH. Usually, N_c can be a non-integer. For instance, $N_c = 1/2$ means that one hop happens in every two bits, implying a slow FH system. On the other hand, if $N_c > 1$, a fast FH system is resultant.

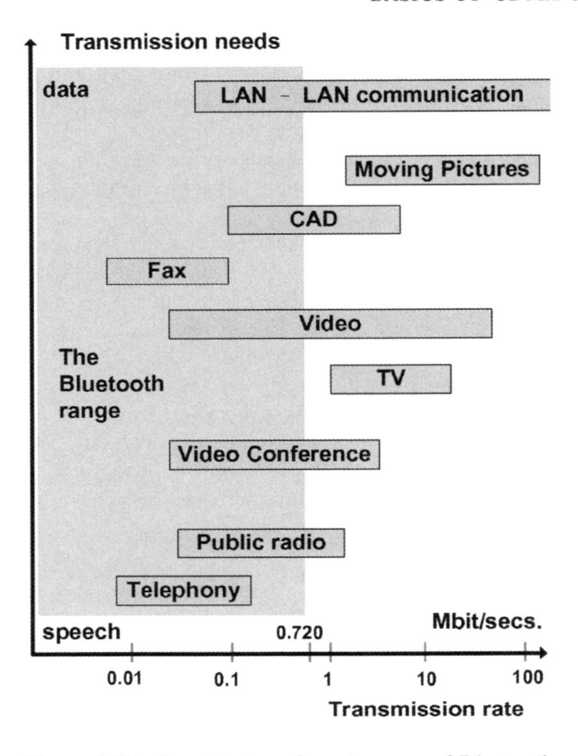

Figure 2.21 The data transfer rate range of Bluetooth.

as a slave in both or as a master of one piconet and a slave in another. A scatternet is the term that describes a network with devices belonging to more than one simultaneously.

The Bluetooth specification has a raw data rate of 1 Mbps. However, because of the overhead reserved for the various Bluetooth protocols, the available net data rate is approximately 723 kbps. Bluetooth can support an asynchronous data channel, up to three simultaneous synchronous voice channels, or a channel which simultaneously supports asynchronous data and synchronous voice. Each voice channel is a synchronous link allowing 64 kbps in each direction. The asynchronous channel can support an asymmetric link of up to 723.2 kbps in either direction while permitting 57.6 kbps in the return direction, or a 433.9 kbps symmetric link. The actual data rates depend on the kind of error correction capability employed, which is a characteristic of the type of packets being transferred [151]. Figure 2.21 shows the spectrum where Bluetooth can be used. The data transfer rate supported by Bluetooth is ideal for voice and other less demanding data rates. However, video data requires a higher bit rate than provided by Bluetooth. Video conferencing is slightly above the Bluetooth data rate, although video conferencing having a bit rate at the lower end of the spectrum could possibly be supported.

The Bluetooth architecture is conceptually segmented into multiple layers. Here we can only briefly describe the functionality of each layer. Figure 2.22 shows an abstract view of the Bluetooth system architecture.

The radio layer acts as the interface to the communication channel allowing various computing devices to communicate. To allow universal communication, the Industrial, Scientific and Medical (ISM) band has been selected for Bluetooth. In most countries the ISM band extends from 2400 to 2483.5 MHz. The radio layer sends information using FH-CDMA, based on the FH-SS technique.

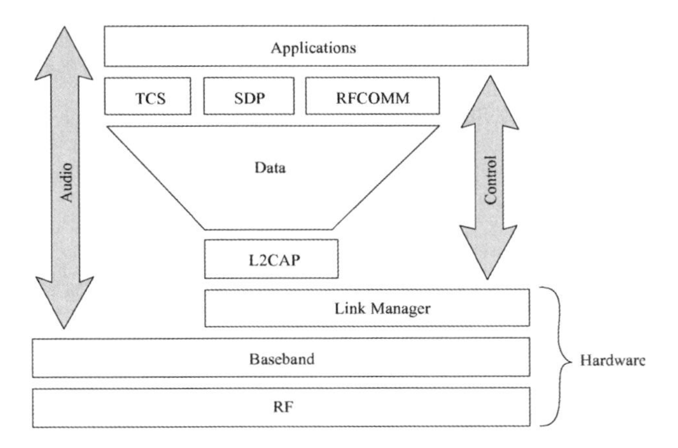

Figure 2.22 A conceptual view of the Bluetooth architecture [152].

Multiple access capabilities are paramount in a standard such as Bluetooth, because multiple users can transmit a spread-spectrum signal at the same frequencies. CDMA is popular because it provides privacy, protection against multipath interference, anti-jamming capabilities, and a low probability of interception (LPI). In CDMA each sender is assigned a unique code sequence that encodes the information-bearing signal. The receiver also knows the code sequence of the sender and decodes the received signal upon reception, recovering the original data. This is possible since the cross-correlation between the codes of the desired senders and the codes of the other senders is small. The encoding process enlarges or spreads the spectrum of the original signal. Therefore, the bandwidth of the coded signal is chosen to be much larger than the bandwidth of the information-bearing signal, which is the reason that CDMA is also known as spread-spectrum modulation [153], as mentioned in the previous subsections.

FH-CDMA is used in Bluetooth. In contrast to DS-CDMA, FH-CDMA transmits short bursts of data over one frequency and then 'hops' to another frequency and sends another data burst. Thus, FH-CDMA uses only a small part of the bandwidth when it transmits, but the frequency location of the transmission differs in time. Bluetooth uses fast frequency hopping (FFH), which implies that the hopping rate is much greater than the symbol rate [154]. FFH is opposed to slow frequency hopping (SFH), in which the hopping rate is less than the symbol rate. The synchronization disadvantage of DS-CDMA is reduced in FH-CDMA, because FH-CDMA synchronization has to be within a fraction of the hop time. Since spectral spreading does not use a very high hopping frequency but rather a large hop-set, the hop time will be much longer than the chip time of a DS-CDMA system. Thus, an FH-CDMA system allows a larger synchronization error [153]. The Bluetooth standard employs frequency hops fixed at $2402 + k$ MHz, where $k = 0, 1, \ldots, 78$, with a nominal hop rate of 1600 hops per second. This results in a single hop slot of 625 microseconds. The modulation scheme is Gaussian pre-filtered binary FSK [154].

The baseband layer is a state machine that controls the radio layer. These control duties include providing the frequency hop sequences to the radio, encrypting for secure links, and packet handling over the wireless link. The baseband controller can establish two types of links. A Synchronous Connection Oriented (SCO) link synchronously transfers data, e.g. voice. The other type of link is called Asynchronous Connectionless (ACL), which is intended for data transfer applications that do not require a synchronous link. The baseband layer provides the functionalities required for devices to synchronize their clocks and establish connections. It also provides inquiry procedures for discovering the addresses of devices within close proximity. Five different channel types are provided for control information, link management information, user synchronous data, user asynchronous data, and isosynchronous (real-time) data. Additional functions of the baseband layer include generating encryption keys, link keys, and controlling error correction for packets.

There are many other layers defined in the Bluetooth specifications, such as Link Manager layer, Logical Link Control and Adaptation Protocol (L2CAP) layer, SDP, TCS, and RFCOMM layers. Due to space limitations, we will not introduce them here. For more information, readers are advised to refer to [151–162], given at the end of this book.

2.4 TIME HOPPING CDMA TECHNIQUES

Having discussed the two CDMA techniques, DS-CDMA and FH-CDMA, we would like also to take a look at another CDMA technique, time hopping (TH) CDMA, in this section. Again we would like to start with the TH-SS technique.

The TH technique, in fact, works in a very similar way as a digital modulation scheme called pulse position modulation (PPM). In other words, the time hopping is nothing but a type of pulse position modulation in the sense that a code sequence is used to key the transmitter on and off, as shown in Figure 2.23, where the times for the transmitter to switch on and off follow a specific pseudo-random code sequence. The average duty cycle of 'on' and 'off' in the transmitter can go as large as 50%. The major difference between PPM and TH-SS lies in the fact that the former uses pulse position patterns to represent the data information symbols, whereas the latter denotes a particular code sequence, which acts as a secret key to decode the data information hidden therein. A conceptual block diagram of a TH system is shown in Figure 2.23, where the on-off switch logic unit in the transmitter is used to control the positions that the sent pulse will hop from one position to another. The receiver in Figure 2.23(b) should also follow exactly the same time-hopping pattern to capture all transmitted power from the transmitter. Obviously, an unintended receiver will not be able to receive all power from a particular TH transmitter due to its unknown pulse-hopping pattern. On the other hand, if a hostile jammer wants to block the communications established in a TH-SS system, it is impossible to block all transmitted pulses without knowledge of the transmitter's hopping patterns unless it keeps on transmitting all the time, requiring a much higher transmission power than that of a TH transmitter.

The TH-SS technologies are not as popular as the other two SS techniques, i.e. the DS-SS and FH-SS techniques. The main reason is implementation difficulties, especially for the pulse generator,

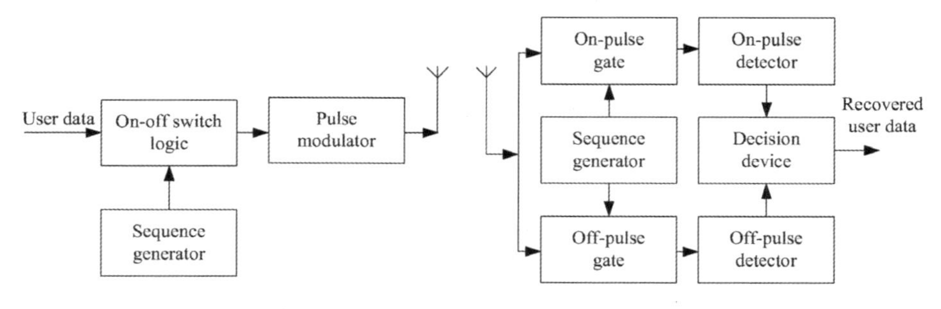

Figure 2.23 Block diagram of a TH-SS transceiver, where the sequence generator is used to control the TH patterns that should change synchronously in both the transmitter and the receiver.

which is the core of a TH-SS system and should be able to produce a train of very narrow impulses with its width being at an order of nano-seconds.[11] The pulse generator should also provide very good timing accuracy, such that the PPM can be effectively applied to code different SS sequences for multiple access. However, it is still a challenging task to make such a pulse generator even today.

Usually, the TH-SS technique seldom works independently in an SS system (except for the case of an ultra-wideband (UWB) system, a technology developed based on the TH technique). Instead, it usually works with some other SS modulation schemes, in particular the FH technique, which has been discussed in the previous section, to result in a time-frequency hopping SS scheme.

The difference separating time-frequency hopping and pure FH lies in the fact that in FH systems the transmitted frequency changes at each code chip time, whereas a time-frequency hopping system may change frequency and/or amplitude only at one or zero transition instants in the code sequence. It is seen from Figure 2.23 that the simplicity of the modulator is obvious. Any pulse-modulating signal source that is capable of following code sequences is eligible as a time-hopping modulator. TH may be used to aid in reducing interference between systems in time-division multiplexing (TDM). However, stringent timing requirements must be placed on the overall system to ensure minimum overlap between transmitters. This is one of the reasons that make a TH system much harder to implement than other SS systems. Also, as in any other SS systems, the spreading sequences must be designed or selected carefully from the viewpoint of their cross-correlation properties. As mentioned earlier, a simple TH-SS system can be blocked by a jammer that uses a continuous carrier at the signal center frequency. The primary advantage offered is in the reduced duty cycle. In other words, to be an effective jammer an interfering transmitter has to be forced to transmit continuously (assuming that the TH sequence used by the time-hopper is unknown to the interferer). The power required by a legitimate time-hopper will be less than that of an interfering transmitter by a factor that should be equal to the PG of this TH system. Because of this relative vulnerability to interferences, a simple TH transmission should not be used for anti-jamming unless combined with FH to prevent single-frequency interferers from causing significant losses.

However, the TH-SS techniques have been found useful in ranging, multiple access, or other special applications. A typical example for such applications is the UWB technologies, which have gained tremendous attention recently, due to their many attractive properties, such as their unique capability to mitigate multipath propagation problems based on their very high time resolution. The issues relating to UWB technologies will not be addressed in more detail in this book due to space limitations.

In the TH-CDMA system, a pseudo-noise sequence defines the transmission moments, rather than the transmission frequency as FH does. As mentioned earlier, the data signal in time-hopping CDMA is transmitted in rapid bursts at time intervals determined by the code assigned to the user. The time axis is divided into frames, and each frame is divided into M, time slots. During each frame the user will transmit in one of the M time slots. Which of the M time slots is transmitted depends on the code signal assigned to the user. Since a user transmits all of its data in one, instead of M, time slots, the frequency it needs for its transmission has increased by a factor M. A block diagram of a TH-CDMA system is given in Figure 2.24.

Figure 2.25 shows the time-frequency plot of the TH-CDMA system. We see from Figure 2.25 that TH-CDMA uses the whole wideband spectrum for short periods instead of parts of the spectrum all of the time.

[11] A nano-second is equal to 10 to minus nine power, or 10^{-9} seconds, whose bandwidth spans at least as wide as 1 GHz or more.

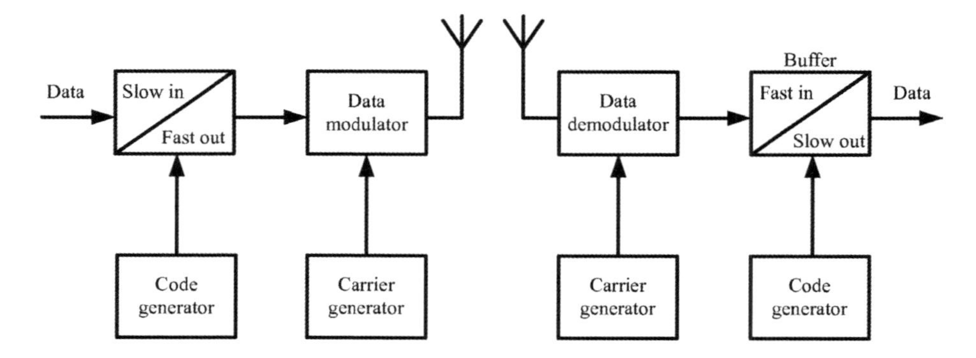

Figure 2.24 Block diagram of a TH-CDMA transmitter and receiver.

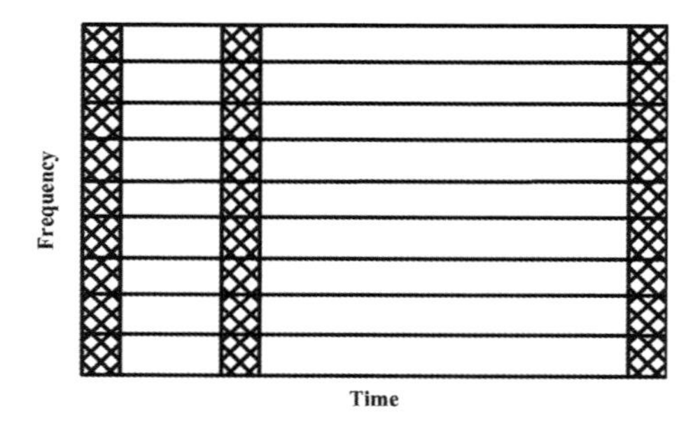

Figure 2.25 Time-frequency plot of the TH-CDMA.

2.5 SPREAD SPECTRUM OR TIME?

Having discussed three major CDMA schemes, we would like to give a brief summary of them as follows.

- The DS-CDMA technique by far is the most popular technique in all CDMA communication applications. There are several reasons for its popularity. First, it is the simplest form of the CDMA technique and can be implemented at a relatively low cost compared with other techniques, such as FH-CDMA and TH-CDMA. All currently available major 3G mobile communication standards are based on the DS-CDMA technique with almost no exception. Second, a DS-CDMA system can also be designed to operate compatibly with many existing communication networks that operate based on other multiple access technologies, such as time-division multiple access (TDMA), to achieve so called smooth upgrading. The proposal of a 3G system or WCDMA [83, 89] was based mainly on this clause as an effort to achieve smooth upgrading from the 2G system named Global System for Mobile Communications (GSM) [105–129] working on TDMA technology.[12]

[12]It should be noted that this claim may not be true to some people because upgrading from GSM to W-CDMA requires in fact a substantial change in network equipment.

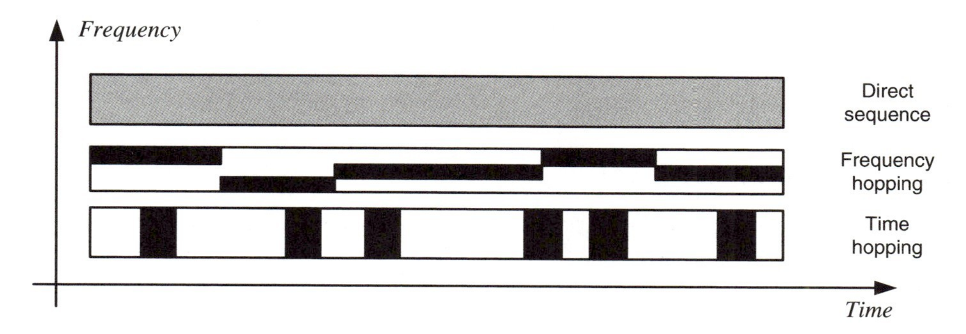

Figure 2.26 Comparison of three basic SS techniques, DS-SS, FH-SS, and TH-SS in terms of their time and frequency occupancy.

- The FH-CDMA technique is less widely used in civilian communication systems compared to the DS-CDMA technique. The FH technique has been applied to GSM networks [105–129] as an option to mitigate the frequency selective fading that may happen in some places, especially in urban and downtown areas, where multipath interference may be serious. The FH-CDMA technique has been applied to the Bluetooth standard [151–162], which requires relatively low data rates. The implementation of an FH-CDMA system relies on an accurate and fast-settling frequency synthesizer, which can be costly if using current state-of-the-art RF and microelectronics technology. However, FH techniques have been widely used in military communication systems, such as battlefield command systems.
- The TH-CDMA technique is even less widely used than either DS-CDMA or FH-CDMA. The reason is partly because it may suffer from serious interference problems if there exists a continuous transmission in the coverage area, as the TH system only works in an on-and-off fashion in a frame. For this reason, the TH technique usually works together with other SS techniques, in particular the FH technique, forming a hybrid TH-FH system. In particular, TH technique has been the basis of UWB technologies.

Figure 2.26 gives a comparison of three basic SS techniques, DS-SS, FH-SS, and TH-SS, in terms of their time and frequency occupancy.

Before ending this section, we would like also to throw several questions to readers in an effort to stimulate some more research interest. All current CDMA technologies were developed based on spread spectrum techniques. Then, what can we get from 'spread time'? As time and frequency form a dual with each other, can a 'spread time' system provide us with similar processing gain to mitigate various impairing factors in the channels, just as SS does? If yes, then how to implement such a 'spread time' system? There have been some research works reported in this particular area, which unfortunately remain very preliminary studies.

2.6 CHARACTERISTIC FEATURES OF CDMA SYSTEMS

The characteristic features of a CDMA system are associated with the properties of several important elements to construct a CDMA system. The major elements to form a CDMA system include CDMA codes, spreading spectrum scheme, spreading modulation technique, carrier modulation, signal processing technique used in both transmitter and receiver, channel condition, and so forth.

In the previous section, we have discussed extensively about the properties of CDMA codes in terms of their correlation functions, including auto-correlation function and cross-correlation function. Therefore, we will not touch the same issues here again. Instead, we would like to take a closer look at how a CDMA system can operate with its unique features and why.

2.6.1 PROCESSING GAIN

To understand various properties of a CDMA system, a very important parameter, called processing gain (PG), for any CDMA system should be well understood. As mentioned earlier, in a DS-SS system, the data is modulated by a spreading signal which occupies much a wider bandwidth than the original data signal. Since multiplication in the time domain corresponds to convolution in the frequency domain, a narrowband signal multiplied by a wideband signal ends up as a wideband signal. One way of doing this is to use a binary waveform as a spreading function, at a higher rate than the data signal, as shown in Figure 2.6. To make it clearer, we plot Figure 2.27.

Here the three signals shown in Figure 2.27 corresponds to the data signal $x(t)$, the spreading signal $c(t)$, and the message signal $m(t)$ as discussed above. The first two signals are multiplied together in a spreading modulator (just a multiplier with two inputs) to give the third waveform. Bits of the spreading signal $c(t)$ are called chips. In Figure 2.27, T_b represents the period of one data bit and T_c represents the period of one chip. The chip rate, $1/T_c$, is often used to characterize a spread spectrum transmission system and also a CDMA system alike. The PG, or sometimes called the spreading factor (SF), is defined as the ratio of the information bit duration over the chip duration, or

$$PG = SF = T_b/T_c \tag{2.29}$$

Therefore, the PG or SF simply represents the number of chips contained in one complete data bit duration. Higher PG means more spreading and thus a wider bandwidth. In general, a high PG also means that more codes can be allocated on the same frequency channel.

2.6.2 PSEUDO-NOISE SEQUENCES

Until now we have not discussed what properties of the spreading signals we would like to have. This depends on the type of CDMA system we want to implement. Let us first consider a system where we want to use spread spectrum to avoid jamming or narrowband interference.

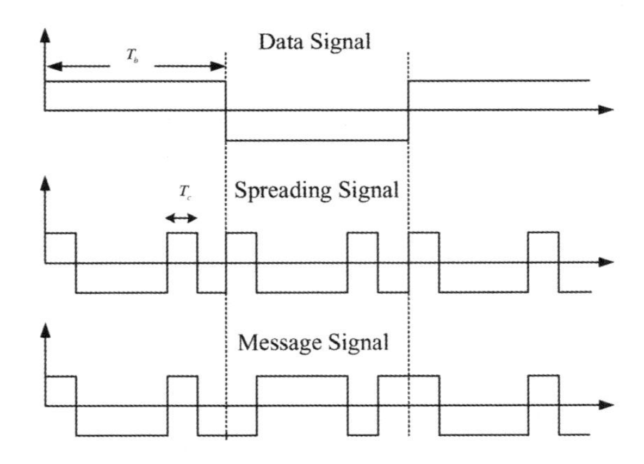

Figure 2.27 Illustration of direct-sequence spreading signal.

We can easily show that, if we want the signal to overcome narrowband interference (which is usually a kind of external interference), the spreading function needs to behave like noise. Random binary sequences (also called pseudo-noise (PN) sequences) are such functions. They have the following important properties:

- They usually have balanced chips, or they have an equal number of 1s and 0s.
- They always give a single peak auto-correlation function.

In fact, the auto-correlation function of a random binary sequence is a triangular waveform as shown in Figure 2.28, where T_c is the period of one chip. Hence, the power spectral density of such a waveform is a sinc function squared, with first zeros at $\pm 1/T_c$.

PN sequences are periodic sequences that have noise-like behavior. They can be conveniently generated using shift registers, modulo-2 adders (i.e. XOR gates) and feedback logic circuitry. Figure 2.29 illustrates a simple block diagram for a PN sequence generator.

The maximum length of a PN sequence is determined by the length of the register and the configuration of the feedback network. An N-bit register can take up to 2^N different combinations of zeros and ones. Since the feedback network performs linear operations, if all the inputs (i.e. the content of the flip-flops) are zero, the output of the feedback network will also be zero. Therefore, the all-zero combination will always give zero output for all subsequent clock cycles, and thus we do not include it in the sequence. Therefore, the maximum length of any PN sequence is $2^N - 1$ and sequences of that length are called maximum-length sequences or m-sequences. They are useful because longer sequences have better properties. Feedback configurations for m-sequences are tabulated and can be found in the literature [7].

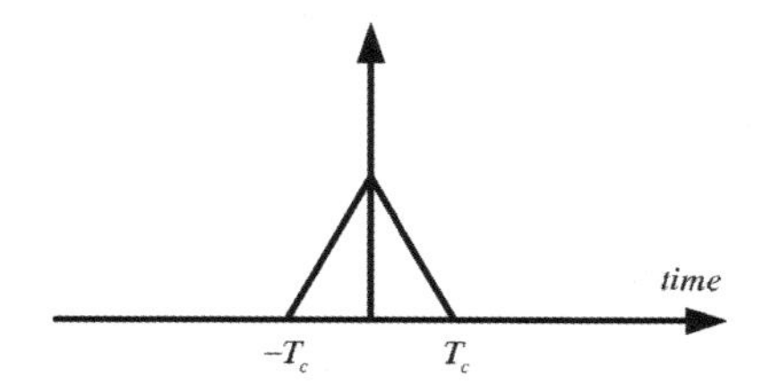

Figure 2.28 Auto-correlation function for a pseudo-noise sequence.

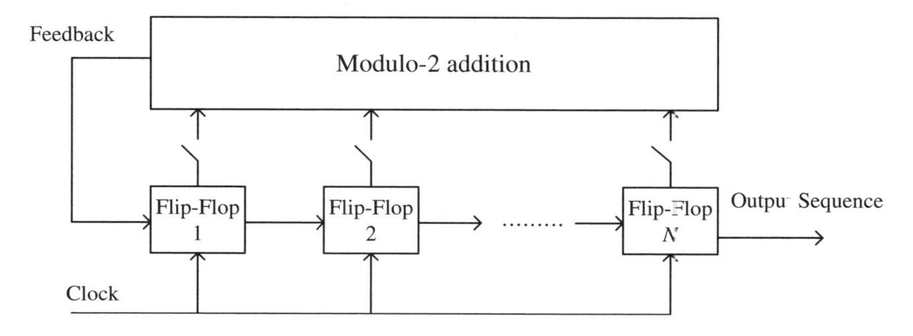

Figure 2.29 Block diagram for a pseudo-noise sequence generator using a shift register and feedback logic circuitry.

In summary, PN sequences are therefore periodic noise-like binary functions generated by a network of feedback loops, modulo-2 adders, and flip-flops. Maximum length PN functions have a period of $2^N - 1$.

2.6.3 MULTIPLE ACCESS CAPABILITY

One of the most important functions for CDMA technologies is to provide multiple access to many different users which share the same air-link infrastructure. In the case of a cellular mobile system, a CDMA-based system can accommodate different mobile terminals in the same cell and they can operate together without introducing much interference with each other.

The fundamental concept behind multiple access is to permit a number of users to share a common channel. As mentioned earlier, two traditional forms of multiple access are frequency division multiple access (FDMA) and time division multiple access (TDMA). In the FDMA scheme, the frequency band is divided into channels. Each user gets one frequency channel assigned that is used at will. It could be compared to AM or FM broadcasting radios where each station has a frequency channel assigned. FDMA demands good inter-channel filtering in order to avoid mutual interference. To play safe, in most FDMA system a guide band should be used in between two neighboring channels. This guide band will inevitably introduce inefficiency in frequency utilization.

On the other hand, in the TDMA scheme, the frequency band is not partitioned but users are allowed to use it only in predefined intervals of time or time slots, one at a time. Thus, a TDMA system demands very good synchronization among the users and it can be very complicated to ensure such a strict time synchronization among all users working in the same TDMA system.

CDMA, for code division multiple access, is different from those traditional ways (either FDMA or TDMA schemes) in that it does not allocate frequency or time in user slots but gives the right to use both to all users simultaneously. To do this, it needs the help of spread spectrum techniques, such as direct-sequence, frequency-hopping, and time-hopping techniques. In effect, each user is assigned a code which spreads its signal bandwidth in such a way that only the same code can recover it at the receiver end. This method has the property that the unwanted signals with different codes get spread even more by the process, making them appear like noise to the receiver.

As discussed earlier, spread spectrum is a means of transmission where the data occupies a larger bandwidth than necessary. Bandwidth spreading is accomplished before the transmission through the use of a spreading code which is independent of the transmitted data. The same code is used to demodulate the data at the receiving end. A simple DS spreading unit consists of a multiplier with two inputs, one being the data signal $x(t)$ and the other the spreading signal $c(t)$. The output from the multiplier is the message signal to be transmitted, or $m(t)$. In the same way, we can have FH spreading or TH spreading, which we will not discuss further here, for brevity.

Originally developed for military use to avoid jamming (interference created on purpose to make a communication channel unusable), spread spectrum modulation now has found wide application in civilian wireless communication systems for its superior performance in an interference-dominated environment.

Obviously, the multiple access property of a CDMA system is directly connected with the properties of the CDMA codes it uses. We have discussed more on the CDMA code properties in terms of their correlation functions in the previous section, but we have not discussed the detail of the CDMA codes used in a CDMA system. We only give a brief discussion on them here (in the next subsection), and a very detailed description of CDMA codes and their properties will be given in Chapter 4.

The advantage of CDMA for wireless communication services is its ability to accommodate many users on the same frequency at the same time. As we mentioned earlier, a specific code is assigned to each user and only that code can demodulate the transmitted signal. Based on the technologies used in current 2G and 3G CDMA-based systems, we know that there are usually two ways of separating users in a CDMA system: (1) using orthogonal codes to implement multiple access, or (2) using non-orthogonal multiple access techniques, which usually happens in asynchronous CDMA channels.

Orthogonal Multiple Access

In a CDMA system using orthogonal multiple access, each user is assigned one or many orthogonal waveforms derived from an orthogonal code.[13] Since the waveforms are orthogonal, users with different codes do not interfere with each other, if they operate in a channel without multipath propagation and all users share a synchronous channel, in which all signals will transmit their bit streams aligned in time in terms of their bit duration. Orthogonal CDMA (O-CDMA) requires strict synchronization among the users, since the waveforms are orthogonal only if they are aligned in time.

An important set of orthogonal codes is the Walsh set. Walsh functions are generated using an iterative process of constructing a Hadamard matrix, starting with $H_1 = [0]$. The Hadamard matrix is built by

$$H_{2n} = \begin{pmatrix} H_n & H_n \\ H_n & \bar{H}_n \end{pmatrix} \tag{2.30}$$

For instance, the Walsh-Hadamard codes of length 2 and 4 can be given, respectively, as follows

$$H_2 = \begin{pmatrix} 0 & 0 \\ 0 & 1 \end{pmatrix} \tag{2.31}$$

and

$$H_4 = \begin{pmatrix} 0 & 0 & 0 & 0 \\ 0 & 1 & 0 & 1 \\ 0 & 0 & 1 & 1 \\ 0 & 1 & 1 & 0 \end{pmatrix} \tag{2.32}$$

From the corresponding matrix, the Walsh-Hadamard codewords are given by the rows. Note that we usually map the binary data to vector form such that we can use real number arithmetic when computing the correlations. Thus, 0s are mapped to 1s and 1s are mapped to -1s. Walsh-Hadamard codes are important because they form the basis for orthogonal codes with different spreading factors. This property becomes useful when we want signals with different spreading factors to share the same frequency channel. The codes that possess this property are called orthogonal variable spreading factor (OVSF) codes. To construct such codes, it is better to use a different approach than matrix manipulation. Using a tree structure allows better visualization of the relation between different code lengths and orthogonality between them.

To illustrate the orthogonality between two Walsh-Hadamard codewords, we can test them as follows. For example, let us see if the second codeword of W_2, which is the second column of the matrix of (2.32), and the third codeword of W_3, which is the third column of the matrix of (2.32), are orthogonal. Hence we get the following two codewords, in vector form, as

$$\begin{cases} W_2 \Rightarrow (1, -1, 1, -1) \\ W_3 \Rightarrow (1, 1, -1, -1) \end{cases} \tag{2.33}$$

Computing the orthogonality, we get (multiplying element by element):

$$W_2 \otimes W_2 = (1 \times 1) + ((-1) \times 1) + (1 \times (-1)) + ((-1) \times (-1)) = 1 - 1 - 1 - 1 = 0 \tag{2.34}$$

[13]Here, we should be very careful to understand the meaning of the word 'orthogonal'. As a matter of fact, the word 'orthogonal' is a bit misleading because so-called orthogonal codes (such as Walsh-Hadamard codes and orthogonal variable spreading factor (OVSF) codes) used in all current CDMA systems will not be orthogonal at all if they are used in an asynchronous channel.

where the operator ⊗ denotes the element-by-element multiplication operation between two vectors or sequences. Therefore, we can see that the Walsh-Hadamard codewords W_2 and W_3 are indeed orthogonal with each other.

However, the auto-correlation function of Walsh-Hadamard codewords does not have good characteristics. It can have more than one peak, therefore it is not possible for the receiver to detect the beginning of the codeword without an external synchronization scheme. The cross-correlation function of the Walsh-Hadamard codewords can also be non-zero for a number of time shifts and unsynchronized users can interfere with each other. This is why Walsh-Hadamard codes can only be used in synchronous CDMA systems. Therefore, the Walsh-Hadamard codes do not have the best spreading behavior or correlation property.

Non-Orthogonal CDMA

The concept behind a non-orthogonal CDMA system is to give up orthogonality among users to reduce the interference by using spread spectrum techniques. PN sequences are used to spread the spectrum. The family of PN sequences, called Gold sequences, are in particular popular for non-orthogonal CDMA. Gold sequences have only three cross-correlation peaks, which tend to get less important as the length of the code increases. They also have a single auto-correlation peak at zero, just like many other ordinary PN sequences.

Gold sequences (codes) can also be constructed from the modulo-2 addition of two maximum length preferred PN sequences. By shifting one of the two PN sequences, we get a different Gold sequence. This property can be use to generate codes which will permit multiple access on the channel.

The use of Gold sequences permits the transmission to be asynchronous. On the other hand, a receiver can be synchronized using the good auto-correlation property of the Gold sequences. For more information on CDMA codes, please refer to Chapter 4.

2.6.4 PROTECTION AGAINST MULTIPATH INTERFERENCE

Another salient feature of CDMA technology is its unique capability to mitigate one of the most formidable impairing factors in mobile channels, i.e. the inter-symbol interference (ISI) caused by multipath propagation through a wireless channel. Once again, we would like to stress that here we are only talking about the CDMA technology based on the DS spreading technique. Of course, the CDMA technologies based on other SS techniques, such as FH or TH, also possess certain capability to mitigate the multipath problem.

The capability of CDMA technology to overcome multipath interference is based on the theory of spreading signal bandwidth to a much wider spectrum than the coherent bandwidth[14] of the multipath channel.

In order to understand how a CDMA scheme can overcome the multipath effect, we should provide a simple introduction to some related channel propagation mechanisms as follows.

In a radio link, the RF signal from the transmitter may be reflected from objects such as hills, buildings, or vehicles. This gives rise to multiple transmission paths at the receiver. Figure 2.30 shows some of the possible ways in which multipath signals can occur. The relative phase of multiple reflected signals can cause either constructive or destructive interference at the receiver. This is experienced over very short distances (typically at half wavelength distances), and thus is given the term fast fading. These variations can vary from 10 to 30 dB over a short distance. Figure 2.31 shows the level of attenuation that can occur due to the fading (normally it obeys the Rayleigh distribution). The Rayleigh distribution is commonly used to describe the statistical time varying nature of the

[14]The coherence channel of a multipath channel is determined by the reciprocal of the maximal delay spread of the channel.

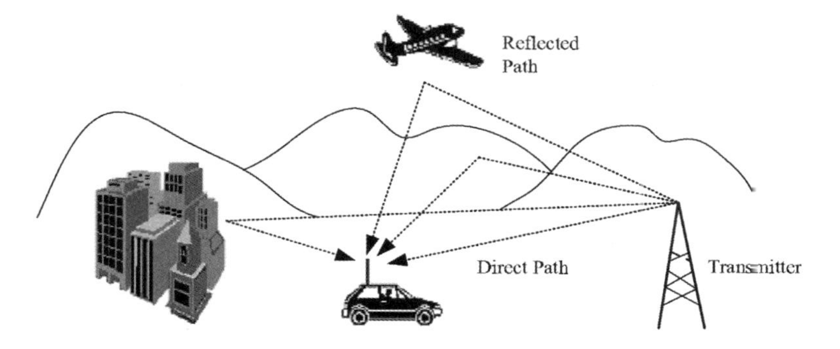

Figure 2.30 Multipath propagation in a mobile channel.

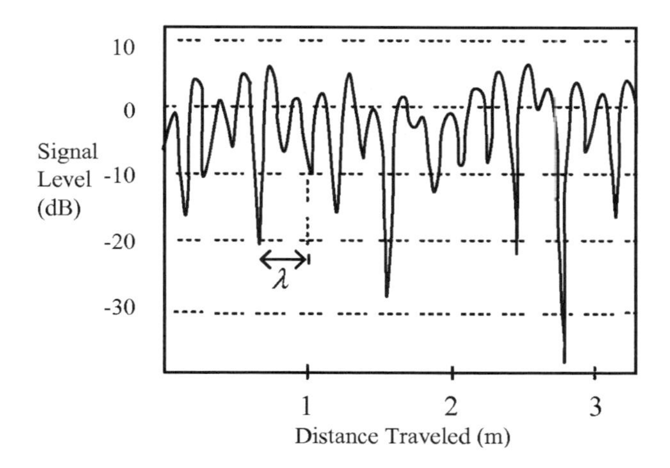

Figure 2.31 Typical Rayleigh fading while a mobile unit is moving (for a carrier frequency at about 900 MHz).

received signal power. It describes the probability of the signal level being received due to fading. Table 2.4 shows the cumulative distribution function of the signal level for the Rayleigh distribution.

The received radio signal from a transmitter consists of typically a direct signal, plus reflections off objects such as buildings, mountains, and other structures. The reflected signals arrive at a later time than the direct signal because of the extra path length, giving rise to slightly different arrival times, spreading the received energy in time. Delay spread is the time spread between the arrival of the first and last significant multipath signals seen by the receiver. In a digital system, the delay spread can lead to ISI. This is due to the delayed multipath signal overlapping with the following symbols. This can cause significant errors in high bit rate systems, especially when using TDMA technology. Figure 2.32 shows the effect of ISI due to delay spread on the received signal. As the transmitted bit rate is increased the amount of ISI can also become more severe. The effect starts to become very significant when the delay spread is greater then about 50% of the bit duration. Table 2.5 shows the typical delay spread for various environments. The maximum delay spread in an outdoor environment is approximately 20 μs, thus significant ISI can occur at bit rates as low as 25 kbps.

Table 2.4 Cumulative distribution function for Rayleigh distribution.

Signal level (dB about median)	% Probability of signal level being less then the value given
10	99
0	50
−10	5
−20	0.5
−30	0.05

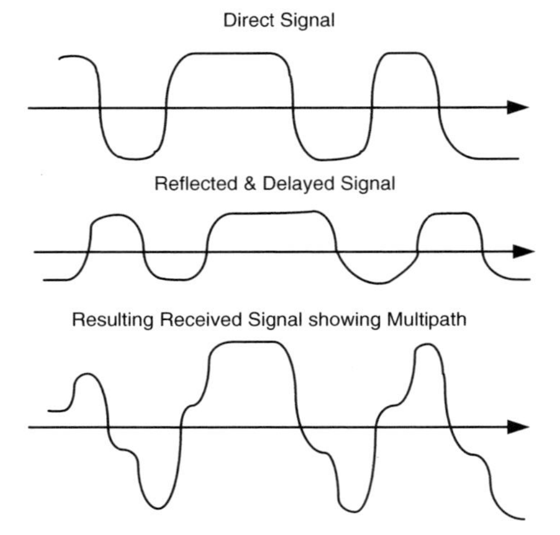

Figure 2.32 Delay spread caused by multipath propagation effect.

Table 2.5 Typical delay spread in different operational environments.

Environment	Delay spread	Maximum path length difference (m)
Indoor (room)	40 nsec–200 nsec	12–60
Outdoor	1 μsec–20μ sec	300–6000

ISI can be minimized in several ways. One method is to slow down the symbol rate by reducing the data rate for each channel (i.e. split the bandwidth into more channels using OFDM). Another is to use a coding scheme that is tolerant to ISI such as CDMA.

A multipath channel will cause frequency-selective fading if the signal sent to the channel has a bandwidth wider than the coherence bandwidth of the channel. In any radio transmission channel, the channel spectral response is usually not flat. It has dips or fades in the response due to reflections causing cancelation of certain frequencies at the receiver. Reflections off near-by objects (e.g. ground, buildings, trees, etc.) can lead to multipath signals of similar signal power as the direct signal. This can result in deep nulls in the received signal power due to destructive interference.

For narrow bandwidth transmissions if the null in the frequency response occurs at the transmission frequency then the entire signal can be lost. This can be partly overcome in two ways. By transmitting a wide bandwidth signal or spread spectrum as CDMA, any dips in the spectrum only result in a small loss of signal power, rather than a complete loss. Another method is to split the wideband transmission into many narrow bandwidth carriers, as is done in a COFDM/OFDM transmission. The original signal is spread over a wide bandwidth and so nulls in the spectrum are likely to affect only a small number of carriers rather than the entire signal. The information in the lost carriers can be recovered by using forward error correction techniques working jointly with an interleaver.

Therefore, it is clearly seen that CDMA uses a totally different approach to mitigate the multipath interference, which is very difficult to overcome in FDMA and TDMA. On one hand, we can explain that CDMA overcomes multipath interference by spreading its signal in a much wider bandwidth than its original baseband signal. Thus, if part of the signal bandwidth unfortunately falls into the nulls in the channel frequency response function, the loss is still manageable due to the much narrower width of nulls compared to the width of the CDMA signal spectrum. On the other hand, the capability of CDMA technology to overcome multipath interference can also be clearly explained in terms of time domain signal processing. In the time domain, a CDMA receiver always makes use of a RAKE receiver structure, which looks like a rake used by farmers to collect grains in a field. A RAKE always consists of several fingers. Thus, a RAKE receiver just mimics the function of a RAKE to describe how it works in a CDMA receiver. A typical configuration of a RAKE receiver is illustrated in Figure 2.33.

Obviously, the word RAKE is not an acronym and, in fact, it is not always capitalized as it is in this book. RAKE derives its name from its inventors Price and Green in 1958. When a wideband signal is received over a multipath channel, the multiple delayed replicas appear at the receiver as depicted in Figure 2.34. By attaching a 'handle' to the plot of the multipath returns, a picture of an ordinary garden rake is created. It is from this picture that the RAKE receiver gets its name.

A RAKE receiver is a CDMA receiver technique which uses several baseband correlators to individually process several multipath signal components. The correlator outputs are coherently (or non-coherently) combined to achieve improved communications reliability and performance.

As to be shown in the next chapter, which introduces the IS-95 standard, both the base station and the mobile receivers use RAKE receiver techniques. Each correlator in a RAKE receiver is

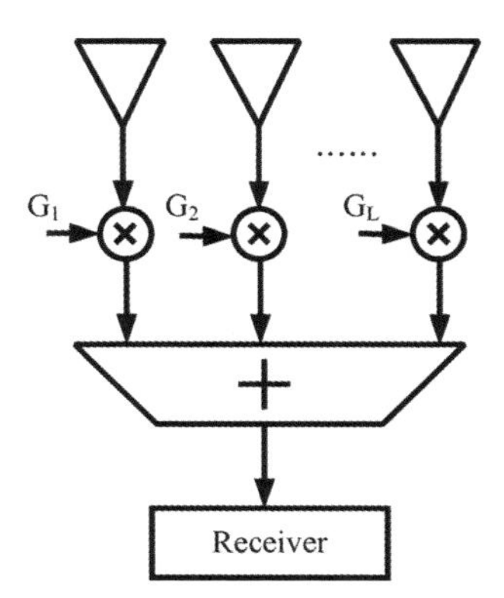

Figure 2.33 Generic RAKE receiver structure, where $G_1, G_2, \ldots$ and G_L are weight coefficients.

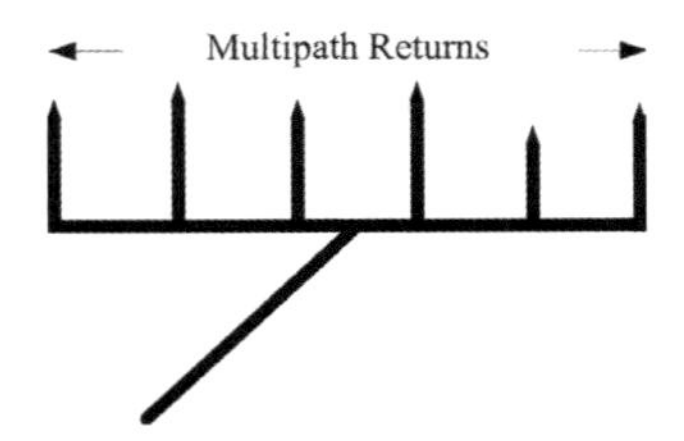

Figure 2.34 How does RAKE mimic a rake in a farm field?

called a RAKE-receiver finger. The base station combines the outputs of its RAKE-receiver fingers non-coherently i.e. the outputs are added in power. The mobile receiver combines its RAKE-receiver finger outputs coherently, i.e. the outputs are added in voltage. The mobile receivers usually have three RAKE-receiver fingers and base station receivers have four or five, depending on the equipment manufacturer. There are two primary methods used to combine the RAKE-receiver finger outputs. One method weights each output equally and is therefore called equal-gain combining (EGC). The second method uses the data to estimate weights which maximize the signal-to-noise ratio (SNR) of the combined output. This technique is known as maximal-ratio combining (MRC). In practice, it is not unusual for both combining techniques to perform about the same, if the multipath returns have the same energy.

Therefore, we can say that a RAKE receiver is a radio receiver designed to counter the effects of multipath fading. It does this by using several 'sub-receivers' (or fingers), each delayed slightly in order to tune in to the individual multipath components. Each component is decoded independently, but at a later stage combined in order to make the best use of the different transmission characteristics of each transmission path. This could very well result in higher SNR (or Eb/No) in a multipath environment than in a 'clean' environment.

The use of a RAKE receiver in a CDMA system in fact realizes multipath diversity in the time domain. This means that a CDMA system can overcome frequency-selective fading due to the fact that it spreads its signal to a much wider bandwidth such that the loss in the nulls of the channel response becomes insignificant compared with the total CDMA signal power. Therefore, this unique capability of CDMA technology has made it extremely attractive in all 3G wireless communications.

2.6.5 *INTERFERENCE/JAMMING REJECTION*

A CDMA system can also offer a very effective means to reject unwanted interference in the same channel. This capability stems from the spread spectrum technique, or the direct sequence spreading technique. It is noted that here we will discuss two different issues, one being 'interference rejection' and the other 'jamming rejection'. The difference between the two lies only in the fact that the former concerns interference in general (both malacious and or unintentional), while the latter refers only to malicious or hostile interference generated on purpose. To make the discussions simpler, we will not differentiate the two below. In addition, it is noted that, if we talk about interference, then it can be either wideband or narrowband.

Now let us look at a narrowband interference signal to a particular CDMA receiver. A typical form of such narrowband interference can be a single-carrier signal whose frequency is just the same as the center frequency of the CDMA system, whose system model is shown in Figure 2.35 and whose generic receiver model is illustrated in Figure 2.36. It is noted that the DS-CDMA system model shown in Figure 2.35 has K users and the receiver should be tuned to one particular user signal (or the first one, as shown in Figure 2.36). It is further assumed that in addition to the channel noise ($n(t)$) there are no other impairing factors in the channel to simplify our discussion here.

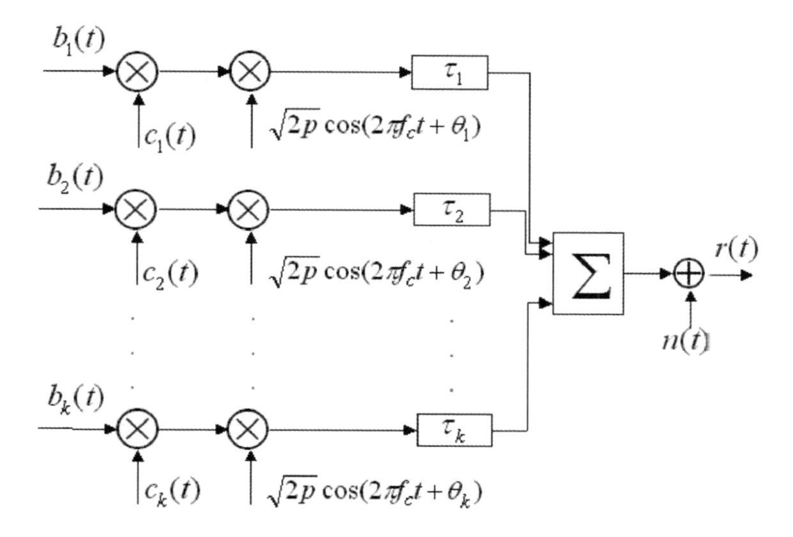

Figure 2.35 DS-CDMA system model with K users.

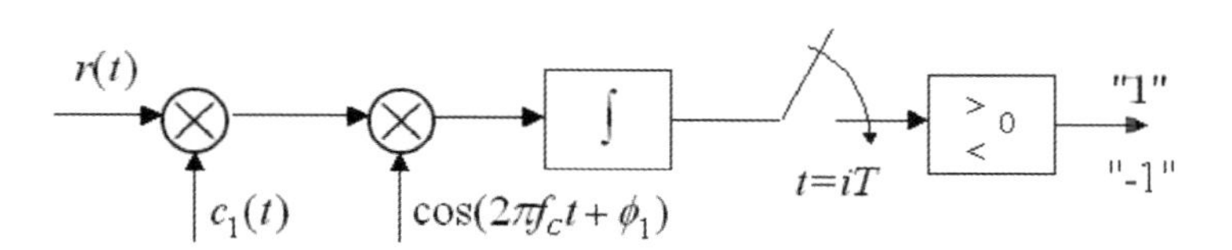

Figure 2.36 A generic CDMA receiver.

In Figure 2.35 $b_k(t)$, where $k = 1, 2, \ldots, K$, is the data signal from the kth user, $c_k(t)$ is the signature code for the kth user, and $r(t)$ is the combined signal viewed at a receiver, with useful signals and noise integrated. If there is no external interference, a CDMA receiver as shown in Figure 2.36 will be able to decode wanted information from $r(t)$, after the received signal is processed by despreading, carrier demodulation, low-pass filtering, sampling, and decision making. Now, consider the case where there is a narrowband interference, such as a narrowband modulated signal (e.g. an AM modulated signal) with its carrier frequency being exactly the same as the center frequency of the CDMA system of concern. Therefore, the input signal to the CDMA receiver will be the addition of $r(t)$ and the narrow band interfering signal, say $i(t)$, whose bandwidth should be much narrower than that of the CDMA signal. Obviously, if $i(t)$ goes into the first stage of the CDMA receiver, it will multiply with the local reference signature code $c_1(t)$. The output from the despreading unit of the receiver will consist of three major components, one being the useful part, which is the result of the despreading operation against the transmission of the first user signal, the second part being the multiple access interference component, which is generated from the despreading process against all other (not the first) users' transmissions, and the last part being the result from multiplication of $i(t)$ and the local reference signature code $c_1(t)$. The last part, or $i(t)c_1(t)$, clearly is not a narrowband signal any more, and thus it will not be able to go through the following low-pass filter. Therefore, the narrowband interference has been rejected.

Next, let us examine the case of wideband interference. A typical type of wideband signal is a spread digital modulated signal, such as the one generated in Figure 2.35. Therefore, in fact, such a wideband interference signal can be viewed as the signal sent from one of the K users, for instance, the second user or $\sqrt{2p}b_2(t)c_2(t)\cos(2\pi f_c t + \theta_k)$. We already know that the second user signal will not be able to pass through the low-pass filter, due to the fact that this signal will still be a wideband signal after going through the despreading unit at the receiver. Therefore, the capability to reject multiple access interference will make it possible to allow the CDMA receiver to reject the wideband interference, which resembles any one of the signals sent from the kth transmitter. For any other forms of wideband interference we can also draw the same conclusion that a CDMA receiver is able to reject most wideband interference.

The superior capability to reject both interference and jamming signals is one of the most desirable technical features a CDMA system can provide. This has made CDMA a very popular choice in all 3G wireless communications, due to the possible coexistence with many other wireless systems in the same or very close neighbor bands.

2.6.6 PRIVACY

Privacy is another important property for a CDMA system. Compared to FDMA and TDMA, CDMA provides a much greater degree of freedom to offer better security in a wireless communication system.

Take a CDMA-based cellular mobile system as an example. The communications between a mobile terminal and a base station in a CDMA cellular mobile system are under full control of the base station or mobile terminal switching office (MTSO). All air-link channels in such a CDMA-based mobile cellular system are encoded by different signature codes, which are known only to the mobile terminal of concern and the base station.A third party needs to know a particular signature code on a real-time basis before it can intercept the communications between the mobile terminal and the base station. It is admitted that the number of signature codes used for a specific CDMA-based mobile cellular standard is limited. For instance, the IS-95 standard uses 64-ary Walsh-Hadamard sequences as the downlink channel signature codes. Therefore, it is understandable that a third party can use trial-and-error to decode the information carried in the air-link of a certain mobile terminal and the base station. However, it usually is very difficult to do so if the signature codes used by a particular pair of mobile terminal and base station are changing in a predetermined way known only to the mobile terminal and its base station. Therefore, it can be said that the real-time tracking of the use of the signature codes for a particular pair of mobile terminal and base station is not an easy task.

On the other hand, an intercepter may easily follow the pattern of the frequency or time slots once it gets to know the way in which all frequency or time slots are assigned to different mobile terminals in an FDMA- or TDMA-based mobile cellular system. In this sense, CDMA-based wireless communication systems can offer much better privacy to the end-customers compared with systems using other multiple access technologies (such as FDMA and TDMA). In addition, a CDMA mobile cellular system can work with many different encryption algorithms to further enhance the privacy of the communications.

2.6.7 LOW PROBABILITY OF INTERCEPTION

The low probability of interception is another salient feature for spread spectrum techniques, which first found applications in military communication systems due to this important technical feature.

It should be noted that the low probability of interception has something to do with privacy, but they function in a slightly different way. The low probability of interception property in a

CDMA-based wireless communication system is due mainly to the fact that the average transmission power level in a CDMA system can be made much lower that that of an FDMA or TDMA system, which will make it very difficult for a third party to detect the existence of its transmission signals under a noisy and interference-prone environment. Therefore, the low probability of interception property of a CDMA wireless system is a direct consequence of the application of spread spectrum techniques (such as direct sequence, frequency-hopping, and time-hopping techniques). On the other hand, the privacy of the system implies security of the communications and has nothing to do with the low average emission level.

The capability of the low probability of interception in a CDMA-based wireless communication system is achieved according to Shannon's channel capacity theory, which specifies that the capacity of a communication channel can be traded between transmission power and bandwidth of the transmitted signal. In other words, in order to achieve a certain channel capacity, we can either raise the transmission power level under the condition that a fixed bandwidth is used, or spread the signal bandwidth with the transmission power level fixed. Therefore, in the context of CDMA communications, a much wider spectrum has been used to transmit the user's information than necessary, and thus a certain transmission rate can still be achieved using a relatively low transmission power level, compared to that used in either FDMA or TDMA systems. In this way, a third party will find it very difficult to detect the presence of a CDMA signal spectrum due to its very low power spectrum density function because of the use of spread spectrum technique.

This property of a CDMA wireless communication system has made it possible to allow its operation to overlay with other existing radio systems, as to be discussed in the next subsection.

2.6.8 OVERLAY WITH EXISTING RADIO SYSTEMS VERSUS COGNITIVE RADIO

When CDMA was first introduced to civilian communication applications in the 1980s, mobile cellular operators in North America faced a serious problem of spectrum shortage because most suitable spectrum sections had already been allocated to other radio operators, such as power line transmission systems in the continent. Obviously, it would be very difficult and expensive to reallocate a new spectrum section for those existing radio users within a short time.

Therefore, the proposal for a CDMA mobile cellular standard (later called IS-95) by Qualcomm Inc. just provided a reasonable solution to the spectrum reallocation dilemma because a CDMA air-link architecture can operate at a much lower power spectrum density level than any of the competing technologies, such as TDMA-based mobile cellular design, such that it may easily be deployed without introducing serious interference to all existing primary radio users, which had been operating for many years there. Also, partly for this reason, the IS-95 standard received tremendous attention in the community when it was first proposed for discussion.

The possibility to overlay with existing radio systems is still a very attractive feature of a CDMA-based radio system even today. Many radio system designers are still very much interested in this unique feature of CDMA technology to design a wireless system which operates within a very crowded radio spectrum.

It is noted that the possibility to overlay with existing radio systems has given CDMA technology a new role to play in the newly emerging cognitive radio technology, which gives a radio terminal high intelligence to understand, know, detect, and adapt to the radio environment in which it operates on a real-time or nearly real-time fashion. The fundamental requirement for the operation of a cognitive radio terminal is not to interfere with existing radio users or so-called incumbent users or licensed users.

Cognitive radio has been proposed to implant some kind of intelligence into a radio terminal that can automatically sense, recognize, and make wise use of any available radio frequency spectrum at a given time. The use of the available frequency spectrum is purely on an opportunity driven basis.

In other words, it can utilize any idle spectrum sector for information exchange and stop using it immediately if the primary user of the spectrum sector needs it. Thus, cognitive radio sometimes also is called 'smart radio', or 'frequency agile radio', 'police radio', 'adaptive software radio', etc. For the same reason, cognitive radio techniques can in many cases make possible licensed exempt use of spectrum that is otherwise not in use or lightly used without infringing upon the rights of licensed users or causing harmful interference to licensed operations.

The core of cognitive radio is the cognitive algorithms it may use. Therefore, it is not difficult to know that a fully functional cognitive radio should have the ability to do the following tasks:

1. Tune to any available channel in the target band.

2. Establish network communications and operate in all or part of the channel.

3. Implement channel sharing and power control protocols which adapt to spectrum occupied by multiple heterogeneous networks.

4. Implement adaptive transmission bandwidths, data rates, and error correction schemes to obtain the best throughput possible.

5. Implement adaptive antenna steering to focus transmitter power in the direction required to optimize received signal strength.

The most important part of a cognitive radio is its inherent intelligence, which makes it different from any normal wireless terminal available toady, in either 2G or 3G systems. This intelligence will allow a cognitive radio to scan all possible frequency spectra before it makes an intelligent decision on how and when to make use of a particular sector of the spectrum for communications. Therefore, it is inevitable that a cognitive radio needs a very great signal processing power to deal with the huge amount of data it captures from various radio channels. Thus, the capability to process all those enormous amount of data on a real-time or quasi-real-time basis is a must for any cognitive radio terminal.

It is still too early to specify exactly the algorithms a cognitive radio should use at the moment when this book is being written. However, we would like to show some evidence of how a primitive cognitive radio may behave. A cognitive radio has to use the following two protocols for its very basic operation: (1) dynamic frequency selection (DFS), and (2) transmit power control (TPC). DFS was originally used to describe a technique to avoid radar signals by 802.11a networks which operate in the 5 GHz unlicensed national information infrastructure (U-NII) band. Now, it has been generalized to refer to an automatic frequency selection process intended to achieve some specific objective (like avoiding harmful interference to a radio system with a higher regulatory priority). On the other hand, TPC was originally a mechanism for 802.11a networks to lower aggregate transmit power by 3 dB from the maximum regulatory limit to protect earth exploration satellite systems (ESSS) operations. Now it has been generalized to a mechanism that adaptively sets transmitter power based on the spectrum or regulatory environment. These two protocols will become a must for all cognitive radios. In addition, a cognitive radio should have incumbent profile detection (IPD) capability, which is another key cognitive radio behavior. IPD is the ability to detect an incumbent user (one with regulatory priority) based on a specific spectrum signature. The operation of IPD bears the following characteristics:

1. DFS requires an IPD protocol to identify unoccupied or lightly used frequencies.

2. IPD includes detection schemes focused on the characteristics of the specific incumbents in the band, or bands, that the cognitive radio is designed to support.

3. IPD eliminates the need for geolocation techniques (GPS, etc.) to determine the location of the radio and identify unused channels using a database.

After having reviewed the basic concept of cognitive radio, we should be able to see the important role that CDMA technology may play in the design of a real operational cognitive radio system/network in the near future. Obviously, it is noted that a successful cognitive algorithm used in a cognitive radio terminal relies on its capability to sense and communicate with the environment without introducing unnecessary interference to the incumbent users working in the same bandwidth or time. Any cognitive radio terminal should work only on an opportunity basis. That is to say, it should grab the opportunity if the bandwidth or time slot is unused and stop using it immediately if it senses that the incumbent user is starting to transmit. No matter how fast a system can respond, any overlapping between the transmission of a cognitive radio terminal and the incumbent users can be dangerous and harmful, as it may interfere with the communications among the incumbent users, which may have a high priority. Therefore, the best strategy to minimize the interference is to allow cognitive radio terminals to work on a transmission power level as low as possible, such that any overlapping in transmissions between a cognitive radio terminal and the incumbent users will not introduce harmful interference with each other.

In this perspective, CDMA is the best choice as an air-link multiple access technology for cognitive radio systems/networks. The main reason is that CDMA is the only technology which can provide users with a sufficiently high data throughput without using too high transmission power due to its powerful processing gain. In this sense, both FDMA and TDMA may not be best suited for application in a cognitive radio system/network as the primary multiple access technology. On the other hand, orthogonal frequency division multiple access (OFDMA) technology may not be a suitable choice either, due to the fact that it uses amplitude modulation in its RF carrier modulation and demodulation, and thus it needs a relatively high transmission power (due to the unavailability of processing gain) to maintain the same data transmission rate compared to that of a CDMA system. Therefore, we can see that CDMA technology will play an extremely important role in the design of a cognitive radio architecture in the near future.

2.6.9 LOW POWER EMISSION TO REDUCE HEALTH RISK

In modern society, people are becoming ever more cautious to any possible hazards to their health. Radio transmission is one of the possible hazards that people are worrying about very much.

It is often said in many reports in the media that there is no clear evidence to show a direct linkage between radio frequency transmissions in a mobile cellular system or any other wireless application and danger to human health. I am not an expert in the field of research on radio environmental pollution, and thus I should not say that all such reports are groundless. However, the increasing use of radio appliances in our daily life, such as wireless LAN, mobile cellular phone, cordless phone, and so on, has obviously increased the danger of certain mutations in our human cells, which has been shown in many research reports. Clearly, it is not possible to ask people to reduce the use of these wireless appliances, which have become a part of their daily life. However, how to reduce the average amount of such radio frequency exposure is what we need to consider. In this sense, CDMA technology is still a superior choice if human health is a priority to think about, let alone many other technological advantages of the technology as discussed earlier.

It is technically true that a CDMA mobile phone will effectively reduce the amount of radio frequency power emission which may directly affect us when we use the mobile cellphone in our daily life. Some reports have shown that the average RF emission from a CDMA mobile cell phone is only about one third of that from a TDMA-based cell phone with almost the same technical features. We will be very interested to see any new reports on the comparison between CDMA radio terminals and OFDMA terminals in the future. Unfortunately, I deeply doubt that an OFDMA mobile terminal will not exceed the CDMA terminal in terms of its average radio frequency power emission if operating under similar technical specifications. The reason is simple and straightforward, because the OFDMA technology does not provide any processing gain and thus is unable to lower its transmission power to retain a certain transmission rate.

2.7 MULTI-CODE AND M-ARY CDMA TECHNIQUES

Multi-code CDMA is a derivative of traditional CDMA architecture. As we know, a traditional CDMA system operates on a one-code-per-user basis, in which every user is assigned one code as its signature code to distinguish itself from others. The separation among all users is implemented by the orthogonality or quasi-orthogonality among the signature codes used in a CDMA system. Therefore, the spreading coding in a traditional CDMA system fulfills two functions. One is to spread the signal bandwidth to achieve certain processing gain and the other is to give a unique identity to each user for code division multiple access.

Therefore, one-code-per-user operation is one of the most important characteristics of a traditional CDMA architecture. The design of one-code-per-user architecture has its limitations. First, in this scheme the user transmission rate can hardly change if all signature codes used in the same system have the same code length.[15] This is not what we want if we want to design a wireless communication system pertaining to multimedia applications, where different users may request different transmission rates at different times. Another limitation of a traditional one-code-per-user design of a CDMA system is that, if we assume that each cell can accommodate in total M users, the cell-wise capacity will not be fully used if there are fewer than M active users. The reason is simple as every user can only use one signature code to encode its data and the use of any other codes is not allowed.

A multi-code CDMA system was proposed to solve the problem. In a multi-code CDMA system, a user can be assigned multiple signature codes to encode its information data whenever needed. In this way, every user has a greater degree of freedom in terms of the number of signature codes used to encode its data information according to its requirements on data transmission rate. In such a system, it is at least theoretically possible to allow a user to use all M codes to encode its information if all of the other users are not active at the moment, and thus maximum cell-wise capacity is always achievable no matter how many active users are present. Of course, this is only the extreme case and may not happen very often in a real multi-code CDMA system. However, one of the most attractive functions of a multi-code CDMA system is to allow each user to send its required data transmission rate at any time by choosing an appropriate number of signature codes for multimedia communications.

It should be noted that each CDMA system (no matter whether or not it is a multi-code CDMA system) has its own limitation on how many signature codes can be used simultaneously in the same CDMA channel, and this limitation is determined by the average multiple access interference caused by non-ideal cross-correlation functions among different signature codes. Therefore, it is clear that the introduction of multi-code CDMA will not increase the total number of active users supportable by a CDMA system. Nevertheless, the unique capability of multi-code CDMA to support multimedia signal (such as voice, data, and video) transmission has made it a very attractive CDMA system architecture, which will find wide application in the future.

Several different multi-code CDMA schemes have been proposed in the literature [163–169]. More specifically, they include (1) orthogonal code system, (2) multi-code system, (3) parallel combinatorial system, and (4) BPSK M-ary CDMA system, which are introduced below.

2.7.1 *ORTHOGONAL CODE SYSTEM*

The first multi-code CDMA scheme was named the orthogonal code system and was proposed in 1987 by Enge and Sarwate [163]. The basic idea for an orthogonal code system is described as follows. Assume that there are H different spreading codes available to a particular user in the system. Each transmitter will first group the original information data stream into symbols, each of which will consist of m bits such that $m = \log_2 H$. The transmission from the transmitter will choose one from

[15] In most cases, we assume that the signature code length is just equal to the processing gain of a CDMA system if short-code spreading is presumed.

H different codes, depending on the bit patterns of a symbol (which contains m bits). The mapping strategy of the orthogonal code system is shown in Figure 2.37, where there are eight spreading codes (A, B, C, D, E, F, G, H) and each code will be selected to denote a particular data bit pattern of (d_0, d_1, d_2). More spreading codes could be used to encode more bits in each symbol. Obviously, in this scheme a transmitter will only send one particular spreading code each time to denote a particular data bit pattern in a symbol. The states number, spreading code number, and bits per symbol (m) in the orthogonal code system are shown in Table 2.6.

The block diagrams for the transmitter and receiver in a orthogonal code system are shown in Figures 2.38 and 2.39, respectively. In the transmitter diagram, there are in total H spreading codes, which are denoted $(V_1, V_2, \ldots, V_H)$. The transmitter should select one of the H spreading codes for a particular input data bit pattern in each symbol. In the receiver block diagram, the received signal $r(t)$ should be despread by all H spreading codes, as the receiver has no way of knowing in advance which spreading code will appear at a given time. The output from the H despreading units will be fed into a processor to select the largest value among the H inputs. This largest value will be considered as the code the transmitter sent and will then be sent into a de-mapping unit to decode the sent information using the mapping relation shown in Figure 2.37.

Performance analysis of an orthogonal code system was carried out by Enge and Sarwate in 1988, as shown in reference [164], where the authors studied the bit error probability of an orthogonal code

Data			Spread code							
d_0	d_1	d_2	A	B	C	D	E	F	G	H
0	0	0	+1	0	0	0	0	0	0	0
0	0	1	0	+1	0	0	0	0	0	0
0	1	0	0	0	+1	0	0	0	0	0
0	1	1	0	0	0	+1	0	0	0	0
1	0	0	0	0	0	0	+1	0	0	0
1	0	1	0	0	0	0	0	+1	0	0
1	1	0	0	0	0	0	0	0	+1	0
1	1	1	0	0	0	0	0	0	0	+1

Figure 2.37 Mapping table for orthogonal code system with $H = 8$, where there are eight spreading codes (A, B, C, D, E, F, G, H) and each code will be selected to denote a particular data bit pattern of (d_0, d_1, d_2).

Table 2.6 The states number, spreading code number, and bits per symbol (m) in an orthogonal code system.

H	States	Bits/symbol (m)
2	2	1
4	4	2
8	8	3
16	16	4

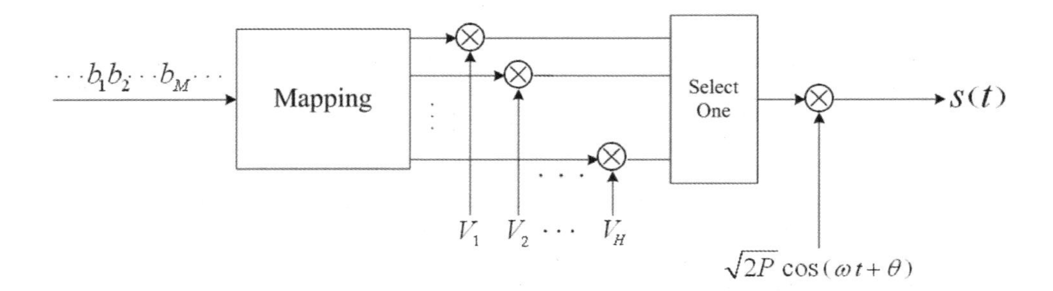

Figure 2.38 Orthogonal code system transmitter with $H = 8$, where there are eight spreading codes ($V_1, V_2, \ldots, V_H$) and each code will be selected to denote a particular data bit pattern.

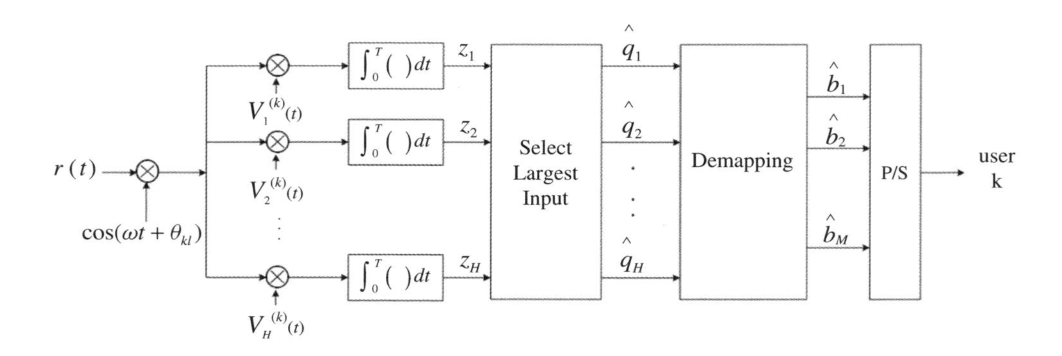

Figure 2.39 Orthogonal code system receiver with $H = 8$.

system with different decoding strategies and receiver architectures under an impulsive noisy and multiple access interference environment. Another publication appearing in 1990 [165] tackled the issues of an orthogonal code system under an indoor environment. The authors used the squared detection law and a RAKE receiver to detect the multi-code CDMA signals and analyzed the bit error rate performance of an orthogonal code system under additive white Gaussian noise (AWGN) and multiple access interference.

2.7.2 MULTI-CODE SYSTEM

The second popular multi-code CDMA system, which just called multi-code system, was proposed by Chih-Lin I in 1995 [166]. In this scheme, each user is allowed to use multiple spreading codes simultaneously to offer a much faster data transmission rate than possible in the orthogonal code system (in which only one spreading code is allowed to send into the channel at a given time from a particular transmitter). The most attractive feature of this scheme is that multiple bit streams can be encoded by different spreading codes and then be sent into the channel at the same time, effectively enlarging the bit pipe formed between the transmitter and the receiver. A fundamental schematic diagram of the multi-code system transmitter is shown in Figure 2.40. The receiver in a multi-code system can use the same structure as that used in an orthogonal code system, as shown in Figure 2.39,

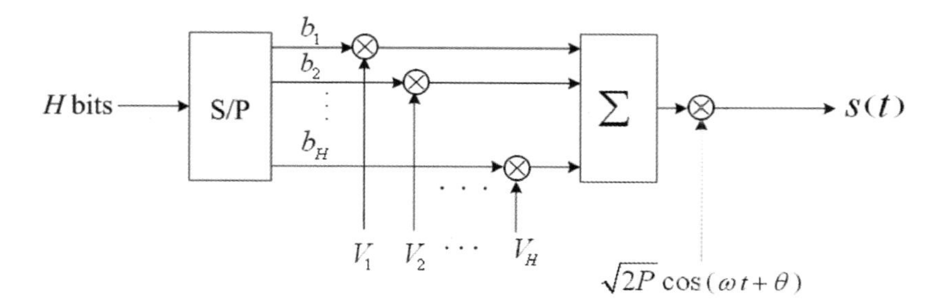

Figure 2.40 Multi-code system transmitter with $H = 8$, where all H spreading codes will be used to encode H different bits or $(b_1, b_2, \ldots, b_H)$.

with the only difference being that the processor following H despreading units will detect all possible signals sent using H spreading codes, instead of only one, as is the case in the orthogonal code system.

The bit error probability (BEP) performance of a multi-code system was analyzed in 2001 by Lee *et al.* [167], in which the authors used the maximal likelihood method to study the BEP performance under AWGN and multiple access interference. The receiver made use of a traditional correlator receiver. A study on the multi-code system can also be seen in [168], in which the authors conducted a performance study for a multi-code system under AWGN, multiple access interference, and multipath interference. The results were also compared to those for a traditional single-code CDMA system, where each user uses only one code for CDMA transmission.

2.7.3 PARALLEL COMBINATORIAL SYSTEM

Yet another multi-code CDMA scheme is the parallel combinatorial system, which was proposed in 1999 by Guo and Milstein [169]. In the parallel combinatorial system, each user will also be allocated H different spreading codes. However, the user can encode its data bit symbol by using different possible combinations of the number of sent spreading codes and their signs. In this way, the parallel combinatorial system will have a greater degree of freedom in controlling the data transmission rate, compared with the multi-code system discussed above.

The block diagrams for a transmitter and a receiver for the parallel combinatorial system are shown in Figures 2.41 and 2.42, respectively. In the parallel combinatorial system transmitter, there are three important parameters, which are the number of spreading codes assigned to a transmitter H, the number of actual codes sent each time u, and the number of bits in each symbol $m = \lfloor \log_2 2^u C_u^H \rfloor$, where the notation $\lfloor x \rfloor$ denotes the largest integer smaller than x, and C_u^H stands for calculation of combinations to select u samples from H elements. Figure 2.43 illustrates the mapping algorithm to encode the original data bit patterns in each symbol to the spreading codes which are to be sent into the channel. Table 2.7 shows the relation among the three key parameters H, u, and m.

2.7.4 BPSK M-ARY CDMA SYSTEM

As the most general multi-code CDMA system, the BPSK M-ary CDMA system is capable of offering the highest data transmission rate among all multi-code CDMA schemes. In addition, it is the most flexible multi-code CDMA scheme in terms of its suitability for multimedia transmissions, where the data rate from a user may change from time to time according to its real-time data rate requirements. As we will have in-depth coverage of the M-ary CDMA system in Chapter 9, we will only give a brief introduction to its basic ideas here.

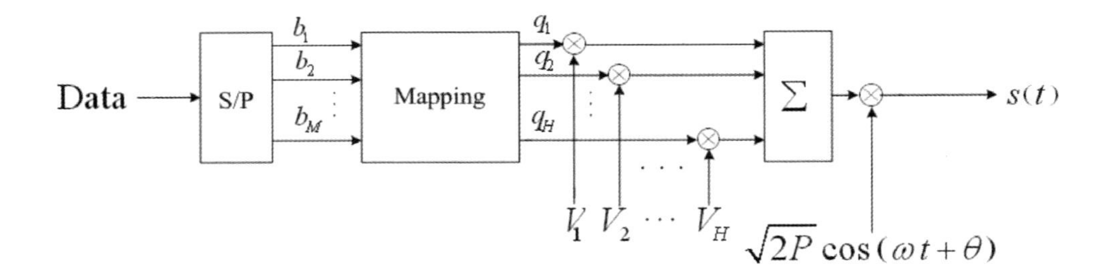

Figure 2.41 parallel combinatorial system transmitter.

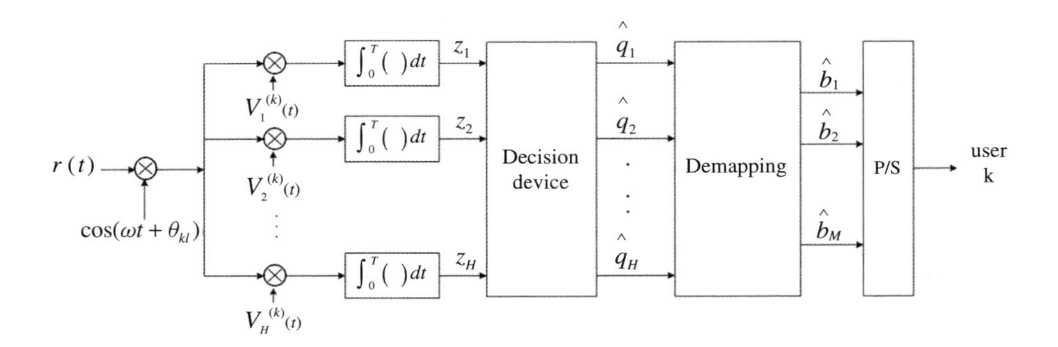

Figure 2.42 Parallel combinatorial system receiver.

Data	Spread code
d_0 d_1 d_2	A B C
0 0 0	+1 +1 0
0 0 1	+1 −1 0
0 1 0	−1 −1 0
0 1 1	−1 +1 0
1 0 0	+1 0 +1
1 0 1	+1 0 −1
1 1 0	−1 0 −1
1 1 1	−1 0 +1
	0 +1 +1
	0 +1 −1
	0 −1 −1
	0 −1 +1

Figure 2.43 Mapping relation for a parallel combinatorial system with $H = 3$ and $u = 2$.

Table 2.7 The relationship among three key parameters (H, u, and m) in the parallel combinatorial system.

H	u	States	Bits/symbol (m)
2	1	4	2
	2	4	2
3	1	6	2
	2	12	3
	3	8	3
4	1	8	3
	2	24	4
	3	32	5
	4	16	4

In the BPSK M-ary CDMA system, each user is given H (>1) spreading codes, each of which can be encoded by three different values, +1, −1, and 0. It is noted that when a code is encoded by 0, it simply means the transmitter will send nothing. Let us use an example to illustrate how it works. Letting $H = 3$ (there are three codes A, B, and C), we can have the following possible states for a transmitter to send different spreading codes. It can send one code (which has six possibilities: $+A$, $-A$, $+B$, $-B$, $+C$, $-C$); it can send two codes (which has 12 possibilities: $+A+B$, $+A+C$, $+B+C$, $-A+B$, $-A+C$, $-B+C$, $+A-B$, $+A-C$, $+B-C$, $-A-B$, $-A-C$, $-B-C$); and it can also send three codes (which has eight possibilities: $+A+B+C$, $-A+B+C$, $+A-B+C$, $+A+B-C$, $-A-B+C$, $-A+B-C$, $+A-B-C$, $-A-B-C$). Therefore, altogether we have $6+12+8=26$ different states, which is just equal to $3^H - 1 = 3^3 - 1 = 26$. In this way, we conclude that such a BPSK M-ary CDMA system can have $3^H - 1$ different code states which can be used to encode information data. Figure 2.44 shows the mapping between the data symbol and sent codes for a BPSK M-ary CDMA system with $H = 3$.

However, we usually may not be able to make full use of all $3^H - 1$ states. Take again $H = 3$ as an example. In this case we have altogether 26 code transmission states but we can only use any 16 of the 26 states, as we can only choose a symbol block size of 4 bits, in this case due to the fact that $2^4 = 16$ and $2^5 = 32 > 26$. In general, we have $m = \lfloor \log_2(3^H - 1) \rfloor$, where $\lfloor x \rfloor$ denotes the largest integer less than x. Table 2.8 has shown the relationship between the three key parameters, H, number of states, and m.

Table 2.8 The relation among H, number of code states, and m in a BPSK M-ary CDMA System.

H	States	Bits/symbol (m)
2	8	3
3	26	4
4	80	6
5	242	7
6	728	9

Data				Spread code		
d_0	d_1	d_2	d_3	A	B	C
0	0	0	0	+1	0	0
0	0	0	1	−1	0	0
0	0	1	0	0	+1	0
0	0	1	1	0	−1	0
0	1	0	0	0	0	+1
0	1	0	1	0	0	−1
0	1	1	0	+1	+1	0
0	1	1	1	+1	−1	0
1	0	0	0	−1	−1	0
1	0	0	1	−1	+1	0
1	0	1	0	+1	0	+1
1	0	1	1	+1	0	−1
1	1	0	0	−1	0	−1
1	1	0	1	−1	0	+1
1	1	1	0	0	+1	+1
1	1	1	1	0	+1	−1
				0	−1	−1
				0	−1	+1
				+1	+1	+1
				−1	+1	+1
				+1	−1	+1
				+1	+1	−1
				−1	−1	+1
				−1	+1	−1
				+1	−1	−1
				−1	−1	−1

Figure 2.44 Mapping relation between data bit pattern and sent codes for a BPSK M-ary CDMA system with $H = 3$.

Obviously, we can also view an orthogonal code system as a parallel combinatorial system if we let $u = 1$. On the other hand, a multi-code system can also be viewed as a parallel combinatorial system when $u = H$. We can easily obtain the bandwidth utilization efficiency (η) for all four different multi-code systems using the following equation:

$$\eta = \frac{\text{Number of bits per symbol}}{\text{Symbol duration}} \tag{2.35}$$

Assume that the symbol duration is T and H spreading codes are assigned to each user. We can readily have the bandwidth utilization efficiency (η) for all four different multi-code systems as follows:

$$\text{Orthogonal code system:} \quad \eta = \frac{\log_2 H}{T} \tag{2.36}$$

$$\text{Multi-code system:}\quad \eta = \frac{H}{T} \tag{2.37}$$

$$\text{Parallel combinatorial system:}\quad \eta = \frac{\lfloor \log_2(2^u C_u^H) \rfloor}{T} \tag{2.38}$$

$$\text{BPSK } M\text{-ary CDMA system:}\quad \eta = \frac{\lfloor \log_2(3^H - 1) \rfloor}{T} \tag{2.39}$$

The bandwidth utilization efficiency for all four different multi-code systems are also listed in Table 2.9.

Table 2.9 Comparison of bandwidth utilization efficiency for four different multi-code CDMA schemes.

	Orthogonal code system	Multi-code system	Parallel combinatorial system	BPSK M-ary CDMA system
$H = 1$	NA	1	1	1
$H = 2$	1	2	$2(u = 1)$ $2(u = 2)$	3
$H = 3$	NA	3	$2(u = 1)$ $3(u = 2)$ $3(u = 3)$	4
$H = 4$	2	4	$3(u = 1)$ $4(u = 2)$ $5(u = 3)$ $4(u = 4)$	6
$H = 5$	NA	5	$3(u = 1)$ $5(u = 2)$ $6(u = 3)$ $5(u = 4)$ $5(u = 5)$	7
$H = 6$	NA	6	$3(u = 1)$ $5(u = 2)$ $7(u = 3)$ $7(u = 4)$ $7(u = 5)$ $6(u = 6)$	9
$H = 7$	NA	7	$3(u = 1)$ $6(u = 2)$ $8(u = 3)$ $9(u = 4)$ $9(u = 5)$ $8(u = 6)$ $7(u = 7)$	11

2.8 MULTI-CARRIER CDMA SYSTEMS

Multi-carrier CDMA technology [170–185] was developed from single-carrier CDMA technology. The motivation to propose the multi-carrier CDMA can be explained as follows.

The basic idea for a CDMA system to overcome the multipath problem is built on the fact that a spread spectrum signal has a much wider bandwidth occupancy than its original baseband signal, such that the spread spectrum signal will offer a unique property that multipath diversity can be made possible due to its excellent time resolution derived from its very wide bandwidth signaling. A RAKE receiver can then be used to resolve different multipath returns individually and combine them again in a coherent or non-coherent way, to achieve the multipath diversity capability, which has become one of the most important technical features that make CDMA a very attractive multiple access technology for 2G and 3G wireless communications.

In the frequency domain, the capability of a CDMA system to overcome the multipath interference problem can be explained using the following text. Also, due to the very wide spectral width of CDMA signaling, the loss of energy owing to the nulls of a frequency-selective fading channel will not cause too much damage to the total energy of the CDMA signal due to its relatively wide spectral occupancy in the frequency domain. Therefore, it can be shown that the resilience of a CDMA system against frequency-selective fading is much stronger than any of the other traditional multiple access technologies, such as FDMA and TDMA, which are also called narrowband technologies.

The loss of energy in the CDMA signaling due to the nulls of a frequency-selective fading channel will still have some negative impact on the signal detection efficiency at a CDMA receiver. Therefore, some methods were proposed to recover or minimize the energy loss of CDMA signaling due to frequency-selective fading. Multi-carrier CDMA is one of them.

One possible implementation of a multi-carrier CDMA system can be described as follows. The input bit stream should first go through a serial to parallel conversion to form N sub-channels and then the N different parallel data streams will be spreading modulated by N different spreading codes, followed by carrier modulation by N distinct carrier frequencies. Of course, another possible scheme to implement multi-carrier CDMA is that after the serial to parallel conversion the N different sub-channels will be spreading modulated by using one spreading code. In this case, the N modulated multi-carrier signals are not separable in the CDMA code space, but only separated by different carrier frequencies.

Nevertheless, for both cases the serial to parallel conversion has converted the wideband single data stream into N narrowband sub-streams, making the signal more resilient against frequency-selective fading in the channel. It is noted that the capability of a multi-carrier CDMA system to mitigate frequency-selective fading is acquired through a very different approach compared to the way that normal wideband CDMA overcomes the same problem. The multi-carrier CDMA system splits up a wideband signal into many narrowband sub-streams, and thus if a few sub-streams unfortunately fall into the nulls of a frequency-selective fading channel, the rest will not be seriously affected. Those sub-streams that fall into the nulls can be recovered by using some kind of interleaving and error correction coding schemes.

In addition, multi-carrier CDMA can overlap the spectra for different sub-streams to forma so-called orthogonal multi-carrier CDMA system to further improve its bandwidth efficiency. The spectra shapes for traditional wideband direct sequence CDMA, multi-carrier CDMA, and orthogonal multi-carrier CDMA are shown in Figure 2.45. It is seen from the figure that inter-symbol interference will be a problem if an orthogonal multi-carrier CDMA system can not maintain the orthogonality among all sub-channels, whereas the normal multi-carrier CDMA system will not have such a problem as long as a sufficient guide band is used to separate neighboring sub-channels. Another salient feature for the orthogonal multi-carrier CDMA system is that it can be implemented by inverse fast Fourier transform (IFFT) and fast Fourier transform (FFT) in the transmitter and receiver, respectively, to greatly reduce the implementation complexity of the radio transceiver.

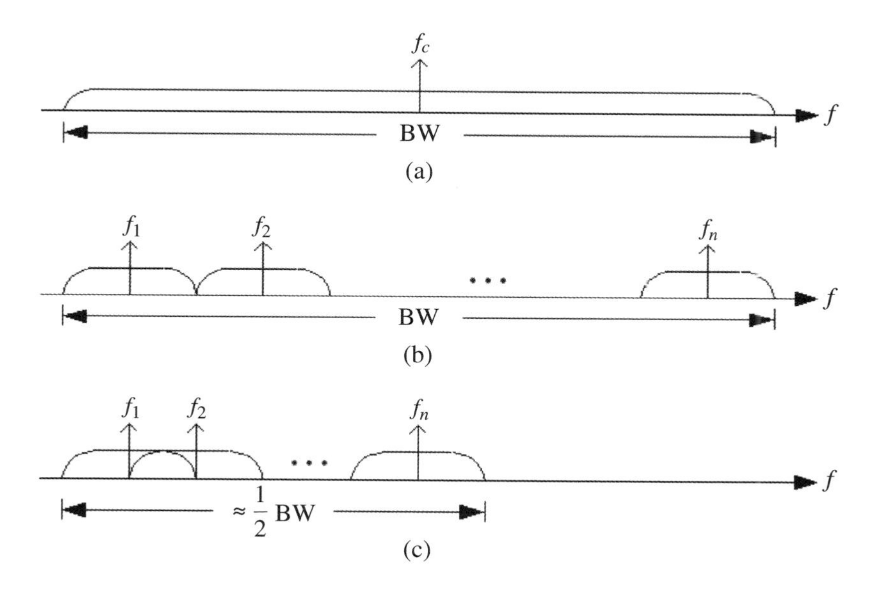

Figure 2.45 The spectra shapes for (a) traditional wideband direct sequence CDMA, (b) multi-carrier CDMA, and (c) orthogonal multi-carrier CDMA signals.

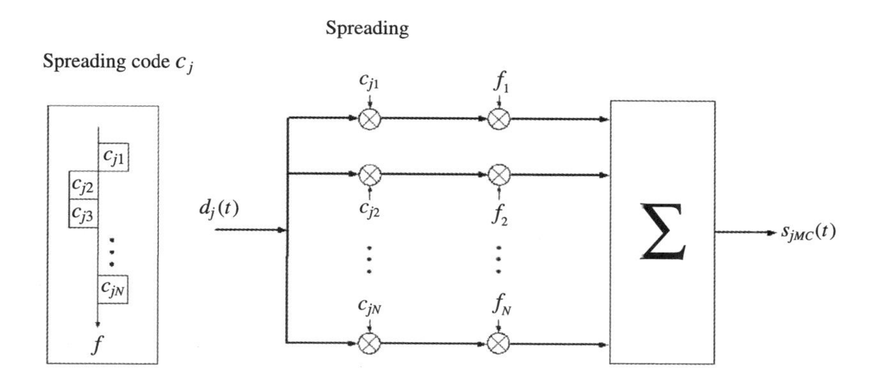

Figure 2.46 A multi-carrier CDMA transmitter with frequency domain spreading.

Another scheme to implement a multi-carrier CDMA system is to use frequency domain spreading instead of time domain spreading, as discussed earlier. Conceptual diagrams for a transmitter and a receiver are shown in Figures 2.46 and 2.47, respectively. It is seen from the figures that in this case one spreading code will be assigned to each user and its transmitter will spread the same bit signal (i.e. $d_j(t)$) with different chips of a spreading code (i.e. $c_{j1}, c_{j2}, \ldots, c_{jN}$), followed by carrier modulation by N different sub-carriers or $f_1, f_2, \ldots, f_N$.

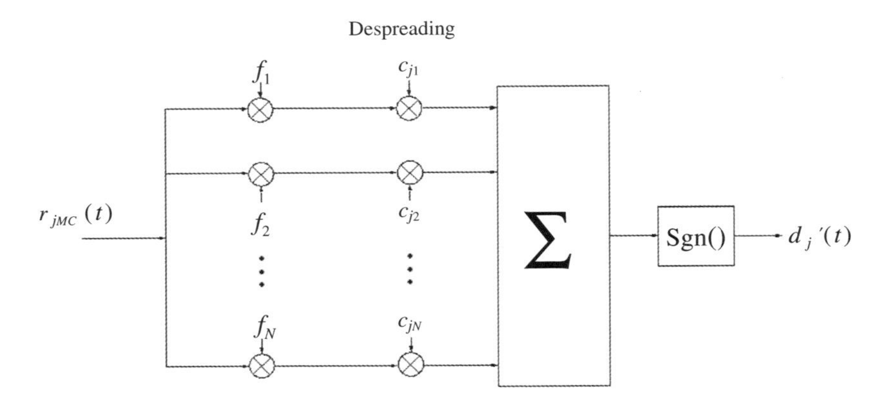

Figure 2.47 A multi-carrier CDMA receiver with frequency domain spreading.

In the multi-carrier CDMA scheme shown in Figures 2.46 and 2.47, no time domain spreading is used, and user data is spread only in the frequency domain. Thus, the reverse process should be applied to the receiver to decode sent information, as shown in Figure 2.47.

It is clear that a multi-carrier CDMA system can also use both time domain spreading and frequency domain spreading at the same transmitter to meet particular application and implementation requirements. Obviously, time domain spreading can offer an extra separation among different sub-carriers and thus it is in particular well suited for a non-orthogonal multi-carrier CDMA system. On the other hand, the frequency domain spreading provides no division among sub-carriers and thus it can only be used in an orthogonal multi-carrier CDMA system.

More discussions on multi-dimensional spreading techniques, which will also be an important part of next generation CDMA technology, will be given in Chapter 5.

2.9 OFDM CDMA TECHNIQUES

To understand OFDM CDMA techniques we have to start with the orthogonal frequency division multiplex (OFDM) technique. As a matter of fact, with knowledge of OFDM, it is very easy to understand the working principle of OFDM CDMA, as OFDM CDMA is nothing but a combination of the OFDM and CDMA techniques.

The OFDM technique is a natural extension of the multi-carrier modulation system.[16] With the help of the OFDM technique, a complex multi-carrier modulator structure in a multi-carrier transmitter can be replaced by a baseband signal processing unit based on an IFFT module. In the same way, the multi-carrier demodulator in a multi-carrier receiver can be replaced by an FFT unit. Both IFFT and FFT can be implemented easily by either hardware or software. In particular, if a software approach is used to implement an IFFFT/FFT module, the use of a powerful digital signal processor is required and nothing more than that is needed. This can greatly reduce the implementation cost compared to the implementation of a multi-carrier modem, which may require N different oscillators if N sub-carriers are used. The implementation of N sub-carrier oscillators is never a trivial task, as all oscillators have to use RF circuitry, which is always costly in terms of power and space.

[16]Here, we have to admit that we are only concerned with a general multi-carrier system, which will not necessarily use CDMA.

The development of the the OFDM technique can be traced back to the mid-1960s, when some people started to think about the way to transmit signals in parallel through multiple sub-channels. In fact, this type of multiple sub-channel transmission technique had been used in some military communication systems in the 1970s. In 1971, Wienstein and Ebert in their research proposed a parallel data transmission system, where all carrier modulation and demodulation processes were introduced to the parallel transmission system. In addition, it was they who first suggested use of a 'guard space' and time-domain raised-cosine-window shaping to reduce the inter-symbol interference induced in the signal propagation process of the channel.

The milestone of the development of OFDM technique came in 1980, when Peled and Ruiz in their research activities proposed the use of cyclic prefix (CP) to overcome the problems associated with signal propagation over time-dispersive channels (later people also used the same technique to mitigate the Doppler effect in frequency-dispersive channels). The use of CP can effectively solve the serious problem caused by multipath interference by using a guard interval to make sure the IFFT or FFT over a complete symbol duration will be implemented by cyclic convolution. This development was a breakthrough, which made the OFDM technique a practical means to be used in many real-world communication systems. As long as the channel delay spread is shorter than the CP, multipath interference will not cause any harmful effect to the signal detection process in an OFDM system. Of course, we should acknowledge that the use of CP has paid the price in reducing the effective transmission rate of an OFDM system as well as wasting some transmission power but this price is meaningful and worthwhile.

The later 1990s is the primary age for the OFDM technique, which has gained great attention, and many communication applications have been found for commercial services. The most important examples OFDM technique applications include high-speed data transmission systems, high-speed digital subscriber loop (HDSL), asynchronous data subscriber loop (ADSL), digital audio broadcasting (DAB), digital video broadcasting (DVB), and IEEE 802.11a/g wireless local area networks (WLANs).

The fundamental idea behind OFDM is quite similar to that proposed for multi-carrier systems, where a wideband data stream will be chopped into many narrowband sub-streams, which are orthogonal with each other in the frequency domain. In this manner, a frequency-selective fading channel with respect to the wideband signal will be converted into many flat-fading sub-channels viewed by those sub-streams. However, we have to note that some data sub-streams may also possibly fall into the nulls of a frequency-selective channel, but they can be recovered by using joint application of frequency-domain interleaving and error-correction coding schemes.

It is noted that the major difference between OFDM and traditional frequency division multiplex (FDM) lies in the fact that all sub-channels in an OFDM system are orthogonal, whereas those in FDM are not. Therefore, the OFDM has the advantage of high bandwidth efficiency compared to that of FDM and other multiplex techniques. In order to make all sub-channels orthogonal with each other, we have to make sure that the frequency spacing between two consecutive sub-carriers (Δf) is equal to the reciprocal of the symbol duration (T_s), or in other words we have $\Delta f = 1/T_s$.

In the following text, we will show how a multi-carrier system will be used to derive an OFDM architecture, thus proving that an OFDM system is a natural extension developed from a multi-carrier system.

A block diagram for a generic multi-carrier transmitter is shown in Figure 2.48, in which a modulated QAM symbol stream is fed into a serial to parallel converter, forming N sub-streams denoted by $X_1, X_2, \ldots, X_N$. The output N sub-streams will then be fed into N sub-carrier modulators with each sub-carrier signal represented by $e^{j2\pi(k\Delta f)n\Delta t}$.

If we assume that the data transmission rate of such a multi-carrier system is $1/\Delta t$, the date rate in each sub-stream will become $1/(N\Delta t)$. Also, assume that the information data are complex and can be expressed by $a(k) + jb(k)$. Then, the output from a complex modulator with its carrier being

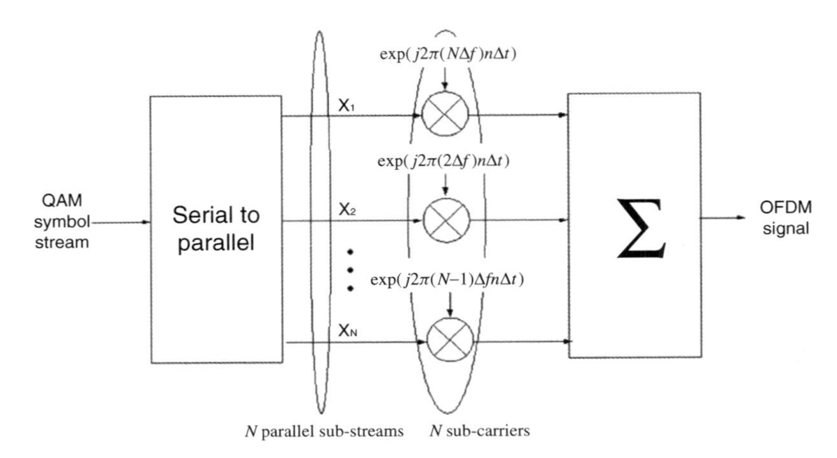

Figure 2.48 A multi-carrier CDMA receiver with frequency domain spreading.

$\cos(\omega_k t) + j\sin(\omega_k t)$ will become

$$D(t) = \sum_{k=0}^{N} [a(k)\cos(\omega_k t) + b(k)\sin(\omega_k t)] \tag{2.40}$$

where we have assumed $\omega_k = 2\pi f_k$, $f_k = f_0 + k\Delta f$, and $\Delta f = 1/(N\Delta t)$. It is noted that the results shown in Equation (2.40) are obtained due to the fact that the modulation of a complex data $a(k) + jb(k)$ with a complex carrier $\cos(\omega_k t) + j\sin(\omega_k t)$ will proceed such that their real parts and imaginary parts will be multiplied with each other, respectively. In general, a complex sub-carrier can be written as

$$S_k(t) = A_k(t)e^j[\omega_k t + \phi_k(t)] \tag{2.41}$$

Therefore, we can combine all sub-carriers into one equation as

$$S(t) = \frac{1}{N}\sum_{k=0}^{N-1} A_k(t)e^j[\omega_k t + \phi_k(t)] \tag{2.42}$$

where $\omega_k = \omega_0 + k\Delta\omega$, and $A_k(t)$, $\phi_k(t)$, and ω_k are the amplitude, phase, and carrier frequency for the kth sub-carrier. If we sample the signal under a sampling frequency of $1/\Delta t$, then we have $A_k(t) = A_k$, $\phi_k(t) = \phi_k$, and the sampled signal becomes

$$S_s(n\Delta t) = \frac{1}{N}\sum_{k=0}^{N-1} A_k e^{j\phi_k} e^{j2\pi(k\Delta f)n\Delta t} \tag{2.43}$$

where we have assumed $\omega_0 = 0$ without losing generality.

Now let us take a look at the definition of inverse discrete Fourier transform (IDFT), which can be expressed by

$$f(n\Delta t) = \frac{1}{N}\sum_{k=0}^{N-1} F(k\Delta f)e^{j\frac{2\pi nk}{N}} \tag{2.44}$$

By comparing Equations (2.43) and (2.44), we can see that the two will become exactly the same under the condition of $\Delta f = \frac{1}{N\Delta t}$. Under this circumstance, $A_k E^{j\phi_k}$ will be the complex signal in the frequency domain, $S(n\Delta t)$ will be the signal in the time domain, Δf is the space of two consecutive sub-carriers, and $N\Delta t$ is the symbol duration in each sub-stream of the multi-carrier system.

Therefore, it can be seen from the above discussion that a multi-carrier system (no matter whether CDMA is used or not) can be easily implemented by an OFDM structure, which offers a great reduction in hardware implementation complexity due to its baseband signal processing capability.

It is also noted that the difference between an OFDM CDMA system and an ordinary OFDM system lies in the fact that the former should use a spreading modulator before or after the serial to parallel converter, and thus the later signal processing will proceed on a chip basis, instead of a bit basis. The rest of the structure of an OFDM CDMA system becomes virtually the same as that of a normal OFDM system, and thus we will not discuss it in more detail here.

Next, we will show the orthogonality among the sub-carriers in an OFDM system. It is observed from Figure 2.48 that, no matter what symbol modulation scheme (i.e. QAM, PAM, QPSK, etc.) is used, N input symbol elements $X_1, X_2, \ldots, X_N$ can always be written in their complex format, such that

$$\begin{cases} X_1 = A_1 e^{j2\pi\theta_1} \\ X_2 = A_2 e^{j2\pi\theta_2} \\ \vdots \\ X_N = A_N e^{j2\pi\theta_N} \end{cases} \tag{2.45}$$

where A_k and θ_k $(k = 1, 2, \ldots, N)$ are the amplitude and the phase of the symbol elements, respectively. Now, let us multiply all sub-carriers with the sub-streams to obtain

$$\begin{cases} x_1 = A_1 \exp(j\omega_1)\exp(-j\pi N(t-t_s)/T_s) \\ x_2 = A_2 \exp(j\omega_2)\exp(-j\pi (N-2)(t-t_s)/T_s) \\ \vdots \\ x_N = A_N \exp(j\omega_N)\exp(j\pi (N-2)(t-t_s)/T_s) \end{cases} \tag{2.46}$$

where we have assumed that, to ensure all sub-carriers are orthogonal, the frequency space between two consecutive sub-carriers Δf should be equal to the reciprocal of the symbol duration T_s, or $\Delta f = \frac{1}{T_s}$. Now, we sum over all symbol elements $X_1, X_2, \ldots, X_N$ to have

$$s(t) = \sum_{i=-\frac{N}{2}+1}^{\frac{N}{2}} x_{i+\frac{N}{2}} = \sum_{i=-\frac{N}{2}+1}^{\frac{N}{2}} X_{i+\frac{N}{2}} \exp\left(j2\pi\frac{i}{T_s}(t-t_s)\right), \qquad t_s \le t \le t_s + T_s \tag{2.47}$$

or

$$s(t) = 0, \qquad t < t_s \quad \text{or} \quad t > t_s + T_s \tag{2.48}$$

Then, we can proceed to see if all sub-carrier modulated signals from an OFDM system are orthogonal or not by performing demodulation to a particular sub-carrier to obtain

$$\begin{aligned} &\int_{t_s}^{t_s+T_s} \exp\left(-j2\pi\frac{j}{T_s}(t-t_s)\right) \sum_{i=-\frac{N}{2}+1}^{\frac{N}{2}} X_{i+\frac{N}{2}} \exp\left(j2\pi\frac{i}{T_s}(t-t_s)\right) \mathrm{d}t \\ &= \sum_{i=-\frac{N}{2}+1}^{\frac{N}{2}} X_{i+\frac{N_s}{2}} \int_{t_s}^{t_s+T_s} \exp\left(j2\pi\frac{i-j}{T_s}(t-t_s)\right)\mathrm{d}t = X_{i+\frac{N}{2}} T_s \end{aligned} \tag{2.49}$$

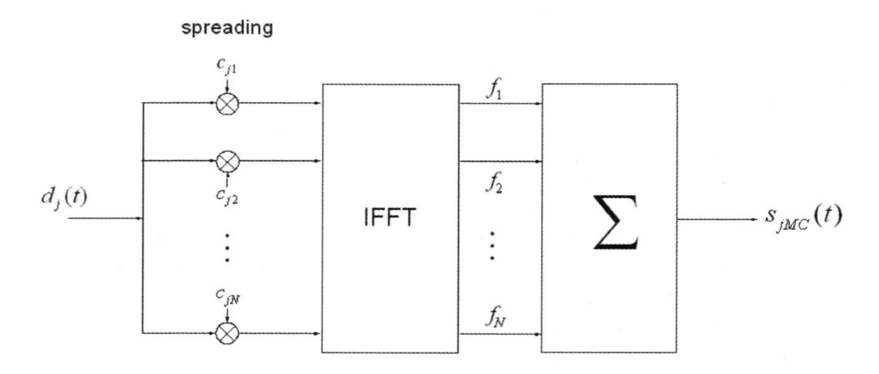

Figure 2.49 Conceptual block diagram for an OFDM CDMA transmitter with frequency domain spreading.

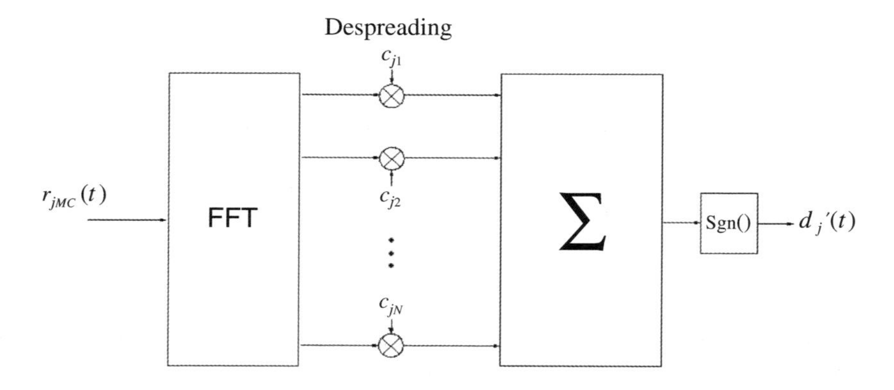

Figure 2.50 Conceptual block diagram for an OFDM CDMA receiver with frequency domain spreading.

where the above results are true for all N different indexes of

$$i = \left\{-\frac{N}{2}, -\frac{N}{2}+1, -\frac{N}{2}+2, \ldots, 0, 2, \ldots, \frac{N}{2}-3, \frac{N}{2}-1\right\} \tag{2.50}$$

Typical transmitter and receiver architectures for an OFDM CDMA system are shown in Figures 2.49 and 2.50, where we again assume that frequency domain spreading is considered.

3

CDMA-Based 2G and 3G Systems

In this chapter, two major 2G and 3G mobile cellular standards (namely, IS-95 and W-CDMA), both of which are based on CDMA technology, will be discussed. The IS-95 standard as a 2G technology undoubtedly stands as a milestone that marks the first ever successful application of CDMA technology in civilian communication systems. Therefore, the making of the IS-95 standard is really a great achievement of the industry with its profound historical impact, which will stay for ever. Qualcomm has made a great contribution to the development of IS-95 systems. Without its significant pioneering works, the success of the IS-95 standard would have been impossible. The company's large number of intellectual property rights (IPRs) ownership of many core CDMA technologies explains how important its position is in the entire R&D process of IS-95 technology, later also called the CdmaOne system, which has been successfully deployed in many countries around the world.

It is noted that cdma2000[1] is a 3G technology, which was developed by 3GPP2. It is in fact an evolutionary version of IS-95 and thus it can work together with legacy IS-95 in a fully downward compatible fashion. This is a very strong point in favour of 3GPP2's cdma2000 technology compared to its rival technology, W-CDMA, which is also a 3G technology but was promoted by 3GPP, which follows the European footprint after the GSM standard. However, the most important similarity between cdma2000 and W-CDMA lies in the fact that they both make use of CDMA as their multiple access technology. As to be seen from the introduction given in the sections following, both cdma2000 and W-CDMA share many common CDMA technologies.

As a matter of fact, not much difference can be found between those CDMA-based 2G and 3G mobile cellular standards in terms of their fundamental system operation schemes, although their individual signalings and architectures may differ. That is why many people are disappointed by the fact that almost no innovative CDMA technology can be seen from the designs of all CDMA-based 3G mobile cellular standards (including cdma2000, W-CDMA, TD-SCDMA,[2] etc), which appeared at least ten year after the first CDMA technology or IS-95 was deployed in the world. However, we can make our own judgment to see if this claim is correct or not after having read through the three sections following.

[1]We will not discuss cdma2000 in particular in this book also due to the fact that it bears many similarities to its predecessor, the IS-95 standard. Therefore, in this book we will only discuss the IS-95 and WCDMA standards, which are very good representatives for CDMA-based 2G and 3G mobile cellular systems.

[2]The TD-SCDMA technology is also a 3G technology, proposed by CATT China.

The Next Generation CDMA Technologies Hsiao-Hwa Chen

3.1 EIA/TIA IS-95 SYSTEM

As mentioned earlier, the importance of the IS-95A standard is underlined by its successful application of CDMA technology in civilian mobile cellular communication systems for the first time in history, marking the beginning of a new era of CDMA.

The TIA/EIA IS-95 standard[3] was first released in July 1993. The IS-95A revision was published in May 1995 and is the basis for many of the later commercial 2G CDMA systems around the world. IS-95A describes the structure of the wideband 1.25 MHz CDMA channels, power control, call processing, handoffs, and registration techniques for system operation. In addition to voice services, many IS-95A operators provide circuit-switched data connections at 14.4 kbps. IS-95A was first deployed in September 1996 by Hutchison, Hong Kong.

The IS-95B revision, also termed TIA/EIA-95, combines IS-95A, ANSI JSTD-008, and TSB-74 into a single standard. The ANSI JSTD-008 specification, published in 1995, defines a compatibility standard for 1.8–2.0 GHz CDMA PCS systems. TSB-74 describes interaction between IS-95A and CDMA PCS systems that conform to ANSI JSTD-008. Many operators that have commercialized IS-95B systems offer 64 kbps packet-switched data, in addition to voice services. Due to its data speeds, IS-95B was also categorized as a 2.5G technology. A mobile cellular network based on IS-95B was first deployed in September 1999 in Korea and since then has been adopted by several other operators in Japan and Peru.

People are often confused by the many different names of CDMA-based mobile cellular technologies, such as cdmaOne, IS-95A, ANSI JSTD-008, and IS-95B. As a matter of fact, cdmaOne is a collective name given to IS-95A, ANSI JSTD-008 and IS-95B. IS-95A and ANSI JSTD-008, are very much the same except for the band of operation and a few messaging protocols; IS-95A operates on the cellular band (i.e. 824–894 MHz), and ANSI JSTD-008 operates on the PCS band (i.e. 1850–1990 MHz), as shown in Figure 3.1. Several other differences exist in the technical specifications of the IS-95A and ANSI JSTD-008 standards. IS-95A uses 45 MHz spacing for both forward and reverse channels, whereas ANSI JSTD-008 uses 80 MHz spacing for both its forward and reverse

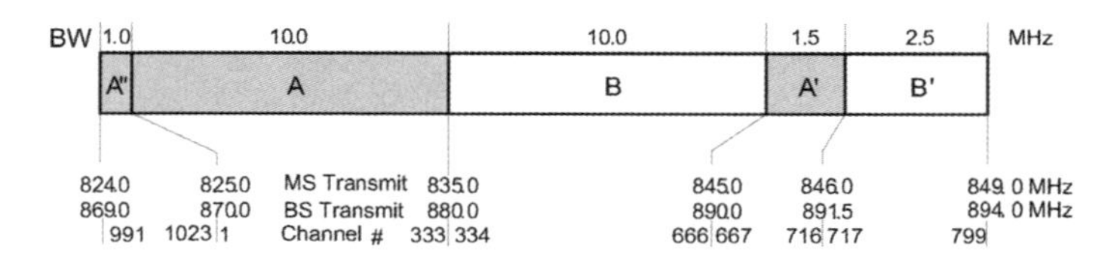

(a) Frequency plan of IS-95A, also called cellular bands

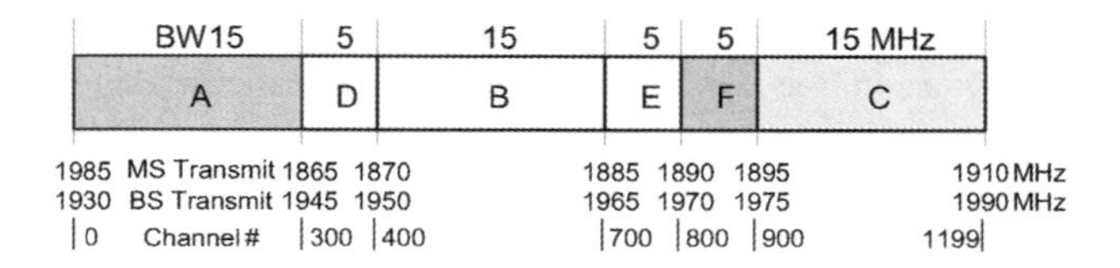

(b) Frequency plan of JSTD-008, also called PCS bands

Figure 3.1 Cellular bands and PCS bands allocated for IS-95A and JSTD-008 standards in the US.

[3]TIA stands for Telecommunications Industry Association, and EIA stands for Electronic Industries Association. IS-95 is the short form of Interim Standard 95.

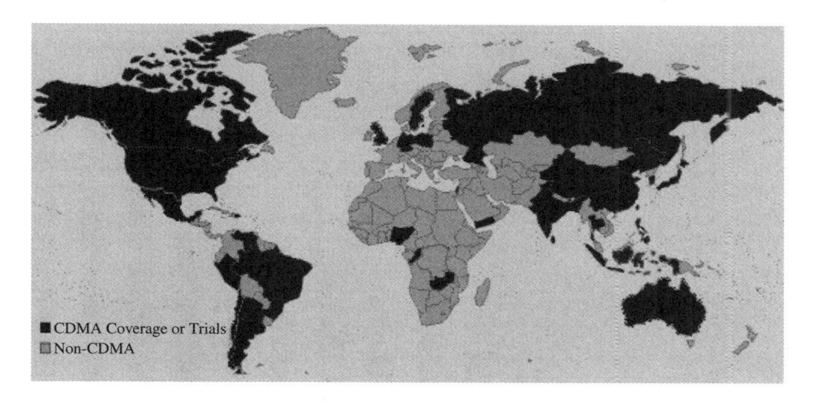

Figure 3.2 cdmaOne service coverage around the world.

channels. Permissible frequency assignments in IS-95A and JSTD-008 are 30 kHz and 50 kHz, respectively. IS-95B is an upgrade version to both these standards and operates in both bands. Therefore, the discussions given in this section will be concentrated only on the IS-95A system, and I give explanations on the differences among IS-95A, IS-95B, and JSTD-008 wherever necessary.

The release of the IS-95 standard in 1993, as well as IS-95A and IS-95B later, was due mainly to the pioneering research and development activities on CDMA-based mobile communication technology carried out in the US as well as many other countries in the world, especially by Qualcomm Inc. [25–52]. The research on CDMA-based mobile cellular technologies was motivated by the fact that an even higher capacity may be possible by using CDMA than achievable using TDMA technology, which had been applied to the IS-54 standard or DAMPS system. Due to the limited capacity of AMPS systems in the US, overlay CDMA on AMPS became commercially attractive to explore the benefits of spread spectrum (SS) techniques, which were mainly used in military communications before. This gave an AMPS operator in the US the option of increasing its network capacity in specific areas by replacing a number of 30 kHz AMPS carriers with one or more 1.25 MHz IS-95 carriers. Dual-mode AMPS/IS-95 handsets can take advantage of IS-95 networks wherever available, and revert to the old AMPS operating mode in those areas where no CDMA coverage is available. Qualcomm started to produce the first commercial AMPS/IS-95 dual-mode phones in 1994.

Since the first commercial IS-95A network was launched by Hutchison in Hong Kong on September 28, 1995, commercial IS-95 networks have been deployed in many countries around the world, including Korea and the US, as shown in Figure 3.2. IS-95 reached 100 million subscribers after only six years of commercial deployment. The numbers of service subscribers for IS-95 systems in terms of different regions in the world are shown in Figure 3.3.

When implemented in a cellular network, IS-95 technology offers several benefits to the cellular operators and their subscribers, compared with its predecessor, AMPS system:

1. Capacity increases eight to ten times compared to that of an AMPS system and four to five times that of a GSM system.

2. Call quality has been improved with better and more consistent sound as compared to AMPS systems.

3. System planning can be simplified due to the use of the same frequency in every sector of every cell.

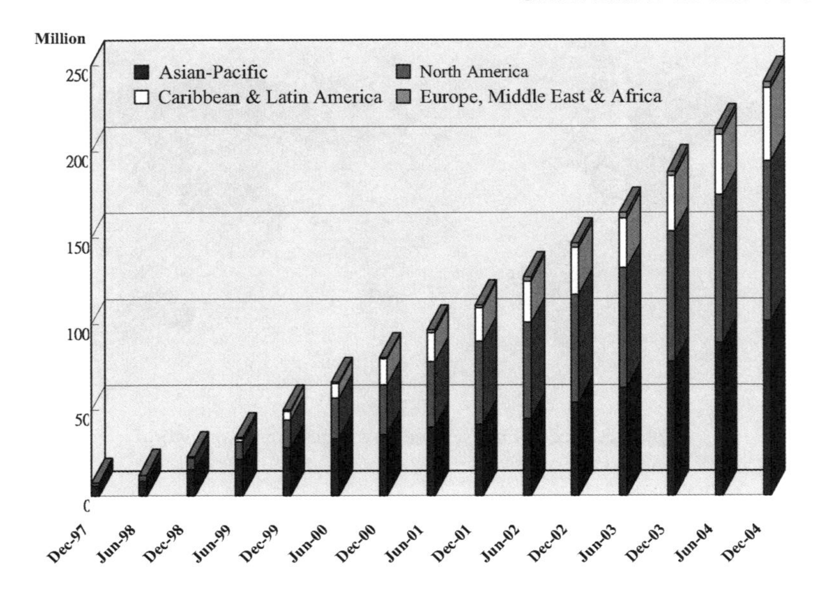

Figure 3.3 The numbers of cdmaOne service subscribers up to December 2004 around the world.

4. It offers enhanced privacy and improved coverage characteristics, allowing for the possibility of fewer cell sites.

5. It offers increased talk time for mobiles.

3.1.1 IS-95A NETWORK CONFIGURATION

Before discussing the details of the IS-95 system, we would like to take a look at the whole IS-95 system from its network perspective, as illustrated in Figure 3.4. It is seen from the figure that an IS-95 system consists of the following basic functional blocks: mobile station, base stations, base station controller, switch, etc.

A mobile station (MS) is defined in the IS-95 specifications [33,34]. An MS is more often called a mobile terminal, mobile handset or cell phone, which is usually very small in physical dimensions and light in weight in order to be carried around easily. A mobile handset usually uses a single antenna for signal transmission and reception and thus it is very hard to use multiple antennas for space diversity. For this reason, a MIMO or transmitter-diversity system based on space-time coding algorithms can hardly be applied to the uplink channel where the handset acts as the transmitter. In addition, an MS will be powered by battery in most of the operating hours, and thus its transmitting power is greatly constrained by power consumption of the mobile terminal.

A base station (BS) is defined in IS-95 as a place where radio signals from and to MSs will be processed. Different from an MS, a BS does not have strict constraints in its physical dimensions. It can be installed on the top of a building or a lamp pole and can be powered by a 100 $\sim$ 240 volt AC power source. A BS is always placed in the center of a cell to provide even coverage in the entire cell. In order to enhance the capacity of the whole network, IS-95 also allows a cell to be divided into three or more sectors. In this case, a BS should use several antennas to cover different sectors in the same cell.

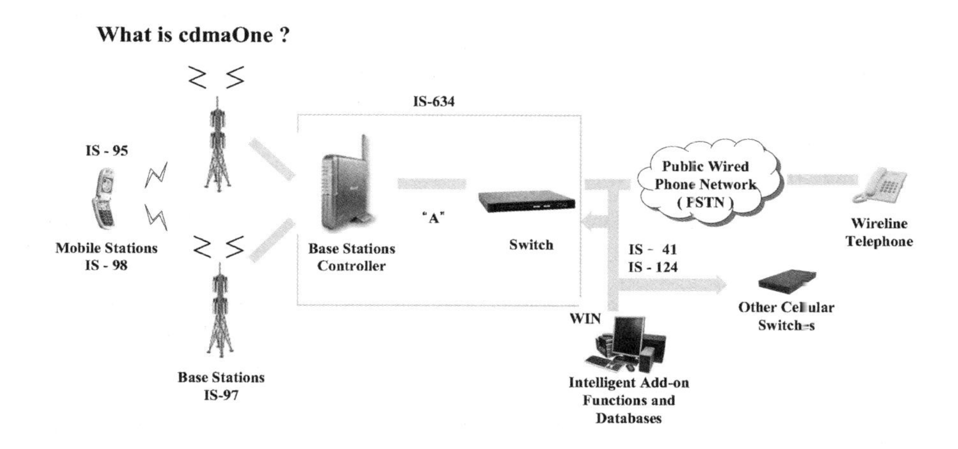

Figure 3.4 The operation and basic configuration of the IS-95 system.

In addition to MS and BS, the operation of an IS-95 system needs a base station controller (BSC), which ought to work jointly with a switch to perform various important functions of a mobile cellular network, such as mobile location registration, soft handoff, mobile paging, dynamic resource allocation, traffic load shedding, etc. The detailed functions of the BSC and the switch have been specified in TIA/EIA IS-634 [40]. The combination of BSC and switch is often called the master switching center (MSC) in an IS-95 system.

The linkage between the MSC and the public subscriber telephone network (PSTN) as well as other cellular switches, which may operate under other mobile service providers, is also shown in Figure 3.4. The technical specifications for this part of the IS-95 network can be found from IS-41 and IS-124 [27,28,37].

Next, we would like also to talk about the cell configuration in an IS-95 system. In a mobile cellular network the whole geographical area is divided into small chunks called cells. A single BS services each of these cells. The cells are often grouped into a cluster of three to seven cells. Groups of clusters are put under a single BSC, as shown in Figure 3.4. All BSCs in a mobile cellular network are controlled by an MSC. By repeating the seven-cell cluster over a city we can cover the entire area by planning just one cluster. However, it can never be considered an easy task to plan cells for a city, as it needs a lot of practical experience in mobile communication engineering. Nevertheless, this hierarchy of cells and clusters helps to increase capacity and facilitate call routing, apart from many other operational advantages. But the final number of cells in an area is usually a compromise on various factors like the density of calls, number of BSs, size of each cell, network capacity, and most importantly the budget available. Planning cells is quite an interesting study topic and good planning can save a lot of money and yet provide good coverage.

CDMA has many advantages over other multiple access schemes, such as FDMA and TDMA, in terms of cell planning. In the case of FDMA or TDMA, a given spectrum should be divided into smaller chunks, each of which is uniquely assigned to a cell in a cluster. By doing so repeatedly a big city is serviced by the limited spectrum. Smaller bandwidth for transmission compromises quality of signal. This planning of spectrum is called frequency planning. The distribution of frequencies in a cluster is important to reduce co-channel interference and adjacent channel interference between repeating clusters. On the other hand, frequency planning in CDMA can be reduced to minimal. The entire spectrum can be assigned to all the seven cells in a cluster. This is possible because of the orthogonality of the signature codes used for transmissions. As a direct consequence, usage of the entire spectrum enhances the quality of transmissions. Adjacent channel interference is controlled by power control, which will be explained in Subsection 3.1.5.

3.1.2 WALSH, SHORT AND LONG PN CODES

In order to make it easy to understand the operation of an IS-95A system, we introduce three important codes, i.e. Walsh codes, short pseudo-noise (PN) codes (or short codes), and long PN codes (or long codes).

Walsh codes originate from orthogonal Walsh-Hadamard functions or orthogonal Walsh-Hadamard matrices. In fact, each Walsh code comes from a row or a column of a Walsh-Hadamard matrix. A matrix is said to be an orthogonal matrix if all its rows or columns are mutually orthogonal with one another. Two functions are said to be orthogonal if and only if the cross-correlation functions between them are zero in all relative time shifts. For more information about the correlation properties of CDMA codes, please refer to Chapter 2. The Walsh codes offer an ideal orthogonal property if they are used in a synchronous channel (e.g. the forward link or downlink in an IS-95A system), such that normalized auto-correlation of a code is one and cross-correlation between any two codes is zero. It is noted that this ideal orthogonality does not exist if they are applied to an asynchronous channel (e.g. the reverse link or uplink in an IS-95A system). IS-95A and IS-95B use 64-ary Walsh codes and cdma2000 uses 128-ary Walsh codes. These codes are used for spreading in a forward link in the IS-95 standard. Therefore, we should understand why the forward link can be divided into 64 code channels, which are equal to the number of different 64-ary Walsh codes. On the reverse link in the IS-95A standard the Walsh codes are not used to differentiate users but for 64-ary modulation, in which every 6 incoming raw data bits will be mapped into one of 64 Walsh codes, depending on the particular 6-bit pattern. Therefore, although both forward and reverse links use Walsh codes, they are used in different ways due to the different properties the Walsh codes possess in synchronous and asynchronous transmission modes.

In addition to the Walsh codes, the IS-95A standard also uses a short PN code, or simply short code, which is a 16-bit short PN code used to identify the different BSs and thus the cells where the BSs are located. Different BSs are distinguished in IS-95A by assigning an offset of this code to each BS under a common time reference in the whole network. On the reverse link the mobiles will use the short codes for extra signal robustness but without any phase offset. Global Positioning System (GPS) services should be used in synchronizing the different offsets of BSs in the mobile cellular network. On the other hand, the long PN code, or simply long code, which is a 42-bit PN code, is used in the reverse link for spreading modulation as a signature code to identify different mobile stations. The long code is used in the forward link for data scrambling in the IS-95A standard.

3.1.3 FORWARD CHANNEL

A forward channel in an IS-95A mobile cellular system is defined as a physical channel whose transmission direction is from BS to MS. Sometimes, the forward channel is also called the downlink channel. Major channel parameters for the forward link are listed in Table 3.1.

Figure 3.5 shows a block diagram of the forward code channel generator, which should be placed right before the QPSK carrier modulator as illustrated in Figure 3.6. It can be seen from the forward code channel generator that time diversity gain is achieved through spreading coding and block interleaving. In addition, IS-95A can provide path diversity through soft handoff and RAKE receiver.

The forward channels may carry normal traffic as well as overhead information. The overhead information in the forward channels is used to establish the system-wise timing and station

Table 3.1 Major channel parameters for the IS-95A forward link.

Channel	Sync	Paging		Traffic			
Data rate (bits/s)	1200	4800	9600	1200	2400	4800	9600
Code repetition	2	2	1	8	4	2	1
Modulation symbol rate (syms/s)	4800	19200	19200	19200	19200	19200	19200
PN chips/ modulation symbol	256	64	64	64	64	64	64
PN (chips/bit)	1024	256	128	1024	512	256	128

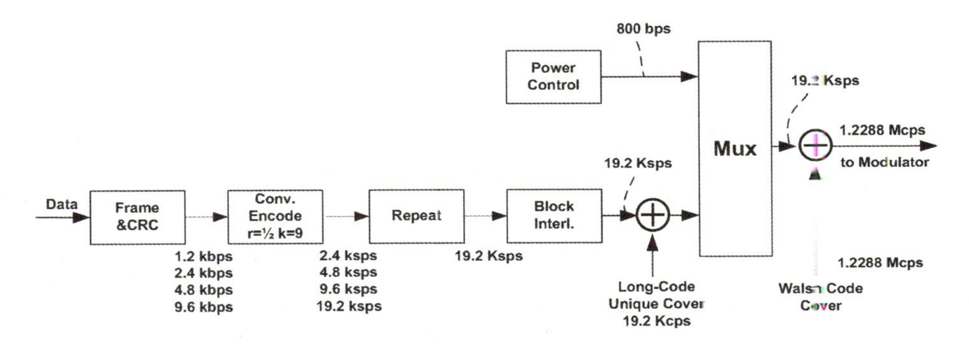

Figure 3.5 The forward CDMA code generator in an IS-95A system, where a long PN code (with period $2^{42} - 1$) is used to scramble the data.

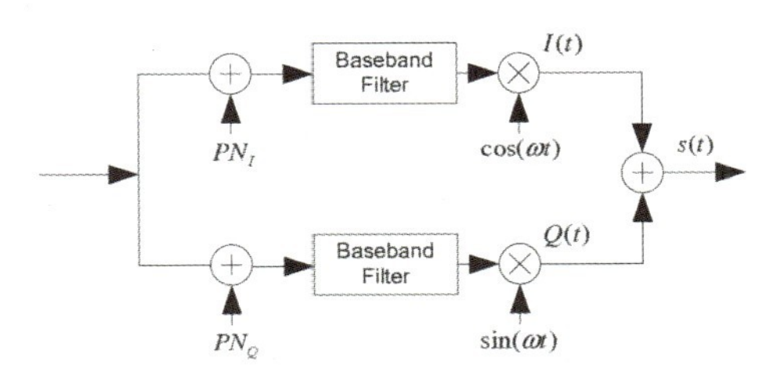

Figure 3.6 The QPSK forward channel carrier modulator in an IS-95A system, where two short PN codes (with period $2^{15} - 1$ or 26.67 ms) are used to identify different base stations with different phase offsets ($64 \times n$ chips).

identity, etc. There are mainly three types of overhead information in the forward channels: pilot, synchronization, and paging signals. It is to be noted that the pilot signal in the forward link channels is a spread but unmodulated DS-SS signal. The pilot channel is also used in the mobile-assisted handoff (MAHO) process as a signal strength reference. The forward error correction (FEC) rate is 1/2 and the PN code rate is 1.2288 MHz, yielded from 128×9600 bits/s.

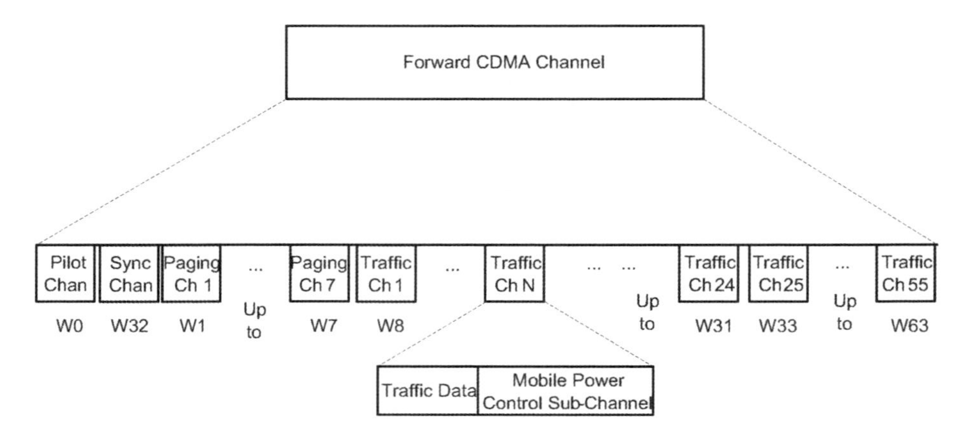

Figure 3.7 The logical forward CDMA channels in an IS-95A system.

The forward link in IS-95A consists of 64 logical channels or code channels, each of which uses one of a set of 64-ary Walsh codes. The use of Walsh sequences for channelization in the forward link is due to the synchronous nature of the forward link, in which all 64 Walsh codes are ideally orthogonal with one another, making the channels completely separable in the receiver. Each forward code channel is spread by the short code, which consists of I and Q components, as shown in Figure 3.6. These two coded, covered, and spread streams are vector-modulated on to the RF carrier. Thus, the spreading modulation is QPSK, superimposed on a BPSK code symbol stream. The I and Q channels in the QPSK modulator of IS-95A carry the same data stream, but use two different short PN codes, or PN_I and PN_Q, as shown in Figure 3.6. Transmitting power in IS-95A should be reduced by 3, 6, or 9 dB for different data rates at 9600, 4800, or 2400 bps.

The correspondence between IS-95A forward CDMA logical channels and code channels is shown in Figure 3.7, where we can see that there are four different logical forward channels: (1) pilot channel, (2) synchronization channel, (3) paging channels, and (4) traffic channels.

It is to be noted that there is only one pilot channel in the forward link, which corresponds to the Walsh code channel W0, or the first Walsh code in the Walsh orthogonal matrix. There is also only one synchronization channel and it uses Walsh code W32 for channelization. However, there are seven paging channels, corresponding to the Walsh codes from W1 to W7, and 55 traffic channels corresponding to the Walsh codes from W8 to W63. Each traffic channel consists of a traffic data sub-channel and a mobile power control sub-channel, as shown in Figure 3.7. Very often, pilot, synchronization, and paging channels are also called overhead channels, due to the fact that they deliver some non-traffic information. In the following text, we will take a look at both the traffic channels and the overhead channels to explain how they work in the IS-95A standard.

Pilot Channel

There is only one pilot channel in IS-95A. This channel sends all zeros, carrying no data information. Thus, the pilot channel contains short code only with no additional cover or information content. The pilot channel sends the beacon signal that defines the radius of a cell and hence is transmitted with the highest power. The signal level in the pilot channel is about 4–6 dB higher than normal traffic channels. Also, it is used as a timing source or a demodulation reference for the mobile receivers and as a measurement device during handoffs, especially in mobile assisted handoffs (MAHOs). The pilot channel is assigned the W0 code for channelization. The period of the pilot short code is

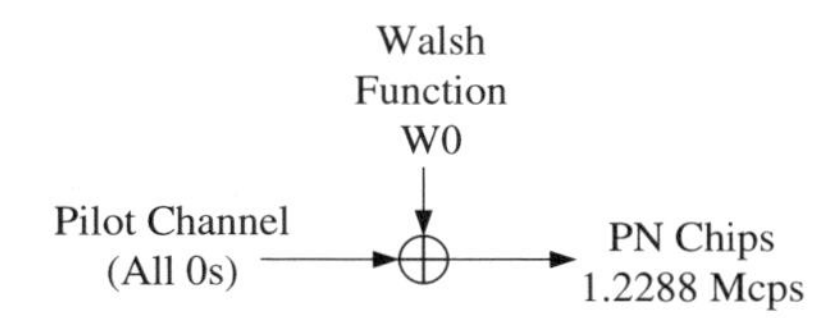

Figure 3.8 The pilot signal generator in the IS-95A system.

$2^{15} = 26.667$ ms at the 1.2288 MHz chip rate. The pilot channels in different base stations use the same short code and are distinguished by their phases. Therefore, the pilot signal is unique per sector or cell. The pilot phases are assigned to BS in multiples of 64 chips, giving a total of $2^{15}/64 = 512$ possible phase offset patterns. Therefore, this 9-bit number (512 different assignments) identifying the pilot phase assignment is called the pilot offset. The generation of the pilot signal is shown in Figure 3.8.

Synchronization Channel

Another overhead channel in IS-95A is the synchronization channel or sync channel, which is used by mobiles during system acquisition to receive reference time, system identification, system parameter information, the state of the long code, as well as other system configuration information. The synchronization channel uses Walsh code W32, which operates at 1200 bps and conveys pilot PN code offset, system clock time, and long code state information to allow an MS to be synchronized immediately to the cellular network. The interleaving period of the synchronization channel is $80/3 = 26.667$ ms, which helps to simplify the procedure to find frame boundaries, after the mobile has located the pilot signal. The code period ambiguity can be resolved by the long code state and system clock reference.

The signal generation process in the synchronization channel is illustrated in Figure 3.9. The Synchronization channel message has a length of $93 \times N$ bits, and will be sent in N superframes with its parameters being defined as follows: (1) one superframe (96 bits, 80 ms) is equal to three

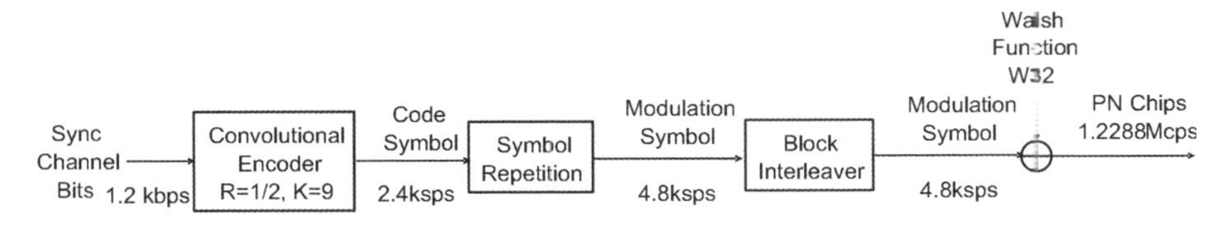

(a) Signal generation process in sync channel

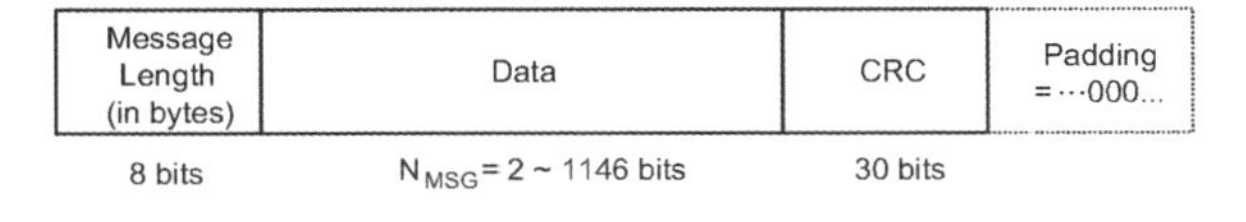

(b) Sync channel frame structure

Figure 3.9 (a) The synchronization signal generator in the IS-95A system. (b) Signal frame structure in the sync channel.

sync channel frames, (2) one frame (32 bits, 26.667 ms) is equal to the 1-bit start-of-message (SOM) plus 31-bit data, and (3) the SOM bit is set to +1.

The message part of the sync frame contains the following information: (1) system identification (SID) and network identification (NID), (2) PN sequences offset and long code state, (3) system time, leap seconds, offset from UTC, etc., and (4) paging channel data rate.

Paging Channels

The paging channel are another type of overhead channel. As shown in Figure 3.7, there are in total seven different paging channels, assigned by code channels from W1 to W7, whereas the other two overhead channels, pilot and synchronization channels, are assigned only one code channel each.

Paging channels carry some overhead information, such as pages, call setup messages, and so on. The first paging channel is called the primary paging channel and its overhead messages are always sent on the primary paging channel. Paging channels operate in a slotted mode, in which mobiles sleep and wake up when it is time to listen. Each paging slot lasts 80 ms, which is just equal to eight paging channel half-frames. One half-frame has a length of 10 ms and contains 48 or 96 bits, which consists of 1-bit start-of-paging-channel-message plus 47 or 95 data bits. A paging channel message contains the following information: (1) system parameters, (2) access parameters for access channel, (3) channel assignment information, and (4) temporary MS identification (TMSI) assignment.

The block diagram for generation of the paging channel signal is shown in Figure 3.10. The paging channel transmits at a data rate of 4800 or 9600 bps. The paging channels always have the same time alignment as the traffic channels. The paging channels are used to page MS and to pass other messages. As shown in Figure 3.10(b), the synchronized paging channel message has length of $47 \times N$ or $95 \times N$ bits.

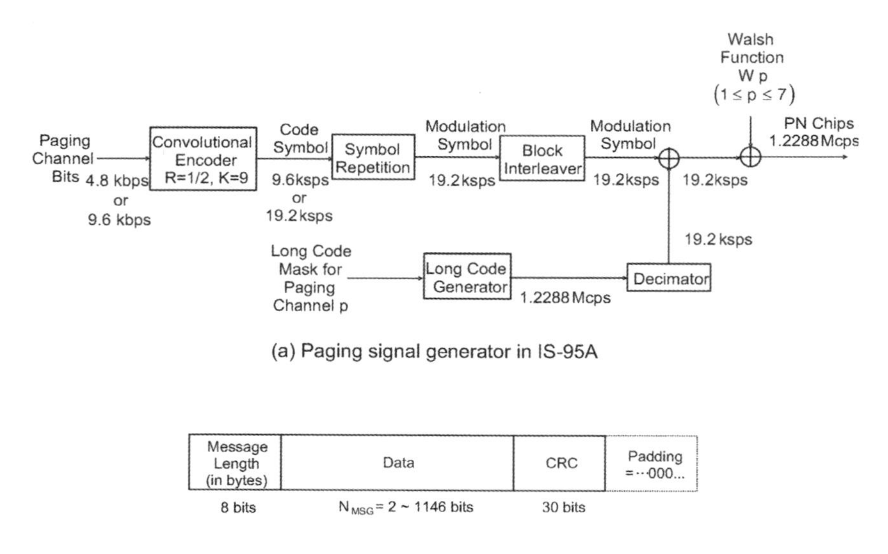

Figure 3.10 (a) The paging signal generator for paging channels in the IS-95A system. (b) Paging channel frame structure in IS-95A.

Traffic Channels

The traffic channels in the forward link in IS-95A are assigned to individual users to carry call traffic. All the remaining Walsh codes, after being assigned to all overhead channels, can be used as traffic channels as long as the noise and interference allow admission to the cell. Therefore, in total the number of available traffic channels is equal to 55,[4] as shown in Figure 3.7. However, it has to be noted that the actual number of active traffic channels that can be used simultaneously in an IS-95A system is very much interference-limited, and can be equal to only about half of the total available traffic channels, depending on the operational environment of a particular cell.

Obviously, the assignment of traffic channels to all MSs changes dynamically in response to all MS access requirements. Each traffic channel is always used to carry user data in 20 ms frames. The data rate in traffic frames can be 1, 1/2, 1/4, or 1/8 of 9600 bps. The change in data rate is accomplished by 1, 2, 4, or 8 times repetition of code symbols, but the energy per bit should be kept approximately constant through the adjustment of symbol levels for different rates. The rate change can be altered on a frame-by-frame basis in each 20 ms time (frame length). In other words, it is impossible to change the transmission rate inside a traffic frame. It should be pointed out also that the 800 bps reverse link power control sub-channel is carried on the forward traffic channels by puncturing 2 out of every 24 symbols transmitted, as shown in Figure 3.5. Another important point we have to mention here is that all base stations in an IS-95A system must be synchronized in time within an accuracy of a few micro-seconds. This is made possible by making use of the information carried on the sync channel in the forward link.

Figure 3.11 shows the block diagram of signal generation in the forward traffic channels. Figure 3.12 gives the forward traffic channel frame structure in the IS-95A system.

As a summary, we can conclude from the above descriptions about the forward link channels of IS-95A system that a forward channel is identified by the following key parameters: (1) its CDMA RF carrier frequency, (2) the unique short PN code pilot offset of the sector or cell, and (3) the unique Walsh code of the user.

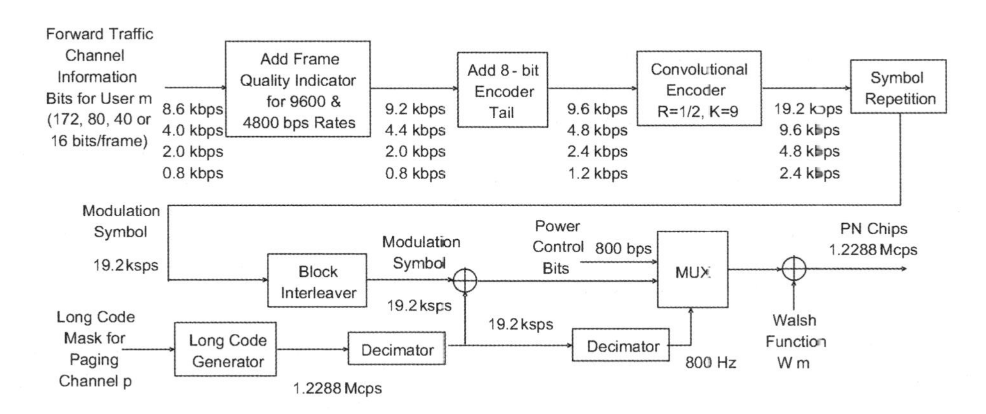

Figure 3.11 The signal generator for forward traffic channels in the IS-95A system.

[4]In practice, the number of active users in each sector/cell in the IS-95A system can range from a few to a bit more than ten, depending on the environment in which the system is operating. This problem is called the 'interference-limited capacity' of the CDMA technology used in all 2G/3G wireless systems.

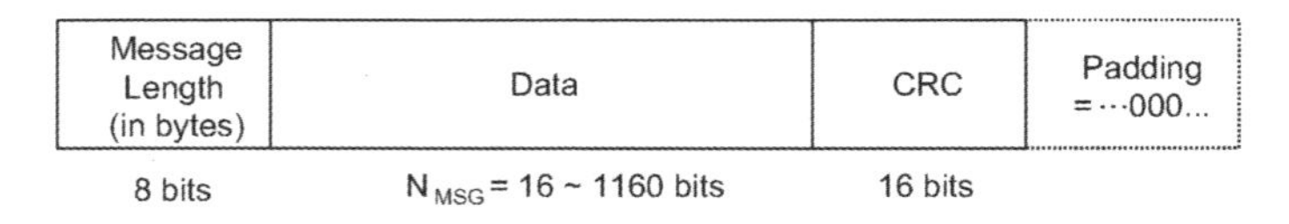

(a) Forward traffic channel frame structure

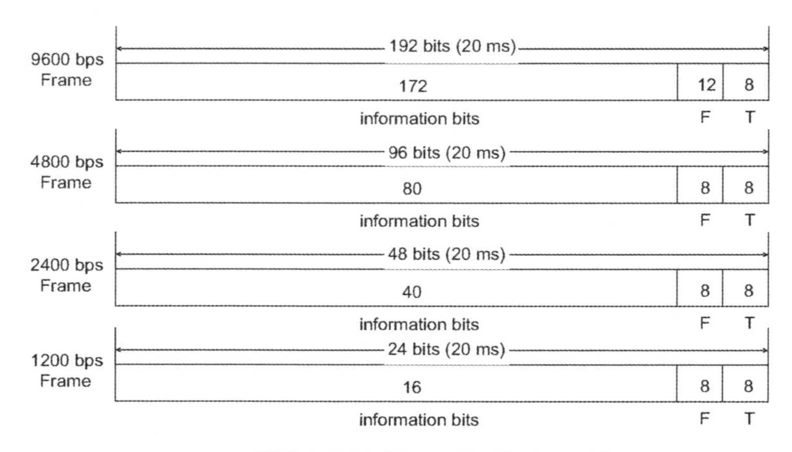

(b) Data field of forward traffic channel frame.
(F: Frame Quaity Indicator (CRC); T: Encoder Tail Bits.)

Figure 3.12 (a) The forward traffic channel frame structure in the IS-95A system. (b) Data field of forward traffic channel frame with different data rates.

3.1.4 REVERSE CHANNEL

A reverse channel in an IS-95A mobile cellular system is defined as a physical channel whose transmission direction is from MS to BS. Sometimes, the reverse channel is also called the uplink channel in contrast to the downlink channel explained in Subsection 3.1.3. The reverse channels carry both traffic and signaling information.

The reverse channel in IS-95A uses an FEC coding scheme with its code rate being 1/3 (in contrast to 1/2 in the forward link), and the code symbol rate is equal to 28 800 symbols per second. Each modulation symbol carries six code symbols and the PN rate is 1.2288 MHz, which is the same as the forward link channels. Different from the forward channels in IS-95A, which need to differentiate different MSs in the sector or cell, the reverse CDMA channels consist of $2^{42} - 1$ logical channels, which can be used to differentiate almost all MSs in the whole mobile cellular network operated by one particular service provider. Each of the logical channels is permanently and uniquely associated with one particular MS. The channel assignment does not change even upon handoffs, in which the MS travels across the boundary of cells or sectors. Reverse link addressing is accomplished through manipulation of the initial phase of the long PN code (whose period is $2^{42} - 1$) as part of the spreading process in a mobile transmitter. It should also be pointed out that the reverse CDMA channel does not observe the strict orthogonal rule in any sense, and rather it uses a very long period spreading

Table 3.2 Major parameters for the IS-95A reverse CDMA channels.

Channel	Access	Traffic			
Data rate (bps)	4800	1200	2400	4800	9600
Code rate	1/3	1/3	1/3	1/3	1/3
Symbol rate before repetition (sps)	14400	3600	7200	14400	28800
Symbol repetition	2	8	4	2	1
Symbol rate after repetition (sps)	28800	28800	28800	28800	28800
Transmit duty cycle	1	1/8	1/4	1/2	1
Code symbols/modulation symbol	6	6	6	6	6
PN chips/modulation symbol	256	256	256	256	256
PN chips transmitted/bit	256	128	128	128	128

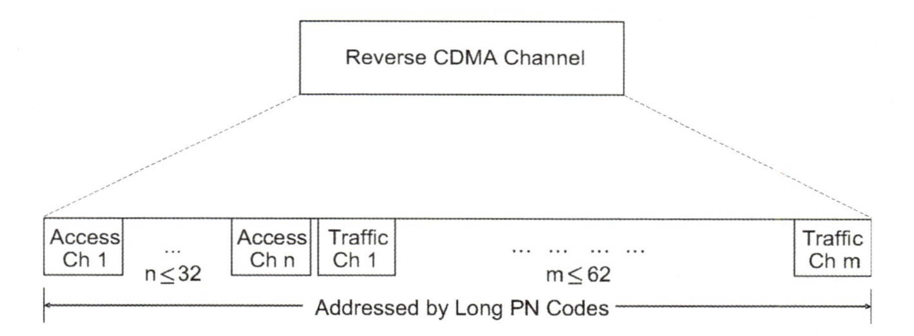

Figure 3.13 The logical reverse CDMA channels in the IS-95A system. There are two types of logical reverse CDMA channels, one being access channels (less than 32) and the other traffic channels (less than 62).

code or long PN code with different phases. The cross-correlation functions between different MSs are not necessarily zero, but are kept sufficiently small to control the interference.[5]

The modulation is carried out using 64-ary orthogonal Walsh functions, each period of which is repeated for four chips of the PN code. Therefore, the Walsh symbol rate can be calculated by 1.2288 MHz/(four chips per Walsh chip)/(64 Walsh chips per Walsh symbol) = 4800 modulation symbols per second. The major parameters of reverse CDMA channels in IS-95A are shown in Table 3.2. Figure 3.13 shows the conceptual diagram for all reverse logical CDMA channels. It is seen from the figure that there are two types of reverse CDMA channels, one being access channels (less than 32) and the other being traffic channels (less than 62).

[5]This non-zero cross-correlation property in the reverse channel in the IS-95A system contributes directly to the serious interference-limited capacity, which is far less than half of the processing gain of the system (64/2 = 32). In fact, this is one of the bottlenecks for capacity improvement in the IS-95A system.

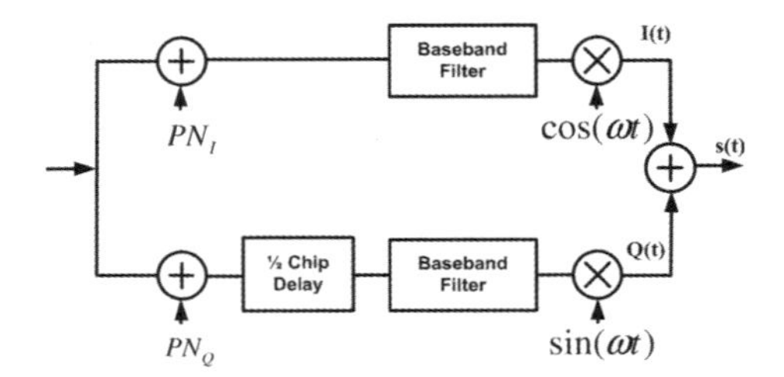

Figure 3.14 The OQPSK carrier modulation used in IS-95A reverse CDMA channels.

The carrier modulator used in the reverse CDMA channels is shown in Figure 3.14, where I and Q channels carry identical data information, but use different PN sequences (both long code and short code are used). Different from the carrier modulation (that is QPSK) used in the forward CDMA channels, the carrier modulation used in the reverse channels is OQPSK. The difference between QPSK and OQPSK lies in the fact that an extra 1/2 chip delay element is introduced in an OQPSK modulator. The benefit of using OQPSK instead of QPSK includes significant reduction in the amount of possible carrier phase change as chip value changes in time. The reduction in the possible carrier phase change can contribute to the improvement in sensitivity to the hard-limiting effect in RF power amplifiers working in class-C state, in which the transmitter operates in a non-linear region for better power efficiency. The rest of the carrier modulator in the reverse link looks very similar to that used in the forward link, as shown in Figure 3.6.

In addition, the reverse link has also the following major modulation parameters. First, it uses 20 ms interleaver for time diversity, as shown in Figure 3.19 (which will be explained later). Second, orthogonal 64-ary Walsh codes are used to modulate symbol through mapping every 6-bit data block into a particular Walsh code. Third, the long PN code is used for channel identification and is concatenated with the short PN code of length 2^{15}. Fourth, non-coherent OQPSK demodulation is used in the mobile receiver for carrier demodulation. Finally, three different diversity schemes are used in the reverse link to improve the signal quality at a receiver: time diversity through coding and interleaving, path diversity through RAKE receiver, and spacial diversity by using two receiver antennas at each BS. In fact, the signal processor after each antenna has four fingers in its RAKE receiver unit, such that altogether eight path-searchers will be available for signal detection. It is also noted that up to four receiver antennas will be available during the mobile handoff process. All of these measurements taken in the reverse link channels serve as extra protection against adverse effects present in a hostile asynchronous transmission environment in the reverse link channels.

As mentioned earlier, there are two different types of logical channels in the reverse link of IS-95A, i.e. access channels and traffic channels, which are discussed next.

Access Channels

The access channels in the reverse link are used by mobiles which are not yet in a call to transmit registration requests, call setup requests, page responses, order responses, and other data bursts, as well as to process other messages between the MS and BS. An access channel is really just a public long code offset unique to the BS sector. It is to be noted that the access channels in the reverse link are always paired to paging channels in the forward link. Each paging channel can have up to 32 access channels, as shown in Figure 3.13. All of these access channels operate as slotted random access channels with a fixed data rate of 4800 bps (or 20 ms frame duration), as listed in Table 3.2.

Figure 3.15 shows the block diagram of the signal generation process in the access channels, which operate in a packet switching mode. Figure 3.16 gives the reverse access channel long code

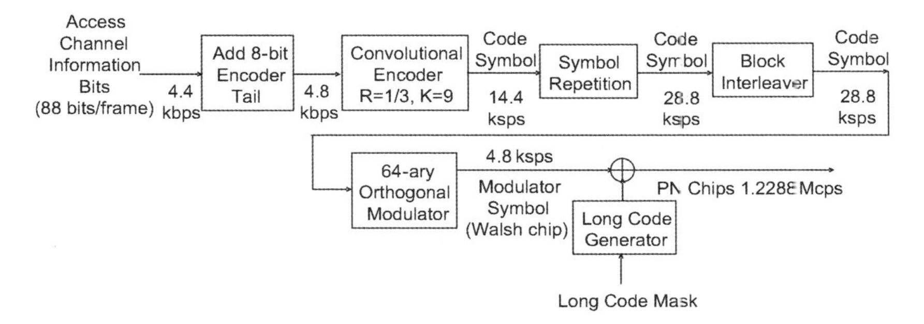

Figure 3.15 The generation of reverse link access channels in the IS-95A system.

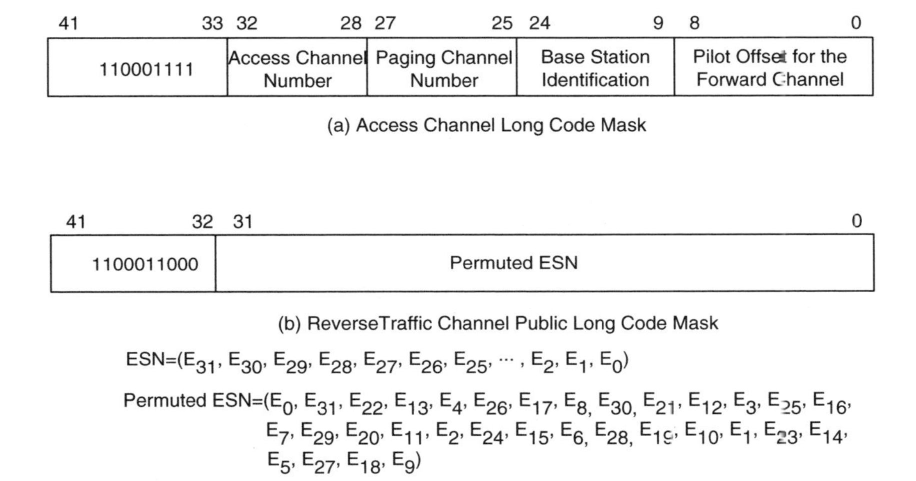

Figure 3.16 (a) The reverse access channel long code mask contents, and (b) reverse traffic channels public long code mask in the IS-95A system.

mask contents and reverse traffic channels public long code mask in the IS-95A system, where the Electronics Serial Number (ESN) is a 32-bit-long unique number assigned by the manufacturer of the mobile terminal. Therefore, it is seen that the ESNs are different for different mobile terminals, and thus they are always distinguishable even when they appear in different mobile networks throughout the world.

The reverse access channel message format, reverse access channel slot structure (which is also plotted in detail in Figure 3.18), and the basic frame structure for reverse access channels are shown in Figure 3.17. Figure 3.18 shows the reverse access channel slot structure, where an access channel slot always begins and ends on an access channel frame boundary. Each access frame has a length of 96 bits or 20 ms. Each access channel slot typically consists of $(3 + \text{MAX\_CAP\_SZ}) + (1 + \text{PAM\_SZ})$ access frames in length, as shown in Figure 3.18, where 'MAX_CAP_SZ' stands for maximal capsule size, and 'PAM_SZ' stands for preamble size, and their values should always be given to mobiles through forward paging channels. In IS-95A, it is also specified that the access channel slot should begin at access channel frames such that $t = 0 \sim \text{mod}(4 + \text{MAX\_CAP\_SZ} + \text{PAM\_SZ})$, where t is

Message Length (in bytes)	Data	CRC	Padding =... 000...
8 bits	N_{MSG}= 2 ~ 842 bits	16 bits	

(a) Reverse access channel message (with length of 88×N bits)

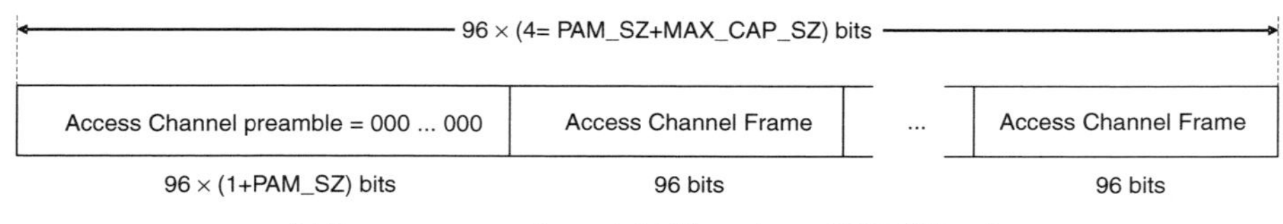

(b) Reverse access channel slot.The values of PAM_SZ and MAX_CAP_SZ are received on the forward link paging channel

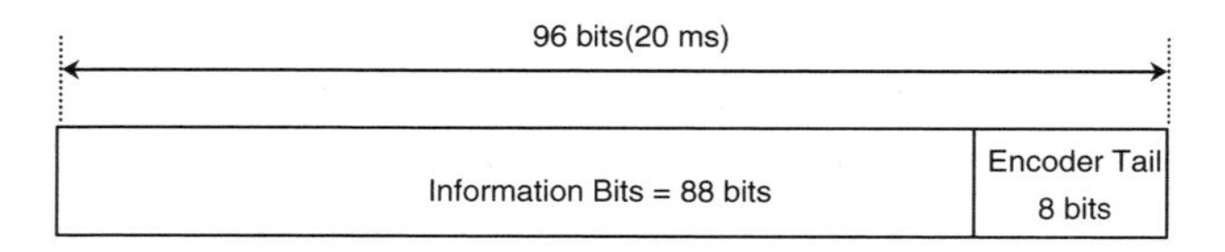

(c) Basic frame structure in reverse access channels. The length of an access channel frame = 88 bits + 8-bit tail bits (all zeros)

Figure 3.17 (a) The reverse access channel message format, (b) reverse access channels slot structure, and (c) basic frame structure of reverse access channels.

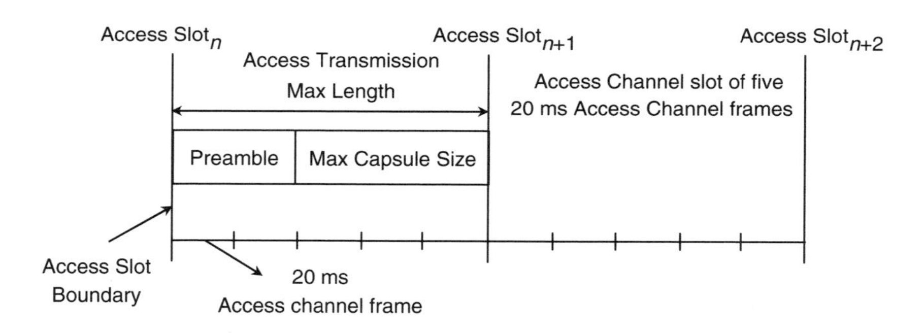

Figure 3.18 The reverse access channel slot structure.

the known system time given in frames. More detailed information about the reverse channel structure is shown in Figure 3.19.

There are many different types of information that the access channel should convey to perform variable functions. Therefore, the access channel field or access channel message, as shown in Figure 3.19, can vary from one to another. Due to the limited space, we will not give more detailed discussions on them; detailed information on the contents of access channel messages can be found in the IS-95A standard document [25–52].

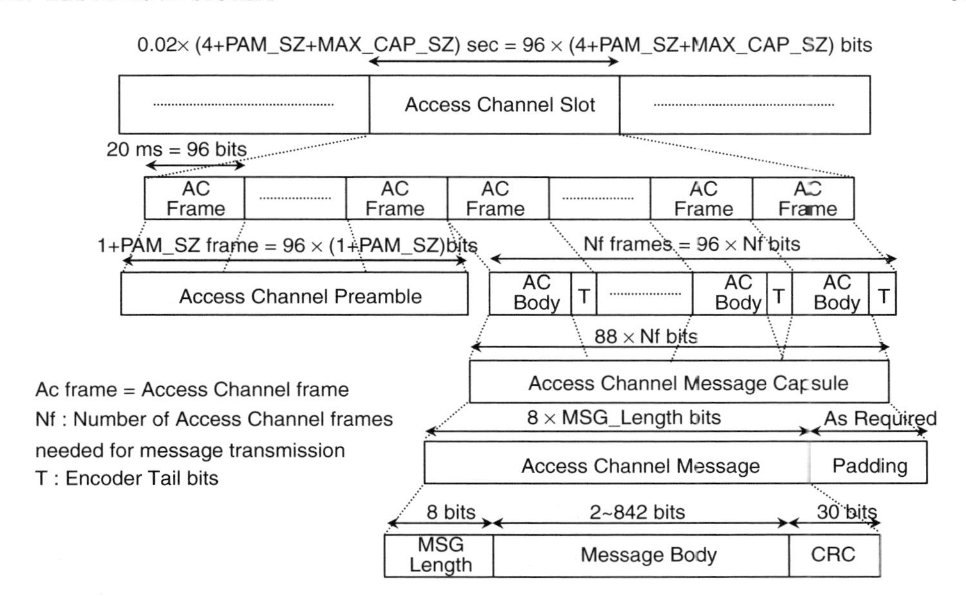

Figure 3.19 The reverse access channel structure details.

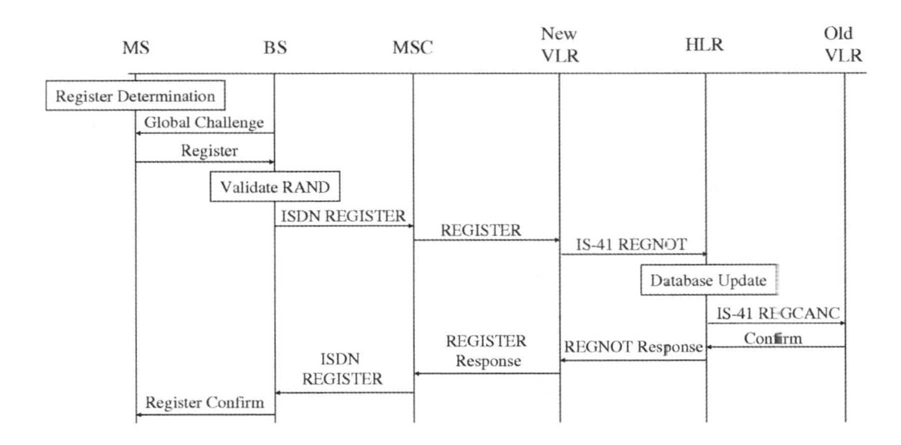

Figure 3.20 Signaling flow of registration process in IS-95A standard.

As an important function of reverse access channels, the registration process deserves some special attention. The registration process starts from the time when an MS notifies a BS of its presence in the system, giving information about its location, status, identification, etc. IS-95 supports the following registration types: (1) power-up registration, (2) power-down registration, (3) time-based registration, (4) distance-based registration, (5) zone-up registration, (6) parameter-change registration, (7) ordered registration, and (8) implicit registration. Figure 3.20 shows the signal flow happening in the registration process.

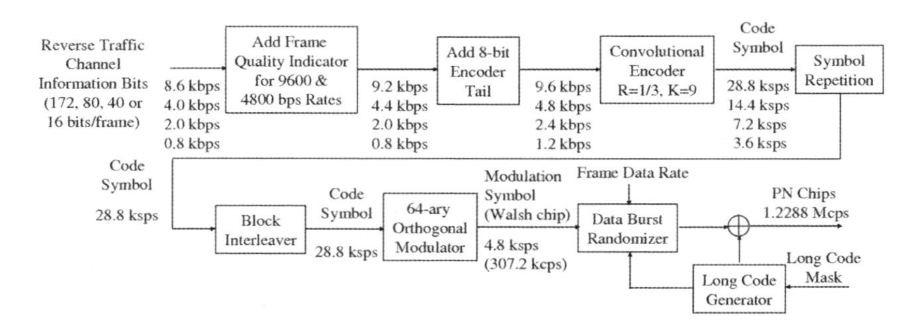

Figure 3.21 The signal generation process in reverse traffic channel in IS-95A.

Traffic Channels

The reverse traffic channels are used by individual users during their actual calls to transmit traffic to the BS. A reverse traffic channel uses a user-specific public or private long code mask. The reverse traffic channels can carry variable data rates and formats, just as forward traffic channels. However, the transmission duty cycle is reduced compared to forward link channels. Usually, bits are not repeated in the reverse traffic channels. The message format in the reverse traffic channels is identical to the forward traffic channels.

The signal generation process in the reverse traffic channels is shown in Figure 3.21, where 64-ary Walsh codes are used to encode every 6-bit code symbol into different Walsh codes, forming a 307.2 kcps stream before the data burst randomizer. Therefore, the spreading modulation in the reverse traffic channels undergoes two stages; the first partial spreading is accomplished when every 6-bit block is mapped into one of the 64-chip Walsh codes to yield a spreading factor of 10.667; the second partial spreading is carried out when the 307.2 kcps chip stream is further spread by the long code with a spreading factor of four, implying that each chip is spread into four chips of the long PN code. Thus, the total spreading factor in the reverse traffic channels is $10.667 \times 4 = 42.667$.

We can conclude from the above discussions on the reverse channels in IS-95A that a reverse channel is identified by the two major parameters: (1) its CDMA RF carrier frequency, and (2) a unique long PN code offset of the individual mobile terminal.

3.1.5 POWER CONTROL

To discuss in detail the power control technique used in an IS-95A system, we have to explain the reason why we need power control in a CDMA system. The reason is because of the near-far problem presented in all traditional CDMA systems. This issue has been addressed briefly in Chapter 2.

The near-far problem stems from the fact that in a CDMA system a user close to a cell site or BS would saturate the BS receiver and swamp the signals coming from all other users further away, unless the transmitting power is controlled. The near-far effect can be better explained in terms of the properties of auto-correlation and cross-correlation functions of a CDMA code set, which is used in a CDMA system for channelization, just like the 64-ary Walsh codes in the forward link of an IS-95A system. To make a good CDMA system, the difference between the auto-correlation peak and out-of-phase auto-correlation functions (also called auto-correlation side lobes) should be made as big as possible. Assume that there are two mobiles inside a cell, with mobile A being very close to the BS and mobile B being located further away from the BS. Now, the BS receiver wants to receive the signal from mobile B. What will happen if no power control is in effect or equivalently the two mobiles are transmitting at equal power? It is understandable that the BS receiver will never detect the auto-correlation peak of the signal from mobile B because the cross-correlation function generated by the strong signal from mobile A will overwhelm the auto-correlation peak of mobile B. Therefore, we have to find a way to control the transmitting power of all mobiles in a cell to make their signals almost the same level seen from the BS receiver.

Due to the imperfect cross-correlation functions of the CDMA codes, any currently available CDMA system, including IS-95A, is an interference-limited system. Having made this statement, let us dwell on this. The success of an IS-95 system lies in controlling the total power in the system.

In a CDMA environment every MS is a source of interference to the others. At the MS receiver we see the radio environment around it as a cumulative addition of information for itself and interference. The interference is actually the information for other MSs plus noise from others sources. Hence if the interference is higher, the information signal cannot be retrieved. A mobile has a special receiver called a RAKE receiver that can make estimates of multipath fading and retrieve the information for a particular mobile.

An IS-95A system applies the power control technique to both forward link and reverse link channels. There are basically two different types of power control techniques, i.e. open-loop power control and closed-loop power control, which are to be explained as follows.

Open-loop power control is used during access attempts. It increases the transmitting power during each attempt. We do not need to worry that such attempts may disturb the communications by increasing power at the BS receiver, since the BS has already informed the MSs of the power increment step on each attempt on a broadcast channel.

In closed-loop power control there is a feedback procedure. This type of power control is used when the MS is using the traffic channel resources, i.e. it is active. The BS is continuously monitoring the reverse link. If it finds that the quality of the reverse link is poor then it will instruct the mobile to increase its power by inserting power control bits in the traffic data. This insertion of power bits for power control is called bit puncturing of the forward link. The BS does this every 800 bits/s, as shown in Figure 3.5.

3.1.6 HANDOVER

The handoff procedure in IS-95A has a lot to do with the forward link pilot channel. Therefore, please recap all your knowledge about the pilot channel discussed earlier in Subsection 3.1.3.

The pilot channel acts just like a lighthouse to a ship. It acts as a beacon for the mobiles and identifies the BS. When the MS powers on it gets latched to a BS by searching for the pilots. Now the question is, which pilot does it latch on to or which BS does it latch to? Obviously it has to latch to the BS that is nearest to it. Then how does the MS know which BS is the nearest? In fact, the MS will scan for the strongest (in terms of power) pilot channel and latch to it. Now comes a question on why handoff? When the MS goes from one cell to another, handover or handoff occurs. As the MS goes away from the BS the power level of the pilot channel may decrease and hence it looks for a pilot of stronger strength to latch on to. This is to ensure that when an MS is using traffic channel resources, the information flow does not stop when control goes from one BS to another. Thus we prevent call dropping. At this point there may be another question. Does the MS scan all the short PN offsets to search for a pilot during handoff? In fact, pilot channels having the largest power, the MS from its position will receive varying power levels of different offsets. Here let us introduce the term pilot database. The pilots are divided into sets, which are used to search for pilots during handoff. The mobile maintains four sets:

1. **Active set**: pilots associated with forward traffic channels assigned by the base station.

2. **Candidate set**: Pilots not currently in the active set, but whose level is high enough to be there (but others are stronger).

3. **Neighbor set**: pilots that are not currently in the active set or candidate set and are likely candidates for handoff. The initial neighbor list is sent to the mobile in the system parameters message on the paging channel.

4. **Remaining set**: includes all pilots in the system which are not in another set.

Next, we want to discuss handoff procedures. There are several different types of handoff procedure depending on the different situations, explained as follows:

1. **Soft handoff**: when the mobile goes from one cell to another but uses the same frequency. We also can have softer handoff, when the mobile goes to a different sector within a cell.

2. **Hard handoff**: when (i) the MS is transferred between disjoint active sets, (ii) CDMA frequency assignment changes, (iii) the frame offset changes, or (iv) the mobile is sent from CDMA channels to analog voice channels.

3. **Idle handoff**: when the paging channel is transferred from one BS to another.

4. **Access handoff**: when the mobile sends the access attempts to another BS.

The following steps should proceed in a handoff:

1. starting in a state where only one cell is supporting the call in question,
2. informing the candidate cell of the imminent handoff,
3. signaling the mobile to begin executing the handoff,
4. new cell beginning to service the mobile,
5. mobile beginning to use the new cell,
6. entering the mid-handoff state (prolonged only in CDMA),
7. mobile discontinuing use of the old cell,
8. old cell stopping service to the mobile,
9. ending in a state where the new cell is supporting the call in question.

The IS-95A system uses CDMA technology that is specifically designed not only to reduce handoff failures but also to provide seamless services. We can see the differences between the traditional 1G AMPS system and IS-95A system in terms of their handoff characteristic features from Table 3.3.

Table 3.3 Differences between the traditional 1G AMPS system and IS-95A system in terms of their handoff characteristic features.

IS-95A	AMPS
CDMA handoffs do not normally require frequency tuning	Hard handoff (communication is interrupted briefly)
CDMA requires change of the code channel in the forward CDMA channel	No simultaneous communication with more than one BS
No tuning, either frequency or code channel, is required in the reverse CDMA channel at any time.	BS does the signal quality measurement

3.2 ETSI WCDMA SYSTEM

UMTS is one of the 3G mobile systems being developed within the ITU's IMT-2000 framework. It is a realization of a new generation of wideband multi media mobile telecommunications technology. The coverage area of service provision is to be worldwide in the form of Future Land Mobile Telecommunications Services (FLMTS), also called IMT-2000. The coverage will be provided presumably by a combination of cell sizes ranging from in-building pico-cells to global cells covered by satellites, giving services to the remote regions of the world. It was expected that the UMTS would not be a replacement for 2G technologies (e.g. GSM, DCS1800, CDMA, DECT, etc.), which will continue to evolve to their full potential.

UMTS was developed mainly for countries with 2G GSM networks, because these countries have agreed to free new frequency ranges for UMTS networks. As a matter of fact, UMTS is a new technology and operates in a new frequency band, and thus a whole new radio access network has to be build. This is obviously a disadvantage compared to the relatively smooth upgrading path from IS-95 to cdma2000 1X, as discussed in Section 3.1. The advantage of the UMTS system is that the new frequency range gives plenty of new capacity for operators. 3GPP has been overseeing the standard development and has wisely kept the core network as close to the GSM core network as possible. It is noted that UMTS phones are not meant to be backward compatible with GSM systems, but subscriptions (e.g. SIM cards) can be. Dual-mode phones will solve the compatibility problems, hopefully. UMTS also has two flavors, or FDD (which is also named UMTS FDD) and TDD (which is also named UMTS TDD). It is quite certain that the former has gained much attention and will be implemented first. Some harmonization has been done between systems, such as chip rate and pilot issues, etc.

The CDMA technology used by UMTS systems is commonly called wideband CDMA or simply WCDMA. 3G WCDMA systems have 5 MHz bandwidth (in either uplink or downlink channel). In fact, a 5 MHz bandwidth is neither wide nor narrow; it is just a bandwidth. Nevertheless, the new 3G WCDMA systems have indeed a wider bandwidth than existing 2G CDMA systems (i.e. 1.25 MHz bandwidth in IS-95), and that is why it is called wideband. It should be noted that the name WCDMA is true in a relative sense, as there are commercially available CDMA systems operating over a 20 MHz bandwidth.

It is meaningful at this moment for us to take a brief look at different 3G standards in the world. There are five major 3G air interface technologies specified by ITU Recommendation ITU-R M.1457:

- IMT-2000 CDMA Direct Spread, also known as UTRA FDD including WCDMA in Japan, ARIB/DoCoMo recommendation. UMTS is developed by 3GPP.

- IMT-2000 CDMA Multi-carrier, also known as CDMA2000 (3X), developed by 3GPP2. IMT-2000 CDMA2000 includes 1X components, like CDMA2000 1X EV-DO.

- IMT-2000 CDMA TDD, also known as UTRA TDD and TD-SCDMA. TD-SCDMA is developed by China and supported by the TD-SCDMA Forum.

- IMT-2000 TDMA Single Carrier, also known as UWC-136 (EDGE), supported by UWCC.

- IMT-2000 DECT, supported by the DECT Forum.

3G is a generic name for a set of mobile technologies which are designed for multimedia communications. Defined by the International Telecommunications Union (ITU), 3G systems must provide: (1) backward compatibility with 2G systems,x (2) multimedia support, (3) improved system capacity compared to 2G and 2.5G cellular systems, and (4) high-speed packet data services ranging from 144 kbps in wide-area mobile environments to 2 Mbps in fixed or in-building environments. The standardization of 3G systems was conducted in several regions through their respective standard organizations:

- ETSI: European Telecommunications Standards Institute
- T1: Standardization Committee—Telecommunications (US)
- TIA: Telecommunications Industry Association (North America)
- ARIB: Association of Radio Industries and Business (Japan)
- TTC: Telecommunications Technology Committee (Japan)
- TTA: Telecommunications Technology Association (Korea)
- CWTS: China Wireless Telecommunications Standard group.

International Mobile Telecommunications-2000 (IMT-2000), initiated by the ITU, is presumably the global standard for 3G wireless communications, defined by a set of interdependent ITU Recommendations. IMT-2000 provides a framework for worldwide wireless access. Out of the ITU's IMT-2000 initiative, the 3rd Generation Partnership Project (3GPP) and the 3GPP2 were born.

3GPP is a collaboration agreement that was established in December 1998. The collaboration agreement brings together a number of telecommunications standards bodies, which are known as organizational partners. The current organizational partners are ARIB and TTC (Japan), CCSA (China), ETSI (Europe), T1 (USA), and TTA (Korea). 3GPP is focused on WCDMA-based technology and its derivative and upgraded versions. Refer to the website at `http://www.3gpp.org` for more information.

On the other side, 3GPP2 is another collaborative effort between five officially recognized standards bodies (ARIB, CCSA, TIA, TTA, and TTC) (see `http://www.3gpp2.org/`), whose activities are focused on cdma2000-based technologies.

The proposal of ETSI that was submitted to 3GPP is called UMTS (Universal Mobile Telecommunications System). The terrestrial version of UMTS is called UMTS Terrestrial Radio Access (UTRA). The proposal of 3GPP is also called UTRA, which stands for Universal Terrestrial Radio Access. UTRA has two modes: (1) frequency division duplex (FDD) and (2) time division duplex (TDD). There are salient features for FDD and TDD operation modes, which can be summarized as follows.

FDD operation mode provides simultaneous radio transmission channels for mobiles and base stations. At the base station, separate transmit and receive antennas are used to accommodate separate uplink and downlink channels. At the mobile unit, a single antenna is used for both transmission to and reception from the base station, and a duplexer is used to enable the same antenna to be used for simultaneous transmission and reception. It is necessary to separate the transmit and receive frequencies so that the duplexer can be given sufficient isolation while being inexpensively manufactured. It is noted that FDD has been used exclusively in earlier analog mobile radio systems.

On the other hand, TDD mode shares a single radio channel in time so that a portion of the time is used to transmit from the base station to the mobile, and the remaining time is used to transmit from the mobile to the base station. If the data transmission rate is much greater than the end-user's data rate, it is possible to store information bursts and provide the appearance of full duplex operation to a user, even though there are not two simultaneous radio transmissions at any instant of time. TDD is only feasible with digital transmission formats and digital modulation, and is very sensitive to timing.

Table 3.4 compares the differences in major air interface parameters for the UMTS UTRA-FDD, UMTS UTRA-TDD, and TD-SCDMA systems. Table 3.5 gives major system parameters for the UMTS WCDMA and cdma2000 systems. Table 3.6 makes a comparison among different 2.5G and 3G technologies in terms of their capabilities.

Table 3.4 Comparison of major system parameters for the UMTS UTRA-FDD, UMTS UTRA-TDD, and TD-SCDMA systems.

	FDD scheme	TDD schemes	
Multiplex technology	WCDMA	TD-CDMA	TD-SCDMA
Bandwidth	2×5 MHz paired	1×5 MHz unpaired	1×1.6 MHz unpaired
Frequency reuse	1	1	1 (or 3)
Handover	Soft, softer (interfreq.: hard)	Hard	Hard
Modulation	QPSK	QPSK	QPSK and 8-PSK
Receiver	RAKE	Joint detection RAKE (mobile station)	Joint detection RAKE (mobile station)
Chip rate	3.84 Mcps	3.84 Mcps	1.28 Mcps
Spreading factor	4–256	1, 2, 4, 8, 16	1, 2, 4, 8, 16
Power control*	Fast: every 667 μs**	Slow: 100 cycles/s***	Slow: 200 cycles/s***
Frame organization	0.667/10 ms	0.667/10 ms	0.675/5 ms
Timeslots/frame	N.A.	15	7

*Range: 80 dB (UL), 30 dB (DL) in steps of **0.25 to 1.5 dB; ***1, 2 or 3 dB.

Table 3.5 Comparison of major system parameters for the UMTS WCDMA and cdma2000 systems.

Parameter	WCDMA	cdma2000
Carrier spacing	5 MHz	3.75 MHz
Chip rate	4.096 MHz	3.6864 MHz
Data modulation	BPSK	FW–QPSK RV–BPSK
Spreading	Complex (OQPSK)	Complex (OQPSK)
Power control frequency	1500 Hz	800 Hz
Variable data rate implement.	Variable SF; multicode	Repeat., puncturing, multicode
Frame duration	10 ms	20 ms (also 5, 30, 40)
Coding	Turbo and convolutional	Turbo and convolutional
Base stations synchronized?	Asynchronous	Synchronous
Base station acquisition/detect	3 step; slot, frame, code	Time-shifted PN correlation
Forward link pilot	TDM dedicated pilot	CDM common pilot
Antenna beam forming	TDM dedicated pilot	Auxiliary pilot

Table 3.6 Comparison of capabilities for different 2.5G and 3G technologies.

	Peak network downlink speed	Average user throughput for file downloads	Capacity	Other features
GPRS	115 kbps	30–40 kbps		
EDGE	473 kbps	100–130 kbps	Double of that for GPRS	GPRS backward compatible
UMTS-WCDMA	2 Mbps	220–320 kbps	Increased over EDGE for high-bandwidth applications	Simultaneous voice and data operation, enhanced security, QoS, multimedia support, and reduced delay
UMTS-HSDPA*	14 Mbps	550–1100 kbps	Two and a half to three and a half times that of WCDMA	Backward compatible with WCDMA
CDMA2000 1xRTT	153 kbps	50–70 kbps		
CDMA2000 1xEV-DO	2.4 Mbps	300–500 kbps		Optimized for data, VoIP in development

*High Speed Downlink Packet Access (HSDPA) is in fact an extension of UMTS and can offer a data rate up to 10 Mbps on the downlink channel. HSDPA is a new 3GPP standard in order to increase the downlink throughput by replacing QPSK in UMTS by 16QAM in HSDPA. It works to offer a combination of channel bundling (TDMA), code multiplex (CDMA) and improved coding (adaptive modulation and coding). It also introduces a separate control channel to facilitate the data transmission speed. Similar techniques will be available later for uplink with HSUPA.

A UMTS network consists of three interacting domains: core network (CN), UMTS Terrestrial Radio Access Network (UTRAN), and user equipment (UE). The main function of the CN is to provide switching for user traffic. The CN also contains the databases and network management functions. The basic CN architecture for UMTS is based on GSM network with GPRS. All equipment has to be modified for UMTS operation and services. The UTRAN provides the air-interface access method for UE. A base station is referred to as a Node-B and the control equipment for Node-Bs is called the radio network controller (RNC).

The spectrum allocation in Europe, Japan, and Korea for the FDD mode is 1920-1980 MHz for the uplink and 2110–2170 MHz for the downlink, with the bands 1980–2010 MHz and 2170–2200 MHz intended for the satellite part of future systems. The UTRA TDD mode utilizes two frequency bands in Europe, the 1900–1920 MHz and 2010–2025 MHz bands. In both modes each carrier has a bandwidth of approximately 5 MHz. In the FDD mode, separate 5 MHz carrier frequencies are used for the uplink and downlink. On the other hand, in TDD only one 5 MHz is shared between uplink and downlink. Each operator, subject to its offered license, can deploy multiple 5 MHz carriers in order to increase capacity. Figure 3.22 shows the UMTS frequency spectrum allocation after the World Radio Conference (WRC) 2002.

Figure 3.23 compares voice capacity per 5 MHz spectrum for different 2G and 3G systems. It is seen from the figure that WCDMA offers a performance that still lags behind cdma2000 1X and

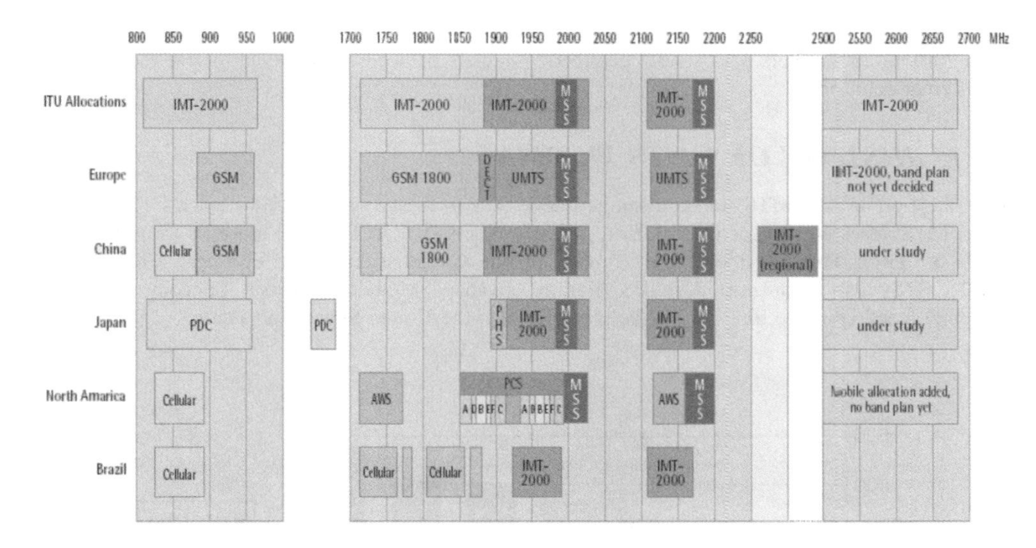

Figure 3.22 IMT-2000/UMTS spectrum allocation for different regions in the world, which was decided at the World Radio Conference (WRC) in 2002.

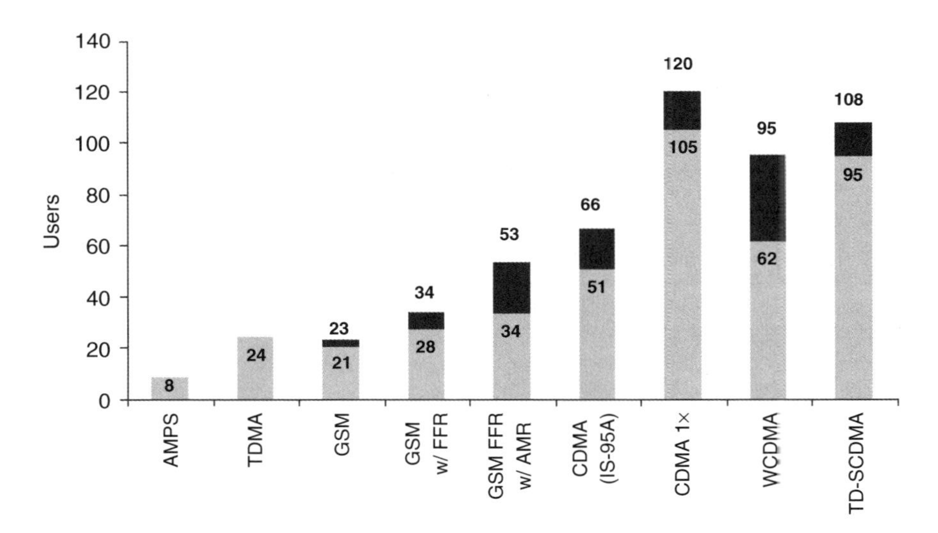

Figure 3.23 Voice capacity comparison in terms of per 5 MHz spectrum for different 2G and 3G systems, where the dark regions on the bars show the capacity variation among applications with variable link setups.

TD-SCDMA systems. Figure 3.24 shows handset sale comparison for different 2G and 3G systems from 2001 to 2007. Table 3.7 lists 3G networks, numbers of licenses, and deployment requirements in different countries.

3.2.1 HISTORY OF UMTS WCDMA

The inception of the UMTS standard can be traced back to as early as the 1990s when ETSI initiated one UMTS research project in RACE 1, seven projects in RACE2, and fourteen projects in the ACTS Program. RACE projects were funded by the Commission of the European Communities (CEC). ETSI also organized FAMOUS (Future Advanced MObile Universal Telecommunications Systems) meetings three times a year between Europe, the United States, and Japan.

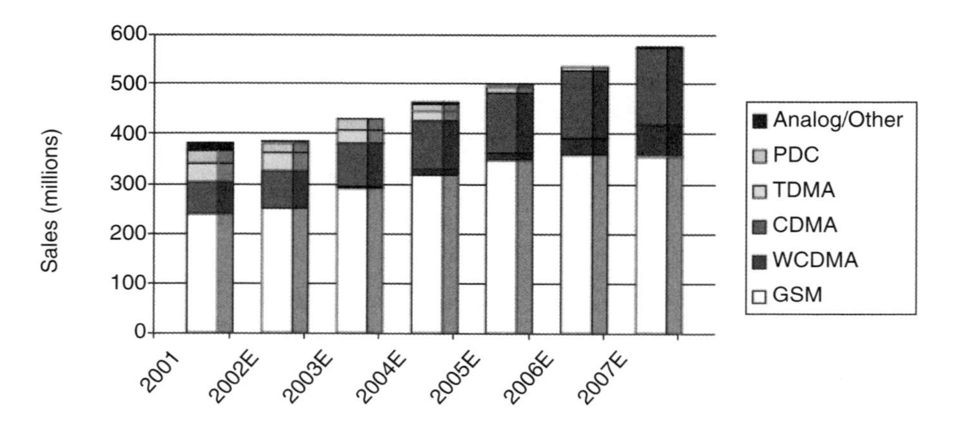

Figure 3.24 Handset sale comparison for different 2G and 3G systems.

Table 3.7 3G networks, numbers of licenses, and deployment requirements in different countries.

Country	3G Network	Number of licenses	Government requirements
Australia	CDMA2000 and WCDMA	6	No coverage obligations
Austria	WCDMA	4	25% coverage by end-2003; 50% by end-2005
Belgium	WCDMA	3	30% coverage in 3 years, 40% in 4 years, 50% in 5 years; 85% by end-2006. In 2/02 delayed launch from 9/02 to 9/03
Brazil	CDMA2000 and WCDMA	NA	No special 3G requirements/policies announced
Canada	CDMA2000 and WCDMA	NA	No special 3G license requirements. Operators use regular spectrum licenses

Table 3.7 *(continued)*

Country	3G Network	Number of licenses	Government requirements
China	CDMA2000 and WCDMA	NA	No special 3G requirements/policies announced
Denmark	WCDMA	4	30% of population by end-2004; 80% by end-2008; then sharing allowed for next 20%
Finland	WCDMA	4	No coverage requirements but ministry may ensure implementation
France	WCDMA	2 (+1 pending)	25% voice coverage and 20% data 2 years after launch; 80% voice and 60% data 8 years after launch
Germany	WCDMA	6	25% of population covered by end-2003, 50% by end-2005, does not allow mergers of 3G license holders
Greece	WCDMA	3	Cover 25% population by 12/03, Olympic Games facilities 02/04, 50% population by 12/06
Hong Kong	WCDMA	4	50% coverage by end-2006; keep 30% available for MVNOs
Ireland	WCDMA	4	License A (more spectrum): 80% coverage by 2008 License B: cover 5 major cities (58% population) by 2008
Italy	WCDMA	5	Cover regional capitals within 30 months, provincial capitals within 60 months
Japan	CDMA2000 and WCDMA	3	Licenses have temporary status, awarded permanent licenses when ministry is satisfied with 3G status of each operator
The Netherlands	WCDMA	5	5 licenses with specific coverage requirements. Infr sharing allowed in 9/01, but service separately
Norway	WCDMA	4	3 licenses (4th given back): 90% coverage to largest cities within 5 years from launch; may fine if buildout not on track
Portugal	WCDMA	4	20% coverage in 1 year, 40% in 3 years, 60% in 5 years; each license holder committed $ 768.4 million to infrastructure

(continued overleaf)

Table 3.7 (*continued*)

Country	3G Network	Number of licenses	Government requirements
Singapore	NA	3	Provisional deadline of 12/31/04 for nationwide network
South Korea	CDMA2000 and WCDMA	2 (+1 pending)	Government warned operators in 02/02 not to switch 3G technologies
Spain	WCDMA	4	Launch by mid-2003, previously cover 23 cities by 06/02 (postponed from 08/01); pre-postponement required 90% coverage by 2005
Sweden	WCDMA	4	Access by 1/1/02; licenses pay 0.15% of revenues; 99.% overall coverage by end-2003. Telia suing for license
Switzerland	WCDMA	4	50% coverage by end-2004; launch 12/31/02. Government said in 8/01 willing to push back launch
UK	WCDMA	5	80% coverage by end-2007
USA	CDMA2000 and WCDMA	NA	No special 3G license requirements, can use regular spectrum; doubts of enough spectrum being available for WCDMA

From 1991 to 1995, two CEC-funded research projects called Code Division Testbed (CODIT) and Advanced Time Division Multiple Access (ATDMA) were carried out by the major European telecommunication manufacturers and network operators. The CODIT and ATDMA projects investigated the suitability of wideband CDMA- and TDMA-based radio access technologies for 3G systems. This work was later continued in the Future Radio Wideband Multiple Access System (FRAMES) project and became the basis of further ETSI UMTS work until decisions were made in 1998.

In February 1992 the World Radio Conference (WRC) in Malaga allocated frequencies for future UMTS use. Frequency bands 1885–2025 and 2110–2200 MHz were identified for IMT-2000 use. The UMTS Task Force was established in February 1995, issuing the report entitled 'The Road to UMTS'.

The UMTS Forum was established at the inaugural meeting, held in Zurich, Switzerland, in December 1996. Since then, the planned 'European' WCDMA standard has been known as the Universal Mobile Telecommunications System (UMTS). In June 1997 the UMTS Forum produced its first report, entitled 'A Regulatory Framework for UMTS'. UMTS core band was decided in October 1997.

In the January 1998 ETSI SMG meeting in Paris, both the WCDMA and TD-CDMA proposals were combined into the UMTS air-interface specification. In June 1998, the terrestrial air-interface proposals (UTRAN, WCDMA(s), cdma2000(s), EDGE, EP-DECT, TD-SCDMA) were handed to the ITU-R as possible IMT-2000 candidate proposals. The first call using a Nokia WCDMA terminal in DoCoMo's trial network was completed in September 1998 at Nokia's R&D unit near Tokyo in Japan.

On December 4, 1998, ETSI SMG, T1P1, ARIB, TTC, and TTA created 3GPP in Copenhagen, Denmark, and the first meeting of the 3GPP Technical Specification Groups was held at Sophia Antipolis, France, on December 7 and 8, 1998.

On April 27 and 28, 1999 Lucent Technologies, Ericsson, and NEC announced that they had been chosen by NTT DoCoMo to supply WCDMA equipment for NTT DoCoMo's next generation wireless commercial network in Japan. This was the first announced WCDMA 3G infrastructure deal.

3GPP approved the UMTS Release 4 specification in March 2001 at the Palm Springs meeting.

NTT DoCoMo launched a trial 3G service, an area-specific information service for i-mode, on June 28, 2001. On September 25, 2001, NTT DoCoMo announced that three 3G phone models were commercially available. NTT DoCoMo launched the first commercial WCDMA 3G mobile network on October 1, 2001.

On March 14, 2002, UMTS Release 5 was issued. It is noted that the initial target date was December 2001. UMTS Release 6 was issued on December 16, 2004, which was delayed from its initial target date of June 2003.

Ericsson demonstrated 9 Mbps with WCDMA, High Speed Downlink Packet Access (HSDPA) phase 2, on February 14, 2005. Ericsson and several operators in three Scandinavian countries demonstrated 1.5 Mbps enhanced uplink in a live WCDMA system on May 10, 2005. HSDPA is a new 3GPP standard in order to increase the downlink throughput by changing the radio modulation (QPSK to 16QAM). Later HSUPA for uplink will also be available. HSDPA and HSUPA have also been known as 3.5G technology.

The milestones of development of UMTS are summarized as:

- February 1992 (Malaga): ITU-R World Radio Conference identified IMT2000 frequency bands.
- January 1998 (Paris): ETSI selected WCDMA for paired (FDD) and TD-CDMA for unpaired (TDD) UMTS-operation out of five competing modes.
- November 1999 (Helsinki): ITU approved IMT-2000 Radio Interface specifications including FDD and TDD modes at ITU meeting (M.1457).
- December 1999 (Nice): 3GPP approved UMTS Release 99 specifications for both FDD and TDD.
- March 2001 (Palm Springs): 3GPP approved UMTS Release 4 specifications for both FDD and TDD

To understand better where the UMTS standard stands in ITU IMT-2000 proposals, we provide Figure 3.25, where we have plotted all major ITU-endorsed IMT-2000 candidate proposals and later called 3G standards. From all those listed proposals or standards, we can classified them into:

1. Two core technologies (TDMA and CDMA).
2. Three systems (UMTS, CDMA2000, and UWC-136 or EDGE).
3. Five radio interfaces:

   i. IMT-DS (direct spread), used in UTRA-FDD
   ii. IMT-MC (multi-carrier), used in CDMA2000 system
   iii. IMT-TC (time code), used in UTRA-TDD and TD-SCDMA
   iv. IMT-SC (single carrier), used in UWC-136 or EDGE technology
   v. IMT-FT (frequency time), used in DECT system.

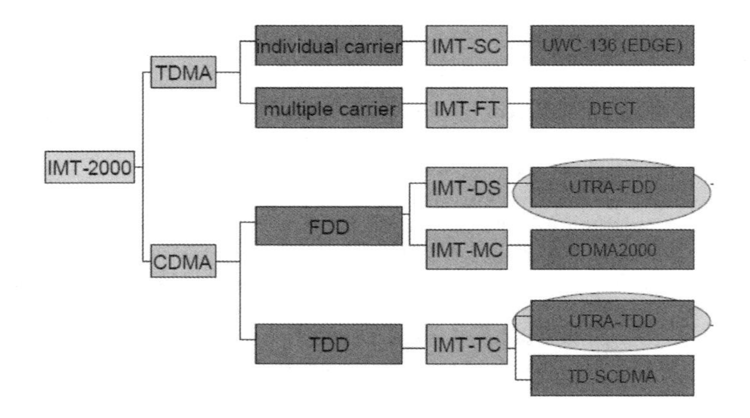

Figure 3.25 Family tree of all major ITU IMT-2000 candidate proposals.

3.2.2 ETSI UMTS VERSUS ARIB WCDMA

In this section, we will focus our discussions on ETSI UMTS WCDMA [83] technology because it has been a standard release issued by 3GPP. All parties of 3GPP should make their best efforts to commit to full compatibility with the 3GPP releases, whose major versions are listed in Table 3.8. On the other hand, the Japanese version of WCDMA launched by NTT DoCoMo in October 2001, also called ARIB WCDMA system [89] or FOMA service, has some technical differences compared to the UMTS standard, and we will give some explanation in this subsection.

Table 3.8 Key aspects of major 3GPP UMTS standard Releases.

Release	Key Aspects
Release 99* (March 2000)	1. New radio interface (UTRAN) 2. SMS, EMS, and MMS 3. FDD and TDD at 3.84 Mcps 4. Handover 5. CAMEL Phase 2 and 3 used for pre-paid services and access charge in GSM and GPRS networks 6. EDGE 7. GSM-UMTS interworking 8. Call forwarding enhancement
Release 4 (March 2001)	1. New TDD mode at 1.28 Mcps 2. Data synchronization (SyncML) 3. Evolution of UTRA to support IP 4. SMS and EMS enhancements 5. MMS Release 4 6. Evolution of core transport to IP 7. UTRAN repeater

Table 3.8 (*continued*)

Release	Key Aspects
Release 5 (March 2002)	1. Multi-rate speech codec 2. Provision of IP-based multimedia services 3. Wideband AMR 4. Security enhancements 5. IP multimedia services 6. MMS Release 5 7. CAMEL Phase 4-optimal routing of mobile-mobile calls 8. Evolution of UTRAN transport to IP

*Release 99 of the UMTS standards is the first version to be deployed. Release 4 and Release 5 specify enhancements and optional features. A key philosophy of the 3GPP standards is that the first release specifies all mandatory features, while later releases add optional features only.

Members of the 3GPP include organizations, such as ETSI of Europe and ARIB of Japan, etc., individual members, and market representatives, such as the GSM Association, etc. Included with the organizational and market representation partners are virtually every major OEM. The 3GPP, similar to 3GPP2, is divided into several technical standards for their respective areas. Once these standards are written, the 3GPP endorses the standards and submits them to the ITU.

One aspect of 3G standard development that is often misunderstood by the public is the concept of releases, a concept that also applies to 2G and 2.5G networks. 3G, in this case UMTS, does not consist of only one release, but a series of releases that were built upon previous releases. Initially, releases were noted by year; for instance, Release 99, Release 00, and so on. However, releases are no longer tied to the year in which they are finalized. Instead, the 3GPP has defined the requirements in Release 4, pretty much finalized Release 5, and has recently begun working on Release 6, all of which are subsequently releases of the UMTS Release 99 standard.

What makes it even more complex is that, within each release (e.g. Release 99) there are multiple versions. For instance, Release 99 began with the March 2000 version of the Release 99 standard, and has since evolved every three months since that time in conjunction with the quarterly 3GPP plenary meetings. Although the basic functionality of Release 99 does not change each quarter, the technical definition of how the functionality is implemented does change. Specifically, 3GPP members submit change requests (CRs), which identify changes to the baseline documentation. CRs can include anything from typographical or grammatical errors to additional/changed text that is inserted/replaced to clear up an ambiguity or correct an error, both of which could prevent a successful launch, in the documentation.

Now let us go back to the topic of interest in this subsection. An interesting issue on the evolutional path of WCDMA technology is the compatibility between ARIB WCDMA technology [89] developed by NTT DoCoMo and UMTS-FDD [83] proposed by ETSI.

In October 2001, NTT DoCoMo launched commercial Freedom of Mobile Multimedia Access (FOMA) services. A lot has been reported on the launch, but we still believe there are some widely held misunderstandings about what happened in Japan in comparison to European activities on UMTS-FDD.

FOMA networks use WCDMA, like the UMTS standard being deployed in Europe. However, FOMA uses an earlier version of the UMTS release. Without going into details, NTT DoCoMo got tired of waiting for Release 99 to become a standard. Instead, it elected to pursue 3G on its own, based upon the pre-finalized Release 99, to meet its own particular technical requirements and subsequently required its suppliers to provide equipment (infrastructure and handsets) that met those requirements.

However, NTT DoCoMo was publicly committed to bring its FOMA into line with UMTS Release 99 in mid-2003.

FOMA has two operational modes: a dedicated 64 kbps circuit connection and a 384 kbps downlink and 64 kbps uplink best-effort connection. The dedicated 64 kbps circuit is currently intended for real-time services like video conferencing. It is important to note that UMTS and cdma2000 do not have this 64 kbps circuit switched mode. Thus, subscribers should not expect real-time high-data-rate services when the launches first occur. We believe UMTS carriers will begin providing real-time services when they deploy a later release of the UMTS standard. For cdma2000 carriers, we believe that they will not provide real-time services until they deploy 1xEV-DO (which is a data only scheme) or 1xE-DV, which support higher peak data rates and have an all-IP core.

FOMA is a hybrid version of Release 99, but it will evolve to become compatible with the UMTS standard. Before we move on to the details of the UMTS standard, we need to clear up any misunderstanding there may be about NTT DoCoMo's FOMA service and its relationship with UMTS Release 99. NTT DoCoMo is a member of 3GPP and it is still involved in the 3GPP process. However, it decided to deploy a 3G service before the Release 99 standard was frozen for its own commercial consideration. Its FOMA service, therefore, was based on a pre-release version of the Release 99 standard. Since DoCoMo went it alone, its 3G solution has evolved, and was not fully compatible with Release 99. However, DoCoMo promised to make its FOMA service fully compatible with the UMTS standard in the next two years after its launch in October 2001. Therefore, the reality is that the compatibility is an issue to be resolved and even today roaming between Japanese WCDMA networks and European UMTS networks is not a commonplace.

3.2.3 UMTS CELL AND NETWORK STRUCTURE

UMTS can offer different coverage-scales to different users. There are in total four different UMTS hierarchical cell structures, which are (1) pico-cell, which covers only a small area such as one office room, (2) micro-cell, which can cover a vicinity of several buildings to provide local UMTS services, (3) macro-cell, which will span an area as large as a few kilometers in radius as a regional service provider, and finally (4) global cell, which will be covered by satellites and will be available to any place around the world. Under such a hierarchical cell structure, UMTS can provide services to users located in various geographical regions on the earth. It is to be noted that formation of a global cell needs to use other technology rather than UTRAN due to the nature of the long propagation delay in a satellite air-link sector.

Figure 3.26 shows a conceptual diagram of the UMTS hierarchical cell structure, which include all four different cells.

A very basic UMTS network structure consists of three fundamental components: (1) access network, in which base stations play a key role in managing the air interface access between the UMTS network and user equipment (UE), (2) core network (CN), also called fixed network, which is responsible for handling all internal connections, and (3) intelligent network (IN), which is in charge of billing, subscriber location registration, roaming, handover, etc.

Figure 3.27 shows the UMTS basic network structure and UMTS general reference architecture in a UMTS network. It can be seen from the figure that a UMTS Terrestrial Radio Access Network (UTRAN) contains several radio subsystems, so called Radio Network Subsystems (RNS), and contains functions for mobility management. The RNS controls handover whenever a mobile changes cell, implements functions for encoding, and administers the resources of the UMTS radio interface. The U_u interface connects UTRAN with mobile end devices, so-called user equipment (UE), and is comparable with U_m in a GSM network. UTRAN is connected over the I_u interface with the CN, comparable with the A interface in GSM between BSC and MSC. The CN contains the interfaces to other networks and mechanisms for connection handover to other systems.

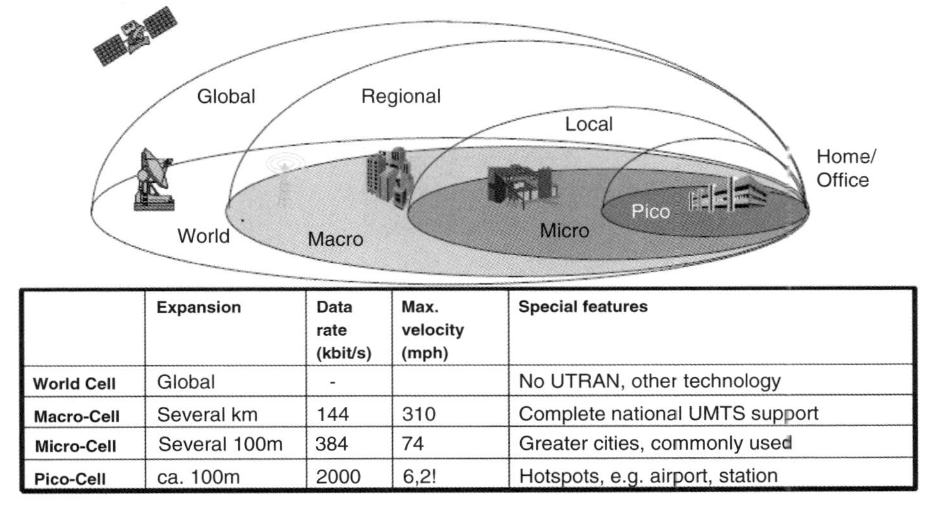

	Expansion	Data rate (kbit/s)	Max. velocity (mph)	Special features
World Cell	Global	-		No UTRAN, other technology
Macro-Cell	Several km	144	310	Complete national UMTS support
Micro-Cell	Several 100m	384	74	Greater cities, commonly used
Pico-Cell	ca. 100m	2000	6,2!	Hotspots, e.g. airport, station

Figure 3.26 UMTS hierarchical cell structure, which includes four different cells: pico-cell, micro-cell, macro-cell, and global cell.

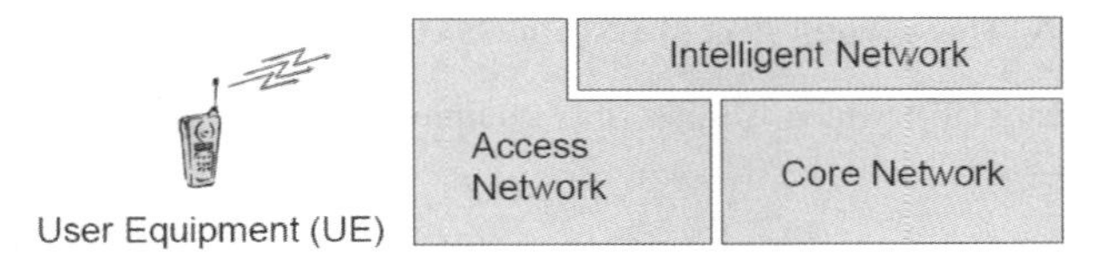

Figure 3.27 (a) UMTS basic network structure and (b) UMTS general reference architecture.

A UMTS network can also be explained using a method commonly referred to in the literature, as shown in Figure 3.28, where there are four basic components, explained as follows:

1. Cell: which specifies a basic coverage area. Hardware associated with the cell includes antenna system, high power amplifier (HPA), transmitter, receiver, etc. A cell in UMTS is equivalent to a sector in GSM or cdma2000.

2. Node-B: which is a common equipment at a cell site to control the cell, and thus equivalent to RBS, BTS, or base station in GSM or cdma2000.

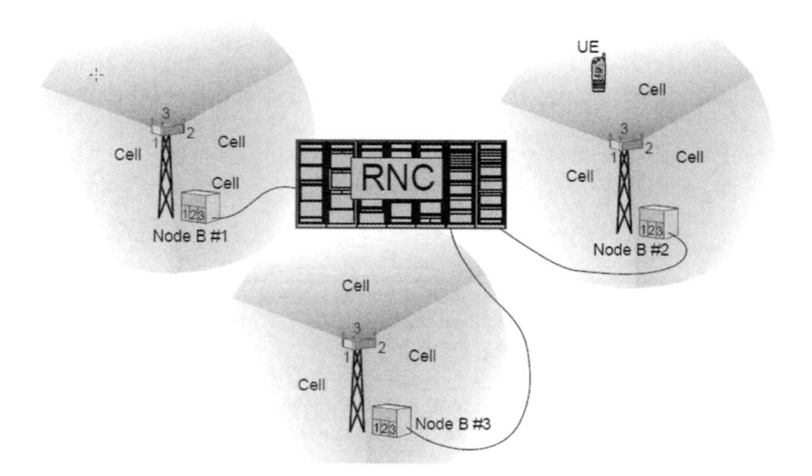

Figure 3.28 A typical UMTS network architecture, which consists of a radio network controller, cells, node-Bs, and user equipment.

3. Radio Network Controller (RNC): which is equipment to control the Node-Bs and interface them to the CN. This is equivalent to BSC in GSM or cdma2000.
4. User Equipment (UE): which is subscriber equipment and equivalent to a mobile station in GSM or cdma2000.

All of these naming conventions or acronyms will be extensively used in this and the sections following whenever the UMTS standard is discussed. Therefore, it is of extreme importance to remember them well from now on.

3.2.4 UMTS RADIO INTERFACE

The radio interface technology used in UMTS is called UMTS Terrestrial Radio Access (UTRA), in which two operating modes have been defined: UTRA frequency division duplex (FDD) and UTRA time division duplex (TDD) modes. We explain the two operation modes defined in UMTS in more detail as follows.

In general, the UTRA-FDD operation mode is mainly suitable for suburban areas where symmetrical transmission of speech and video is required. The data transmission rate in UTRA-FDD mode can go up to 384 kbps. The UTRA-FDD mode can also work for circuit- and packet-switched services in urban areas. On the other hand, the UTRA-TDD operation mode works mainly in households and other restricted areas, such as a company's premises, similar to DECT. The UTRA-TDD is in particular suitable for broadcast of speech and video, in both symmetrical (up to 384 kbps) and asymmetrical (up to 2 Mbps) ways.

Figure 3.29(a) is a simple diagram to illustrate how the UTRA-FDD operation mode works in terms of its carrier frequency allocation. Similarly, Figure 3.29(b) depicts the operation principle for the UTRA-TDD scheme in the time and frequency domains.

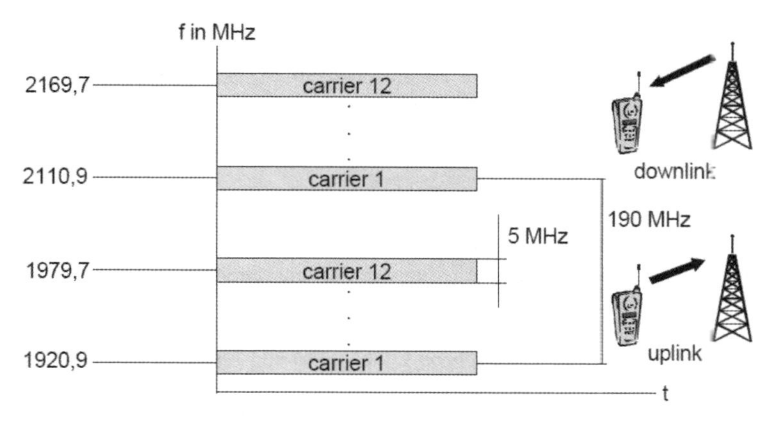

(a) UMTS UTRA-FDD operation mode

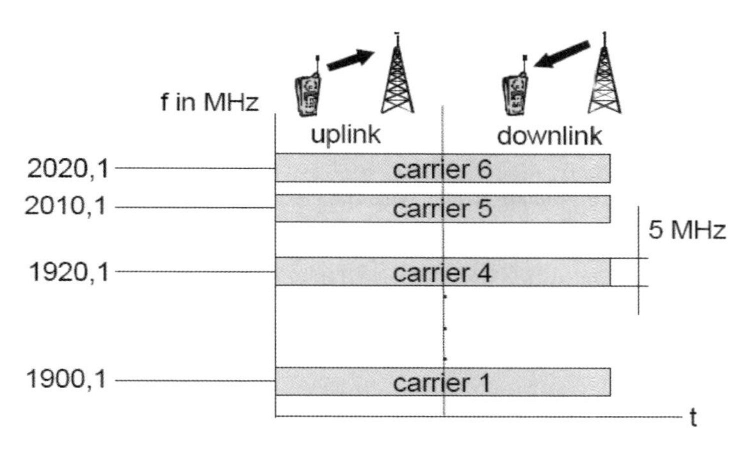

(b) UMTS UTRA-TDD operation mode

Figure 3.29 (a) UMTS UTRA-FDD operation mode and (b) UMTS UTRA-TDD operation mode.

It is seen from Figure 3.29 that UTRA-FDD puts wideband CDMA (WCDMA) together with direct sequence spread spectrum (DS-SS) as a bandwidth expansion technique.[6] All channels are separated by carrier frequencies, spreading codes, and phase positions (only for uplink). There are in total 250 channels created for user data transmission, whose transmission rates can be up to 384 kbps for high mobility. It has to be admitted that UTRA-FDD needs a relatively complex performance control mechanism due to the nature of FDD, which is in particular suitable for coverage driven roll-out.

[6]Therefore, the core CDMA technology used in WCDMA has no big difference compared to that used in IS-95, which has been discussed in the previous section in this chapter.

UTRA-TDD makes use of wideband TDMA/CDMA techniques together with DS-SS scheme. Data signals can be sent and received on the same carrier due to the use of TDD to separate uplink and downlink transmissions. There are in total 120 channels created for user data transactions, whose rates can be up to 2 Mbps, which is higher than the UTRA-FDD scheme. Channel separation is implemented through using different spreading codes and timeslots, and thus a lower spreading factor is required than that in the UTRA-FDD scheme. However, UTRA-TDD operation needs cell-wise precise synchronization in order to keep the same reference time among different UEs. UTRA-TDD technology is suitable for small cells with more asymmetric traffic, as well as for unlicensed cordless and public wireless local loops. TDD technology is best suited to indoor use where interference from base stations is manageable and the lower transmission range does not matter.

It is noted that FDD and TDD Node-Bs can operate at the same radio network controller (RNC), as shown in Figure 3.30. The differences between UTRA-FDD and UTRA-TDD are listed in Table 3.9. Figure 3.31 depicts the frame structure used in UTRA-TDD.

To understand better the UTRAN specifications, one can find much more information from the 3GPP RAN documentations. One can easily download all of those specifications free of charge, with only the need to register once.

The general description of UTRAN specifications of 3GPP RAN has been given in the following document:

1. 3GPP TS 25.301: Radio Interface Protocol Architecture
2. 3GPP TS 25.302: Services Provided by the Physical Layer
3. 3GPP TS 25.304: UE Procedures in Idle mode and Procedures for Cell Reselection in Connected Mode.

The Layer 3 (RRC) protocols for both FDD and TDD operation modes are discussed in:

1. 3GPP TS 25.331: Description of the RRC Protocol.

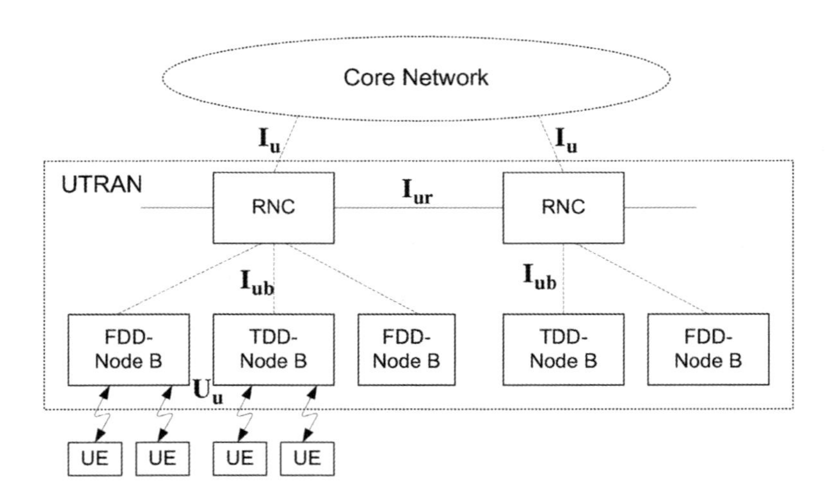

Figure 3.30 Illustration of co-existence of FDD and TDD Node-Bs under the same UTRAN RNC in a UMTS system.

Table 3.9 Different parameters used in UTRA-FDD and UTRA-TDD modes.

	UTRA-FDD	UTRA-TDD
Duplex method	FDD	TDD
Channel spacing (MHz)	5	5
Carrier chip rate (Mcps)	7.68 (HCR)	3.84 (LCR)
Time slot structure(slots/frame)	15	15
Frame length (ms)	10	10
Modulation	QPSK	QPSK
Detection	Based on pilot symbols	Based on midamble
Intra-frequency handover	Soft	Hard (cell reselection)
Inter-frequency handover	Hard	Hard
Spreading factors	4...512	1...15

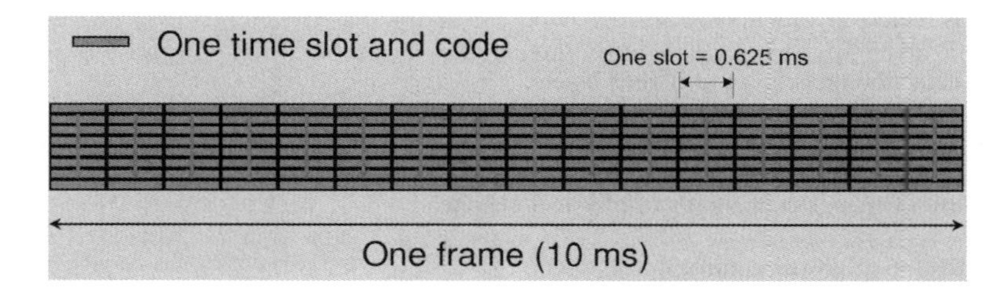

Figure 3.31 UMTS UTRA-TDD frame structure.

The Layer 2 (MAC/RLC) specifications for both FDD and TDD operation modes are given in:

1. 3GPP TS 25.321: MAC Protocol Specification
2. 3GPP TS 25.322: Description of the RLC protocol.

The Layer 1 (physical layer) specifications for FDD operation mode are given in the documents:

1. 3GPP TS 25.211: Transport Channels and Physical Channels (FDD)
2. 3GPP TS 25.212: Multiplexing and Channel Coding (FDD)
3. 3GPP TS 25.213: Spreading and Modulation (FDD).

The Layer 1 (physical layer) specifications for TDD operation mode are given in the documents:

1. 3GPP TS 25.221: Transport Channels and Physical Channels (TDD)
2. 3GPP TS 25.222: Multiplexing and Channel Coding (TDD)
3. 3GPP TS 25.223: Spreading and Modulation (TDD).

Therefore, 3GPP UTRA specifications cover both FDD and TDD operation modes. However, due to the limited space in this book, we will not discuss all of them in a very detailed way. Therefore, in the following discussions given in this section we focus on the FDD operation mode of UTRA. We discuss the TDD mode only if very necessary, or if we need to compare the FDD mode with the TDD mode in terms of their performance and complexity, and so forth.

3.2.5 UMTS PROTOCOL STACK

The UMTS protocol stack architecture is shown in Figure 3.32, in which three main layers–the physical layer (L1), the Medium Access Control (MAC) and the Radio Link Control (RLC) sublayers (L2), and the Radio Resource Control (RRC) layer (L3)–are illustrated. The Control Plane and the User Plane in the UMTS protocol stacks are also shown.

Layer 1: Physical Layer

Of course, the frontmost part of the protocol stack in the UMTS layered architecture is the physical layer, which offers information transfer services to the MAC layer. These services are denoted as transport channels. There are also physical channels, which comprise the following major functions:

1. various handover functions;
2. error detection and report to higher layers;
3. multiplexing of transport channels;
4. mapping of transport channels to physical channels;
5. fast close loop power control;

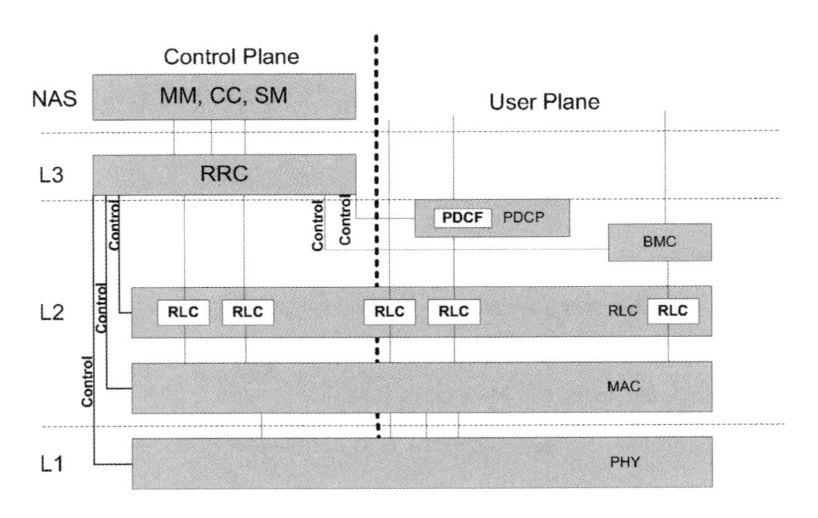

Figure 3.32 UMTS protocol stack architecture (MM: mobility management; CC: call control; SM: service management).

6. frequency and time synchronization; and

7. other responsibilities associated with transmitting and receiving signals over the radio media.

The complete list and explanations on all physical channels are given in the following subsections.

Layer 2-1: MAC Sub-Layer

The MAC sub-layer offers data transfer to RLC and higher layers. The MAC sub-layer comprises the following functions:

1. selection of appropriate TF (basically bit rate), within a predefined set, per information unit delivered to the physical layer;

2. service multiplexing on RACH, FACH, and dedicated channels;

3. priority handling between 'data flows' of one user as well as between data flows from several users (the latter being achieved by means of dynamic scheduling);

4. access control on RACH

5. address control on RACH and FACH; and

6. contention resolution on RACH.

It is to be noted that the term 'sub-layer' is used here to distinguish the full layers, such as 'physical layer' etc., as the MAC sub-layer itself does not form a full layer. Instead, it is only a part of Layer 2 in the UTRA protocol stack structure. The MAC sub-layer and the RLC sub-layer together form Layer 2.

Layer 2-2: RLC Sub-Layer

The RLC sub-layer offers the following services to the higher layers:

1. Layer 2 connection establishment/release;

2. transparent data transfer, i.e. no protocol overhead is appended to the information unit received from the higher layer; and

3. assured and unassured data transfer.

The RLC sub-layer comprises the following functions:

1. segmentation and assembly;

2. transfer of user data;

3. error correction by means of retransmission optimized for the WCDMA physical layer;

4. sequence integrity (used by at least the control plane);

5. duplicate detection;

6. flow control; and

7. ciphering.

Layer 3: RRC Layer

The RRC layer is also called Layer 3 in the UMTS protocol stack and offers the core network the following services:

1. general control service, which is used as an information broadcast service;
2. notification service, which is used for paging and notification of a selected UEs; and
3. dedicated control service, which is used for establishment/release of a connection and transfer of messages using the connection.

The RRC layer comprises the following functions:

1. broadcasting information from the network to all UEs;
2. radio resource handling (e.g. code allocation, handover, admission control, and measurement reporting/control);
3. QoS control;
4. UE measurement reporting and control of the reporting; and
5. power control, encryption and integrity protection.

3.2.6 UTRA CHANNELS

UTRA FDD radio interface has logical channels, which are mapped to transport channels, which are again mapped to physical channels. Logical to transport channel conversion happens in the MAC layer, which is a lower sub-layer in the Data Link Layer (Layer 2).

There are in total six different logical channels, which are listed as follows:

1. Broadcast Control Channel (BCCH) (DL)
2. Paging Control Channel (PCCH) (DL)
3. Dedicated Control Channel (DCCH) (UL/DL)
4. Common Control Channel (CCCH) (UL/DL)
5. Dedicated Traffic Channel (DTCH) (UL/DL)
6. Common Traffic Channel (CTCH) (broadcasting),

where the abbreviations of DL and UL stand for downlink and uplink, respectively.

There are in total seven different transport channels:

1. Dedicated Transport Channel (DCH) (UL/DL), mapped to DCCH and DTCH
2. Broadcast Channel (BCH) (DL), mapped to BCCH
3. Forward Access Channel (FACH) (DL), mapped to BCCH, CCCH, CTCH, DCCH and DTCH
4. Paging Channel (PCH) (DL), mapped to PCCH
5. Random Access Channel (RACH) (UL), mapped to CCCH, DCCH and DTCH

6. Uplink Common Packet Channel (CPCH) (UL), mapped to DCCH and DTCH
7. Downlink Shared Channel (DSCH) (DL), mapped to DCCH and DTCH.

There are in total thirteen different physical channels:

1. Primary Common Control Physical Channel (PCCPCH), mapped to BCH
2. Secondary Common Control Physical Channel (SCCPCH), mapped to FACH, PCH
3. Physical Random Access Channel (PRACH), mapped to RACH
4. Dedicated Physical Data Channel (DPDCH), mapped to DCH
5. Dedicated Physical Control Channel (DPCCH), mapped to DCH
6. Physical Downlink Shared Channel (PDSCH), mapped to DSCH
7. Physical Common Packet Channel (PCPCH), mapped to CPCH
8. Synchronization Channel (SCH)
9. Common Pilot Channel (CPICH)
10. Acquisition Indicator Channel (AICH)
11. Paging Indication Channel (PICH)
12. CPCH Status Indication Channel (CSICH)
13. Collision Detection/Channel Assignment Indication Channel (CD/CA-ICH).

Figure 3.33 shows all physical, transport, and logical channels, as well as the mapping relations among them, in UMTS UTRA. It is noted that the mapping from logical channels to transport channels happens on the MAC layer (L2), while the mapping from transport channels to physical channels happens on the physical layer (L1). Figure 3.34 shows the general architecture of Layers 1, 2, and 3 in the 3GPP UTRA standard. It is also shown clearly that the mapping from the logical channels to the transport channels happens on the MAC layer.

Transport Channels

More detailed information about the transport channels is given as follows. As shown in Figure 3.34, all transport channels carry services offered by Layer 1 to the higher layers. A transport channel is defined by how and with what characteristics data is transferred over the air interface.

There are two groups of transport channels: dedicated transport channels, and common transport channels. Only one dedicated transport channel exists; called the Dedicated Channel (DCH). The DCH is a downlink or uplink transport channel, and is transmitted over the entire cell or over only a part of the cell using, e.g. beam-forming antennas. DCH carries both the service data, such as speech frames, and higher layer control information, such as handover commands or measurement reports from the terminal.

The content of the information carried on the DCH is not visible to the physical layer, and thus higher layer control information and user data are treated in the same way. The physical layer parameters set by UTRAN may vary between control and data. DCH supports possible fast rate change (every 10 ms) and fast power control, as well as soft handover.

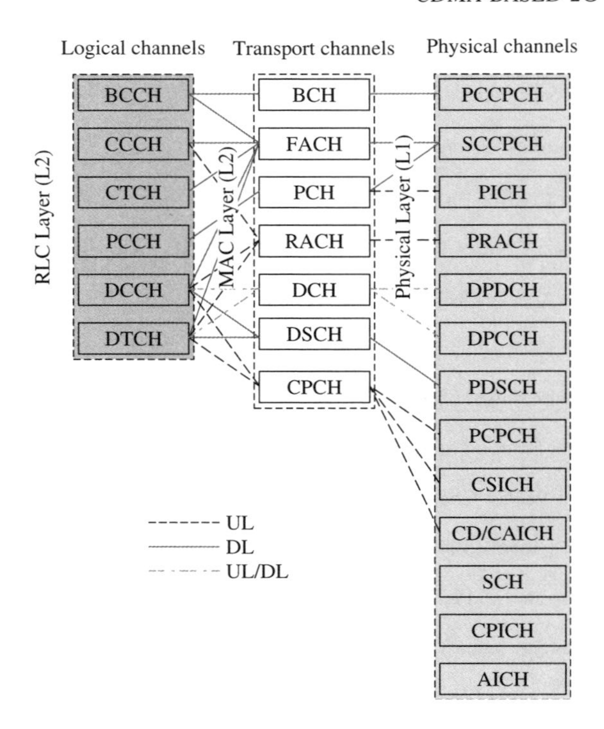

Figure 3.33 Physical, transport, and logical channels in UMTS standard and the mapping relations among them.

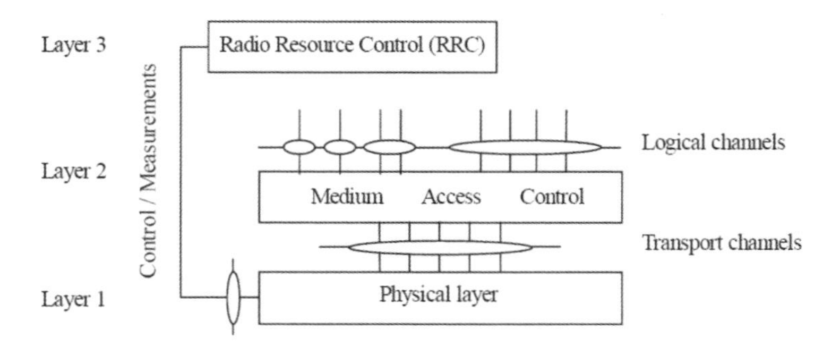

Figure 3.34 Layer 1, 2, and 3 architecture in 3GPP UTRA specifications.

However, there are six common transport channels, which are divided between all or a group of users in a cell. It is noted that they do not support soft handover, but some of them can support fast power control. They are:

1. BCH: Broadcast Channel

2. FACH: Forward Access Channel

3. PCH: Paging Channel

4. RACH: Random Access Channel

5. CPCH: Common Packet Channel; and

6. DSCH: DL Shared Channel.

BCH is a downlink transport channel that is used to broadcast system and cell-specific information. BCH is always transmitted over the entire cell. The most typical data needed in every network is the available random access codes and access slots in the cell, or the types of transmit diversity. BCH is transmitted with relatively high power. Single transport format offers a low and fixed data rate for the UTRA broadcast channel to support low-end terminals.

PCH is also a downlink transport channel. PCH is always transmitted over the entire cell. PCH carries data relevant to the paging procedure, that is, when the network wants to initiate communication with the terminal. The identical paging message can be transmitted in a single cell or in up to a few hundreds of cells, depending on the system configuration.

RACH is an uplink transport channel. RACH is intended to be used to carry control information from the terminals, such as requests to set up a connection. RACH can also be used to send small amounts of packet data from the terminal to the network. The RACH is always received from the entire cell. The RACH is characterized by a collision risk and is transmitted using open loop power control.

FACH is a downlink transport channel. FACH is transmitted over the entire cell or over only a part of the cell using beam-forming antennas. FACH can carry control information, e.g. after a random access message has been received by the base station. FACH can also transmit packet data. FACH does not use fast power control. FACH can be transmitted using slow power control. There can be more than one FACH in a cell. The messages transmitted need to include in-band identification information.

CPCH is an optional uplink transport channel. CPCH is an extension to the RACH channel that is intended to carry packet-based user data. CPCH is associated with a dedicated channel on the downlink which provides power control and CPCH Control Commands (e.g. Emergency Stop) for the uplink CPCH. The CPCH is characterized by initial collision risk and by using inner loop power control. CPCH may last several frames.

DSCH is an optional downlink transport channel shared by several UEs to carry dedicated user data and/or control information. The DSCH is always associated with one or several downlink DCH. The DSCH is transmitted over the entire cell or over only a part of the cell using beam-forming antennas. DSCH supports fast power control and variable bit rate on a frame-by-frame basis.

Physical Channels

The mapping between the transport channels and physical channels happens in the UTRA physical layer, as shown in Figure 3.33. We would like to discuss the uplink physical channels, followed by the downlink physical channels.

There are two dedicated uplink physical channels: Uplink Dedicated Physical Data Channel (UL DPDCH) and Uplink Dedicated Physical Control Channel (UL DPCCH). Also, there are two common uplink physical channels, which are Physical Random Access Channel (PRACH) and Physical Common Packet Channel (PCPCH).

The uplink Dedicated Physical Data Channel (UL DPDCH) carries the DCH transport channel (generated at Layer 2 and above). There may be zero, one, or several uplink DPDCHs on each radio link.

The UL DPCCH carries control information generated at Layer 1. One and only one UL DPCCH exists on each radio link.

PRACH is used to carry the RACH. The random access transmission is based on a slotted ALOHA approach with fast acquisition indication. The UE can start the random-access transmission at the beginning of a number of well-defined time intervals, denoted access slots. There are 15 access slots per two frames and they are spaced 5120 chips apart. Information on what access slots are available for random-access transmission is given from higher layers.

PCPCH is used to carry the CPCH, whose transmission is based on the Digital Sense Multiple Access-Collision Detection (DSMA-CD) approach with fast acquisition indication. The UE can start transmission at the beginning of a number of well-defined time intervals.

There is only one type of downlink dedicated physical channel, that is the Downlink Dedicated Physical Channel (DL DPCH). Within one downlink DPCH, dedicated data generated at Layer 2 and above, i.e. the dedicated transport channel (DCH), is transmitted in time-multiplex with control information generated at Layer 1 (known pilot bits, TPC commands, and an optional TFCI).

Common Pilot Channel (CPICH) is a fixed rate (30 kbps, SF = 256) downlink physical channel that carries a predefined bit/symbol sequence. In case transmit diversity (open or closed loop) is used on any downlink channel in the cell, the CPICH should be transmitted from both antennas using the same channelization and scrambling code. There are two types of common pilot channels: the Primary CPICH and the Secondary CPICH.

The Primary Common Pilot Channel (P-CPICH) has the following characteristics: the same channelization code is used for the P-CPICH; the P-CPICH is scrambled by the primary scrambling code. There is one and only one P-CPICH per cell. The P-CPICH is broadcast over the entire cell. The Primary CPICH is a phase reference for the following downlink channels: SCH, Primary CCPCH, AICH, PICH APAICH, CD/CA-ICH, CSICH, DL-DPCCH for CPCH, and the S-CCPCH. By default, the P-CPICH is also a phase reference for downlink DPCH and any associated PDSCH. The P-CPICH is always a phase reference for a downlink physical channel using closed loop TX diversity.

A Secondary Common Pilot Channel (S-CPICH) has the following characteristics: an arbitrary channelization code of SF = 256 is used for the S-CPICH; an S-CPICH is scrambled by either the primary or a secondary scrambling code; there may be zero, one, or several S-CPICHs per cell; an S-CPICH may be transmitted over the entire cell or only over a part of the cell; an S-CPICH may be a phase reference for a downlink DPCH. The S-CPICH can be a phase reference for a downlink physical channel using open loop TX diversity, instead of the P-CPICH being a phase reference.

The Primary Common Control Physical Channel (P-CCPCH) bears the following characteristics: it has a fixed rate: 30 kbps, SF = 256, and is used to carry the BCH transport channel. Neither TPC commands, nor TFCI, nor pilot bits will be sent in P-CCPCH.

The Secondary Common Control Physical Channel (S-CCPCH) is used to carry the FACH and PCH. Two types of S-CCPCHs exist: those that include TFCI and those that do not include TFCI. It is the UTRAN that determines if a TFCI should be transmitted, hence making it mandatory for all UEs to support the use of TFCI.

Synchronization Channel (SCH) is a downlink signal used for cell search. The SCH consists of the Primary and Secondary SCH. The 10 ms radio frames of the Primary and Secondary SCH are divided into 15 slots, each having a length of 2560 chips.

Physical Downlink Shared Channel (PDSCH) is used to carry the Downlink Shared Channel (DSCH). A PDSCH corresponds to a channelization code below or at a PDSCH root channelization code. A PDSCH is allocated on a radio frame basis to a UE. Within one radio frame, UTRAN may allocate different PDSCHs under the same PDSCH root channelization code to different UEs based on code multiplexing. Within the same radio frame, multiple parallel PDSCHs, with the same spreading factor, may be allocated to a single UE. All PDSCHs are operated with radio frame synchronization.

The Acquisition Indicator Channel (AICH) is a fixed rate (SF = 256) physical channel used to carry Acquisition Indicators (AI). AIs correspond to signatures on the PRACH.

CPCH Access Preamble Acquisition Indicator Channel (AP-AICH) is a fixed rate (SF = 256) physical channel used to carry AP acquisition indicators (API) of CPCH. APIs correspond to AP signatures transmitted by UE.

CPCH Collision Detection/Channel Assignment Indicator Channel (CD/CA-ICH) is a fixed rate (SF = 256) physical channel used to carry CD Indicator (CDI) only if the CA is not active, or to carry CD Indicator/CA Indicator (CDI/CAI) at the same time if the CA is active.

Paging Indicator Channel (PICH) is to provide terminals with efficient sleep mode operation. For detection of the PICH, the terminal needs to obtain the phase reference from the CPICH, and, as with the AICH, the PICH needs to be heard by all terminals in the cell and thus needs to be sent at high power level without power control. The PICH is a fixed rate (SF = 256) physical channel used to carry the paging indicators. The PICH is always associated with an S-CCPCH to which a PCH transport channel is mapped.

CPCH Status Indicator Channel (CSICH) is a fixed rate (SF = 256) physical channel used to carry CPCH status information. The CSICH bits indicate the availability of each physical CPCH channel and are used to tell the terminal to initiate access only on a free channel but, on the other hand, to accept a channel assignment command to an unused channel. A CSICH is always associated with a physical channel used for transmission of CPCH AP-AICH and uses the same channelization and scrambling codes.

3.2.7 UTRA MULTIPLEXING AND FRAME STRUCTURE

The frame structure is associated with specific transport or physical channels. In the UTRA specifications, different transport channels are in general given distinct frame structures to fit particular requirements in terms of their contents, data rates, multiple access schemes, duplex techniques (FDD or TDD), downlink or uplink, etc. Therefore, to give detailed information about the frame structures for all different transport and physical channels would take too much space, and thus I will not do so in this section. Instead, I provide only some discussions on generic UTRA multiplexing and frame structures, together with some examples here.

As shown in Table 3.9, the detailed operational parameters used in both FDD and TDD modes are given and then compared. In January 1998, ETSI decided that the UMTS should be given an option to operate on two different duplex modes, FDD and TDD modes. The WCDMA technology was chosen for wide-area services and will use paired FDD bands: 1920–1980 MHz for uplink and 2110–2170 MHz for downlink. On the other hand, TD/CDMA was chosen for private, indoor services in unpaired TDD bands, i.e., 1900–1920 MHz and 2010–2025 MHz.

UTRA FDD

The UTRA FDD operation mode is the multiplexing scheme, which separates downlink and uplink transmissions in different carriers, to implement full-duplex operation in radio link.

Several salient features exist in the UTRA FDD operation mode, as listed below:

1. It uses wideband Direct Sequence CDMA technology.
2. It uses 4.096 Mcps chip rate, which is expandable to 8.192/16.384 Mcps.
3. It supports asynchronous base stations.
4. It uses variable spreading and multi-code operation.
5. It enables coherent detection in both uplink and downlink.
6. It offers optimized packet access on common or dedicated channels.

Figure 3.35 illustrates the frame structure used for UTRA FDD operation mode. It is seen from the figure that the frame length is 10 ms with in total 16 slots. The downlink and uplink will use different multiplexing schemes. The downlink uses time multiplexed control and data frames, which will be transmitted via the I and Q channels in quadrature digital modem. On the other hand, the uplink will use I/Q code multiplexed control and data frames, in which the data signal will be sent in the Dedicated Physical Data Channel (DPDCH) channel, and the control signal will be sent in the Dedicated Physical Control Channel (DPCCH) channel.

UTRA FDD can select various ways to map transport channels into physical channels, depending on the different operational requirements, as shown in Figure 3.36. Figure 3.37 shows the super-frame, frame, and slot structures for UTRA FDD physical channels, DPCCH and DPDCH.

UTRA TDD

TDD mode should use harmonized parameters to UTRA FDD, such as chip rate, frame length and slot size, modulation, etc. Due to the relatively low average transmission power level, UTRA TDD is dedicated primarily for private, uncoordinated systems, which can be deployed in unpaired UMTS

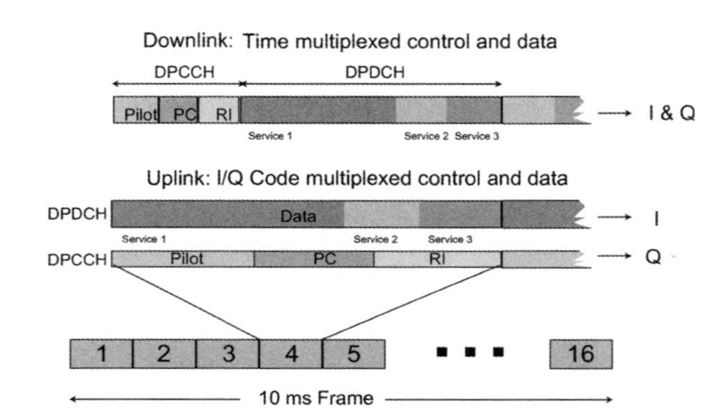

Figure 3.35 UTRA FDD downlink and uplink time multiplexing schemes.

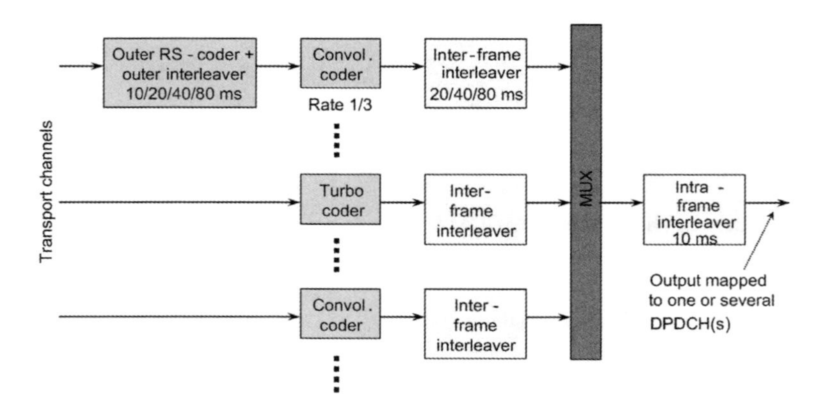

Figure 3.36 Mapping from transport channels to physical channels using different coding and multiplexing schemes in UTRA FDD.

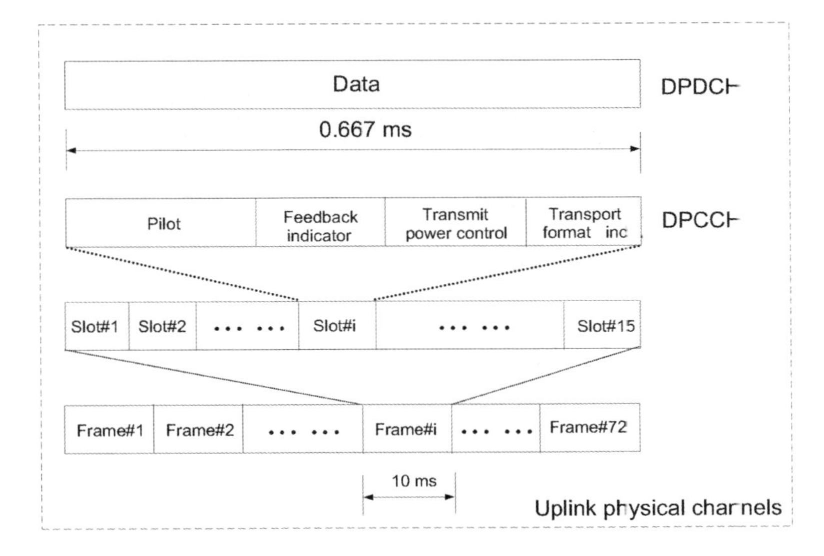

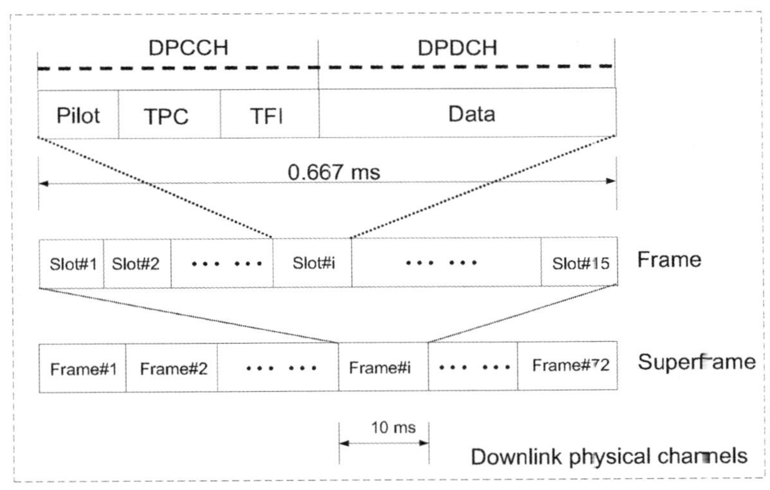

Figure 3.37 Super-frame, frame, and slot structures for UTRA FDD physical channels, DPCCH, and DPDCH.

bands. This is of extreme importance especially for those countries where the spectral allocation has become very difficult.

Each 0.625 ms slot in UTRA TDD can be allocated to either uplink or downlink transmission, as shown in Figure 3.38. However, at least one slot should be assigned to downlink (BCCH) and one to uplink (RACH). The same asymmetry and frame synchronization is needed within a continuous area in coordinated systems.

In each slot up to 8 codes are used for multiple access. It allows multi-code transmission. Different users can share the same timeslot. Since only a few codes are used in each timeslot, joint detection is supported easily in UTRA TDD operation mode.

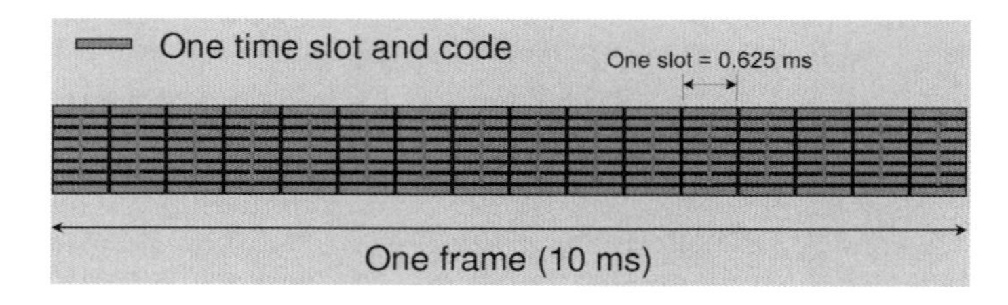

Figure 3.38 UTRA FDD time slot and frame structure.

3.2.8 SPREADING AND CARRIER MODULATIONS

Like any CDMA system, UMTS also needs spreading and carrier modulations. The carrier modulation is for sending baseband signals into air through radio frequency carrier. The spreading modulation functions as a vehicle to span the spectrum and implement the multiple access, achieving processing gain.

OVSF Codes

Both UTRA FDD and UTRA TDD use orthogonal variable spreading factor (OVSF) codes[7] for spreading modulation. The chip duration of the OVSF codes is T_c. The OVSF codes can be generated from a unique tree structure. The OVSF codes in UTRA systems perform the following functionalities:

1. widen the band from $1/T_b$ to $1/T_c$;
2. characterize the users and user services in downlink; and
3. characterize the user services in uplink.

The tree structure of OVSF codes is shown in Figure 3.39. There are several important characteristic features for the OVSF codes:

1. they can maintain orthogonality among all leaves (which is defined as the end of each branch);
2. the number of available OVSF codes is equal exactly to the spreading factor (SF); and
3. they can be completely orthogonal if they operate in exactly synchronized channels.

In the discussions given in this part, we will concentrate on the UTRA FDD operation mode due to the constraint on space.

Uplink Spreading and Modulation

Figure 3.40 illustrates the spreading and modulation for the case of a single uplink DPDCH. Data modulation is carried out by dual-channel QPSK, where the uplink DPDCH and DPCCH are mapped to the I and Q branches, respectively. The I and Q branches are then spread to the chip rate with two different channelization codes c_D/c_C and subsequently complex scrambled by a mobile-station-specific complex scrambling code c_{scramb}.

For multi-code transmission, each additional uplink DPDCH may be transmitted on either the I or the Q branch. For each branch, each additional uplink DPDCH should be assigned its own

[7]More discussions on OVSF codes can be found in Subsection 4.1.1.

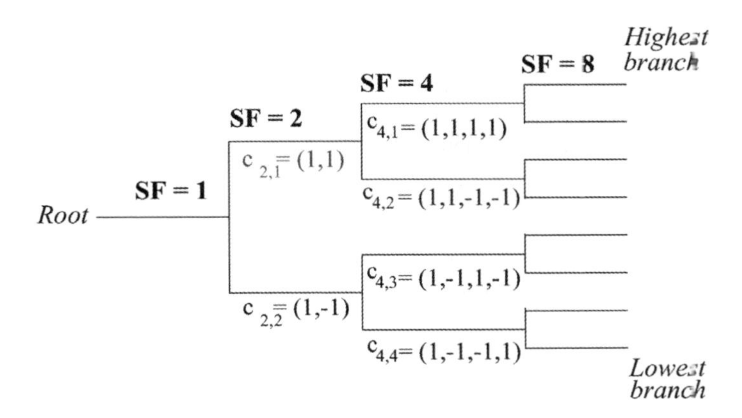

Figure 3.39 OVSF code generation tree in UTRA.

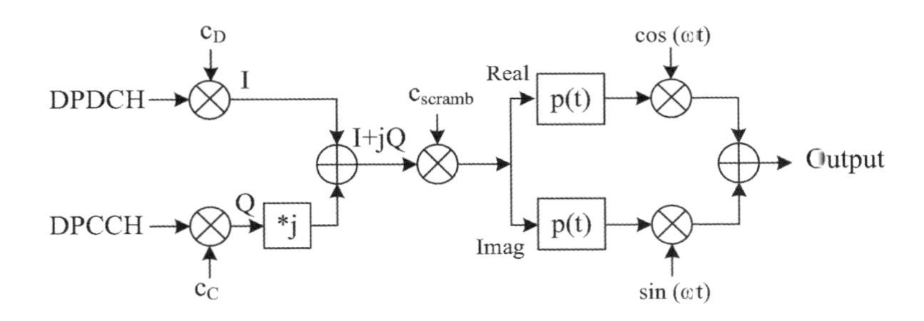

Figure 3.40 Spreading/modulation for uplink DPDCH/DPCCH in UTRA.

channelization code. Uplink DPDCHs on different branches may share a common channelization code.

The spreading and modulation of the message part of the random-access bursts is basically the same as for the uplink dedicated physical channels, as shown in Figure 3.40, where the uplink DPDCH and uplink DPCCH are replaced by the data part and the control part, respectively. The scrambling code for the message part is chosen based on the base-station-specific preamble code, the randomly chosen preamble sequence, and the randomly chosen access slot (random-access time-offset). This guarantees that two simultaneous random-access attempts that use different preamble codes and/or different preamble sequences will not collide during the data part of the random-access bursts.

The channelization codes of Figure 3.40 are the same type of OVSF codes as that for the downlink, as shown in Figure 3.41. Each connection is allocated at least one uplink channelization code, to be used for the uplink DPCCH. In most cases, at least one additional uplink channelization code is allocated for a uplink DPDCH. Further uplink channelization codes may be allocated if more than one uplink DPDCH is required. As different mobile stations use different uplink scrambling codes, the uplink channelization codes may be allocated without coordination between different connections. The uplink channelization codes are therefore always allocated in a predefined order. The mobile station and network only need to agree on the number and length (spreading factor) of the uplink channelization codes.

Either short or long scrambling codes should be used on uplink. The short scrambling code is a complex code $c_{scramb} = c_I + jc_Q$, where c_I and c_Q are two different codes from the extended Very Large Kasami set of length 256. The network decides the uplink short scrambling code. The mobile station is informed about what short scrambling code to use in the downlink Access Grant message that is the base-station response to an uplink Random Access Request. The short scrambling code may, in rare cases, be changed during a connection.

The long uplink scrambling code is typically used in cells without multi-user detection in the base station. The mobile station is informed if a long scrambling code should be used in the Access Grant message following a Random Access Request and in the handover message. What long scrambling code to use is directly given by the short scrambling code. No explicit allocation of the long scrambling code is thus needed. The scrambling code sequences are constructed as the position-wise modulo-2 sum of 40 960 chip segments of two binary m-sequences generated by means of two generator polynomials of degree 41. Let x and y be the two m-sequences. The x sequence is constructed using the primitive (over GF(2)) polynomial $1 + X^3 + X^{41}$. The y sequence is constructed using the polynomial $1 + X^{20} + X^{41}$. The resulting sequences thus constitute segments of a set of Gold sequences. The scrambling code for the quadrature component is a 1024-chip shifted version of the in-phase scrambling code. The uplink scrambling code word has a period of one radio frame of 10 ms.

For random access channels, the spreading code for the preamble part is cell specific and is broadcast by the base station. More than one preamble code can be used in a base station if the traffic load is high. The preamble codes must be code planned, since two neighboring cells should not use the same preamble code. The code used is a real-valued 256-chip orthogonal Gold code. All 256 codes are used in the system. The preamble codes are generated in the same way as the codes used for the downlink synchronization channel.

The modulating chip rate is 4.096 Mcps. This basic chip rate can be extended to 8.192 or 16.384 Mcps. The pulse-shaping filters are root-raised cosine (RRC) with roll-off factor of $\alpha = 0.22$ in the frequency domain. QPSK modulation is used.

Downlink Spreading and Modulation

Figure 3.41 illustrates the spreading and modulation for the downlink DPCH. Data modulation is QPSK, in which each pair of two bits are serial-to-parallel converted and mapped to the I and Q branches, respectively. The I and Q branches are then spread to the chip rate with the same channelization code c_{ch} (real spreading) and subsequently scrambled by the same cell-specific scrambling code c_{scramb} (real scrambling).

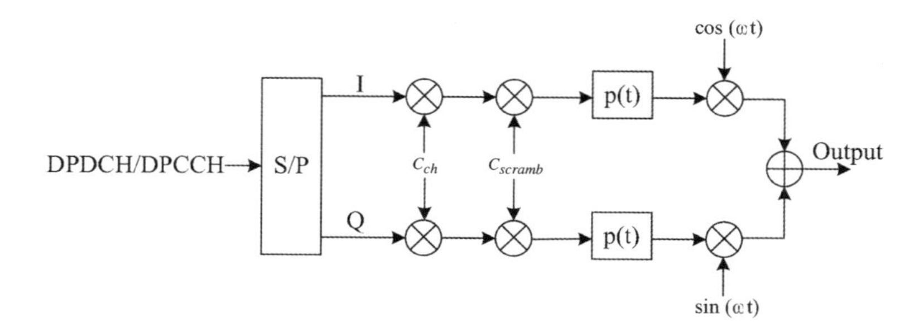

Figure 3.41 Spreading/modulation for downlink DPDCH/DPCCH in UTRA, where c_{ch}: channelization code; c_{scramb}: scrambling code; and $p(t)$: pulse-shaping filter (root raised cosine, roll-off rate of 0.22).

For multi-code transmission, each additional downlink DPCH should also be spread or modulated according to Figure 3.41. Each additional downlink DPCH should be assigned its own channelization code.

The channelization codes of Figure 3.41 are OVSF codes that preserve the orthogonality between downlink channels of different rates and spreading factors. The OVSF codes have been defined in Figure 3.39. Each level in the code tree defines channelization codes of length SF, corresponding to a spreading factor of SF, as shown in Figure 3.41. All codes within the code tree cannot be used simultaneously within one cell. A code can be used in a cell if and only if no other code on the path from the specific code to the root of the tree or in the sub-tree below the specific code is used in the same cell. This means that the number of available channelization codes is not fixed but depends on the rate and spreading factor of each physical channel. The channelization code for the BCCH is a predefined code which is the same for all cells within the system. The channelization code(s) used for the Secondary Common Control Physical Channel is broadcast on the BCCH. The channelization codes for the downlink dedicated physical channels are decided by the network. The mobile station is informed about what downlink channelization codes to receive in the downlink Access Grant message that is the base-station response to an uplink Random Access request. The set of channelization codes may be changed during a connection, typically as a result of a change of service or an inter-cell handover. A change of downlink channelization codes is negotiated over a DCH.

The total number of available scrambling codes is 512, divided into 32 code groups with 16 codes in each group. The grouping of the downlink codes is done in order to facilitate a fast cell search. The downlink scrambling code is assigned to the cell (sector) at the initial deployment. The mobile station learns about the downlink scrambling code during the cell search process. The scrambling code sequences are constructed as the position wise modulo-2 sum of 40 960 chip segments of two binary m-sequences generated by means of two generator polynomials of degree 18. Let x and y be the two sequences. The x sequence is constructed using the primitive (over GF(2)) polynomial $1 + X^7 + X^{18}$. The y sequence is constructed using the polynomial $1 + X^5 + X^7 + X^{10} + X^{18}$. The resulting sequences thus constitute segments of a set of Gold sequences. The scrambling codes are repeated for every 10 ms radio frame.

The modulating chip rate is 4.096 Mcps. This basic chip rate can be extended to 8.192 or 16.384 Mcps. The pulse-shaping filters are root raised cosine (RRC) with roll-off factor $\alpha = 0.22$ in the frequency domain.

3.2.9 PACKET DATA

In UMTS WCDMA systems, data applications are expected to dominate the overall traffic volume. The salient feature of 3G wireless is its capability to support packet switched traffic compared to the 2G systems (e.g. IS-95). The need for all-IP wireless services makes this feature of UMTS WCDMA system even more important.

In the packet switched operation mode, the traffic generated by data applications is inherently bursty and asymmetric by nature, with higher data rates in the downlink than those in the uplink. In addition, the reverse transmissions of all users in one cell share the same set of OVSF channelization codes. Therefore, optimal resource utilization is essential in the downlink.

In (UTRA FDD) WCDMA, there are three types of downlink transport channels that can be used to transmit bursty packet data: common, dedicated, and shared transport channels. Among these, the dedicated and shared channels are suited for the transmission of medium to large data amounts, while the common channels are suited for the transmission of small data amounts, such as signaling data or small IP packets. Consequently, for the transmission of bursty packet data we have to select between the dedicated and the shared channels

The Downlink Dedicated Channel (DCH) has a fixed spreading factor that does not vary on a frame-by-frame basis and is determined by the highest transmission rate of the source. The variable data rate transmission may be implemented with discontinuous transmission (DTX) by gating the transmission on and off or by the use of flexible positions. If flexible positions are used, the transmission is continuous and the DTX is implemented by repeating the transmitted bits. Therefore the use of the DCH channel for the transmission of bursty data is not efficient, as it results in low OVSF code utilization and decreased system capacity.

On the other hand, shared channels are made for the transmission of bursty data as they allow a single OVSF channelization code to be shared among several users. However, the use of the shared channels has some restrictions. The most important restrictions are:

1. The OVSF codes must be allocated from the same branch of the code tree. The root code of the DSCH subtree defines the spreading factor for maximum data rate transmission, while the rest codes are used when lower rates are needed.

2. The DSCH does not support spreading factor 512 (the highest spreading factor of the OVSF code tree).

3. HS-DSCH has a fixed spreading factor.

A comparison among the DCH, DSCH, HS-DSCH, and FACH channels in terms of their packet data operation parameters is shown in Table 3.10. Figure 3.42 shows the time diagrams for small/infrequent and large/frequent packet access in UTRA FDD operation mode.

Table 3.10 A comparison on packet data operation among the DCH, DSCH, HS-DSCH, and FACH channels.

Channel	HS-DSCH	DSCH	DCH (DL)	FACH
Spreading factor	Fixed 16	Variable (4-256) frame-by-frame	Fixed (4-512)	Fixed (4-256)
Power control	Fixed/slow power setting	Fast, based on associated DCH	Fast at 1500 kHz	Fixed/slow power setting
Interleaving	2 ms	10–80 ms	10–80 ms	10–80 ms
Soft-handover	For associated DCH	For associated DCH	Yes	No
3GPP Release Version	Release 5	Release 99	Release 99	Release 99

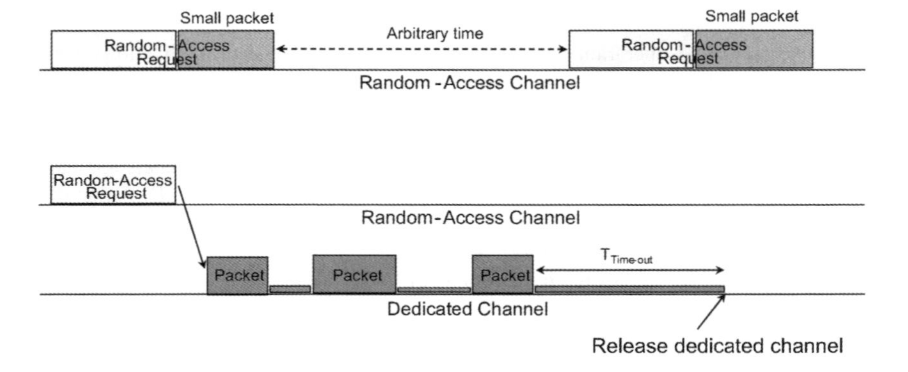

Figure 3.42 UTRA FDD packet access with small-infrequent and large-frequent packet transactions.

3.2.10 POWER CONTROL

Different from the power control schemes used in IS-95 (discussed in Section 3.1), UMTS defines three main different power control mechanisms: (1) open-loop power control, (2) inner-loop power control, and (3) outer-loop power control, which will be introduced here as follows.

Open-Loop Power Control

In the UMTS standard, open-loop power control is defined as the ability of the UE transmitter to set its output power to a specific value. It is used for setting initial uplink and downlink transmission powers when a UE is accessing the network. The open loop power control tolerance is ±9 dB (under normal conditions) and ±12 dB (under extreme conditions).

Figure 3.43 shows the major functional blocks involved with the power control mechanism in a UTRA transceiver.

Inner Loop Power Control

In UMTS, inner-loop power control, also called fast closed-loop power control, in the uplink is defined as the ability of the UE transmitter to adjust its output power in accordance with one or more Transmit Power Control (TPC) commands received in the downlink, in order to keep the received uplink signal-to-interference ratio (SIR) at a given SIR target. The UE transmitter is capable of changing the output power with a step size of 1, 2, and 3 dB, in the slot immediately after the TPC_cmd can be derived. Inner-loop power control frequency is 1500 Hz. The serving cells estimate the SIR of the received uplink DPCH, generate TPC commands (TPC_cmd), and transmit the commands once per slot according to the following rule: if $SIR_{est} > SIR_{target}$ then the TPC command to transmit is '0', while if $SIR_{est} < SIR_{target}$ then the TPC command to transmit is '1'.

Upon reception of one or more TPC commands in a slot, the UE derives a single TPC command for each slot, combining multiple TPC commands if more than one is received in a slot. Two algorithms are supported by the UE for deriving a TPC_cmd. Which of these two algorithms is used should be

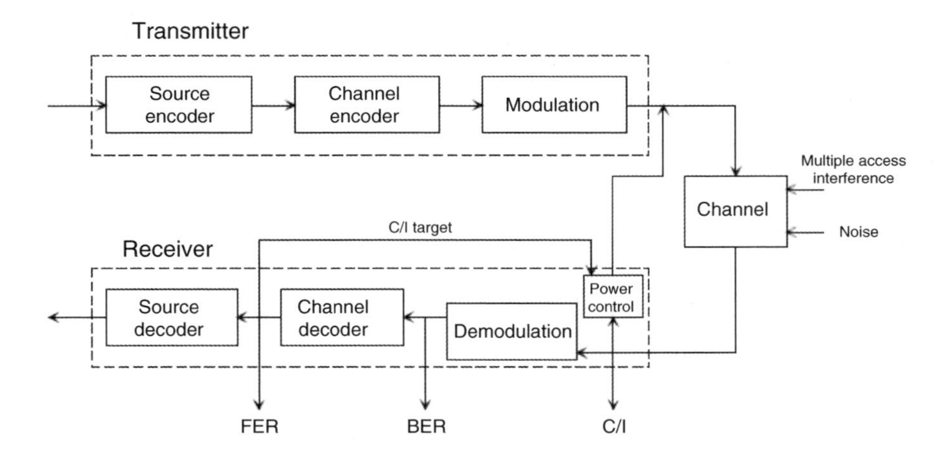

Figure 3.43 Major functional blocks involved with power control mechanism in a UTRA transceiver.

determined by a UE-specific higher-layer parameter, or Power_Control_Algorithm. More specifically, the two algorithms used in the inner-loop power control can be explained as follows:

- Algorithm 1: The power control step is the change in the UE transmitter output power in response to a single TPC command.
- Algorithm 2: If all five estimated TPC command are 'down', the transmit power is reduced by 1 dB; if all five estimated TPC command are 'up', the transmit power is increased by 1 dB; otherwise the transmit power is not changed.

As a matter of fact, the transmit power of the downlink channels is determined by the network. The power control step size can take four different values: 0.5, 1, 1.5, or 2 dB. It is specified in the UMTS standard that it is mandatory for UTRAN to support a step size of 1 dB, while support of other step sizes is optional. The UE generates TPC commands to control the network transmit power and send them in the TPC field of the uplink DPCCH. Upon receiving the TPC commands UTRAN adjusts its downlink DPCCH/DPDCH power accordingly.

To summarize, the inner-loop power control fulfil the following three functions:

- it can mitigate fast fading effect at a rate of 1.5 kbps;
- it functions in both downlink and uplink; and
- it works based on a fixed quality target set in MS or BS, depending on downlink or uplink.

Outer Loop Power Control

Outer-loop power control is used to maintain the quality of communications at the level of bearer service quality requirements, while using as low power as possible. The uplink outer-loop power control is responsible for setting a target SIR in the Node-B for each individual uplink inner-loop power control. This target SIR is updated for each UE according to the estimated uplink quality (e.g. Block Error Rate, Bit Error Rate, etc.) for each Radio Resource Control connection. The downlink outer-loop power control is the ability of the UE receiver to converge to required link quality (BLER) set by the network (RNC) in downlink.

It is to be noted that in the UMTS standard the power control in the downlink common channels is determined by the network. In general the ratio of the transmit power between different downlink channels is not specified in the 3GPP specifications and may change with time, even dynamically. Additional special situations of power control are power control in compressed mode and downlink power control during handovers, and so on.

To summarize, the outer-loop power control works for the following functionalities:

- it can compensate changes in the environment;
- it can adjust the SIR target to achieve the required FER/BER/BLER;
- it depends on MS mobility and multipath diversity; and
- In the case of soft handover it comes after frame selection.

3.2.11 HANDOVERS

The handover defined in the UMTS standard is always associated with the cell reselection process, which occurs when a UE moves away from a cell under the control of one Node-B to another. Defined in the UMTS standard, Cell Reselection is the process of selecting a new cell when the UE is not in traffic (e.g. Idle, cell FACH, cell PCH, cell URA). Although carried out autonomously by the UE, a number of system parameters carried in System Information Block (SIB) types 3 and 11 influence the procedure.

There are three types of handovers specified in UMTS: (1) intra-frequency handovers, (2) inter-frequency handovers, and (3) inter-RAT handovers, where RAT stands for Radio Access Technologies.

Intra-frequency handover occurs between cells on the same UMTS radio frequency. The UE can measure the signal strength of other cells without interrupting connectivity with the current cell. Inter-frequency handover occurs between cells on different UMTS radio frequencies. To measure the signal strength of an inter-frequency neighbor cell, the UE must tune away from the serving cell's frequency and tune to the neighbor cell's frequency without losing data. Inter-RAT cell handover occurs between cells on different RATs. For example, handover to a GSM, cdma2000, or UMTS TDD is considered an inter-RAT handover (Release 99 considers only GSM). This requires significant reconfiguration of hardware and software in the UE. Measurements must be taken when it is possible to tune the radio away from the serving cell without losing data. For obvious reasons, the Inter-RAT cell handover is very important especially during the initial deployment of UMTS in a region where it might happen that GSM or any other mobile cellular systems are operating, and thus dual-mode UE will be dominant in the 3G handset market.

Figure 3.44 shows cell reselection signal thresholds used for the three different types of handovers defined in UMTS, i.e, intra-frequency handovers, inter-frequency handovers and inter-RAT handovers.

When an inter-frequency cell reselection is in progress, it is recommended that the condition $S_{intersearch} < S_{intrasearch}$ must be satisfied. In this case, the inter-frequency searching will only begin when the quality of the pilot signal is worse than the value at which intra-frequency searching begins. When an inter-RAT cell reselection is carried out, it is recommended that the requirement of $S_{search,RAT} = 0$ dB must be satisfied. In this case, the inter-RAT reselection will be considered only when the quality of the pilot signal is below the level of $Q_{qualmin}$ (as shown in the figure), thereby maximizing WCDMA coverage. The goal in setting these parameters is to ensure that intra-frequency neighbors are considered first for reselection, followed by the inter-frequency neighbors. These parameters attempt to maximize WCDMA usage, only considering other technologies when

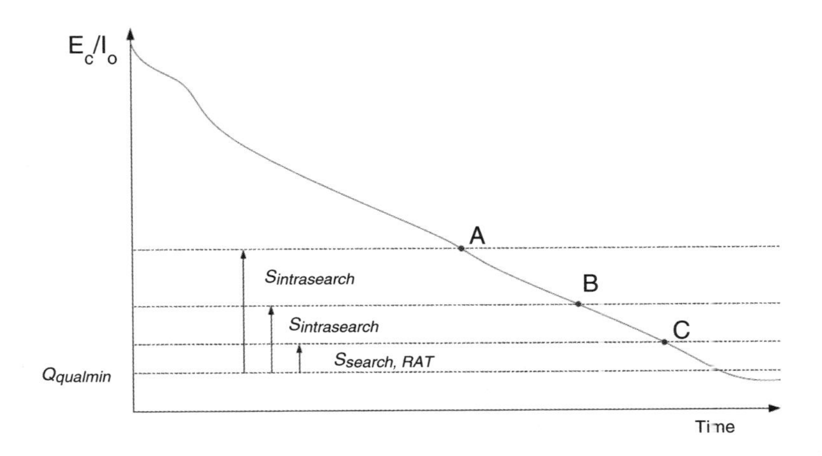

Figure 3.44 Cell reselection signal thresholds used for three different types of handovers in UMTS: intra-frequency handovers, inter-frequency handovers and inter-RAT handovers.

WCDMA is unsuitable. $Q_{qualmin}$ is defined as the minimum CPICH E_c/N_o for the cell to be suitable, where E_c stands for chip energy, and N_o is the two-sided power spectral density of noise.

There are in total six reported events as a result of inter-frequency measurement, which are listed below.

- Event 2a: Change of best frequency.
- Event 2b: The estimated quality of the currently used frequency is below a certain threshold and the estimated quality of a non-used frequency is above a certain threshold.
- Event 2c: The estimated quality of a non-used frequency is above a certain threshold.
- Event 2d: The estimated quality of the currently used frequency is below a certain threshold.
- Event 2e: The estimated quality of a non-used frequency is below a certain threshold.
- Event 2f: The estimated quality of the currently used frequency is above a certain threshold.

It is noted that the frequency quality estimate used in all aforementioned events is defined as:

$$Q_{frequency\,j} = 10W_j \log\left(\sum_{i=1}^{N_{A\,j}} M_{i,j}\right) + 10(1 - W_j)\log M_{Best\,j} \tag{3.1}$$

where $Q_{frequency\,j}$ is the estimated quality of the active or virtual Active Set on frequency j, $M_{i,j}$ is a measurement result of cell i in the active or virtual Active Set on frequency j, $N_{A\,j}$ is the number of cells in the active or virtual Active Set on frequency j, $M_{Best\,j}$ is the measurement result of the cell in the active or virtual Active Set on frequency j with the highest measurement result, and W_j is a parameter sent from UTRAN to UE and used for frequency j.

Similarly, we can also have four different events reported from inter-RAT measurements:

- Event 3a: The estimated quality of the currently used UTRAN frequency is below a certain threshold and the estimated quality of the other system's frequency is above a certain threshold.
- Event 3b: The estimated quality of the other system's frequency is below a certain threshold.
- Event 3c: The estimated quality of the other system's frequency is above a certain threshold.
- Event 3d: Change of the best cell in the other system.

The frequency quality estimate for the serving UTRAN frequency used in Event 3a is the same as that used for inter-frequency events. The triggering conditions for Event 3a are defined as:

$$\begin{cases} Q_{used} \le T_{used} - \frac{H_{3a}}{2} \\ M_{other\,RAT} + CIO_{other\,RAT} \ge T_{other\,RAT} + \frac{H_{3a}}{2} \end{cases} \tag{3.2}$$

where Q_{used} is the quality measurement of the serving UTRAN frequency, T_{used} is the absolute threshold that applies for the UTRAN system in that measurement, $M_{other\,RAT}$ is the measurement quantity for the cell of the other system, $CIO_{other\,RAT}$ is the cell individual offset for the cell of the other system, $T_{other\,RAT}$ is the absolute threshold that is applied for the other system in that measurement, and H_{3a} is the hysteresis parameter for Event 3a. Similar equations can be defined for all other events.

3.3 DISCUSSION:LESSONS TO LEARN

In this chapter, we have discussed two major 2G and 3G mobile cellular standards, IS-95 and W-CDMA, one being the 2G technology and the other the 3G standard, both of which work based on traditional CDMA technology, characterized by largely similar technical features, such as unitary code spreading modulation,[8] precious power control, RAKE receiver, etc. Obviously, they share many similar core intellectual property rights (IPRs), most of which are owned by Qualcomm Inc., which was the major company developing those key technologies in the 1990s.

As to be seen from the time-frame when these CDMA-based 2G and 3G mobile cellular standards were developed, CDMA technology was first introduced into cellular mobile systems as the IS-95 standard in 1993.[9] Therefore, more than ten years have elapsed since the first CDMA cellular mobile technology was developed. Ten years is not a long time, but it should not be considered as a short period of time either, if we think about how fast modern IT technologies advance. Unfortunately, CDMA technology is still in its first generation stage. Even looking at the air-link technologies used in all major 3G mobile cellular standards, such as WCDMA, cdma2000, and so forth, we can hardly see any big difference from its predecessor, IS-95.

Now the first question we might ask here is why CDMA technology could not advance as fast as other technologies have done. The history of making CDMA cellular mobile systems can tell us some of the reason. As mentioned earlier, Qualcomm has undoubtedly made a great contribution to the development of practical CDMA technology, which paved the way for its wide application in cellular mobile phone systems. All will remember the extremely important role it has played. However, we also have to acknowledge the fact that Qualcomm owns the majority of the IPRs related to CDMA technologies. The ownership of those core CDMA IPRs has brought the company a great deal of profits, as witnessed by its share price on the Wall Street. Nowadays any company, regardless of its size, that would like to use CDMA technologies in its products needs to think about the cost of paying huge amount of licensing fees to Qualcomm if the products are commercialized in the market. As a matter of fact, Qualcomm is enjoying a great deal from the status quo of CDMA technology, without much incentive to push for the evolution toward second-generation CDMA technology.

The sluggish nature of CDMA technology's evolution may also be contributing to the premature emergence of the 3G mobile cellular standards, pushed mainly by Japanese telecommunication giant NTT DoCoMo. The birth of WCDMA on October 1, 2001[10] was less than seven years after IS-95, whose standard was first published for public access in July 1993. The introduction of 3G mobile cellular was too hurried to have enough time to make it possible to carry out serious research and development on next generation CDMA technologies. Japan was worrying very much about the fact that Korea had obtained state-of-the-art CDMA technology from Qualcomm, and was very eager to team up with the Europeans to develop a 3G mobile cellular system, namely the WCDMA standard, by taking advantage of the success of European's second generation mobile cellular technology, the GSM system. Some people suggested that WCDMA could make use of much more sophisticated CDMA technology if its emergence could happen a few more years later than October 2001. The race between Japan and Korea made a premature WCDMA, using virtually almost the same CDMA technology as IS-95. This is clearly reflected in the fact that the birth of WCDMA never changed

[8]As to be explained later, the unitary codes stands for the spreading/channelization codes used in all traditional CDMA systems, where each user/channel is spread by one single code.

[9]TIA/EIA IS-95 standard was first published in July 1993. TIA stands for Telecommunications Industry Association, and EIA stands for Electronic Industries Association. IS-95 is the short form of Interim Standard 95

[10]NTT DoCoMo launched a trial 3G service, an area-specific information service for i-mode on June 28, 2001. on September 25, 2001, NTT DoCoMo announced that three 3G phone models were commercially available. NTT DoCoMo launched the first commercial WCDMA 3G mobile network on October 1, 2001.

Qualcomm's worldwide domination of CDMA IPRs. For instance, the OVSF codes used in WCDMA are basically the Walsh-Hadamard sequences used in the IS-95 standard. Sadly to say, we can hardly find many different core CDMA technologies used in WCDMA compared with those employed in the IS-95 system. Therefore, too rapid development of a technology may severely compromise its novelty. This is a serious lesson we should learn from the evolution from 2G to 3G mobile cellular systems.

After having discussed the possible causes for the slow evolution of CDMA technology, naturally we will turn to the question of when the new generation CDMA will come, or even if it will come at all. It is well known that all major 3G mobile cellular systems are using CDMA as their multiple access solution. Therefore, it is justified to say that the worldwide application of 3G mobile cellular has brought CDMA into its prime age. However, it is also suggested that traditional CDMA technology was originated from the need for voice-centric applications, characterized by circuit-switched traffic and continuous-time slow speed communications. On the other hand, Beyond 3G (B3G) wireless communications will mainly carry packet-switched signals dominated by high-speed short-burst traffic. Thus, the traditional CDMA technology may not be well suited to B3G communications. Orthogonal frequency division multiplex (OFDM) and orthogonal frequency division multiple access (OFDMA) have been proposed for their applications in B3G wireless communications. It should be noted that OFDM is only a multiplexing technology, which gives a way to transmit multiple data streams via different tones in parallel and works in a single-user scenario, whereas OFDMA is capable of dividing users in terms of different tone subsets for multiple access. Therefore, the comparison should be made between CDMA and OFDMA (rather than OFDM) in the context of the discussion here.

The operational advantages of OFDMA can be summarized as follows. It offers an efficient way to overcome many channel impairment factors, such as multipath interference (or inter-symbol interference induced by multipath propagation), channel frequency-dispersion caused by mobility and so on, with the help of cyclic prefix (CP) insertion in each OFDMA symbol. The overlapping in consecutive sub-channels in the OFDMA also makes the scheme very bandwidth efficient, compared with other multiple access schemes, such as CDMA and so forth. In addition, OFDMA is suitable in particular for high-speed burst traffic due to the fact that the signal detection there proceeds based on symbols instead of frames, thus making it especially suitable for high-speed bursty signal detection. On the other hand, traditional CDMA technology designed for slow-speed voice-centric applications may not work well in detection of very short packets (for instance, a packet containing only a few bits) as the signature codes were not designed to work for signal detection of packet edges, where partial auto-correlation and partial cross-correlation functions, instead of complete auto-correlation and complete cross-correlation functions,[11] govern the detection efficiency. Unfortunately, design of all spreading codes or signature codes used in traditional CDMA systems does not take into account the fact that both partial and complete correlation functions will determine the CDMA system performance. Therefore, in this sense the evolution of CDMA technology should begin with the innovation of the CDMA code design approach, which should be the core of the next generation CDMA technologies.

What, then, is the next generation CDMA technology? First of all, the next generation CDMA should be developed based on revolutionary design concepts of a CDMA system, which include the innovation in the CDMA code design approach, spreading modulation technique, and so forth. Second, the next generation CDMA technologies should be able to provide agile system architecture to suit the applications of B3G wireless communications, characterized by high-speed burst traffic and all-IP wireless core. Third, the next generation CDMA ought to offer system-level performance at least similar to that of OFDMA technology, if not better. Of course, it is not appropriate for us to reveal clearly what next generation CDMA will look like at this moment, and this issue will be

[11]Partial correlation functions are defined as correlations that involve only partial length of spreading sequences. On the other hand, complete correlation functions involve the full length of spreading sequences.

the topic of interest covering the whole book. Obviously, one thing can be ascertained: CDMA has a long way to go along its evolutionary path, no matter how fast it goes. There is no reason to say that CDMA technology is already dead and thus it should be dumped in favor of some newcomers, such as OFDMA technology. We will make this point very clear from the discussions given throughout this book.

4

Technical Limitations of Traditional CDMA Technology

In this chapter, we will address the issues related to the technical limitations of traditional CDMA technology. The discussions carried out here will pave the way for the definition and proposal of next generation CDMA technology, which should be built up based on a thorough understanding of the problems associated with traditional CDMA systems.

As mentioned in the previous chapter, the currently available CDMA technology was proposed basically for slow-speed voice-centric applications, and thus bears many features pertaining to second generation mobile cellular systems, in particular the IS-95 standard. On the other hand, the birth of all CDMA-based 3G mobile cellular standards, such as WCDMA, cdma2000, and TD-SCDMA, was driven primarily by the competition between Japan and Korea for development of CDMA-based mobile cellular technologies, and thus sacrificing some great opportunities to bring sufficient technological innovations to the CDMA-based 3G mobile cellular standards.

A careful review of current CDMA technology will be done in the sections followed. The review will start from the CDMA code design approach used in all traditional CDMA systems. Then, the discussion will continue to cover spreading modulation schemes, and other related techniques used in CDMA-based 2G and 3G mobile cellular standards.

4.1 PROBLEMS WITH TRADITIONAL CDMA CODES

Before we start our discussion on problems associated with traditional CDMA systems, we should first define the terminologies we will use to distinguish two major categories of spreading codes used in all CDMA systems of concern, one being the unitary codes and the other the complementary codes. If we say a CDMA system is based on unitary codes, it means that the system works on a one-code-per-user or one-code-per-channel basis. In this sense, it is clear that all traditional CDMA-based 2G and 3G mobile cellular systems are using unitary codes without exception. These unitary codes can be further divided into two sub-categories, one being the orthogonal codes,[1] such

[1]Here, so-called orthogonal codes are not necessarily truly orthogonal if they are used in an asynchronous operation mode, such as in the uplink channels of a mobile cellular system. Thus, the term 'orthogonal codes' is quite misleading. As this term has been used since CDMA was first introduced, we will still use it to represent those codes, such as Walsh-Hadamard codes, OVSF codes, etc.

The Next Generation CDMA Technologies Hsiao-Hwa Chen

as Walsh-Hadamard codes (mainly used in IS-95 and cdma2000 systems) and orthogonal variable spreading factor (OVSF) codes (used in WCDMA and TD-SCDMA systems), and the other being quasi-orthogonal codes, such as Gold codes, Kasami codes, m-sequences, and so forth. Therefore, the quasi-orthogonal codes comprise a fairly large group of spreading codes, while orthogonal codes currently have only two types, namely Walsh-Hadamard sequences [186–189] and OVSF codes [83–89].

4.1.1 ORTHOGONAL CDMA CODES

The design of orthogonal CDMA codes came from a simple idea based on matrix operations. We can define an orthogonal matrix as follows. Let $\mathbf{X}$ be a square matrix. Let us consider a matrix $\mathbf{X}$, with all its elements taking complex values without loss of generality, and we say that it is an orthogonal matrix if and only if

$$\mathbf{X}\mathbf{X}^H = \mathbf{I}, \tag{4.1}$$

where $\mathbf{X}^H$ is the Hermitian transform of the matrix $\mathbf{X}$ and $\mathbf{I}$ is an identity matrix with its diagonal line elements being identically unit and all other elements being zero. It is noted that the Hermitian transform of a matrix $\mathbf{X}$ is always equivalent to performing joint transpose and conjugate operations of the same matrix. If all elements in a matrix are real, then its Hermitian transform is just equal to applying the transpose operation to it. It is noted also that most cases we will consider here are involved with real matrices, and thus the transpose operation will be more common than the Hermitian transform operation.

If we indeed can find a matrix $\mathbf{X}$, which satisfies Equation (4.1), then either its row vectors or column vectors are orthogonal with each other. Those row or column vectors can form a set of orthogonal sequences or orthogonal codes, which can be used for CDMA applications. For instance, the Walsh-Hadamard sequences are derived from Walsh-Hadamard matrices, as shown in the example below,

$$\begin{pmatrix} 1 & 1 & 1 & 1 & 1 & 1 & 1 & 1 \\ 1 & -1 & 1 & -1 & 1 & -1 & 1 & -1 \\ 1 & 1 & -1 & -1 & 1 & 1 & -1 & -1 \\ 1 & -1 & -1 & 1 & 1 & -1 & -1 & 1 \\ 1 & 1 & 1 & 1 & -1 & -1 & -1 & -1 \\ 1 & -1 & 1 & -1 & -1 & 1 & -1 & 1 \\ 1 & 1 & -1 & -1 & -1 & -1 & 1 & 1 \\ 1 & -1 & -1 & 1 & -1 & 1 & 1 & -1 \end{pmatrix} \tag{4.2}$$

which gives an 8×8 Walsh-Hadamard matrix. Either eight rows or eight columns can be used to form eight Walsh-Hadamard sequences to support eight different users in a CDMA system. As a matter of factor, the IS-95 system just uses a 64×64 Walsh-Hadamard matrix to form 64-ary Walsh-Hadamard sequences for downlink channelization and uplink 64-ary modulation, as discussed in Section 3.1.

On the other hand, the OVSF codes can be generated using a tree structure. The OVSF code generation algorithm generates an OVSF code from a set of orthogonal codes. OVSF codes were first introduced for 3G communication systems, such as WCDMA and TD-SCDMA. OVSF codes are primarily used to preserve orthogonality between different channels in a communication system.

The OVSF codes are defined as the rows of an $N \times N$ matrix, $\mathbf{C}_N$, which is defined recursively as follows. First, define $\mathbf{C}_1 = [1]$. Next, assume that $\mathbf{C}_N$ is defined and let $\mathbf{C}_N(\mathrm{k})$ denote the kth row

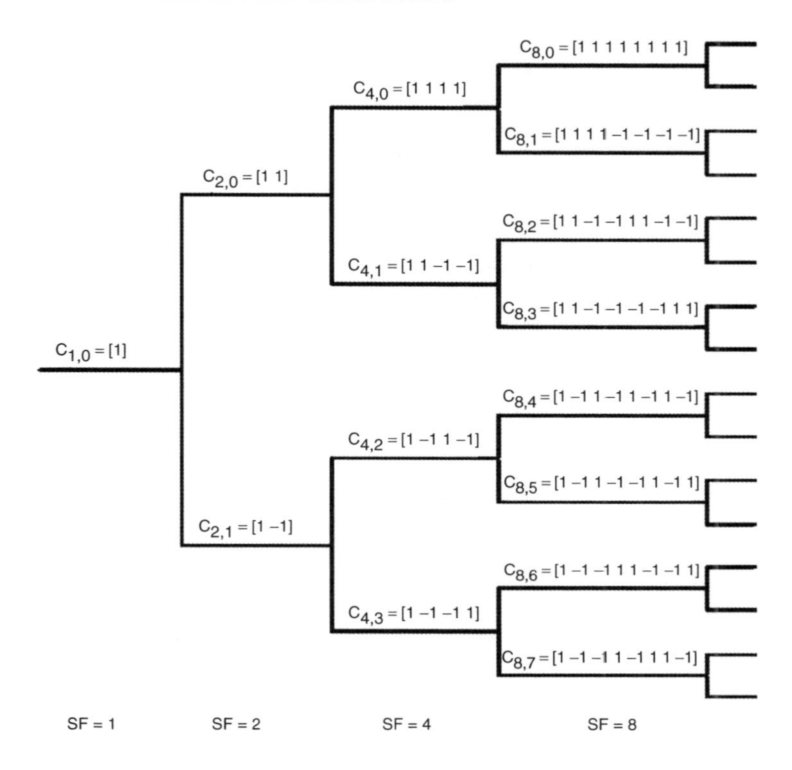

Figure 4.1 Generation tree structure for OVSF codes, where SF denotes spreading factor or processing gain of a CDMA system using the OVSF codes, and only four layers or depths are shown for simplicity.

of $\mathbf{C}_N$. Define $\mathbf{C}_{2N}$ by

$$\mathbf{C}_{2N} = \begin{pmatrix} \mathbf{C}_N(0) & \mathbf{C}_N(0) \\ \mathbf{C}_N(0) & -\mathbf{C}_N(0) \\ \mathbf{C}_N(1) & \mathbf{C}_N(1) \\ \mathbf{C}_N(1) & -\mathbf{C}_N(1) \\ \cdots & \cdots \\ \mathbf{C}_N(N-1) & \mathbf{C}_N(N-1) \\ \mathbf{C}_N(N-1) & -\mathbf{C}_N(N-1) \end{pmatrix}. \tag{4.3}$$

It has to be noted that $\mathbf{C}_N$ is only defined for N a power of two. It follows by induction that the rows of $\mathbf{C}_N$ are orthogonal with each other. In fact, the OVSF codes are more often defined recursively by a tree structure, as shown in Figure 4.1, where only four different layers or 'depths' of the tree structure are shown for description simplicity. The layer can continue until a specific spreading factor is obtained. For instance, in the WCDMA system the maximal spreading factor allowed is 256, forming a fairly big tree structure.

It is also noted that if [**C**] (for example, $\mathbf{C}_{4,1} = [1\ 1\ \ -1\ \ -1]$ in Figure 4.1) is a code length 2^r at depth r in the tree structure, where the root has depth 0, the two branches leading out of **C** are labeled by the sequences [**C C**] and [**C** −**C**], which have length 2^{r+1}. The codes at depth r in the tree are the rows of the matrix $\mathbf{C}_N$, where $N = 2r$. Note that two OVSF codes are orthogonal if and only if neither code lies on the path from the other code to the root. Since codes assigned to different users in the same cell must be orthogonal, this restricts the number of available codes for a given cell. For example, if the code $\mathbf{C}_{4,1}$ in the tree is assigned to a user, the codes $\mathbf{C}_{1,0}$, $\mathbf{C}_{2,0}$, $\mathbf{C}_{8,2}$, $\mathbf{C}_{8,3}$, and so on, cannot be assigned to any other user in the same cell. In general, if a code $\mathbf{C}_{2^r,n}$ (where $n = 0, \ldots, 2^r - 1$) is used in a cell, all its parent codes ($\mathbf{C}_{2^s,m}$, where $s = 0, \ldots, r-1$ and $m = 0, \ldots, 2^s - 1$) and all its child codes ($\mathbf{C}_{2^t,l}$, where $t > r$ and $l = 0, \ldots, 2^t - 1$) should not be used in the same cell to maintain orthogonality among the users. We call this restriction of the OVSF codes the code assignment blocking (CAB) problem.

We can specify the two code parameters from the OVSF code generator block: the spreading factor, which is the length of the code, and the code index, which must be an integer in the range $(0, 1, \ldots, N-1)$, where N is the spreading factor. If the code appears at depth r in the preceding tree, the spreading factor is simply 2^r. The code index specifies how far down the column of the tree at depth r the code appears, counting from 0 to $N-1$. For $\mathbf{C}_{N,k}$, in, N is the spreading factor and k is the code index.

We can recover the code from the spreading factor and the code index as follows. Convert the code index to the corresponding binary number, and then add 0s to the left, if necessary, so that the resulting binary sequence $x_1, x_2, \ldots, x_r$ has length r, where r is the logarithm base 2 of the spreading factor. This sequence describes the path from the root to the code. The path takes the upper branch from the code at depth i if $x_i = 0$, and the lower branch if $x_i = 1$.

To reconstruct the code, recursively define a sequence of codes $\mathbf{C}_i$ as follows. Let $\mathbf{C}_0$ be the root code of [1]. Assuming that $\mathbf{C}_i$ has been defined, for $i < r$, define $\mathbf{C}_{i+1}$ by

$$\mathbf{C}_{i+1} = \begin{cases} \mathbf{C}_i \mathbf{C}_i, & x_i = 0 \\ \mathbf{C}_i(-\mathbf{C}_i), & x_i = 1 \end{cases} \tag{4.4}$$

The code $\mathbf{C}_N$ has the specified spreading factor and code index. For example, to find the code with spreading factor 16 and code index 6, do the following:

- Convert 6 to the binary number 110.
- Add one zero to the left to obtain 0110, which has length $4 = \log_2 16$.
- Construct the sequences $\mathbf{C}_i$ according to the procedure shown below.

The OVSF codes can be generated using the following procedure:

1. Let $i = 0$ to form $\mathbf{C}_0 = [1]$.
2. Let $i = 1$ and $x_1 = 0$ to form $\mathbf{C}_1 = \mathbf{C}_0\mathbf{C}_0 = [1][1]$.
3. Let $i = 2$ and $x_2 = 1$ to form $\mathbf{C}_2 = \mathbf{C}_1(-\mathbf{C}_1) = [1\ 1][-1\ -1]$.
4. Let $i = 3$ and $x_3 = 1$ to form $\mathbf{C}_3 = \mathbf{C}_2(-\mathbf{C}_2) = [1\ 1\ -1\ -1][-1\ -1\ 1\ 1]$.
5. Let $i = 4$ and $x_4 = 0$ to form $\mathbf{C}_4 = \mathbf{C}_3(\mathbf{C}_3) = [1\ 1\ -1\ -1\ -1\ -1\ 1\ 1][1\ 1\ -1\ -1\ -1\ -1\ 1\ 1]$;
6. Continue until the desirable spreading factor is reached.

It is noted that the code $\mathbf{C}_4$ has spreading factor 16 and code index 6. As a matter of fact, the OVSF codes are not completely new CDMA codes, and they are virtually the same codes as the Walsh-Hadamard sequences. This fact can be easily verified if we look at Figure 4.1, where we stagger all codes in the same depth (for instance, the depth $SF = 8$) to form a square matrix, obtaining

$$\begin{pmatrix} \mathbf{C}_{8,0} \\ \mathbf{C}_{8,1} \\ \mathbf{C}_{8,2} \\ \mathbf{C}_{8,3} \\ \mathbf{C}_{8,4} \\ \mathbf{C}_{8,5} \\ \mathbf{C}_{8,6} \\ \mathbf{C}_{8,7} \end{pmatrix} = \begin{pmatrix} 1 & 1 & 1 & 1 & 1 & 1 & 1 & 1 \\ 1 & 1 & 1 & 1 & -1 & -1 & -1 & -1 \\ 1 & 1 & -1 & -1 & 1 & 1 & -1 & -1 \\ 1 & 1 & -1 & -1 & -1 & -1 & 1 & 1 \\ 1 & -1 & 1 & -1 & 1 & -1 & 1 & -1 \\ 1 & -1 & 1 & -1 & -1 & 1 & -1 & 1 \\ 1 & -1 & -1 & 1 & 1 & -1 & -1 & 1 \\ 1 & -1 & -1 & 1 & -1 & 1 & 1 & -1 \end{pmatrix}$$

$$\overset{\text{Row interchanges}}{\Longrightarrow} \begin{pmatrix} 1 & 1 & 1 & 1 & 1 & 1 & 1 & 1 \\ 1 & -1 & 1 & -1 & 1 & -1 & 1 & -1 \\ 1 & 1 & -1 & -1 & 1 & 1 & -1 & -1 \\ 1 & -1 & -1 & 1 & 1 & -1 & -1 & 1 \\ 1 & 1 & 1 & 1 & -1 & -1 & -1 & -1 \\ 1 & -1 & 1 & -1 & -1 & 1 & -1 & 1 \\ 1 & 1 & -1 & -1 & -1 & -1 & 1 & 1 \\ 1 & -1 & -1 & 1 & -1 & 1 & 1 & -1 \end{pmatrix} \quad (4.5)$$

which is exactly an 8×8 Walsh-Hadamard matrix, as shown in Equation (4.2). Therefore, the OVSF codes are basically the same as the Walsh-Hadamard sequences. The only difference between the two is that the OVSF codes allow combinational use of Walsh-Hadamard sequences with different lengths. However, the combinational use of Walsh-Hadamard sequences with different lengths brings in a new problem, which is the CAB problem as mentioned earlier.

It has to be admitted that the CAB problem poses a great challenge to the code assignment process in any CDMA system which uses OVSF codes as its spreading codes. The use of any one OVSF code with a particular spreading factor ($\mathrm{SF} = 2^r$) will effectively block all its parent codes and all its child codes, as shown in Figure 4.2, where the code $\mathbf{C}_{4,1} = [1\ 1\ -1\ -1]$ is the assigned code, which will block all its parent codes and child codes, as shown by the thick solid lines in the figure. As described in Section 3.2, every user in a WCDMA or TD-SCDMA system requires a particular spreading factor due to its certain data rate demand, and the change in the date rate will need a change of spreading factor and thus of OVSF code. This requires the reassignment of a code to meet the particular data rate requirement of the user. However, the reassignment of spreading codes in a WCDMA or TD-SCDMA system will inevitably waste a lot of network resources due to the reshuffling of the code assignments in a cell. Also due to this CAB problem, a WCDMA or TD-SCDMA system can support far fewer users than the maximal spreading factor used by the system. For example, if a WCDMA system uses the OVSF codes whose maximal spreading factor is 256, the total number of users allowed in a particular cell will definitely be much lower than 256, even if all users require only a low-rate voice service. From many measurement reports revealed by many 3G mobile operators around the world, it is very unlikely that the systems could support low-rate service users whose number was close to one fifth of the maximal spreading factor, depending very much on the environment where the cell is located.

Another non-trivial problem with the OVSF codes is the difficulty of supporting variable rate services, which are in particular important in all multimedia mobile services, where the data rate may change from time to time to fit a particular data rate and meet its quality of service (QoS) requirements. It is seen from Figures 4.1 and 4.2 that the spreading factors in the OVSF code tree structure can change only in multiples of two. In other words, the end users can only be given a data rate of 2^r, where r is an integer. Therefore, it is impossible to provide a user with a continuous rate change. For example, if a user requires a data rate of 5 units, the system has to assign it the nearest data rate of 8 units, which is equal to 2^3, but not exactly 5. This will cause a huge waste

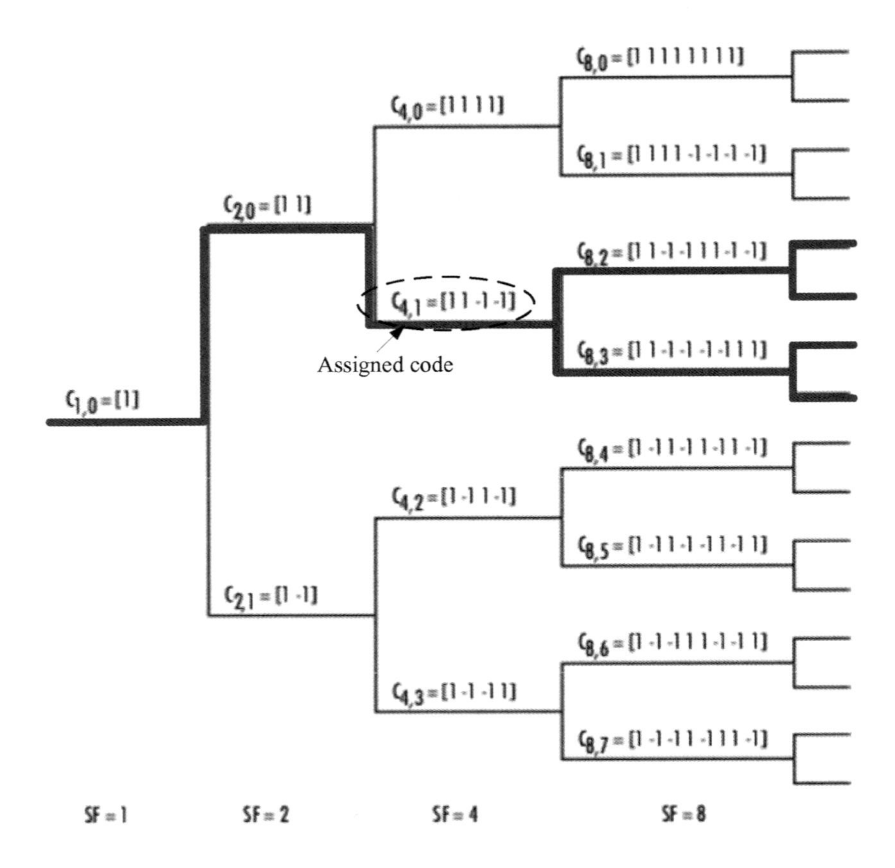

Figure 4.2 The code assignment blocking (CAB) problem with OVSF codes, where the code $\mathbf{C}_{4,1}$ is the assigned code, which will block all its parent codes and child codes, as shown by the thick solid lines.

of precious bandwidth (about 37.5 % of assigned bandwidth has been wasted in this case). The most serious problem is that this kind of waste always exists for all users in WCDMA and TD-SCDMA systems, no matter what data rates they ask for, as long as the rates are not exactly equal to multiples of two. A simple estimation of the bandwidth waste due to this problem can be done by the following method.

Assume that we consider only the data rate change from 1 to 16 units, as shown in Table 4.1, where the estimated percentage of bandwidth wasted in WCDMA and TD-SCDMA systems due to the use of OVSF codes is illustrated. For this simple case, assuming that it is equally likely that a user will take any of the data rates listed in the table, the average estimated percentage of wasted bandwidth can be equal to about 17.19%, which is not a trivial figure at all.

Yet another problem to be noted with orthogonal codes, such as Walsh-Hadamard and OVSF codes, is that so-called orthogonal codes will not be orthogonal at all if they are used in asynchronous transmission channels, such as the uplink channels in a mobile cellular environment. This raises another serious concern on the legitimacy of the use of Walsh-Hadamard codes and OVSF codes in

Table 4.1 The estimated percentage of bandwidth wasted in WCDMA and TD-SCDMA systems due to the use of OVSF codes, resulting in an average percentage of wasted bandwidth equal to about 17.19%.

Real rate	Rate assigned	Wasted bandwidth (%)
1	1	0
2	2	0
3	4	25
4	4	0
5	8	37.5
6	8	25
7	8	12.5
8	8	0
9	16	43.75
10	16	37.5
11	16	31.25
12	16	25
13	16	18.75
14	16	12.5
15	16	6.25
16	16	0

a mobile or wireless communication application, as long as there exist asynchronous transmissions in the direction from mobile stations to the base stations. The asynchronous transmissions are always there due to the fact that all mobile terminals will always be stationed in various positions within a cell or a cell-like region, and thus their communication with the base station of the cell will always take place along different propagation paths (and therefore different distances). In this way, the signals from different mobile terminals will arrive at the same base station at different times randomly, and they will sum up at the receiver of the base station in an asynchronous fashion. This situation can be typically illustrated in Figure 4.3, which shows one particular detection frame in an asynchronous CDMA system of concern.

It is to be noted from Figure 4.3 that there are K different bit streams received from different mobile terminals located in different places in a cell. The bits from different users may take either $+1$ or -1 with equal likelihood, if the sent signals are binary signals for illustration simplicity. We consider only the cases in which each bit will be covered exactly by one spreading code, which can be either a Walsh-Hadamard sequence or an OVSF code. Then, we can easily imagine that detection of a particular bit from a particular mobile terminal, say $b_k[1]$ (where $k \in (1, \ldots, K)$), as shown in Figure 4.3, will be interfered with by all other $K-1$ bits coming from different users, which thus use different signature codes for spreading purposes. Even worse, the duration of the bit of interest, say $b_k[1]$, will in general overlap with two consecutive bits of all other asynchronous transmissions, as shown in Figure 4.4. As each bit from all other users may take either $+1$ or -1 with equal likelihood, the cross-correlation function between $b_k[1]$ and any signal from other users may be made up of two partial cross-correlation functions. Therefore, the detection efficiency in this case will be mainly governed by partial cross-correlation functions (rather than complete cross-correlation functions) between any two codes. Unfortunately, the partial cross-correlation functions of all orthogonal codes, such as Walsh-Hadamard sequences and OVSF codes, are virtually totally out of control. In fact, the partial cross-correlation functions of any orthogonal codes can be even worse than those of known quasi-orthogonal codes, such as Gold codes, m-sequences, Kasami codes, etc.

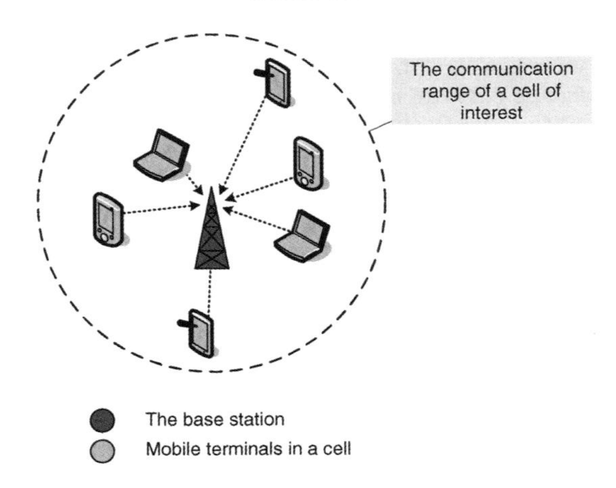

Figure 4.3 Asynchronous transmission channels in a CDMA system, where different mobile terminals are located in different places in a cell and send their signals to the base station along different propagation paths.

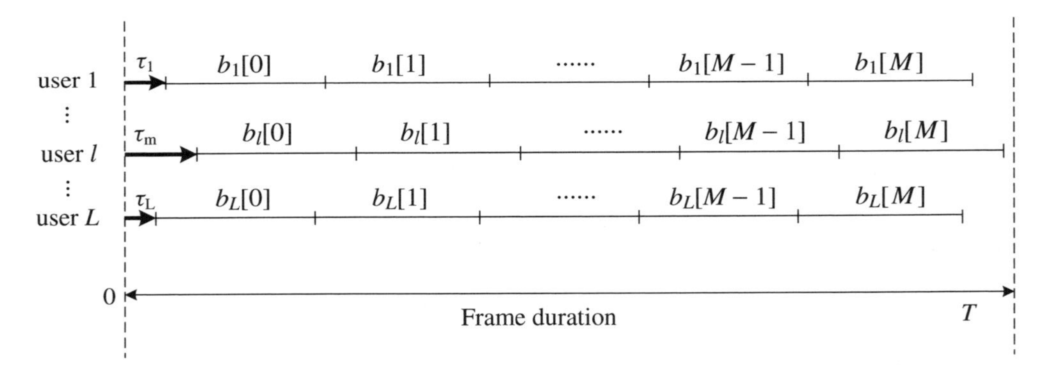

Figure 4.4 K asynchronous bit streams in an asynchronous CDMA system, where the bits sent from K different mobile terminals are not aligned in time, forming aperiodic cross-correlations among different user signals.

Obviously, if this problem can not be overcome, the detection efficiency of an asynchronous CDMA receiver can be seriously undermined.

In fact, many reported studies have suggested that the partial cross-correlation functions for all orthogonal codes, such as Walsh-Hadamard sequences and OVSF codes, are very different from their complete cross-correlation functions. Unfortunately, the partial auto-correlation functions and partial cross-correlation functions of the signature codes or sequences have very much to do with the detection performance of all future wireless applications, in which packet-switched bursty traffic (in contract to circuit-switched continuous traffic) will dominate. The reason is obvious, as a packet in all IP-based wireless systems can be as short as a few bits, making traditional CDMA signal detection very difficult if the partial correlation properties of the signature codes are not under control.

It is noted that the problems with poor partial correlation functions exist not only in traditional orthogonal spreading codes, such as Walsh-Hadamard sequences and OVSF codes, but also in all quasi-orthogonal codes, such as Gold codes, Kasami codes, and so forth. As there will be a dedicated section in the following text to discuss the impact of poor partial correlation properties on bursty traffic signal detection, we will not discuss it in much detail here.

Based on the discussions given in this section, we can summarize the limitations of orthogonal codes as follows. There are three inherent problems with orthogonal spreading codes currently used by 2G and 3G mobile cellular systems:

1. The CAB problem, which comes from the fact that the OVSF codes are generated by a tree structure.
2. The difficulty of supporting continuously variable data rate services due to the fact that the OVSF codes can change their spreading factors only at multiples of two, not at any number.
3. The problem with their poor partial cross-correlation functions resulting from their application in asynchronous CDMA transmission channels.

The aforementioned technical limitations of orthogonal codes have to be resolved in the development of the next generation CDMA technologies.

4.1.2 QUASI-ORTHOGONAL CDMA CODES

Other important spreading codes or sequences used in traditional CDMA systems are quasi-orthogonal codes. The name 'quasi-orthogonal codes' indicates that all of these spreading codes or sequences are not exactly orthogonal even if they are used in a synchronous transmission CDMA system. However, their cross-correlation functions are somehow still quite acceptable in the sense that at different chip offsets they are under control, usually being below a certain level to limit multiple access interference (MAI).

As a matter of fact, the quasi-orthogonal CDMA codes comprise a fairly large number of codes, and thus the majority of the work on the subject of CDMA coding in the literature was dedicated to them. The most widely used quasi-orthogonal CDMA codes include Gold codes [190–191], Kasami codes [192–195], m-sequences [196–204], etc. Other less widely cited spreading codes include Kronecker sequences [205–209], GMW codes [210–212], 4-CCL and 5-CCL codes [213], Bent codes [214], No codes [215], and Wavelets-code [216], which are all unitary codes.[2] One of the most important characteristics of quasi-orthogonal CDMA codes is that they offer relatively more uniform performance for their operation in both synchronous and asynchronous channels, compared with orthogonal CDMA codes, such as Walsh-Hadamard sequences and OVSF codes, which usually perform extremely badly if they are used for asynchronous channel transmission in a wireless communication system.[3] Also for this reason, quasi-orthogonal CDMA codes have been found more suitable for their applications in many wireless applications. In the following text we will take a look at different quasi-orthogonal spreading codes.

m-Sequences

m-Sequences probably are the quasi-orthogonal CDMA codes which can be generated using the simplest hardware structure. m-Sequences can be generated from output of a linear feedback shift

[2]Unitary codes are defined as spreading codes which work on a one-code-per-user basis. In contrast, complementary codes can work on a one-flock-per-user basis.

[3]Any wireless communication application, such as wireless local area networks or mobile cellular systems, rely on both uplink and downlink transmissions to implement full-duplex communication. Thus both uplink and downlink need to use spreading codes for CDMA purposes. The uplink transmissions are always asynchronous, whereas the downlink channel transmissions happen in an synchronous way.

register (LFSR) with certain feedback logic, and they have the property that if the shift register is set to any non-zero state and then cycled, a pseudo-random binary sequence of a maximum of $n = 2^r - 1$ bits will be generated, where r is the number of stages, i.e. the number of bit positions in the register, before the shift register returns to its original state and the n-bit output sequence repeats.[4]

In fact, various pseudo-random codes can be generated using an LFSR. The generator polynomial governs all major characteristics of the generator. For a given generator polynomial, there are two ways [7] of implementing LFSR. A Galois feedback generator uses only the output bit to add (in Galois field) several stages of the shift register and is desirable for high-speed hardware implementation as well as software implementation. The other way, known as a Fibonacci feedback generator, can generate several delays of sequences without any additional logic.

Shift-register sequences having the maximum possible period for an r-stage shift register are called maximal length sequences or m-sequences. A primitive generator polynomial [7] always yields an m-sequence. The maximum period of an r-stage shift register can be proven to be $2^r - 1$. The m-sequences have three important properties, i.e. balance property, run-length property, and shift-and-add property. The periodic auto-correlation function of an m-sequence $R_n(k)$ is two-valued and is given by

$$R_n(k) = \begin{cases} 1.0 & k = lN \\ -\frac{1}{N} & k \neq lN \end{cases} \tag{4.6}$$

where l is an integer and N is the period of the m-sequence. The excellent auto-correlation property comes from the first and third properties. Refer to [7] for more detailed information about the properties. m-Sequences have good auto-correlation properties and have been used in many applications, including the IS-95 standard. As the cross-correlation property of these sequences is relatively poor compared to Gold codes, the same sequence with different offsets is usually used for different users or for different base stations. With this method, the discrimination property between different spreading codes only depends on the partial auto-correlation property.

The software implementation of Galois construction for a given generator polynomial is straightforward and well known [217]. A typical configuration of an LFSR for m-sequence generation is shown in Figure 4.5, where there are five stages or $r = 5$, and its length can be $32 - 1 = 31$. Table 4.2 gives the number of different primitive polynomials which can be used to generate m-sequences, where r is the degree of the primitive polynomials and N_p is the number of different primitive polynomials available.

Gold Codes

Gold codes were first developed by Robert Gold [190, 191]. These codes possess pseudo-random properties ensuring both favorable auto-correlation and cross-correlation characteristics. Such codes are used in direct sequence spread spectrum systems such as UMTS in order to distinguish different channels across the radio interface.

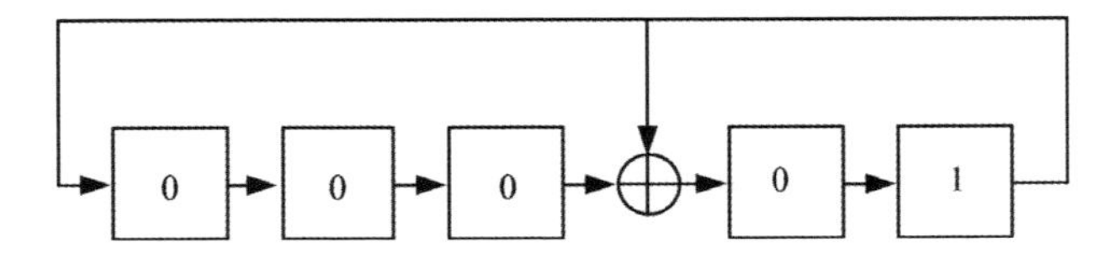

Figure 4.5 A typical configuration of an LFSR for m-sequence generation, where there are five stages or $r = 5$, and its length can be $2^r - 1 = 31$.

---

[4] A typical example for m-sequences is that the register may be used to control the sequence of frequencies for a frequency-hopping spread spectrum transmission system.

Table 4.2 Number of different primitive polynomials which can be used to generate m-sequences, where r is the degree of the primitive polynomials and N_p is the number of different primitive polynomials available.

r	N_p	r	N_p
2	1	11	176
3	2	12	144
4	2	13	630
5	6	14	756
6	6	15	1800
7	18	16	2048
8	16	17	7710
9	48	18	8064
10	60	19	27594

Some pairs of m-sequences with the same degree can be used to generate Gold codes by linearly combining two m-sequences with different offsets in the Galois field. Not all pairs of m-sequences can be used to yield Gold codes, and those which do yield Gold codes are called preferred pairs. Gold codes have three-valued auto-correlation and cross-correlation functions with their values being $\{-1, -t(m), t(m) - 2\}$, where

$$t(m) = \begin{cases} 2^{(m+1)/2} + 1 & \text{for odd } m \\ 2^{(m+2)/2} + 1 & \text{for even } m \end{cases} \tag{4.7}$$

Some references (e.g. [7]) have shown the dramatic decrease of the peak cross-correlations of Gold codes compared to m-sequences. Furthermore, the cross-correlation values show many -1s, which is very desirable, whereas Kasami sequences show much fewer -1s while providing half-peak cross-correlations.

The generation of Gold codes is very simple. Using two preferred m-sequence generators of degree r, with a fixed non-zero initial loading state in the first generator, Gold codes are obtained by changing the initial loading state of the second generator from 0 to $2^r - 1$. Another Gold sequence can be obtained by setting all zero to the first generator, which is the second m-sequence itself. In total, $2^r + 1$ Gold codes are available.

Kasami Codes

Another type of important quasi-orthogonal CDMA code is the Kasami code [192, 193], which was proposed by Professor Kasami in the 1960s.

Kasami sequences have optimal cross-correlation values touching the Welch lower bound [218]. A lower bound on the cross-correlation between any pair of binary sequences of period n in a set of M sequences is given by

$$\phi_{n,M} \geq n\sqrt{\frac{M-1}{Mn-1}} \tag{4.8}$$

For an m-sequence $\mathbf{a}$, $\mathbf{a}'$ is obtained by taking every qth bit of $\mathbf{a}$ and denoted by $\mathbf{a}[q]$. $\mathbf{a}'$ is called a decimated sequence of $\mathbf{a}$. By choosing $q = 2^{m/2} + 1$, where m is the degree of sequence $\mathbf{a}$, $\mathbf{a}'$ is periodic with period $2^{m/2} - 1$. By repeating $\mathbf{a}'q$ times, a new sequence $\mathbf{b}$ is obtained. With $\mathbf{a}$ and $\mathbf{b}$, we form a new set of sequences by adding **a** and $2^{m/2} - 2$ cyclically shifted **b**s. Including **a** and **b**, we get $2^{m/2}$ sequences. These sequences are called Kasami sequences. The hardware implementation of Kasami sequences is painful because the decimation process requires a much faster clock. Luckily, the decimated sequence itself is an m-sequence [5] of order $m/2$ and we can exploit this fact to implement a Kasami code generator.

It should be noted that not all pairs of polynomials with degree m and $m/2$ can be used to generate Kasami sequences, and we should check their cross-correlation values using some testing calculations.

As a matter of fact, there are two classes of Kasami sequences: the small set and the large set. The large set contains all the sequences in the small set. Only the small set is optimal in the sense of matching Welch's lower bound [218] for correlation functions. Kasami sequences have period $N = 2^n - 1$, where n is a non-negative and even integer. Let **u** be a binary sequence of length N, and let **w** be the sequence obtained by decimating **u** by $2^{n/2} + 1$. The small set of Kasami sequences is defined by the following formulas, in which T denotes the left shift operator, m is the shift parameter for **w**, and $\oplus$ denotes modulo addition-2.

A small set of Kasami sequences for an even n can be defined as

$$K_s(\mathbf{u}, n, m) = \begin{cases} \mathbf{u}, & m = -2 \\ \mathbf{u} \oplus T_m \mathbf{w}, & m = 0, \ldots, 2^{n/2} - 2 \end{cases} \tag{4.9}$$

It is noted that a small set Kasami code set contains $2^{n/2}$ sequences.

For $\mathrm{mod}(n, 4) = 2$, a large set of Kasami sequences can be defined as follows. Let **v** be the sequence formed by decimating the sequence **u** by $2^{n/2+1} + 1$. The large set is defined as follows, in which k and m are the shift parameters for the sequences **v** and **w**, respectively. Thus, in this way, a large set of Kasami sequences can be defined:

$$K_L(\mathbf{u}, n, k, m) = \begin{cases} \mathbf{u}, & k = -2, m = -1 \\ \mathbf{v}, & k = -1, m = -1 \\ \mathbf{u} \oplus T^k \mathbf{v}, & k = 0, \ldots, 2^n - 2, m = -1 \\ \mathbf{u} \oplus T^m \mathbf{w}, & k = -2, m = 0, \ldots, 2^{n/2} - 2 \\ \mathbf{v} \oplus T^m \mathbf{w}, & k = -1, m = 0, \ldots, 2^{n/2} - 2 \\ \mathbf{u} \oplus T^k \mathbf{v} \oplus T^m \mathbf{w}, & k = 0, \ldots, 2^n - 2, m = 0, \ldots, 2^{n/2} - 2 \end{cases} \tag{4.10}$$

The sequences described in the first three rows above correspond to the Gold sequences for mod $(n, 4) = 2$. See the previous discussion on Gold sequence generation for a description of Gold sequences. However, the Kasami sequences form a larger set than the Gold sequences. The correlation functions for the sequences take on the values $\{-t(n), -s(n), -1, s(n) - 2, t(n) - 2\}$, where $t(n) = 1 + 2^{(n+2)/2}$ for even n and $s(n) = (t(n) + 1)/2$.

LFSR Generation of PN Codes

This section gives the irreducible connection polynomials of the LFSR generators of optimized PN code sets in octal form. Polynomials are taken from table C.2 in appendix C of the book [145]. In the case of m-sequences note that the second half of each code set is generated using the reciprocal feedback polynomials of the first half of a code set, i.e. codes having order numbers from $N/2 + 1$ to N are time-reversed reciprocals of codes having order numbers from 1 to $N/2$ in the code matrix (in an optimized initial phase, of course), where N is the total number of sequences in a set. The sets of Gold and Kasami sequences do not contain reciprocals. Codes were generated by changing the modulo-2 summation phase of several linear component code generators. A small Kasami set always contains as the last sequence the longer m-sequence from which the set was generated. Also, in the case of Gold codes of lengths 31, 63, and 127 the last two codes in a code matrix are the sequences corresponding to the preferred pairs.

The following data are the feedback polynomials in the successive order of rows of the code matrix, which can be used to generate m-sequences of different lengths:

1. Codes of length 15:
   23, 31.

2. Codes of length 31:
   23, 31.

3. Codes of length 63:
   103, 147, 155, 141, 163, 133.

4. Codes of length 127:
   211, 217, 235, 367, 277, 325, 203, 313, 345, 221, 361, 271, 357, 375, 253, 301, 323, 247.

5. Codes of length 255:
   435, 551, 747, 453, 545, 543, 537, 703, 561, 455, 717, 651, 515, 615, 765, 607.

6. Codes of length 511:
   1021, 1131, 1461, 1423, 1055, 1167, 1541, 1333, 1605, 1751, 1743, 1617, 1553, 1157, 1715, 1563, 1713, 1175, 1725, 1225, 1275, 1773, 1425, 1267, 1041, 1151, 1063, 1443, 1321, 1671, 1033, 1555, 1207, 1137, 1437, 1707, 1533, 1731, 1317, 1473, 1517, 1371, 1257, 1245, 1365, 1577, 1243, 1665.

7. Codes of length 1023:
   2011, 2415, 3771, 2157, 3515, 2773, 2033, 2443, 2461, 3023, 3543, 2745, 2431, 3177, 3525, 2617, 3471, 3323, 3507, 3623, 2707, 2327, 3265, 2055, 3575, 3171, 2047, 3025, 3337, 3211, 2201, 2605, 2377, 3661, 2627, 3375, 3301, 3045, 2145, 3103, 3067, 2475, 2305, 3763, 2527, 3615, 2347, 3133, 3427, 3117, 3435, 3531, 2553, 2641, 2767, 2363, 3441, 2503, 3733, 2213.

8. Codes of length 2047:
   4005, 4445, 4215, 4055, 6015, 7413, 4143, 4563, 4053, 5023, 5623, 4577, 6233, 6673, 7237, 7335, 4505, 5337, 5263, 5361, 5171, 6637, 7173, 5711, 5221, 6307, 6211, 5747, 4533, 4341, 6711, 7715, 6343, 6227, 6263, 5235, 7431, 6455, 5247, 5265, 4767, 5607, 4603, 6561, 7107, 7041, 4251, 5675, 4173, 4707, 5463, 5755, 6675, 7655, 5531, 7243, 7621, 7161, 4731, 4451.

9. Codes of length 4095:
   10123, 15647, 16533, 16047, 11015, 14127, 17673, 13565, 15341, 15053, 15621, 15321, 11417, 13505, 13275, 11471, 16237, 12515, 12255, 11271, 17121, 14227, 12117, 14135, 14711, 13131, 16521, 15437, 12067, 12147, 14717, 14675, 10663, 16115, 12247, 17675, 10151, 14613, 11441, 10321, 11067, 14433, 12753, 13431, 11313, 13425, 16021, 17025, 15723, 11477, 14221, 12705, 14357, 16407, 11561, 17711, 13701, 11075, 16363, 12727.

The following data are the feedback polynomials in the successive order of rows of the code matrix, which can be used to generate Gold codes of different lengths:

1. Codes of length 31:
   Polynomial 1 : 45
   Polynomial 2: 75

2. Codes of length 63:
   Polynomial 1: 103
   Polynomial 2: 147

3. Codes of length 127:
   Polynomial 1: 211
   Polynomial 2: 217

4. Codes of length 511:
   Polynomial 1: 1021
   Polynomial 2: 1131

5. Codes of length 1023:
   Polynomial 1: 2011
   Polynomial 2: 2415

6. Codes of length 2047:
   Polynomial 1: 4005
   Polynomial 2: 4445

The following data are the feedback polynomials in the successive order of rows of the code matrix, which can be used to generate small set Kasami codes of different lengths:

1. Codes of length 15:
   Polynomial 1: 23
   Polynomial 2: 7

2. Codes of length 63:
   Polynomial 1: 103
   Polynomial 2: 15

3. Codes of length 255:
   Polynomial 1: 435
   Polynomial 2: 23

4. Codes of length 1023:
   Polynomial 1: 2011
   Polynomial 2: 75

5. Codes of length 4095:
   Polynomial 1: 10123
   Polynomial 2: 141

The following data are the feedback polynomials in the successive order of rows of the code matrix, which can be used to generate large set Kasami codes of different lengths:

1. Codes of length 15:
   Polynomial 1: 23
   Polynomial 2: 37
   Polynomial 3: 7

2. Codes of length 63:
   Polynomial 1: 103
   Polynomial 2: 147
   Polynomial 3: 15

3. Codes of length 255:
   Polynomial 1: 435
   Polynomial 2: 675
   Polynomial 3: 23

4. Codes of length 1023:
   Polynomial 1: 2011
   Polynomial 2: 2415
   Polynomial 3: 75

5. Codes of length 4095:
   Polynomial 1: 10123
   Polynomial 2: 13311
   Polynomial 3: 141

4.1.3 OTHER CDMA CODES AND SEQUENCES

There are many other CDMA codes or sequences, which are less widely used in real CDMA systems. Some of them have their unique properties, which may be found useful in future wireless applications. Those less widely used CDMA codes include Kronecker sequences [205–209], GMW sequences [210–212], 4-CCL and 5-CCL codes [213], Bent sequences [214], multi-band wavelet packet spreading codes [216], and No codes [215].

The application of Kronecker sequences in a CDMA system was studied in an article written by K. H. A. Karkkainen, who is my former colleague when I worked at the University of Oulu. In this article, the author proposed six conjectures for the linear complexity (LC) of some Kronecker sequences of two- and three-component codes based on the application of the Berlekamp-Massey algorithm. In this study, the components of interest were chosen from the families of Gold codes, Kasami codes, Barker codes, Golay complementary codes, and m-sequences. It was indicated in the study that typically the LC value is a large part of the code length. The LC value of the outermost code influences most on the LC value.

GMW sequence is another interesting code, which possesses 2-valued periodic auto-correlation functions. This property of GMW sequences is very similar to that of m-sequences, and thus it helps to reduce the multipath interference due to its good auto-correlation function. On the other hand, however, the uncontrollable periodic cross-correlation levels of GMW sequences undermine their successful application in CDMA systems when compared with other spreading codes such as Gold codes or Kasami codes, which maintain controllable cross-correlations but do not possess 2-valued periodic auto-correlation functions.

GMW sequences have relatively large families. However, in many practical CDMA applications, such as indoor wireless networks, micro-/pico-cell wireless communication systems, etc., the average number of simultaneously active users is usually limited. In a paper published in 1997 [211], we proposed some useful algorithms to construct some desirable sub-families from the original GMW sequence family by deliberately eliminating those sequences which do not meet certain criteria, such that the constructed sub-families can ensure improved performance. Three different optimization algorithms–reciprocal code deleting, highest-peak-deleting, and most-peak-deleting–were introduced and the performance of their resultant sub-families was compared in terms of bit error rate (BER) under CDMA co-channel interference.

The 4-CCL and 5-CCL codes in fact can be generated in a very similar dual LFSR structure as used for generating Gold codes but with different initial loading. As discussed in the previous sections, m-sequences and Gold codes can be generated by using some primitive polynomials (also called irreducible polynomials), while 4-CCL and 5-CCL codes can be generated by using some non-primitive polynomials. The cross-correlation functions for 4-CCL and 5-CCL codes have four and five different levels, respectively, and this is the reason that their names were given. An extensive study of 4-CCL and 5-CCL codes was published in 1992 [213]. It was noted in the paper [213] that both 4-CCL and 5-CCL codes in general offer quite good auto-correlation and cross-correlation functions, which in many cases are better than those for Gold codes, although the cross-correlation functions for the Gold codes have only three different levels. It was also found in the paper that the performance of a CDMA system is basically determined by the frequency, at which the most significant cross-correlation level appears, rather than the number of different cross-correlation levels. Therefore, under this judgment the 4-CCL and 5-CCL codes in general offer better performance than Gold codes if they are used as signature codes in a CDMA system.

Another interesting spreading code is the multi-band wavelet packet spreading code, which I proposed in a paper published in 2001 [216]. The work revealed in this paper was based on the application of wavelet functions, which contain multi-band signal in the codes. With those multi-band signals, the wavelet spreading codes can perform better under certain circumstances in a CDMA system, especially when the channel suffers a severe frequency-selective fading.

Bent code [214] is another useful spreading code for CDMA applications due to its uniquely large linear span compared to many other spreading codes reported in the literature. However, due to the problem of the complexity of the code generation process, Bent codes have not found wide application in CDMA systems. It should be noted that No code [215] is another spreading code, which has also quite a large linear span in its code structure.

As a concluding remark for this section, we should stress that all aforementioned spreading codes or sequences for CDMA applications are called unitary codes, which work on a one-code-per-user basis. Many important properties (such as their auto-correlation and cross-correlation functions) of the unitary codes stem from the nature of one-code-per-user operation. It is quite understandable that one-code-per-user operation constraints the code design process in optimization of their auto-correlation and cross-correlation functions. As a consequence, it is noted that all unitary codes, including orthogonal codes and quasi-orthogonal codes, can never offer ideal auto-correlation and cross-correlation functions. We say the auto-correlation function is ideal if and only if the peak of the auto-correlation is non-zero, and all the side lobes of the auto-correlation function should be zero. We say the cross-correlation functions are ideal if and only if all cross-correlation levels are zero for all possible time shifts. It should be mentioned again that so-called orthogonal codes, such as Walsh-Hadamard codes and OVSF codes, are not orthogonal at all if they are used in asynchronous transmission channels, such as the uplink channels of a mobile cellular system.

Therefore, it can be easily concluded from the discussions given above that there is a long way to go in the search for better spreading codes for future wireless or wired communication systems.

4.2 SPREADING MODULATIONS

Before we start to discuss spreading modulations, we should stress again that our discussion will focus only on direct sequence spreading and we will not discuss the details of other spreading schemes (such as frequency hopping and time hopping) in this section.

Spreading modulation is a unique modulation process pertaining to a spread spectrum or CDMA system, and it performs several important functions in a spread spectrum or a CDMA system alike. First, spreading modulation is a process to modulate the original data signal with the wideband spreading signal, which is usually a noise-like sequence or pseudo-random (PN) signal. All users will use a different PN signal in order to be distinguished from one another for user signal separation. Second, spreading modulation is to spread the original baseband signal to a bandwidth much wider than necessary to transmit the baseband signal itself. In this way, many operational advantages are possible due to the powerful processing gain achieved in the spreading modulation process. Third, spreading modulation schemes will also determine the hardware complexity of a CDMA transceiver, and thus its selection is a primary concern for system implementation complexity and cost. Different wireless communication applications should use a suitable spreading modulation scheme to meet the cost and physical size requirements for a particular application.

4.2.1 DS SPREADING MODULATION

Direct sequence spreading is the best well-known spreading modulation technique, and has been widely used in many commercial mobile cellular radio systems. The data signal is multiplied by a pseudo-random noise-like code (PN code) and every user will be assigned a different PN code for CDMA purposes.

Direct sequence spread spectrum, also known as direct sequence code division multiple access (DS-CDMA), is one of two common approaches (DS and FH) for spread spectrum modulation for digital signal transmission over the airwaves. In direct sequence spread spectrum, the stream of information to be transmitted is divided into small pieces, each of which is allocated across to a frequency channel across the spectrum. A data signal at the point of transmission is combined with a higher data-rate bit sequence (also known as a chipping code) that divides the data according to a spreading factor or processing gain. The redundant chipping code helps the signal resist interference and also enables the original data to be recovered if data bits are damaged during transmission.

Direct sequence contrasts with the other spread spectrum process, known as frequency hopping spread spectrum, or frequency hopping code division multiple access (FH-CDMA), in which a broad slice of the bandwidth spectrum is divided into many possible broadcast frequencies. In general, frequency-hopping devices use less power and are cheaper, but the performance of DS-CDMA systems is usually better and more reliable.

Spread spectrum first was developed for use by the military because it uses wideband signals that are difficult to detect and that resist attempts at jamming. In recent years, researchers have turned their attention to applying spread spectrum processes for commercial purposes, especially in local area wireless networks and mobile cellular systems, such as the IS-95, W-CDMA, and TD-SCDMA standards.

Bandwidth spreading by direct modulation of signals by a wideband spread signal (also called code) is called the direct sequence spreading technique. The DS spreading modulated signal is then modulated by a carrier before final transmission. In DS spreading modulation, the baseband signals are usually called bits, and the code bits are called chips. Typically, the baseband signal bandwidth is multiplied several times by the spreading signals. In other words, the chip rate is much higher than the bit rate. The spreading signal sequence is unique for a transmitter, and the same chip sequence is used at the receiver to reconstruct the signals (data bits). A mechanism, called correlation, is used to synchronize the received spread signals (that contain data) with the locally generated code. At maximum received signal strength, correlation occurs. The receiver then enters the tracking mode, such that the spreading modulated signals are received without interruption. A simple DS spreading modulation system is shown in Figure 4.6, where $d(t)$ is the input data bits, $c(t)$ is the code bits, and $x(t)$ is the carrier frequency modulated signal, ready for transmission.

A note about why frequency up-conversion is required for radio transmission in a CDMA system can be given as follows. Baseband and very low frequencies are susceptible to heavy attenuation during transmission. In addition, imagine a scenario where every transmitter transmits in the baseband frequencies. It is practically impossible for everyone to transmit in baseband frequencies (a baseband frequency is the frequency spectrum that is occupied by the unmodulated signals). Hence up-conversion of frequency is normally required to comply with the transmission requirements. However, it is to be noted that in some applications signal transmission can happen without a specific carrier modulation. The newly emerging ultra-wideband (UWB) technology usually does not require carrier modulation, and the wideband signal will transmit without carrier up-conversion.

In the DS spreading modulation transmitter, a code generator is a pseudo-random generator that generates a known pseudo-noise code sequence. Normally, the code has finite length (for instance,

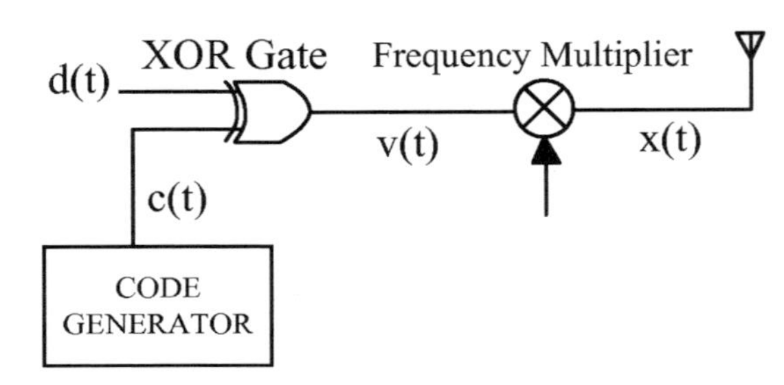

Figure 4.6 A typical configuration of a DS-CDMA transmitter using direct sequence spreading modulation.

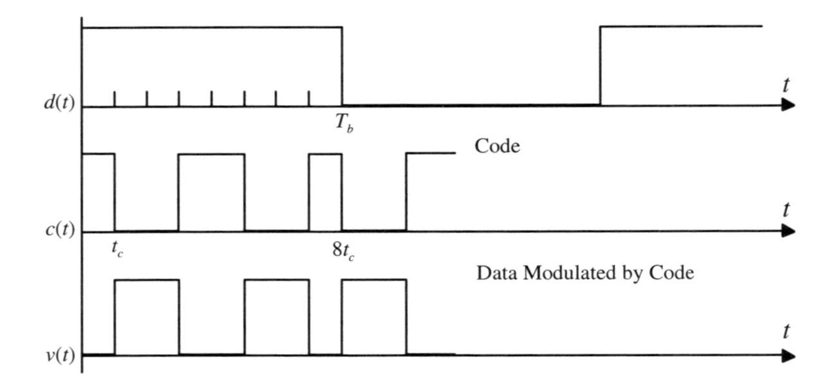

Figure 4.7 Signaling format generated from a DS-spreading modulated CDMA transmitter.

1024 chips), and repeats periodically. The requirements for a good PN code were discussed in the previous section.

As shown in Figure 4.6, an XOR gate is used for spreading the data bits. The input code and the resulting output are displayed in Figure 4.7, where each data bit is coded with eight chips. In practice, this would be much higher, of the order of 1024 or even more. The higher the number of chips per bit, the higher will be the processing gain.

One important parameter of a DS spreading modulation CDMA system is the processing gain. Consider a data rate of 10 kbps, and chip rate of 1 mcps. The processing gain is given by $10\log(R_c/R_b)$, where R_c is the chip rate and R_b is the data rate. For a chip rate of 1 mcps and a data rate of 1 kbps, the processing gain is $10\log(1000)$ or 30 dB. The processing gain is a measure of immunity to noise and jamming signals. The higher the processing gain, the greater the band spread of the signals.

A simplified DS spreading modulation receiver block diagram is shown in Figure 4.8. It consists of a PN generator that feeds the matching chip sequence to an XOR gate to reproduce the original bit sequence. The PN generator is driven by an error signal from the output of the low pass filter (LPF), so that chip timing is adjusted to produce the maximum signal threshold. Normally, the acquisition of the data is done through a two-step process. The first is acquisition and the second is tracking. Acquisition refers to acquiring the chip timing of the received signals. This may be further subdivided into coarse acquisition and fine acquisition. The two are differentiated by the amount of chip timing adjustment. Once the acquisition is achieved, then the received signals must be tracked properly. Otherwise, you

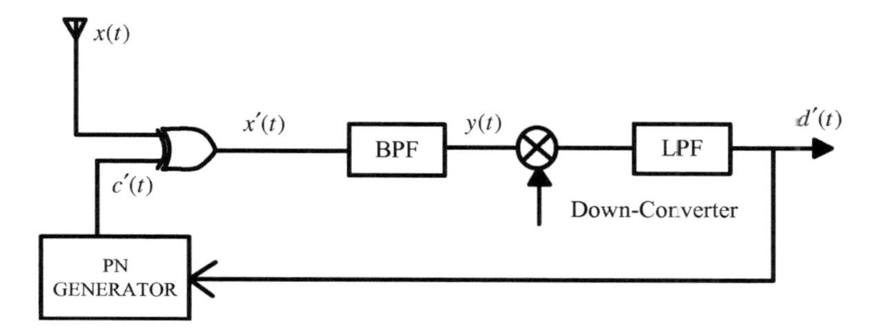

Figure 4.8 A typical configuration of a DS-CDMA receiver using direct sequence spreading modulation.

may lose the lock, resulting in loss of data bits. As with conventional receiver operation, an error voltage at the output of the LPF(or an integrator) provides necessary correction to the PN generator.

4.2.2 PROBLEMS WITH DS SPREADING

Most currently available (2G and 3G) wireless or mobile cellular standard use CDMA technology as their multiple access scheme and almost all of them use direct sequence (DS) modulation as their spreading technique. On the other hand, the other two spreading techniques, frequency hopping (FH) and time hopping (TH),[5] are much less widely used in wireless local area networks and mobile cellular standards. This is the major reason why we will treat DS spreading as the focus of discussion in this book.

It is recalled that when DS spreading was first proposed in the literature in the 1960s people always wondered why a baseband signal should be spread artificially into a frequency bandwidth which can be much wider than necessary to carry the baseband information in a radio frequency channel. It seemed to many people then that it wastes the precious spectrum resource, instead of being considered as a bandwidth-efficient air-link architecture. The confusion is removed when we realize that DS and other spread spectrum techniques (such as FH and TH, etc.) empower a digital communication system with many other operational advantages due to its unique processing gain, which is roughly equal to the ratio between the bandwidth occupied by the RF spread spectrum and that for its original baseband signal. As a matter of fact, the relationship between signal bandwidth and power has been given in the famous Shannon's theory,[6] in which it is clearly indicated that we can trade signal power with its spectral occupancy to keep the same maximal data transmission rate or channel capacity. Of course, at that time, Shannon did not specifically point out the advantages of using spectral occupancy in exchange for signal power in his theory.

Another milestone in the application of spread spectrum techniques is the time when Qualcomm Inc. successfully developed the first CDMA-based civilian mobile cellular communication standard,

[5]It is noted that there is another form of spreading modulation scheme, which is called chirp modulation. The chirp spreading scheme was initially used as a form of pulse compression pulse waveform in many traditional radar applications and it needs to use a surface acoustic wave (SAW) analog filter to carry out the matched filtering process. Chirp spreading is not used in any digital communication systems due to its bulky size and difficulties in digital implementation.

[6]In the late 1940s Claude Shannon, a research mathematician at Bell Telephone Laboratories, invented a mathematical theory of communication that gave the first systematic framework in which to optimally design telephone systems. The main questions motivating this were how to design telephone systems to carry the maximum amount of information and how to correct for distortions on the lines.

also called IS-95, in the 1990s. Since then, it has been successfully demonstrated in theory as well as in practice that a CDMA system based on the DS spreading technique can in fact offer a higher bandwidth efficiency than its predecessors, such as frequency division multiple access (FDMA) and time division multiple access (TDMA) techniques, in addition to many other extremely useful technical features, such as low probability of interception, privacy, protection against multipath interference, overlay operation with existing radio systems, etc., as discussed in Chapter 2 Since then, DS-based CDMA technology has been the prime multiple access radio technology for many wireless networks and mobile cellular standards, such as CDMA2000, W-CDMA, and TD-SCDMA. Therefore, CDMA technology reached its climax at the beginning of this century. As a direct beneficiary from the success of CDMA, Qualcomm enjoys a huge amount of royalty income from the applications of the technology even from other companies in the industry. Thus, the use of CDMA technology becomes very expensive and it is in a company's best interest not to use any of the CDMA-related technologies such that the company can effectively reduce the cost of wireless communication products.

Under such a scenario, the technological evolution of CDMA itself has been seriously affected and most of the industry do not want to touch CDMA any more. Instead, they would like very much to find some other competing technology which can offer equally good performance for wireless applications. Orthogonal frequency division multiple access (OFDMA) technology was proposed and developed under this scenario.

It is not the time for us to discuss here which technology is better, CDMA or OFDMA, but rather it has to be mentioned that the monopoly or high concentration of CDMA-related intellectual property rights (IPRs) in one company in effect has dumped the will to push CDMA technology into its second and even third generation from its current technological stage (it is only in its first generation).

It is one of the main ideas behind the birth of this book that CDMA technology should be encouraged to move on to its next generation, just like many other important technologies in the industry. CDMA as a core air-link technology used in 2G and 3G wireless communications is only at its first generation stage and we need to work hard to break the barriers in order to make it better for its future applications in 4G and 5G wireless applications. We still have a long way to go.

Under this context, we have now understood that there are many problems with existing CDMA technology and we need to find some ways to overcome these problems. For this reason, let us revisit the principle that a direct sequence spreading modulation takes places. To make it easier to define its performance benchmark, we introduce a parameter called spreading efficiency (SE), which can be used to compare the effectiveness of different spreading modulation schemes. The SE can be defined as

$$\text{Spreading efficency} = \frac{\text{Number of bits covered by each spreading code}}{\text{Number of chips used in each spreading code}} \tag{4.11}$$

which has a unit of number of bits carried by each chip. It is noted that SE can be easily translated into bandwidth efficiency due to the fact that the bandwidth is largely determined by the chip width of the spreading codes, which is usually a fixed parameter. Therefore, SE will be exactly equal to bandwidth efficiency if the chip width is known.

In this context, a CDMA system using an N-chip long spreading code to spread one complete user information bit (as only the short codes will be discussed here) has an SE of $\frac{1}{N}$ bits per chip, which is still a rather low value. If we can find some ways to effectively increase the SE, say approaching one, under the condition that it should not sacrifice other operational advantages of a CDMA system, we will succeed in improving the overall bandwidth efficiency of CDMA technology.

It is to be noted that the problem of a relatively low SE with all CDMA systems (such as IS-95, cdma2000, W-CDMA, and TD-SCDMA) currently available has a lot to do with unitary spreading codes (such as Walsh-Hadamard sequences, Gold codes, Kasami codes, m-sequences, etc.) and the DS spreading modulation scheme. Most CDMA-related IPRs held by Qualcomm today have also been built up based largely on them. Therefore, innovation on the spreading modulation and spreading code design approaches will also help to break the IPR barrier. We will give more discussions on how to improve the SE of a CDMA system using some innovative spreading techniques in Chapter 5.

Another problem with traditional CDMA technology is its inherent difficulties in facilitating multimedia signal transmission, which may need to change the data transmission rate from time to time on the fly, to reflect the real-time rate changes in source information, such as video, data, and voice, which can come together or separately, very much depending on the nature of the wireless application. A typical example which can be used to illustrate the problem is the W-CDMA or TD-SCDMA standards, which were developed by ETSI of the EU and ARIB of Japan, and CATT of China, respectively. Both standards are considered to be 3G standards and use orthogonal variable spreading factor (OVSF) codes as their major spreading or channelization sequences.

As shown in Figure 4.1, OVSF codes are generated from a tree-like structure and thus their spreading factors have to change in multiples of two. In other words, the spreading factors should not change by an arbitrary value, such as 3, 5, etc. If a user requires a spreading factor of 5 to fit its particular application, the system can only give the lowest possible spreading factor 8, as the system could not give 4, which is lower than 5 as required. Therefore, in this simple example, the system wastes bandwidth of about $(8-5)/8 = 37.5\,\%$.

In addition to the possible waste of bandwidth due to the inability of a W-CDMA or TD-SCDMA system to match exactly to the rate demand from a user, the transmission rate change is always a painful process which involves almost all active terminals in a cell, also due to the nature of the tree-like code generation process for OVSF codes. The reason is because the change of spreading factor in one terminal may affect the code assignments plan as a whole at a cell-wise scale, and thus a code reshuffling is always necessary due to the request for a spreading factor change from one single terminal. Obviously, the reshuffling is a time consuming and bandwidth consuming process and it wastes a lot of network resources as a whole.

It is noted that in fact the cell-wise reshuffling of the code assignment plan is also needed if a new user enters the cell due to its location being nearer to the cell, or a mobile inside the cell receives a new call, and so forth. Thus, it is understandable that the whole system or network is kept busy all the time due only to the code assignment reshuffling problem with the application of OVSF codes.

The question is why not using a better code instead of OVSF codes at the very beginning of the system design process? The answer lies with economic and political issues, which in most cases have much more leverage power than technical issues in the decision making process of a company or country that supports the development of mobile cellular standards. Clearly, the OVSF codes technically are not a very good choice for the W-CDMA and TD-SCDMA standards, both as 3G solutions. However, politically, the process was so hurried that Japan virtually did not have time to think about some other better spreading codes as its 3G solution when it was pushed hard by Korea, which had acquired key CDMA technology from Qualcomm at that time. The use of OVSF codes has been considered a failure because it represents a lost opportunity to further boost the W-CDMA system's performance by using some other better spreading code in its air-link sector.

Therefore, we can see that the problems (such as the difficulties in supporting rate-on-demand in multimedia applications) with the W-CDMA and TD-SCDMA standards stem from the joint application of poor spreading codes and the DS spreading modulation scheme. Therefore, there is a big room for improvement in terms of spreading code design or generation approaches under DS spreading modulation or other spreading schemes. As a matter of fact, all spreading codes used in 2G and 3G standards, such as Walsh-Hadamard codes, OVSF codes, Gold codes, Kasami codes, m-sequences, etc., were designed and proposed without considering the real applications scenarios in wireless channels, where many impairment factors exist, such as multipath propagation, asynchronous transmission in uplink channels, and random bit streams. Most spreading codes, such as Gold codes, m-sequences, Kasami codes, etc., were proposed many years ago and they were studied not necessarily for mobile cellular applications. More specifically, they were designed mainly for their applications in radar applications based on their good auto-correlation functions. Therefore, their poor performance when used in a mobile cellular system is never a surprise to us.

Innovation in the CDMA code design approach is a major issue of concern in this book. We have been working extensively on this issue for many years and some very exciting results have been obtained from the research. One of the most important steps forward in innovation of CDMA code design is the use of a new design methodology, called the real environment adapted linearization (REAL) approach, which can be used to design CDMA codes by taking into account almost all real application scenarios, such as multipath propagation, random bit signs in the data stream, bursty traffic in packet-switched networks, asynchronous transmission in uplink channels, etc. Some interesting conclusions have been made from the REAL approach, such as that we have shown that a CDMA system can be made interference-free if it uses CDMA codes obtained from the REAL approach. There will be more discussions on this topic in Chapter 6, and we will not go further here.

4.3 SCRAMBLING TECHNIQUES

As mentioned many times earlier, this book is mainly concerned with the next generation CDMA system and its performance under the context that it may use short codes to perform spreading modulation. Thus, long code scrambling is not an issue of major concern. However, in this section, we touch on some issues on scrambling techniques that may be associated with other design issues of our concern, such as spreading code design approaches, spreading modulation schemes, and so forth.

There are mainly two functions for a scrambling technique to perform in a mobile cellular or a generic wireless communication application, which supports both uplink and downlink transmissions.

Table 4.3 Synchronization codes, spreading codes, and scrambling codes used in the W-CDMA standard.

	Synch. code	Spreading code	Scrambling code (uplink)	Scrambling code (downlink)
Code type	Gold codes	OVSF codes	Complex-valued Gold code segments (long) or complex-valued (2) codes (short)	Complex-valued Gold code segments
Code length	256 chips	4-512 chips	38400 chips	38400 chips/ 256 chips
Code number	1 major code, 16 minor codes	SF(uplink) $= 4, 8, 16, \ldots, 256$ SF(downlink) $= 4, 8, 16, \ldots, 512$	16, 777, 216	512 major codes, each having 15 minor codes
Spreading	No	Yes	No	No
Purpose	Synchronization between BS and mobiles	Uplink: separate channels from a mobile Downlink: separate different mobiles from same BS	Separate different mobiles	Separate different BSs

One is to achieve signal separation and the other to achieve signal synchronization, for both mobile terminals and base stations alike. Usually, scrambling techniques use a relatively long code compared to those used for data spreading. Therefore, most commonly used scrambling codes include m-sequences, Gold codes, Kasami codes, etc.

As a typical example for the applications in signal separation, scrambling techniques can be used in different base stations to scramble spreading modulated signals (for instance, the pilot signals) using a unique long code for each base station, allowing mobile stations to distinguish the pilot or other signals from different base stations. It is very important when performing handoff and other processes at a mobile terminal.

The scrambling techniques can also find wide applications for synchronization among terminals. Synchronization refers to the issues of how a receiving node synchronizes its scrambler to the one located in the transmitting node. Scrambler synchronization requires two conditions to be met at both the transmitting and receiving nodes: first, that each node start scrambling at the same place in the data stream, and second, that each node use the same seed at the start.

Table 4.3 shows the synchronization codes, spreading codes, and scrambling codes used in both uplink and downlink transmissions in the W-CDMA system. It is seen from the table that different codes are used to perform spreading modulation (to achieve processing gain), the synchronization process (to achieve terminal synchronization), and user separation (for code division multiple access), and thus the system implementation can be very complex due to the different layers of codes used in both base station and mobile terminals. Also, the need for multiple layer codes to achieve spreading modulation, terminal synchronization, and base station or mobile terminal separation using scrambling codes can be reduced to a minimum if better CDMA codes can be designed and the code design process can be done by taking into account the system operation requirements under a unified framework.

4.4 NEAR-FAR EFFECT

When working with the first CDMA standard for civilian mobile cellular communications, Qualcomm spent a great deal of effort on devising a way to overcome the near-far effect by introducing a sophisticated and fast power control mechanism, which consists of open-loop power control for an initial coarse power adjustment and closed-loop power control for power level fine tune.

The near-far effect is explained below. In fact, the near-far effect became a major obstacle for the successful application of CDMA technology in a multiple user wireless system. The near-far effect problem with applying direct sequence spreading CDMA is illustrated in Figure 4.9. This effect is present when an interfering transmitter (terminal B) is much closer to the receiver than the

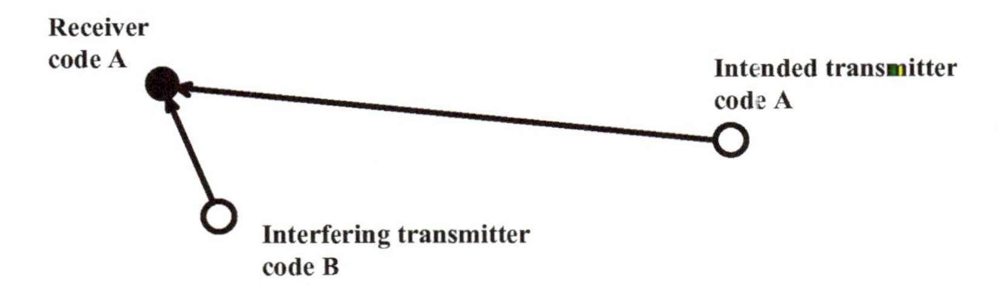

Figure 4.9 Illustration of the near-far effect in a DS-CDMA system, where an intended transmitter A is much further a way from the receiver than the unwanted interfering transmitter B, making the detection of signal from A impossible at the receiver.

intended transmitter (terminal A). Although the cross-correlation between codes A and B is low, the correlation between the received signal from the interfering transmitter and code A can be higher than the correlation between the received signal from the intended transmitter and code A. The result is that proper data detection is not possible. It is noted that, on the other hand, the frequency-hopping technique is less affected by the near-far effect than direct sequence. Frequency-hopping sequences have only a limited number of 'hits' with each other. This means that if a near-interferer is present, only a number of frequency-hops will be blocked instead of the whole signal. From the hops that are not blocked it should be possible to recover the original data message. However, we do not discuss FH spreading in this book. For more information, readers are advised to check the references on the topic, such as [7–10].

The mechanism of the near-far effect problem can also be explained as follows. Consider a receiver and two transmitters (one close to the receiver, and the other far away). If both transmitters transmit simultaneously and at equal powers, then due to the inverse square law, the receiver will receive more power from the nearer transmitter. This makes the farther transmitter more difficult, if not impossible, to hear. Since one transmission's signal is the other's noise, and the signal-to-noise ratio (SNR) for the farther transmitter is much lower, if the nearer transmitter transmits a signal that is orders of magnitude higher than the farther transmitter, then the SNR for the farther transmitter may be below the detectable level and the farther transmitter may just as well not transmit. This effectively jams the communication channel.

In a CDMA system or some other cellular phone-like network, this problem is commonly solved by dynamic output power adjustment of the transmitters. That is, the closer transmitters use lower power so that the SNR for all transmitters at the receiver is roughly the same. This sometimes can have a noticeable impact on battery life, which can be dramatically different depending on distance from the base station.

To place this problem in a more common context, imagine you are talking to someone 20 feet away. If the two of you are in an anechoic chamber then a conversation is quite easy to hold at normal voice levels. Instead, now imagine that two people are yelling into your friend's ears. Your friend will not be able to understand anything you say. In order for your friend to hear you, you would have to use loudspeakers and radically increase the volume such that you overpower the other two. Taking this analogy back to a wireless communication system, the farther transmitter would have to drastically increase transmission power, which simply may not be possible.

This analogy could also be extended to headlights on an automobile, though much more loosely. On a dark night and facing directly into the headlights, it is very difficult to read the license plate, the hood ornament, or any other detail that is dimly lit. In this case, the brightness of the headlights is many times brighter than the light reflecting off other parts of the car. Seeing stars during the daytime is similar: the background sky light signal drowns out the weaker light of the star. In summary, the near-far effect problem is one of detecting and receiving a weaker signal among stronger signals.

It is very costly to mitigate the near-far problem in a CDMA system, such as what has been done in the IS-95 standard, as discussed in Chapter 3. The most commonly used approach to overcome the near-far effect problem in a CDMA system is to use a precision power control algorithm, which is normally initiated by a receiver of interest. The power control algorithms used in a CDMA system often consist of open loop power control and closed-loop power control, which make a CDMA transceiver very complicated in both its hardware and software implementation.

In the UMTS UTRA standard, three different level power control schemes are defined: open loop power control, inner loop power control, and outer loop power control.

Open loop power control is the ability of the user equipment (UE) transmitter to set its output power to a specific value. It is used for setting the initial uplink and downlink transmission powers when a UE is accessing the network. The open loop power control tolerance is ± 9 dB (normal conditions) or ± 12 dB (extreme conditions).

On the other hand, the fast closed loop power control (also called inner loop power control) in the uplink is the ability of the UE transmitter to adjust its output power in accordance with one or more

Transmit Power Control (TPC) commands received in the downlink, in order to keep the received uplink signal-to-interference ratio (SIR) at a given SIR target. The UE transmitter is capable of changing the output power with a step size of 1, 2, and 3 dB, in the slot immediately after the TPC_cmd can be derived. The inner loop power control frequency is 1500 Hz, which is very fast indeed.

Finally, the outer loop power control is used to maintain the quality of communication at the level of bearer service quality requirement, while using as low power as possible. The uplink outer loop power control is responsible for setting a target SIR in the Node B for each individual uplink inner loop power control. This target SIR is updated for each UE according to the estimated uplink quality (Block Error Rate, Bit Error Rate) for each Radio Resource Control connection. The downlink outer loop power control is the ability of the UE receiver to converge to the required link quality (BLER) set by the network (RNC) in the downlink. More detailed descriptions of the power control algorithms were given in Chapter 3.

However, after having a much closer look at the reason that causes the near-far effect in all traditional CDMA systems, such as IS-95, CDMA2000, W-CDMA, and TD-SCDMA, we can find that the near-far effect is also a by-product that stems from the non-ideal auto-correlation and cross-correlation functions of the spreading codes used by the CDMA system. The near-far effect is in fact caused by the fact that the cross-correlation function side lobes from unwanted signals (such as the terminal B in Figure 4.9) overwhelm the auto-correlation function peak of wanted signal (such as the terminal A in Figure 4.9). Therefore, it is natural to ask: what if the spreading codes used by a CDMA system possess ideal correlation properties? The answer is clear: there will be no near-far effect as all cross-correlation functions will be zero in any possible chip shifts. If so, then complex power control will also not be necessary.

4.5 ASYNCHRONOUS TRANSMISSIONS IN UPLINK CHANNELS

Asynchronous transmissions in the uplink channels of a CDMA system pose another serious thread to successful signal detection. Asynchronous transmissions exist in the uplink channels of almost all wireless systems, such as wireless local area networks, and mobile cellular systems. The cause of asynchronous transmissions is the different separation distances between mobile transmitters and the receiver (usually the base station), due to the different locations of different mobile terminals. Therefore, the signals from distinct mobile terminals will arrive at the same receiver (of the base station) at different times, and their information bit streams are often not aligned in time (or the starting times and the ending times of bits are not synchronized in time).

Usually, the mobile-to-base links can hardly be precisely coordinated, particularly due to the mobility of the handsets, and require a somewhat different modulation approach in a CDMA system, compared to the downlink transmissions, which are synchronous. Since it was not mathematically possible to create signature sequences that are orthogonal with each other for arbitrarily random starting points, usually pseudo-random or pseudo-noise (PN) sequences are used in the asynchronous channels of all traditional CDMA systems. These PN sequences are statistically uncorrelated, and the sum of a large number of PN sequences results in multiple access interference (MAI) that is approximated by a Gaussian noise process (via the theorem of the 'law of large numbers' in statistics). If all of the users are received with the same power level, then the variance (e.g. the noise power) of the MAI increases in direct proportion to the number of users.

All currently available CDMA systems use spread spectrum processing gain to allow receivers to partially discriminate against unwanted signals. Signals with the desired chip code and timing are received, while signals with different chip codes (or the same spreading code but a different timing offset) appear as wideband noise reduced by the processing gain.

Since each user generates MAI, controlling the signal strength is an important issue with CDMA transmitters, just as discussed in the previous section. A CDMA (or synchronous CDMA), TDMA or FDMA receiver can in theory completely reject arbitrarily strong signals using different codes, time slots or frequency channels due to the orthogonality of these systems. However, this is not true for an asynchronous CDMA; rejection of unwanted signals is only partial. If any or all of the unwanted signals are much stronger than the desired signal, they will overwhelm it. This leads to a general requirement in any asynchronous CDMA system to approximately match the various signal power levels as seen at the receiver. In a CDMA mobile cellular system, the base station uses a fast closed-loop power control scheme to tightly control each mobile's transmit power, just as discussed earlier.

Having mentioned the major properties of an asynchronous CDMA transmission system, we can easily understand that its performance will not be better than a synchronous CDMA system due to less controllable correlation properties of the spreading codes in asynchronous operation mode. However, an asynchronous CDMA can also have some useful features compared to synchronous CDMA, FDMA or TDMA systems, as explained below.

Asynchronous CDMA's main advantage over synchronous CDMA, TDMA, and FDMA is that it can use the spectrum more efficiently in mobile cellular applications. TDMA systems must carefully synchronize the transmission times of all the users to ensure that they are received in the correct timeslot and do not cause interference. Since this cannot be perfectly controlled in a mobile environment, each timeslot must have a guard-time, which reduces the probability that users will interfere, but decreases the spectral efficiency. Similarly, FDMA systems must use a guard-band between adjacent channels, due to the random doppler shift of the signal spectrum which occurs due to the user's mobility. The guard-bands will reduce the probability that adjacent channels will interfere, but decrease the utilization of the spectrum.

Most importantly, asynchronous CDMA offers a key advantage in the flexible allocation of resources. There are a fixed number of orthogonal codes, timeslots or frequency bands that can be allocated for synchronous CDMA, TDMA, and FDMA systems, which remain underutilized due to the bursty nature of packetized data transmissions (which will be the major traffic in B3G wireless systems). There is no strict limit to the number of users that can be supported in an asynchronous CDMA system, only a practical limit governed by the desired bit error probability, since the signal to interference ratio (SIR) varies inversely with the number of users. In a bursty traffic environment like a mobile cellular system, the advantage afforded by asynchronous CDMA is that the performance (bit error rate) is allowed to fluctuate randomly, with an average value determined by the number of users times the percentage of utilization. Suppose there are $2N$ users who only talk half of the time, then $2N$ users can be accommodated with the same average bit error probability as N users who talk all of the time. The key difference here is that the bit error probability for N users talking all of the time is constant, whereas it is a random quantity (with the same mean) for $2N$ users talking half of the time.

In other words, asynchronous CDMA is ideally suited to a mobile cellular network where large numbers of transmitters each generate a relatively small amount of traffic at irregular intervals. Synchronous CDMA, TDMA, and FDMA systems cannot recover the underutilized resources inherent to bursty traffic due to the fixed number of orthogonal codes, timeslots or frequency channels that can be assigned to individual transmitters. For instance, if there are N timeslots in a TDMA system and $2N$ users who talk half of the time, then half of the time there will be more than N users needing to use more than N timeslots. Furthermore, it would require significant overhead to continually allocate and deallocate the orthogonal code, timeslot or frequency channel resources. By comparison, asynchronous CDMA transmitters simply send when they have something to say, and go off the air when they don't, keeping the same PN signature sequence as long as they are connected to the system.

It should be noted that all of the above operational advantages for an asynchronous CDMA are based on the assumption that no ideally orthogonal CDMA codes exist for their applications in an asynchronous channels. Now, let us think about the same issues in a different direction. If we can find a CDMA code set which can offer ideally orthogonal correlation property for its asynchronous transmission just as the orthogonal codes do in a synchronous CDMA channel, what can we expect in terms of the performance gain compared to the case where no such ideally orthogonal codes exist for their applications in asynchronous channels? This is another interesting question we would like to keep open at this moment, as we will offer some more intensive discussions in Chapter 6.

4.6 RANDOM SIGNS IN CONSECUTIVE SYMBOLS

All traditional spreading codes were designed and generated based basically on their seemingly acceptable periodic auto-correlation functions and periodic cross-correlation functions.

The periodic auto-correlation function and periodic cross-correlation function are defined as follows. Let us look at Figure 4.10, which illustrates the formation of the periodic cross-correlation function[7] between two codes (or $c_1(t)$ and $c_k(t)$) in an asynchronous DS-CDMA system, where a local correlator is tuned to code $c_1(t)$, the incoming interference from the kth transmitter is encoded by code $c_k(t)$, and presumably two consecutive bits of the interference carry the same sign '+1'.

It is seen from Figure 4.10 that the periodic cross-correlation function between the first and the kth users' codes (or $c_1(t)$ and $c_k(t)$) is yielded due to the same sign carried in the two consecutive bits of the kth interfering signal. We can easily show that the periodic cross-correlation function for two Gold codes is well controlled and has only three different levels, as indicated in the discussions given in Section 4.1.2, or $\{-1, -t(m), t(m) - 2\}$, where

$$t(m) = \begin{cases} 2^{(m+1)/2} + 1, & \text{for odd } m \\ 2^{(m+2)/2} + 1, & \text{for even } m \end{cases} \quad (4.12)$$

Unfortunately, the aperiodic cross-correlation functions of the same Gold code normally will be much more difficult to predict. The aperiodic cross-correlation function is the result if the signs for the two consecutive bits become different, as shown in Figure 4.10. In general, this conclusion (the aperiodic cross-correlation functions of a unitary code are always more difficult to predict than its periodic cross-correlation functions) is applicable to all unitary codes, such as Gold codes, Kasami codes, Walsh-Hadamard sequences, OVSF codes, m-sequences, and so forth, which have been applied to all 2G and 3G CDMA-based wireless standards.

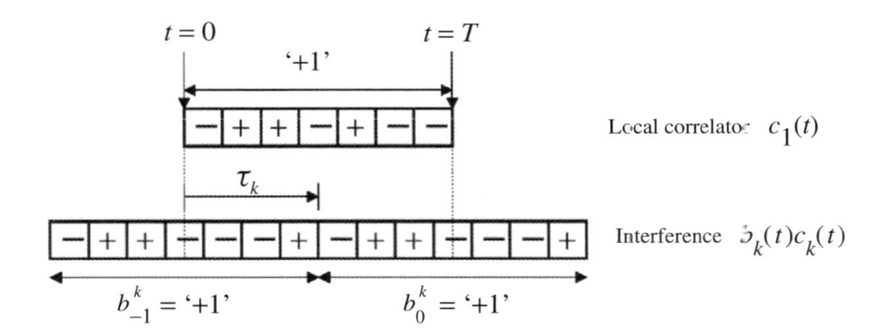

Figure 4.10 Illustration of periodic cross-correlation function in an asynchronous DS-CDMA system, where a local correlator is tuned to code $c_1(t)$, the incoming interference from the kth transmitter is encoded by code $c_k(t)$, and two consecutive bits of the interference carry the same sign '+1'.

[7]Some literature also refers to it as the even cross-correlation function. In this way, the odd cross-correlation functions denote the aperiodic cross-correlation functions used throughout this book. The same can be applied to the auto-correlation functions.

Therefore, the detection efficiency for a series of bits with the same sign will in general be better than for a random sign changing bit stream, causing an uneven detection efficiency which is dependent on bit patterns. Obviously, this problem is there due to the fact that all existing unitary CDMA codes used in all 2G and 3G CDMA systems were designed and proposed based only on their acceptable periodic correlation properties. The aperiodic correlation properties of a code were very rarely taken into account in the traditional sequence/code design process. This is the problem which should be improved.

4.7 MULTIPATH INTERFERENCE

Multipath propagation is a serious impairment factor which can cause serious damage to the signal detection process of a receiver in a mobile cellular or wireless communication system. The cause of multipath propagation is related to the carrier frequency used in most mobile cellular and wireless network standards. Most of these wireless applications operate at the 800–900 MHz, 2 GHz, and 5 GHz spectrum sectors, in which the RF signal transmissions become much more like a light ray than a wave, which can circumambulate buildings and other objects on the earth when the signal impinges on it. Therefore, the reflections from different objects on the propagation path will arrive at the same receiver with different multipath returns, which have different delays. If the delays are longer than symbol duration, intersymbol interference (ISI) will occur and can cause severe damage to the signal detection efficiency.

The major causes of multipath interference also include atmospheric ducting, ionospheric reflection and refraction, and reflection from terrestrial objects, such as mountains and buildings, etc.

The effects of multipath interference include constructive and destructive interference, and phase shifting of the signal. This causes Rayleigh fading, named after Lord Rayleigh. The standard statistical model of this gives a distribution known as the Rayleigh distribution. Rayleigh fading with a strong line of sight content is said to have a Rician distribution, or to be Rician fading.

The most commonly used techniques to combat the multipath effect include equalization algorithms, RAKE receiver, and OFDM technique. Equalization is a technique that was used in many earlier wireless applications to overcome the multipath induced ISI problem. A RAKE receiver is in particular useful to mitigate multipath interference in a CDMA system, and thus it has found very wide applications in all 2G and 3G CDMA-based wireless systems, including IS-95, cdma2000, W-CDMA, and TD-SCDMA systems. Also for this reason, the RAKE receiver has become one of the most profitable IPRs in Qualcomm's IPR collections, and any products using RAKE receiver or its related techniques will be liable for IPR transfer fees, which are usually not a small amount.

A RAKE receiver in fact is a digital radio receiver designed to counter the effects of multipath fading. It does this by using several sub-receivers (also called fingers), each delayed slightly in order to tune in to the individual multipath components. Each component is decoded independently, but at a later stage combined in order to make the most use of the different transmission characteristics of each transmission path. This could very well result in a higher signal-to-noise ratio (or Eb/No) in a multipath environment than in a 'clean' environment, to make multipath diversity possible. There are two common methods on how to combine the individually collected paths. One is to combine them with a proper weight for each path corresponding to its signal power strength, and it is thus called maximal ratio combining (MRC). The other way is to simply give the same weight to all fingers, to result in equal gain combining (EGC), with a lower implementation complexity than that for the MRC scheme.

The RAKE receiver is so named because of its analogous function to a garden rake, each finger collecting bit or symbol energy similarly to how a rake collects leaves. RAKE receivers are common in a wide variety of CDMA and W-CDMA radio devices such as mobile phones and wireless LAN equipment.

The last solution to combat multipath-induced ISI is to use the OFDM technique, which emerged very recently, and its derivative, OFDMA, which has been widely viewed as a strong contender to CDMA technology as a major air-link architecture for B3G wireless systems.

OFDM, also sometimes called discrete multitone modulation (DMT), is based upon the principle of frequency division multiplexing (FDM), but is utilized as a digital modulation scheme. The bit stream that is to be transmitted is first split up into several parallel bit streams, typically dozens to thousands, depending on the data rate requirements. The available frequency spectrum is divided into several sub-channels, and each low-rate bit stream is transmitted over one sub-channel by modulating a sub-carrier using a standard modulation scheme, for example PSK, QAM, PAM, and so on. The sub-carrier frequencies are chosen such that the modulated data streams are orthogonal to each other, meaning that cross-talk between the sub-channels is eliminated.

Channel equalization is simplified by using many slowly modulated narrowband signals instead of one rapidly modulated wideband signal. The primary advantage of the OFDM technique is its ability to cope with severe channel fading conditions, for example, multipath[8] and narrowband interference, without using complex equalization filters. OFDM has been developed into a popular wireless air-link scheme for many wideband digital communication systems. Examples of those applications include:

1. ADSL and VDSL broadband access via telephone network copper wires.
2. IEEE 802.11a and 802.11g Wireless LANs.
3. The digital audio broadcasting systems EUREKA 147, Digital Radio Mondiale, HD Radio, T-DMB, and ISDB-TSB.
4. The terrestrial digital TV systems DVB-T, DVB-H, T-DMB, and ISDB-T.
5. The IEEE 802.16 or WiMax Wireless MAN standard.
6. The IEEE 802.20 or Mobile Broadband Wireless Access (MBWA) standard
7. The Flash-OFDM cellular system.
8. Some ultra wideband (UWB) systems.
9. Power line communication (PLC).
10. Point-to-point (PtP) and point-to-multipoint (PtMP) wireless applications.

After having discussed the multipath interference problem and its countermeasures, we should think about the fundamental cause of multipath interference. Of course, multipath propagation is a natural phenomenon, which will always be there, and we can not change it. However, the damage caused by multipath propagation will be very much different if we can use a CDMA code family with all codes having ideal auto-correlation and cross-correlation functions. The ideal auto-correlation function of a code is defined as there being no auto-correlation side lobes and only its peak auto-correlation level being non-zero. The ideal cross-correlation function is defined as the cross-correlation function between any two codes being zero irrespective of their relative chip shift.

The ideal auto-correlation function can eliminate ISI caused by self-interference due to reflections from different objects but originating from the same transmitter, if the wanted signal can be sampled correctly. The ideal cross-correlation functions can ensure there will be no ISI caused by non-zero

[8]To overcome multipath-induced ISI, an OFDM system has to use a specially designed signaling structure, called a cyclic prefix (CP), to be inserted in each symbol to ensure that no distortions will be added if multipath propagation is present.

cross-correlation functions between wanted and unwanted signals. Therefore, the multipath interference problem will not be a problem if we can use a CDMA code set with all its codes having ideal correlation properties. Thus, we see once again that the mitigation mechanism in a CDMA system against multipath interference should be initiated from the innovation of the code design approach, not merely using some external sub-systems, such as equalization, RAKE receiver, etc.

4.8 HIGH-SPEED BURSTY-TYPE TRAFFIC

Bursty traffic is one of the most salient features which make a B3G wireless system very much different from a 2G or 3G system, where the majority of traffic is still voice-centric and continuous-time transmission based largely on circuit-switching techniques. All B3G wireless systems will carry packet data as the major payload in their air-link traffic. Thus, the design of a next generation CDMA system should take the burst traffic into account as well.

It is admitted that all currently available CDMA wireless applications for both 2G and 3G standards, such as IS-95, cdma2000, W-CDMA, TD-SCDMA, etc., share the same IPRs, which were devised by Qualcomm when it developed the first CDMA mobile cellular technology, IS-95. Obviously, the time when Qualcomm developed the IS-95 standard was too early to think about its ability to support burst traffic. As a matter of fact, the IS-95 standard was designed mainly for voice communications, which are dominated by continuous-time transmission characterized by very long frames in its transmission sessions. A relatively long frame is useful to facilitate the synchronization and tracking process, followed by data detection, if using traditional spreading codes, whose partial auto-correlation functions and partial cross-correlation functions can be very bad. The partial cross-correlation functions are defined as the correlation between two different codes that are not completely overlapped in time with each other. Thus, only partial chips of the two codes are involved in the cross-correlation function calculation. The partial auto-correlation function is defined as the correlation process between a code and its delayed version with only their partial chips being involved. Figure 4.11 illustrated the two sections of a partial cross-correlation function in an asynchronous DS-CDMA system, where a local correlator is tuned to code $c_1(t)$, the incoming

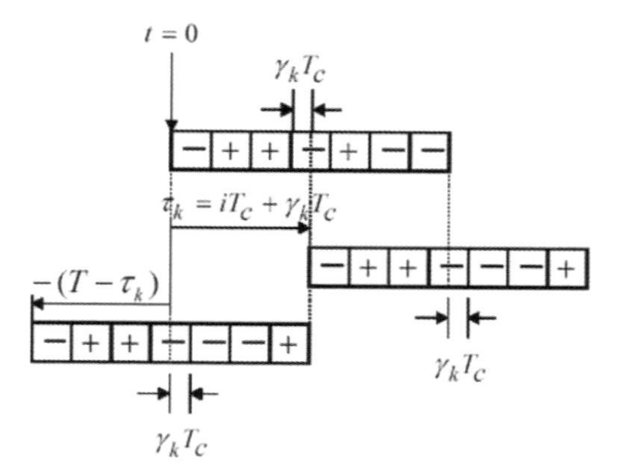

Figure 4.11 Formation of two sections of a partial cross-correlation function in an asynchronous DS-CDMA system, where a local correlator is tuned to code $c_1(t)$, the incoming interference from the kth transmitter is encoded by code $c_k(t)$, and two consecutive bits of the interference carry the same sign '+1'.

interference from the kth transmitter is encoded by code $c_k(t)$, and two consecutive bits of the interference carry the same sign '+1'.

Obviously, in order to ensure a satisfactory operational performance for a CDMA system working in B3G wireless applications (where bursty traffic will dominate), the CDMA codes designed for the system should have good partial auto-correlation functions and good partial cross-correlation functions. This is another challenge we should tackle in the design of next generation CDMA technologies.

4.9 RATE-MATCHING PROBLEMS

Rate matching is a very important issue in a wireless system which is designed for supporting multimedia services. Rate-matching algorithms are required when the source information stream will have different data rates from time to time during a communication session. Rate matching is critical to meet certain quality of service (QoS) requirements in wireless communication systems. The rate-matching process is carried out so that the block size matches the radio frame(s). It will either repeat bits to increase the rate or puncture bits to decrease the rate. A typical example of a wireless system that needs rate matching is a mobile terminal switching from previous video-conference mode to a voice-only mode, asking to slow down the data rate.

Therefore, it is clear that a CDMA technology should be able to support the need for data rate change at any time whenever needed. This is also called rate-on-demand capability in the literature dealing with the issues of data rate-matching algorithms.

Now, let us revisit the several existing mobile cellular standards. In particular, we would like to take a look at the W-CDMA standard to see how it will work to perform its rate-matching algorithm in the communication process. It is well known that W-CDMA uses orthogonal variable spreading factor (OVSF) to channelize its traffic load. The OVSF codes, as discussed in Subsection 4.1.1, are generated from a tree-like structure and they can offer only spreading factors as multiples of two, being $2, 4, \ldots, 512$ for downlink channels and $2, 4, \ldots, 256$ for uplink channel transmissions, as shown in Table 4.3. Therefore, it will not be a surprise if we see that the rate matching conducted in W-CDMA systems can proceed only at multiples of two, and it is not possible to have exact rate matching for a particular rate demand which is not a multiple of two. In other words, if a terminal asks for a rate of 5 units, the system can only offer the lowest possible figure of 8 units. As a consequence, the system wastes about $(8-5)/8 = 37.5\%$ bandwidth in this particular case. If other rates are needed, the percentage of bandwidth waste will be varying, and it can be even larger, as shown in Table 4.1.

Since the first release of the UMTS W-CDMA standard, many research papers have been published on how to optimize rate-matching algorithms to fit multimedia QoS requirements in 3G mobile cellular services. This was once a very hot topic in many wireless communication related technical journals and conferences.

The rate changes in a mobile cellular system operating based on UMTS W-CDMA can be a very complicated process, initiated by either mobile terminals or base station, whenever the channel condition is changed or the nature of the information sources is changed, and so forth. Let us look at a typical example of the rate-matching process that may happen in a UMTS W-CDMA system. We assume that the rate-change demand is initiated from a mobile terminal, which should first issue the need to the base station with which it is working and tell the base station what spreading factor it is asking for. Then, the base station should check whether or not a code with a specific spreading factor is available. If it is indeed available, the process can be relatively simple, as the code with the spreading factor can be allocated to the mobile terminal right away, as long as the use of this particular OVSF code will not block others' ongoing communication sessions in the same cell. However, if a code with the specific spreading factor is not available, the process can be much more complicated. Usually, the base station should first check whether some in-use codes can be reallocated to other branches such that the needed code can become available to the demanding mobile terminal. If yes code assignment reshuffling will take place, which will take some time and waste a lot of network resources for the reshuffling. Otherwise, the base station has to wait until an available code becomes available at the end of others' communication session. Therefore, it can be seen from this example that the rate-matching process used in a W-CDMA-based mobile cellular system is a time-consuming and bandwidth-consuming process, which could be avoided if some other better spreading codes could be used.

In addition to the time-consuming nature of the rate-matching process conducted in the UMTS W-CDMA system, it is usually impossible to perform exact rate matching, and waste of precious bandwidth is commonplace in W-CDMA systems. It has to be admitted that the difficulties in performing exact rate matching in W-CDMA and TD-SCDMA, or any other system using OVSF codes for spreading modulation, stem from the OVSF codes used by these systems. In this case, we have seen again that the use of wrong spreading codes will really affect the whole system's performance for ever, as long as the system is still in use.

4.10 ASYMMETRIC DATA RATE IN UP- AND DOWN-LINKS

It is a well-known fact that the traffic loads in uplink and downlink channels in a mobile cellular or any other wireless system can be very different due to the operational modes of the users. In most cases, we can always say that the traffic in uplink and downlink channels is usually not symmetric, depending very much on the types of services a wireless system can offer and the operational modes in which a user can be. For instance, if a mobile user is using his/her laptop to brows the Internet, the traffic in the downlink can be very much more intensive than what happens in the uplink channel, as he or she only needs to press a few keys to download a fairly large data file, containing web pages and other files. On the other hand, if a user is using FTP services to upload a big file to a server, the traffic will go to another extreme, in which the uplink channel is heavily used while the downlink channel is completely idle, as he or she may just wait there for the file transfer to come to an end. Therefore, asymmetry in the uplink and downlink channels is very common in all wireless applications, and thus a CDMA system should be designed carefully to cater for the asymmetrical nature of the uplink and downlink channels.

To make it easier to support the asymmetric traffic in the uplink and downlink channels in a wireless system, the time division duplex (TDD) technique has been proposed and used in many wireless systems, such as the UMTS UTRA-TDD and TD-SCDMA standards, both of which use TDD to separate uplink and downlink channel transmissions. In a TDD system, uplink transmissions and downlink transmissions will happen in different time segments, and usually the downlink transmissions are arranged to happen before the uplink channel transmissions in a time frame structure.

The use of TDD in a wireless system offers several attractions. First, the agility in spectrum allocation for mobile services is a great advantage of the TDD operation mode, compared with frequency division duplex (FDD), which requires pair-wise spectrum allocation for the uplink and downlink, causing a big burden for countries where spectrum resources are already very tight, such as the US and Japan. Second, the use of the same carrier in both up the links and downlinks helps to implement smart antenna and other technologies that rely on identical propagation characteristics in both uplinks and downlinks. Third, the TDD scheme facilitates asymmetric traffic support in uplinks and downlinks, associated with the increasing popularity of the Internet services. The transmission rates in two links can be adjusted dynamically according to the specific traffic requirements, such that the overall bandwidth utilization efficiency can be maximized. Fourth, the TDD technique is responsible for the lower implementation cost of an RF transceiver, which does not require high isolation for transmit and receive multiplexing as needed in a FDD transceiver, and thus an entire TDD-based RF transceiver can be integrated into a single IC chip. On the contrary, an FDD transceiver requires two independent sets of RF electronics for uplink and downlink signal loops. The cost saving can be as much as 20–22% compared with FDD solutions. Due to the aforementioned merits, some people expected TDD technology to become more attractive for future wireless communications, especially for those applications that cover relatively small areas.

However, it is to be noted also that the use of the TDD technique bears some technical limitations compared to FDD mode. The relatively high peak-to-average power (PTAP) ratio is one problem. Because a CDMA transceiver requires a good linearity, a relatively high PTAP ratio will limit the effective transmission range and thus the coverage area of a cell. Also, the discontinuity of slotted signal transmission in the TDD mode reduces its capability to mitigate fast fading and the Doppler effect in mobile channels, thus limiting the highest terminal mobility supported by TDD systems.

The capability of a CDMA system to support asymmetric traffic loads in uplink and downlink channels has a lot to do with the capability of supporting rate-matching algorithms, as both are needed to solve the problem with transmission rate change to fit different types of traffic requirements. To make a CDMA system better at supporting asymmetrical traffic in uplink and downlink channels, we should also pay a lot of attention to the CDMA codes and spreading modulation schemes. The spreading codes and spreading modulation schemes used in all existing CDMA systems are not good enough in this respect, and thus it leaves us an exciting research topic to explore further.

4.11 SENSITIVITY TO TIME-SELECTIVE FADING

Time-selective fading is caused by the time-variant property of a mobile communication channel, where terminals move from place to place, sometimes at a very high speed.

Before going further on the solutions to combat time-selective fading, we would first like to explain clearly what time-selective fading is. Time-selective fading is a terminology of channel theory, which is just the opposite to frequency-selective fading. Frequency-selective fading is caused by multipath propagation of the channel due to the fact that the frequency response function of the channel exhibits uneven gains in different frequencies, hence the name 'frequency-selective fading'. Therefore, frequency-selective fading can be simply interpreted as uneven gains at different frequencies.

On the other hand, time-selective fading is caused by the Doppler effect in the mobile communication channel, where terminals are in motion at a certain speed relative to the base station which is the receiver of the signals from the mobile terminals. Therefore, time-selective fading occurs due to the time-variant properties of a mobile channel. In this respect, we can also describe time-selective fading as the effect that uneven gains occur at different times. Depending on how fast the time-selective fading occurs, we can define the severity of the time-selective fading to the signal detection process. For instance, if the cycle (which is also called the 'coherent time' of the channel) of time-selective fading is shorter than a symbol duration, we say that it is fast fading; otherwise it is viewed as slow fading. Fast fading is always more difficult to overcome than slow fading, as it makes it more difficult for the receiver to predict or adapt to the changing signal.

It should be noted that the coherent time of the fading channel is roughly equal to the reciprocal of the maximal Doppler shift observed in the channel. For example, if we observe 1 kHz Doppler shift in the channel, the maximal moving velocity of the terminal will be about 300 km/h, which is about the speed of a bullet train. Therefore, we see that the Doppler shift (say 1 kHz) is much lower than the carrier frequency used in most current wireless applications (which can be as high as 5 GHz), but its damage to the signal detection process should never be underestimated. Imagine if the received signal is changing at a frequency of 1 kHz, which simply means that the signal can change 1000 times within a one second time period. This is a lot indeed, and it makes a receiver work very hard to catch up with the signal's change. All channel estimation work should be updated within 1 ms. We can see that this is not an easy task to fulfill.

Now, let us go back to a CDMA system and its design requirements associated with the time-selective fading effect in the channel. Obviously, a CDMA system always works based on a simple principle, in which different users' transmissions should be kept as orthogonal with one another as possible in order not to induce or at least to minimize multiple access interference (MAI). The signature codes and in particular their correlation properties play a key role in this respect. Careful examination of how all the commonly used spreading codes work in 2G and 3G wireless systems shows that the orthogonality or quasi-orthogonality among the signature codes used by different users/channels is built completely on the time-domain correlation properties. Thus, we can say that their orthogonality or quasi-orthogonality is built in the time domain only and they of course will be very sensitive to time-selective fading. The different gains at different instants of time will sensitively affect the orthogonality reconstruction process at a receiver, possibly inducing serious multiple access interference.

Therefore, the fact that time-domain orthogonality or time-domain quasi-orthogonality is susceptible to time-selective fading tells us that we have to work out some other better spreading codes, which should provide us with inherent immunity against the time-variant effect of the channels. The design of next generation CDMA technology should also take this important requirement into account, for its successful application in a mobile channel with a relatively high Doppler spread, such as the case in vehicle-to-vehicle communications, which has been in the process of standardization as a new IEEE 802.11p standard.

4.12 IMPAIRED POWER-EFFICIENCY DUE TO MAI

Multiple-access interference (MAI) is a major source of interference in a CDMA system. The fundamental concern in the design of CDMA codes is to optimize the cross-correlation functions such that the MAI can be kept under an acceptable level. Therefore, we can understand that the cross-correlation function among the CDMA codes will be a critical design parameter that should be taken into account seriously in the entire system design process. As a matter of fact, MAI is the major impairment factor to limit the performance of all currently available CDMA systems. This is also a main reason why everybody considers CDMA to be an interference-limit system. Take IS-95 or cdmaOne as an example, where the system uses 64-ary Walsh-Hadamard sequences as the channelization codes for its downlink transmissions. Unfortunately, in most cases the system can only support a capacity per cell no more that half of its processing gain (or $64/2 = 32$ users).

The interference-limited capacity of a traditional CDMA system has motivated us to think about the question of why CDMA should always be an interference-limited system. Can we make it better? Or can we make a CDMA with some interference-resistant characteristics, and how?

To make the discussions on the issues clearer, we would like to show some of the results obtained from our study on the correlation properties of widely used signature codes, which have been applied to 2G and 3G wireless applications. Due to space limitations, we show the results only for two CDMA codes, Walsh-Hadamard sequences and Gold codes. Walsh-Hadamard sequences can be considered as a typical orthogonal code and have the same correlation properties as OVSF codes. On the other hand, Gold codes can be viewed as representative of quasi-orthogonal codes due to their good cross-correlation properties, which possess three controllable cross-correlation levels.

To make the presentation of the results more informative, we have used two plotting systems in all figures to be shown. One type of plot shows maximal (a positive level) and minimal (a negative level) auto-correlation function (ACF) and cross-correlation function (CCF), together with their variance. This plot can tell us how those codes will perform in terms of the worst ACF and CCF levels as well as their variance. With this information, we can judge how they will perform under multipath interference and MAI jointly within their ACF and CCF dynamic ranges.

Another plotting system will show us the peak CCF level together with their appearance frequency. This type of plot will be able to tell us how the codes will perform in terms of average MAI.

Code lengths for Walsh-Hadamard sequences of 32, 64, and 128 will be considered for illustration simplicity. It is easy to show that a similar trend will exist if longer codes are considered. Code lengths for Gold codes of 31, 63, and 127 are presented for the same purpose. It is noted that the code set sizes for the Walsh-Hadamard sequences are the same as their code lengths, while the code set sizes for the Gold codes are 33, 65, and 129, respectively.

Figures 4.12, 4.13, and 4.14 show the auto-correlation functions for Walsh-Hadamard sequences of lengths 32, 64, and 128, where both highest and lowest ACF levels are shown, together with their variances.

Figures 4.15, 4.16, and 4.17 show the cross-correlation functions for Walsh-Hadamard sequences of lengths 32, 64, and 128, where both highest and lowest CCF levels are shown, together with their variances.

Figures 4.18, 4.19 and 4.20 show the cross-correlation functions for Walsh-Hadamard sequences of lengths 32, 64, and 128, where both CCF peaks and their appearance frequencies are shown.

Figures 4.21, 4.22 and 4.23 show the auto-correlation functions for Gold codes of lengths 31, 63, and 127, where both highest and lowest ACF levels are shown, together with their variances.

Figures 4.24, 4.25 and 4.26 show the cross-correlation functions for Gold codes of lengths 31, 63, and 127, where both highest and lowest CCF levels are shown, together with their variances.

Figures 4.27, 4.28 and 4.29 show the cross-correlation functions for Gold codes of lengths 31, 63, and 127, where both CCF peaks and their appearance frequencies are shown.

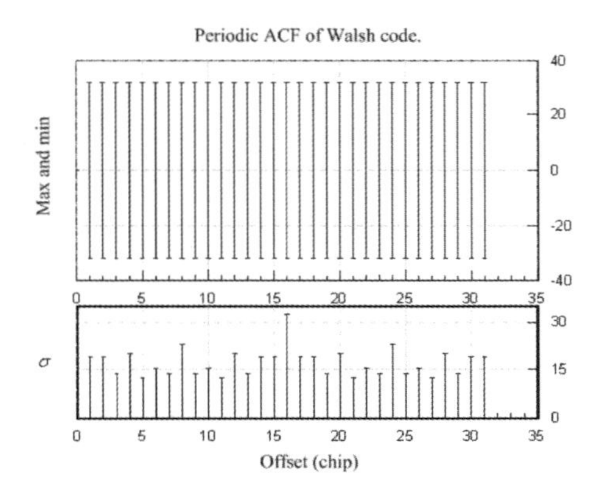

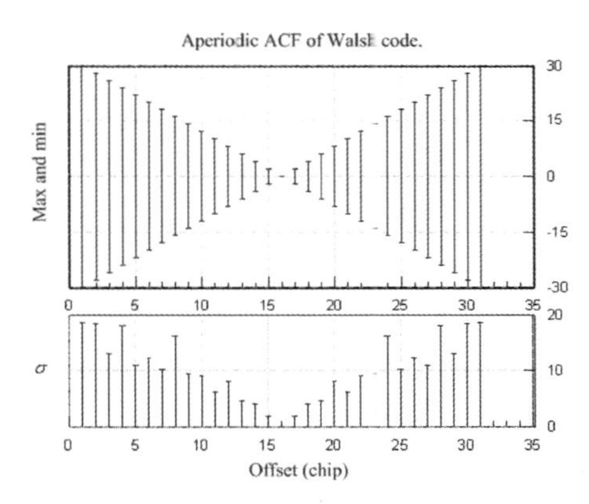

Figure 4.12 Auto-correlation functions for Walsh-Hadamard sequences of length 32, where both highest and lowest ACF levels are shown, together with their variances.

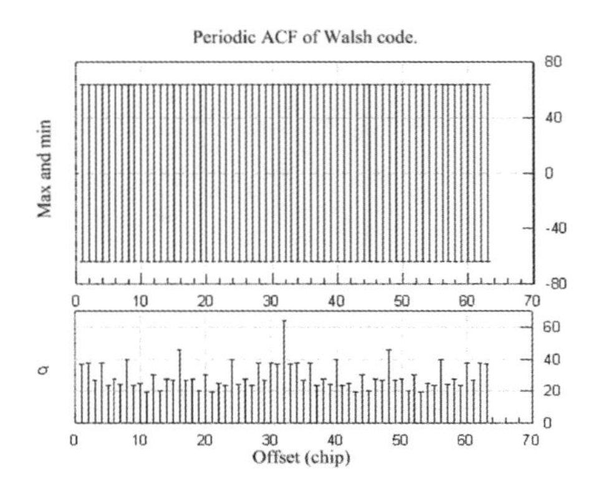

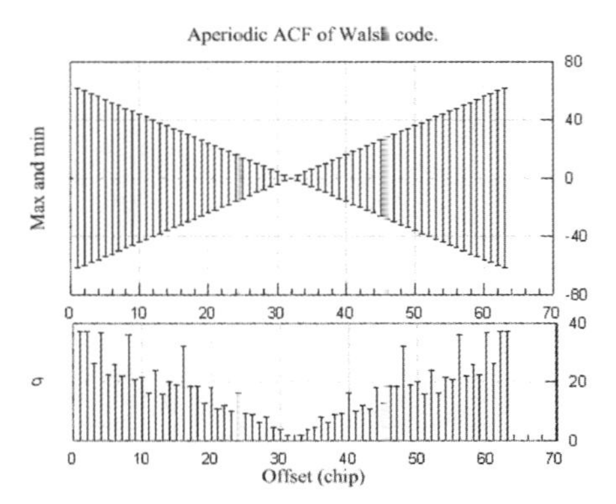

Figure 4.13 Auto-correlation functions for Walsh-Hadamard sequences of length 64, where both highest and lowest ACF levels are shown, together with their variances.

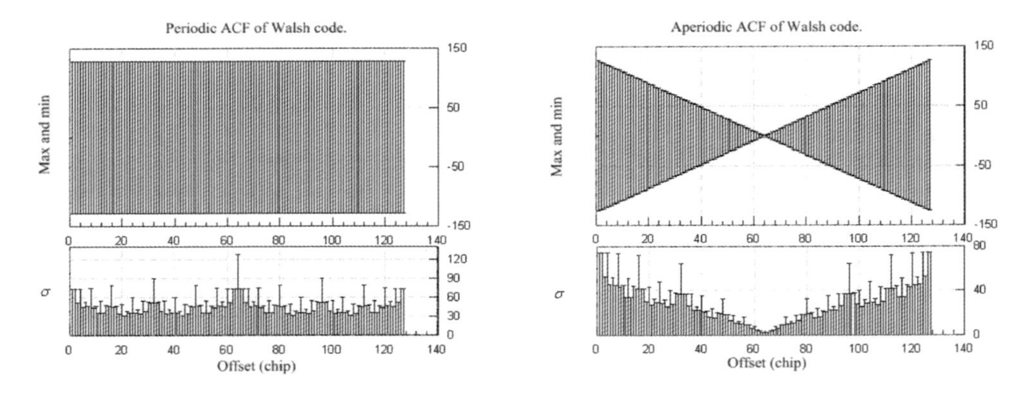

Figure 4.14 Auto-correlation functions for Walsh-Hadamard sequences of length 128, where both highest and lowest ACF levels are shown, together with their variances.

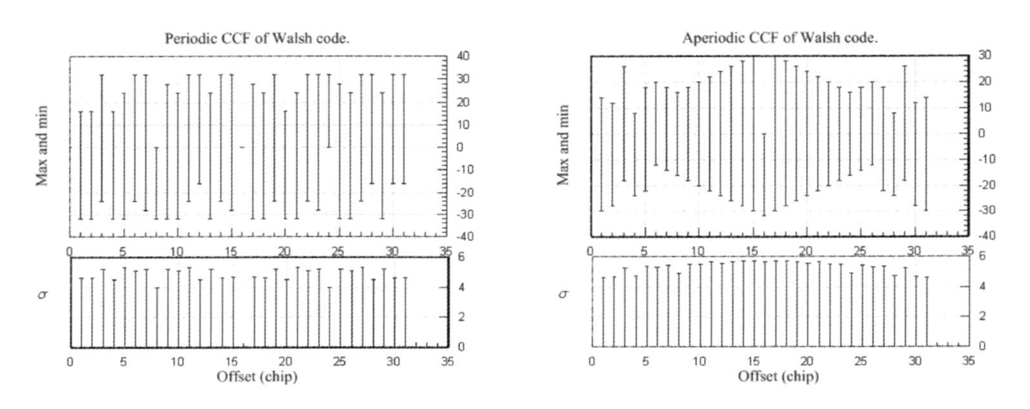

Figure 4.15 Cross-correlation functions for Walsh-Hadamard sequences of length 32, where both highest and lowest CCF levels are shown, together with their variances.

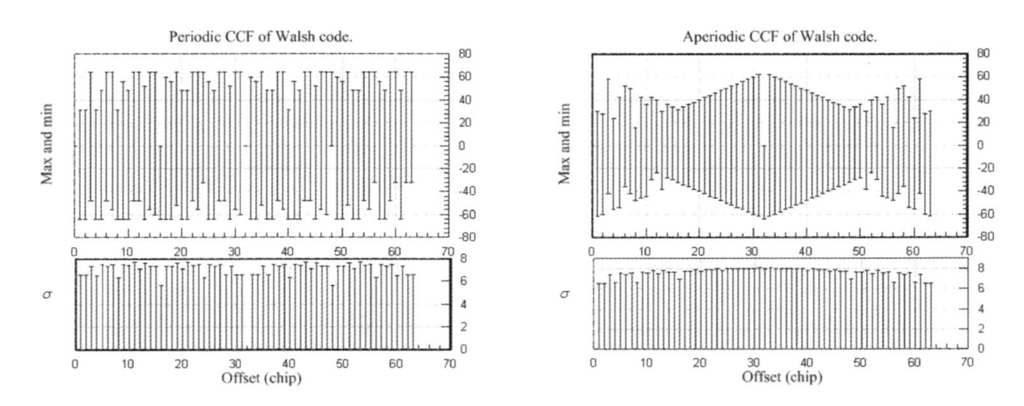

Figure 4.16 Cross-correlation functions for Walsh-Hadamard sequences of length 64, where both highest and lowest CCF levels are shown, together with their variances.

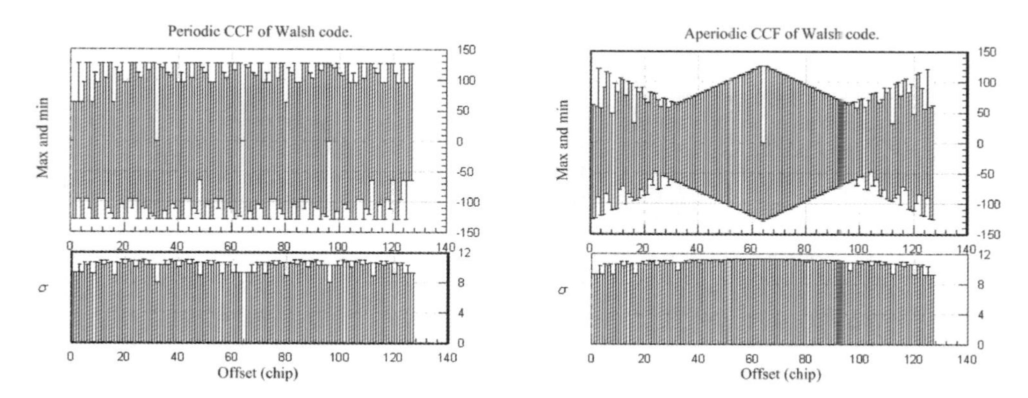

Figure 4.17 Cross-correlation functions for Walsh-Hadamard sequences of length 128, where both highest and lowest CCF levels are shown, together with their variances.

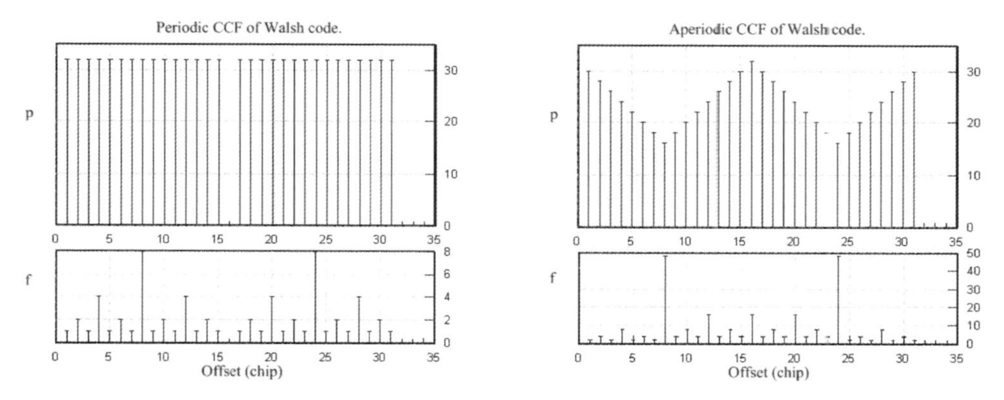

Figure 4.18 Cross-correlation functions for Walsh-Hadamard sequences of length 32, where both CCF peaks and their appearance frequencies are shown.

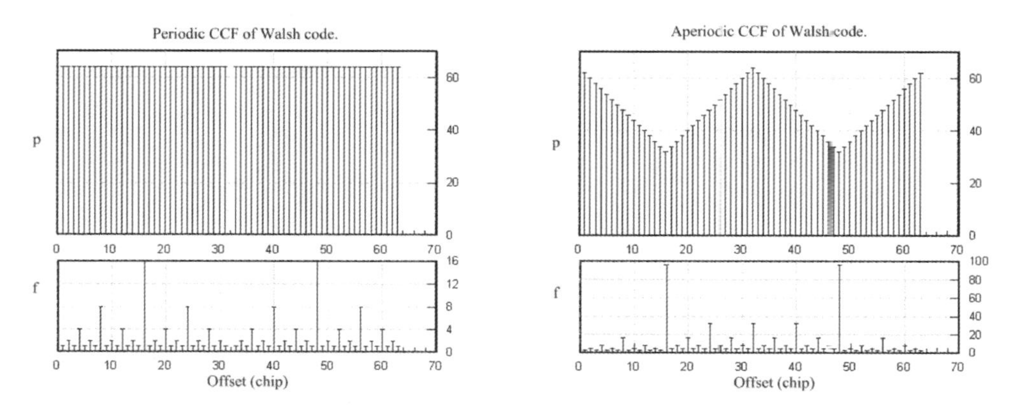

Figure 4.19 Cross-correlation functions for Walsh-Hadamard sequences of length 64, where both CCF peaks and their appearance frequencies are shown.

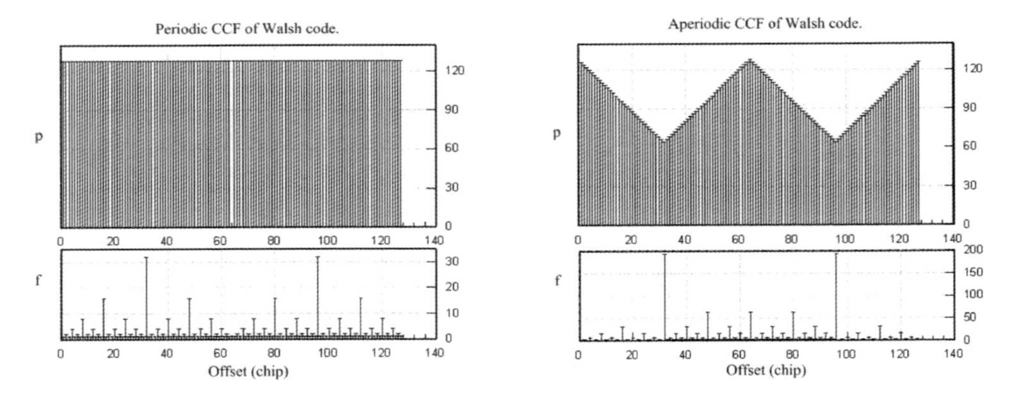

Figure 4.20 Cross-correlation functions for Walsh-Hadamard sequences of length 128, where both CCF peaks and their appearance frequencies are shown.

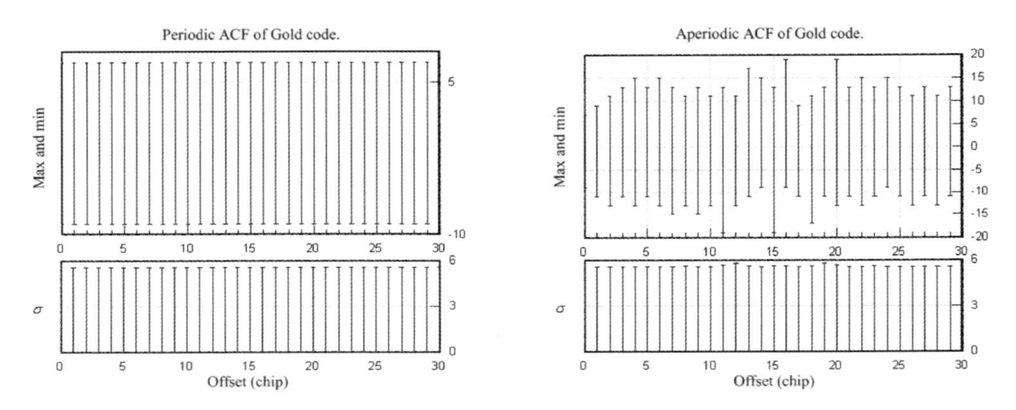

Figure 4.21 Auto-correlation functions for Gold codes of length 31, where both highest and lowest ACF levels are shown, together with their variances.

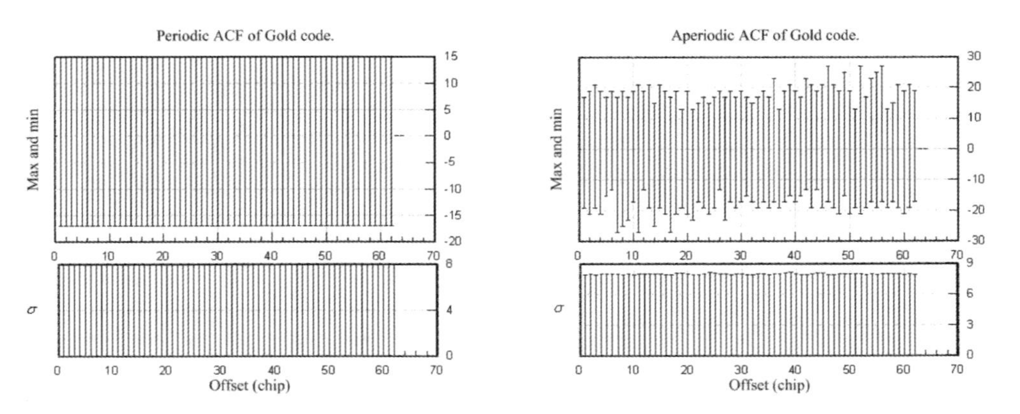

Figure 4.22 Auto-correlation functions for Gold codes of length 63, where both highest and lowest ACF levels are shown, together with their variances.

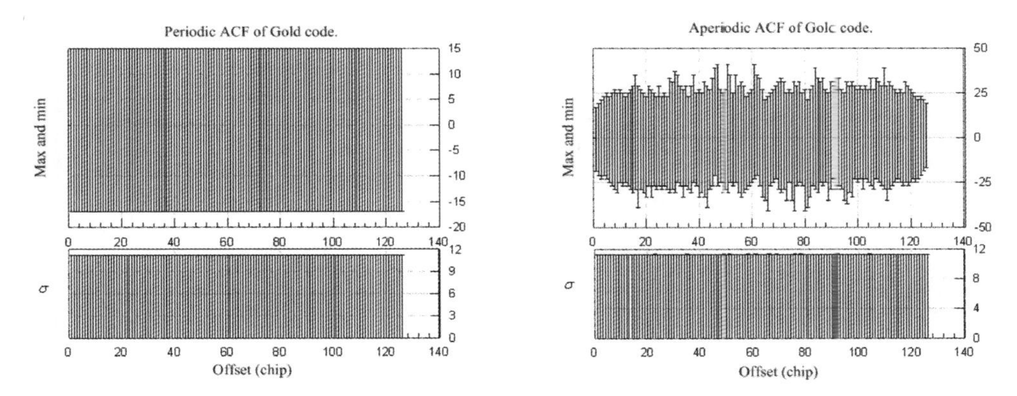

Figure 4.23 Auto-correlation functions for Gold codes of length 127, where both highest and lowest ACF levels are shown, together with their variances.

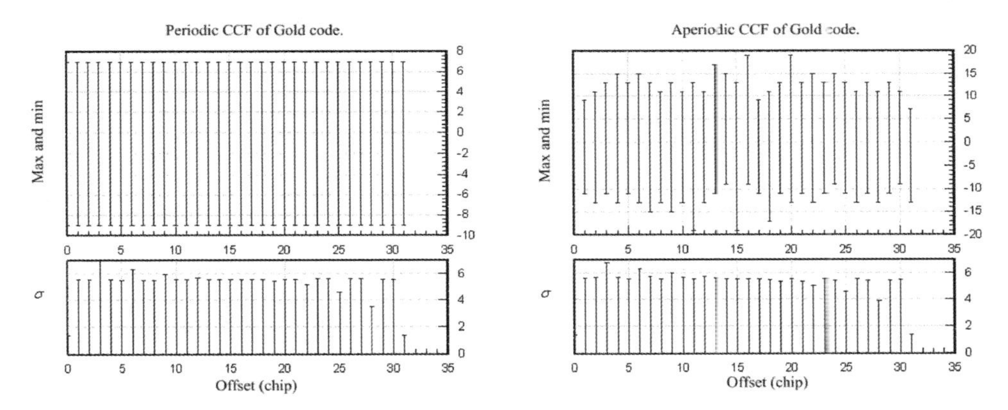

Figure 4.24 Cross-correlation functions for Gold codes of length 31, where both highest and lowest ACF levels are shown, together with their variances.

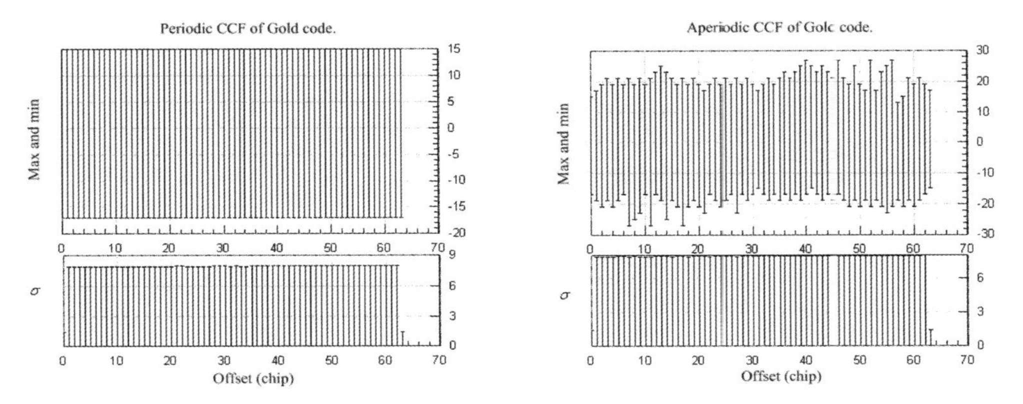

Figure 4.25 Cross-correlation functions for Gold codes of length 63, where both highest and lowest ACF levels are shown, together with their variances.

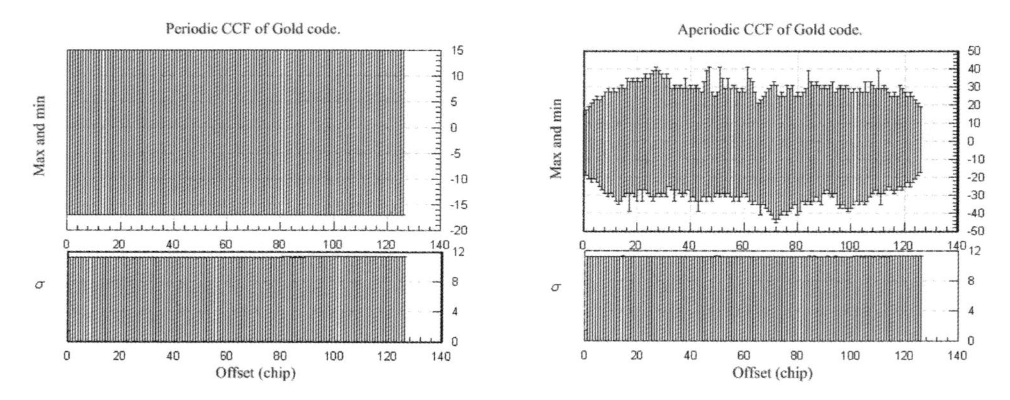

Figure 4.26 Cross-correlation functions for Gold codes of length 127, where both highest and lowest ACF levels are shown, together with their variances.

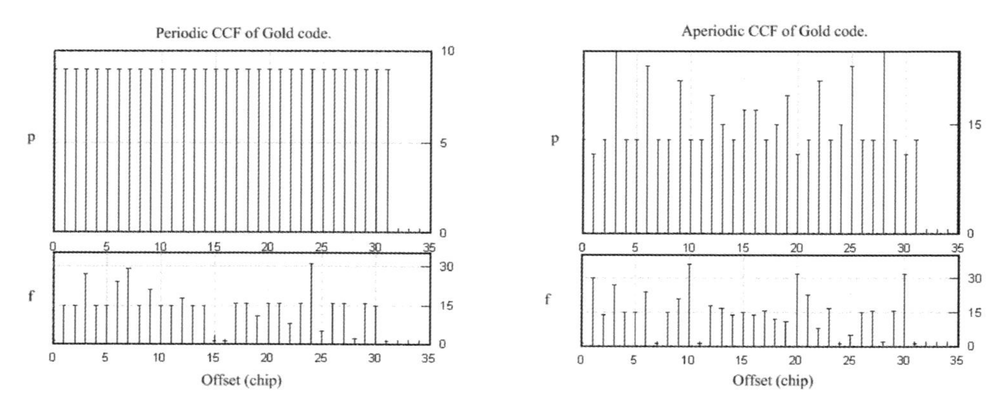

Figure 4.27 Cross-correlation functions for Gold codes of length 31, where both CCF peaks and their appearance frequencies are shown.

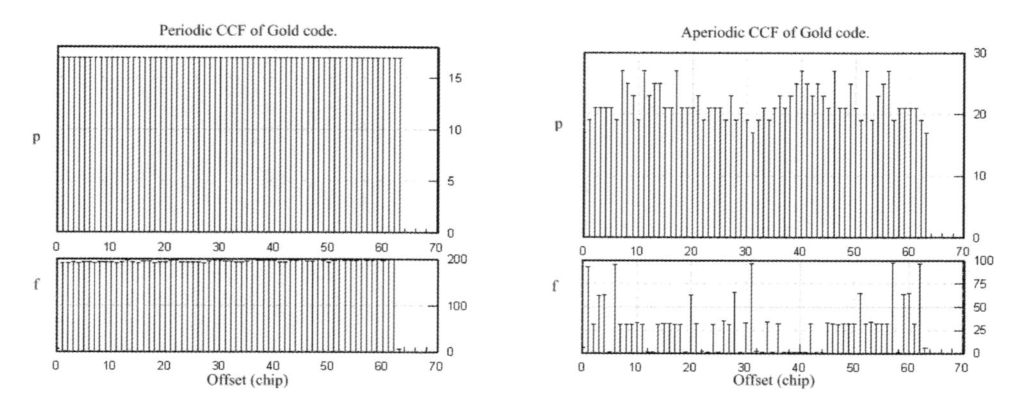

Figure 4.28 Cross-correlation functions for Gold codes of length 63, where both CCF peaks and their appearance frequencies are shown.

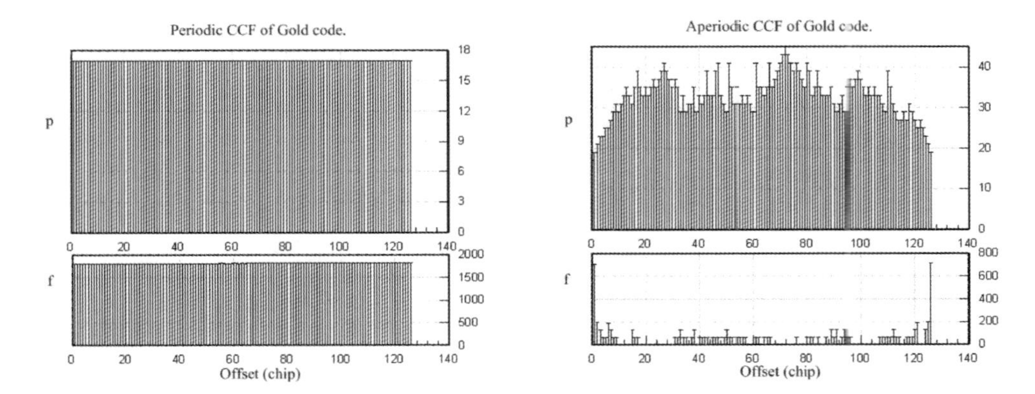

Figure 4.29 Cross-correlation functions for Gold codes of length 127, where both CCF peaks and their appearance frequencies are shown.

From the figures we can see that the Walsh-Hadamard sequences are virtually not orthogonal at all in asynchronous channels, as shown by their extremely poor aperiodic ACFs and aperiodic CCFs. On the other hand, Gold codes can offer much better performance in asynchronous transmission mode than Walsh-Hadamard sequences.

5

What is Next Generation CDMA Technology?

In the previous chapters, we have gone through some detailed discussions on existing code division multiple access (CDMA) technologies, which have been applied to the second and third generation (2G and 3G) wireless communication systems.

It is noted that the CDMA technologies used in all 2G and 3G systems are basically the same and they share many core IPRs in their respective system architectures. Those common core IPRs include direct sequence spreading modulation, unitary spreading codes or sequences, closed-loop and open-loop power control schemes, RAKE receiver, soft handover techniques, and so forth. Therefore, it is unfortunately true to say that not many new techniques can be seen in all 3G CDMA-based mobile cellular standards compared to those used in IS-95, which was introduced more than ten years ago. Therefore, it is reasonable to call all those CDMA technologies used in both 2G and 3G mobile cellular systems 'first generation CDMA technology', which should evolve along its own evolutionary path into next generation.

Then comes the important question we are supposed to discuss in this book: what is the next generation CDMA technology? Of course, it is very difficult to give a good answer at the moment, as we are still working hard toward this goal. However, we can ascertain that the next generation CDMA technologies should provide technical advantages that the first generation does not have. All major technical limitations with the first generation CDMA technology should be overcome in the new generation CDMA technology. Also, a large number of technological innovations should be introduced to make the next generation CDMA technology a very strong candidate for (if not dominating in) 4G and 5G wireless applications.

To better understand the development of CDMA technology, we would like to revisit the history of CDMA, from which we can learn a lot for its future technological evolution. The major milestones in the development of the first generation CDMA technology are illustrated as follows, in chronological order:

- November 1988: CDMA cellular concept was proposed.
- November 1989: A CDMA open demonstration conducted in San Diego.
- February 1990: NYNEX and QUALCOMM successfully demonstrated CDMA in New York City.

The Next Generation CDMA Technologies Hsiao-Hwa Chen

- 1991: QUALCOMM successfully performed large-scale capacity tests in San Diego.
- 1992:
  - US West ordered the first CDMA network equipment
  - CDMA soft handoff patent granted.
- 1993:
  - IS-95A standard completed
  - US Telecommunications Industry Association (TIA) adopted CDMA as North American digital standard
  - First commercial CDMA market trial
  - South Korea adopted CDMA.
- 1994:
  - Sprint PCS adopted CDMA
  - QUALCOMM formed QUALCOMM Personal Electronics (QPE), a joint venture with Sony Electronics to develop and manufacture CDMA phones.
- 1995:
  - CDMA standardized for US PCS
  - First commercial launch of cdmaOne (Hutchison Telecom, Hong Kong)
  - QUALCOMM launched first commercial cdmaOne handset.
- 1996:
  - cdmaOne was commercially launched in South Korea
  - PrimeCo launched cdmaOne in 14 US cities (now Verizon Wireless)
  - CDMA Development Group (CDG) announced more than one million cdmaOne subscribers.
- 1997: IS-95B standard completed (including 64 kbps data transmission capability):
  - Commercial service available in 100 US cities
  - CDMA chosen in Japan
  - QUALCOMM unveiled the Q Phone
  - QPE ships one million CDMA phones.
- 1998:
  - TIA endorsed CDMA2000 as a 3G solution for International Telecommunication Union (ITU)
  - LG Telecom launched first CDMA data services
  - cdma2000 submitted to ITU as part of the IMT-2000 process for global 3G standards
  - More than 12.5 million cdmaOne subscribers in 30 countries
  - First 1xEV-DO demonstration.

- 1999:
  - QUALCOMM and Ericsson reached agreement to support single 3G CDMA standard
  - QUALCOMM sold wireless infrastructure division to Ericsson
  - China Unicom joined CDG and announced plans for commercial services
  - 83 CDMA operators in 35 countries
  - CDG announced CDMA is fastest growing mobile technology with nearly 42 million subscribers
  - QUALCOMM introduced the Thin Phone and the pdQ phone
  - QUALCOMM exceeded 14 million CDMA phones shipped since production began.
- 2000:
  - Japan's IDO and DDI started nationwide 64 kbps CDMA packet data service
  - DDI announced they will use cdma2000 for 3G wireless service
  - IUSACELL became first Latin American operator to offer wireless Internet services
  - QUALCOMM, Samsung, and Sprint PCS made the first 3G cdma2000 voice call. Lucent and QUALCOMM completed the first 153 kbps 3G CDMA2000 data call
  - QUALCOMM and Sprint commenced US trials for 3G cdma2000 solution
  - SK Telecom launched world's first 3G cdma2000 commercial service
  - QUALCOMM sold CDMA handset business to Kyocera Wireless Corp.
- 2001:
  - More than 100 million CDMA subscribers globally
  - More than 22 million cdmaOne Internet and data users
  - cdma2000 surpassed three million subscribers
  - QUALCOMM shipped cumulative total of more than 500 million chips
  - QCT and Nortel Networks conducted industry's first mobile IP call
  - QCT, SchlumbergerSema, and Samsung demonstrated CDMA/GSM roaming using R-UIM-enabled CDMA handsets
  - QUALCOMM introduced BREW system
  - KDDI announced successful completion of CDMA2000 1xEV-DO trial with QUALCOMM, Hitachi, Sony, and Kyocera
  - Telesp Cellular in Brazil was the first Latin American operator to deploy 3G cdma2000
  - Romania launched commercial cdma2000 services
  - First commercial gpsOne deployment by SECOM in Japan.
- 2002:
  - 3G CDMA subscribers surpassed 27 million
  - QUALCOMM's CMX multimedia software debuted in US as Sprint launched CMX with 'Sprint PCS Ringers & More' service

- China Unicom launched nationwide cdmaOne network in China
- SK Telecom launched cdma2000 1xEV-DO in South Korea
- QUALCOMM shipped cumulative total of more than 725 million chips
- More than five million gpsOne subscribers
- 14 countries launched commercial cdma2000 services (Australia, Canada, Chile, Columbia, Ecuador, India, Israel, Japan, Moldova, New Zealand, Panama, Russia, United States, and Venezuela)
- Monet Mobile launched the first cdma2000 1xEV-DO commercial network in the United States
- BREW subscribers top 2.5 million.

- 2003:
  - 3G CDMA subscribers surpassed 73 million
  - 18 countries launched commercial cdma2000 services (Argentina, Belarus, Bermuda, Brazil, Canada, China, Dominican Republic, Guatemala, Indonesia, Kazakhstan, Mexico, Nicaragua, Nigeria, Peru, Puerto Rico, Taiwan, Thailand, Vietnam)
  - BREW subscribers top 16 million
  - 19 more operators deployed commercial BREW services worldwide
  - MSM7xxx chipset solution series for high-performance multimedia wireless devices was released
  - MSM6250 chipset solution for WCDMA (UMTS), GSM and GPRS was released
  - QUALCOMM's WCDMA solutions selected by 13 infrastructure vendors and 13 leading device manufacturers
  - More than 25 million gpsOne subscribers
  - Verizon Wireless launched cdma2000 1xEV-DO in the United States
  - KDDI launched cdma2000 1xEV-DO in Japan.
- 2004:
  - 240.2 million CDMA subscribers worldwide
  - 146.8 million cdma2000 subscribers
  - CDMA2000 1xEV-DO Revision A approved by Third Generation Partnership Project 2 (3GPP2)
  - Eurotel Praha (Czech Republic) launched world's first cdma2000 1xEV-DO network at 450 MHz (CDMA450).
- 2005:
  - More than 200 million commercial cdma2000 subscribers worldwide
  - 143 cdma2000 operators commercially deployed in 67 countries on 6 continents
  - 950 cdma2000 devices offered commercially since 2000
  - 64 cdma2000 device manufacturers
  - Number of commercial cdma2000 1xEV-DO operators almost doubled from 16 to 29

- Number of commercial CDMA450 operators increased to 31 in 22 countries
- Average growth rate of nearly 4.9 cdma2000 subscribers per month.

- 2006:
  - QUALCOMM acquired Flarion Technologies in January 2006 to expand its already extensive portfolio of orthogonal frequency division multiple access (OFDMA) intellectual property and enhance QUALCOMM's industry-leading R&D organization with expertise in OFDMA technology and products
  - QUALCOMM signed OFDM/OFDMA License and FLASH-OFDM Design Transfer License Agreements with AnyDATA
  - QUALCOMM and SOMA Networks signed OFDM/OFDMA Subscriber and Infrastructure Equipment License.

It is noted that, when this book was written, the worldwide subscribers for CDMA2000 services (which is one of the 3G standards developed directly from the IS-95 standard) had reached about 280 million, as shown in Figure 5.1.

From the major milestones of CDMA technological development outlined above, it is seen that the first generation CDMA technology has been successfully used in 2G and 3G mobile cellular communication systems around the world. The success of the first generation CDMA technology has made a new technological giant, Qualcomm Inc., which is the inventor and initiator of first generation CDMA technology.

About 12 years have elapsed since the CDMA cellular concept was first proposed in 1988. We have to note that, during the same period of time, worldwide mobile cellular technology has experienced three generations, from its first generation, such as TACS, NMT, and AMPS, to its second generation, such as IS-95, GSM, and DAMPS, to its third generation, such as the cdma2000, UMTS W-CDMA, and TD-SCDMA standards.

Today, it is very interesting to recall that there was fierce competition between CDMA and TDMA (or between IS-95 and GSM) technologies when Qualcomm first proposed CDMA for the second

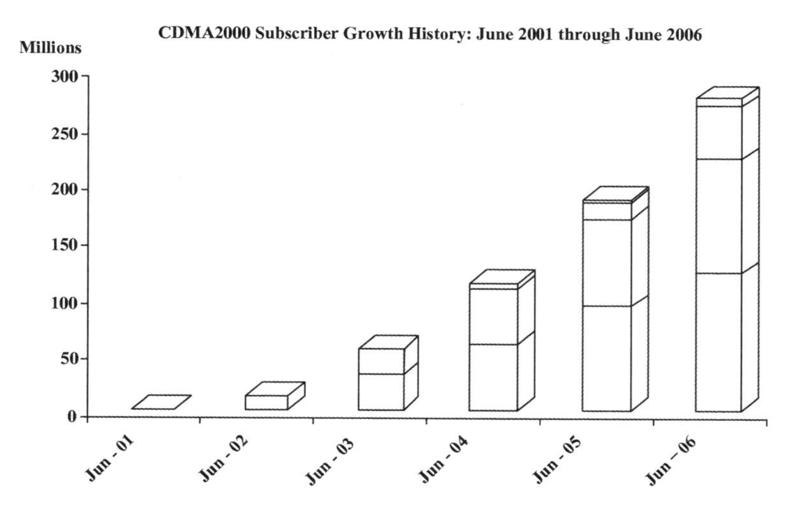

Figure 5.1 Worldwide CDMA2000 subscriber growth history: June 2001 through June 2006.

generation civilian mobile cellular applications. At that time, the major players in the competition were Qualcomm and Ericsson.

When the CDMA cellular concept was first proposed, many people thought that it could not be made to work. Indeed, at least one European company deeply involved with GSM, Ericsson, went through the three classic stages of 'Not Invented Here' syndrome: (1) it is impossible, (2) it is infeasible, (3) actually, we thought of it first. At first, the most vocal top brains at Ericsson tried to claim that CDMA violated information theory.

In IS-95 CDMA, a single carrier frequency has a bandwidth of 1.2288 MHz, and up to 30–40 cell phones in a given sector can all be transmitting chips at that rate on the same carrier frequency, which seemed on first examination to assume that it was possible to send fifty million bits through a one-and-a-quarter MHz band, which would indeed violate Shannon's Theory. The mistake they made was that chips are not 'information' based on Shannon's definition, and though those phones were sending chips that fast, they were actually sending bits (real data) at no more than 14 400 bits per second each. Unfortunately for Ericsson, Qualcomm did a field test in New York City where several prototype phones mounted in vans were able to operate at once on the same frequency talking to multiple cells all of which also operated on the same frequency.

The next argument from Ericsson was that though it seemed technically possible, it would be too expensive. Everyone knew that the electronics required to make CDMA work was a lot more complicated than what TDMA used, and Ericsson's loud voices claimed that it could never be reduced in price enough to make it competitive. And shortly thereafter Qualcomm proved Ericsson wrong again, by beginning to produce both infrastructure and phones at very competitive prices.

After that, Ericsson suddenly decided that it had applicable patents and took Qualcomm to court. Over the long drawn out process of litigation, every single preliminary court judgment went in favor of Qualcomm, and it became obvious that Ericsson did not have a case and that Qualcomm was not going to be intimidated. Ultimately, the entire case was settled in a massive omnibus agreement where Ericsson became the last of the large companies in the industry to license Qualcomm's patents (on the same royalty terms as everyone else) while taking a large money-losing division off Qualcomm's hands and assuming all the liabilities associated with it, and granting Qualcomm a full license for GSM technology. The industry consensus was that this represented a full-scale surrender by Ericsson. On the other hand, Nokia was not anything like this and it had licensed several years before.

History is just like a mirror we have to watch from time to time, in order for us to become smarter. Today, we are facing some more serious challenges to develop 4G and 5G wireless communication systems, whose data rate can be up to 1 Gbps. The major contenders for being selected as a prime air-link technology are CDMA (probably we should name it as the next generation CDMA here, as the first generation CDMA technology will most likely not be a winner for technological, economic, and political reasons) and OFDMA technologies. Compared with the time when the competition between CDMA and TDMA for the 2G mobile cellular systems happened in the late 1990s, most of the major players are still there but their roles and influence have been different. Qualcomm has become a super player in terms of its CDMA technological IPRs and it is eager to keep its superior position in the new technological competitions. It seems clear that nobody can avoid the IPR transfer problem if they want to use CDMA and its related technologies in the development of 4G and future wireless applications. Therefore, the rule of the games in this commercial world have pushed people away from CDMA and toward some other replacement technologies, such as OFDMA technology.

The 3GPP LTE E-UTRAN standardization process has already shown the trend, which tends to reluctantly use any of the CDMA technology not merely because of technical reasons. As most participants of 3GPP are from the GSM/W-CDMA camp, they have suffered a lot in developing their 3G mobile cellular standards and have paid a huge amount of license fees or were forced to cross-license many of their own patents with Qualcomm because of the use of CDMA and its related technologies in W-CDMA systems.

We see a parallel between what happened due to the competition between CDMA and TDMA about ten years ago and what is happening now in the 4G standardization process: CDMA versus

OFDMA. Qualcomm today is acting somehow differently compared with Ericsson about ten year ago, although Qualcomm has used a different strategy, which is reflected in its recent (in January 2006) move on purchasing Flarion at a cost of 600 million USD, a small spinoff company from Lucent that owns many of the original IPRs for OFDM and OFDMA technologies.

Flarion has been testing the next-generation OFDMA wireless IP technology since 2003. In addition to OFDMA, Flarion also pioneered Flash-OFDMA for mobile IP-based broadband services. Flash-OFDM is a proprietary cellular broadband technology that network operators can use to link laptops to act as a fixed wireless access system, bridging the 'last mile' to connect computers in homes and small offices. It means an all-IP architecture and high speeds. The technology means users traveling at 250 kilometers per hour can download data at up to 1.5 Mbit/s and upload at 500 Kbit/s.

As a key supplier of CDMA mobile communications technology, Qualcomm also aims to support operators that prefer an OFDMA or hybrid OFDM/CDMA offering to differentiate their control on key technologies, which might be used in 4G wireless.

This also tells us what role politics and economics can play in those seemingly very technical matters. I do not make any conclusion here and it is better to leave the issue open to all for your own judgment. Nevertheless, what I should say is that technological monopoly can be a big barrier to the evolution of any technology, just like CDMA technology. I, and maybe many other people as well, fully believe that CDMA technology would not have stayed at its first generation for such a long time if its major IPRs had not been so concentrated in the hands of only one company.

At least, one of the reasons for me to write this book was to push for faster technological evolution of CDMA technology. The motivation to develop the next generation CDMA technologies is there. However, we have to go back to the fundamentals and ask ourselves honestly the question of what is the next generation CDMA technology. In this book, I will try my best to figure out the framework for the next generation CDMA, and maybe it will not be as complete as everybody wants, but at least it can offer some ideas which can be used as the bricks to build the whole mansion: next generation CDMA technologies.

CDMA codes will still be the core of the next generation CDMA technology. There needs to be a revolution in the design approaches for CDMA codes. The design process should take into account as many real operational conditions as possible, and the code selection criteria should not be built only on the periodic auto-correlation function and periodic cross-correlation function. At least the following operational conditions should be taken into account, such as the multipath propagation effect, asynchronous transmissions, random bit signs in the data stream, bursty traffic, etc. At the same time, we should pay enough attention to maintain a sufficiently large code set size when searching for suitable CDMA codes for next generation CDMA systems.

Innovation in the spreading modulation scheme is another focal point when we carry out research on next generation CDMA technology. The improvement in spreading efficiency, which is defined as the information bits carried in each chip (as shown in Equation (4.11)), is an interesting topic for further investigation. A higher spreading efficiency implies a higher bandwidth efficiency, which is an extremely important concern in the design of B3G wireless communication systems. One of the most important advantages for OFDMA technology is its relatively high bandwidth efficiency due to the fact that it can use overlapping sub-carriers to reduce the spectral occupancy requirement. Therefore, to compete with OFDMA in terms of spectral efficiency, the next generation CDMA technology ought to excel in this respect as well.

Another important claim for OFDMA technology for its suitability in B3G wireless applications is that it can use relatively cheap and simple baseband signal processing techniques to replace otherwise complicated RF digital modulation units in many other multiple access schemes, such as traditional CDMA systems. Therefore, the system architecture of the next generation CDMA technology should be implemented using simple hardware structure. The next generation CDMA technology should work satisfactorily without relying heavily on many complex sub-systems, which have been widely used in first generation CDMA systems, such as multi-user detection, open-loop and closed-loop power control, RAKE receiver, smart-antenna systems, etc., just to name a few as examples.

Inability to support high-speed bursty traffic is a serious concern that makes people believe that CDMA technology is only suitable for slow-speed continuous-time transmissions relevant to voice-centric applications. Many supporters of OFDMA have been using this excuse to justify their conclusions against CDMA technology for its possible applications in B3G wireless systems. Therefore, the next generation CDMA technology has to be able to support high-speed burst-type traffic to meet the requirements of all-IP wireless architecture that will be used in all B3G wireless services. To achieve this, innovations in the whole CDMA architecture, from CDMA code design to other signaling structure, should be encouraged.

In addition, the overall capacity of the next generation CDMA systems should be greatly enhanced compared to that available in the current first generation CDMA systems, such as IS-95, cdma2000, W-CDMA, etc. Obviously, the capacity can be greatly enhanced if the next generation CDMA technology can operate in an interference-free or at least an interference-resistant mode. To make it happen we have to break the myth that a CDMA system is always interference-limited.

We have to admit that it is an extremely challenging task to develop next generation CDMA technology and we have a lot to do to achieve our ultimate goal.

5.1 APPLICATION SCENARIOS

Obviously, the next generation CDMA technologies should be developed to fit different types of future wireless applications, which include mobile cellular, wireless local area networks, wireless personal area networks, etc.

Futuristic wireless applications can be built on a heterogenous infrastructure, under which many different layers of wireless networks/systems will be integrated under a unified giant umbrella. To facilitate the network deployment and future upgrading, it will be a plus if the different layers of wireless networks use similar, if not exactly the same, multiple access air-link technology. This will be proved to be a vital requirement from the point of view of service operators to provide cost-effective communication solutions to different end-users. In this sense, we should be very careful when designing the next generation CDMA technology to make it suitable for its applications in various wireless application scenarios or environments. The next generation CDMA technology should be made as versatile as possible and it should not be defined in a narrow sense, for instance only being suitable for mobile cellular systems.

Of course, we have to admit that it will be an extremely difficult job to design an advanced air-link technology, which at the same time should be suitable for applications in many different wireless communication systems. The same may happen to the design of the next generation CDMA technology, which involves many cutting-edge research topics as a whole in the entire design process. If this case is true, then we have to make sure that the fundamental goals we have set should be achieved. Those fundamental goals include enhanced bandwidth efficiency through the revolution of spreading modulation schemes, improved system capacity requirements to break the interference-limited myth, and so forth. Therefore, the enhancement in bandwidth efficiency and system capacity are our two fundamental goals for the applications of next generation CDMA technology in any wireless systems.

In the following subsection we will discuss several major application scenarios for next generation CDMA technology. These major scenarios include mobile cellular (also called wireless wide area networks (WWANs)), wireless local area networks (WLANs), wireless personal area networks (WPANs), and cognitive radio networks. The relatively extensive discussions on these application scenarios are based on the fact that they will form the most important part of the B3G wireless communication infrastructure under the heterogeneous wireless network umbrella. Of course, it should be noted that there are many other wireless networks/systems scenarios, such as wireless metropolitan area networks (WMANs), wireless regional area networks (WRANs), etc. Due to space limitations in this book, we will not include them as application scenarios in our discussions. However, readers may easily extend our discussions to cover them as well.

As a special note, we would like to emphasize that next generation CDMA could play an extremely important role in the newly emerging cognitive radio wireless networks and cooperative wireless networks. A cognitive radio network may consist of a large number of service nodes, which can be divided mainly into three types: incumbent nodes (which are the primary licensed users of the spectrum), non-cooperative nodes (which are unlicensed users of the spectrum, whose operations may not be able to be cooperative with others), and cooperative nodes (which are again unlicensed users of the spectrum, whose operations can be cooperative with the nodes of interest). Therefore, a cognitive radio node should operate in the spectrum in a purely opportunity-driven fashion, or it should stop immediately if it senses that the bandwidth is in use by the incumbent node(s). Taking into account the fact that time is also needed to switch off the transmissions from a cognitive radio node, no matter how fast the response time of the node can be, we would like to know which type of multiple access technology can cause the least interference if the switching-off time inevitably overlaps with the transmission of the incumbent node(s). Obviously, CDMA is the only natural choice due to its much lower average power emission level than any other multiple access scheme, such as FDMA, TDMA, and OFDMA. In particular, OFDMA may not be suitable in this sense, as it uses an amplitude modulation scheme in its carrier frequency up-conversion unit, which is not a power-efficient carrier modulation scheme and requires a relatively high transmission power to achieve a certain signal-to-noise-ratio at the receivers.[1] We will offer more detailed discussions on this issue in Subsection 5.1.4.

At the end of this section, we also offer some discussions on multi-carrier CDMA and OFDM-CDMA system architecture, which might be used as a cost-effective implementation structure for next generation CDMA technology due to the possible use of multi dimensional spreading modulation schemes.

5.1.1 MOBILE CELLULAR

The first generation CDMA technology was initially proposed by Qualcomm in the late 1990s for its applications in cellular radio systems. The standard developed for this purpose is called IS-95 (also called cdmaOne), which has been extensively discussed in Chapter 3. Therefore, we can see that the initial applications for CDMA technology were aimed primarily at mobile cellular systems, which constitute a huge market in the wireless service business sector due to the great demand for mobile communications in the world. This demand will not be diminished in many years to come, as seen from Figure 5.2, which shows only the CDMA mobile cellular services.

To understand better the historical background of first generation CDMA and its applications in mobile cellular systems, we would like to have some more discussions as follows, although more detailed information can also be found in Chapter 3.

Original targets for the development of first generation CDMA technology (such as the IS-95 standard) were set for voice-centric applications, which are characterized by relatively slow speed continuous-time transmissions with a signaling structure dominated by very long frames.

The 2G CDMA networks support Internet applications such as email, web browsing, and access to corporate intranets as well as digital fax and other 'mobile office' functions at 14.4 kb/s. The evolution of cdmaOne networks to support cdma2000 Phase I capabilities provides enhanced network performance in many ways, including speed and efficiency of data transmission.

The voice and data services supported by first generation CDMA technology are provided under the IS-95-A infrastructure and subscriber equipment. CdmaOne networks provide both packet- and

[1] The advantages for OFDMA's application in a cognitive radio network include that it can make use of any fractional bandwidth currently empty such that some part of the sub-channels can be allocated to them in a dynamic fashion.

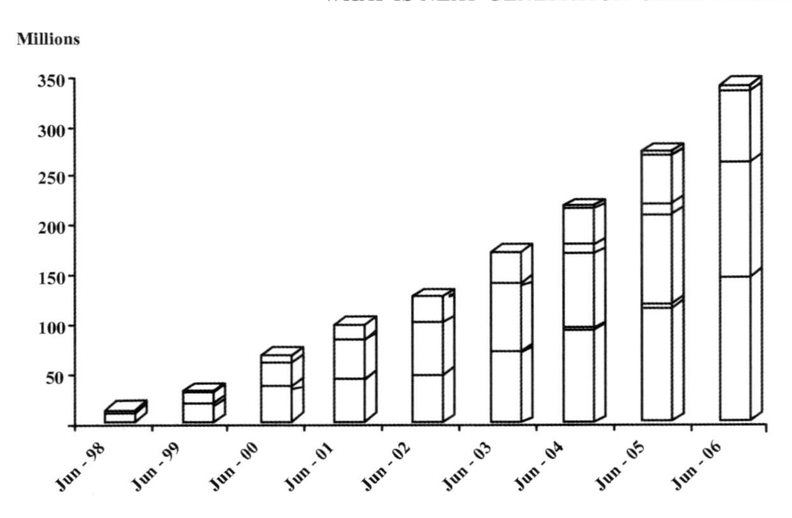

Figure 5.2 Worldwide CDMA subscriber growth history: June 1998 through June 2006.

circuit-switched data services at speeds up to 14.4 kb/s along with a choice of 8K, 13K, and 18K EVRC voice services from the same platform. This is one of the primary benefits of the network design: carriers do not have to make significant investments to add data services to voice. Typical circuit-switched applications include both analog and digital fax, file transfers, etc. Internet connections and email delivery are popular packet-based applications.

High data rate (HDR) and cdma2000 Phase I are complementary technology developments that offer higher capacities, better use of spectrum, higher speeds, and additional services. As the first part of the development and standards effort for cdma2000, cdma2000 Phase I gives end-users the ability to double the capacity, or spectral efficiency, of current cdmaOne networks. Also, it expands capabilities to allow packet and circuit data services at speeds up to 144 kb/s.

Carriers that want to offer peak data rates of greater than 1.8 Mb/s or that want to offer data services to a large number of customers without diminishing voice capacity soon will be able to implement this data feature that uses a standard-band 1.25 MHz carrier. HDR was first demonstrated publicly in September 1998, and field trials were done with US West and Cisco Systems in early spring of 1999. HDR is designed to optimize packet-data services. It separates data service from voice service because the two have fundamentally different requirements. Voice services are real-time, delay-sensitive applications that aim to provide equal service to all users, regardless of their location in the cell. This results in power-sharing schemes, where weaker users are allocated more power than stronger users, being the optimal solution for voice. A relatively modest data rate is sufficient for high-quality voice service, and voice users cannot benefit substantially from higher data rates. Physical layer designs for both voice and data services must make compromises in their design features (including choice of frame sizes, control and signaling methods, and delay budgets) to accommodate both services.

On the other hand, packet-data systems are aimed at maximizing the sector throughput. Because data users have various data-rate requirements, the goal is to allocate each user the maximum data rate that he can accept, based on his application needs and the wireless channel conditions.

Circuit-switched networks establish an end-to-end channel capacity for data transmission just as they do for a voice call. While the channel is in use, no other user's data can be transmitted. The user has exclusive control of the channel.

Packet-switched networks break messages into small data packets and send each packet over the network with a destination address. These networks are referred to as 'connectionless' because there is no end-to-end fixed connection. The Internet is a packet-switched network that shares the channel capacity. Packet-switched networks allow multiple users to share the same channel at the same time. This allows a smooth transition for packet-based networks, such as HDR, to offer high-speed Internet access.

In early 1999, TIA published a draft standard for its cdma2000 (also known as 1XRTT submission, cdma2000 Phase I or 95-C) proposal, supported by the CDMA Development Group (CDG) and a coalition of wireless equipment manufacturers. Cdma2000 uses a standard-band 1.25 MHz carrier but doubles the capacity of current cdmaOne systems and offers other enhancements such as even wider coverage and longer standby times. It also provides a high degree of backward compatibility with today's cdmaOne infrastructure, handsets, and services.

The enhanced cdma2000 standard (also known as 3XRTT) is being defined by a broad-based group that includes the TIA and other industry organizations. This standard will satisfy the objectives of IMT-2000, including the following data requirements: 144 kb/s for mobile users, 384 kb/s for pedestrian users, and 2 Mb/s for fixed users. Enhanced cdma2000 can be deployed in new or current frequency bands with much backward compatibility to cdmaOne and IS-41-based networks.

From the aforementioned 2G and 3G CDMA services (based on 3GPP2 standards, such as IS-95/cdmaOne and cdma2000, etc.), we can see clearly what is the current status of the application of first generation CDMA technology. It is expected that B3G mobile cellular systems will be operating on an all-IP wireless platform, which is built based on fully packet-switched techniques. Therefore, the situation of mixed circuit-switching and packet-switching design existing in 2G and 3G CDMA mobile cellular will be switched to completely packet-switching architecture. Therefore, all channel traffic will be dominated by short bursts/packets, and the long signaling frame structure existing in 2G and 3G CDMA standards will never appear in future B3G wireless systems.

Based on this thought flow, we can understand that it is of the ultimate importance for the next generation CDMA technology to be able to support high-speed packet-switching transmissions to meet the needs of all-IP-based wireless network architecture.

In order to design the next generation CDMA technology in particular for its applications in mobile cellular networks, we should make sure that it can offer coverage in a relatively large area. This consideration is very much different if the technology is developed to support only a small area, such as a local area network or personal area network. To this end, the design of next generation CDMA air-link technology should be done jointly with uplink and downlink channel duplex techniques. It is widely believed that a time division duplex (TDD) technique is only suitable for applications for a relatively small area due to its low average transmission power, compared with that of the frequency division duplex (FDD) technique. In the TDD scheme, the signal transmissions for uplink and downlink channels happen in different time segments in a time frame, and usually the downlink channel transmission happens before the uplink channel transmission. The transmissions in the uplink and downlink channels should be separated by a guide time interval to prevent possible interference between the two. Therefore, it is understandable that the transmission for either downlink or uplink can only happen about half of a time frame, reducing the average transmission power achievable to either downlink or uplink channel transmissions. On the other hand, the FDD scheme will use two separate bandwidths for the uplink and downlink channel transmissions and thus either direction can send its signal all the time. For this reason, FDD can offer much higher average transmission power than TDD. Therefore, most mobile cellular standards use FDD as their uplink and downlink duplex scheme due to the consideration to cover a relatively large cell size in a mobile cellular system. Understanding this point is important for design of the next generation CDMA technology suitable for mobile cellular applications, as the channel duplex techniques will effectively affect the design of spreading modulation schemes and spreading codes. Thus, they should be taken into account jointly.

Frequency reuse factor is another important concern for the design of an air-link technology for B3G mobile cellular networks/systems, as it will have a strong impact on the cellular network deployment plan and thus the implementation/service cost. A mobile cellular network is always planned in clusters of cells. The number of cells per cluster is restricted by the requirement that the clusters must fit together like jig saw pieces. The possible cell clusters are 4-, 7-, 12-, and 21-cell clusters. To support more users using a limited number of carrier frequencies, initial research undertaken by Bell Labs has found a solution which involves reusing the same channel frequency in many different clusters. The cells in these different clusters using the same channel frequency must be sufficiently separated so that co-channel interference would not occur.

The hexagonal-shaped cells shown in network planning are artificial and cannot be generated in the real world. However, this shape is chosen to simplify the planning and design of a cellular system as hexagons fit together without any overlap or gap in between them. Another advantage of using a hexagon is that it approaches a circular shape, which is the ideal power coverage area. The real cell shape can be arbitrary and will keep changing due to prevailing conditions.

The size of the cell largely depends on the area in which it is located. Generally, rural areas have fewer subscribers than urban areas. Thus, in an urban area more channels are needed to accommodate the larger number of subscribers. If each cell in a given rural and urban area had a fixed number of channels, the cell size in the urban area would have to be smaller to allow more channels in the given area. Reducing the cell size would result in cells using similar channel frequency to be located closer to each other. Therefore, reducing the size too much would cause an increase in co-channel interference. The size of the cell can be varied by varying the power and sensitivity of the base station. An alternative way to change the size of the cell is to split it into sectors.

It is widely believed that a CDMA mobile cellular can make the frequency reuse factor one, to greatly simplify the cell planning process, which is always a headache for a service provider. Figure 5.3 shows the typical cell planning scheme (7-cell cluster) in an FDMA mobile cellular network with a frequencies reuse factor of 7. Interference may occur if the frequency reuse factor is chosen

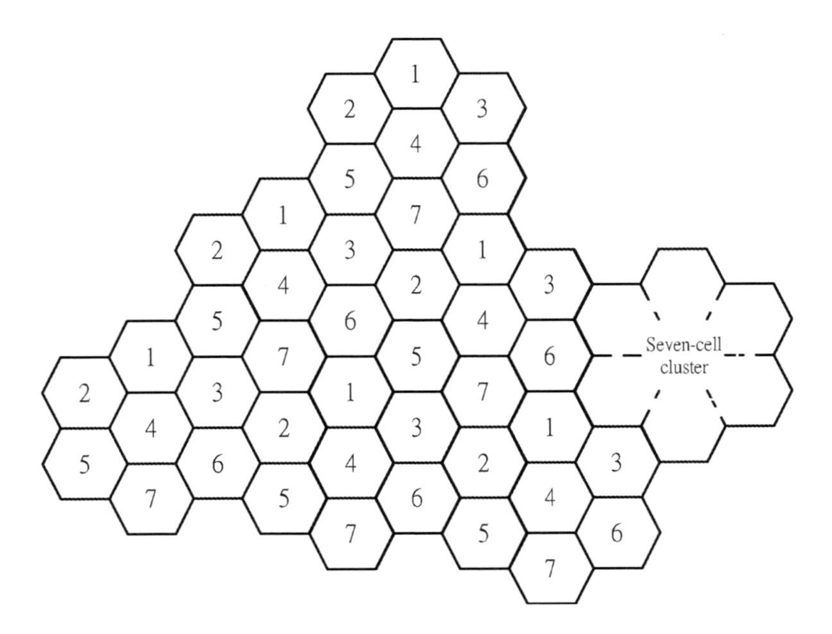

Figure 5.3 A 7-cell cluster cell planning scheme in an FDMA mobile cellular network.

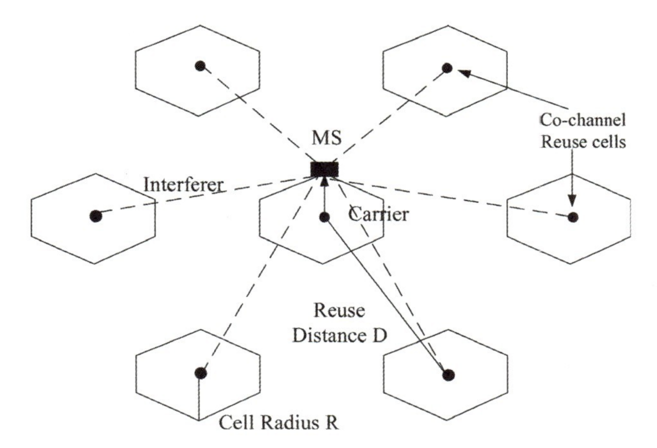

Figure 5.4 Co-channel interference caused by insufficient distance between two cells using the same frequency in a mobile cellular network.

wrongly, usually being too small, which makes two cells using the same frequency too close to each other, as shown in Figure 5.4, which illustrates co-channel interference caused by insufficient distance between two cells using the same frequency in a mobile cellular network.

In fact, the unit frequency reuse factor in a CDMA system has been considered as an important plus, compared to other multiple access technologies, such as FDMA, TDMA, and even OFDMA. However, in reality the situation may be different from what has been assumed in theoretical or simulation studies. A CDMA system may still need to use a frequency reuse factor larger than one to overcome the problems with an insufficient number of signature codes in neighboring cells, as a practical CDMA code set has always limited size and it is impossible to accommodate all users in the entire mobile cellular network. The same will apply to next generation CDMA systems. Therefore, these issues on the code design approaches should be raised to make sure that the CDMA code set can be made as large as possible to ease the frequency reuse planning process.

5.1.2 *WIRELESS LANs*

WLAN is another thriving wireless service arena dedicated mainly for data communications, although voice-over-IP (VOIP) services have also been developed for WLANs recently. As most traffic carried in initial WLANs is data and their coverage areas are relatively small and localized (compared to those of a mobile cellular network), the design concern can be very much different.

The standardization process for WLANs has gone through a very different path compared to that for mobile cellular standards, which failed to be unified into one single standard throughout the world. The standardization process for WLANs went along a much smoother path and was mainly coordinated by IEEE under its IEEE 802.11 working groups. Also for this reason we can see very uniform worldwide standards for WLANs, which are very good for fast proliferation of the technology in the world, as seen in recent years. There are several popular WLAN standards, which include IEEE 802.11b, IEEE 802.11a, IEEE 802.11g, and IEEE 802.11n.

Family of IEEE 802.11 Standards

In order to give us a whole picture on IEEE 802.11 series standards, we would like to list all IEEE standards and their task groups within the IEEE 802.11 working group as follows. Please note that some 802.11 standards may not necessarily fall into the category of WLANs. For instance, the IEEE 802.11p standard was developed mainly for vehicle-to-vehicle (V2V) communications.

- IEEE 802.11: The original 1 Mbit/s and 2 Mbit/s, 2.4 GHz RF and IR standard (1999)
- IEEE 802.11a: 54 Mbit/s, 5 GHz standard (1999, shipping products in 2001)
- IEEE 802.11b: Enhancements to 802.11 to support 5.5 and 11 Mbit/s (1999)
- IEEE 802.11c: Bridge operation procedures; included in the IEEE 802.1D standard (2001)
- IEEE 802.11d: International (country-to-country) roaming extensions (2001)
- IEEE 802.11e: Enhancements: QoS, including packet bursting (2005)
- IEEE 802.11F: Inter-Access Point Protocol (2003), and withdrawn February 2006
- IEEE 802.11g: 54 Mbit/s, 2.4 GHz standard (backward compatible to IEEE 802.11b) (2003)
- IEEE 802.11h: Spectrum Managed 802.11a (5 GHz) for European compatibility (2004)
- IEEE 802.11i: Enhanced security (2004)
- IEEE 802.11j: Extensions for Japan (2004)
- IEEE 802.11k: Radio resource measurement enhancements (proposed—2007)
- IEEE 802.11l: (reserved and will not be used)
- IEEE 802.11m: Maintenance of the standard; odds and ends (ongoing)
- IEEE 802.11n: Higher throughput improvements using MIMO antennas (pre-draft-2007)
- IEEE 802.11o: (reserved and will not be used)
- IEEE 802.11p: WAVE: Wireless Access for the Vehicular Environment (such as ambulances and passenger cars) (working–2008)
- IEEE 802.11q: (reserved and will not be used, can be confused with 802.1Q VLAN trunking)
- IEEE 802.11r: Fast roaming (working 'Task Group r'—2007)
- IEEE 802.11s: ESS Mesh Networking (working—2008)
- IEEE 802.11T: Wireless Performance Prediction (WPP)—test methods and metrics Recommendation (working—2008)
- IEEE 802.11u: Interworking with non-802 networks (for example, cellular) (proposal evaluation)
- IEEE 802.11v: Wireless network management (early proposal stages)
- IEEE 802.11w: Protected Management Frames (early proposal stages—2008)
- IEEE 802.11x: (reserved and will not be used)
- IEEE 802.11y: 3650-3700 Operation in the US (early proposal stages).

It is noted that there is no standard or task group named 'IEEE802.11x'. Rather, this term is used informally to denote any current or future 802.11 standard, in cases where further precision is not necessary.[2] 802.11F and 802.11T are stand-alone documents, rather than amendments to the 802.11 standard and are capitalized as such.

Next Generation CDMA for WLANs

We can understand that various multiple access technologies have been proposed and eventually applied to different IEEE 802.11 WLAN standards. For instance, 802.11b uses CSMA/CA with direct sequence spread spectrum as its air-link design. On the other hand, both 802 11a and 802.11g choose OFDM as their air-link architecture. MIMO technology has been used in 802.11n to improve its data transmission rate (with its multiplexing capability) or transmission range (with its spatial diversity capability).

We should also aware of that the operational environment for WLANs is very much different from that in a mobile cellular system. The former usually has a relatively small coverage area, and sometimes it only needs to cover indoor areas. However, the latter should always cover a relatively wide or large area, which can be as large as a few kilometers in radius. This is partly the reason why mobile cellular systems are also called wireless wide area networks (WWANs). In addition, a mobile cellular system needs to support the communications between mobile terminals and base station with high mobility, which is at least at the speed of cars (normally lower than 250 km/h). On the other hand, WLANs usually do not support such high mobility in their coverage area, and thus only pedestrian handheld or stationary terminals will be supported. A typical WLAN application scenario is the wireless data connection between laptop PCs and an access point in a coffee-shop such as Starbucks, etc.

Another different technical feature between a mobile cellular system and a WLAN is that the average data rate supported in WLANs is much higher than that for a mobile cellular system. This is true even if we compare currently available 3G data services with 802.11g networks, the former being at about 2 Mbps at its peak rate and the latter being at 54 Mbps. Their difference is more than ten times. Even if we use a 3.5G cellular data service (such as HSDPA[3]), the WLAN still has a speed advantage.

Obviously, due to the nature of their data payload, all WLANs operate in packet switched mode and thus the next generation CDMA technology designed for WLANs should also be able to operate fully compatible with the high-speed packet transmission scheme.

5.1.3 *WIRELESS PANs*

Another important application scenario for next generation CDMA technology is wireless personal area networks (WPANs), which has been flourishing recently due to the increasing demands for high-speed wireless data transfer between various personal wireless devices, such as PDAs, digital cameras, laptop PCs, cell phones, portable storage units, and so forth. Compared to WLANs, a WPAN normally operates in an even smaller coverage area. A WPAN can also be considered as a computer network used for communication among computer devices (including telephones and personal digital

---

[2]The IEEE 802.1X standard for port-based network access control is often mistakenly called '802.11x' when used in the context of wireless networks.

[3]High-Speed Downlink Packet Access (HSDPA) is a mobile telephony protocol, a 3.5G technology, which provides a smooth evolutionary path for UMTS-based 3G networks allowing for higher data transfer speeds. Current HSDPA deployments support 1.8 MBit/s or 3.6 MBit/s in downlink. Further steps to 7.2 MBit/s and beyond are planned for the future. As an evolution of the W-CDMA standard, HSDPA achieves the increase in the data transfer speeds by defining a new W-CDMA channel: a high-speed downlink shared channel (HS-DSCH) that operates in a different way from existing W-CDMA channels and is used for downlink communications to the mobile.

assistants (PDAs)) close to one person. The devices may or may not belong to the person in question. The reach of a WPAN is typically a few meters. WPANs can be used for communication among the personal devices themselves (intra-personal communication), or for connecting to a higher level network and the Internet (an uplink).

PANs may be wired with computer buses such as USB and FireWire. A WPAN can also be made possible with wireless networking technologies such as IrDA, Bluetooth, and ultra-wideband (UWB) technology.

The WPAN standardization process has been coordinated under IEEE 802.15 Working Groups. IEEE 802.15 is the 15th working group of the IEEE 802 which specializes in WPAN standards. It currently includes five task groups (numbered from 1 to 5):

- Task group 1 (WPAN/Bluetooth): IEEE 802.15.1 specifies a WPAN standard based on the Bluetooth v1.1 specifications.
- Task group 2 (Coexistence): IEEE 802.15.2 addresses the issue of coexistence of WPANs with other wireless devices operating in ISM frequency bands which are also shared by WLANs.
- Task group 3 (High Rate WPAN): IEEE 802.15.3 is a MAC and PHY standard for high data rate (11 to 55 Mb/s) WPANs.
  - IEEE 802.15.3 (High Rate WPAN)
  - IEEE 802.15.3a (WPAN High Rate Alternative PHY)
  - IEEE 802.15.3b (MAC Amendment)
  - IEEE 802.15.3c (WPAN Millimeter Wave Alternative PHY)
- Task group 4 (Low Rate WPAN):
  - IEEE 802.15.4 (Low Rate WPAN)
  - IEEE 802.15.4 (WPAN Low Rate Alternative PHY)
  - IEEE 802.15.4b (Revisions and Enhancements)
- Task group 5 (Mesh Networking): Standardization work for mesh networking of WPANs.

For more detail information on Bluetooth technology, please refer to Section 2.3.

Next Generation CDMA for WPANs

Most WPANs use spread spectrum techniques as their major air-link architecture. In particular, Bluetooth technology (IEEE 802.15.1; see Section 2.3) uses frequency hopping as its multiple access technology. UWB or Wireless USB, specified in the IEEE 802.15.3a standard, can use either multi-band OFDM or DS spreading for its air-link technology. The ZigBee technique has adopted DS-SS as its physical layer multiple access scheme and CSMA/CA as its media access control layer protocol, as specified in the IEEE 802.15.4 standard. All three major WPAN standards have opted for CDMA (or more specifically, the first generation CDMA technology) as their primary air-link architecture.

To work out the next generation CDMA technology suitable for WPANs, it is very important to overcome the shortcomings existing in all currently available CDMA technology, which uses unitary codes and DS spreading modulation, etc.

Another issue for the application of CDMA technology in WPANs is to avoid possible interference with other wireless applications, as all currently available major WPAN standards have been allowed to operate in ISM bands, where many other wireless services could also operate, such as WLANs, etc. Therefore, the capability for the next generation CDMA technology to reduce the average interference should be enhanced, compared to existing CDMA technology.

5.1.4 COGNITIVE RADIO

Cognitive radio is a newly emerging technology, which can be viewed as a solution to solve the severe worldwide spectral shortage problem, due to the rapid proliferation of wireless applications, which include mobile cellular, wireless LANs, wireless PANs, wireless MANs, wireless RANs, and so forth.

Radio frequency (RF) bandwidth is a natural resource that is very limited in terms of the total capacity it can support as a whole. Due to the fast development of wireless communication technologies, radio frequency spectrum has been treated as a non-replacement resource that everybody wants to access for their daily life needs. The allocation of the RF bandwidth has in general gone through three different ways in history. In the early days of wireless applications, which were dominated by broadcasting services, aviation communications, power transmission line control systems, microwave relay towers, and so forth, the radio spectra were usually allocated to certain users through government assignment or authorization, which were usually done on a need (instead of money) basis. However, with the rapid advancement of radio technologies in the past 20 years, the spectra allocation in many developed countries has been at a stage where only a very small amount of free bandwidth suitable for certain wireless applications is available and everybody wants to get the spectrum to provide some wireless communication services. Therefore, auction became the only way to allocate this scarce resource in a relatively fair way. This practice has been done for several years in some countries, such as the US and many other nations when in particular the licenses for 3G mobile cellular services were issued. The auction of the radio spectrum has brought some negative consequence due to the huge business interest involved in the value of limited RF bandwidth. This has created a lot of speculations on possible increase in the value of RF spectra with many years to come. Some business entity can spend the money to get the bandwidths and keep them there without using them, just waiting for a future increase in value before selling them. Obviously, this is not in the best interests of the end-users as a whole.

Cognitive radio technology has found its way into the fore front of wireless communication research just to solve the dilemma caused by the spectrum allocation/auction practice. The basic idea for cognitive radio is to allow the use of radio spectrum on an opportunity basis under the condition that it should not interfere with the existing incumbent or licensed users. More specifically, cognitive radio is a paradigm for wireless communication in which either the network or the wireless node itself changes particular transmission or reception parameters to execute its tasks efficiently without interfering with licensed users. This parameter alteration is based on observations of several factors from the external and internal cognitive radio environment, such as radio frequency spectrum, user behavior, and network state.

The idea of cognitive radio was first presented in an article written by Mitola and Maguire [219]. It was a novel approach for wireless communication that Mitola later described as 'the point in which wireless personal digital assistants (PDAs) and the related networks are sufficiently computationally intelligent about radio resources and related computer-to-computer communications to detect user communications needs as a function of use context, and to provide radio resources and wireless services most appropriate to those needs' [220]. It was thought of as a final point towards which a software-defined radio platform should evolve or a fully reconfigurable wireless black-box that changes its communication functions depending on network and/or user demands.

On the other hand regulatory bodies in various countries (such as the Federal Communications Commission in the United States) found that most of the radio frequency spectrum was inefficiently utilized. For example cellular network bands are overloaded in most parts of the world but amateur radio or paging frequencies are not (independent studies performed in some countries confirmed that observation [221,222], and concluded that spectrum utilization depends strongly on time and place). Moreover fixed spectrum allocation meant that rarely used frequencies, assigned to specific services, cannot be accessed by non-licensed users, even if non-licensed user transmission does not introduce any interference to the preempted service. This was the reason for allowing non-legitimate users to utilize licensed bands, assuming it would not cause any interference (thus deferring from licensed bands whenever legitimate user presence was sensed). Such a paradigm for wireless communication is also called cognitive radio.

Depending on historical reasons and the set of parameters taken into account when while making decisions on transmission or/and reception alteration, we can have two major types of cognitive radio, as follows:

- Full cognitive radio ('Mitola radio'): in which every possible parameter observed by a wireless node and/or network is taken into account when making decisions on transmission and/or reception parameter change.

- Spectrum sensing cognitive radio: this is a special case of full cognitive radio in which only radio frequency spectrum is observed.

Also depending on the parts of the spectrum available for cognitive radio, we can distinguish them into the following two types:

- Licensed band cognitive radio: when cognitive radio is capable of using bands assigned to licensed users, apart from utilization of unlicensed bands such as UNII band or ISM band. One of the licensed band cognitive radio-like systems is the IEEE 802.15 Task group 2 [223] specification.

- Unlicensed band cognitive radio: when cognitive radio can only utilize unlicensed parts of radio frequency spectrum. An example of unlicensed band cognitive radio is IEEE 802.19 [224].

Although cognitive radio was initially thought of as an extension to software-defined radio (or full cognitive radio), most of the research work currently is focusing on spectrum sensing cognitive radio, particularly on the utilization of TV bands for communication. The essential problem of spectrum sensing cognitive radio is a design based on high quality spectrum sensing devices and algorithms for exchanging spectrum sensing data between nodes. It has been shown in the literature that a simple energy detector cannot guarantee accurate detection of signal presence. This calls for more sophisticated spectrum sensing techniques and requires that information about spectrum sensing must be exchanged between nodes regularly. (The authors in [225] showed that increasing the number of cooperating sensing nodes decreases the probability of false detection.)

Some applications of spectrum sensing cognitive radio [226] are emergency networks and WLANs with higher throughput and transmission distance extensions.

To adaptively fill free radio frequency bands OFDM seems to be a perfect candidate. Indeed, in [222] Timo A. Weiss and Friedrich K. Jondral from the University of Karlsruhe proposed a spectrum pooling system in which free bands sensed by nodes were immediately filled up by OFDM subbands. It is totally justified if an OFDM receiver is used in a cognitive radio terminal to fill up the sub-channel holes in the bandwidth of interest if the terminal recognizes that these sub-channels are free. Thus, opportunity-based communications can be initiated immediately. However, our question is: as the cognitive radio terminal finds out that the incumbent licensed users are starting to use the radio spectrum, can the terminal give back all of these sub-channels immediately in time, quickly enough such that no serious interference will be made to the incumbent users. Or in other words, whether OFDM is a suitable choice in the sense that the interference could be minimized if a cognitive radio terminal stopped its transmission immediately because of the commencement of transmission from a licensed user?

This is a very interesting question to be explored further in our study on the design of the next generation CDMA technology for its possible applications in cognitive radio systems/networks. We may not be able to give a complete answer at this moment, but what we can be certain of is that, in order to minimize damage or interference due to the transmission commencement of incumbent users, spread spectrum based CDMA technology will definitely be more suitable for cognitive radio terminals due to its relatively low average transmission power emission level compared to any other multiple access air-link technologies, such as FDMA, TDMA, and OFDM (OFDMA). In particular, OFDM- or OFDMA-based technologies use amplitude modulation as their RF front-end architecture, which requires much higher average transmission power than that of a CDMA transmitter. Therefore, it can be anticipated that on average the possible interference can be reduced if a CDMA-based transceiver is used in a cognitive radio terminal.

5.1.5 COOPERATIVE COMMUNICATIONS

Cooperative communications have been an active research topic, which emerged only very recently. Study of the cooperative nature of a communication system originated mainly from wireless ad hoc and sensor networks, where all communication nodes must work in a cooperative (instead of a greedy/selfish) way to optimize the overall performance of the ad hoc or sensor network as a whole. The major motivation for the research is that most terminals in a wireless ad hoc or sensor network are power and other capability constrained, and thus they rely heavily on cooperation among many other nodes to route or deliver data information from the sources to destinations successfully.

Study of cooperative communications has created many new focal points of research from the point of view of wireless network design. First, research activities on cooperative communications have successfully brought together two different issues, network and channel. Thus, some joint optimization in terms of both network and channel issues can be obtained to maximize the capacity of a communication system or network as a whole. On the other hand, traditional researches on either the network level or the system level were not combined in the way that is being done in the research on cooperative communications. This concept is indeed a revolutionary breakthrough in the sense that it allows us to see communication problems at an even higher level than ever before and thus optimized cooperative communication system can offer a much higher efficiency in using scarce communication resources. Second, at least theoretically speaking, joint optimization taking both network and channel into account can always yield a much better solution than treating the two issues separately. We have seen many similar cases in previous communication research. For instance, in order to improve the signal detection efficiency in a CDMA system, it has been shown that a joint CDMA signal detection scheme can always outperform any individual signal detection scheme. In a joint CDMA signal detection algorithm, all received signals from different transmitters will be taken into account jointly in the signal detection process, and thus the detection outcomes will be results in which information from all signals has been utilized to make the best possible decision. The same conclusion can be obtained if it is interpreted from the point of view of information theory: a decision made using more information available is always better than that made using less information. Therefore, it is quite understandable that the application of the cooperative communication concept has great potential to improve the communication quality in many different wireless communication systems or networks.

The nature of cooperative communications is very similar to the way in which a soccer team can play in a competition. One can imagine that if every individual player in a soccer team wants to be a hero and plays in a very selfish way, never passing the ball to his or her team mates, the team will definitely lose the game, due to the fact that there is no cooperation in the way they play. On the other hand, the competition result will be very different if all players play in a way that cherishes team-spirit, and cooperate with each other closely throughout the entire game. This soccer team can beat any rivals. This simple example mimics the performance of a cooperative communication system compared to that of a traditional communication system without using cooperative algorithms.

The research on cooperative communications spans several different topics [226–234]. In other words, many cooperative issues are proposed in the context of cooperative communication research in terms of different wireless applications and their operational modes.

Cooperative Relay Transmission

One interesting issue in cooperative communications is to achieve cooperative diversity in the perspective of networking. This topic sometimes is also called cooperative relay transmission. Nowadays, most emerging mobile devices are equipped with some form of embedded wireless radio which makes them able to receive and transmit data. The cooperative communication paradigm pools distributed resources of different nodes, such that the nodes act like a collaborative system instead of greedy adversarial participants. Cooperative communication has been shown to be an effective way of creating diversity (more reliability) in wireless fading networks. In particular in the slow fading scenario, once a channel is weak due to deep fade, channel coding no longer helps the transmission. In this situation, cooperative transmission can dramatically improve the performance by creating diversity using the antennas available at the other nodes of the network. This observation has led to recent interest in the design and analysis of efficient cooperative transmission protocols. Here, the cooperative scenario is modeled by a slow fading relay channel where the transmission from the source to the destination is helped by another node (relay). We can design different cooperative protocols that the relay can pick to improve the outage performance of this network. In the low SNR regime where the transmission power is low, some reported works [229,235,236] have shown that the use of some simple cooperative protocols can significantly improve the outage capacity of such a network.

In the context of cooperative relay communications, there are several different relay channel problems. Figure 5.5 shows the simplest form of cooperative relay communication, in which the source terminal (S) wants to send information to its destination (D). The information is sent via two routes, one being directly to the destination and the other through a relay terminal (R).

Another form of cooperative relay communication is shown in Figure 5.6, in which two branches from the source to destination terminals will use their respective relay terminals (R) to relay information to the destination (D). This form of cooperative relay communication is called the dual-relay scenario.

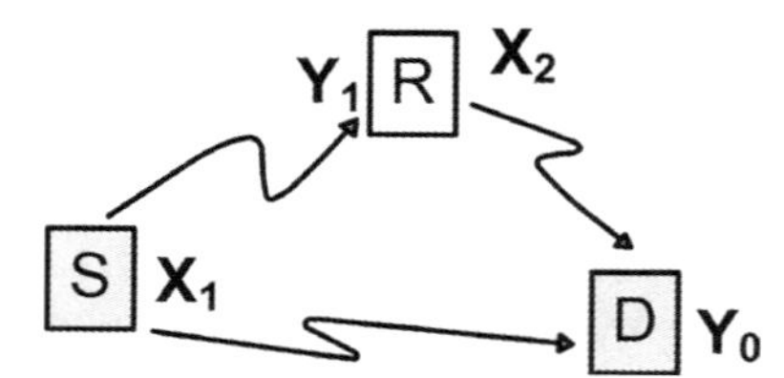

Figure 5.5 Single branch relay cooperative communication, where the source terminal (S) wants to send information to its destination (D). The information is sent in two ways, directly to the destination or via a relay terminal (R).

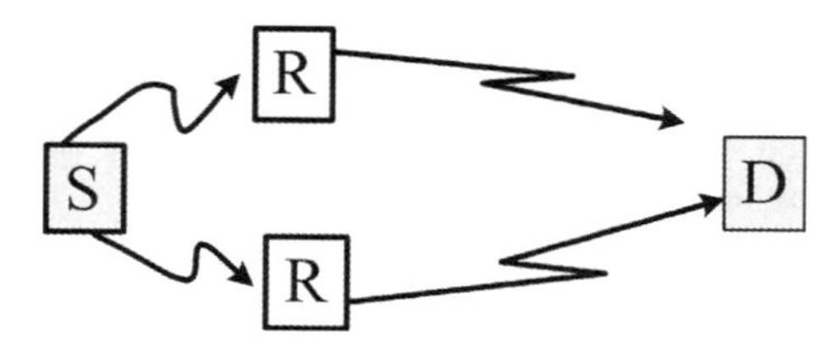

Figure 5.6 Dual-relay cooperative communication, where the source terminal (S) wants to send information to its destination (D). The information is sent via two branches, which will relay the information to the destination.

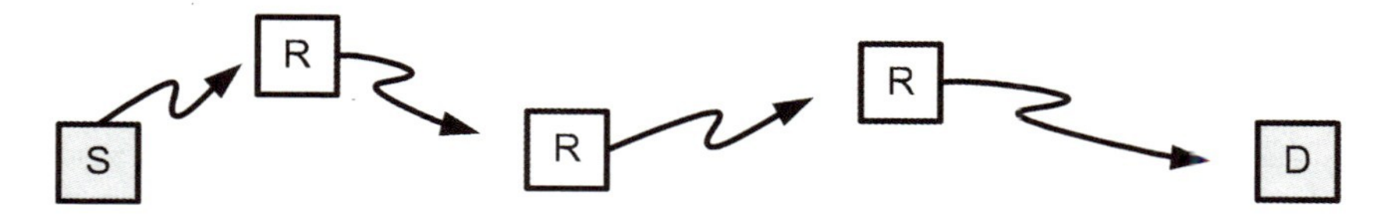

Figure 5.7 A multiple hop relay cooperative communication scheme, where the source terminal (S) wants to send information to its destination (D) via three relay terminals.

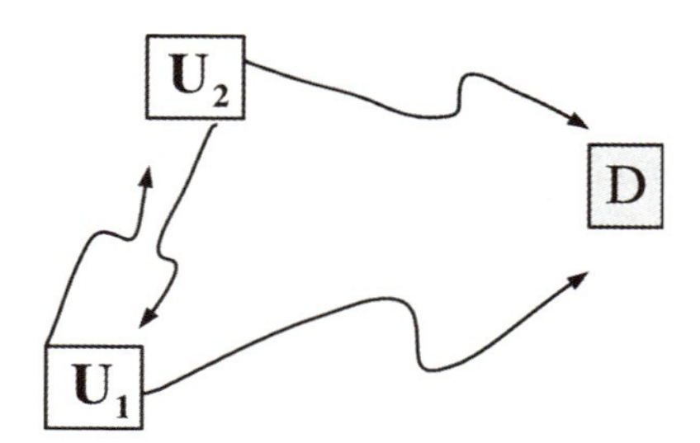

Figure 5.8 A generic cooperative communication scenario, where two terminals (U_1 and U_2) want to send their information to their destination (D).

Yet another form of cooperative relay communication is to use multiple hop relay from the source to destination, as shown in Figure 5.7, where three relay terminals are used to deliver the information from the source to the destination. Of course, in general an arbitrary number of relay terminals can be used for this configuration.

A more general type of cooperative relay communication is illustrated in Figure 5.8, in which two terminals (U_1 and U_2) are cooperating in sending their information to their destination, terminal D.

Decentralized Space-Time Coding

Another topic in cooperative communications, decentralized space-time coding, aims to achieve spatial diversity gain. In this study, similar to the space-time coding scheme used in a single MISO system, orthogonal space-time codes are proposed to achieve diversity in a distributed MISO system, in which

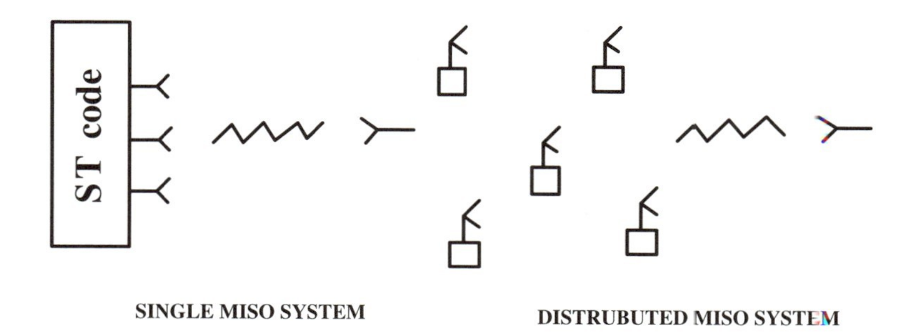

Figure 5.9 Basic configuration of a distributed MISO system in contrast to a single MISO system.

space-time codes have to be assigned to each node and each node will use a linear combination of the columns of an underlining space-time code. Figure 5.9 illustrates the basic configuration of a single MISO system in contrast to a distributed MISO system. The results of study in this area have shown that cooperative ST coding can truly decentralize the use of space-time coding in distributed networks. Also, different designs of decentralized space-time coding schemes can provide the diversity order ($\min(N; L)$), when number of nodes N is different from the number of virtual antennas L. For $N = L$, diversity order can be fractional. Finally, it has been shown that the use of a randomized ST coding scheme can achieve the performance of a centralized space-time code in terms of coding gain as the number of nodes increases.

CDMA versus Cooperative Communications

There are many relay operation schemes currently under investigation, depending on the network topology and service requirements. Here, the relay operation scheme defines the way a relay terminal may operate.

Relay cooperative communication can happen in a full duplex fashion, in which the relay terminal is able to receive and transmit at the same time using the same frequency band. To make this happen, we need to carefully design a transceiver with a sufficiently high isolation between the transmitter and receiver RF modules. In addition, we have to make sure that the strong transmitting power will not overwhelm the weak received signal, which can be 100–150 dB lower than the transmitting signal level.

On the other hand, a half duplex scheme is a relatively simple hardware configuration to implement relay cooperative terminals compared with a full duplex terminal. A half duplex relay terminal will not receive and transmit at the same time using the same frequency band. Therefore, there can be three different half duplex methods: (1) time division duplex (TDD), (2) frequency division duplex (FDD), and (3) code division duplex (CDD). Figure 5.10 shows a TDD-based cooperative relay operation, in which the relay terminal will receive the signal from the source terminal (S) in the first time slot, and will send the signal to the destination (D) in the next time slot. Also, it is easy to show the other two half duplex relay operation schemes.

There are several different relay algorithms that have been proposed in the literature. The most commonly accepted relay algorithms include two major categories, one being fixed relaying algorithms, and the other being adaptive relaying algorithms:

- Fixed relaying algorithms
  - amplify and forward
  - estimate and forward
  - decode and forward.

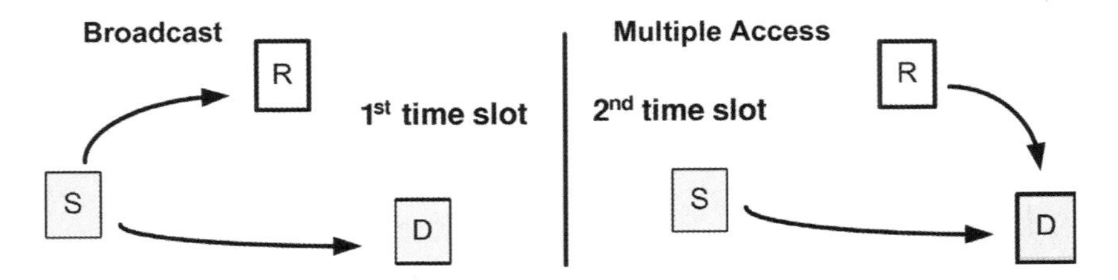

Figure 5.10 TDD-based cooperative relay operation, in which the relay terminal receives the signal from the source terminal (S) in the first time slot, and sends the signal to the destination (D) in the next time slot.

- Adaptive relaying algorithms
  - selection adaptive relay
  - incremental adaptive relay.

It is understandable that the TDD half duplex implementation of a cooperative relay system needs two time slots for receiving and transmitting signal at a relay terminal, and thus its signal throughput will be seriously constrained. Therefore, a TDD half duplex scheme is not a good choice for a high-speed wireless network. The FDD half duplex relaying scheme can receive and send information at a relaying terminal at the same time, but requires two separate frequency bands for this purpose. Therefore, it may not be the best solution for the application scenario where frequency spectrum allocation has been a problem.

On the other hand, CDD is the most versatile duplex scheme for a relaying terminal in a cooperative communication system. Under the CDD scheme, each relaying terminal should be assigned two signature codes, one being used for signal reception from the source terminal and the other for sending signal to the destination. If the signal flow from the source terminal is continuous, reception and transmission at a relaying terminal can proceed simultaneously, resulting in a very high signal throughput in the cooperative communication system. Therefore, basically, the CDD scheme for relay operation in a cooperative communication system can be used to implement either half or full duplex relaying operation, depending on the requirements of network throughput and the number of signature codes available to the system.

To implement a CDD relaying operation in a cooperative communication system, there are a few serious challenging issues which should be tackled before the scheme can be viewed as practical.

First, we should design a proper (either half or full) duplex scheme for a cooperative communication system based on code division multiplex, in which there are many relaying terminals. To make the issue clearer, let us consider a generic cooperative communication system with many terminals, which can be further categorized as the source node and destination node, together with many relaying nodes in between, as shown in Figure 5.11. Assume that all relaying terminals will perform the relaying function only and do not have their own information to send. Also assume that each relaying

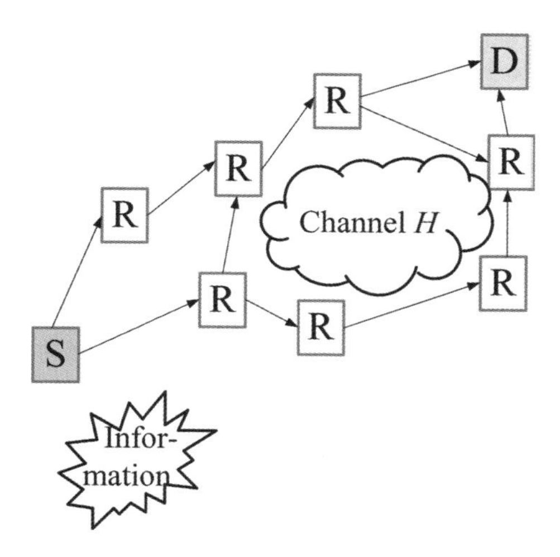

Figure 5.11 A generic relaying scenario in a cooperative communication system, where the information should be sent to the destination via many relaying terminals, each of which operates on a CDD-based full duplex relaying function.

terminal will be assigned two codes, one for signal reception and the other for sending relayed signal. If the two codes are allowed to be used simultaneously, the CDD full duplex relaying operation will result. On the other hand, if each relaying terminal is assigned only one code (which is unique among all terminals), a relaying terminal will not be able to receive and send information at the same time. Therefore, the only solution is that it has to use a TDD scheme to perform a half duplex relaying operation. Obviously, a one-code-per-node cooperative communication system bears many similar characteristics to a normal CDMA system, but it allows cooperative diversity to gain substantially from its operational advantages.

Second, for either a two-code-per-node or one-code-per-node scheme, we need to propose an efficient code protocol to maximize the overall performance of a cooperative communication system. Possible collisions should be avoided due to the fact that several relaying terminals may contend for the same node to deliver the information on its way to the destination. Excessive collisions in the relaying process will effectively reduce the reliability of the cooperative communication system.

Third, to implement either a one-code-per-node or two-code-per-node scheme in a cooperative communication system, we should choose a suitable signature code family for its application. As the number of nodes or terminals can be fairly large, we have to find a signature code family sufficiently large to accommodate all relaying nodes as well as other nodes.

There are many new issues that should be taken into account if a cooperative communication system is using code division multiplex (or multiple access) as its relaying signal coding scheme. Many problems with traditional CDMA technologies need to be overcome before they can be used for cooperative communication systems. This also gives us a great opportunity to find a completely new direction to design the next generation CDMA technology for its applications in cooperative communications.

5.2 INNOVATIVE SPREADING MODULATIONS

Now that we have discussed the issues on different application scenarios for next generation CDMA technology, we would like to go further to address the issues on what are the fundamental technological elements that can be used to implement next generation CDMA systems.

Undoubtedly, if a technology is considered as a new generation compared to its predecessor, the technology itself has to offer something special, different and superior from its previous generation. The same applies to the next generation CDMA technology, which is the focal point of this book and should be clearly defined here, although the definition may not be exhaustively complete. Therefore, based on this consideration we would like to discuss several ideas that we proposed to form the most important parts of next generation CDMA technology. Innovation in spreading modulation scheme is the main subject to be covered in this section.

As a well known fact, all currently available mainstream CDMA technology uses direct sequence spreading spectrum (DS-SS) as the foundation of its multiple access air-link architecture. Although some wireless applications, such as Bluetooth technology, have opted for other spreading techniques, such as frequency hopping, as their major air-link access scheme, most mobile cellular and wireless networking applications still use direct sequence as their spreading modulation scheme for bandwidth spreading and multiple user separation purposes. On the other hand, it should also be noted that the DS-SS technique has become one of the most important core IPRs in Qualcomm and it has brought the company many millions of dollars through IPR transfer. Therefore, the development of the next generation CDMA technology has to work out something very different from what belongs to this core IPR.

The fundamental objectives to develop a new spreading modulation scheme are given as follows. First of all, the new spreading modulation scheme should offer a much better spreading efficiency than all existing spreading techniques, such as direct sequence, frequency hopping, time hopping, etc. Better spreading efficiency means better bandwidth efficiency, which is what we want to design the next generation CDMA technology. Second, the new spreading technique developed for the next generation CDMA technology should work with relatively simple hardware to make it easy to implement in a terminal with small physical dimensions, such as a handheld portable unit. Third, the new spreading technique should be fully compatible with the requirement to operate in an all-IP wireless platform, which will work on high-speed short-packets instead of continuous-time transmissions. Fourth, the new spreading modulation scheme should also be able to meet the requirements of multimedia transmissions, in which rate-on-demand is a must and the transceiver should be agile enough to perform on-the-fly rate change at any time.

It is noted that all of the above four requirements on the new spreading modulation scheme suitable for the next generation CDMA technology are somehow related with one another. Thus, the design of the new spreading schemes should be very careful to reach a good balance where conflicts exist.

5.2.1 OS SPREADING MODULATION

As an effort to improve the spreading efficiency of a traditional CDMA system, a novel spread modulation scheme, called offset stacking (OS) spreading modulation scheme, was proposed in [237].

As mentioned previously, the spreading efficiency of the direct sequence spreading modulation scheme used in most 2G and 3G wireless communication systems is equal to the reciprocal of the processing gain of the systems. For example, if a 64-ary Walsh-Hadamard code is used to spreading the data information, as is the case in the IS-95 standard, the spreading efficiency of the system is equal to only 1/64, which is much less than one. Therefore, methods to improve the spreading efficiency will be very important to enhance the bandwidth efficiency of a CDMA system as a whole. The offset stacking spreading modulation scheme is such a method to be proposed to work with orthogonal complementary codes in a CDMA system. Imagine the improvement in bandwidth efficiency a CDMA system could achieve if the spreading efficiency of the system were as high as one. The improvement in the bandwidth efficiency would be as high as the processing gain of the CDMA system, or 64 times for a CDMA system with a processing gain of 64, which is really a significant improvement.

To introduce the way in which the OS spreading modulation works, let us consider a generic multiuser CDMA system model, as shown in Figure 5.12, in which only downlink transmission is shown for description simplicity. It is noted that in this generic CDMA system model each user is assigned a flock of M element codes as its signature code and the spreading modulator can use any type of spreading modulation scheme, such as direct sequence spreading or offset stacking spreading, without loss the generality. Each transmitter will use M different sub-carriers to send M different element codes and the whole CDMA system will share the same M sub-carriers, and no more sub-carriers will be required. Of course, the transmitter looks like a multi-carrier CDMA transceiver, which can also be implemented using an OFDM architecture for hardware implementation simplification. We have to emphasize that the multi-carrier structure used here is different from the normal multi-carrier CDMA as all sub-carriers here carry the same data information but encoded by different element codes belonging to the same code flock. Therefore, the different sub-carriers will not be able to provide any redundancy for frequency diversity purpose. As a matter of fact, diversity for such a system should be obtained through other feasible ways, such as multi-dimensional spreading or space-time coding schemes, etc.

The block diagram of the receiver tuned to the first user's transmission is shown in Figure 5.13. It is noted that the receiver also needs to use M different sub-carrier demodulators to decode all M element codes of the flock assigned to the first user.

Now, we can introduce the basic principle of offset stacked spreading modulation. The spreading modulator used by every user, as shown in Figure 5.12, is illustrated in Figure 5.14, in which the input

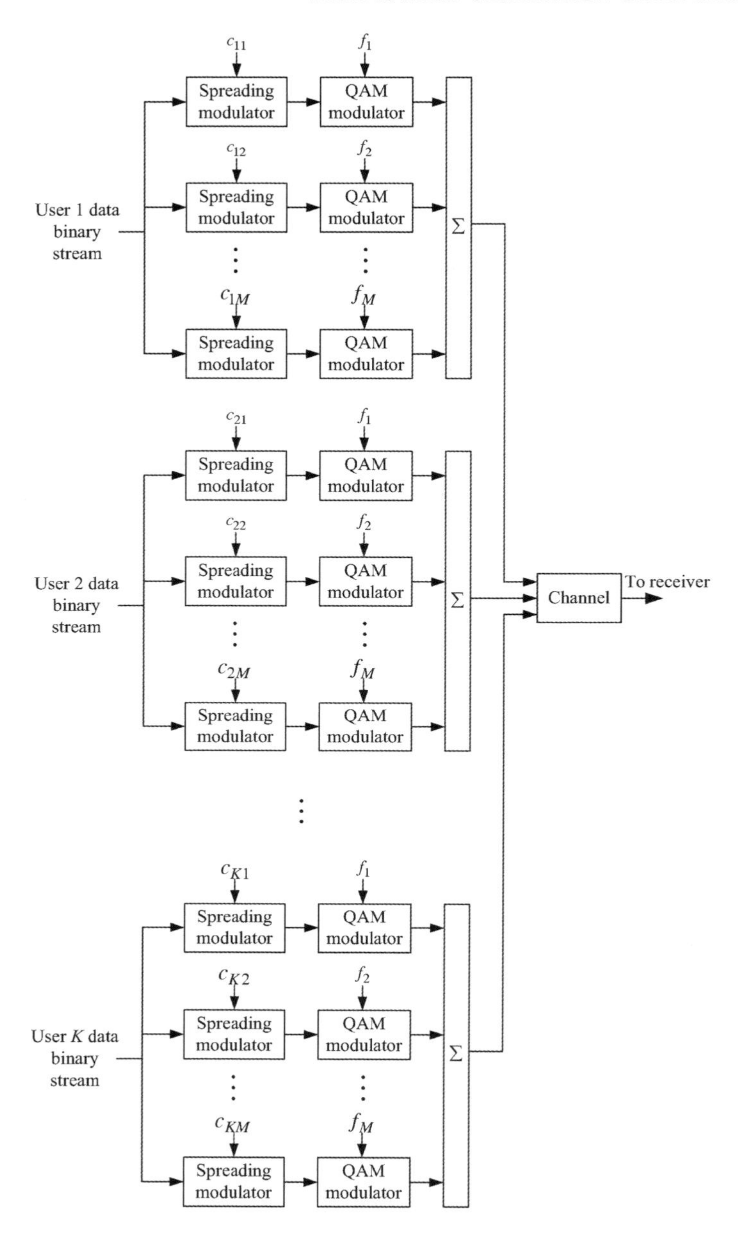

Figure 5.12 A generic multi-user CDMA system, where K users exist.

data stream is from the kth user $\mathbf{b}_k = (b_{k1}, b_{k2}, \ldots, b_{kj} \ldots,)$. The OS spreading modulator works on the element code $\mathbf{c}_{km} = (c^1_{km}, c^2_{km}, \ldots, c^N_{km})$, which is N chips long.[4] Therefore, the spreading

[4]In this book, we should pay attention to the variables used to denote the parameters of an orthogonal complementary code set. We always use (if no further explanations are given) the three variables M, N, and K to

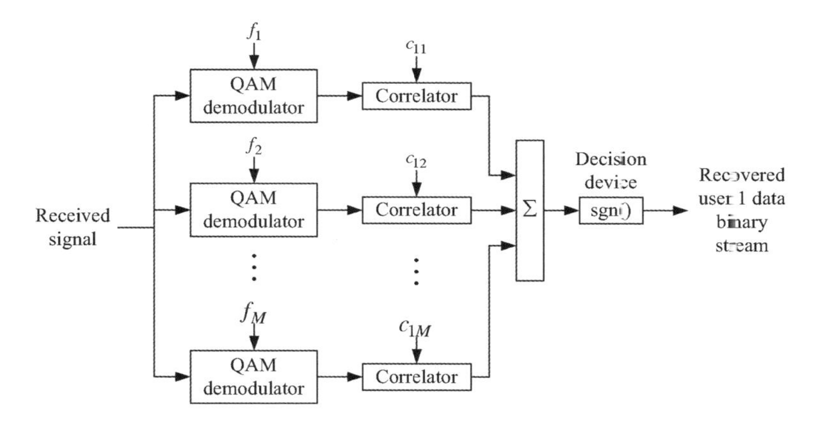

Figure 5.13 A generic multi-user CDMA system receiver, tuned to the first user transmission.

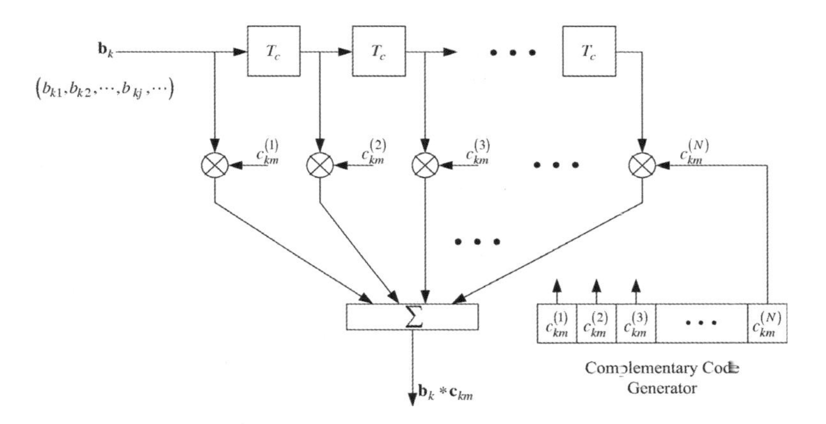

Figure 5.14 Illustration of offset stacking spreading modulator.

modulator illustrated in Figure 5.14 is to implement OS spreading modulation for the mth element code of the kth user, where $m = 1, 2, \ldots, M$ and $k = 1, 2, \ldots, K$.

The difference between the conventional direct sequence spreading and OS spreading schemes is shown in Figure 5.15. It is to be noted that the OS spreading is considered as a general spreading modulation scheme, which also includes DS spreading as a special case. This observation is significant as study of a more general scheme is always more attractive to us, in order to reveal some more important results due to the observation from a broader perspective. It is seen from Figure 5.15 that the OS spreading can use any offset chips to overlap two consecutive bits, if we let the number of offset chips be n, and thus n can take any value from 1 to N, which is the length of the element code. If indeed $n = N$, the OS spreading scheme is reduced to a normal DS spreading modulation scheme.

It should also be pointed out that the baseband signal after OS spreading modulation is not necessarily binary if the original input bit stream is binary. Therefore, the carrier modulation scheme followed should use some schemes which can handle multiple level baseband signal modulation,

denote the flock size, element code length, and set size of the orthogonal complementary code set, and m, n, and k denote their indexes, respectively.

(a) Generally CDMA spreading

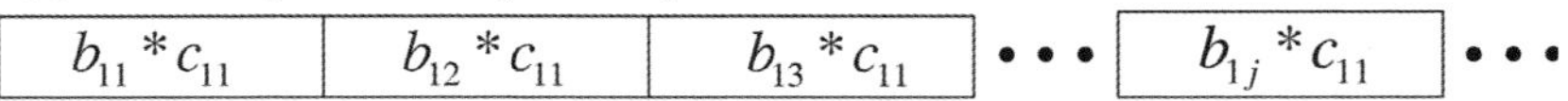

(b) Spreading of Complete Complementary Code System

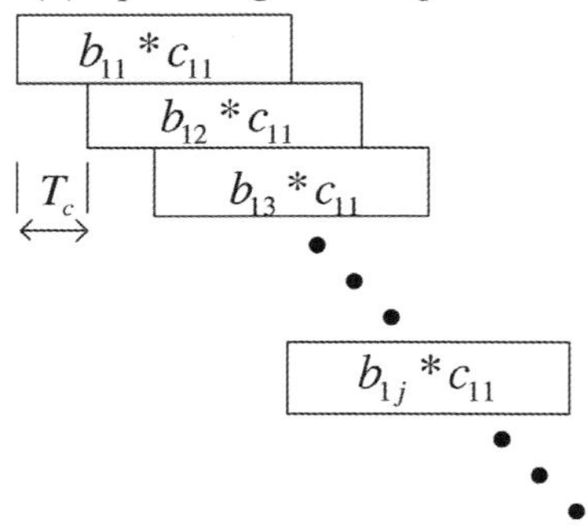

Figure 5.15 Comparison between offset stacking spreading modulation and conventional DS spreading modulation. It is seen that DS spreading is only a special case of OS spreading when $n = N$.

such as M-ary QAM, as shown in Figure 5.12. Therefore, we see that the carrier modulation scheme should be selected together with the spreading modulation scheme and it should not be chosen independently.

The use of OS spreading is significant not only in terms of its research value (due to the fact that it is a rather general form of spreading modulation scheme, which effectively includes the DS spreading modulation scheme as its special case when $n = N$), but also in terms of its many operational advantages (in addition to its greatly improved spreading efficiency), which will never be made possible if only the DS spreading modulation scheme is used. We discuss this issue in a much more detailed way in the following sections.

It should be mentioned that the OS spreading modulation scheme should not be considered for applications with most unitary codes, which are used in a CDMA system operating on a one-code-per-user basis. The OS spreading modulation scheme should only be used with orthogonal complementary codes due to their special orthogonal properties, which will be realized with all element codes jointly.

5.2.2 TWO-DIMENSIONAL SPREADING MODULATION

As discussed in the previous subsection, complementary codes can be used to formulate a CDMA system based on the OS spreading modulation technique to construct a unique OS-CDMA system, which can have many operational advantages compared to a traditional CDMA system based on direct sequence spreading. The OS-CDMA system uses different sub-carriers to transmit different element codes in a flock of complementary codes in parallel from a transmitter, and thus a multi-carrier CDMA architecture is given as a result.

The block diagram of the CDMA system using complementary codes is shown in Figures 5.12 and 5.13. Figure 5.12 shows a CDMA system based on complementary codes with M users, each of which uses a multi-carrier CDMA transmitter to send its information into the channel. Figure 5.12 in fact illustrates the scenario of asynchronous uplink channel transmissions, in which each user will send its information from a mobile terminal to the base station, and thus different delays exist for different mobile transmission channels. All M uplink transmission signals will be combined in the channel, added by noise to form a received signal at a receiver, whose diagram is shown in Figure 5.13. The receiver is tuned to the transmission from the first user.

In the CDMA system based on complementary codes illustrated in Figures 5.12 and 5.13, there are M users. The kth (where $k = 1, 2, \ldots, M$) user is assigned a flock of M element codes, or $(\mathbf{c}_{k1}, \mathbf{c}_{k2}, \ldots, \mathbf{c}_{kM})$, and each element code (say $\mathbf{c}_{km}$, where $m = 1, 2, \ldots, M$) is carrier modulated with sub-carrier f_m, where $m = 1, 2, \ldots, M$. Therefore, viewed at each user's transmitter, the same information bit stream will be spread modulated by M element codes and sent in parallel via M different sub-carriers. Thus, in effect, the user data information has been spread in two dimensions, or in both the time and frequency domains. The time domain spreading is carried out by each individual element code, while the frequency spreading is fulfilled across different sub-carriers in different carrier frequencies.

Viewed at the receiver, the two-dimensional spread signal will be carrier demodulated by different sub-carrier frequencies and then is fed into individual correlators to perform matched filtering. Only wanted information (for instance, the transmission from the first transmitter as shown in Figure 5.13) will be reproduced at the output of the summation unit of the receiver and all unwanted signals will contribute as multiple access interference (MAI). Therefore, the receiver structure shown in Figure 5.13 is in fact a two-dimensional despreading and signal detection unit.

In such a two-dimensional spreading CDMA system based on complementary codes, the orthogonality of the spreading codes is based on both the time domain and the frequency domain simultaneously. The time domain matched filtering happens in each individual element code correlator in Figure 5.13, while the frequency domain matched filtering is carried out at the summation unit of the receiver shown in Figure 5.13. In particular, two-dimensional spreading can yield much better orthogonality than any possible one-dimensional spreading technique (which have been used in all traditional CDMA systems based on unitary codes). The reason is that two-dimensional spreading can offer much more degrees-of-freedom to achieve orthogonality of the spreading codes in both the time and frequency domains.

Figures 5.16 and 5.17 show the orthogonality reconstruction process which happens to a complementary code set. They actually give both the ACF and CCF of a particular complementary code with its PG equal to $16 \times 4 = 64$, as shown in Figures 5.16 and 5.17, respectively, from which it can be clearly seen that, although the ACF of each individual element code is not ideal (there are many non-zero side lobes, as shown in Figure 5.16), the sum of them yields $R_A(\tau)$, an ideal ACF, which is just what we want. The same observation can be made with regard to the CCF of some particular complementary codes or $R_{A,B}(\tau)$, as shown in Figure 5.17. Similar characteristic features can also be found to any other complementary codes, as given in Appendix E. It has to be noted that this desirable feature of complementary codes has not been found in any other conventional or unitary CDMA code (with one-dimensional spreading), including all so-called orthogonal codes, such as Walsh-Hadamard sequences, OVSF codes, etc.

Therefore, we have seen the way to construct a CDMA system using the two-dimensional spreading technique, which can be made possible only with the help of complementary codes. If an orthogonal code set is able to construct perfect ACF and CCF under the two-dimensional spreading scheme, as shown in Figures 5.16 and 5.17, we say that this complementary code set is an orthogonal complementary code set, which can be used in a CDMA system to offer greatly improved performance. In the later part of this book, we will also show that a two-dimensional spreading technique can be used to construct an interference-free CDMA architecture with the help of orthogonal complementary codes.

The application of two-dimensional spreading technique also allows us to unveil many new facets of the orthogonality reconstruction mechanisms that exist in an orthogonal complementary code set. In this way, some new approaches can be found to design some orthogonal complementary code sets for some particular applications.

As the orthogonality of spreading codes can be realized based on two dimensions, in both the time domain (through the chip-wise correlation process in a correlator) and the frequency domain

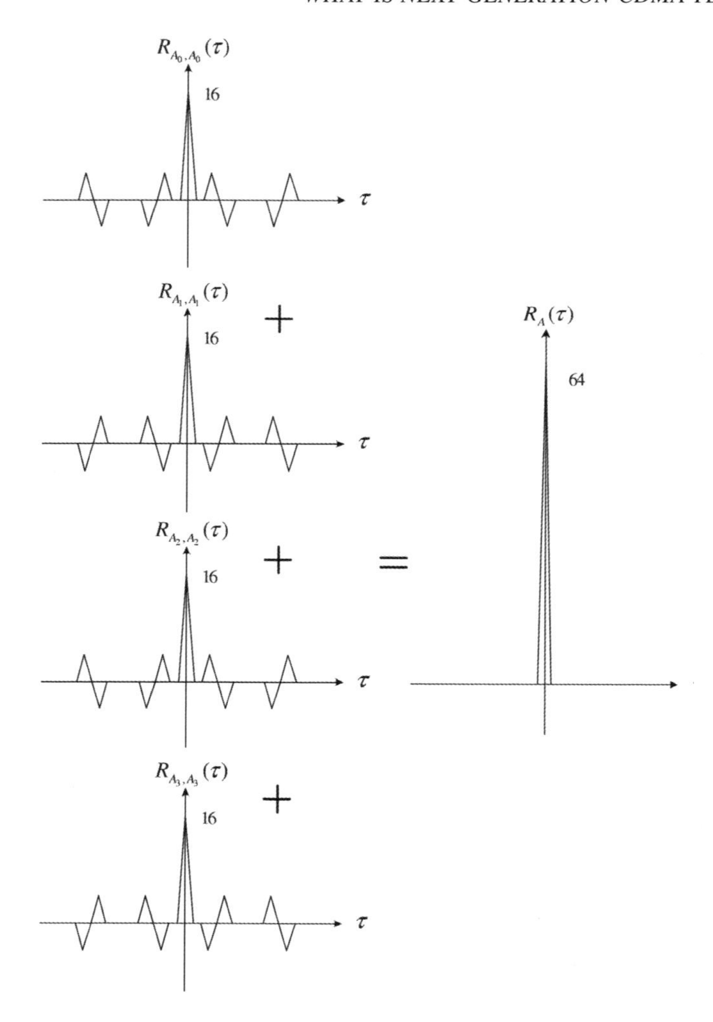

Figure 5.16 Auto-correlation function of an orthogonal complementary code with its element code length being 16 and PG $16 \times 4 = 64$, where $R_{A_0,A_0}(\tau)$ is the ACF of element code A_0, $R_{A_1,A_1}(\tau)$ is the ACF of element code A_1, $R_{A_2,A_2}(\tau)$ is the ACF of element code A_2, $R_{A_3,A_3}(\tau)$ is the ACF of element code A_3, and $R_A(\tau)$ is the sum of $R_{A_0,A_0}(\tau)$, $R_{A_1,A_1}(\tau)$, $R_{A_2,A_2}(\tau)$, and $R_{A_3,A_3}(\tau)$.

(through the summation operation in the receiver, as shown in Figure 5.13), we can have several different ways to achieve orthogonality among the orthogonal complementary codes in a set. The following conclusions have been obtained from our study on two-dimensional spreading techniques.

- Orthogonality among complementary codes in a set can never be achieved solely in the time domain. First of all, we would like to ask whether ideal orthogonality among all complementary codes in a set could be achieved only in the time domain. In fact, it is not difficult to understand that to achieve ideal orthogonality among all complementary codes in the time domain simply means that all ACF side lobes in $R_{A_0,A_0}(\tau)$, $R_{A_1,A_1}(\tau)$, $R_{A_2,A_2}(\tau)$, and $R_{A_3,A_3}(\tau)$, shown in

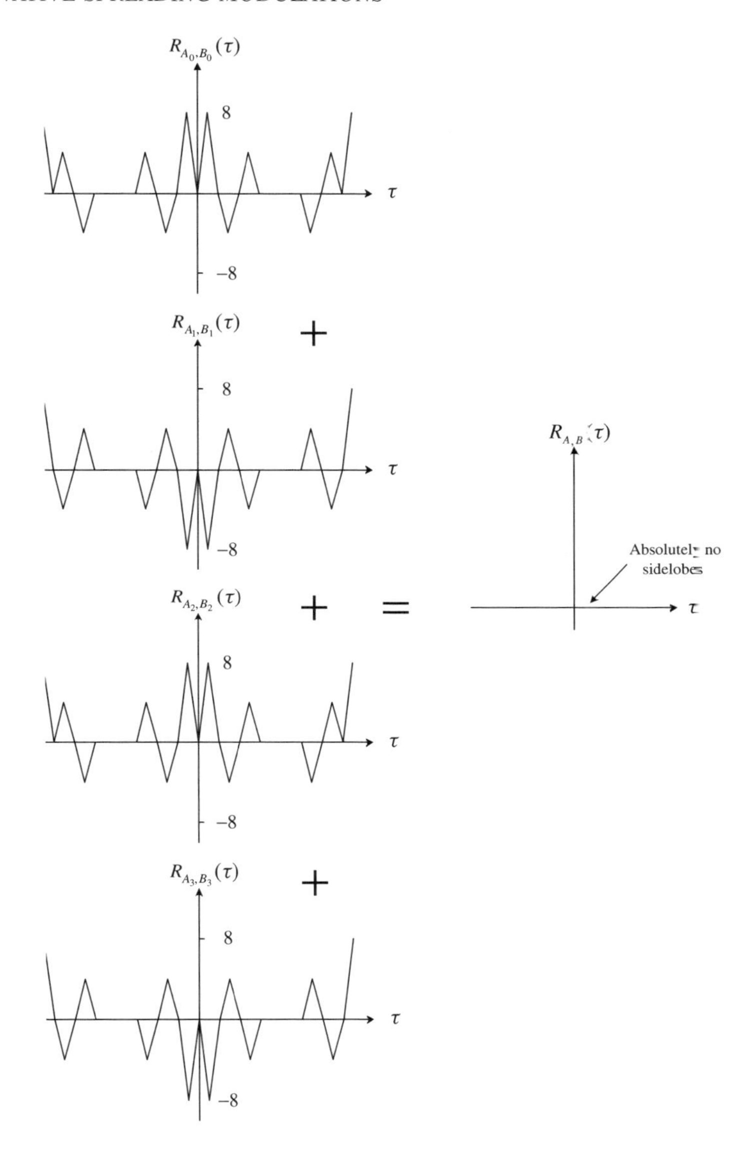

Figure 5.17 Cross-correlation function of an orthogonal complementary code with its element code length being 16 and PG $16 \times 4 = 64$, where $R_{A_0,B_0}(\tau)$ is the CCF of element codes A_0 and B_0, $R_{A_1,B_1}(\tau)$ is the CCF of element codes A_1 and B_1, $R_{A_2,B_2}(\tau)$ is the CCF of element codes A_2 and B_2, $R_{A_3,B_3}(\tau)$ is the CCF of element codes A_3 and B_3, and $R_{A,B}(\tau)$ is the sum of $R_{A_0,B_0}(\tau)$, $R_{A_1,B_1}(\tau)$, $R_{A_2,B_2}(\tau)$, and $R_{A_3,B_3}(\tau)$.

Figure 5.16, and all non-zero CCF levels in $R_{A_0,B_0}(\tau)$, $R_{A_1,B_1}(\tau)$, $R_{A_2,B_2}(\tau)$, and $R_{A_3,B_3}(\tau)$, shown in Figure 5.17, should disappear. If this indeed happens, then we could have found already a unitary code set with an ideal ACF and an ideal CCF at the same time. Unfortunately, the community has spent much time but has failed to find such a unitary code set for CDMA purposes. Therefore, the conclusion is that it is not possible to achieve ideal orthogonality among all complementary codes in a set solely in the time domain.

- Orthogonality among complementary codes in a set can possibly be achieved solely in the frequency domain. To achieve orthogonality among complementary codes only in the frequency domain implies that ACF side lobes in $R_{A_0,A_0}(\tau)$, $R_{A_1,A_1}(\tau)$, $R_{A_2,A_2}(\tau)$, and $R_{A_3,A_3}(\tau)$, as shown in Figure 5.16, and CCF levels in $R_{A_0,B_0}(\tau)$, $R_{A_1,B_1}(\tau)$, $R_{A_2,B_2}(\tau)$, and $R_{A_3,B_3}(\tau)$, as shown in Figure 5.17, have non-zero values, but they all will be canceled at the summation unit of the receiver, as shown in Figure 5.13. In this case, the orthogonality among the complementary codes is established solely in the frequency domain and has nothing to do with the time domain.

- Orthogonality among complementary codes in a set can possibly be achieved in both the time and the frequency domains jointly. Finally, it is also reasonable to deduce that orthogonality among the complementary codes in a set can be achieved partially from the time domain and partially from the frequency domain. In fact, this case can happen very often and has been illustrated in Figures 5.16 and 5.17, as we can see that at certain chip offset positions, we have $R_{A_0,A_0}(\tau) = 0$, $R_{A_1,A_1}(\tau) = 0$, $R_{A_2,A_2}(\tau) = 0$, and $R_{A_3,A_3}(\tau) = 0$, as shown in Figure 5.16, and $R_{A_0,B_0}(\tau) = 0$, $R_{A_1,B_1}(\tau) = 0$, $R_{A_2,B_2}(\tau) = 0$, and $R_{A_3,B_3}(\tau) = 0$, as shown in Figure 5.17. However, at some other chip offset positions, the ACF side lobes in $R_{A_0,A_0}(\tau)$, $R_{A_1,A_1}(\tau)$, $R_{A_2,A_2}(\tau)$, and $R_{A_3,A_3}(\tau)$, as shown in Figure 5.16, and CCF levels in $R_{A_0,B_0}(\tau)$, $R_{A_1,B_1}(\tau)$, $R_{A_2,B_2}(\tau)$, and $R_{A_3,B_3}(\tau)$, as shown in Figure 5.17, yield non-zero values, but they all will be canceled at the summation unit of the receiver, as shown in Figure 5.13. Therefore, the orthogonality among complementary codes in a set is achieved through the joint effect in both the time and frequency domains.

Study of the above properties for two-dimensional spreading techniques is significant as we can make wise use of them for a particular wireless communication application. For instance, if we would like to design a CDMA system in particular for its application in a scenario where the Doppler effect is strong, then we could choose orthogonal complementary codes whose orthogonality is based solely on the frequency domain. The reason is that a Doppler channel is a frequency-dispersive channel, and it is also a time-variant channel. High Doppler spread means fast changing speed in channel state information (CSI). Therefore, the channel will exhibit different gains at different times. If the channel changes at a cycle which is comparable to the chip width of a CDMA system, the variable gains in different chips (caused by Doppler spread) will destroy the orthogonality among CDMA codes if their orthogonality based on the time domain only. Therefore, the solution is simple: to use an orthogonal complementary code set with its orthogonality is based only on the frequency domain. In this way, the time varying property of the Doppler channel will not affect the orthogonality among the complementary codes, ensuring good detection efficiency.

Many wireless applications with a strong Doppler effect can be found, such as mobile communications in bullet trains, where a speed of about 300 km/h can effectively block all currently available mobile cellular services. Therefore, some special anti-Doppler wireless technologies should be considered. Also, research on vehicle-to-vehicle (V2V) communications has become a hot topic recently. CDMA technology based on orthogonal complementary codes with their orthogonality built solely on the frequency domain can be an attractive solution.

Dynamic Two-Dimensional Spreading

The basic idea behind two-dimensional spreading is the use of different signature code matrices to identify different users. Therefore, the signature code should be spread in both the time and frequency domains at the same time.

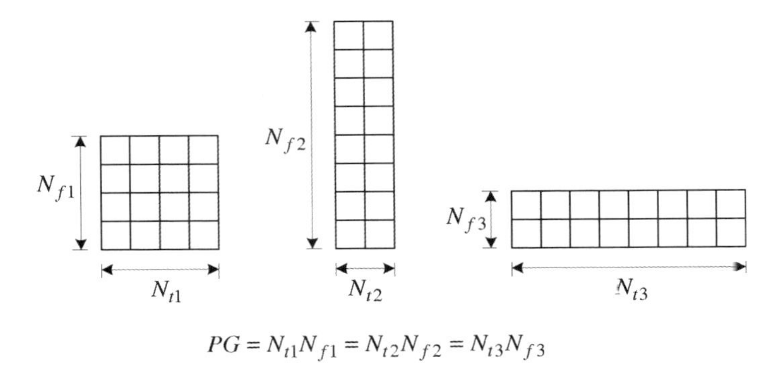

Figure 5.18 Two-dimensional spreading with the same processing gain but different time and frequency spreading dimensions.

Assume that the time and frequency domains' spreading dimensions are N_t and M_f, respectively. The spreading code for each user has a processing gain of $N_t \times N_f$. With the same processing gain or the product of $N_t \times N_f$, we can change the values for N_t and M_f to form different spreading dimensions in both time and frequency domains, as shown in Figure 5.18, where three different time and frequency domains' spreading dimensions are shown.

The capability to allow dynamic change in time and frequency domain spreading factors can help to exploit different characteristic features in the time and frequency domain spreading. For instance, if we use a relatively large time domain spreading factor (or N_t) and relatively small frequency domain spreading factor (or N_f), the orthogonality of the spreading codes will relay mainly on the time domain orthogonality of the spreading codes. Therefore, in this case the orthogonality of the codes is relatively sensitive to the time-variant properties of the channel but less susceptible to the frequency-selective fading effect in the channel. On the other hand, if we choose to use two-dimensional spreading codes with a relatively small time domain spreading factor but relatively large frequency domain spreading factor, the orthogonality of the two-dimensional spreading codes will be more sensitive to the frequency-selectivity of the channel than to the time-selective fading caused by the Doppler effect. Therefore, dynamically choosing the right N_t and N_f will give us another degree-of-freedom in designing a wireless communication system for many future applications.

The dynamic changes in the time and frequency domain spreading dimensions allow us to implement some new air-link (in particular for multiple access and CDMA code assignment) schemes for a wireless communication system. In the following text, we will give some more detailed discussions on the issues.

2D-Spreading CDMA-OFDMA Scheme

The concept of the two-dimensional spreading technique based on OFDM can be extended to a more general and more appealing application scenario, in which CDMA can be combined with OFDMA to form a new scheme for multiple access, called the 2D-spreading CDMA-OFDMA scheme.

The basic concept of the 2D-spreading CDMA-OFDMA scheme is illustrated in Figures 5.19 and 5.20, where the same processing gain of 16 is applied to the scheme but with different time and frequency domain spreading factors. It is seen from Figure 5.19 that there are in total $2K$ sub-carriers that are used to accommodate K users. Each user is assigned an $N_{t1} \times N_{f1}$ code matrix for CDMA purposes. Therefore, the users' signals are orthogonal in both the frequency space and code space, providing very good isolation between the users' signals for multiple access interference suppression.

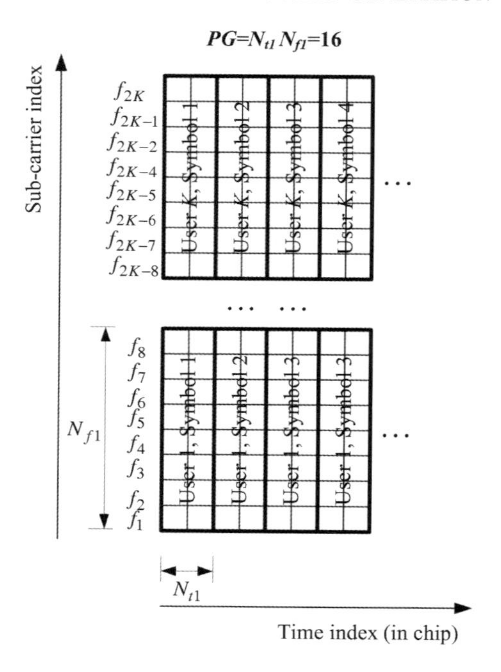

Figure 5.19 2D-spreading CDMA-OFDMA scheme, where the time and frequency spreading factors are $N_{t1} = 2$ and $N_{f1} = 8$ and the processing gain is $N_{t1} \times N_{f1} = 16$.

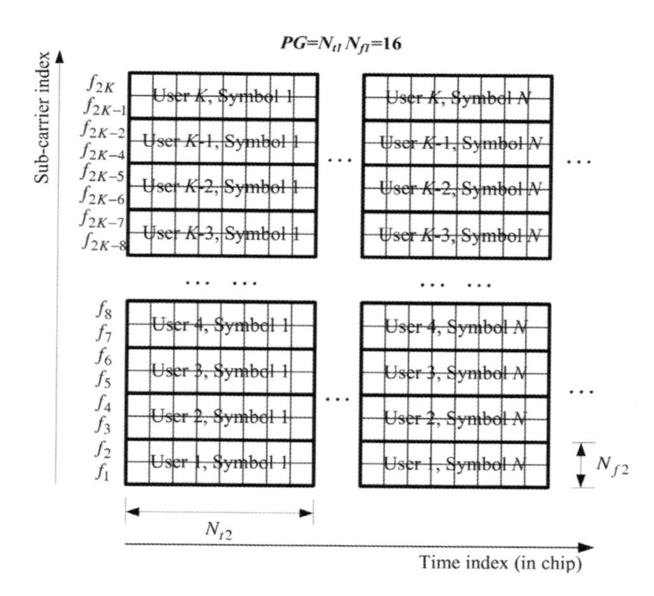

Figure 5.20 2D-spreading CDMA-OFDMA scheme, where the time and frequency spreading factors are $N_{t2} = 8$ and $N_{f2} = 2$ and the processing gain is $N_{t2} \times N_{f2} = 16$.

Also, the same can be observed from Figure 5.20, where the only difference from Figure 5.19 is that it uses a relatively large time domain spreading factor compared to that used in Figure 5.19.

The choice of the 2D-spreading CDMA-OFDMA schemes, as shown in Figures 5.19 and 5.20, is determined by the characteristics of the channel on which a wireless system is operating. If the channel has relatively high Doppler spread and thus the time-varying effect is a problem, we should use the scheme shown in Figure 5.19 due to its relatively small time domain spreading factor. On the other hand, if the channel of concern has a strong multipath propagation effect, then we should choose the scheme shown in Figure 5.20, in which a relatively low frequency domain spreading factor is used, as an effort to reduce the impact of frequency-selective fading on the orthogonality of the two-dimensional (2D) spreading codes.

The bandwidth efficiency of the 2D-spreading CDMA-OFDMA scheme can be further improved if we use both sub-carrier sets and two-dimensional spreading codes to separate different users. For example, let us again look at Figure 5.19, where we have used $N_{f1} = 8$ and $N_{t1} = 2$. To implement this two-dimensional spreading CDMA system, it can be shown that we can find an orthogonal complementary code set with its flock size[5] being equal to $N_{f1} = 8$ and its element code length being $N_{t1} = 2$. It will also be shown in the analysis given in Chapter 6 that an orthogonal complementary code set with its flock size being N_{f1} can always have in total N_{f1} different flocks, to support in total N_{f1} users in a CDMA system. Therefore, we can stack $N_{f1} = 8$ codes on each $N_{f1} \times N_{t1}$ code matrix to allow $N_{f1} = 8$ simultaneous transmissions over the same sub-carrier set (with eight sub-carriers in the set). Therefore, Figure 5.21 shows the way such a 2D-spreading CDMA-OFDMA system is

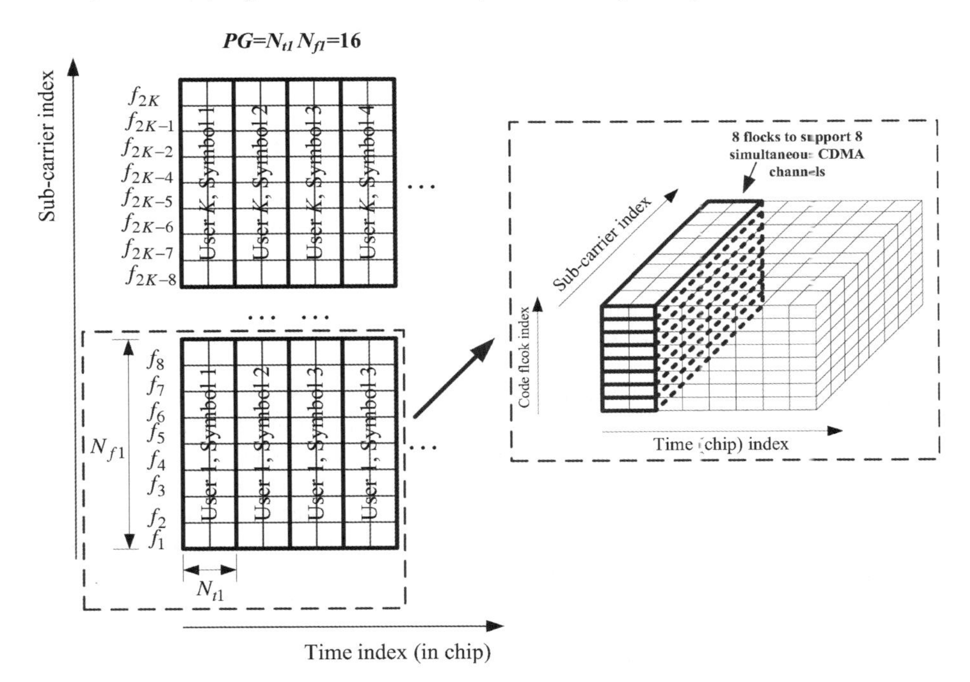

Figure 5.21 High-bandwidth efficient 2D-spreading CDMA-OFDMA scheme, where the time and frequency spreading factors are $N_{t2} = 2$ and $N_{f2} = 8$ and the processing gain is $N_{t2} \times N_{f2} = 16$. There are in total $N_{f2} = 8$ stacked 2D spreading code matrices for simultaneous CDMA transmissions.

[5]The flock size of an orthogonal complementary code set is defined as the number of element codes assigned to one user for CDMA purposes.

formulated and in total $N_{f2} = 8$ 2D-spreading code matrices are stacked for simultaneous CDMA transmissions. This scheme has a very high bandwidth efficiency as well as very agile implementation in terms of its capability to choose different time and frequency domain spreading factors, and thus the number of stacked 2D spreading code matrices.

There are several salient features of the proposed 2D-spreading CDMA-OFDMA system compared to an ordinary OFDMA system. First, the 2D-spreading CDMA-OFDMA system suggested here offers a processing gain, which is achieved through the time and frequency two-dimensional spreading, while the OFDMA system usually does not provide any processing gain to the users. Therefore, the performance of the proposed 2D-spreading CDMA-OFDMA system can be made much more robust than that of an ordinary OFDMA system. Second, the proposed 2D-spreading CDMA-OFDMA system allows stacking of multiple user signals over the two-dimensional time-frequency indexes to implement multiple access, while the OFDMA scheme only separates users in the time-frequency indexes and no code division multiple access is integrated in the scheme. Therefore, the proposed 2D-spreading CDMA-OFDMA system can make joint use of three different dimensions, the time, frequency, and code spaces, to achieve a much more flexible resource allocation as a whole in the system.

For this reason, as an important part of the next generation CDMA technology, the proposed 2D-spreading CDMA-OFDMA system can be found extremely useful in future wireless applications.

It is to be noted that the performance of the proposed 2D-spreading CDMA-OFDMA system is still under investigation at present. We are currently in particular interested to know how much benefit can be obtained compared with a traditional OFDMA scheme which does not provide any processing gain. We would like to obtain bandwidth efficiency analysis as well as bit error rate study over a multipath propagation channel.

5.2.3 SPACE-TIME-FREQUENCY SPREADING MODULATION

Space-time coding has been proposed to provide spatial diversity gain or spatial multiplex capability in a multiple-input-multiple-output (MIMO) system [238–244]. Spatial diversity gain can be exploited to improve the detection efficiency of a receiver and the spatial multiplex capability can be an effective means to enhance the overall data throughput in the channel. Therefore, MIMO technology based on space-time coding has become an important performance enhancing technology for all B3G wireless communications, which can be mobile cellular system, wireless local area networks, and any other forms of wireless applications.

Many space-time coding schemes have been proposed in the literature, which include space-time block coding (STBC), space-time trellis coding (STTC), space-time differential coding (STDC), and so on. In general, it is noted that space-time coding provides a way to separate the transmitted signals through different antennas. Therefore, in this sense any approach, as long as it can provide an effective way to separate signals sent from different antennas, can be used in a space-time coding scheme for MIMO system implementation. From this concept, we can expand our imagination based on the discussions made in the previous subsection on two-dimensional spreading techniques.

It is known that in a CDMA system based on the two-dimensional spreading scheme, as mentioned in the preceding subsection, each user is assigned a flock of element codes as its signature code for CDMA purposes. In other words, we can also say that, in a CDMA system based on two-dimensional spreading techniques, each user will be assigned a code matrix to fulfill two functions: one being data information spreading in both time time and frequency domains, and the other user separation. Therefore, it is natural to propose that, in a CDMA system based on a three-dimensional spreading modulation scheme, each user will be assigned a three-dimensional code with its three dimensions expanding the space, time, and frequency domains. Assume that each user will use a code volume (which extends the three dimensions: space, time, and frequency), with its three parameters being N_s, N_t, and N_f, to denote the number of antennas, the length of element code in the time domain, and

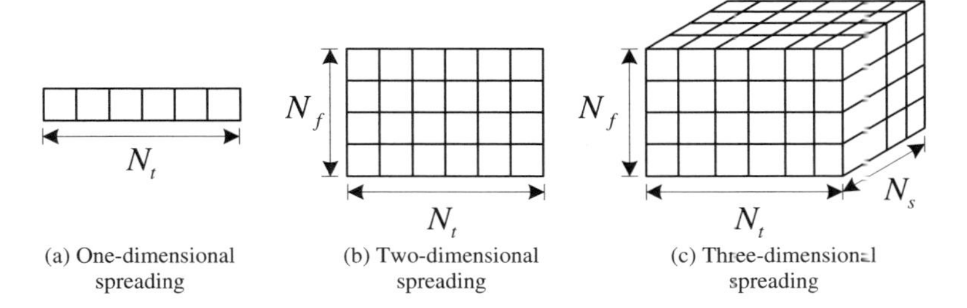

Figure 5.22 Signature codes used in one-, two-, and three-dimensional spreading techniques, where N_t, N_f, and N_s denote the element code length in the time domain, the number of sub-carriers, and the number of antennas.

the number of sub-carriers (to be used to spread the N_f element codes in the flock), respectively. Therefore, for this CDMA system based on the three-dimensional spreading technique the processing gain will become $N_s \times N_t \times N_f$.

Figure 5.22 shows a conceptual plot to illustrate the idea of three-dimensional spreading for a new generation CDMA system and comparison with the one- and two-dimensional spreading techniques.

The signature codes for the three-dimensional spreading technique can be generated based on the signature codes obtained for two-dimensional spreading. A straightforward way is to extend the two-dimensional code set, such as an orthogonal complementary code set (as discussed in the previous subsection), into a three-dimensional code volume with the help of a random coding approach, in which N_s code matrices can be generated based on each orthogonal complementary code flock with N_f element codes and N_t chips in each element code. It can be shown that, as the value of $N_t \times N_f$ approaches infinity, the cross-correlation functions among all N_s code matrices (each with dimension $N_t \times N_f$) approaches zero.

A CDMA system using the space-time-frequency spreading modulation scheme can be easily constructed based on the two-dimensional spreading architecture shown in the previous subsection, with the only difference being that each transmitter now will use multiple antennas to send N_f carrier modulated signals.

5.3 ISOTROPIC MAI-FREE AND MI-FREE OPERATION

It is a well-known fact that, in a multi-user CDMA system, the uplink and downlink channel transmissions are very different in terms of their transmission timing relationship. The downlink channel transmissions take place in the transmission direction from a base station to mobiles and they are always synchronous, as all signals are sent from the base station transmitter at the same time and will arrive at a receiver also at the same time. Thus, all signals are aligned bit-to-bit in time. This can greatly simplify and facilitate the correlation reconstruction process at the receiver.

On the other hand, the transmissions in the uplink channels happen in the direction from mobiles to a base station, and thus they always are asynchronous. This is because different mobiles can be located at any place in a cell and their transmissions will arrive at the same base station receiver at different times. Therefore, the signal detection with respect to the asynchronous transmissions is always more challenging compared to that with respect to the synchronous transmissions.

Therefore, it is common that different detection efficiency may exist for synchronous and asynchronous transmissions in all 2G and 3G wireless systems based on traditional CDMA technology, such as IS-95, W-CDMA, cdma2000, etc. The asymmetry in signal detection efficiency for synchronous and asynchronous channels also has a great impact on system design in all 2G and 3G systems. For example, the IS-95 system uses 64-ary Walsh-Hadamard codes for downlink channel spreading and channelization due to the good orthogonality of the codes in synchronous channels. However, the same 64-ary Walsh-Hadamard codes are used only for 64-ary modulation in the up link channels of IS-95 systems due to the concern that the Walsh-Hadamard codes will not be orthogonal at all in asynchronous transmission channels.

Another asymmetry may exist in a CDMA system to mitigate the inter-symbol interference (ISI) caused by the multipath propagation effect, very much depending on the spreading codes used by the CDMA system. Let us again take a look at the IS-95 standard, in which 64-ary Walsh-Hadamard codes are used to channelize its downlink transmissions. Due to very good auto-correlation and cross-correlation functions of the Walsh-Hadamard codes in the synchronous channels, the ISI problem caused by the multipath propagation in downlink signal detection is much less severe compared to the uplink transmission channels of the same system, which may use some other quasi-orthogonal codes to separate the transmissions from different mobiles. The same observation can be made on any other mobile cellular system based on CDMA, and different detection efficiency happens due mainly to the problems stemming from the asymmetrical correlation properties of the spreading codes used.

Obviously, the detection efficiency of the CDMA air-interface technologies designed for current 3G systems should be greatly improved to cater for future wireless systems. In particular, we should target the bottlenecks of the current systems, the most obvious of which is the uplink sector of today's CDMA-based 2G and 3G systems (i.e. IS-95, W-CDMA, cdma2000, etc.). The uplink channels are usually asynchronous, being a vulnerable sector of the entire air-interface because orthogonality amid the user signature codes is utterly destroyed, resulting in a much more severe multiple access interference (MAI) than that in the downlink sector. The asynchronous transmission in the uplink is also responsible for a much lower data rate than that in the downlink channels, thus limiting the overall throughput of the system. As mentioned earlier, in a mobile cellular system it is inevitable for the uplink to operate in an asynchronous mode due to the mobility of mobile terminals in a cell and the varying distances between different mobiles and their base stations. The asynchronous nature of the uplink channels makes it impossible to use orthogonal codes, such as Walsh-Hadamard codes and OVSF codes, etc., for uplink channelization. Therefore, it is common practice in existing 2G and 3G systems to use quasi-orthogonal codes, such as Gold codes or Kasami codes, in uplink channels, as they are more robust than the orthogonal codes in asynchronous channels. However, the use of non-orthogonal codes in the uplink channels does not solve the problem, because it is still difficult to control the cross-correlation functions among Gold or Kasami codes in an asynchronous uplink channel and sometimes it may result in unacceptably high even/odd periodic cross-correlations, making the air-link extremely unreliable. This problem has to be fully addressed in the next generation CDMA technology and it has motivated us to propose the isotropic air-interface technologies, which involve the application of various advanced technologies such as spreading code design, spreading modulation innovation, precision uplink synchronization control, etc.

In traditional CDMA-based systems (such as IS-95, cdma2000, W-CDMA, etc.), little effort has been made so far to ensure an isotropic or uniform link performance in both synchronous downlink and asynchronous uplink. The uplink (also called reverse link) in a mobile cellular system is the transmission direction from a mobile to its base station (or cell site) and usually suffers much more impairing factors than the opposite direction, called downlink (also called forward link). First, the transmitting signal power in the uplink is always limited owing to the battery capacity constraint in mobile handsets, whose effective range is further reduced due to the fact that users often place mobile phone calls indoors. Thus, most radiating power is trapped inside a room and only a very small fraction of the radiating energy can get through windows. Second, the difference in antenna elevation heights between a mobile phone and a base station also determines that the uplink channels will suffer more from shadowing, local scattering, propagation/penetration loss, etc. than the downlink. Third, the uplink is always asynchronous in the sense that data symbols from different mobiles in a cell

Table 5.1 Different operational environments for uplink and downlink signal transmission in the mobile cellular scenario.

	Downlink (forward link)	Uplink (reverse link)
Transmission mode	Synchronous	Asynchronous
Power source	AC power	Battery
RF Tx power	$\geq$10 W	$\leq$0.5 W
Antenna beam pattern	Directional/omni-directional	Omni-directional
Antenna elevation	High (roof-mounted)	Low (handheld)
Propagation/penetration loss	Medium	High
SNIR* at Rx	Medium$\sim$high	Very low$\sim$Low
Possible Rx diversity	Time/frequency	Space/time/frequency
Pilot/signal multiplexing	DPTI†/DPCC‡	DPTI†
Rx joint detection	Difficult	Possible

*SNIR: Signal to Noise and Interference Ratio.
†DPTI: Data Pilot Time Interleaving.
‡DPCC: Dedicated Pilot Code Channel.

arriving at the base station are not necessarily aligned in time, and thus partial aperiodic correlations, rather than periodic correlations, of user signature codes govern the link-performance. Unfortunately, the former is often much more difficult to control than the latter, causing a much greater MAI than that in downlink channels. Table 5.1 compares different operational environments for uplink and downlink channels in a mobile cellular system. It should be stressed that we can not alter all the existing imparity in uplinks and downlinks, which is a reality we have to deal with. However, what we can do is to work out some new CDMA air-interface technologies that could hopefully counteract some of these disadvantages, which is also one of the focal points of this book.

A slower uplink in 2G systems is tolerable simply because the high-speed upload demand in most 2G applications is much less than the download demand. However, the situation in future wireless applications may change and high-speed upload applications will become commonplace when a mobile terminal with a high resolution video camera wants to send signals to someone else. Future wireless systems need to support ‘mobile server’ applications as a mobile terminal has a much higher computational power and battery capacity than today’s mobile phones. A typical scenario is a news journalist using a mobile unit with a high resolution video camera to capture scenes of interest and simultaneously transmit the video signals through a broadband wireless uplink to an all-IP backbone network, from where the scenes will be viewed in real time by the worldwide Internet users. In this case, the journalist’s mobile unit will act as a mobile server for the information source.

In general, there are two methods to address the link-performance imparity problem in uplink and downlink channels. One possible solution is to use isotropic spreading techniques, which consists of two elements, isotropic spreading codes and isotropic spreading modulation, to offer uniform link performance regardless of channel operating modes: either synchronous or asynchronous. The other way is to transform the uplink into a synchronous channel with the help of sophisticated open- and closed-loop synchronization control techniques, as proposed in the TD-SCDMA standard.

Therefore, the design of the next generation CDMA technology should be based on sufficient improvement of the detection efficiency for both synchronous and asynchronous transmissions. We would like to describe this attractive property of next generation CDMA technology as the capability to provide isotropic MAI-free and MI-free operation, where MI stands for multipath interference.

As we will give much more detailed discussions on the isotropic MAI-free and MI-free properties of the next generation CDMA technology in the following chapters, we here only offer some discussions on another interesting way to convert an asynchronous channel into a synchronous one using uplink synchronization control.

Uplink synchronization control is an important technique to transform an otherwise asynchronous uplink channel into a synchronous one, paving the way for successful application of existing orthogonal spreading codes, such as OVSF codes and Walsh-Hadamard sequences. To realize the uplink synchronization, a precision synchronization control technique, consisting of both open-loop and closed-loop control algorithms, should be applied in order to accomplish a fast and adaptive uplink synchronization with an accuracy of a fraction of one chip. In a mobile cellular system with uplink synchronization control, all mobiles in a cell, coordinated by the base station, should carry out transmission timing adjustment constantly, such that all their burst signals arriving at the base station should be synchronous in time. After achieving synchronization, mobiles and the base station should also work jointly to track the synchronization adaptively.

The uplink synchronization technique has been successfully integrated into the TD-SCDMA standard [91, 245], China's 3G technology for mobile cellular, which was proposed to the ITU in 1998 as one of the IMT-2000 candidate proposals, and approved in 2000. One of the salient features of the TD-SCDMA standard is its TDD-based synchronous CDMA technology, where OVSF codes are used in both uplink and downlink channels, due to its ideal orthogonality.[6] In other words, it will be impossible for TD-SCDMA to use orthogonal OVSF codes in its uplink if uplink synchronization control is not used. Figure 5.23 illustrates the detailed procedure of uplink synchronization control algorithm proposed in the TD-SCDMA, where burst signals in time slots received at both base station and mobiles, such as UE1, UE2, and UE3, are plotted on a step-by-step basis. It is assumed that UE1 and UE2 are existing mobiles having already achieved up-link synchronization with the base station, whereas UE3 is a newly arrived mobile that wants to connect to the base station and thus initiates an open-loop and closed-loop synchronization control process, as shown in Figure 5.23 from steps (a) to (h). Due to space constraints in this section, it is impossible for us to give a full description of all relevant technical contents of the TD-SCDMA standard, such as its frame structure and time slots/channels assignments, etc. Those who are interested in detailed information about the TD-SCDMA standard should refer to [91, 245]. Here, we would like to give only a rather brief introduction to the uplink synchronization control algorithms used in TD-SCDMA, as depicted in Figure 5.23.

During a cell-search procedure, the mobile UE3 should first establish downlink synchronization with the base station in a nearby cell by looking for the DwPTS burst from the cell site, as shown in steps (a) and (b) in Figure 5.23, based on which it will proceed to the uplink synchronization procedure, as shown in steps (c) and (d). At the beginning a mobile can estimate the propagation delay from a base station according to the received power level of the DwPTS burst. Its first transmission in the uplink is performed in the UpPTS time slot, which falls within the 'Searching Window' at the base station receiver, to reduce interference existing in normal time slots, such as TS0 to TS6, which are dedicated for link-established users' data traffic. The timing used for the SYNC_UL burst (in the UpPTS slot) is set according to the received power level of DwPTS, accomplishing the open-loop synchronization. On detecting the SYNC_UL burst, the base station will evaluate the received power level as well as timing and reply to the mobile UE3 by sending adjustment information, such as Synchronization Shift (SS) and Power Control (PC) commands, for UE3 to modify its uplink transmission timing and power level in the next burst transmission, executing the closed-loop synchronization control, as shown in steps (e) and (f) in Figure 5.23. The initial uplink synchronization should be achieved within four consecutive subframes, as specified in the TD-SCDMA standard [91, 245].

To maintain the uplink synchronization (its procedure is not shown in Figure 5.23), the midamble field of every uplink burst will be used. In each uplink time slot, such as TS1 or TS2 in Figure 5.23, the midamble codes from different mobiles in the cell are distinct. The base station can estimate

[6]It is hard to say if it is an advantage or a disadvantage to use OVSF code in both uplink and downlink channels of TD-SCDMA systems, as it requires a very complicated uplink synchronization scheme to curb otherwise excessive MAI in the uplink transmissions.

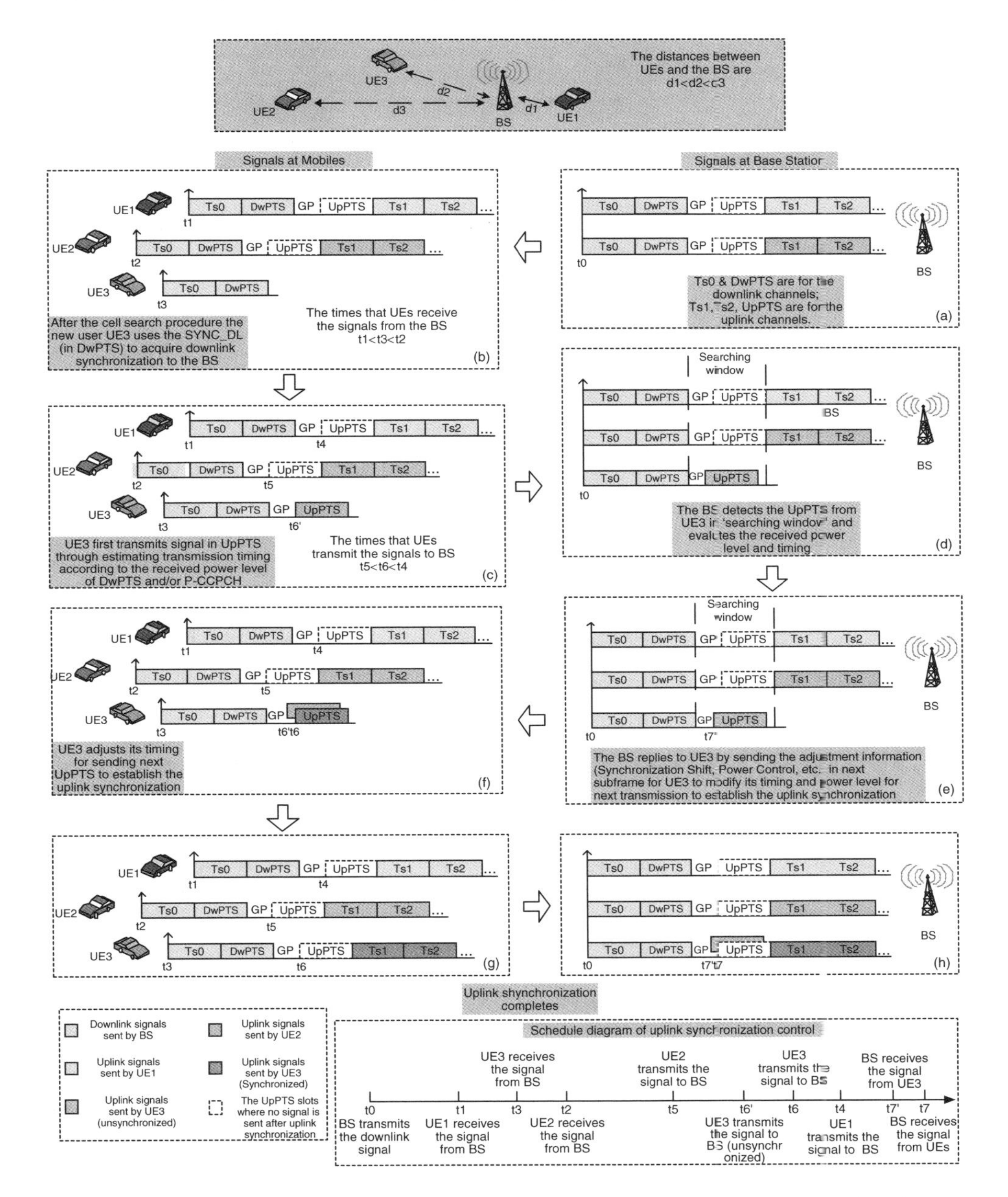

Figure 5.23 Step-by-step procedure of open-loop and closed-loop uplink synchronization algorithms proposed in the TD-SCDMA standard for a new mobile UE3 with two existing mobiles UE1 and UE2.

the power level and timing by measuring the midamble field from each mobile in the same time slot. Then in the next available downlink time slot, such as TS0 in Figure 5.23, the base station will signal the SS and PC commands, which occupy part of the midamble field, to enable the mobile to properly adjust its transmission timing and power level, respectively. The uplink synchronization can be checked once per TDD sub-frame and the step size in the uplink synchronization can be adapted from 1/8 chip to 1 chip duration, which is sufficient to maintain orthogonality of the OVSF codes from different mobiles.

To facilitate uplink synchronization control, the use of the mobile positioning technique is the key. The TD-SCDMA standard uses smart antenna technology to track direction-of-arrivals (DOAs) of incoming signals from different mobiles in order to improve the estimation accuracy of propagation delay for each mobile and thus the accuracy in the uplink synchronization control. It should also be pointed out that the uplink synchronization control needs a precision common synchronization clock to make the propagation delay estimation possible. GPS is a ready technique for this purpose, as suggested by the TD-SCDMA standard. However, the shortcoming for using GPS as the synchronization clock is that it will not be available indoors, unless a GPS repeater is installed in every building, tunnel, and so forth, adding a great infrastructural cost to the mobile core network deployment. Therefore, some other measures are also under consideration, such as a solution based on a synchronization beacon (which can be integrated into some common transport channel of TD-SCDMA) sent at a fixed interval by the core network operator itself. Every active mobile in a cell can acquire the common clock information via listening to the synchronization beacon for uplink synchronization control purposes.

As a summary remark, we would like to stress that the next generation CDMA technology ought to provide an isotropic MAI-resistant and MI-resistant operation (no matter which approach is used), which is an important benchmark for next generation CDMA technology, making its performance truly noise-limited rather than interference-limited.

5.4 BANDWIDTH EFFICIENCY VERSUS POWER EFFICIENCY

Bandwidth efficiency and power efficiency are two important performance requirements for any communication system, no matter whether it is wireless or wired. At the same time, it is very unfortunate to say that bandwidth efficiency and power efficiency usually act in a contradictory way. In most cases, to achieve high bandwidth efficiency may sacrifice power efficiency.

For instance, to improve the bandwidth efficiency of a digital modem we can use multiple-level digital modulation schemes, such as M-QAM, M-PSK, and so forth, to pack more information in to each symbol, thus enhancing the overall bandwidth efficiency of the digital modulation scheme. However, at the same time to successfully detect more information in each symbol requires that we have to raise the transmission power at the transmitter to achieve the same bit error rate at a receiver. Otherwise, too many errors will make the bandwidth efficiency meaningless, as no information can be delivered to the destination successfully. The only exception is the transformation from BPSK to QPSK modulation, which is an exceptional case in terms of their performance under the same operational condition. It is a well-known fact that BPSK and QPSK modulation schemes have the same power efficiency, which means that they will offer exactly the same bit error rate under the same transmitting power. The reason is simple, as a QPSK modem can be viewed as two BPSK arms working in parallel. Thus, the bit error rate for two BPSK arms should be the same as one BPSK modem, resulting in the same bit error performance for both BPSK and QPSK schemes. However, QPSK can offer twice bandwidth efficiency that of a BPSK modem under the same data transmission rate requirement, because the QPSK modem can use a symbol duration twice as long as that in a BPSK modem to have the same data rate, thus effectively reducing the bandwidth occupancy. Therefore, the bandwidth efficiency of QPSK is twice as high as that of the BPSK modem. For this reason, most digital communication systems prefer to use QPSK is an effort to enhance the bandwidth efficiency.

Now, let us look at the issues relating to next generation CDMA technology. We will face the same problem in trading bandwidth efficiency with power efficiency in the CDMA system architecture design. In many cases, they are somehow related with each other and should be treated very carefully in order to achieve the best balance whenever possible.

We can illustrate the issue with the help of an example in the design of OS-spreading-based CDMA systems. There is also a contradiction between bandwidth efficiency and power efficiency in a CDMA system using the OS spreading modulation scheme. As mentioned earlier, an OS spreading modulation scheme can offer a relatively high spreading efficiency, which can approach one if the next bit can be sent right after one chip offset delay relative to the previous bit. Therefore, it is obviously a very big improvement compared to traditional DS-CDMA systems in terms of their spreading efficiency. The improvement in spreading efficiency is also equal to the improvement in bandwidth efficiency. Therefore, there is no question that the OS spreading modulation scheme is indeed a big step forward in terms of bandwidth efficiency improvement compared to the DS spreading technique, which is used in almost all CDMA-based 2G and 3G mobile cellular standards.

Now, the problem we would like to discuss with the OS spreading modulation scheme is the choice of its carrier modulation scheme, as given in the next subsection.

5.4.1 OS-SPREADING-BASED CDMA

The OS modulation/demodulation scheme is the core of the OS-CDMA architecture. The name offset stacking comes from the fact that the spreading codes are shifted by one or more chips relative to one another. The offset chips between two consecutive data bits can be chosen on the fly, which is important for multirate processing in all future wireless applications, in particular multimedia communications.

In the following discussions, we would like to assume that the complementary codes can be constructed by using a $\sqrt{L}$-dimensional orthogonal matrix where L is the length of the element code of the complementary codes that will be generated. Let $\mathbf{A}$ be a K-dimensional orthogonal matrix such that its elements satisfy $|a_{ij}|=1$ and $\sum_{i=1}^{K} a_{ji}a_{ki}^* = 0$, where $j \neq k$.

Let $\mathbf{B}$ be another $K \times K$ orthogonal matrix. Then, K sequences of length K^2 can be constructed in the following way:

$$\begin{cases} \mathbf{C}_1 = (b_{11}\mathbf{A}_1, b_{12}\mathbf{A}_2, \ldots, b_{1K}\mathbf{A}_K) \\ \mathbf{C}_2 = (b_{21}\mathbf{A}_1, b_{22}\mathbf{A}_2, \ldots, b_{2K}\mathbf{A}_K) \\ \vdots \\ \mathbf{C}_K = (b_{K1}\mathbf{A}_1, b_{K2}\mathbf{A}_2, \ldots, b_{KK}\mathbf{A}_K) \end{cases} \tag{5.1}$$

where $\mathbf{A}_i$ is the ith row of matrix $\mathbf{A}$.

Each $\mathbf{C}_i$ $(i = 1, 2, \ldots, K)$ has an auto-correlation of zero for any shifts (except the zero shift) and the cross-correlation of any two of them is zero for all shifts. Now, let $\mathbf{D}$ be another K-dimensional orthogonal matrix. Then the final spreading sequences can be built up by

$$\begin{aligned} \mathbf{E}_{ij} = (&c_{i1}d_{j1}, c_{i2}d_{j2}, \ldots, c_{iK}d_{jK}, c_{i(K+1)}d_{j1}, c_{i(K+2)}d_{j2}, \ldots, \\ &c_{i(K^2-K+1)}d_{j2}, \ldots, c_{iK^2}d_{jK}) \end{aligned} \tag{5.2}$$

where $\mathbf{E}_{ij}$ means the jth sequence in the ith flock, and c_{ij} and d_{ij} are the (i, j)th elements of matrices $\mathbf{C}$ and $\mathbf{D}$, respectively.

The OS spreading modulation scheme uses complementary codes (CCs), which do not function like other conventional unitary spreading codes, such as Gold codes, Walsh-Hadamard codes, etc. The number of channels required by an OS spreading modulation system depends on the length of the element code and is equal to the flock size. The spreading efficiency (SE), which is defined as the number of bits conveyable in each chip, is not equal to 1/PG as in all traditional CDMA systems. Instead, it is equal to the reciprocal of the flock size (if only one chip offset is used) due to the fact that the bit stream is not aligned in time one bit after another. Instead, a new bit will start right after another chip delay relative to the previous bit. Therefore, CDMA systems using the OS spreading modulation scheme are more bandwidth efficient on a single link basis than conventional CDMA systems. For example, the SE of conventional CDMA and CC-based CDMA (CC-CDMA) for a PG of 64 is 1/64 and 1/4 (if the flock size is 4 here), respectively.

An example with element codes of length $L = 4$ (two users) applied to the downlink channel can be found in [237]. Each bit of the input sequence of each user is spread with the corresponding two spreading codes. The two spread sequences are not aligned in time like in traditional CDMA systems but shifted by one chip and then added. Therefore, the signal sent over channel f_1 consists of data from both users, spread with the first element code of the first flock, while the signal sent over channel f_2 consists of data from both users spread with the second element code of the second flock.

Assume that we only consider one particular type of complementary code, called complete complementary codes,[7] in this section. The maximum and minimum level values for any spreading code length is $\pm L\sqrt{L}$. The theoretical number of possible baseband signal levels $(0, \pm 2, \pm 4, \ldots, \pm L\sqrt{L})$ generated by the OS spreading modulation is $L\sqrt{L} + 1$. However, the probability that higher levels occur is very small due to the fact that the levels have a Gaussian-like distribution. The probability that either the maximum or the minimum number of levels occurs is $2 \times (0.5)^{L\sqrt{L}}$.

It can be shown that the probability of occurrence of level x for any L is

$$P(x) = \frac{\left(L \cdot \sqrt{L}\right)!}{\left(\frac{L\cdot\sqrt{L}-x}{2}\right)! \cdot \left(L \cdot \sqrt{L} - \left(\frac{L\cdot\sqrt{L}-x}{2}\right)\right)! \cdot 2^{L\cdot\sqrt{L}}} \tag{5.3}$$

Figure 5.24 shows the distribution (histogram) of the levels for an element code length $L = 16$. It is obvious that the levels have a Gaussian-like probability distribution. It can also be readily shown that the average symbol energy of a pulse amplitude modulation (PAM) signal constellation is

$$E_s = E\{A^2\} = \sum_{i=1}^{M-1} |a_i|^2\, p_i \tag{5.4}$$

where A is the random amplitude level that takes on the values a_i with probability p_i.

The OS spreading modulation scheme output signal is actually a PAM signal with Gaussian distributed levels. The symbol energy of this signal depends on the element code length L and on the number of users. In the case where not all users in the system are transmitting data, the level distribution and therefore the symbol energy on each channel may vary.

Symbol times and rates defined in this OS-spreading-based CDMA system have slightly different meanings from those given in conventional CDMA systems. Let T_{bit} be the duration of the input bit and R_{bit} the input bit rate $1/T_{bit}$. Let the input bit sequence be spread with the CC spreading codes. Then the 'chip' sequence after the spreading procedure is of rate R_{chip} which is L times higher than R_{bit}. This chip rate is relevant only for the OS spreading modulation process and must not be confused with the chip rate of conventional CDMA systems. The OS spreading modulator converts these chips into symbols at a rate of R_{symbol}, which are sent through the channel. The symbol rate

[7] It is noted that there are many more complementary codes to be introduced in the later part of this book. Complete complementary codes are one type of them with a relatively small set size, and thus able to support fewer users in a CDMA system.

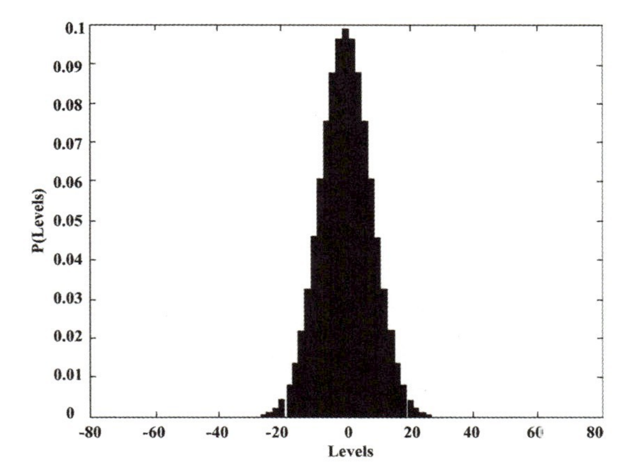

Figure 5.24 Distribution of baseband signals after OS spreading modulation in an OS-CDMA system transmitter.

is comparable to the chip rate of conventional CDMA systems, where these levels are produced by superposition of several non-return zero (NRZ) streams.

On the other hand, the demodulation process of OS spreading modulated signals can be divided into four parts:

1. The signal of each channel is correlated with the corresponding element code of each user, e.g. the signal transmitted over the first channel is correlated with the first spreading codes $A_0, B_0, \ldots$. After this process, there are $C \times K$ signals: $C_0A_0, C_0B_0, \ldots, C_1A_1, C_1B_1, \ldots, C_KA_K, C_KB_K, \ldots$, where C is the number of channels and K is the number of users.

2. The resulting signals of this correlation process are added user-wise, e.g. the first signal after this process will consist of $C_0A_0 + C_1A_1 + \cdots + C_KA_K$. The number of signals after this process is N; now each signal carries the data of the corresponding user.

3. The signals are normalized by applying an attenuation factor $L\sqrt{L}$, where L is the length of the spreading code. The normalization is only necessary if channel estimation will be applied, otherwise it can be skipped.

4. Hard decision is applied to recover the transmitted data bit.

Based on the discussions earlier, we can understand that an OS-spreading-based CDMA system should use a multiple-level carrier modulation to up-covert the baseband OS spreading modulated signal into the radio frequency band and then send it over the channel. The signal detection process at a receiver will perform an inverse process to demodulate the carrier signal and then make the decision on the signal levels. After that, a de-mapping algorithm should be used to recover the data information hidden in the multiple-level baseband signal.

The detection on the multiple-level digital modulated signals is usually more difficult than the detection on binary digital modulated signals. Also, it is more likely that there will be detection errors as the signal constellation points in a multiple-level modulation scheme are close to each other. In order to reduce the detection error, it is usually required that more transmission power should be used to allow us to distinguish more easily one constellation point from others. This will introduce some loss in power efficiency, at the price of improved bandwidth efficiency obtained by using OS spreading modulation scheme. We will conduct more analysis on this issue in Chapter 7.

5.5 HIGH SPEED BURST DATA ACCESS AND NEXT GENERATION CDMA

The development of mobile cellular communications has already gone through three generations. First generation mobile cellular systems were basically based on analog technology and were developed only for low capacity voice communications. Those first generation mobile cellular technologies include TACS (for the UK), NMT (for the Nordic countries), AMPS (for North America), and so forth. Second generation mobile cellular communications are based on digital technologies and were developed initially also for voice communications, but later were modified to cater for low speed data communications as well. Typical examples of second generation mobile cellular systems include GSM (Europe), DAMPS and IS-95 (US), and JDC (Japan). The 3G mobile cellular technologies have been deployed in many countries in the world only very recently. The introduction of 3G mobile cellular technologies was mainly targeted to multimedia communications, but most of the traffic carried in 3G mobile phones is still not based on all-IP wireless architecture. Therefore, high-speed all-IP wireless communications will be one of the most important characteristic features carried in future wireless communication systems, including mobile cellular and wireless data networks as a whole. For this reason, the development of next generation CDMA technology should definitely take it into account as a major design consideration.

Also, it is noted that much criticism of traditional CDMA technology is based largely on the fact that first generation CDMA technology was developed mainly for slow-speed continuous-time transmissions, which are characterized by very long frame structure and relatively low data transmission rate. Slow-speed transmissions with relatively long frames help to achieve synchronization and tracking for CDMA signal detection, as happened in all transceiver designs based on first generation CDMA technology. Some have also pointed out in a very straightforward way that all CDMA systems based on first generation CDMA technology are not suitable for 4G wireless communications, which should support high-speed bursty traffic, due to the fact that their signaling structure is too long to fit packet-switched signaling.

On the other hand, orthogonal frequency division multiple access (OFDMA) technology has become a new important player for 4G wireless system platforms, also due to its capability to support high-speed packet transmissions in wireless channels. Transmission of short packets with a few bits can be done easily in an OFDMA transceiver. The detection process in an OFDMA receiver can proceed with the detection of tones at the output of an FFT block, which can easily be made to work with short packets.

Now, we approach a key question on whether or not next generation CDMA technology can support high-speed bursty traffic as well, and if so, how.

Let us examine carefully why traditional CDMA technology could not support high-speed bursty traffic communications. A relatively long frame structure is necessary to allow a traditional CDMA transceiver to lock on to the header and then to identify the beginning of each bit. Based only on this, follow-up signal detection on the bit stream can be made possible, as the transceiver will know the starting point of spreading codes or scrambling codes to reconstruct a good correlation peak before the decision device. Why does a traditional CDMA transceiver need a long header to process the signal detection algorithms that follow? It has a lot to do with the spreading codes used by all traditional CDMA systems. These traditional spreading codes include Walsh-Hadamard codes, Gold codes, Kasami codes, m-sequences, OVSF codes, etc., all of which were proposed for CDMA applications based mainly on their seemingly acceptable even periodic correlation functions (which include both even periodic auto-correlation functions and even cross-correlation functions). In other words, all of these codes can work properly only under the condition that the complete code should be fed into a local correlator to reconstruct correlation functions. Otherwise, the correlation functions based only on a segment of code will be very bad for all traditional spreading codes used in 2G and 3G CDMA-based systems. The problem is that they have very bad partial correlation properties, which make them unsuitable for applications in high-speed bursty traffic communications.

Therefore, we need a new code design approach to find some new CDMA codes for next generation CDMA technology. All existing spreading codes, which are also called unitary codes,[8] will not be considered as suitable candidates for applications in next generation CDMA technology. Therefore, we urgently need some revolution in code design methodology here.

In Chapter 6 of this book, we will in particular address the issues on the design of innovative spreading codes for next generation CDMA technology. Therefore, in this sense, we can say that the application of innovative spreading codes is significant to make the CDMA technology discussed in this book completely different from all existing CDMA systems. It should be pointed out that we will show that an interference-free CDMA system is possible if we use orthogonal complementary codes, which have been one of the cores in next generation CDMA technology. We will also propose several different types of orthogonal complementary codes for their applications in different wireless systems by exploiting their particular properties. One of the most salient features in all these orthogonal complementary codes is their very good partial correlation properties, which make them very suitable to reduce MAI and MI when they are used in a CDMA system that supports high-speed bursty type traffic.

As an example to show the ideal partial correlation property of these orthogonal complementary codes, we use Figure 5.25, where a complementary-code-based CDMA system with DS spreading is illustrated to overcome MAI and MI without the help of a RAKE receiver. In Figure 5.25, a 3-bit data burst is sent into a 2-user and 2-path uplink asynchronous CDMA channel. The inter-path and inter-user delays are only 1 chip for illustration simplicity (the same result will be given if any other delays are applied).[9] A simple correlator receiver is used to detect an incoming burst in the presence of both MAI and MI. It is seen from the figure that the correlator can successfully recover the original data information ($+ - +$) (no matter whether it appears in the middle or at the beginning of the packet) without any impairment caused by either MAI or MI. Similarly, we can show the same result for a down-link multipath channel, which usually presents much less difficulty than an uplink channel.

Thus, a simple correlator can solve MI and MAI problems in this CDMA system based on orthogonal complementary codes (OCC). It should be noted that avoiding the use of a RAKE receiver is significant as it paves the way for an OCC-CDMA receiver to work in a truly blind fashion without the need for any prior channel information. On the other hand, a RAKE has to acquire virtually all channel information, such as delays and amplitudes of all multipath returns, for its maximal-ratio-combining (MRC) operation, whose impact on implementation complexity in a mobile handset should never be underestimated. Obviously, for a DS-spreading OCC-CDMA (OCC/DS-CDMA) we can also use a RAKE receiver to further boost signal-to-interference ratio. It is noted that the use of RAKE here can provide a much higher multipath-diversity gain than achievable in a conventional CDMA system due to the fact that the output signal from each finger now contains only the useful auto-correlation peak, as shown explicitly in Figure 5.25. On the other hand, the output from a finger of a conventional CDMA RAKE receiver contains both useful and unwanted signals caused by non-trivial auto-correlation side lobes.

---

[8]The unitary codes are defined in this book as the spreading codes which work on a one-code-per-user basis. The opposite to the unitary codes is the complementary code, which works on an one-flock-per-user basis.

[9]It is noted that the relative inter-path delay should never become multiples of bit duration. If this happens, then multipath interference will cause problem. The same result applies to any CDMA system, no matter what type of spreading code (for both unitary codes or complementary codes) is used.

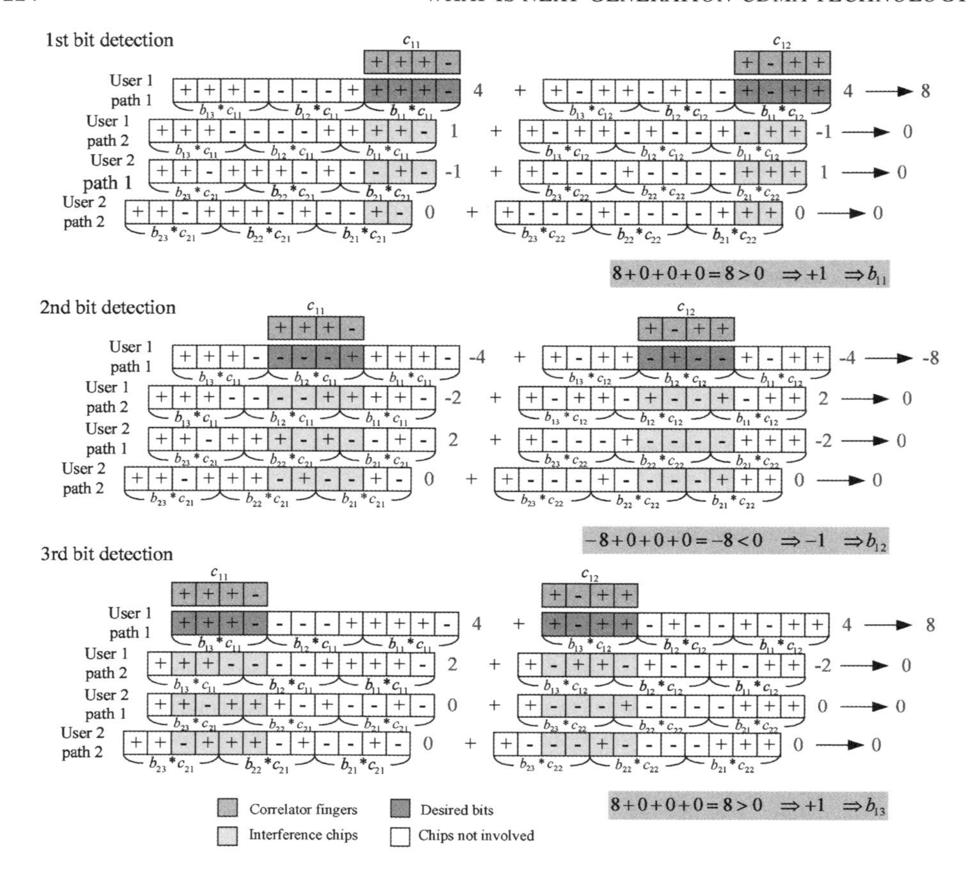

Figure 5.25 MI-free and MAI-free operation for a two-user OCC/DS-CDMA system in asynchronous high-speed bursty uplink channels, where a two-way multipath channel is considered with both inter-path delay and inter-user delay being one chip for illustration simplicity. The parameters of the orthogonal complementary codes used in this figure are $K = M = 2$ and $N = 4$. The 3-bit data information sent is $\{+ - +\}$.

Furthermore, if a RAKE has to be used in an OCC/DS-CDMA receiver, equal gain combining (EGC) is preferred to yield satisfactory detection efficiency. The advantage of using EGC rather than MRC in a RAKE is to avoid having to use a complicated multipath amplitude estimation unit.

As an important requirement for a CDMA technology suitable for future gigabit all-IP wireless communications, we should pay sufficient attention to detection efficiency at the edges of a packet or burst. In this sense, an OCC-CDMA scheme is in particular well suited as shown in the upper parts of Figure 5.25, which illustrates the signal detection process for the first (or the rightmost) bit of a packet. In detecting those bits at the edges of a packet, partial ACFs and partial CCFs of the codes will become extremely important. It is seen from the figure that the OCC/DS-CDMA system yields zero partial CCFs, which ensures an ideal performance for signal detection even on the edges of a frame or packet. This observation is significant due to the domination of burst traffic in future wireless systems.

5.6 INTEGRATION OF MIMO AND CDMA TECHNOLOGIES

Diversity schemes have been widely used in wireless applications to improve the signal detection efficiency. There are three major diversity schemes which have found wide application: time diversity, frequency diversity, and spatial diversity. Multiple-input-multiple-output (MIMO) technology is a type of spatial diversity scheme, which was proposed recently.

It is a well-known fact that any type of diversity technique should satisfy the three requirements stated as follows. First, the signals should be sent through statistical independent channels. Second, the sent signals via different channels should be the same to have some kind of redundancy for coherent or non-coherent recombination. Third, the sent signals via different channels should be separable at a receiver. All three conditions should be met to make diversity possible. MIMO technology exploits the spatial diversity advantages to improve the signal detection efficiency at a receiver. One MIMO channel can be formed by a pair of transmitter (Tx) antenna and receiver (Rx) antenna. If we send the same data through different Tx-Rx channels, which are separated by sufficiently long distance for their independent statistical properties, then a spatial diversity gain can be obtained at a receiver to improve the detection efficiency, especially under some fading environments.

MIMO technology has been widely used in various wireless applications to significantly improve the performance of a wireless communication system as a whole. The performance enhancement due to the use of MIMO technology is reflected mainly in the two facts explained as follows. First, the application of MIMO technology can send data through different channels formed by a pair of antennas (one transmitter antenna and one receiver antenna). If the system uses different Tx-Rx antenna pairs to send the same data stream, a powerful spatial diversity gain is made possible. It is noted that the spatial diversity gain obtained in a MIMO system does not consume any time and frequency resource, which is very limited in any wireless application. Therefore, the popularity of MIMO technology for modern wireless systems is intuitively understandable. In addition to the spatial diversity gain, a MIMO system can also be configured to send different data information streams through different Tx-Rx data pipes, in order to increase the data throughput between the transmitter and receiver. This is also called the multiplexing gain of a MIMO system, which was demonstrated in the V-BLAST system developed by Bell Laboratory several years ago. Therefore, the spatial diversity gain and spatial multiplexing capability are the two major operational advantages for MIMO technology.

Obviously, MIMO technology will also play an important role in next generation CDMA technology. As a matter of fact, MIMO technology can be naturally integrated within the next generation CDMA technology as an indispensable part of it. We can explain why MIMO technology will become an important and natural part of next generation CDMA technology as follows.

As discussed in Subsection 5.2.3, we can extend the frequency-time two-dimensional spreading technique into a three-dimensional spreading scheme, which includes the space, time, and frequency domains. Successful operation of a MIMO system requires some effective space-time coding schemes, which are used to facilitate the signal separation process at a receiver. Many space-time coding schemes have been suggested in the literature and they have been used in different MIMO systems. The most popular space-time coding schemes include space-time block coding (STBC), space-time trellis coding (STTC), and space-time differential coding (STDC). We would like to emphasize here again that a space-time coding scheme is used only for separating signals sent through different Tx-Rx antenna pairs at a receiver, and nothing else. Therefore, it is easy to understand that any ways can be used in a MIMO system, as long as they can help to facilitate the signal separation process at the receiver end.

We have been worked on signal separation schemes for many years (as a CDMA code set is just a coding scheme which is used to separate different signals sent from different users), and many superior

CDMA codes have been found by us. Those CDMA codes can also be used as space-time coding schemes for a MIMO system for antenna separation. As suggested in Subsection 5.2.3, we can use a random coding method to expand the two-dimensional CDMA codes (based on various orthogonal complementary codes already in our hands) into three-dimensional codes for a MIMO-based next generation CDMA system architecture.

On the other hand, we can also use some existing space-time coding schemes in orthogonal complementary code (OCC) CDMA systems to achieve either spatial diversity gain or spatial multiplexing capability. This is discussed further below. To make the issues clearer, let us consider a generic block diagram for a complementary-code-based CDMA system with space-time block coding, as shown in Figures 5.26 and 5.27, which show the transmitter and receiver, respectively.

Assume that the transmitter is designed for transmission of the first user signal and the receiver is tuned also to the first user. It is seen from Figure 5.26 that in this STBC OCC-CDMA system the transmitter uses its signature code flock $(\mathbf{c}_{11}, \mathbf{c}_{12}, \ldots, \mathbf{c}_{1M})$ to spread its bit stream $\bar{\mathbf{b}}_1$, which will be encoded in the STBC unit into two sub-streams $\mathbf{b}_1^{(1)}$ and $\mathbf{b}_1^{(2)}$. It should be noted that the STBC encoding happens before the spreading modulation with M different element codes in the same flock. M different sub-carriers are used to modulate M different element codes in the same code flock with the two ST encoded data sub-streams, $\mathbf{b}_1^{(1)}$ and $\mathbf{b}_1^{(2)}$ (which are orthogonal with each other at bit level). Therefore, the DS spreading modulated signals for the two sub-streams with the M different carrier frequencies are added together before being sent out into the channel. It is noted that the two

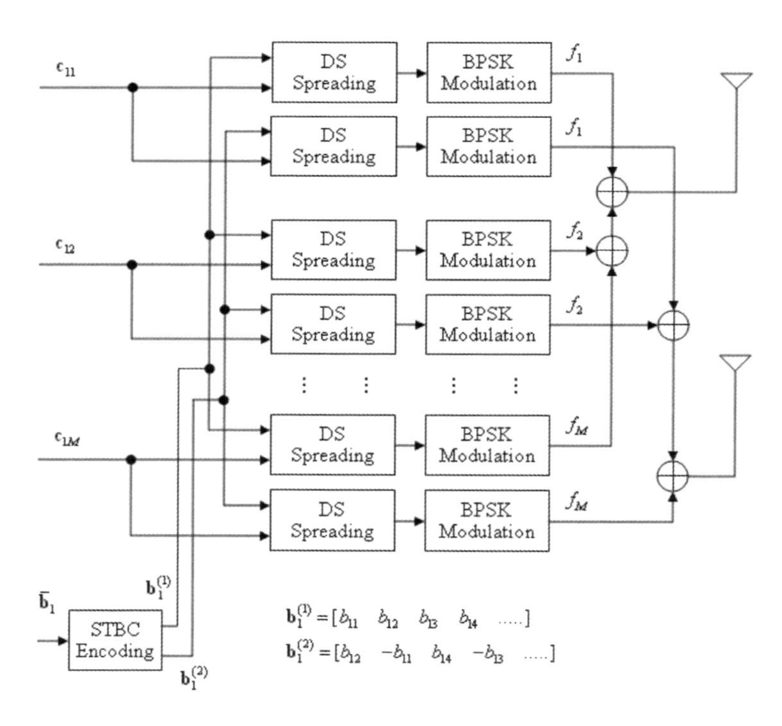

Figure 5.26 Block diagram for the transmitter of an STBC CDMA system based on orthogonal complementary codes.

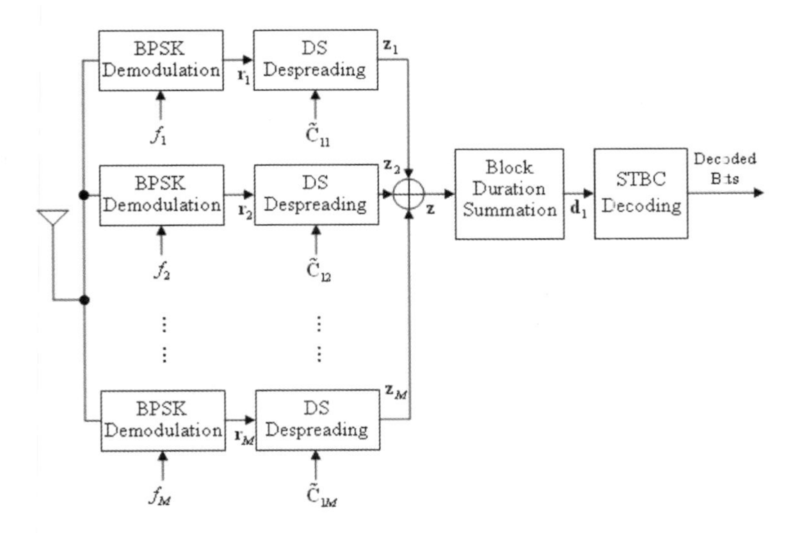

Figure 5.27 Block diagram for the receiver of an STBC CDMA system based on orthogonal complementary codes.

sub-streams after the STBC encoding algorithm are shown as follows

$$\begin{cases} \mathbf{b}_1^{(1)} = (b_{11}, b_{12}, b_{13}, b_{14}, \ldots) \\ \mathbf{b}_1^{(2)} = (b_{12}, -b_{11}, b_{14}, -b_{13}, \ldots) \end{cases} \tag{5.5}$$

which will ensure that they are orthogonal with each other in terms of each block which consists of two bits.

In the receiver module, as shown in Figure 5.27, the received signal from the channel first goes through carrier demodulation and direct sequence despreading, before the STBC decoding.

Therefore, we can see that the OCC-CDMA system can work harmonically with a space-time coding scheme to form different configurations of joint MIMO and OCC-CDMA systems. More detailed discussions on the issues will be covered in Chapter 8.

5.7 *M*-ARY CDMA TECHNOLOGIES

M-ary CDMA [164–169; 246–250] is another possible technique which can be used in next generation CDMA systems. Therefore, we will put it into the discussions in this book.

The idea for *M*-ary CDMA systems is that the CDMA codes can also be used for encoding different information bit array patterns using different states that the spreading codes are sent into the channel. Therefore, in this way source information can be transmitted to a receiver if the receiver knows how to de-map the states of code transmissions into the data bit array patterns. It is noted that in an *M*-ary CDMA system the source information data will not be sent bit by bit, as it is in all traditional DS-CDMA systems. Instead, the source information should first be grouped into a bit array, which may consist of a several bits, and then each bit array should be encoded by a certain code transmission pattern, which is represented by the transmissions of certain number of spreading codes in parallel with their different signs.

Therefore, the working principle of the M-ary CDMA system tells us that the spreading codes or CDMA codes can also be used as an effective means to encode source information for communication purposes. If you have more codes available and they can be used simultaneously (without introducing excessive mutual interference), you can encode (or pack) more information in one particular state of code transmissions. Therefore, the spreading code set size will play an important role in determining the transmission efficiency of an M-ary CDMA system. Therefore, in this sense, the CDMA code space or spreading code space can form another important dimension and it can play the same role in information transmissions, just like the time, frequency, and space dimensions.

For example, if 2^M orthogonal codes are available in an M-ary CDMA system, then we can use them to encode a bit array with its size being M bits, if only one code is allowed to be sent at any time and the sent code has only one sign. In such an M-ary CDMA system, the transmission of one particular code will denote a particular bit pattern in the bit array of M bits. At the receiver side, M correlators should be used to check which code indeed appears in the channel, and the detection algorithm should use maximal likelihood (ML) detection to obtain an optimal detection performance.

On the other hand, we can also control the number of codes to be sent into the channel, plus their signs (either positive or negative). Then, we will have much more degrees-of-freedom to denote different bit array patterns. For example, if we are allowed to use both numbers and signs of codes to represent different bit array patterns, then we will have in total $3^M - 1$ code transmission states if there are M codes available in an M-ary CDMA system. However, in this case, the number of code transmission states will not necessarily be an integer multiple of two. To illustrate this situation, let us assume that $M = 3$ and we will have altogether $3^M - 1 = 3^3 - 1 = 26$ code transmission states, which can only be used to represent $2^4 = 16$ different four-bit array patterns, as $2^5 = 32 > 26$. Thus, we have seen that in this case $26 - 16 = 10$ code transmission states can not be used in an M-ary CDMA system.

This problem should not necessarily be considered as bad. Instead, it gives us a very interesting research topic, on which we would like to see which code transmission state subset can be the best to minimize the error rate if we can only choose 16 states from 26. We will spend a lot of effort to address the issue in Chapter 9.

Due to the limitation of space here, we will not go further to cover every aspect of M-ary CDMA systems. For more information please refer to Section 2.7. Also, more detailed discussions on the performance of M-ary CDMA systems will be presented in Chapter 9.

6

Complementary Codes

As discussed in the previous chapter, orthogonal complementary code plays a very important role in the next generation CDMA technology.

It is well understood that signature codes or spreading codes form the core of a CDMA system or technology. A CDMA system relies on signature codes or spreading codes to divide different users for multiple access purposes. Therefore, the choice of signature codes for a CDMA system will largely determine the fundamental performance of the system. As a matter of fact, the signature codes should be chosen well before the whole CDMA architecture has been formulated and the code selection should preferably be done jointly with the design of the CDMA system architecture.

Many lessons can be learnt from the applications of first generation CDMA technology in 2G and 3G CDMA-based mobile cellular systems. At the time when the 2G mobile cellular system was designed based on spread spectrum techniques, which was later called the IS-95 standard, how to combine the design of spreading codes with the system architecture was not well understood. As a matter of fact, the system design for IS-95 was completely separated from the code design. The engineers could only select some available spreading codes from their database or the literature, most of which were built by mathematicians who were working on information theory. Some typical examples of those spreading codes include Gold codes, Kasami codes, Walsh-Hadamard codes, maximal length sequences (or m-sequences), etc., most of which were studied and proposed by people working in modern algebra. Many of these codes were investigated in the context of radar systems, in which only the auto-correlation functions of the codes are important. The design of these spreading codes was done at a time when channel propagation theory was only at its very preliminary stage: virtually nobody then understood channel fading and the multipath propagation effect. Therefore, it is an undoubted fact that spreading code design was done at least a couple of decades before the first generation CDMA technology was proposed in the later 1990s by Qualcomm. The complete disjunction between the design of spreading codes and the design of the first generation CDMA technology is the main reason why IS-95 and all later 3G CDMA mobile cellular standards are unable to work at a performance better than interference-limited capacity.

It is ironic to note that, as a result of the development of the first generation CDMA technology (which has been used in many 2G and 3G mobile cellular standards), almost everybody is used to saying that all CDMA systems are always interference-limited. We do not know whether or not this conclusion was made on purpose, but it has already damaged the will in the research community to search for better CDMA technology. Therefore, it is the time for us to work out some new and more promising CDMA technology, which is the focus of this book.

The Next Generation CDMA Technologies Hsiao-Hwa Chen

6.1 MAGIC POWER OF COMPLEMENTARY CODES

Due to their many unique desirable properties, orthogonal complementary codes play an extremely important role in the development of next generation CDMA technology.

It should be emphasized here that the choice of orthogonal complementary codes for the next generation CDMA technology is not an accident. As a matter of fact, we did not have any preemptive ideas on which type of signature codes would be suitable for the next generation CDMA technology before our study on the issues, which have been continuing for many years. As shown in the results to be revealed later in this chapter, we will show step by step the process which led us to regard orthogonal complementary codes as the most desirable spreading codes in a natural way, which guides us to conclude that an interference-free CDMA system will be made possible if and only if orthogonal complementary codes are used as the spreading codes (or signature codes) for the system. This process will be introduced when we discuss the real environment adapted linearization (REAL) approach, which was proposed by us as an effective methodology to search for the optimal spreading codes for the next generation CDMA technology. We do not detail the REAL algorithm here, but just summarize the major contributions made by the REAL approach as follows.

1. It is the REAL approach that first time uses the concept of joint code and system design in the code search process. No previous effort was not be able to do this.

2. In the REAL approach, we have taken into account many real operational conditions when searching for the optimal spreading codes for CDMA applications. Those real operational conditions include multipath propagation, synchronous and asynchronous transmissions, random bit signs in the data stream, bursty traffic, and so forth. Therefore, the solution from this optimization problem can ensure an interference-free CDMA operation if a CDMA system operates under the same conditions.

3. Two important conclusions have been obtained from the REAL approach. The first is that the REAL approach has shown that it is possible to implement an interference-free CDMA system, and only orthogonal complementary codes should be used for such an interference-free CDMA system. Second, in such an interference-free CDMA system the set size of the orthogonal complementary codes must be equal to the flock size[1] of the codes.

Therefore, we have to say that the selection of orthogonal complementary codes as the core of the next generation CDMA technology is not an accident or an idea popping up from nowhere. Instead, it is a necessity, and no other type of code should be used. The magic power of the orthogonal complementary codes comes from their unique correlation reconstruction properties, which were built up on much more degrees-of-freedom, compared to the ways of reconstructing their correlation functions in all traditional spreading codes (such as Gold codes, Walsh-Hadamard codes, Kasami codes, m-sequences, OVSF codes, etc.).

We will make a lot of effort to give more detailed explanation on the issues in the sections following in this chapter, as well as in later other chapters.

6.2 DIFFERENT TYPES OF COMPLEMENTARY CODES

It is noted that complementary codes constitute a fairly large group of codes, whose auto-correlation functions and cross-correlation functions are built on their several element codes (whose number

[1]In CDMA using orthogonal complementary codes each user is assigned a particular flock of element codes for CDMA transmissions.

usually is even for obvious reasons) jointly. Therefore, if they are used in a CDMA system, each user in the system has to be assigned a flock of element codes. On the contrary, all traditional spreading codes, such as Gold codes, m-sequences, Walsh-Hadamard codes, OVSF codes, Kasami codes, etc., are unitary codes, which are defined by us to characterize their way of being used in a CDMA system, which works on a one-code-per-user basis.

In the following subsections, we will give a fairly detailed introduction to all complementary codes we know so far. There are six different types as listed below:

1. Primitive complementary codes
2. Complete complementary codes
3. Extended complementary codes
4. Super complementary codes
5. Pair-wise complementary codes
6. Column-wise complementary codes.

It should be noted that the first three types of complementary codes were introduced in the literature, and the last three were proposed by us. Some of the complementary codes are orthogonal complementary codes (OCCs) and some are not. In fact, only the first and fifth types, or the primitive complementary codes and pair-wise complementary codes, are not orthogonal; the rest are orthogonal complementary codes, which at least in theory can be used in CDMA systems. However, even among these orthogonal complementary codes, they may exhibit many different characteristic features in terms of their set sizes, their correlation properties, and so forth. Therefore, we should say that the study of complementary-code-based CDMA technology is only at its beginning and we have a very long way to go to work out the next generation CDMA technology as a whole.

Here we would like to define what we mean by the ideally orthogonal set of the spreading codes as follows. If we say the spreading codes are ideally orthogonal, then both even and odd periodic cross-correlation functions of any two codes in the set are zero for any possible relative chip shifts. The even and odd periodic cross-correlation function for any pair of codes is mathematically expressed as follows.

Assume that codes $\mathbf{A}$ and $\mathbf{B}$ are two codes in an ideally orthogonal set of the spreading codes. They can also be defined as $\mathbf{A} = (\mathbf{A}_1, \mathbf{A}_2, \ldots, \mathbf{A}_M)$ and $\mathbf{B} = (\mathbf{B}_1, \mathbf{B}_2, \ldots, \mathbf{B}_M)$, which states that codes $\mathbf{A}$ and $\mathbf{B}$ are complementary codes, each of which contains M element codes. Each element code consists of N chips, or $\mathbf{A}_m = (a_{m1}, a_{m2}, \ldots, a_{mN})$ and $\mathbf{B}_m = (b_{m1}, b_{m2}, \ldots, b_{mN})$, where $m = 1, 2, \ldots, M$. If $\mathbf{A}$ and $\mathbf{B}$ are two codes from an ideally orthogonal set, they must satisfy the following equations:

$$\rho(\mathbf{A}, \mathbf{B}; \tau) = \sum_{m=1}^{M} \left[\sum_{i=1}^{\tau} a_{mi} b_{m(N-\tau+i)} \pm \sum_{i=\tau+1}^{N} a_{mi} b_{m(i-\tau)} \right] = 0, \quad \text{(for any } \tau) \tag{6.1}$$

where τ is the relative delay between the two codes $\mathbf{A}$ and $\mathbf{B}$. It is noted that the above equation gives the requirements on the even (if the positive sign is used in Equation (6.1)) and odd (if the negative sign is used in Equation (6.1)) periodic cross-correlation functions for the two ideally orthogonal codes. It is noted that Equation (6.1)) is a general form for any code, and letting $M = 1$ implies that we are considering the unitary spreading codes; otherwise it will refer to complementary codes.

It is noted that not all complementary codes listed above are ideally orthogonal codes. For example, the primitive complementary codes and pair-wise complementary codes are not ideally orthogonal codes, as they do not satisfy the requirements on the perfect cross-correlation functions as specified in Equation (6.1) On the other hand, the remaining complementary codes listed above, have perfect cross-correlation functions and thus they are ideally orthogonal codes.

In addition to the cross-correlation function, which defines the ideal orthogonality of the set of spreading codes, we also need to define their auto-correlation function, which will affect the detection efficiency, especially in a multipath channel. The desirable auto-correlation function for an ideal spreading code set should be defined as follows.

$$\rho(\mathbf{A};\tau)=\begin{cases}\sum_{m=1}^{M}\left[\sum_{i=1}^{\tau}a_{mi}a_{m(N-\tau+i)}\pm\sum_{i=\tau+1}^{N}a_{mi}a_{m(i-\tau)}\right]=NM, & \tau=0\\ \sum_{m=1}^{M}\left[\sum_{i=1}^{\tau}a_{mi}a_{m(N-\tau+i)}\pm\sum_{i=\tau+1}^{N}a_{mi}a_{m(i-\tau)}\right]=0, & \text{elsewhere}\end{cases}\tag{6.2}$$

in which τ is the relative chip delay between the local correlator and the incoming code received at a receiver, and M can be any integer without losing generality. If $M = 1$, the above auto-correlation requirement fits a unitary code set; otherwise it refers to a complementary code set. Equation (6.2) includes both even (for the positive sign in (6.2)) and odd (for the negative sign in (6.2)) periodic auto-correlation functions.

Obviously, if the code set size is K, we will have $C_K^2 = \binom{K}{2} = \frac{K!}{2!(K-2)!}$ equations to specify the ideally orthogonal conditions as given in Equation (6.1). We will also have K equations to specify the ideal auto-correlation functions for all K codes, as given in Equation (6.2).

The analysis given here will form the basis for us to propose a new spreading code design approach, called the real environment adapted linearization (REAL) approach, which will be discussed in detail in Subsection 6.3.5.

6.2.1 PRIMITIVE COMPLEMENTARY CODES

As its name implies, the primitive complementary code is the first type of complementary code ever proposed in the literature. The origin of the primitive complementary codes can be traced back to the 1960s, when Golay [251] and Turyn [252] first studied pairs of binary complementary codes whose auto-correlation function is zero for all even shifts except the zero shift.

Therefore, it should be noted that the primitive complementary codes are not orthogonal codes in the sense that the cross-correlation functions between any two codes are not zero in all chip shifts. As a matter of fact, the study of primitive complementary codes was mainly driven by the development of radar technologies in the 1960s.

In 1961, Golay proposed a set of complementary series which were defined as a pair of equal long and finite sequences of two kinds of elements having the property that the number of pairs of like elements with any one given separation in one series is equal to the number of pairs of unlike elements with the same given separation in the other series.

For instance the two series, 1001010001 and 1000000110 have, respectively, three pairs of like and three pairs of unlike adjacent elements, four pairs of like and four pairs of unlike alternate elements, and so forth for all possible separations. These series, which were originally conceived in connection with the optical problem of multi-slit spectrometry, also have possible applications in communication engineering, for when the two kinds of elements of these series are taken to be $+l$ and -1, it follows immediately from their definition that the sum of their two respective auto-correlation series is zero everywhere, except for the center term. Several propositions relative to these series, to their permissible number of elements, and to their synthesis are demonstrated in Golay's paper [251]. This was probably the earliest work ever reported on complementary codes, which were called complementary series then.

In Golay's work, a set of complementary series was defined as a pair of equally long, finite sequences of two kinds of elements which have the property that the number of pairs of like elements with any given separation in one series is equal to the number of pairs of unlike elements with the same separation in the other series. For instance, the two series $\mathbf{A} = 00010010$ and $\mathbf{B} = 00011101$ are complementary, as mentioned earlier.

The basic property of complementary series may be expressed also in auto-correlative terms. Let the various a_i and b_i elements ($i = 1, 2, \ldots, n$) of two n-long complementary series be either $+1$ or -1, and let their respective auto-correlative series be defined by

$$\begin{cases} c_i = \sum_{i=1}^{i-n-j} a_i a_{i+j} \\ d_i = \sum_{i=1}^{i-n-j} b_i b_{i+j} \end{cases} \tag{6.3}$$

where we have $c_i + d_i = 0$ for $j \neq 0$ and $c_0 + d_0 = 2n$.

In Golay's paper, he also mentioned that this auto-correlative property of complementary series may lead to applications in the field of communication in so-called horizontal modulation systems, which permit several communication channels to utilize simultaneously the same frequency bands. These modulation systems are acquiring increasing importance. Of course, at the time when Golay did his work, he clearly did not know about CDMA but his description is very much like what a CDMA system will do.

In this early work, Golay also described some general properties of complementary series (or the primitive complementary codes defined here in this book), as summarized below.

1. The numbers of elements in two complementary series are equal. If it were not so, the pair of extreme elements of the longest series would remain unmatched by an unlike pair of elements with the same spacing in the other series.

2. Two complementary series are interchangeable. It will be noted that this results from the symmetry of the definition with respect to the **A** and **B** series, as discussed earlier.

3. The order of the elements of either or both of a pair of complementary series may be reversed. This results from the circumstance that the order of a pair of elements does not affect the parity of this pair.

4. One or both of a pair of complementary series may be altered without affecting their complementary property. This results from the circumstance that the parity of a pair is invariant under alteration of both elements of that pair.

5. Alternate elements in each of two complementary series may be altered, without affecting their complementary property. Such a transformation results in the change of both or neither elements of an even spaced pair, so that the parity of such pairs remains unaffected. Conversely, the parity of the odd spaced pairs is changed in both series, and this, by virtue of the remarks made in item 2, does not affect the complementary property of the series.

   It is concluded from properties 2 to 5 that a single pair of complementary series can be the basis for the construction of 26 pairs of complementary series (some of which may be identical) by either performing or not performing the following six operations:

   i. interchanging the series
   ii. reversing the first series
   iii. reversing the second series
   iv. altering the first series
   v. altering the second series
   vi. altering the elements of even order of each series.

6. When the complementary property is written explicitly for the two pairs which are $n-1$ elements distant in the **A** and **B** series, we obtain

$$a_1 + a_n + b_1 + b_n = 1 \pmod{2} \tag{6.4}$$

When the complementary property is written for the four pairs which are $n-2$ elements distant, we obtain

$$a_1 + a_2 + a_{n-1} + a_n + b_1 + b_2 + b_{n-1} + b_n = 0 \pmod{2} \tag{6.5}$$

and by addition modulo 2 of Equations (6.4) and (6.5), we have

$$a_2 + a_{n-1} + b_2 + b_{n-1} = 1 \pmod{2} \tag{6.6}$$

The process can continue and we have generally

$$a_r + a_{n-r+1} + b_r + b_{n-r+1} = 1 \pmod{2} \tag{6.7}$$

When $n = 2s + 1$ and $r = s + 1$, substitution in Equation (6.7) yields

$$a_{s+1} + a_{s+1} + b_{s+1} + b_{s+1} = 1 \pmod{2} \tag{6.8}$$

which is self-contradictory. Hence, it is concluded that the number of elements in complementary series must be even.

7. Let us define

$$u(x, y) = (x - y)^2, \quad x = 0 \text{ or } 1, y = 0 \text{ or } 1 \tag{6.9}$$

where the function of x and y thus defined is 0 or 1 depending upon the xy pair being even or odd. Therefore, for complementary series, we have

$$\sum_{s=1}^{s=v} u(a_s, a_{n-v+s}) + u(b_s, b_{n-v+s}) = v \tag{6.10}$$

which indicates that the total number of odd pairs of elements which are $n - v$ elements distant in two complementary series is v. That is also, of course, the total number of even pairs of elements which are spaced likewise. Now let

$$\begin{aligned} t(v) &= \frac{1}{2}\sum_{s=1}^{s=v}(a_s + a_{n-v+s} + b_s + b_{n-v+s}) - \frac{v}{2} \\ &= \frac{1}{2}\sum_{s=1}^{s=v}(a_s + a_{n-s+1} + b_s + b_{n-s+1}) - \frac{v}{2} \end{aligned} \tag{6.11}$$

The terms under each sum appear the same number of times as in the left-hand side (LHS) of Equation (6.10) and since there are v ls among them which must be associated each with a 0 to make v odd pairs, one half of the excess of the ls included under the $\sum$ sign over v represents the number of pairs of ls which are $n - v$ elements distant. The last member of Equation (6.11) may be utilized therefore to determine how many pairs of ls there should be with any given spacing, among two complementary series, and this reduces to approximately one quarter the number of pairs which must be examined in order to verify the complementary property of two series.

8. Let p and q designate the numbers of ls in two complementary series. Since the total number of even pairs in one must be equal to the total number of odd pairs in the other, we have

$$\frac{1}{2}p(p-1)+\frac{1}{2}(n-p)(n-p-1)=q(n-q) \tag{6.12}$$

where we have

$$n=(n-p-q)^2+(p-q)^2 \tag{6.13}$$

which tells us that the number of elements in complementary series must be expressible as a sum of at most two squares. Since this number must also be even, the allowable numbers up to 50 are 2, 4, 8, 10, 16, 18, 20, 26, 32, 34, 36, 40, and 50.

There are altogether 15 properties for complementary series (or what are called primitive complementary codes in this book) discussed in Golay's work and we do not go on to give a full description of them here. For those who are interested, please refer to the reference [251].

Two years after Goley's work, R. Turyn in his paper [252] gave more relevant discussions on primitive complementary codes. The title of his paper is 'Ambiguity function of complementary sequences', which seems very much related to the applications in radar systems, as the term ambiguity function appears only in books or papers dealing with radar systems. This paper published by Turyn would be called a letter in today's terminology as it is only a bit more than one page long.

In Turyn's paper, he defined that a pair of binary sequences (x_i), (y_i) will be called complementary if the sum of their (aperiodic) correlation functions is zero except for zero shift (i.e is a δ function). Turyn pointed out that there exist the following four notions, which are clearly equivalent:

1. A pair of complementary sequences each of length n.

2. A quaternary code of length $2n$, with correlation function a δ function, whose elements with odd index are all of the form $\pm\alpha$ and whose elements with even index are all of the form $\pm\gamma$.

3. A quaternary code of length $2n$, with correlation function a δ function, whose first n elements are all of the form $\pm\alpha$, and whose last n elements are all of the form $\pm\gamma$.

4. A binary sequence of length $2n$ whose correlation function is zero for all even shifts.

where a quaternary code is defined as a sequence whose terms are $\pm\alpha$ or $\pm\gamma$, with the multiplication rules implied by $\alpha\gamma=\gamma\alpha=0$ and $\alpha^2=\gamma^2=1$. In items 2 and 4, the odd and even elements form the two complementary sequences; in item 3 the first n and last n elements form the two complementary sequences.

Turyn also mentioned that it is simple to show that a quaternary code of length n has the property that if the autocorrelation function is a δ function, and if the ith element is $\pm\alpha$, then the $(n+1-i)$th element is $\pm\gamma$; i.e. omitting the signs, the sequence backwards is precisely the sequence with α and γ interchanged. Moreover, given any quaternary code of the type described above in item 2, we may form another such of length k times the original length simply by writing each element k times.

The method of forming long complementary sequences from shorter ones can be described most briefly in terms of 4 above. If (z_i) is a binary sequence of length $2n$ with correlation function equal to zero for all even shifts, then the sequence (z_i); $((-l)^i z_i)$, i.e. the sequence z'_i with

$$\begin{cases} z'_i = z_i, & 1\le i\le 2n \\ z'_{i+n}=(-1)^i z_i, & 2n+1\le i\le 4n \end{cases} \tag{6.14}$$

also has correlation function equal to zero for all even shifts. In terms of complementary sequences, if (x_i) and (y_i) are two complementary sequences of length n, then (x_i); (y_i) and (x_i); $(-y_i)$ are complementary sequences of length $2n$ (where the semicolon denotes juxtaposition).

Before the end of his paper, Turyn also mentioned that the purpose of this work is to point out that there is an analog to the uncertainty principle for pairs of sequences. The mean square value of the sum of the ambiguity functions is at least the sum of the values given by the uncertainty principle, and is equal to it only for pairs of sequences orthogonal in the usual sense, thus automatically for doubled sequences.

From the above description on the primitive complementary codes or sequences, we can see that the initial study on them was mainly for radar applications, not for communication purposes. In addition, the primitive complementary codes or sequences suggested by both Golay and Turyn ensure zero auto-correlation side lobes and zero cross-correlation functions only at even relative chip shifts. This almost excludes the possibility of their application in a CDMA system, in which we want to see good auto-correlation functions and good cross-correlation functions in all possible relative chip shifts. Therefore, in the rest of this book, we will not discuss the issues relating to primitive complementary codes, but rather we may only use them as examples to compare different complementary codes.

6.2.2 *COMPLETE COMPLEMENTARY CODES*

The name complete complementary codes was suggested by Naoki Suehiro [253, 254], in which he extended the concept of primitive complementary codes to the generation of the complete complementary code families whose auto-correlation function is zero for all even and odd shifts except the zero shift and whose cross-correlation function for any pair is zero for all possible shifts. It should be noted that, unlike the primitive complementary codes, the complete complementary codes are ideal orthogonal codes, whose auto-correlation function for any one code and cross-correlation function between any two codes are perfect in the sense that their orthogonality prevails even in asynchronous transmissions. Therefore, we call the complete complementary code a kind of ideal orthogonal code, which is very much different from Walsh-Hadamard codes or OVSF codes, whose orthogonality does not exist if they are used in asynchronous transmission channels, such as uplink channels in a mobile cellular system.

The work carried out in [253, 254] paved the way for practical applications of complementary codes in CDMA systems. It is also noted that a practical architecture for a CDMA system based on the complete complementary codes was first proposed in [237].

Before introducing the complete complementary codes, we would like to summarize what had been achieved before the complete complementary codes were proposed by Naoki Suehiro and Mitsutoshi Hatori in 1988 [254]. As mentioned earlier, Turyn [252] and Taki *et al.* [255] discussed a class of binary sequences whose elements are either 1 or -1 and whose auto-correlation function is zero for all even (multiples of two) shifts except the zero shift. These sequences have been called even-shift orthogonal sequences. A pair of binary sequences whose auto-correlation sum is zero, except for the zero shift, was discussed by Golay [251] and by Turyn [252]. The pair of sequences were then called a complementary code. The concept of complementary codes was further extended by Tseng and Liu [256], Sivaswamy [257], and Frank [258]. Therefore, it is clearly seen that the application of complementary codes for CDMA systems was not possible when the primitive complementary codes were proposed and studied in the 1960s and 1970s.

The concept of even-shift orthogonal sequences was extended by Naoki Suehiro [253] in his paper (written in Japanese) published in 1982. The new codes suggested in [253] were called N-shift cross-orthogonal sequences, and they have the properties of complementary codes. At that time, N-shift cross-orthogonal sequences were still not perfect orthogonal codes, but they offered some desirable properties which might be needed if they were used in a synchronous spread spectrum system. One of the most important characteristics in the N-shift cross-orthogonal sequences is that the cross-correlation function for any two codes in the set is zero if their relative chip shifts are N chips, where usually $N \geq 2$.

Having discussed the N-shift cross-orthogonal sequences, we are now ready to give the detailed definition of the complete complementary sequences.

Let us consider a sequence of L complex numbers or

$$\mathbf{S} = (s_1, \ldots, s_j, \ldots, s_L), \tag{6.15}$$

where we have $|s_j|$ for $j = 1, \ldots, L$.

Let us define the auto-correlation function $\rho(\mathbf{S}; i)$ for sequence $\mathbf{S}$ with i chips shift as

$$\rho(\mathbf{S}; i) \triangleq \begin{cases} \frac{1}{L}\sum_{j=1}^{L-i} s_j s_{j+i}^*, & i = 0, 1, \ldots, L-1 \\ \frac{1}{L}\sum_{j=1-i}^{L} s_j s_{j+i}^*, & i = -1, \ldots, -L+1 \end{cases} \tag{6.16}$$

where s_{j+i}^* is the complex conjugate of s_{j+i}. Let us also define the cross-correlation function $\rho(\mathbf{S}_1, \mathbf{S}_2; i)$ between sequences $\mathbf{S}_1$ and $\mathbf{S}_2$ with the i-chip shift as

$$\rho(\mathbf{S}; i) \triangleq \begin{cases} \frac{1}{L}\sum_{j=1}^{L-i} s_{1,j} s_{2,j+i}^*, & i = 0, 1, \ldots, L-1 \\ \frac{1}{L}\sum_{j=1-i}^{L} s_{1,j} s_{2,j+i}^*, & i = -1, \ldots, -L+1 \end{cases} \tag{6.17}$$

where we have used the definitions for the two sequences as

$$\begin{cases} \mathbf{S}_1 = (s_{1,1}, \ldots, s_{1,L}) \\ \mathbf{S}_2 = (s_{2,1}, \ldots, s_{2,L}) \end{cases} \tag{6.18}$$

An N-shift auto-orthogonal sequence of length L is defined as a sequence $\mathbf{S}$, defined by Equation (6.15), whose auto-correlation function $\rho(\mathbf{S}; i)$, defined by Equation (6.17), is zero for any N-multiple shift except the zero shift, or

$$\rho(\mathbf{S}; i) = 0, \quad i = \pm N, \pm 2N, \ldots \tag{6.19}$$

A pair of N-shift cross-orthogonal sequences of length L is defined as a pair of sequences $\mathbf{S}_1$ and $\mathbf{S}_2$, both of length L and defined by Equation (6.15), whose cross-correlation function $\rho(\mathbf{S}_1, \mathbf{S}_2; i)$, defined by Equation (6.18), is zero for any N-multiple shift (including the zero-shift), or

$$\rho(\mathbf{S}_1, \mathbf{S}_2; i) = 0, \quad i = 0, \pm N, \pm 2N, \ldots \tag{6.20}$$

A set of M sequences $\{\mathbf{S}_1, \ldots, \mathbf{S}_M\}$ is called a set of N-shift cross-orthogonal sequences if any two out of the M sequences form a pair of N-shift cross-orthogonal sequences, or

$$\rho(\mathbf{S}_j, \mathbf{S}_k; i) = 0, \quad i = 0, \pm N, \pm 2N, \ldots, \tag{6.21}$$

where $\mathbf{S}_j, \mathbf{S}_k \in \{\mathbf{S}_1, \ldots, \mathbf{S}_M\}$ for $j \neq k$.

A set of M sequences $\{\mathbf{S}_1, \ldots, \mathbf{S}_M\}$ is called an auto-complementary sequence of order M if the sum of the M auto-correlation functions is zero for any i-chip shift, except the zero shift, or

$$\sum_{j=1}^{M} \rho(\mathbf{S}_j; i) = 0, \quad i \neq 0. \tag{6.22}$$

Two sets of M sequences $\{\mathbf{S}_1, \ldots, \mathbf{S}_M\}$ and $\{\mathbf{T}_1, \ldots, \mathbf{T}_M\}$ are called a cross-complementary code of order M if the sum of M cross-correlation functions between the corresponding sequences $\mathbf{S}_j$ and $\mathbf{T}_j$ is zero for any i-chip shift (including the zero shift), or

$$\sum_{j=1}^{M} \rho(\mathbf{S}_j, \mathbf{T}_j; i) = 0, \quad \text{for any } i. \tag{6.23}$$

M sets of M sequences $\{\mathbf{S}_{1,1}, \ldots, \mathbf{S}_{1,M}\}, \ldots, \{\mathbf{S}_{M,1}, \ldots, \mathbf{S}_{M,M}\}$ are called a complete complementary code of order M if every set is an auto-complementary code and every pair of distinct sets chosen from the M sets is a cross-complementary code, or

$$\begin{cases} \sum_{j=1}^{M} \rho(\mathbf{S}_{k,j}; i) = 0, & i \neq 0;\ k = 1, \ldots, M \\ \sum_{j=1}^{M} \rho(\mathbf{S}_{k,j}, \mathbf{S}_{l,j}; i) = 0, & \text{for any } i;\ k, l = 1, \ldots, M;\ k \neq l \end{cases} \tag{6.24}$$

Therefore, it is seen from the above definitions that the auto-correlation function and cross-correlation functions of complete complementary codes are defined based on the flock, which will be assigned to a user for code division multiple access. Each flock here consists of M element codes, which should be used jointly, instead of individually as is the case for all traditional CDMA systems.

As to be clarified later, any complementary code sets can always be defined by three parameters, set size (K), flock size (M), and element code length (N), which can be denoted in a concise form as (N, M, K). Here in this case, the set size and flock size of the complete complementary codes are made equal, being M, and the element code length is L. This implies that each user should use M element codes and the whole system can only support maximal M users. It is easily seen from Equation (6.24) that the complete complementary code offers ideal orthogonality among the codes.

We will see later that every ideally orthogonal code should bear this unique property that its set size must be equal to its flock size in order to be ideally orthogonal. An ideally orthogonal code set is defined as a code set in which the auto-correlation function for any one code and cross-correlation functions between any two codes are always ideal, such that all auto-correlation function side lobes are zero whatever the relative chip shift and all cross-correlation functions are zero for all possible relative chip shift.

6.2.3 EXTENDED COMPLEMENTARY CODES

As to be shown in the following section, a complete complementary code set can offer ideal orthogonality among all codes in the set. The ideal orthogonality of a code set should be defined as follows:

- A complete complementary code always yields an ideal auto-correlation function for a flock of element codes.
- A pair of complete complementary codes always offers an ideal cross-correlation function between any pair of flocks in the same code set.
- The above mentioned ideal auto-correlation functions and cross-correlation functions are ensured regardless of the transmission mode of a CDMA system. That is to say, their ideal

auto-correlation functions and cross-correlation functions should be ensured in both synchronous and asynchronous transmission channels.

It should be stressed here again that each user in a CDMA system based on the complementary codes should be assigned one flock of M element codes, each of which has a length of N chips. The set size of such complementary codes is K, which is just equal to the number of different flocks (M) in the set.

There are several distinctions between traditional CDMA codes and complementary codes. The major difference is that the orthogonality of the complementary codes is based on a flock of element codes jointly, instead of a single code as in the case for traditional CDMA codes. It means that every user is assigned a flock of element codes as its signature code. Spread sub-streams are transmitted over different channels (via either different carriers or different time slots) and arrive at the correlator receiver at the same time to produce an auto-correlation peak. Unlike traditional CDMA codes, the orthogonality of complementary codes is based on a flock of codes as described above. Therefore the correlation properties of complementary codes have to be discussed on a flock basis. The auto-correlation function of complementary codes is always zero for all possible integer chip shifts except the zero shift. The cross-correlation function of any pair of complementary codes in a complementary code set is zero for any possible integer chip shifts. It can also be shown that the same result will apply to the cases with non-integer chip shifts.

We will also show that the complete complementary code is the only CDMA spreading code which offers such an ideal orthogonality. Therefore, more extensive study of them is significant and we should acknowledge the contributions made by many previous researchers to the generation of the complete complementary codes [237;251–254,259–267]. Based on the work of complete complementary codes, we have proposed many different types of complementary codes, such as super complementary codes, pair-wise complementary codes, column-wise complementary codes, etc., which will be introduced later in this chapter.

However, the complete complementary codes have some severe problems, which makes it impossible to apply them to a practical CDMA system. The major problem with complete complementary codes is associated with their small set-size, or parameter K. The set size (number of flocks) of a complementary code set is always equal to the flock size (number of element codes in one flock) and is $M = \sqrt{L}$. The processing gain (PG) of complete complementary (CC) codes, on the other hand, is equal to $L\sqrt{L}$, where L is the length of the element code. Table 6.1 shows the set sizes (K) and flock sizes (M) for various CC codes with different element code lengths (L). It is seen from Table 6.1 that, with a fixed element code length L, the set size will be equal to $M = \sqrt{L}$, which is too small to support a sufficiently large number of users with a complementary code set.

Therefore, in order to increase the set size to support more users in a CDMA system based on complementary codes, extended complementary codes were introduced in [254]. Clearly, in this sense, the extended complementary codes are nothing new compared with the complete complementary codes. The properties and generation approach for the extended complementary codes will be introduced in Subsection 6.3.2.

Table 6.1 Set sizes and flock sizes for complete complementary codes with various element code lengths L.

Element code length ($L = 4^n$)	4	16	64	256	1024	4096
PG ($L\sqrt{L}$)	8	64	512	4096	32 768	262 144
Family size ($\sqrt{L}$)	2	4	8	16	32	64
Flock size ($\sqrt{L}$)	2	4	8	16	32	64

6.2.4 SUPER COMPLEMENTARY CODES

The super complementary codes were proposed by us to overcome the problems with the complete complementary codes and extended complementary codes, both of which were introduced in [254]. There are several technical limitations with either complete complementary codes or extended complementary codes, as listed below.

- First of all, to maintain a sufficiently large set size, both complete complementary codes and extended complementary codes have to use very long element codes. In this way, a CDMA system based on them will be very complicated due to the fact that a very high sampling rate has to be used to sample at least one sample per chip, given a certain data transmission rate. A high sampling rate will tremendously increase the signal processing load of the central processing unit (CPU), and thus slow down the overall processing speed as a whole.

- Second, the use of very long element codes will also make it very difficult to increase the data transmission rate with a fixed signal processing power. With the increase of data rate, the bit duration will be substantially reduced, thus making the chip width even shorter. This will make it very difficult for the receiver to maintain good synchronization with the sent data streams.

- Yet again, the application of long element codes will also make it very difficult to implement various rate-matching algorithms to support multimedia wireless communications. The complete complementary codes use a fixed flock size, which is equal to the set size or $\sqrt{L}$, if the element code length is equal to L. The ratio between the flock size (or the set size) $\sqrt{L}$ and element code length L is a fixed number, equal to $\frac{\sqrt{L}}{L}$. There is no way to change this ratio, thus leaving the system designer very little degree of freedom. As a matter of fact, we will show later that all ideally orthogonal complementary codes have the property that their set size (K) is always equal to their flock size (M). Therefore, if we can make the flock size larger, we can increase the set size as well, to support more users in a CDMA system. Unfortunately, the rigid ratio between the flock size and element code length makes it impossible to increase the set size under a fixed processing gain, which is equal to the product of flock size and element code length $(M \times N)$.

We have been motivated to find some better complementary codes, which can maintain the same ideal orthogonality and at the same time offer a relatively large set size in order to support more users in a CDMA system. We have been searching for them in several different directions. One direction is to find some ideally orthogonal complementary codes with a relatively large set size under a fixed processing gain. That means that we should try to enlarge the flock size as much as possible in order to have a large set size, due to the fact that the set size is always equal to the flock size for any ideally orthogonal complementary codes.[2]

This work has been done very fruitfully and several different types of ideally orthogonal codes (we will introduce them in the text following) have been found. One group is the super complementary codes, whose name was given due to its superior property that its set size can reach its maximal value, being equal to the processing gain of the super complementary codes. The properties and generation approach for the super complementary codes will be discussed in detail in the next section.

6.2.5 PAIR-WISE COMPLEMENTARY CODES

As discussed earlier, there are two major groups of CDMA codes available for CDMA applications. One is unitary codes, which work on a one-code-per-user basis. Almost all of these popular CDMA

[2]At this moment, we only could deduce this conclusion from our observation for all available complete complementary codes or extended complementary codes. However, we will show in the REAL approach to be discussed in the next section that all ideally orthogonal complementary codes always have this property.

codes adopted in the current 2G and 3G systems are unitary codes. All these unitary codes can be further classified into two sub-groups, one being quasi-orthogonal codes (such as m-sequences, Gold codes, Kasami codes, etc.) and the other being orthogonal codes (such as Walsh-Hadamard sequences, OVSF codes, etc.).

Another group of CDMA codes is complementary codes, which were first studied by Golay and Turyn in the early 1960s for their possible applications in radar systems. Later on, there came some sporadic research on complementary codes but not much serious attention had been given to them, due mainly to their implementation complexity and relatively small set sizes. unlike the unitary codes, the complementary codes work on a flock-per-user basis. It was found in our earlier works that joint application of orthogonal complementary codes and offset stacking (OS) spreading can effectively improve the bandwidth-efficiency of a CDMA system in addition to several other desirable properties, such as isotropic MAI-free operation, agility to implement rate-matching, suitability for burst traffic applications, and so on.

In order for a CDMA system to work on a one-flock-per-user basis, different element codes should be sent through the channel via different carriers or time slots. Obviously, the bigger the flock size, the more complex the transceiver has to be. Therefore, it is natural for us to think about how to generate a complementary code set with relatively small flock size, but still with an acceptable correlation property. This is the second research direction we have been working on as an effort to improve on the complete complementary codes and extended complementary codes. The generalized pair-wise complementary (GPC) code set [268] was proposed if the flock size of a complementary code set is limited to two. In fact, the GPC codes were introduced as a result to combine the advantages of unitary codes (for their implementation simplicity) and complementary codes (for their robustness against MAI and MI). A generation approach based on complete complementary codes and generalized even-shift orthogonal sequences is proposed to generate GPC code sets with desirable interference-free windows (IFWs) that are uniform across the entire code set. The GPC codes work on pairs, allowing us to use highly power-efficient QPSK to implement the carrier modems with a single carrier. The GPC code sets possess sparsely distributed auto-correlation function (ACF) side-lobes and cross-correlation function (CCF) levels outside the IFW, implying that controllable ACF and CCF levels can be ensured even outside the IFWs. The width of the IFWs for the GPC codes can be as wide as $8N$, where N is the element code length of the complete complementary codes used to generate the GPC codes. Therefore, using different N we can change the width of IFWs of GPC code sets. If only half of the GPC code set is used in a CDMA system, the width of CCF IFWs can be further extended to cover the entire code length, ensuring an ideally MAI-free operation of a CDMA system.

6.2.6 COLUMN-WISE COMPLEMENTARY CODES

In the previous subsections, we have introduced super complementary codes and pair-wise complementary codes, which were proposed as an effort to overcome some inherent problems with complete complementary codes, in particular their relatively small set size under a fixed spreading factor (or processing gain).

The super complementary codes can offer a large set size compared to that of complete complementary codes and thus they are able to support more users within the code set. It is noted that both super complementary codes and complete complementary codes are ideally orthogonal codes, meaning that they can offer perfect orthogonality for their applications in both synchronous and asynchronous channels. On the other hand, the pair-wise complementary codes are not ideally orthogonal codes, implying that their cross-correlation functions between any two codes may not necessarily be zero for all possible relative chip shifts. However, their cross-correlation functions are still under control (their maximal level should not exceed a certain value). The most important advantage for the pair-wise complementary codes is that they work on a pair basis, such that a transmitter can use a single-carrier (instead of multi-carrier) modem to send CDMA signals. Therefore, the complexity of a CDMA transceiver can be substantially reduced.

Our research on searching for new CDMA code sets continued, and in this subsection we will propose another type of interesting complementary code, namely column-wise complementary code, which was introduced in order to allow us to understand better the relationship between the code generation process and its orthogonality properties.

The introduction of column-wise complementary codes was largely based on N-shift orthogonal sequences [254,255], which are defined as a sequence always orthogonal to its peer sequences as long as their relative chip shift is equal to exactly N chips. It is noted that the N-shift orthogonal sequences comprise a fairly large group of sequences, which should not be considered as ideally orthogonal sequences as they may not be orthogonal with each other if their relative chip shifts are not exactly equal to N chips. It is important to note that the value of N in an N-shift orthogonal sequence usually is an even number due to the fact that an odd relative shift can make it extremely difficult to cancel all non-zero cross-correlation levels between any two codes. Among all N-shift orthogonal sequences, the most commonly used one is the even-shift orthogonal sequences, which can ensure zero cross-correlation function between any two codes if their relative chip shift is an even number. Therefore, the theory on the N-shift orthogonal sequences forms the basis for the introduction of the column-wise complementary codes.

Study of column-wise complementary codes is significant due to the fact that it gives a very general approach to generate many different types of complementary codes under a unified framework. In addition, study of column-wise complementary codes also helps us to understand much better the relationship between the ideal orthogonality and code structure for any ideally orthogonal complementary codes, such as complete complementary codes and super complementary codes, etc.

The reason that we call this fairly large group of ideally orthogonal complementary codes 'column-wise complementary codes' is that both generation process and the orthogonality of the codes can be nicely presented by the correlation function between different columns of the codes, if we write a code flock (which should be assigned to one user for CDMA and usually consists of M element codes) into a matrix and each element code is written as a column of the matrix. Based on this methodology, we will show that virtually all ideally orthogonal complementary codes can be generated and analyzed via column-wise correlation functions of the codes.

To illustrate the issue more clearly, let us look at an example code set, which in fact is a complete complementary code set with its parameters (M, N, K) being $(2, 4, 2)$, where M, N, and K stand for the flock size, element code length, and set size, respectively. The code set can be written as

$$\begin{aligned}\mathbf{C}^{(1)}_{M\times N} &= \begin{pmatrix} c_1^{(1)}(1) & c_1^{(1)}(2) & c_1^{(1)}(3) & c_1^{(1)}(4) \\ c_2^{(1)}(1) & c_2^{(1)}(2) & c_2^{(1)}(3) & c_2^{(1)}(4) \end{pmatrix} \\ &= \begin{pmatrix} \mathbf{c}^{(1)}(1) & \mathbf{c}^{(1)}(2) & \mathbf{c}^{(1)}(3) & \mathbf{c}^{(1)}(4) \end{pmatrix} \end{aligned} \tag{6.25}$$

$$\begin{aligned}\mathbf{C}^{(2)}_{M\times N} &= \begin{pmatrix} c_1^{(2)}(1) & c_1^{(2)}(2) & c_1^{(2)}(3) & c_1^{(2)}(4) \\ c_2^{(2)}(1) & c_2^{(2)}(2) & c_2^{(2)}(3) & c_2^{(2)}(4) \end{pmatrix} \\ &= \begin{pmatrix} \mathbf{c}^{(2)}(1) & \mathbf{c}^{(2)}(2) & \mathbf{c}^{(2)}(3) & \mathbf{c}^{(2)}(4) \end{pmatrix} \end{aligned} \tag{6.26}$$

We would like to take a look at the auto-correlation function of a code and cross-correlation function between any two codes in the set. In the analysis conducted in this book, we will only use partial or aperiodic correlation functions to study the orthogonality of a CDMA code set. It is proved in Appendix A that if a CDMA code set retains perfect aperiodic auto-correlation and aperiodic cross-correlation functions, it will also give perfect periodic auto-correlation and periodic cross-correlation functions, and vice versa. The aperiodic and periodic auto-correlation and cross-correlation functions are illustrated in Figures A.1 to A.7.

The perfect auto-correlation functions indicate that the auto-correlation function should be zero for all relative chip shifts except at the zero shift, and the perfect cross-correlation functions represent the fact that the cross-correlation functions should be zero for all possible relative chip shifts. The condition of the perfect auto-correlation functions will ensure an ideal multipath interference-free operation in a CDMA system and that of the perfect cross-correlation functions will enable multiple access interference (MAI) free operation in a CDMA system.

In a real CDMA system, the aperiodic auto-correlation functions and cross-correlation functions usually do not take place unless in some very rare situation, such as at the beginning or the end of a packet. On the other hand, the periodic auto-correlation functions and cross-correlation functions will mainly govern the performance of a CDMA system. There are two different types of periodic auto-correlation functions, even periodic auto-correlation functions and odd periodic auto-correlation functions. Similarly, there are two different types of periodic cross-correlation functions, even periodic cross-correlation functions and odd periodic cross-correlation functions.

After having defined the terminologies, we are ready to take a look at a simple complementary code set as shown in Equations (6.25) and (6.26). As aforementioned, we use partial auto-correlation and partial cross-correlation functions to check its orthogonal properties.

The partial auto-correlation function of either code (6.25) or code (6.26) can be written as

$$\rho\left(\mathbf{C}_{M\times N}^{(k)}; i\right) = \frac{1}{MN}\sum_{m=1}^{M}\sum_{n=1}^{N-i} c_m^{(k)}(n)\, c_m^{(k)}(n+i) \tag{6.27}$$

where k takes either 1 or 2.

The partial cross-correlation function between the two codes given in (6.25) and (6.26) can be written as

$$\rho\left(\mathbf{C}_{M\times N}^{(k)}, \mathbf{C}_{M\times N}^{(k')}; i\right) = \frac{1}{MN}\sum_{m=1}^{M}\sum_{n=1}^{N-i} c_m^{(k')}(n)\, c_m^{(k)}(n+i) \tag{6.28}$$

where $k \neq k'$. Now, we can insert the code given in (6.25) into Equation (6.27), and we obtain the peak auto-correlation level as

$$\begin{aligned}
\rho\left(\mathbf{C}_{2\times 4}^{(1)}; i=0\right) &= \frac{1}{8}\sum_{m=1}^{2}\sum_{n=1}^{4} c_m^{(k)}(n)\, c_m^{(k)}(n) \\
&= \frac{1}{8}\left(c_1^{(1)}(1)\, c_1^{(1)}(1) + c_2^{(1)}(1)\, c_2^{(1)}(1)\right) + \frac{1}{8}\left(c_1^{(1)}(2)\, c_1^{(1)}(2) + c_2^{(1)}(2)\, c_2^{(1)}(2)\right) \\
&\quad + \frac{1}{8}\left(c_1^{(1)}(3)\, c_1^{(1)}(3) + c_2^{(1)}(3)\, c_2^{(1)}(3)\right) + \frac{1}{8}\left(c_1^{(1)}(4)\, c_1^{(1)}(4) + c_2^{(1)}(4)\, c_2^{(1)}(4)\right) \\
&= \frac{1}{8}\rho\left(\mathbf{c}^{(1)}(1)\right) + \frac{1}{8}\rho\left(\mathbf{c}^{(1)}(2)\right) + \frac{1}{8}\rho\left(\mathbf{c}^{(1)}(3)\right) + \frac{1}{8}\rho\left(\mathbf{c}^{(1)}(4)\right)
\end{aligned} \tag{6.29}$$

It is seen from (6.29) that $\rho\left(\mathbf{c}^{(1)}(1)\right)$ is nothing but the auto-correlation function of the first column vector in the code matrix given in (6.25). The same happens to the other three terms in (6.29). Therefore, the auto-correlation of a complementary code is equal to the summation of column-wise auto-correlations of all element codes (the column vectors in the code matrix).

Now, let us look at the one-chip offset auto-correlation function of the same code matrix, or

$$\begin{aligned}
\rho\left(\mathbf{C}^{(1)}_{2\times 4}; i=1\right) &= \frac{1}{8}\sum_{m=1}^{2}\sum_{n=1}^{3} c^{(k)}_m(n)\, c^{(k)}_m(n+1) \\
&= \frac{1}{8}\left(c^{(1)}_1(1)\, c^{(1)}_1(2) + c^{(1)}_2(1)\, c^{(1)}_2(2)\right) + \frac{1}{8}\left(c^{(1)}_1(2)\, c^{(1)}_1(3) + c^{(1)}_2(2)\, c^{(1)}_2(3)\right) \\
&\quad + \frac{1}{8}\left(c^{(1)}_1(3)\, c^{(1)}_1(4) + c^{(1)}_2(3)\, c^{(1)}_2(4)\right) \\
&= \frac{1}{8}\rho\left(\mathbf{c}^{(1)}(1); \mathbf{c}^{(1)}(2)\right) + \frac{1}{8}\rho\left(\mathbf{c}^{(1)}(2); \mathbf{c}^{(1)}(3)\right) + \frac{1}{8}\rho\left(\mathbf{c}^{(1)}(3); \mathbf{c}^{(1)}(4)\right) \qquad (6.30)
\end{aligned}$$

which gives nothing but the summation of column-wise correlation functions between the first and second column vectors, that between the second and third column vectors, and that between the third and fourth column vectors, subject to normalization. The same observation can be made about the auto-correlation functions with any other chip offsets.

The fact that the auto-correlation function of a complementary code can always be expressed as a summation of column-wise auto-correlation functions for all different element codes can also be easily seen in many other orthogonal complementary codes, such as complete complementary codes, super complementary codes, and so forth. Let us look at another more general case with two orthogonal complementary codes, whose parameters are $(N, M, K) = (N, M, M)$, as follows

$$\begin{aligned}
\mathbf{C}^{(k)}_{M\times N} &= \begin{pmatrix} c^{(k)}_1(1) & c^{(k)}_1(2) & \dots & c^{(k)}_1(N) \\ c^{(k)}_2(1) & c^{(k)}_2(2) & \dots & c^{(k)}_2(N) \\ \vdots & \vdots & \ddots & \vdots \\ c^{(k)}_M(1) & c^{(k)}_M(2) & \dots & c^{(k)}_M(N) \end{pmatrix} \\
&= \begin{pmatrix} \mathbf{c}^{(k)}(1) & \mathbf{c}^{(k)}(2) & \dots & \mathbf{c}^{(k)}(N) \end{pmatrix} \qquad (6.31) \\
\mathbf{C}^{(k')}_{M\times N} &= \begin{pmatrix} c^{(k')}_1(1) & c^{(k')}_1(2) & \dots & c^{(k')}_1(N) \\ c^{(k')}_2(1) & c^{(k')}_2(2) & \dots & c^{(k')}_2(N) \\ \vdots & \vdots & \ddots & \vdots \\ c^{(k')}_M(1) & c^{(k')}_M(2) & \dots & c^{(k')}_M(N) \end{pmatrix} \\
&= \begin{pmatrix} \mathbf{c}^{(k')}(1) & \mathbf{c}^{(k')}(2) & \dots & \mathbf{c}^{(k')}(N) \end{pmatrix} \qquad (6.32)
\end{aligned}$$

which can be inserted into (6.28) to yield the auto-correlation function with i chips offset as

$$\begin{aligned}
\rho\left(\mathbf{C}^{(k)}_{M\times N}; i\right) &= \frac{1}{MN}\sum_{m=1}^{M}\sum_{n=1}^{N-i} c^{(k)}_m(n)\, c^{(k)}_m(n+i) \\
&= \frac{1}{MN}\sum_{n=1}^{N-i}\sum_{m=1}^{M} c^{(k)}_m(n)\, c^{(k)}_m(n+i) \\
&= \frac{1}{MN}\sum_{n=1}^{N-i}\left(c^{(k)}_1(n)\, c^{(k)}_1(n+i) + c^{(k)}_2(n)\, c^{(k)}_2(n+i) + \dots + c^{(k)}_M(n)\, c^{(k)}_M(n+i)\right) \\
&= \frac{1}{MN}\sum_{n=1}^{N-i}\rho\left(c^{(k)}(n); c^{(k)}(n+i)\right) \qquad (6.33)
\end{aligned}$$

which again shows that the auto-correlation function of a complementary code is equal to the sum of the column-wise correlation functions between the first and $(1+i)$th column vectors, the second and $(2+i)$th column vectors, ..., the $(N-i)$th and Nth column vectors, subject to the normalization.

We can also take a look at the cross-correlation function between the two codes given in (6.31) and (6.32) as

$$\begin{aligned}
\rho\left(\mathbf{C}_{M\times N}^{(k)}, \mathbf{C}_{M\times N}^{(k')}; i\right) &= \frac{1}{MN}\sum_{m=1}^{M}\sum_{n=1}^{N-i} c_m^{(k)}(n)\, c_m^{(k')}(n+i) \\
&= \frac{1}{MN}\sum_{n=1}^{N-i}\sum_{m=1}^{M} c_m^{(k)}(n)\, c_m^{(k')}(n+i) \\
&= \frac{1}{MN}\sum_{n=1}^{N-i}\left(c_1^{(k)}(n)\, c_1^{(k')}(n+i) + c_2^{(k)}(n)\, c_2^{(k')}(n+i)\right. \\
&\qquad \left. + \cdots + c_M^{(k)}(n)\, c_M^{(k')}(n+i)\right) \\
&= \frac{1}{MN}\sum_{n=1}^{N-i}\rho\left(c^{(k)}(n); c^{(k')}(n+i)\right)
\end{aligned} \tag{6.34}$$

which also illustrates that the cross-correlation function between the two orthogonal complementary codes with i chips offset can be obtained by summing all column-wise cross-correlation functions between the nth and $(n+i)$th column vectors of the two code matrices, where $n = 1, \ldots, N-i$.

The analysis given above demonstrates the clear fact that most orthogonal complementary codes can be further studied based on their column-wise correlation properties. This gives us a new way to search for some unknown CDMA codes or explore more in-depth properties of some orthogonal complementary code sets, otherwise not possible. This is the reason why we use the name 'column-wise complementary code' for them in this book. It is stressed here again that the introduction of the name 'column-wise complementary code' is not only for creation of some new orthogonal complementary codes, but also for having a better way to study the properties of orthogonal complementary codes. We do find some unknown orthogonal complementary codes along with the direction given by the column-wise correlation properties of the orthogonal complementary codes, but more precisely we would like to say that the introduction of the concept of 'column-wise complementary code' can help us understand much better the orthogonality formation process in a complementary code set. Therefore, we can safely say that the column-wise complementary codes form a fairly large superset of orthogonal complementary codes, which may encompass most orthogonal complementary codes ever discussed in this book. This is the most interesting feature of column-wise complementary codes.

A column-wise complementary code set can be characterized by four parameters, denoted as $(K, M, N, \mathcal{N})$, where K denotes the set size, M denotes the flock size, N is the length of an element code, and $\mathcal{N}$ indicates the fact that this particular column-wise complementary code set is generated using an $\mathcal{N}$-shift orthogonal sequence. The $\mathcal{N}$-shift orthogonal sequence is defined as a sequence which exhibits ideal orthogonality if and only if at the $\mathcal{N}$ offset chips. Therefore, we can see here that a column-wise complementary code can be uniquely determined by the four parameters $(K, M, N, \mathcal{N})$. For more information on the $\mathcal{N}$-shift orthogonal sequences, readers may refer to [254, 255].

6.3 GENERATION OF COMPLEMENTARY CODES

Up to now, we have introduced six different types of complementary codes, namely primitive, complete extended, super , pair-wise, complementary codes, and column-wise complementary codes.

As we can see from the introduction, some complementary codes are perfectly orthogonal codes (complete, extended, super complementary codes, and column-wise complementary codes) and some are quasi-orthogonal codes (primitive and pair-wise complementary codes). Some may form a subset of the others and vice versa.

In the previous sections, we have only touched on some very basic properties of the complementary codes, and have not given the methods to generate them. In this section we will find some explicit ways to demonstrate that different complementary codes can be generated using different approaches, some of which are very general in the sense that they can be used to generate several different types of complementary codes, and some of which are only valid for one particular complementary code but not others.

In particular, we will also illustrate a very unique code design approach, called the 'real environment adaptation linearization' (REAL) approach, which reveals much important information regarding the architecture of the next generation CDMA system, such as the possibility to implement an interference-free CDMA system, the type of CDMA codes for such an interference-free CDMA system, the relation between the code set size and the flock size, etc. Therefore, understanding the knowledge covered in this section is extremely important to design the next generation CDMA systems.

In this chapter, the discussions on the generation methods for various complementary codes will start with the complete complementary code, which is also the first perfectly orthogonal complementary code ever proposed. We will not discuss primitive complementary codes, as they are only orthogonal at some particular chip offsets, and thus they virtually are not useful for any CDMA applications.

6.3.1 GENERATION OF COMPLETE COMPLEMENTARY CODES

The complete complementary code is the first orthogonal complementary code ever proposed. Therefore, it is of great importance to understand the method by which they can be generated. We will give a method to generate complete complementary codes whose element code length is N^2 as follows.

Step 1

Let $\mathbf{A}$ be an N-dimensional orthogonal matrix which consists of complex elements a_{ij}, whose absolute values are $|a_{ij}| = 1$ for $i = 1, 2, \ldots, N$ and $j = 1, 2, \ldots, N$, such that

$$\mathbf{A} = \begin{pmatrix} a_{11} & a_{12} & \cdots & a_{1N} \\ a_{21} & a_{22} & \cdots & a_{2N} \\ \vdots & \vdots & \vdots & \vdots \\ a_{N1} & a_{N2} & \cdots & a_{NN} \end{pmatrix} \tag{6.35}$$

and for an obvious reason the inner product of any two different rows in the above matrix should be zero, or

$$\sum_{i=1}^{N} a_{ji} a_{ki}^* = 0, \qquad j \neq k \tag{6.36}$$

Then, we can formulate a sequence $\tilde{\mathbf{A}}$ based on the matrix given in (6.35) such that

$$\tilde{\mathbf{A}} = \begin{pmatrix} a_{11}, & a_{12}, & \ldots, & a_{1N}, \\ a_{21}, & a_{22}, & \ldots, & a_{2N}, \\ \cdots & \cdots & \cdots & \cdots \\ a_{N1}, & a_{N2}, & \ldots, & a_{NN} \end{pmatrix} \tag{6.37}$$

It can be easily shown that the auto-correlation function of sequence $\tilde{\mathbf{A}}$ is zero for all N-multiple shifts, except the zero shift. This is because the inner products of any two different rows of the orthogonal matrix $\mathbf{A}$ are always zero, as defined earlier.

Now, let $\mathbf{A}_1, \mathbf{A}_2, \ldots, \mathbf{A}_N$ be the rows of the orthogonal matrix $\mathbf{A}$ as defined earlier, or

$$\begin{pmatrix} \mathbf{A}_1 \\ \mathbf{A}_2 \\ \vdots \\ \mathbf{A}_N \end{pmatrix} \tag{6.38}$$

Therefore, Equation (6.37) can be rewritten into

$$\tilde{\mathbf{A}} = (\mathbf{A}_1, \mathbf{A}_2, \ldots, \mathbf{A}_N) \tag{6.39}$$

Step 2

Let $\mathbf{B}$ be another orthogonal matrix such that

$$\mathbf{B} = \begin{pmatrix} b_{11} & b_{12} & \ldots & b_{1N} \\ b_{21} & b_{22} & \ldots & b_{2N} \\ \vdots & \vdots & \vdots & \vdots \\ b_{N1} & b_{N2} & \ldots & b_{NN} \end{pmatrix} \tag{6.40}$$

where $|b_{ij}| = 1$ for $i = 1, 2, \ldots, N$ and $j = 1, 2, \ldots, N$. Based on the two orthogonal matrices $\mathbf{A}$ and $\mathbf{B}$, we can formulate a set of N sequences of length N^2, or $\mathbf{C}_1, \mathbf{C}_2, \ldots, \mathbf{C}_N$, such that

$$\begin{cases} \mathbf{C}_1 = & (b_{11}\mathbf{A}_1 & b_{12}\mathbf{A}_2 & \ldots & b_{1N}\mathbf{A}_N) \\ \mathbf{C}_2 = & (b_{21}\mathbf{A}_1 & b_{22}\mathbf{A}_2 & \ldots & b_{2N}\mathbf{A}_N) \\ \vdots & \vdots & \vdots & \vdots & \vdots \\ \mathbf{C}_N = & (b_{N1}\mathbf{A}_1 & b_{N2}\mathbf{A}_2 & \ldots & b_{NN}\mathbf{A}_N) \end{cases} \tag{6.41}$$

It can be easily shown that each code $\mathbf{C}_i$ $(i = 1, 2, \ldots, N)$ given in (6.41) satisfies the property of the N-shift auto-orthogonal sequence of length N^2, as defined in Subsection 6.2.2.

Step 3

After having obtained the set of N sequences $\mathbf{C}_i$ $(i = 1, 2, \ldots, N)$, we can proceed as follows. We rewrite the sequence set $\mathbf{C}_1, \mathbf{C}_2, \ldots, \mathbf{C}_N$ as

$$\begin{cases} \mathbf{C}_1 = (c_{11}, c_{12}, \ldots, c_{1N^2}) \\ \mathbf{C}_2 = (c_{21}, c_{22}, \ldots, c_{2N^2}) \\ \vdots \\ \mathbf{C}_i = (c_{i1}, c_{i2}, \ldots, c_{iN^2}) \\ \vdots \\ \mathbf{C}_N = (c_{N1}, c_{N2}, \ldots, c_{NN^2}) \end{cases} \tag{6.42}$$

Now, let us introduce the third $N \times N$ orthogonal matrix $\mathbf{D}$, such that $|d_{jk}| = 1$ for $j = 1, 2, \ldots, N$ and $k = 1, 2, \ldots, N$. We can make a set of N sequence $\mathbf{E}_{ij}$ $(i, j = 1, 2, \ldots, N)$ of length N^2 from each $\mathbf{C}_i$ ($i = 1, 2, \ldots, N$) and $\mathbf{D}$ as follows

$$\mathbf{E}_{ij} = \begin{pmatrix} c_{i1}d_{j1}, & c_{i2}d_{j2}, & \ldots, & c_{iN}d_{jN}, \\ c_{i(N+1)}d_{j1}, & c_{i(N+2)}d_{j2}, & \ldots, & c_{i(2N)}d_{jN}, \\ \vdots & \vdots & \vdots & \vdots \\ c_{i(N^2-N+1)}d_{j1}, & c_{i(N^2-N+2)}d_{j2}, & \ldots, & c_{i(N^2)}d_{jN}) \end{pmatrix} \tag{6.43}$$

We can prove that the sequence set $\{\mathbf{E}_{i1}, \ldots, \mathbf{E}_{iN}\}$ ($i = 1, 2, \ldots, N$) is an auto-complementary code of order N, and that any two of the generated N auto-complementary codes or $\{\mathbf{E}_{11}, \ldots, \mathbf{E}_{1N}\}, \{\mathbf{E}_{21}, \ldots, \mathbf{E}_{2N}\}, \ldots, \{\mathbf{E}_{i1}, \ldots, \mathbf{E}_{iN}\}, \ldots, \{\mathbf{E}_{N1}, \ldots, \mathbf{E}_{NN}\}$ satisfy the property of the cross-complementary code. The detailed definitions for auto-complementary codes and cross-complementary codes have been given in Subsection 6.2.2.

Example

We can follow the method given above to generate a simple example of a complete complementary code set. Here we only consider Walsh-Hadamard matrices as the orthogonal matrices needed for generation of complete complementary codes. Of course, any other orthogonal matrices can be used for this purpose, resulting in possibly different complete complementary codes. Assume that there are three 2×2 orthogonal matrices $\mathbf{A}$, $\mathbf{B}$, and $\mathbf{D}$ as follows:

$$\begin{cases} \mathbf{A} = \begin{pmatrix} + & + \\ + & - \end{pmatrix} \\ \mathbf{B} = \begin{pmatrix} + & + \\ + & - \end{pmatrix} \\ \mathbf{D} = \begin{pmatrix} + & + \\ + & - \end{pmatrix} \end{cases} \tag{6.44}$$

Using (6.41) we can obtain the matrix $\mathbf{C}$ as

$$\begin{cases} \mathbf{C}_1 = (+ + \; + -) \\ \mathbf{C}_2 = (+ + \; - +) \end{cases} \tag{6.45}$$

Then, we can use $\mathbf{C}$ and $\mathbf{D}$ to generate $\mathbf{E}$ as follows

$$\begin{cases} \mathbf{E}_{11} = (+ + \; + -) \\ \mathbf{E}_{12} = (+ - \; + +) \\ \mathbf{E}_{21} = (+ + \; - +) \\ \mathbf{E}_{22} = (+ - \; - -) \end{cases} \tag{6.46}$$

which is a complete complementary code set with $N = 2$ and length of $N^2 = 4$. A comprehensive list of the complete complementary codes (PG $= 8 \sim 512$) is given in Appendix E.

6.3.2 GENERATION OF EXTENDED COMPLEMENTARY CODES

The extended complementary codes were proposed based on the complete complementary codes, as discussed earlier. It is noted that, compared to the complete complementary codes, the extended complementary codes will remain at the same set size but with enlarged length of element codes. Therefore, we should make it very clear that use of the extended complementary codes will not help if we want to support more users in the same CDMA system.

An extended complementary code set can be viewed as a complete complementary code set of order N and length N^n, which can be generated based on a complete complementary code set of order N and length N^{n-1}. In this way, a complete complementary code of order N and length N^n for any integer number n can be obtained by extending the complete complementary code for $n = 2$, as shown in the previous subsection.

Let $\{\mathbf{F}_{11}, \ldots, \mathbf{F}_{1N}\}, \ldots, \{\mathbf{F}_{i1}, \ldots, \mathbf{F}_{iN}\}, \ldots, \{\mathbf{F}_{N1}, \ldots, \mathbf{F}_{NN}\}$ be a complementary code of order N and length N^{n-1}. Let f_{ijk} be the kth element of $\mathbf{F}_{ij}$. Therefore, we have

$$\mathbf{F}_{ij} = (f_{ij1}, f_{ij2}, \ldots, f_{ijN^{n-1}}) \tag{6.47}$$

Let $\mathbf{G}_i$ be a sequence of length N^n, which consists of $\mathbf{F}_{i1}, \ldots, \mathbf{F}_{iN}$ arranged as follows

$$\mathbf{G}_i = \begin{pmatrix} f_{i11}, & f_{i21}, & \cdots & f_{iN1}, \\ f_{i12}, & f_{i22}, & \cdots & f_{iN2}, \\ \vdots & \vdots & \vdots & \vdots \\ f_{i1N^{n-1}}, & f_{i2N^{n-1}}, & \cdots & f_{iNN^{n-1}} \end{pmatrix} \tag{6.48}$$

Therefore, we have

$$\rho(\mathbf{G}_i, \mathbf{G}_i : lN) = \frac{1}{N} \sum_j \rho(\mathbf{F}_{ij}, \mathbf{F}_{ij} : l) \tag{6.49}$$

where l is an integer. Thus, $\mathbf{G}_i$ is an N-shift auto-orthogonal sequence of length N^n, because $\{\mathbf{F}_{i1}, \ldots, \mathbf{F}_{iN}\}$ is an auto-complementary sequence. On the other hand, if $i \neq k$, we have

$$\rho(\mathbf{G}_i, \mathbf{G}_h : lN) = \frac{1}{N} \sum_j \rho(\mathbf{F}_{ij}, \mathbf{F}_{hj} : l) \tag{6.50}$$

which shows that $\mathbf{G}_i$ and $\mathbf{G}_h$ are N-shift cross-orthogonal sequences of length N^n, as $\{\mathbf{F}_{i1}, \ldots, \mathbf{F}_{iN}\}$ and $\{\mathbf{F}_{h1}, \ldots, \mathbf{F}_{hN}\}$ are cross-orthogonal sequences. Thus, $\mathbf{G}_1, \ldots, \mathbf{G}_N$ are N-shift cross-orthogonal sequences of length N^n.

Finally, applying the method given in Step 3 of the previous subsection to the sequences $\mathbf{G}_1, \mathbf{G}_2, \ldots, \mathbf{G}_N$, we can obtain the extended complementary codes of length N^n.

The example of an extended complementary code set generated based on the example complete complementary code set given in the previous subsection is given as follows:

$$\begin{cases} E_{11} = (+ + + - + + - + + + + - - - + -) \\ E_{12} = (+ - + + + - - - + - + + - + + +) \\ E_{21} = (+ + + - + + - + - - - + + + - +) \\ E_{22} = (+ - + + + - - - - + - - + - - -) \end{cases} \tag{6.51}$$

in which we have used $n = 4$ such that its length is equal to $N^n = 2^4 = 16$.

6.3.3 GENERATION OF SUPER COMPLEMENTARY CODES

It is seen from the discussions carried out in the previous subsection that complete complementary codes offer a relatively small code set under a fixed element code length or processing gain (PG). For example, if the length of the element codes is N^2, the set size for the complementary codes is merely equal to N. In this complete complementary code set, each flock consists of N element codes, and each element code has a length of N^2. Therefore, the PG value for such a complementary code set becomes $N^2 \times N$, which is a fairly large number compared to the code set size N. As a typical case, if we need in total 64 flocks of complete complementary codes to support an equal number of users in a cell, we have to make the processing gain of such a CDMA system as large as $64^3 = 262\,144$, which is never a small number!

Therefore, we have seen that the complete complementary codes can hardly be applied to a practical CDMA system, which always has a limit on the maximal PG value it can handle. Therefore, to make a practical complementary-code-based CDMA system, we have to find some other better orthogonal complementary codes, which can support a reasonably large number of users under a relatively low PG value. The super complementary codes were proposed to respond to the call for a better orthogonal complementary code. As a matter of fact, both complete complementary codes and super complementary codes are ideally orthogonal complementary codes. Thus, there is no big difference in terms of their orthogonality properties, as they are both perfectly orthogonal. However, the most important feature of the super complementary code is that it can support much more users in the same code set under a fixed PG value than the complete complementary code. This paves the way to implement a practical CDMA system using orthogonal complementary codes, which could be considered as one of the most important blocks for building up the next generation CDMA technology.

The generation approach for the super complementary codes was also developed based on the generation algorithm for the complete complementary codes in the sense that they all start with three orthogonal matrices, **A**, **B**, and **D**, which are always square matrices and have the same dimensions.

To make the description simpler, we would like to summarize the procedure to generate a complete complementary code set as follows.

Assume **A** is an orthogonal matrix with its dimensions being $N \times N$, and $\mathbf{A}_i$ is the ith row of **A**, where $i = 1, 2, \ldots, N$. Also, we assume that the norm of each element is always unit, and thus we have

$$\mathbf{A} = (a_{ij}) = \begin{pmatrix} A_1 \\ A_2 \\ \vdots \\ A_N \end{pmatrix}, \quad |a_{ij}| = 1, \quad \text{for } i, j = 1, 2, \ldots, N \tag{6.52}$$

Assume that we have another orthogonal matrix **B**, whose dimension is also $N \times N$ and whose element can be complex with its norm also being unit. Then, we have

$$\mathbf{B} = (b_{ij}); \quad |b_{ij}| = 1, \quad \text{for } i, j = 1, 2, \ldots, N \tag{6.53}$$

Based on **A** and **B**, we can generate N sequences $\mathbf{C}_1, \mathbf{C}_2, \ldots, \mathbf{C}_N$, with each having a length of N^2, as follows

$$\begin{cases} \mathbf{C}_1 = (b_{11}\mathbf{A}_1, b_{12}\mathbf{A}_2, \ldots, b_{1N}\mathbf{A}_N) = (c_{11}, c_{12}, \ldots, c_{1N^2}) \\ \mathbf{C}_2 = (b_{21}\mathbf{A}_1, b_{22}\mathbf{A}_2, \ldots, b_{2N}\mathbf{A}_N) = (c_{21}, c_{22}, \ldots, c_{2N^2}) \\ \vdots \\ \mathbf{C}_N = (b_{N1}\mathbf{A}_1, b_{N2}\mathbf{A}_2, \ldots, b_{NN}\mathbf{A}_N) = (c_{N1}, c_{N2}, \ldots, c_{NN^2}) \end{cases} \tag{6.54}$$

Now, let us use the third orthogonal matrix **D** with its dimension being also $N \times N$. Again, we assume

$$D = (d_{ij}); \quad |d_{ij}| = 1, \quad \text{for } i, j = 1, 2, \ldots, N \tag{6.55}$$

Then, we can readily have N^2 sequences, each of which has a length of N^2 chips as follows

$$\begin{aligned}\mathbf{E}_{ij} &= (c_{i1}d_{j1}, \ldots, c_{iN}d_{jN}, c_{i(N+1)}d_{j1}, \ldots, c_{i(2N)}d_{jN}, \ldots, c_{i(N^2-N+1)}d_{j1}, \ldots, c_{iN^2}d_{jN}) \\ &= (e_{ij1}, e_{ij2}, \ldots, e_{ijN^2})\end{aligned} \tag{6.56}$$

where $i, j = 1, 2, \ldots, N$. From Equation (6.56) we can formulate N flocks of complete complementary codes (each flock consists of N element codes), to support N users in a CDMA system. Those N flocks of complete complementary codes can be written as

$$\begin{cases}\mathbf{E}_1 = (\mathbf{E}_{11}, \mathbf{E}_{12}, \ldots, \mathbf{E}_{1N}) \\ \mathbf{E}_2 = (\mathbf{E}_{21}, \mathbf{E}_{22}, \ldots, \mathbf{E}_{2N}) \\ \vdots \\ \mathbf{E}_N = (\mathbf{E}_{N1}, \mathbf{E}_{N2}, \ldots, \mathbf{E}_{NN})\end{cases} \tag{6.57}$$

Based on this obtained complete complementary code set, we can proceed to generate the super complementary code set as follows. For description simplicity, we would like to start with the simplest case with $N = 2$. Therefore, all three orthogonal matrices, $\mathbf{E}$, $\mathbf{B}$, and $\mathbf{D}$, will have dimensions of 2×2. Thus, we will obtain a complete complementary code set as

$$\begin{cases}\mathbf{E}_1 = (\mathbf{E}_{11}, \mathbf{E}_{12}) \\ \mathbf{E}_2 = (\mathbf{E}_{21}, \mathbf{E}_{22})\end{cases} \tag{6.58}$$

Now, let us split $\mathbf{E}_{ij}$ (where $i, j = 1, 2$) up into four segments, $\mathbf{T}_{i1}, \mathbf{T}_{i2}, \mathbf{T}_{i3}, \mathbf{T}_{i4}$, each with equal length. Then, we obtain

$$\begin{cases}\mathbf{E}_{11} = \mathbf{T}_1 = (\mathbf{T}_{11}, \mathbf{T}_{12}, \mathbf{T}_{13}, \mathbf{T}_{14}) \\ \mathbf{E}_{12} = \mathbf{T}_2 = (\mathbf{T}_{21}, \mathbf{T}_{22}, \mathbf{T}_{23}, \mathbf{T}_{24}) \\ \mathbf{E}_{21} = \mathbf{T}_3 = (\mathbf{T}_{31}, \mathbf{T}_{32}, \mathbf{T}_{33}, \mathbf{T}_{34}) \\ \mathbf{E}_{22} = \mathbf{T}_4 = (\mathbf{T}_{41}, \mathbf{T}_{42}, \mathbf{T}_{43}, \mathbf{T}_{44})\end{cases} \tag{6.59}$$

which will be used as a set of the seed codes to generate the whole super complementary code set, as described in the text as follows.

We still use $N = 2$ as an example to simplify the description, and then we will extend it to a general case later. First, based on the results given in (6.59), we can perform the first step extension as

$$\begin{cases}\mathbf{S}_1 = (\mathbf{T}_{11}, \mathbf{T}_{21}, \mathbf{T}_{12}, \mathbf{T}_{22}, \ldots, \mathbf{T}_{14}, \mathbf{T}_{24}) = (\mathbf{S}_{11}, \mathbf{S}_{12}, \ldots, \mathbf{S}_{18}) \\ \mathbf{S}_2 = (\mathbf{T}_{11}, \overline{\mathbf{T}_{21}}, \mathbf{T}_{12}, \overline{\mathbf{T}_{22}}, \ldots, \mathbf{T}_{14}, \overline{\mathbf{T}_{24}}) = (\mathbf{S}_{21}, \mathbf{S}_{22}, \ldots, \mathbf{S}_{28}) \\ \mathbf{S}_3 = (\mathbf{T}_{21}, \mathbf{T}_{11}, \mathbf{T}_{22}, \mathbf{T}_{12}, \ldots, \mathbf{T}_{24}, \mathbf{T}_{14}) = (\mathbf{S}_{31}, \mathbf{S}_{32}, \ldots, \mathbf{S}_{38}) \\ \mathbf{S}_4 = (\mathbf{T}_{21}, \overline{\mathbf{T}_{11}}, \mathbf{T}_{22}, \overline{\mathbf{T}_{12}}, \ldots, \mathbf{T}_{24}, \overline{\mathbf{T}_{14}}) = (\mathbf{S}_{41}, \mathbf{S}_{42}, \ldots, \mathbf{S}_{48}) \\ \mathbf{S}_5 = (\mathbf{T}_{31}, \mathbf{T}_{41}, \mathbf{T}_{32}, \mathbf{T}_{42}, \ldots, \mathbf{T}_{34}, \mathbf{T}_{44}) = (\mathbf{S}_{51}, \mathbf{S}_{52}, \ldots, \mathbf{S}_{58}) \\ \mathbf{S}_6 = (\mathbf{T}_{31}, \overline{\mathbf{T}_{41}}, \mathbf{T}_{32}, \overline{\mathbf{T}_{42}}, \ldots, \mathbf{T}_{34}, \overline{\mathbf{T}_{44}}) = (\mathbf{S}_{61}, \mathbf{S}_{62}, \ldots, \mathbf{S}_{68}) \\ \mathbf{S}_7 = (\mathbf{T}_{41}, \mathbf{T}_{31}, \mathbf{T}_{42}, \mathbf{T}_{32}, \ldots, \mathbf{T}_{44}, \mathbf{T}_{34}) = (\mathbf{S}_{71}, \mathbf{S}_{72}, \ldots, \mathbf{S}_{78}) \\ \mathbf{S}_8 = (\mathbf{T}_{41}, \overline{\mathbf{T}_{31}}, \mathbf{T}_{42}, \overline{\mathbf{T}_{32}}, \ldots, \mathbf{T}_{44}, \overline{\mathbf{T}_{34}}) = (\mathbf{S}_{81}, \mathbf{S}_{82}, \ldots, \mathbf{S}_{88})\end{cases} \tag{6.60}$$

where $\bar{\mathbf{X}}$ denotes the negative value of $\mathbf{X}$ or simply '$-\mathbf{X}$'. The procedure can continue by using the following generic iterating equations:

$$\begin{cases}
\mathbf{S}_1 = (\mathbf{T}_{11}, \mathbf{T}_{21}, \mathbf{T}_{12}, \mathbf{T}_{22}, \ldots, \mathbf{T}_{1L}, \mathbf{T}_{2L}) = (\mathbf{S}_{11}, \mathbf{S}_{12}, \ldots, \mathbf{S}_{1(2L)}) \\
\mathbf{S}_2 = (\mathbf{T}_{11}, \overline{\mathbf{T}_{21}}, \mathbf{T}_{12}, \overline{\mathbf{T}_{22}}, \ldots, \mathbf{T}_{1L}, \overline{\mathbf{T}_{2L}}) = (\mathbf{S}_{21}, \mathbf{S}_{22}, \ldots, S_{2(2L)}) \\
\mathbf{S}_3 = (\mathbf{T}_{21}, \mathbf{T}_{11}, \mathbf{T}_{22}, \mathbf{T}_{12}, \ldots, \mathbf{T}_{2L}, \mathbf{T}_{1L}) = (\mathbf{S}_{31}, \mathbf{S}_{32}, \ldots, \mathbf{S}_{3(2L)}) \\
\mathbf{S}_4 = (\mathbf{T}_{21}, \overline{\mathbf{T}_{11}}, \mathbf{T}_{22}, \overline{\mathbf{T}_{12}}, \ldots, \mathbf{T}_{2L}, \overline{\mathbf{T}_{1L}}) = (\mathbf{S}_{41}, \mathbf{S}_{42}, \ldots, \mathbf{S}_{4(2L)}) \\
\vdots \\
\mathbf{S}_{4j+1} = (\mathbf{T}_{(2j+1)1}, \mathbf{T}_{(2j+2)1}, \mathbf{T}_{(2j+1)2}, \mathbf{T}_{(2j+2)2}, \ldots, \mathbf{T}_{(2j+1)L}, \mathbf{T}_{(2j+2)L}) \\
\qquad = (\mathbf{S}_{(4j+1)1}, \mathbf{S}_{(4j+1)2}, \ldots, \mathbf{S}_{(4j+1)2L}) \\
\mathbf{S}_{4j+2} = (\mathbf{T}_{(2j+1)1}, \overline{\mathbf{T}_{(2j+2)1}}, \mathbf{T}_{(2j+1)2}, \overline{\mathbf{T}_{(2j+2)2}}, \ldots, \mathbf{T}_{(2j+1)L}, \overline{\mathbf{T}_{(2j+2)L}}) \\
\qquad = (\mathbf{S}_{(4j+2)1}, \mathbf{S}_{(4j+2)2}, \ldots, \mathbf{S}_{(4j+2)2L}) \\
\mathbf{S}_{4j+3} = (\mathbf{T}_{(2j+2)1}, \mathbf{T}_{(2j+1)1}, \mathbf{T}_{(2j+2)2}, \mathbf{T}_{(2j+1)2}, \ldots, \mathbf{T}_{(2j+2)L}, \mathbf{T}_{(2j+1)L}) \\
\qquad = (\mathbf{S}_{(4j+3)1}, \mathbf{S}_{(4j+3)2}, \ldots, \mathbf{S}_{(4j+3)2L}) \\
\mathbf{S}_{4j+4} = (\mathbf{T}_{(2j+2)1}, \overline{\mathbf{T}_{(2j+1)1}}, \mathbf{T}_{(2j+2)2}, \overline{\mathbf{T}_{(2j+1)2}}, \ldots, \mathbf{T}_{(2j+2)L}, \overline{\mathbf{T}_{(2j+1)L}}) \\
\qquad = (\mathbf{S}_{(4j+4)1}, \mathbf{S}_{(4j+4)2}, \ldots, \mathbf{S}_{(4j+4)2L}) \\
\vdots \\
\mathbf{S}_{2L-3} = (\mathbf{T}_{(L-1)1}, \mathbf{T}_{L1}, \mathbf{T}_{(L-1)2}, \mathbf{T}_{L2}, \ldots, \mathbf{T}_{(L-1)L}, \mathbf{T}_{LL}) \\
\qquad = (\mathbf{S}_{(2L-3)1}, \mathbf{S}_{(2L-3)2}, \ldots, \mathbf{S}_{(2L-3)2L}) \\
\mathbf{S}_{2L-2} = (\mathbf{T}_{(L-1)1}, \overline{\mathbf{T}_{L1}}, \mathbf{T}_{(L-1)2}, \overline{\mathbf{T}_{L2}}, \ldots, \mathbf{T}_{(L-1)L}, \overline{\mathbf{T}_{LL}}) \\
\qquad = (\mathbf{S}_{(2L-2)1}, \mathbf{S}_{(2L-2)2}, \ldots, \mathbf{S}_{(2L-2)2L}) \\
\mathbf{S}_{2L-1} = (\mathbf{T}_{L1}, \mathbf{T}_{(L-1)1}, \mathbf{T}_{L2}, \mathbf{T}_{(L-1)2}, \ldots, \mathbf{T}_{LL}, \mathbf{T}_{(L-1)L}) \\
\qquad = (\mathbf{S}_{(2L-1)1}, \mathbf{S}_{(2L-1)2}, \ldots, \mathbf{S}_{(2L-1)2L}) \\
\mathbf{S}_{2L} = (\mathbf{T}_{L1}, \overline{\mathbf{T}_{(L-1)1}}, \mathbf{T}_{L2}, \overline{\mathbf{T}_{(L-1)2}}, \ldots, \mathbf{T}_{LL}, \overline{\mathbf{T}_{(L-1)L}}) \\
\qquad = (\mathbf{S}_{(2L)1}, \mathbf{S}_{(2L)2}, \ldots, \mathbf{S}_{(2L)2L})
\end{cases} \tag{6.61}$$

where $j = 0, 1, \ldots, (L/2) - 1$ and again the notation $\bar{\mathbf{X}}$ denotes the value of $-\mathbf{X}$ and we have always $L = 2^R$ (R is an integer and always satisfies $R \geq 2$). From Equation (6.61), we have the following observations:

1. After each iteration using (6.61), we double the number of element codes contained in each flock compared to the number before the iteration. For instance, if we initially have four element codes in each flock, then we will have eight element codes in each flock after one iteration using (6.61). One more iteration will increase the number of element codes to 16, and so forth.

2. After each iteration using (6.61), we will also double the number of flocks in the same code set. For example, we initially have a set of super complementary codes with four flocks. Then, the number of flocks will be increased to eight after the first iteration. The number of flocks will be increased to 16 after the second iteration, and so on.

3. The element code length will never change after the iterations using (6.61).

Assume that the original complete complementary code set has its set size as N, its flock size as N, and its element code length as N^2. We first should formulate a seed set from this complete complementary code set based on Equation (6.59), whose set size is $N^2 = 4$, flock size is $N^2 = 4$, and element code length is $N^2/4 = 1$ chip, if $N = 2$ here. The code set can be used to generate the super complementary code set as mentioned earlier. After one iteration, we will have a super complementary code set with set size $2N^2$, flock size $2N^2$, and element code length $N^2/4$ (which is unchanged compared to that before the iteration). The process continues and we can obtain the parameters for a super complementary code set after any number of iterations, depending on the parameters we defined.

Therefore, we can see from the above description that the element code length of a super complementary code set will never change no matter how many iterations using (6.61) are performed. If we want to generate a super complementary code set with a particular element code length, we should first transfer a complete complementary code set into an extended complementary code set, as discussed in Subsection 6.3.2. As mentioned in Subsection 6.3.2, we can control the element code length in the generation process of an extended complementary code set. Also, it is noted that if we extend a complete complementary code set into an extended complementary code set, their set size and flock size are always unchanged but the element code length increases from N^2 to N^n.

For example, if the element code length of a complete complementary code set is N^2, we can generate an extended complementary code set with its element code length being N^n. Based on this extended complementary code set, we can have a seed code set using Equation (6.59), whose set size is N^2, flock size is N^2, and element code length is $N^n/4$. By controlling the element code length of the extended complementary code set, we can generate a super complementary code set with any arbitrary element code length.

Now, we would like to give two simple examples to show how to generate two super complementary code sets with different element code lengths.

Assume that we have three 2×2 orthogonal matrices as

$$\mathbf{A} = \begin{pmatrix} + & + \\ + & - \end{pmatrix}, \quad \mathbf{B} = \begin{pmatrix} + & + \\ + & - \end{pmatrix}, \quad \mathbf{D} = \begin{pmatrix} + & + \\ + & - \end{pmatrix}. \tag{6.62}$$

We can generate the matrix $\mathbf{C}$ as follows

$$\begin{cases} \mathbf{C}_1 = (++\ +-) \\ \mathbf{C}_2 = (++\ -+) \end{cases} \tag{6.63}$$

Based on $\mathbf{C}$ and $\mathbf{D}$, we can have the seed code set using (6.59) as

$$\begin{cases} \mathbf{E}_{11} = \mathbf{T}_1 = (++\ +-) \\ \mathbf{E}_{12} = \mathbf{T}_2 = (+-\ ++) \\ \mathbf{E}_{21} = \mathbf{T}_3 = (++\ -+) \\ \mathbf{E}_{22} = \mathbf{T}_4 = (+-\ --) \end{cases} \tag{6.64}$$

Obviously, the element code length for this seed code set is one chip and the set size and flock size are 4. After one iteration using (6.61) we obtain a super complementary code set with its set size and flock size being $4 \times 2 = 8$ and its element code length unchanged (still one chip), as

$$\begin{cases} \mathbf{S}_1 = (+,+,+,-,+,+,-,+) \\ \mathbf{S}_2 = (+,-,+,+,+,-,-,-) \\ \mathbf{S}_3 = (+,+,-,+,+,+,+,-) \\ \mathbf{S}_4 = (+,-,-,-,+,-,+,+) \\ \mathbf{S}_5 = (+,+,+,-,-,-,+,-) \\ \mathbf{S}_6 = (+,-,+,+,-,+,+,+) \\ \mathbf{S}_7 = (+,+,-,+,-,-,-,+) \\ \mathbf{S}_8 = (+,-,-,-,-,+,-,-) \end{cases} \tag{6.65}$$

Now, we will demonstrate the process to generate a super complementary code set with its element code length being four and its set size and flock size being eight. Again, assume the same three orthogonal matrices as shown in (6.62). In the same way, we will have the seed code set as

$$\begin{cases} \mathbf{E}_{11} = \mathbf{T}_1 = (++\ +-) \\ \mathbf{E}_{12} = \mathbf{T}_2 = (+-\ ++) \\ \mathbf{E}_{21} = \mathbf{T}_3 = (++\ -+) \\ \mathbf{E}_{22} = \mathbf{T}_4 = (+-\ --) \end{cases} \tag{6.66}$$

which can be extended into a code set with element code length 16, such that

$$\begin{cases} \mathbf{E}_{11} = (+ + + - + + - + + + + - - - + -) \\ \mathbf{E}_{12} = (+ - + + + - - - + - + + - + + +) \\ \mathbf{E}_{21} = (+ + + - + + - + - - - + + + - +) \\ \mathbf{E}_{22} = (+ - + + + - - - - + - - + - - -) \end{cases} \tag{6.67}$$

Then, we can use the iteration algorithm given in (6.61) to generate a super complementary code set with its element code length being four and set size and flock size being four, as follows:

$$\begin{cases} \mathbf{T}_1 = [\mathbf{T}_{11}, \mathbf{T}_{12}, \mathbf{T}_{13}, \mathbf{T}_{14}] = (+ + + -, + + - +, + + + -, - - + -) \\ \mathbf{T}_2 = [\mathbf{T}_{21}, \mathbf{T}_{22}, \mathbf{T}_{23}, \mathbf{T}_{24}] = (+ - + +, + - - -, + - + +, - + + +) \\ \mathbf{T}_3 = [\mathbf{T}_{31}, \mathbf{T}_{32}, \mathbf{T}_{33}, \mathbf{T}_{34}] = (+ + + -, + + - +, - - - +, + + - +) \\ \mathbf{T}_4 = [\mathbf{T}_{41}, \mathbf{T}_{42}, \mathbf{T}_{43}, \mathbf{T}_{44}] = (+ - + +, + - - -, - + - -, + - - -) \end{cases} \tag{6.68}$$

After one iteration, we obtain the following super complementary code set with element code length four and set size and flock size eight

$$\begin{cases} \mathbf{S}_1 = (++\ +-, +-\ ++, ++\ -+, +-\ --, ++\ +-, +-\ ++, --\ +-, +-\ --) \\ \mathbf{S}_2 = (++\ +-, -+\ --, ++\ -+, -+\ ++, ++\ +-, -+\ --, --\ +-, -+\ ++) \\ \mathbf{S}_3 = (+-\ ++, ++\ +-, +-\ --, ++\ -+, +-\ ++, ++\ +-, -+\ ++, --\ +-) \\ \mathbf{S}_4 = (+-\ ++, --\ -+, +-\ --, --\ +-, +-\ ++, --\ -+, -+\ ++, ++\ -+) \\ \mathbf{S}_5 = (++\ +-, +-\ ++, ++\ -+, +-\ --, --\ -+, -+\ --, ++\ -+, +-\ --) \\ \mathbf{S}_6 = (++\ +-, -+\ --, ++\ -+, -+\ ++, --\ -+, +-\ ++, ++\ -+, -+\ ++) \\ \mathbf{S}_7 = (+-\ ++, ++\ +-, +-\ --, ++\ -+, -+\ --, --\ -+, +-\ --, ++\ -+) \\ \mathbf{S}_8 = (+-\ ++, --\ -+, +-\ --, --\ +-, -+\ --, ++\ +-, +-\ --, --\ +-) \end{cases} \tag{6.69}$$

The comparison among three different ideally orthogonal complementary codes, complete extended and super complementary codes, is made in Table 6.2. It can be seen from Table 6.2 that the super complementary code set can provide the largest set size, implying that it can support the most users in the same code set. This is just what we wanted from the very beginning of the proposal for the super complementary codes.

A comprehensive list of super complementary codes (PG $= 4 \sim 64$) is given in Appendix F.

6.3.4 GENERATION OF GENERALIZED PAIR-WISE COMPLEMENTARY CODES

So far, we have discussed the generation approaches for three different types of complementary codes, namely complete extended, and super complementary codes, all of which are ideally orthogonal complementary codes. An ideally orthogonal complementary code should process ideal auto-correlation functions and ideal cross-correlation functions in both synchronous and asynchronous operational

Table 6.2 Comparison among three different types of ideally orthogonal complementary codes, complete, extended, and super complementary codes.

	Orthogonal matrix dimension	Element code length	Flock size	Set size	Processing gain
Complete CC	N	N^2	N	N	N^3
Extended CC	N	N^{2+n}	N	N	N^{3+n}
Super CC	2	2^r	2^R	2^R	2^{r+R}

Note: n, r, R, and N are integers, where $n \geq 3$, $r \geq 0$, $R \geq 2$, and $N = x^2$

modes. Therefore, if they are applied to a CDMA system, they could ensure an interference-free operation, at least from the theoretical point of view.[3]

In this subsection, we are going to discuss how to generate pair-wise complementary codes, which are not ideally orthogonal complementary codes. However, the pair-wise complementary codes do possess some advantages that other complementary codes may not have. Each pair-wise complementary code consists of only two element codes (or its flock size is identically two), which should be assigned to a particular user. Due to its relatively small flock size, a CDMA transceiver can be made much simpler by using a normal quadrature modem, in which the I and Q channels can carry two different element codes of the same flock. On the other hand, a CDMA transceiver in a CDMA system using other complementary codes, such as super complementary code set, etc., has to employ a complex multi-carrier modulator/demodulator to send different element codes in parallel. Therefore, in this sense the pair-wise complementary codes offer a unique opportunity for us to simplify the hardware implementation of a CDMA transceiver. Of course, we have to pay some price in exchange for the low complexity. As mentioned earlier, a pair-wise complementary code set is no longer an ideally orthogonal complementary code set and thus it will not be able to offer interference-free operation to a CDMA system which chooses it as its spreading code set.

Generalized Even Shift Orthogonal Sequences

The generation of generalized pair-wise complementary codes needs the help of a generalized even shift orthogonal (GESO) sequence, which has been briefly discussed earlier. Therefore, we should first give some more detailed discussions about it as follows. The complete complementary (CC) codes again will be used here to formulate GESO sequences, which ensure orthogonality for all even relative time shifts between any two sequences in a set. It should be noted that this approach is different from what was suggested in [255].

Let us consider a CC code set with its element code length and flock size being $N = 2$ and $M = 2$, respectively. The first flock of this CC code set is $\mathbf{c}_{11} = (c_{11}^{(1)}, c_{11}^{(2)}, c_{11}^{(3)}, c_{11}^{(4)})$ and $\mathbf{c}_{12} = (c_{12}^{(1)}, c_{12}^{(2)}, c_{12}^{(3)}, c_{12}^{(4)})$; the second flock is $\mathbf{c}_{21} = (c_{21}^{(1)}, c_{21}^{(2)}, c_{21}^{(3)}, c_{21}^{(4)})$ and $\mathbf{c}_{22} = t(c_{22}^{(1)}, c_{22}^{(2)}, c_{22}^{(3)}, c_{22}^{(4)}t)$. Consider only the first flock or $\mathbf{c}_{11}$ and $\mathbf{c}_{12}$. It can be easily shown that, if first spreading each chip in $\mathbf{c}_{11}$ and $\mathbf{c}_{12}$ by [1, 1] and [1, −1], respectively, we have

$$\begin{aligned}\mathbf{c}_{11}^{[++]} &= \left(c_{11}^{(1)}, c_{11}^{(1)}, c_{11}^{(2)}, c_{11}^{(2)}, c_{11}^{(3)}, c_{11}^{(3)}, c_{11}^{(4)}, c_{11}^{(4)}\right) \\ &= \left(\mathbf{c}_{11}^{(1)}, \mathbf{c}_{11}^{(2)}, \mathbf{c}_{11}^{(3)}, \mathbf{c}_{11}^{(4)}\right) \qquad (6.70)\end{aligned}$$

[3]Again, we should note the fact that multipath interference free operation will never be possible if a multipath return happens to take multiples of the element code length, or xN, where N is the element code length and x is any positive integer. This is true whatever spreading codes are used, either unitary codes or complementary codes.

$$\mathbf{c}_{12}^{[+-]} = \left(c_{12}^{(1)}, -c_{12}^{(1)}, c_{12}^{(2)}, -c_{12}^{(2)}, c_{12}^{(3)}, -c_{12}^{(3)}, c_{12}^{(4)}, -c_{12}^{(4)}\right)$$
$$= \left(\mathbf{c}_{12}^{(1)}, \mathbf{c}_{12}^{(2)}, \mathbf{c}_{12}^{(3)}, \mathbf{c}_{12}^{(4)}\right) \quad (6.71)$$

and then combining $\mathbf{c}_{11}^{[+-]}$ and $\mathbf{c}_{12}^{[+-]}$ to form a one-dimensional sequence or

$$\mathbf{c}_1 = \left(\mathbf{c}_{11}^{[++]}, \mathbf{c}_{12}^{[+-]}\right)$$
$$= \left(\mathbf{c}_{11}^{(1)}, \mathbf{c}_{11}^{(2)}, \mathbf{c}_{11}^{(3)}, \mathbf{c}_{11}^{(4)}, \mathbf{c}_{12}^{(1)}, \mathbf{c}_{12}^{(2)}, \mathbf{c}_{12}^{(3)}, \mathbf{c}_{12}^{(4)}\right) \quad (6.72)$$

which is a new sequence with all its even-shifted auto-correlation function (ACF) side lobes being zero. It is easy to show that, for any odd time shift i and j, we always have

$$\mathbf{c}_{11}^{[++]}(i)\,\mathbf{c}_{12}^{[+-]}(j) + \mathbf{c}_{11}^{[++]}(i+1)\,\mathbf{c}_{12}^{[+-]}(j+1)$$
$$= \mathbf{c}_{11}^{(i+1)/2}\left(\mathbf{c}_{12}^{(j+1)/2}\right)^H = c_{11}^{(i)} c_{12}^{(j)}\left(1, 1\right)\left(1, -1\right)^H = 0 \quad (6.73)$$

where $\mathbf{x}^H$ denotes the Hermitia transpose of $\mathbf{x}$, and $\text{mod}(i, 2) = \text{mod}(j, 2) = 1$. We will use the above property to show ideal even-shifted ACF as shown in Figure 6.1, where ACF between $\mathbf{c}_1$ and its 2-chip shifted version is considered.

Therefore, we can calculate the ACF between $\mathbf{c}_1$ and its 2-chip shifted version (it is noted that the same result will be obtained for any other even relative shifts) as follows:

$$\mathbf{c}_1\left[\mathbf{g}_{11}, \mathbf{g}_{12}\right]^* = \mathbf{c}_1\left[\mathbf{s}_{11}, \mathbf{s}_{12}\right]^* + \mathbf{c}_1\left[\mathbf{v}_{11}, \mathbf{v}_{12}\right]^*$$
$$= \left[\mathbf{c}_{11}^{[++]}, \mathbf{c}_{12}^{[+-]}\right]\left[\mathbf{s}_{11}, \mathbf{s}_{12}\right]^* + \left[\mathbf{c}_{11}^{[++]}, \mathbf{c}_{12}^{[+-]}\right]\left[\mathbf{v}_{11}, \mathbf{v}_{12}\right]^*$$
$$= \mathbf{c}_{11}^{[++]}\mathbf{s}_{11}^* + \mathbf{c}_{12}^{[+-]}\mathbf{s}_{12}^* + \mathbf{c}_{11}^{[++]}\mathbf{v}_{11}^* + \mathbf{c}_{12}^{[+-]}\mathbf{v}_{12}^* \quad (6.74)$$

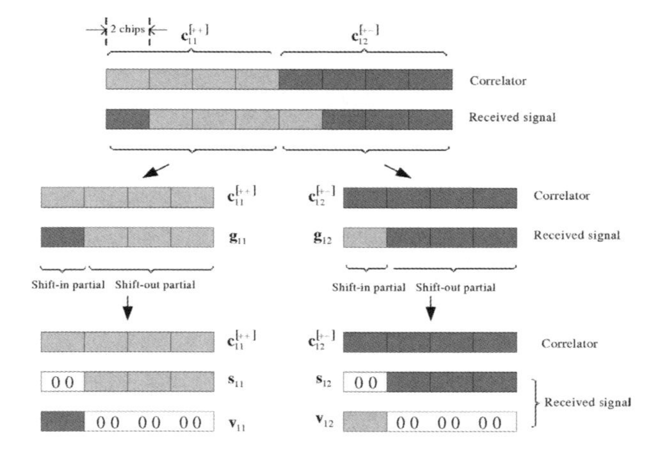

Figure 6.1 Illustration of ideal aperiodic ACF for a GESO sequence, $\mathbf{c}_1 = (\mathbf{c}_{11}^{[++]}\mathbf{c}_{12}^{[+-]})$, whose aperiodic ACF can be decomposed into two components: 'shift-in partial ACF' $(\mathbf{c}_{11}^{[++]}\mathbf{v}_{11}^*, \mathbf{c}_{12}^{[+-]}\mathbf{v}_{12}^*)$ and 'shift-out partial ACF' $(\mathbf{c}_{11}^{[++]}\mathbf{s}_{11}^*, \mathbf{c}_{12}^{[+-]}\mathbf{s}_{12}^*)$.

From the results obtained in Appendix C, it is recognized that the shift-out partial ACFs (as illustrated in Figure 6.1) must be zero, as shown in Equations (C.1) and (C.5):

$$\mathbf{c}_{11}^{[++]}\mathbf{s}_{11}^{*} + \mathbf{c}_{12}^{[+-]}\mathbf{s}_{12}^{*} = 0 \tag{6.75}$$

Using the property shown in (6.73), we can also have

$$\mathbf{c}_{11}^{[++]}\mathbf{v}_{11}^{*} = \mathbf{c}_{12}^{[+-]}\mathbf{v}_{12}^{*} = 0 \tag{6.76}$$

Therefore, it has been proved here that the aperiodic ACF for a GESO sequence is zero. Although the above proof is carried out for a particular relative time shift (or two chips, as illustrated in Figure 6.1), the result is valid in general for any other even shifts. The more rigorous proof can be given as follows.

Assume that the length of $\mathbf{c}_1$ or $\mathbf{c}_2$ is P. The aperiodic ACF of code $\mathbf{c}_1$ with a relative time shift of $i \in (\bmod(i,2) = 0) \cap (i < P/2)$ can be written as

$$\begin{aligned}
\rho(\mathbf{c}_1; i) &= \frac{1}{P}\sum_{j=1}^{P-i}\mathbf{c}_1(j+i)\,\mathbf{c}_1^*(j) \pm \underbrace{\frac{1}{P}\sum_{j=P-i+1}^{P}\mathbf{c}_1(j+i-P)\,\mathbf{c}_1^*(j)}_{\text{Shift-in partial ACF}} \\
&= \frac{1}{P}\sum_{j=1}^{P/2-i/2}\left[\mathbf{c}_1(2j-1+i)\,\mathbf{c}_1^*(2j-1) + \mathbf{c}_1(2j+i)\,\mathbf{c}_1^*(2j)\right] \\
&\quad \pm \frac{1}{P}\sum_{j=P/2-i/2+1}^{P/2}\Big(\mathbf{c}_1(2j-1+i-P)\,\mathbf{c}_1^*(2j-1) + \mathbf{c}_1(2j+i-P)\,\mathbf{c}_1^*(2j)\Big) \\
&= \underbrace{\frac{1}{P}\sum_{j=1}^{P/4-i/2}\left[\mathbf{c}_{11}^{(j+i/2)}\mathbf{c}_{11}^{*(j)}\right]}_{\text{Shift-out partial ACF}} + \underbrace{\frac{1}{P}\sum_{j=p/4-i/2+1}^{P/4}\left[\mathbf{c}_{12}^{(j+i/2-P/4)}\mathbf{c}_{11}^{*(j)}\right]}_{\text{Shift-in partial ACF=0}} \\
&\quad + \underbrace{\frac{1}{P}\sum_{j=p/4+1}^{P/2-i/2}\left[\mathbf{c}_{12}^{(j+i/2-P/4)}\mathbf{c}_{12}^{*(j-P/4)}\right]}_{\text{Shift-out partial ACF}} \\
&\quad \pm \underbrace{\frac{1}{P}\sum_{j=P/2-i/2+1}^{P/2}\left[\mathbf{c}_{11}^{(j+i/2-P/2)}\mathbf{c}_{12}^{*(j-P/4)}\right]}_{\text{Shift-in partial ACF=0}} \\
&= \left[\frac{1}{P}\sum_{j=1}^{P/4-i/2}2c_{11}^{(j+i/2)}c_{11}^{*(j)} + \frac{1}{P}\sum_{j=1}^{P/4-i/2}2c_{12}^{(j+i/2)}c_{12}^{*(j)}\right] + 0 + 0 \\
&= \frac{1}{P}\sum_{m=1}^{2}\sum_{j=1}^{P/4-i/2}2\mathbf{c}_{1m}(j+i/2)\,\mathbf{c}_{1m}^*(j) = 0
\end{aligned} \tag{6.77}$$

where we have used the ideal correlation properties for CC codes given in Equation (B.17) in Appendix B.

Similarly, for $i \in (\bmod(i, 2) = 0) \cap (i > P/2)$ we have

$$\rho(\mathbf{c}_1; i) = \frac{1}{P}\sum_{j=1}^{P-i} \mathbf{c}_1(j+i)\,\mathbf{c}_1^*(j) \pm \underbrace{\frac{1}{P}\sum_{j=P-i+1}^{P} \mathbf{c}_1(j+i-P)\,\mathbf{c}_1^*(j)}_{\text{Shift-in partial ACF}}$$

$$= \frac{1}{P}\sum_{j=1}^{P/2-i/2}\left[\mathbf{c}_1(2j-1+i)\,\mathbf{c}_1^*(2j-1) + \mathbf{c}_1(2j+i)\,\mathbf{c}_1^*(2j)\right]$$

$$\pm\frac{1}{P}\sum_{j=P/2-i/2+1}^{P/2}\left[\mathbf{c}_1(2j-1+i-P)\,\mathbf{c}_1^*(2j-1) + \mathbf{c}_1(2j+i-P)\,\mathbf{c}_1^*(2j)\right]$$

$$= \underbrace{\frac{1}{P}\sum_{j=1}^{P/2-i/2}\left(\mathbf{c}_{12}^{(j+i/2-P/4)}\mathbf{c}_{11}^{*(j)}\right)}_{\text{Shift-in partial ACF}=0}$$

$$\pm\left\{\underbrace{\frac{1}{P}\sum_{j=P/2-i/2+1}^{P/2}\left(\mathbf{c}_{11}^{(j+i/2-P/2)}\mathbf{c}_{11}^{*(j)}\right)}_{\text{Shift-out partial ACF}} + \underbrace{\frac{1}{P}\sum_{j=p/4-i/2+1}^{P/4}\left(\mathbf{c}_{11}^{(j+i/2-P/2)}\mathbf{c}_{12}^{*(j-P/4)}\right)}_{\text{Shift-in partial ACF}=0}\right.$$

$$\left. + \underbrace{\frac{1}{P}\sum_{j=p/4+1}^{P/2-i/2}\left(\mathbf{c}_{12}^{(j+i/2-P/2-P/4)}\mathbf{c}_{12}^{*(j-P/4)}\right)}_{\text{Shift-out partial ACF}}\right\}$$

$$= \left[\frac{1}{P}\sum_{j=P/2-i/2+1}^{P/2} 2c_{11}^{(j+i/2-P/2)}c_{11}^{*(j)} + \frac{1}{P}\sum_{j=P/2-i/2+1}^{P/2} 2c_{12}^{(j+i/2-P/2)}c_{12}^{*(j)}\right] + 0 + 0$$

$$= \frac{1}{P}\sum_{m=1}^{2}\sum_{j=1}^{P/4-i/2} 2\mathbf{c}_{1m}(j+i/2-P/2)\,\mathbf{c}_{1m}^*(j) = 0 \tag{6.78}$$

In this way, it has been shown that the aperiodic ACF for a GESO sequence generated from a flock of CC codes, $\mathbf{c}_{11}$ and $\mathbf{c}_{12}$, is zero for any even relative time shift.

In fact, we can use another flock of the CC codes considered here, or $\mathbf{c}_{21}$ and $\mathbf{c}_{22}$, to generate another GESO sequence $\mathbf{c}_2$ in the same way as the generation of $\mathbf{c}_1$, as illustrated earlier, to yield

$$\mathbf{c}_2 = \left(\ \mathbf{c}_{21}^{[++]} \quad \mathbf{c}_{22}^{[+-]}\ \right) \tag{6.79}$$

where we have

$$\begin{cases} \mathbf{c}_{21}^{[++]} = \left(c_{21}^{(1)}, c_{21}^{(1)}, c_{21}^{(2)}, c_{21}^{(2)}, c_{21}^{(3)}, c_{21}^{(3)}, c_{21}^{(4)}, c_{21}^{(4)}\right) \\ \mathbf{c}_{22}^{[+-]} = \left(c_{22}^{(1)}, -c_{22}^{(1)}, c_{22}^{(2)}, -c_{22}^{(2)}, c_{22}^{(3)}, -c_{22}^{(3)}, c_{22}^{(4)}, -c_{22}^{(4)}\right) \end{cases} \tag{6.80}$$

Using the orthogonality properties of a CC code set, as proved in Appendix B, it can be easily shown that the CCFs of $\mathbf{c}_1$ and $\mathbf{c}_2$ at zero and all even relative time shifts are always zero.

Walsh-Hadamard Matrix Expansion

In the above text, we have demonstrated how to generate a GESO sequence set $\{\mathbf{c}_1, \mathbf{c}_2\}$ from a CC code set $(+++-, +-++)$ and $(++-+, +---)$ with its element code length, flock size, and set size being $N = 4$, $M = 2$, and $K = 2$, respectively. Obviously, the other CC codes can also be used to generate the GESO sequence sets with different lengths and set sizes. Unfortunately, a CC code set usually offers only a relatively small set size compared to its PG value. In fact, we can use Walsh-Hadamard matrix expansion to enlarge its set size as follows.

With the first step expansion, we obtain

$$\mathbf{H}_{1,2} = \begin{pmatrix} \mathbf{c}_1 & \mathbf{c}_1 \\ \mathbf{c}_1 & -\mathbf{c}_1 \end{pmatrix} \tag{6.81}$$

$$\mathbf{H}_{2,2} = \begin{pmatrix} \mathbf{c}_2 & \mathbf{c}_2 \\ \mathbf{c}_2 & -\mathbf{c}_2 \end{pmatrix} \tag{6.82}$$

In general, if matrix $\mathbf{H}_{1,K}$ can be expanded from matrix $\mathbf{H}_{1,K/2}$, we have

$$\mathbf{H}_{1,K} = \begin{pmatrix} \mathbf{H}_{1,K/2} & \mathbf{H}_{1,K/2} \\ \mathbf{H}_{1,K/2} & -\mathbf{H}_{1,K/2} \end{pmatrix} \tag{6.83}$$

$$\mathbf{H}_{2,K} = \begin{pmatrix} \mathbf{H}_{2,K/2} & \mathbf{H}_{2,K/2} \\ \mathbf{H}_{2,K/2} & -\mathbf{H}_{2,K/2} \end{pmatrix} \tag{6.84}$$

The two matrices in (6.83) and (6.84) can be combined into one as

$$\mathbf{H}_K = \begin{pmatrix} \mathbf{H}_{1,K} \\ \mathbf{H}_{2,K} \end{pmatrix} = \begin{pmatrix} \mathbf{H}_{1,K/2} & \mathbf{H}_{1,K/2} \\ \mathbf{H}_{1,K/2} & -\mathbf{H}_{1,K/2} \\ \mathbf{H}_{2,K/2} & \mathbf{H}_{2,K/2} \\ \mathbf{H}_{2,K/2} & -\mathbf{H}_{2,K/2} \end{pmatrix}_{2K \times KP} \tag{6.85}$$

The codes included in matrix $\mathbf{H}_K$ can be viewed as two groups, group 1 being $\mathbf{H}_{1,K}$ and group 2 being $\mathbf{H}_{2,K}$. It can be shown that the inter-group even-shift CCFs are always zero. On the other hand, the intra-group even-shift CCFs will be non-zero at time shifts of nP, where $n = 1, \ldots, \infty$, $P = 2N$, and N is the element code length of the CC codes.

It is seen from (6.85) that with Walsh-Hadamard matrix expansion we have successfully expanded the original GESO sequence set $(\mathbf{c}_1, \mathbf{c}_2)$ into a new set with its set size being $2K$ and its PG value being $PK = 2NK$ after the Kth expansion. The inter-group even-shifted CCFs are always zero and the intra-group even-shifted CCFs are non-zero only at $2nN$ $(n = 1, \ldots, \infty)$, where N is the element code length. However, we still can not control their odd shift CCFs. Therefore, an odd-shift correlation elimination algorithm should be used to suppress all odd-shift correlations of the sequence set generated above.

Odd-Shift Correlation Elimination

Let us define

$$\mathbf{H}_K = \begin{pmatrix} \mathbf{h}_1 \\ \mathbf{h}_2 \\ \vdots \\ \mathbf{h}_{2K} \end{pmatrix} \tag{6.86}$$

and a 2×2 Walsh-Hadamard matrix as

$$\tilde{\mathbf{D}} = \begin{pmatrix} d_{11} & d_{12} \\ d_{21} & d_{22} \end{pmatrix} = \begin{pmatrix} 1 & 1 \\ 1 & -1 \end{pmatrix} = \begin{pmatrix} \mathbf{d}_1 \\ \mathbf{d}_2 \end{pmatrix} \tag{6.87}$$

based on which we can proceed to generate two new matrices as

$$\begin{cases} \mathbf{U}_I(p,q) = \mathbf{H}_K(p,q)\,\tilde{\mathbf{D}}(1,\varsigma) \\ \mathbf{U}_Q(p,q) = \mathbf{H}_K(p,q)\,\tilde{\mathbf{D}}(2,\varsigma) \end{cases} \tag{6.88}$$

where p, q, and ς are defined as

$$\begin{cases} p = 1\ldots\ldots 2K \\ q = 1\ldots\ldots PK \\ \varsigma = -\bmod(q,2) + 2 \end{cases} \tag{6.89}$$

Thus, we can rewrite (6.88) into

$$\mathbf{U}_I = \begin{pmatrix} \mathbf{u}_{I,1} \\ \mathbf{u}_{I,2} \\ \vdots \\ \mathbf{u}_{I,2K} \end{pmatrix}, \qquad \mathbf{U}_Q = \begin{pmatrix} \mathbf{u}_{Q,1} \\ \mathbf{u}_{Q,2} \\ \vdots \\ \mathbf{u}_{Q,2K} \end{pmatrix} \tag{6.90}$$

such that we can combine them into a complex matrix $\mathbf{U}$ as

$$\mathbf{U} = \begin{pmatrix} \mathbf{u}_{I,1} \\ \mathbf{u}_{I,2} \\ \vdots \\ \mathbf{u}_{I,2K} \end{pmatrix} + j\begin{pmatrix} \mathbf{u}_{Q,1} \\ \mathbf{u}_{Q,2} \\ \vdots \\ \mathbf{u}_{Q,2K} \end{pmatrix} = \begin{pmatrix} \mathbf{u}_1 \\ \mathbf{u}_2 \\ \vdots \\ \mathbf{u}_{2K} \end{pmatrix} = \begin{pmatrix} \mathbf{U}_{G1} \\ \mathbf{U}_{G2} \end{pmatrix} \tag{6.91}$$

where group 1 code subset $\mathbf{U}_{G1}$ and group 2 code subset $\mathbf{U}_{G2}$ can be defined as

$$\mathbf{U}_{G1} = \begin{pmatrix} \mathbf{u}_{I,1} \\ \mathbf{u}_{I,2} \\ \vdots \\ \mathbf{u}_{I,K} \end{pmatrix} + j\begin{pmatrix} \mathbf{u}_{Q,1} \\ \mathbf{u}_{Q,2} \\ \vdots \\ \mathbf{u}_{Q,K} \end{pmatrix} = \begin{pmatrix} \mathbf{u}_1 \\ \mathbf{u}_2 \\ \vdots \\ \mathbf{u}_K \end{pmatrix} \tag{6.92}$$

$$\mathbf{U}_{G2} = \begin{pmatrix} \mathbf{u}_{I,K+1} \\ \mathbf{u}_{I,K+2} \\ \vdots \\ \mathbf{u}_{I,2K} \end{pmatrix} + j\begin{pmatrix} \mathbf{u}_{Q,K+1} \\ \mathbf{u}_{Q,K+2} \\ \vdots \\ \mathbf{u}_{Q,2K} \end{pmatrix} = \begin{pmatrix} \mathbf{u}_{K+1} \\ \mathbf{u}_{K+2} \\ \vdots \\ \mathbf{u}_{2K} \end{pmatrix} \tag{6.93}$$

The code set given in (6.91) or (6.92) and (6.93) is the generalized pair-wise complementary (GPC) code set, which offers well-controlled even-and odd-shift CCFs and ACFs, as to be shown below.

It can be seen from (6.91) that the obtained GPC code sets work in pairs as $\mathbf{u}_{I,k}$ and $\mathbf{u}_{Q,k}$, where $k = 1, \ldots, 2K$. The correlation functions of the codes should be constructed in a complementary way, or they should be sent via two orthogonal carriers, perform matched-filtering individually, and then be summed over to form a decision variable at a receiver. The PG value of the GPC codes is $4NK$ with its set size being $2K$.

Figure 6.2 shows a schematic diagram for a CDMA system based on the GPC codes, where a QPSK modem has been used to send a pair of GPC codes over the I and Q channels separately using a single carrier.

Code Generation Example

Here, an example is given to show step by step how to generate a GPC code set with its PG being 64 and set size being 4.

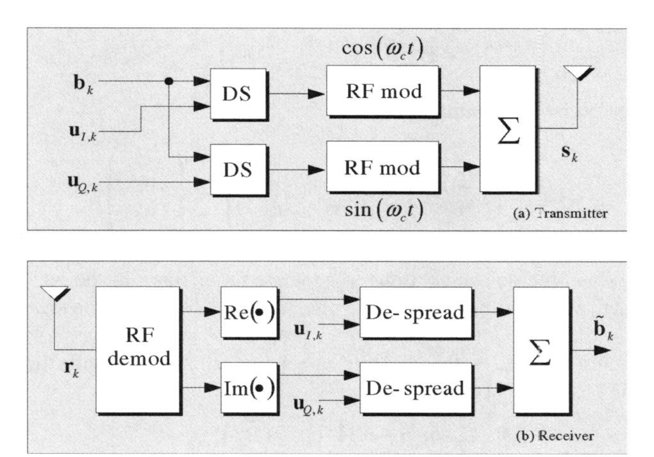

Figure 6.2 Block diagram of a CDMA transceiver using generalized pair-wise complementary (GPC) codes.

First, we need to generate a set of complete complementary codes. Assume that the initial matrices **A**, **B**, and **D** are all equal to a 2 × 2 Walsh matrix as

$$\begin{pmatrix} + & + \\ + & - \end{pmatrix}.$$

According to the definition given in Equation (B.5), we have $\mathbf{Z}_{11} = (1, 1)$, $\mathbf{Z}_{12} = (1, -1)$, $\mathbf{Z}_{21} = (1, -1)$, and $\mathbf{Z}_{22} = (1, 1)$, with which we can obtain the complete complementary code set as

$$\mathbf{E}_1 = \begin{pmatrix} b_{11}\mathbf{Z}_{11} & b_{12}\mathbf{Z}_{21} \\ b_{11}\mathbf{Z}_{12} & b_{12}\mathbf{Z}_{22} \end{pmatrix} = \begin{pmatrix} + & + & + & - \\ + & - & + & + \end{pmatrix} \equiv \begin{pmatrix} \mathbf{c}_{11} \\ \mathbf{c}_{12} \end{pmatrix} \tag{6.94}$$

$$\mathbf{E}_2 = \begin{pmatrix} b_{21}\mathbf{Z}_{11} & b_{22}\mathbf{Z}_{21} \\ b_{21}\mathbf{Z}_{12} & b_{22}\mathbf{Z}_{22} \end{pmatrix} = \begin{pmatrix} + & + & - & + \\ + & - & - & - \end{pmatrix} \equiv \begin{pmatrix} \mathbf{c}_{21} \\ \mathbf{c}_{22} \end{pmatrix} \tag{6.95}$$

Next, we should form a GESO sequence set by spreading each chip of $\mathbf{c}_{11}$ and $\mathbf{c}_{21}$ by [++] and each chip of $\mathbf{c}_{12}$ and $\mathbf{c}_{22}$ by [+−], to yield

$$\begin{cases} \mathbf{c}_{11}^{(++)} = (+ + + + + + - -) \\ \mathbf{c}_{12}^{(+-)} = (+ - - + + - + -) \\ \mathbf{c}_{21}^{(++)} = (+ + + + - - + +) \\ \mathbf{c}_{22}^{(+-)} = (+ - - + - + - +) \end{cases} \tag{6.96}$$

By cascading $\mathbf{c}_{11}^{++}$ and $\mathbf{c}_{12}^{+-}$, $\mathbf{c}_{21}^{++}$ and $\mathbf{c}_{22}^{+-}$, we will obtain $\mathbf{c}_1$ and $\mathbf{c}_2$, respectively, as

$$\begin{cases} \mathbf{c}_1 = (\mathbf{c}_{11}^{[++)}\mathbf{c}_{12}^{[+-]}] = (+ + + + + + - - + - - + + - + -) \\ \mathbf{c}_2 = (\mathbf{c}_{21}^{[++]}\mathbf{c}_{22}^{[+-]}) = (+ + + + - - + + + - - + - + - +) \end{cases} \tag{6.97}$$

Thirdly, we will use Walsh-Hadamard matrix expansion to obtain

$$\mathbf{H}_{1,2} = \begin{pmatrix} \mathbf{c}_1 & \mathbf{c}_1 \\ \mathbf{c}_1 & -\mathbf{c}_1 \end{pmatrix} \tag{6.98}$$

$$\mathbf{H}_{2,2} = \begin{pmatrix} \mathbf{c}_2 & \mathbf{c}_2 \\ \mathbf{c}_2 & -\mathbf{c}_2 \end{pmatrix} \tag{6.99}$$

which can be expressed by combining the two as

$$\mathbf{H}_2 = \begin{pmatrix} \mathbf{H}_{1,2} \\ \mathbf{H}_{2,2} \end{pmatrix} = \begin{pmatrix} \mathbf{c}_1 & \mathbf{c}_1 \\ \mathbf{c}_1 & -\mathbf{c}_1 \\ \mathbf{c}_2 & \mathbf{c}_2 \\ \mathbf{c}_2 & -\mathbf{c}_2 \end{pmatrix} = \begin{pmatrix} \mathbf{h}_1 \\ \mathbf{h}_2 \\ \mathbf{h}_3 \\ \mathbf{h}_4 \end{pmatrix} \tag{6.100}$$

Of course, the expansion process can continue to generate more codes in the set. With more codes in the set, the code length will also increase proportionally. Therefore, we stop with $\mathbf{H}_2$ for illustration simplicity.

Fourthly, we should proceed with odd-shift correlation elimination with the help of matrix $\tilde{\mathbf{D}}$, which was defined in (6.87), to obtain

$$\begin{cases} \mathbf{U}_I(p,q) = \mathbf{H}_2(p,q)\,\tilde{\mathbf{D}}(1,\varsigma) \\ \mathbf{U}_Q(p,q) = \mathbf{H}_2(p,q)\,\tilde{\mathbf{D}}(2,\varsigma) \end{cases} \tag{6.101}$$

where $\varsigma = -\bmod(q,2) + 2$, $p = 1, \ldots, 4$, and $q = 1, \ldots, 32$. The finally obtained GPC code set can be expressed as

$$\mathbf{U}_I = \begin{pmatrix} \mathbf{u}_{I,1} \\ \mathbf{u}_{I,2} \\ \mathbf{u}_{I,3} \\ \mathbf{u}_{I,4} \end{pmatrix} \tag{6.102}$$

$$= \begin{pmatrix} +++++\,+--+--++-+-++++++--+--++-+- \\ ++++++--+--++-+-------++-++--+-+ \\ ++++--+++--+-+-+++++--+++--+-+-+ \\ ++++--+++--+-+-+----++---++-+-+- \end{pmatrix}$$

$$\mathbf{U}_Q = \begin{pmatrix} \mathbf{u}_{Q,1} \\ \mathbf{u}_{Q,2} \\ \mathbf{u}_{Q,3} \\ \mathbf{u}_{Q,4} \end{pmatrix} \tag{6.103}$$

$$= \begin{pmatrix} +-+-+--+++--+++++-+-+--+++--++++ \\ +-+-+--+++--++++-+-+-++---++---- \\ +-+--++-++------+-+--++-++------ \\ +-+--++-++-------+-++--+--++++++ \end{pmatrix}$$

which can be written into a complex form as

$$\mathbf{U} = \begin{pmatrix} \mathbf{u}_{I,1} \\ \mathbf{u}_{I,2} \\ \mathbf{u}_{I,3} \\ \mathbf{u}_{I,4} \end{pmatrix} + j \begin{pmatrix} \mathbf{u}_{Q,1} \\ \mathbf{u}_{Q,2} \\ \mathbf{u}_{Q,3} \\ \mathbf{u}_{Q,4} \end{pmatrix} = \begin{pmatrix} \mathbf{u}_1 \\ \mathbf{u}_2 \\ \mathbf{u}_3 \\ \mathbf{u}_4 \end{pmatrix} = \begin{pmatrix} \mathbf{U}_{G1} \\ \mathbf{U}_{G2} \end{pmatrix} \tag{6.104}$$

or written into two groups, $\mathbf{U}_{G1}$ and $\mathbf{U}_{G2}$, as

$$\begin{cases} \mathbf{U}_{G1} = \begin{pmatrix} \mathbf{u}_{I,1} \\ \mathbf{u}_{I,2} \end{pmatrix} + j \begin{pmatrix} \mathbf{u}_{Q,1} \\ \mathbf{u}_{Q,2} \end{pmatrix} = \begin{pmatrix} \mathbf{u}_1 \\ \mathbf{u}_2 \end{pmatrix} \\ \mathbf{U}_{G2} = \begin{pmatrix} \mathbf{u}_{I,3} \\ \mathbf{u}_{I,4} \end{pmatrix} + j \begin{pmatrix} \mathbf{u}_{Q,3} \\ \mathbf{u}_{Q,4} \end{pmatrix} = \begin{pmatrix} \mathbf{u}_3 \\ \mathbf{u}_4 \end{pmatrix} \end{cases} \tag{6.105}$$

which possess different inter- or intra-group cross-correlation properties.

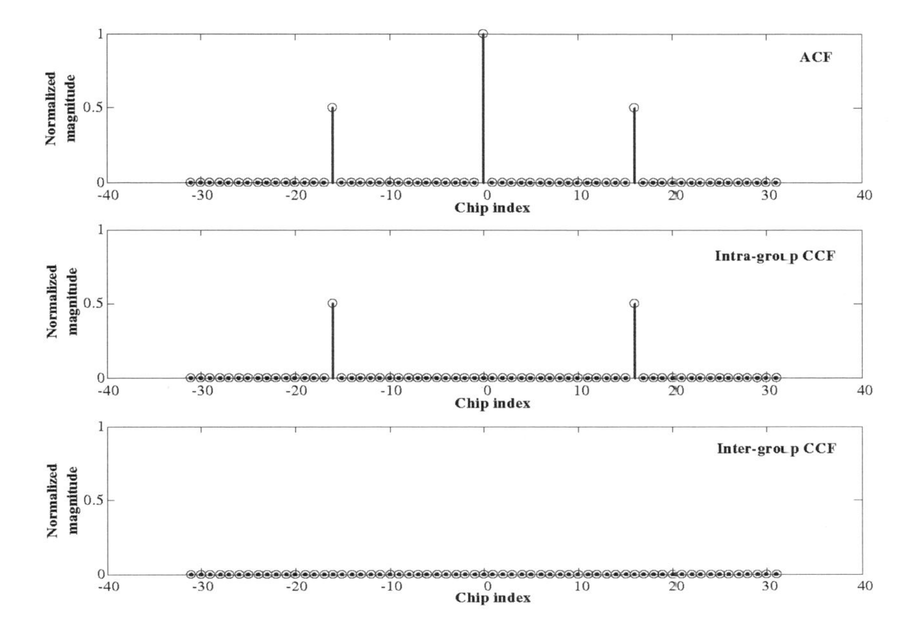

Figure 6.3 ACF and intra-group and inter-group CCFs for GPC codes generated as an example in Section 6.3.4. It is observed that the GPC codes offer an IFW with its width being $8N$ (where $N = 4$ in this case) in both their ACF and CCF, where N is the length of the element code of the complete complementary codes that were used to generate the GPC codes. The PG of the code set is 64. The code length is 32 and the set size is 4.

Figure 6.3 shows the ACF, intra-group CCF, and inter-group CCF of the GPC code set generated as an example above. It is seen from the figure that the inter-group CCF of this GPC code set is perfect, being all zero for all relative time shifts. On the other hand, its intra-group CCF and ACF side lobes are non-zero only at a time shift of $4N$, where N is the element code length of the complete complementary codes.

Figure 6.3 shows the ACF and intra-group and inter-group CCFs of the GPC code set, as to be exemplified in the next section, where $P = 2N = 2 \times 4 = 8$ is assumed. For this particular code set, its processing gain is 64, the code length is 32, and set size is 4, to yield an IFW whose width spans 32 chips. In general, if the element code length of complete complementary codes is N, an IFW equal to $4P = 8N$ will be created. If a longer complete complementary code is used, the IFW will be wider.

6.3.5 ALGEBRA APPROACHES: THE REAL APPROACH

So far we have introduced several different complementary code generation approaches, each of which pertains to the generation of a particular complementary code. Therefore, it is seen from the discussions given in the above subsections that their approaches are in general very different from one another. The advantage of these special approaches is their computational efficiency, while the shortcoming is that they give little information on the connection between the complementary codes generated and their performance. Based on these approaches we have little knowledge to understand why the orthogonal complementary codes play such an important role in the implementation of an interference-free CDMA system.

In this subsection we will turn to another methodology, called the real environment adaptation linearization (REAL) approach. This approach can give us a more general view on the technical requirements to implement an interference-free CDMA, as an effort to understand better the orthogonal complementary codes and the role they will play in the next generation CDMA technology, which is the key issue covered in this book.

In the REAL approach, we should take into account many real operational scenarios in the spreading code design process such that generated spreading codes can effectively address most major adverse operational conditions in their code structure. To make the study as general as possible, the design problem will be formulated initially from complementary codes, which work on a one-user-per-flock basis, because any non-complementary codes (such as the unitary codes, which work on a one-code-per-user basis) are only special cases with the flock size being reduced to one.

Non-Linear Code Design Problem

Let us start with a generic CDMA code set, whose element code length, flock size, and set size are N, M, and K, respectively. We will use the notation (M, N, K) to characterize a generic CDMA code set. If $M = 1$, it reduces to a conventional unitary code set, which works on a one-code-per-user basis. On the other hand, if $M > 1$, it is a complementary code set, in which a flock of element codes should be assigned to a user for CDMA communications. Assume that we have two flocks in a set, $\mathbf{x} = \{\mathbf{x}_1, \mathbf{x}_2, \ldots, \mathbf{x}_M\}$ and $\mathbf{y} = \{\mathbf{y}_1, \mathbf{y}_2, \ldots, \mathbf{y}_M\}$, where $\mathbf{x}_i = \{x_{i1}, x_{i2}, \ldots, x_{iN}\}$, $\mathbf{y}_i = \{y_{i1}, y_{i2}, \ldots, y_{iN}\}$ $(1 \le i \le M)$, and $x_{ij} \in \{\mathcal{C}\}$, $y_{ij} \in \{\mathcal{C}\}$ $(1 \le i \le M, 1 \le j \le N)$, in which $\{\mathcal{C}\}$ denotes the set of complex values. Also, we consider an asynchronous channel (asynchronous one is only a special case here), where both multiple access interference (MAI) and multipath interference (MI) are present.

Thus, this generic spreading code design problem can be expressed as follows: to search for a perfect orthogonal complementary (POC) code set such that its periodic and aperiodic out-of-phase auto-correlation functions, and its periodic and aperiodic cross-correlation functions should be zero, ensuring MAI-free and MI-free operation. To formulate such a design problem mathematically, let us examine in particular two cases with even and odd lengths of the element codes. As all element codes have the same length N, the discussion can be focused on any one of them, say the ith element code, $\mathbf{x}_i = \{x_{i1}, x_{i2}, \ldots, x_{iN}\}$ $(1 \le i \le M)$, without losing generality. For representation simplicity, we assume $N = 5$ and $N = 6$ for odd and even element code lengths, as shown in Figures 6.4 and 6.5, where the i-th element code of a POC code $\mathbf{x} = \{\mathbf{x}_1, \ldots, \mathbf{x}_M\}$ and its j chips shifted versions $(0 \le j \le N - 1)$ are illustrated. It is seen that, if two relatively shifted versions of an element code with their offsets being a and b (which are two integers and satisfy $a + b = N$, as shown by the pairs connected by arrowed lines in the figures) will generate the same periodic and aperiodic auto-correlation functions as the zero-shifted version. Thus, for either even or odd N we can establish an equation set for the code $\mathbf{x} = \{\mathbf{x}_1, \ldots, \mathbf{x}_M\}$ with MN unknown variables and $\lfloor \frac{N}{2} \rfloor + \lfloor \frac{N-1}{2} \rfloor + 1$ independent non-linear equations, such that

$$\begin{cases} \sum_{i=1}^{M} \left[x_{i1}^2 + x_{i2}^2 + \ldots + x_{iN}^2\right] = NM \\ \sum_{i=1}^{M} \left[x_{iN}x_{i1} + x_{i1}x_{i2} + \ldots + x_{i(N-1)}x_{iN}\right] = 0 \\ \sum_{i=1}^{M} \left[x_{i(N-1)}x_{i1} + x_{iN}x_{i2} + \ldots + x_{i(N-2)}x_{iN}\right] = 0 \\ \sum_{i=1}^{M} \left[x_{i(N-2)}x_{i1} + x_{i(N-1)}x_{i2} + \ldots + x_{i(N-3)}x_{iN}\right] = 0 \\ \ldots\ldots \\ \sum_{i=1}^{M} \left[-x_{iN}x_{i1} + x_{i1}x_{i2} + \ldots + x_{i(N-1)}x_{iN}\right] = 0 \\ \sum_{i=1}^{M} \left[-x_{i(N-1)}x_{i1} - x_{iN}x_{i2} + \ldots + x_{i(N-2)}x_{iN}\right] = 0 \\ \ldots\ldots \end{cases} \tag{6.106}$$

Periodic auto-correlation

$x_{i1}\ x_{i2}\ x_{i3}\ x_{i4}\ x_{i5}\ x_{i6}$

Delay 0	$x_{i1}\ x_{i2}\ x_{i3}\ x_{i4}\ x_{i5}\ x_{i6}$
Delay 1	$x_{i6}\ x_{i1}\ x_{i2}\ x_{i3}\ x_{i4}\ x_{i5}$
Delay 2	$x_{i5}\ x_{i6}\ x_{i1}\ x_{i2}\ x_{i3}\ x_{i4}$
Delay 3	$x_{i4}\ x_{i5}\ x_{i6}\ x_{i1}\ x_{i2}\ x_{i3}$
Delay 4	$x_{i3}\ x_{i4}\ x_{i5}\ x_{i6}\ x_{i1}\ x_{i2}$
Delay 5	$x_{i2}\ x_{i3}\ x_{i4}\ x_{i5}\ x_{i6}\ x_{i1}$

(a)

$x_{i1}\ x_{i2}\ x_{i3}\ x_{i4}\ x_{i5}$

Delay 0	$x_{i1}\ x_{i2}\ x_{i3}\ x_{i4}\ x_{i5}$
Delay 1	$x_{i5}\ x_{i1}\ x_{i2}\ x_{i3}\ x_{i4}$
Delay 2	$x_{i4}\ x_{i5}\ x_{i1}\ x_{i2}\ x_{i3}$
Delay 3	$x_{i3}\ x_{i4}\ x_{i5}\ x_{i1}\ x_{i2}$
Delay 4	$x_{i2}\ x_{i3}\ x_{i4}\ x_{i5}\ x_{i1}$

(b)

Figure 6.4 Periodic auto-correlation functions of an even ($N = 6$ in (a)) or odd ($N = 5$ in (b)) length element code $\mathbf{x}_i$ ($1 \le i \le M$) with two consecutive bits having the same sign in an asynchronous channel, where $N - 1 = 5$ (in (a)) or $N - 1 = 4$ (in (b)) delayed multipath returns or periodical cyclic-shifted versions are present. The shaded sequences represent the local element code of $\mathbf{x}_i$ ($1 \le i \le M$) generated at a correlator and the arrowed lines indicate the two relatively cyclic-shifted versions of an element code that will generate the same out-of-phase auto-correlation functions with the local element code.

Aperiodic auto-correlation

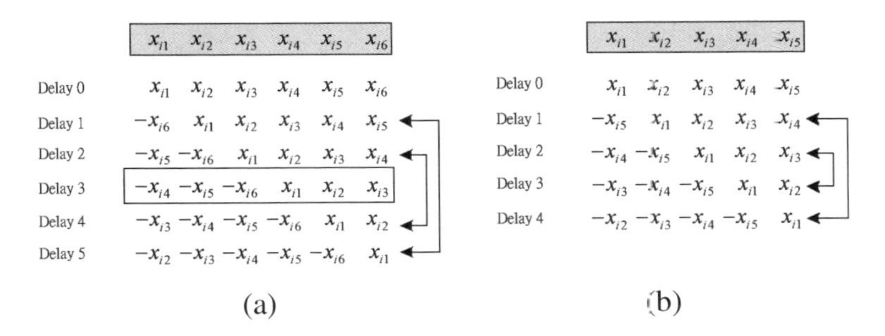

(a) (b)

Figure 6.5 Aperiodic auto-correlation functions of an even ($N = 6$ in (a)) or odd ($N = 5$ in (b)) length element code $\mathbf{x}_i$ ($1 \le i \le M$) with two consecutive bits having different signs in an asynchronous channel, where $N - 1 = 5$ (in (a)) or $N - 1 = 4$ (in (b)) delayed multipath returns or aperiodically cyclic-shifted versions are present. The shaded sequences represent the local element code of $\mathbf{x}_i$ ($1 \le i \le M$) generated at a correlator. The arrowed lines indicate the two relatively cyclic-shifted versions of an element code that will generate the same out-of-phase auto-correlation functions with the local element code and the code inside the solid-line frame yields zero out-of-phase auto-correlation function.

where the first $\lfloor \frac{N}{2} \rfloor + 1$ equations are from ideal periodic auto-correlation functions, the last $\lfloor \frac{N-1}{2} \rfloor$ ones from ideal aperiodic auto-correlation functions, and $\lfloor \alpha \rfloor$ stands for the largest integer less than α.

Now, let us introduce the second code $\mathbf{y} = \{\mathbf{y}_1, \ldots, \mathbf{y}_M\}$ into the set, which consists of $\mathbf{x}$ and $\mathbf{y}$ only, such that they should satisfy all conditions as a pair of POC codes. Similarly, we can establish the equation sets based on the ideal periodic and aperiodic cross-correlation functions between the codes $\mathbf{y} = \{\mathbf{y}_1, \ldots, \mathbf{y}_M\}$ and $\mathbf{x} = \{\mathbf{x}_1, \ldots, \mathbf{x}_M\}$ as follows. For either even or odd N, there are $2N - 1$ non-linear homogenous equations such that they satisfy

$$\begin{cases} \sum_{i=1}^{M} \left[y_{i1}x_{i1} + y_{i2}x_{i2} + \ldots + y_{iN}x_{iN}\right] = 0 \\ \sum_{i=1}^{M} \left[y_{iN}x_{i1} + y_{i1}x_{i2} + \ldots + y_{i(N-1)}x_{iN}\right] = 0 \\ \sum_{i=1}^{M} \left[y_{i(N-1)}x_{i1} + y_{iN}x_{i2} + \ldots + y_{i(N-2)}x_{iN}\right] = 0 \\ \sum_{i=1}^{M} \left[y_{i(N-2)}x_{i1} + y_{i(N-1)}x_{i2} + \ldots + y_{i(N-3)}x_{iN}\right] = 0 \\ \ldots\ldots \\ \sum_{i=1}^{M} \left[-y_{iN}x_{i1} + y_{i1}x_{i2} + \ldots + y_{i(N-1)}x_{iN}\right] = 0 \\ \sum_{i=1}^{M} \left[-y_{i(N-1)}x_{i1} - y_{iN}x_{i2} + \ldots + y_{i(N-2)}x_{iN}\right] = 0 \\ \ldots\ldots \end{cases} \tag{6.107}$$

where the upper N equations are due to ideal periodic cross-correlation functions and the lower $N - 1$ ones from the ideal aperiodic cross-correlation functions. In addition, $\mathbf{y} = \{\mathbf{y}_1, \ldots, \mathbf{y}_M\}$ itself should satisfy the ideal auto-correlation functions as

$$\begin{cases} \sum_{i=1}^{M} \left[y_{i1}^2 + y_{i2}^2 + \ldots + y_{iN}^2\right] = NM \\ \sum_{i=1}^{M} \left[y_{iN}y_{i1} + y_{i1}y_{i2} + \ldots + y_{i(N-1)}y_{iN}\right] = 0 \\ \sum_{i=1}^{M} \left[y_{i(N-1)}y_{i1} + y_{iN}y_{i2} + \ldots + y_{i(N-2)}y_{iN}\right] = 0 \\ \sum_{i=1}^{M} \left[y_{i(N-2)}y_{i1} + y_{i(N-1)}y_{i2} + \ldots + y_{i(N-3)}y_{iN}\right] = 0 \\ \ldots\ldots \\ \sum_{i=1}^{M} \left[-y_{iN}y_{i1} + y_{i1}y_{i2} + \ldots + y_{i(N-1)}y_{iN}\right] = 0 \\ \sum_{i=1}^{M} \left[-y_{i(N-1)}y_{i1} - y_{iN}y_{i2} + \ldots + y_{i(N-2)}y_{iN}\right] = 0 \\ \ldots\ldots \end{cases} \tag{6.108}$$

where the first $\left\lfloor \frac{N}{2} \right\rfloor + 1$ equations reflect ideal periodic auto-correlation functions and the last $\left\lfloor \frac{N-1}{2} \right\rfloor$ equations specify the ideal aperiodic auto-correlation functions In general, to generate a POC code set $\{N, M, K\}$ $(M \geq 1, K > 1)$, regardless of even or odd N, we can have the following non-linear equations:

1. $\left(\left\lfloor \frac{N}{2} \right\rfloor + 1\right) K$ non-linear equations from perfect periodic auto-correlation conditions.
2. $\frac{NK(K-1)}{2}$ non-linear equations from zero periodic cross-correlation conditions.
3. $\left(\left\lfloor \frac{N-1}{2} \right\rfloor + 1\right) K$ non-linear equations from perfect aperiodic auto-correlation conditions.
4. $\frac{(N-1)K(K-1)}{2}$ non-linear equations from zero aperiodic cross-correlation conditions, where $K > 1$.

Thus, we have altogether $K\left(\left\lfloor \frac{N}{2} \right\rfloor + \left\lfloor \frac{N-1}{2} \right\rfloor + 1\right) + (2N - 1)\frac{K(K-1)}{2}$ non-linear equations, which contain *MNK* unknown variables for K unknown POC codes. It is possible that these non-linear equations can be solved under the condition that the following inequality is held

$$K\left(\left\lfloor \frac{N}{2} \right\rfloor + \left\lfloor \frac{N-1}{2} \right\rfloor + 1\right) + (2N - 1)\frac{K(K-1)}{2} \geq MNK \tag{6.109}$$

However, the solutions to a non-linear equation set are not guaranteed even if (6 109) is satisfied. Therefore, we would like to turn to a methodology based on the linear equation sets. It is noted that (6.107) in fact can be transformed into a homogenous linear equation set if $\mathbf{x} = \{\mathbf{x}_1, \ldots, \mathbf{x}_M\}$ is already known, where $\mathbf{x}_i = \{x_{i1}, x_{i2}, \ldots, x_{iN}\}, x_{ij} \in \{\mathcal{C}\}$ ($\mathcal{C}$ is the set of all complex values), $1 \le i \le M$ and $1 \le j \le N$. We will take this known code $\mathbf{x}$ as a seed code, which itself should satisfy all conditions that both its periodic and aperiodic out-of-phase auto-correlation functions should be zero. If so, we readily have a linear equation set from (6.107) to solve the second code $\mathbf{y} = \{\mathbf{y}_1, \ldots, \mathbf{y}_M\}$ (where $\mathbf{y}_i = \{y_{i1}, y_{i2}, \ldots, y_{iN}\}$, $y_{ij} \in \{\mathcal{C}\}$, $1 \le i \le M$ and $1 \le j \le N$) jointly with (6.108), which is not a linear equation set but can help to determine all unknown variables in (6.107). The solution to these equation sets or code $\mathbf{y}$ must satisfy all MAI-free and MI-free conditions, or Equations (6.106) to (6.108). The same procedure can be repeated until all codes in the set are found.

It should be emphasized that the aforementioned design problem has never imposed any restrictions on the values each chip may take, which can be either real or complex. We have chosen the complementary codes (or $M \ge 1$) as our starting point simply because they represent general cases, including all spreading codes currently available. If $M = 1$, the obtained results are applicable to traditional unitary spreading codes; otherwise, if $M > 1$ complementary codes will be yielded. In other words, the solutions obtained from (6.106) to (6.108) should include all possible codes satisfying the MAI-free and MI-free conditions. If these equations are not solvable, we then can conclude that no such POC codes exist to ensure MAI-free and MI-free operation.

Necessary Conditions to Generate POC Codes

Consider an $m \times n$ coefficient matrix $\mathbf{A} = \{a_{ij}\}$ $(1 \le i \le m,\ 1 \le j \le n)$, a variable vector $\mathbf{x} = \{x_i\}$ $(1 \le i \le m)$, and an $m \times 1$ constant vector $\mathbf{b} = \{b_i\}$ $(1 \le i \le m)$, which are related by an equation $\mathbf{Ax} = \mathbf{b}$, where we have $a_{ij} \in \{\mathcal{C}\}$ $(1 \le i \le m,\ 1 \le j \le n)$, $x_i \in \{\mathcal{C}\}$ $(1 \le i \le m)$, and $b_i \in \{\mathcal{C}\}$ $(1 \le i \le m)$, and $\{\mathcal{C}\}$ is the set of complex values. It is noted that m is the number of linear equations and n gives the number of unknown variables in $\mathbf{x}$. The following well-known linear algebra theorems will be useful in our code design problem.

> **Theorem:** The solution to the linear equation set $\mathbf{Ax} = \mathbf{b}$ has only the following three possible outcomes:
>
> (a) If $rank(\mathbf{A}|\mathbf{b}) > rank(\mathbf{A})$, $\mathbf{Ax} = \mathbf{b}$ has no solution.
>
> (b) If $rank(\mathbf{A}|\mathbf{b}) = rank(\mathbf{A})$ and $rank(\mathbf{A}) = n$, $\mathbf{Ax} = \mathbf{b}$ has only one solution.
>
> (c) If $rank(\mathbf{A}|\mathbf{b}) = rank(\mathbf{A})$ and $rank(\mathbf{A}) < n$, $\mathbf{Ax} = \mathbf{b}$ has more than one solution.
>
> $(\mathbf{A}|\mathbf{b})$ stands for an extended matrix for the linear equation set $\mathbf{Ax} = \mathbf{b}$. For a homogenous linear equation set, we always have $rank(\mathbf{A}|\mathbf{b}) = rank(\mathbf{A})$ as $\mathbf{b} = \mathbf{0}$. Therefore, it will have at least one solution or all-zero solution as long as $rank(\mathbf{A}) \le n$.

Assume that there is a flock of codes, $\mathbf{x} = \{x_{11}, x_{12}, \ldots, x_{2N}; x_{21}, x_{22}, \ldots, x_{2N}; \ldots; x_{M1}, x_{M2}, \ldots, x_{MN}\}$, which satisfies the conditions for ideal auto-correlation functions, as defined in (6.106), where $x_{ij} \in \{\mathcal{C}\}$ (where $1 \le i \le M, 1 \le j \le N$, and $\mathcal{C}$ denotes the set of all complex values) represents the jth chip in the ith element code of the flock. To determine the second flock of element codes, $\mathbf{y} = \{y_{11}, y_{12}, \ldots, y_{1N}; y_{21}, y_{22}, \ldots, y_{2N}; \ldots; y_{M1}, y_{M2}, \ldots, y_{MN}\}$, where $y_{ij} \in \{\mathcal{C}\}$ $(1 \le i \le M, 1 \le j \le N)$, we can obtain a $(2N - 1) \times (NM)$ coefficient matrix of a homogeneous linear equation set from (6.107), which specifies ideal periodic and aperiodic cross-correlation functions between $\mathbf{x}$ and

$\mathbf{y}$ and can be written into $\mathbf{A}_y\mathbf{y}^T = \mathbf{0}$, as

$$\mathbf{A}_y = \begin{pmatrix} x_{11} & x_{12} & \cdots & x_{1N}, & x_{21} & x_{22} & \cdots & x_{2N}, & \ldots, & x_{M1} & x_{M2} & \cdots & x_{MN} \\ x_{12} & x_{13} & \cdots & x_{11}, & x_{22} & x_{23} & \cdots & x_{21}, & \ldots, & x_{M2} & x_{M3} & \cdots & x_{M1} \\ \cdots & \cdots & \cdots & \cdots & \cdots & \cdots & \cdots & \cdots & \cdots & \cdots & \cdots & \cdots & \cdots \\ x_{1N} & x_{11} & \cdots & x_{1(N-1)}, & x_{2N} & x_{21} & \cdots & x_{2(N-1)}, & \ldots, & x_{MN} & x_{M1} & \cdots & x_{M(N-1)} \\ x_{12} & x_{13} & \cdots & -x_{11}, & x_{22} & x_{23} & \cdots & -x_{21}, & \ldots, & x_{M2} & x_{M3} & \cdots & -x_{M1} \\ x_{13} & x_{14} & \cdots & -x_{12}, & x_{23} & x_{24} & \cdots & -x_{22}, & \ldots, & x_{M3} & x_{M4} & \cdots & -x_{M2} \\ \cdots & \cdots & \cdots & \cdots & \cdots & \cdots & \cdots & \cdots & \cdots & \cdots & \cdots & \cdots & \cdots \\ x_{11} & -x_{12} & \cdots & -x_{1N}, & x_{21} & -x_{22} & \cdots & -x_{2N}, & \ldots, & x_{M1} & -x_{M2} & \cdots & -x_{MN} \end{pmatrix} \tag{6.110}$$

where the upper half of $\mathbf{A}_y$ is the result of the periodic cross-correlation functions between $\mathbf{x}$ and $\mathbf{y}$ and the lower half is from the aperiodic cross-correlation functions between the two codes. Because $\mathbf{x}$ itself must satisfy all conditions for ideal auto-correlation functions specified by (6.106), all row vectors in the upper half of (6.110) have to be mutually independent. To illustrate it more clearly, let us examine a matrix formed by all possible relatively cyclic-shifted versions of the ith element code $\mathbf{x}_i = \{x_{i1}, \ldots, x_{iN}\}$ (where $x_{ij} \in \{\mathcal{C}\}$, $1 \le i \le M$ and $1 \le j \le N$) in a flock of codes $\mathbf{x} = \{\mathbf{x}_1, \ldots, \mathbf{x}_M\}$, where $N = 4$ for illustration simplicity, or

$$\begin{pmatrix} x_{i1} & x_{i2} & x_{i3} & x_{i4} \\ x_{i2} & x_{i3} & x_{i4} & x_{i1} \\ x_{i3} & x_{i4} & x_{i1} & x_{i2} \\ x_{i4} & x_{i1} & x_{i2} & x_{i3} \end{pmatrix}, \qquad (1 \le i \le M) \tag{6.111}$$

which in fact is just one of the M sub-matrices in the upper half of (6.110). Since $\mathbf{x} = \{\mathbf{x}_1, \ldots, \mathbf{x}_M\}$ has a perfect periodic auto-correlation function, the following equations must be true:

$$x_{i1}x_{i2} + x_{i2}x_{i3} + x_{i3}x_{i4} + x_{i4}x_{i1} = 0 \tag{6.112}$$

$$x_{i1}x_{i3} + x_{i2}x_{i4} + x_{i3}x_{i1} + x_{i4}x_{i2} = 0 \tag{6.113}$$

$$x_{i1}x_{i4} + x_{i2}x_{i1} + x_{i3}x_{i2} + x_{i4}x_{i3} = 0 \tag{6.114}$$

which merely states the fact that three pairs of the row vectors in (6.111) are mutually independent. Now let us establish the inner products of the other three pairs of relatively cyclic-shifted versions of the element code $\mathbf{x}_i = \{x_{i1}, x_{i2}, x_{i3}, x_{i4}\}$ $(1 \le i \le M)$, which are

$$x_{i2}x_{i3} + x_{i3}x_{i4} + x_{i4}x_{i1} + x_{i1}x_{i2} \tag{6.115}$$

$$x_{i3}x_{i4} + x_{i4}x_{i1} + x_{i1}x_{i2} + x_{i2}x_{i3} \tag{6.116}$$

$$x_{i2}x_{i4} + x_{i3}x_{i1} + x_{i4}x_{i2} + x_{i1}x_{i3} \tag{6.117}$$

where (6.115) is the inner product between the second and third rows of (6.111), (6.116) is that between the third and fourth rows, and (6.117) is that between the second and fourth rows. By comparing (6.115) and (6.112), (6.116) and (6.114), and (6.117) and (6.113), we can find that they are all equal. Therefore, we have

$$x_{i2}x_{i3} + x_{i3}x_{i4} + x_{i4}x_{i1} + x_{i1}x_{i2} = 0 \tag{6.118}$$

$$x_{i3}x_{i4} + x_{i4}x_{i1} + x_{i1}x_{i2} + x_{i2}x_{i3} = 0 \tag{6.119}$$

$$x_{i2}x_{i4} + x_{i3}x_{i1} + x_{i4}x_{i2} + x_{i1}x_{i3} = 0 \tag{6.120}$$

or all rows in matrix (6.111) are mutually independent with each other, implying that the matrix has a rank equal to $N = 4$, or the length of the element code $\mathbf{x}_i = \{x_{i1}, x_{i2}, x_{i3}, x_{i4}\}$ ($x_{ij} \in \{\mathcal{C}\}$, $1 \leq i \leq M$ and $1 \leq j \leq N = 4$). Although the above conclusion is obtained from the observation of a single element code $\mathbf{x}_i$ with a particular length $N = 4$, it can be easily shown that it is equally applicable to the cases for any M and N.

In general, it can be shown that the rank of a matrix formed by all possible cyclic-shifted versions of an element code $\mathbf{x}_i$, where $\mathbf{x}_i = \{x_{i1}, x_{i2}, \ldots, x_{iN}\}$ ($x_{ij} \in \{\mathcal{C}\}$, $1 \leq i \leq M$ and $1 \leq j \leq N$), must be equal to the length of the element code or N, as long as the element code itself has a perfect auto-correlation function specified by (6.106).

Observed from the upper half of the $(2N-1) \times (NM)$ matrix $\mathbf{A}_y$ shown in (6.110), formed by all possible cyclic-shifted versions of $\mathbf{x} = \{\mathbf{x}_1, \ldots, \mathbf{x}_M\}$, where $\mathbf{x}_i = \{x_{i1}, x_{i2}, \ldots, x_{iN}\}$, $x_{ij} \in \{\mathcal{C}\}$, $1 \leq i \leq M$, and $1 \leq j \leq N$, it is concluded that the rank of this coefficient matrix of the homogeneous linear equation set $\mathbf{A}_y\mathbf{y} = \mathbf{0}$ is also equal to N. If we let $M = 1$, we will get a homogeneous linear equation set with its rank equal to the number of variables or $NM = N$, resulting in an all-zero solution or $\mathbf{y} = \mathbf{0}$ from **Theorem** (b). It means that, when $M = 1$, we can not find the second code $\mathbf{y} = \{\mathbf{y}_1, \ldots, \mathbf{y}_M\}$ ($\mathbf{y}_i = \{y_{i1}, y_{i2}, \ldots, y_{iN}\}$, $y_{ij} \in \{\mathcal{C}\}$, $1 \leq i \leq M$, and $1 \leq j \leq N$) such that it will ensure perfect periodic and aperiodic cross-correlation functions with the existing code $\mathbf{x} = \{\mathbf{x}_1, \ldots, \mathbf{x}_M\}$, where $\mathbf{x}_i = \{x_{i1}, x_{i2}, \ldots, x_{iN}\}$ ($x_{ij} \in \{\mathcal{C}\}$, $1 \leq i \leq M$, and $1 \leq j \leq N$). However if we let $M > 1$, the solution of $\mathbf{A}_y\mathbf{y} = \mathbf{0}$ may possibly exist according to **Theorem** (c). Therefore, we have the first important corollary obtained in this subsection as follows.

Corollary 6.3.1 *To ensure perfect periodic and aperiodic auto- and cross-correlation functions of a POC code set, the flock size* M *of the code set must be greater than one (or* M > 1*). In other words, only complementary codes can possibly achieve the perfect periodic and aperiodic auto-correlation and cross-correlation functions. All traditional unitary codes, such as Gold, Kasami, Walsh-Hadamard, and OVSF codes, can never yield such perfect periodic and aperiodic auto-correlation and cross-correlation functions.*

It should be noted that the above conclusion was obtained without imposing any constraints on the chip values ($x_{ij} \in \{\mathcal{C}\}$ and $y_{ij} \in \{\mathcal{C}\}$, where $1 \leq i \leq M$ and $1 \leq j \leq N$) the POC codes might take. Therefore, the conclusion should be valid for any type of POC code, either real or complex valued.

Capacity of a CDMA System Based on POCs

If $M > 1$, we then know from **Theorem** (c) that it is impossible for us to readily determine a unique solution to the homogeneous linear equation set $\mathbf{A}_y\mathbf{y} = \mathbf{0}$, which specifies zero periodic and aperiodic cross-correlation functions between the codes $\mathbf{x} = \{\mathbf{x}_1, \ldots, \mathbf{x}_M\}$ and $\mathbf{y} = \{\mathbf{y}_1, \ldots, \mathbf{y}_M\}$, where $\mathbf{x}_i = \{x_{i1}, x_{i2}, \ldots, x_{iN}\}$ and $\mathbf{y}_i = \{y_{i1}, y_{i2}, \ldots, y_{iN}\}$ ($x_{ij} \in \{\mathcal{C}\}$, $y_{ij} \in \{\mathcal{C}\}$, $1 \leq i \leq M$, and $1 \leq j \leq N$). Fortunately, the relations specifying the ideal periodic and aperiodic auto-correlation functions of the code $\mathbf{y}$ itself have never been used so far, and thus they can help us to find the solutions of the second code $\mathbf{y}$. The number of suitable solutions may not necessarily be one, as in the example shown in Subsubsection 6.3.4.

With the two codes obtained, we can proceed to find the third code or $\mathbf{z} = \{\mathbf{z}_1, \mathbf{z}_2, \ldots, \mathbf{z}_M\}$, where $\mathbf{z}_i = \{z_{i1}, z_{i2}, \ldots, z_{iN}\}$, $z_{ij} \in \{\mathcal{C}\}$, $1 \leq i \leq M$ and $1 \leq j \leq N$. Similarly, we can have $2(2N-1) \times (NM)$ homogeneous linear equations to specify the zero periodic and aperiodic cross-correlation functions between the new code $\mathbf{z} = \{\mathbf{z}_1, \mathbf{z}_2, \ldots, \mathbf{z}_M\}$ and the existing ones, or $\mathbf{x} = \{\mathbf{x}_1, \mathbf{x}_2, \ldots, \mathbf{x}_M\}$

and $\mathbf{y} = \{\mathbf{y}_1, \mathbf{y}_2, \ldots, \mathbf{y}_M\}$, as follows.

$$
\mathbf{A}_z = \begin{pmatrix}
x_{11} & x_{12} & \cdots & x_{1N}, & x_{21} & x_{22} & \cdots & x_{2N}, & \ldots, & x_{M1} & x_{M2} & \cdots & x_{MN} \\
x_{12} & x_{13} & \cdots & x_{11}, & x_{23} & x_{23} & \cdots & x_{21}, & \ldots, & x_{M2} & x_{M3} & \cdots & x_{M1} \\
\cdots & \cdots & \cdots & \cdots & \cdots & \cdots & \cdots & \cdots & \cdots & \cdots & \cdots & \cdots & \cdots \\
x_{1N} & x_{11} & \cdots & x_{1(N-1)}, & x_{2N} & x_{21} & \cdots & x_{2(N-1)}, & \ldots, & x_{MN} & x_{M1} & \cdots & x_{M(N-1)} \\
x_{12} & x_{13} & \cdots & -x_{11}, & x_{22} & x_{23} & \cdots & -x_{21}, & \ldots, & x_{M2} & x_{M3} & \cdots & -x_{M1} \\
x_{13} & x_{14} & \cdots & -x_{12}, & x_{23} & x_{24} & \cdots & -x_{22}, & \ldots, & x_{M3} & x_{M4} & \cdots & -x_{M2} \\
\cdots & \cdots & \cdots & \cdots & \cdots & \cdots & \cdots & \cdots & \cdots & \cdots & \cdots & \cdots & \cdots \\
x_{11} & -x_{12} & \cdots & -x_{1N}, & x_{21} & -x_{22} & \cdots & -x_{2N}, & \ldots, & x_{M1} & -x_{M2} & \cdots & -x_{MN} \\
y_{11} & y_{12} & \cdots & y_{1N}, & y_{21} & y_{22} & \cdots & y_{2N}, & \ldots, & y_{M1} & y_{M2} & \cdots & y_{MN} \\
y_{12} & y_{13} & \cdots & y_{11}, & y_{23} & y_{23} & \cdots & y_{21}, & \ldots, & y_{M2} & y_{M3} & \cdots & y_{M1} \\
\cdots & \cdots & \cdots & \cdots & \cdots & \cdots & \cdots & \cdots & \cdots & \cdots & \cdots & \cdots & \cdots \\
y_{1N} & y_{11} & \cdots & y_{1(N-1)}, & y_{2N} & y_{21} & \cdots & y_{2(N-1)}, & \ldots, & y_{MN} & y_{M1} & \cdots & y_{M(N-1)} \\
y_{12} & y_{13} & \cdots & -y_{11}, & y_{22} & y_{23} & \cdots & -y_{21}, & \ldots, & y_{M2} & y_{M3} & \cdots & -y_{M1} \\
y_{13} & y_{14} & \cdots & -y_{12}, & y_{23} & y_{24} & \cdots & -y_{22}, & \ldots, & y_{M3} & y_{M4} & \cdots & -y_{M2} \\
\cdots & \cdots & \cdots & \cdots & \cdots & \cdots & \cdots & \cdots & \cdots & \cdots & \cdots & \cdots & \cdots \\
y_{11} & -y_{12} & \cdots & -y_{1N}, & y_{21} & -y_{22} & \cdots & -y_{2N}, & \ldots, & y_{M1} & -y_{M2} & \cdots & -y_{MN}
\end{pmatrix} \tag{6.121}
$$

As shown earlier, the row vectors in (6.121) generated from all possible cyclic-shifted versions of element codes $\mathbf{x} = \{\mathbf{x}_1, \mathbf{x}_2, \ldots, \mathbf{x}_M\}$ and $\mathbf{y} = \{\mathbf{y}_1, \mathbf{y}_2, \ldots, \mathbf{y}_M\}$ must be mutually independent. Therefore, the rank of the matrix $\mathbf{A}_z$ should be at least $2N$, where N is the length of the element codes. Now if the flock size M becomes two, we face again the situation that the rank of the homogeneous linear equation set $\mathbf{A}_z\mathbf{z} = \mathbf{0}$ or $rank(\mathbf{A}_z)$ is greater than the number of its unknown variables or $MN = 2N$, yielding either all-zero solution $\mathbf{z} = \mathbf{0}$ or no solution according to **Theorem** (a) and **Theorem** (b), which is not the result we want. Therefore, we have to let $M > 2$, say $M = 3$ (it is noted that we have $K = 3$ here), to ensure possible non-zero solutions for the homogeneous linear equation set $\mathbf{A}_z\mathbf{z} = \mathbf{0}$. Similar results can also be obtained for other values of M, N, and K. Thus, we obtain another interesting corollary as follows.

Corollary 6.3.2 *To generate a POC code set with its set size being* K*, in which all its codes should have perfect periodic and aperiodic auto-correlation and cross-correlation functions, the flock size* M *of all codes must not be less than the set size* K*, or* $M \geq K$*. Usually,* $M = K$ *suffices.*

Corollaries 6.3.1 and 6.3.2 should be used jointly to generate the POC code sets. The major steps for the code generation in the REAL approach are summarized in Table 6.3. Figure 6.6 illustrates the differences between the traditional code design approach and the REAL approach proposed here. It is seen from the figure that the new approach considers many real working conditions that never were before, such that POC codes so obtained will be more relevant to the actual working environment.

Example of a POC Code Set $\{M, N, K\} = \{4, 2, 4\}$

As an example, we would like to show how to generate a POC code set with $M = 4$ and $N = 2$. Assume that we already have a seed code, which can be found through exhaustive search, as $\mathbf{x} = \{x_{11}, x_{12}; x_{21}, x_{22}; x_{31}, x_{32}; x_{41}, x_{42}\} = \{-1, -1; -1, -1; -1, 1; -1, 1\}$, where the semi-colon separates the two element codes, each of which is 2 chips long. We will only consider the binary POC codes for illustration simplicity. Starting from $\mathbf{x}$, we can proceed to find the second code $\mathbf{y} = \{y_{11}, y_{12}; y_{21}, y_{22}; y_{31}, y_{32}; y_{41}, y_{42}\}$, which should satisfy cross-correlation and auto-correlation relations specified in (6.107) and (6.108), respectively, as

$$
\begin{cases}
-y_{11} - y_{12} - y_{21} - y_{22} - y_{31} + y_{32} - y_{41} + y_{42} = 0 \\
-y_{11} - y_{12} - y_{21} - y_{22} + y_{31} - y_{32} + y_{41} - y_{42} = 0 \\
y_{11} - y_{12} + y_{21} - y_{22} - y_{31} - y_{32} - y_{41} - y_{42} = 0
\end{cases} \tag{6.122}
$$

Table 6.3 Procedure to generate POC codes in the REAL approach.

Step	Operation
1	Specify K, M, and N, where the conditions for $M > 1$ and $M = K$ must be satisfied.
2	Generate a *seed code* $\mathbf{x}$ using conditions for $\mathbf{x}$ to maintain ideal PACFs[1] and AACFs.[2]
3	Proceed to search for second code $\mathbf{y}$ using homogeneous linear equation set specifying perfect PCCFs[3] and ACCFs[4] between $\mathbf{x}$ and $\mathbf{y}$. If $M > 1$ and $M = K$, the homogeneous linear equation set should give some suitable solutions, which take the forms as expressions of some undetermined variables.
4	Valid solutions can be obtained by solving the undetermined variables using the equations specifying ideal PACF and AACF for $\mathbf{y}$ itself.
5	Repeating steps 1-4 above to generate all K codes in the set.

[1] PACFs: Periodic auto-correlation functions.
[2] AACFs: Aperiodic auto-correlation functions.
[3] PCCFs: Periodic cross-correlation functions.
[4] ACCFs: Aperiodic cross-correlation functions.

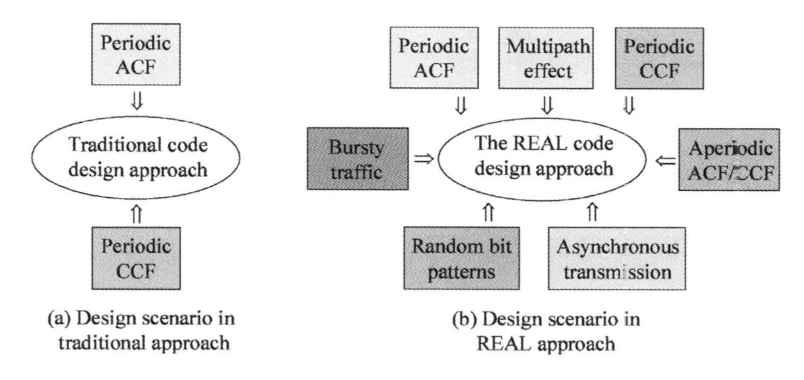

Figure 6.6 Comparison of traditional code design approach and the REAL approach, where (a) shows factors taken into account in the traditional code design approach and (b) shows factors taken into account in the REAL approach.

and

$$\begin{cases} y_{11}^2 + y_{12}^2 + y_{21}^2 + y_{22}^2 + y_{31}^2 + y_{32}^2 + y_{41}^2 + y_{42}^2 = 8 \\ y_{11}y_{12} + y_{12}y_{11} + y_{21}y_{22} + y_{22}y_{21} + y_{31}y_{32} + y_{32}y_{31} + y_{41}y_{42} + y_{42}y_{41} = 0 \end{cases} \tag{6.123}$$

(6.122) can be rewritten into

$$\begin{cases} y_{11} + y_{12} + y_{21} + y_{22} = 0 \\ y_{12} + y_{22} + y_{32} + y_{42} = 0 \\ y_{12} + y_{22} + y_{31} + y_{41} = 0 \end{cases} \tag{6.124}$$

From (6.123) and (6.124) we can easily solve for the second code as $\mathbf{y} = \{y_{11}, y_{12}; y_{21}, y_{22}; y_{31}, y_{32}; y_{41}, y_{42}\} = \{-1, 1; -1, 1; -1, -1; -1, -1\}$.

The third code $\mathbf{z} = \{z_{11}, z_{12}; z_{21}, z_{22}; z_{31}, z_{32}; z_{41}, z_{42}\}$ should satisfy the cross-correlation relation with two existing codes $\mathbf{x}$ and $\mathbf{y}$ as

$$\begin{cases} -z_{11} - z_{12} - z_{21} - z_{22} - z_{31} + z_{32} - z_{41} + z_{42} = 0 \\ -z_{11} - z_{12} - z_{21} - z_{22} + z_{31} - z_{32} + z_{41} - z_{42} = 0 \\ z_{11} - z_{12} + z_{21} - z_{22} - z_{31} - z_{32} - z_{41} - z_{42} = 0 \\ z_{11} - z_{12} + z_{21} - z_{22} - z_{31} - z_{32} - z_{41} - z_{42} = 0 \\ -z_{11} + z_{12} - z_{21} + z_{22} - z_{31} - z_{32} - z_{41} - z_{42} = 0 \\ z_{11} - z_{12} + z_{21} - z_{22} + z_{31} + z_{32} + z_{41} + z_{42} = 0 \end{cases} \tag{6.125}$$

and its own auto-correlation relation as

$$\begin{cases} z_{11}^2 + z_{12}^2 + z_{21}^2 + z_{22}^2 + z_{31}^2 + z_{32}^2 + z_{41}^2 + z_{42}^2 = 8 \\ z_{11}z_{12} + z_{12}z_{11} + z_{21}z_{22} + z_{22}z_{21} + z_{31}z_{32} + z_{32}z_{31} + z_{41}z_{42} + z_{42}z_{41} = 0 \end{cases} \tag{6.126}$$

We can obtain $\mathbf{z} = \{z_{11}, z_{12}; z_{21}, z_{22}; z_{31}, z_{32}; z_{41}, z_{42}\} = \{-1, -1; 1, 1; -1, 1; 1, -1\}$ using (6.125) and (6.126). As in this case we have $M = 4$, we know that there is another code $\mathbf{v}$ in this code set. It has to comply with the cross-correlation relation with all other known codes, $\mathbf{x}$, $\mathbf{y}$ and $\mathbf{z}$, as

$$\begin{cases} -v_{11} - v_{12} - v_{21} - v_{22} - v_{31} + v_{32} - v_{41} + v_{42} = 0 \\ -v_{11} - v_{12} - v_{21} - v_{22} - v_{31} + v_{32} - v_{41} + v_{42} = 0 \\ v_{11} - v_{12} + v_{21} - v_{22} - v_{31} - v_{32} - v_{41} - v_{42} = 0 \\ v_{11} - v_{12} + v_{21} - v_{22} - v_{31} - v_{32} - v_{41} - v_{42} = 0 \\ -v_{11} + v_{12} - v_{21} + v_{22} - v_{31} - v_{32} - v_{41} - v_{42} = 0 \\ v_{11} - v_{12} + v_{21} - v_{22} + v_{31} + v_{32} + v_{41} + v_{42} = 0 \\ -v_{11} - v_{12} + v_{21} + v_{22} - v_{31} + v_{32} + v_{41} - v_{42} = 0 \\ -v_{11} - v_{12} + v_{21} + v_{22} + v_{31} - v_{32} - v_{41} + v_{42} = 0 \\ v_{11} - v_{12} - v_{21} + v_{22} - v_{31} - v_{32} + v_{41} + v_{42} = 0 \end{cases} \tag{6.127}$$

together with its own auto-correlation relation as

$$\begin{cases} v_{11}^2 + v_{12}^2 + v_{21}^2 + v_{22}^2 + v_{31}^2 + v_{32}^2 + v_{41}^2 + v_{42}^2 = 8 \\ v_{11}v_{12} + v_{12}v_{11} + v_{21}v_{22} + v_{22}v_{21} + v_{31}v_{32} + v_{32}v_{31} + v_{41}v_{42} + v_{42}v_{41} = 0 \end{cases} \tag{6.128}$$

We can find two solutions from (6.127) and (6.128), which differ only from their element code appearance order, as $\mathbf{v} = \{-1, 1; 1, -1; -1, -1; 1, 1\}$ and $\mathbf{v}' = \{1, -1; -1, 1; 1, 1; -1, -1\}$. The final POC code set $\{M, N, K\} = \{4, 2, 4\}$ is obtained as

$$\begin{cases} \mathbf{x} = (-, -; -, -; -, +; -, +) \\ \mathbf{y} = (-, +; -, +; -, -; -, -) \\ \mathbf{z} = (-, -; +, +; -, +; +, -) \\ \mathbf{v} = (-, +; +, -; -, -; +, +) \end{cases} \tag{6.129}$$

Significance of REAL Approach

The introduction of REAL is significant in terms of the following:

1. This is the first time in the literature to show that a CDMA system can be made interference-free (i.e. MAI-free and MI-free operation) by using orthogonal complementary code sets. The performance of a CDMA system is not always necessarily interference-limited. The REAL approach has told us that a CDMA system can be made interference-free if it can use some properly designed orthogonal complementary codes.

2. This is also the first time in the literature to take into account so many real operational parameters, such as multipath propagation, asynchronous transmission, random bit signs, bursty traffic, etc., in the spreading code design process. As a consequence, the obtained orthogonal complementary code sets can have some inherent immunity against these impairing factors.

3. The REAL approach tells us that the orthogonal complementary codes are the only option to implement an interference-free CDMA system. Therefore, it has been clearly shown that all traditional unitary codes, such as Gold codes, Kasami codes, Walsh-Hadamard codes, OVSF codes, etc., will never yield an interference-free performance if they are used in a CDMA system. Therefore, just forget about all unitary codes.

4. Interference-free operation is one of the most important characteristic features of the next generation CDMA technology. Therefore, the orthogonal complementary codes will definitely play a critical role in the next generation CDMA technology.

When Does MI Become Harmful to DS/CC-CDMA?

It should be pointed out here that the orthogonal complementary codes generated using the REAL approach and any other methods can work with two different types of spreading modulation schemes, one being traditional direct sequence (DS) spreading modulation and the other offset stacking (OS) spreading modulation. Both the DS and OS spreading modulation schemes have been discussed in Chapter 5.

If the orthogonal complementary codes (OCCs) work with DS spreading modulation, a DS/CC-CDMA system results. On the other hand, if the orthogonal complementary codes work with OS spreading modulation, an OS/CC-CDMA system is yielded. It is noted that a DS/CC-CDMA system will offer the widest opening in its auto-correlation interference-free window (ACIFW), whose width is equal to $N - 1$ if N is the element code length. In other words, a DS/CC-CDMA system can resist any multipath interference as long as the delay of a multipath return is not equal to the multiples of N or xN, where x is any positive integer. The multipath returns will still cause harmful inter-symbol interference to the detection of signals if their delays are equal to exactly xN. As a matter of fact, this problem is not new to the orthogonal complementary codes and the same problem occurs in all existing unitary CDMA codes, such as Gold codes, m-sequences, Kasami codes, Walsh-Hadamard sequences, OVSF codes, etc.

The OS/CC-CDMA system has a relatively small ACIFW compared to that of DS/CC-CDMA. More specifically, an OS/CC-CDMA system has an ACIFW whose length is equal to $N - (N - n) - 1$ chips, where n is the number of offset chips in two consecutive bits. Obviously, if $n = N$, the OS spreading will be reduced to the DSspreading and the ACIFW reaches the maximal value of $N - 1$. On the other hand, if $n = 1$, the OS spreading will give the smallest ACIFW, whose length is zero, meaning that it can not offer any multipath interference resistance at all.

As mentioned earlier, a DS/CC-CDMA system based on orthogonal complementary codes has flexibility in choosing different N and M, while keeping their product or processing gain $N \times M$ unchanged. As shown earlier, the number of available codes in an orthogonal complementary code set is K, which should be always equal to M, as required by the REAL approach. Therefore, to maximize $K = M$, we have to minimize N, whose smallest value is $N = 1$. However, we should be very careful here because the use of $N = 1$ will effectively make the ACIFW length shrink to zero even in a DS/CC-CDMA system because its ACIFW is equal to $N - 1$. Zero ACIFW implies that the DS/CC-CDMA system with $N = 1$ is not able to resist any multipath interference induced by any multipath returns, whose relative delay can be any arbitrary values (due to the fact that now $xN = x$, and any x value will cause harmful multipath interference to signal detection). Therefore, we see that even for a DS/CC-CDMA, we should not use $N = 1$ to maximize the set size of an orthogonal complementary code set.

Therefore, for the sake of enhancement of the multipath interference resistance capability in a DS/CC-CDMA system, we are encouraged to use a relatively large N value such that the probability that the relative delay of a multipath return happens to be xN chips will be relatively small, thus minimizing the harmful effect of the multipath interference to a DS/CC-CDMA system.

Again, we would like to stress that any CDMA system (either traditional or newly proposed) will not be able to resist multipath interference if the relative delay of a multipath return happens to be exactly xN, where N is the element code length (if unitary codes are concerned here, then N will just be equal to the code length) and x is any positive integer. Therefore, if we let $N = 1$, then xN will take $\{1, 2, \ldots\}$ chips, implying that every chip will be subject to multipath interference. It is the worst case scenario. One method can be used to overcome the multipath interference caused by the multipath returns whose relative offset delays are equal exactly to xN. That is to use the long code scrambling technique, by which different multipath returns will be distinguished by different offset phases of the same scrambling code, which has a length to cover at least several bits. In fact, any methods which have been used to solve the problem associated with the multipath interference caused by the multipath returns whose relative offset delays are equal exactly to xN in a traditional CDMA system, can also be used for a CDMA system based on orthogonal complementary codes.

Having explained the problem with harmful multipath returns to a DS/CC-CDMA system, we should not be confused by the claim that a CDMA system using orthogonal complementary codes can be made interference-free. Nevertheless, we have to admit that a CDMA system using orthogonal complementary codes does have the strongest multipath resistance capability.

7

CDMA Systems Based on Complementary Codes

After having discussed the generation and properties of various complementary codes, we are ready to go further to take a look at the issues on how to implement the CDMA architecture with the help of those innovative spreading codes.

It has to be admitted that the design of spreading codes has a lot to do with the formation of CDMA system architecture. For instance, while introducing the generation process of the pair-wise complementary codes in Subsections 6.2.5 and 6.3.4, we have already made it very clear that the basic architecture of a CDMA system based on pair-wise complementary codes will use a quadrature digital modem to modulate two element codes assigned to the same user with the binary data bit streams. Therefore, the architecture of a CDMA system based on pair-wise complementary codes has already been proposed in the code design process, as shown in Figure 6.2. The same conclusion can be made from the discussions on the code generation process for many other complementary codes. Therefore, the CDMA system architecture has a lot to do with the spreading code design approach, and they come together in many cases.

In particular, the code generation or design approach is closely associated with the spreading modulation scheme[1] suitable for some particular spreading codes generated from the approach. Here, the spreading modulation schemes are defined as the ways to modulate source data bit streams with the spreading sequences, and in most cases a multiplier or modulo two addition unit should be used for implementation of the spreading modulation.

However, it should also be noted in some cases that some new spreading techniques can also be devised based on some existing spreading codes due to their very special correlation properties. In other words, new spreading modulation schemes can be proposed by exploiting some unique characteristic features of the new spreading codes. A very good example of this is the introduction of the offset stacking (OS) spreading modulation scheme. In contrast with the traditional direct sequence (DS) spreading modulation scheme, the OS spreading scheme will start to modulate a new bit only after one or a few chips offset, while the DS scheme will modulate a new bit only after having finished modulating the whole preceding bit such that it goes in a bit-by-bit fashion. The motivation to propose the OS spreading modulation scheme is to improve the bandwidth efficiency of the spread modulated

[1]Here we only discuss direct sequence spreading and we do not consider other spreading methods, such as frequency-hopping and time-hopping spreading methods.

The Next Generation CDMA Technologies Hsiao-Hwa Chen

signal, which can be defined by a parameter called spreading efficiency, measured by number of bits conveyable in one chip, or in an unit of bits/chip. Therefore, in this chapter we will in particular discuss the CDMA system architecture based on two major spreading modulation schemes: DS and OS. It should be noted that OS spreading modulation is a general case of DS spreading modulation, or the DS spreading scheme is only a special case of OS spreading with the number of offset chips being equal to the length of element code. Therefore, the discussions on the OS-spreading-based CDMA system is significant in terms of its generality. In addition, many other salient features of the OS spreading modulation scheme have made it a vital component in the next generation CDMA technology.

It is also noted that carrier modulation in a traditional CDMA system based on DS spreading modulation is always designed independent of the spreading modulation scheme. In other words, the selection of carrier modulation scheme has little to do with the spreading modulation scheme. The independence in selection of spreading modulation and carrier modulation in all traditional DS-CDMA systems is due to the fact that the output baseband signal from the spreading modulator is a binary signal, and thus any type of digital modulation scheme (such as BPSK, QPSK, M-QAM, etc.) can be used as the carrier modulator. However, if some other spreading modulation scheme (such as OS spreading modulation) is used, the situation may be very different, depending very much on the nature of the output signal from a spreading modulator. In many cases, joint design of spreading and carrier modulations is preferred in order to optimize the overall system performance. This case will apply to an OS spreading-modulation-based CDMA system, where the output signal from an OS spreading modulator is no longer binary, and thus the selection of carrier modulation scheme is no longer arbitrary. We will discuss the issue in detail in Section 7.2.

7.1 DIRECT SEQUENCE SPREADING AND DS/CC-CDMA SYSTEMS

The most straightforward way to use orthogonal complementary codes (CC) in a CDMA system is to use DS spreading modulation in a CC-based CDMA architecture. This CDMA system architecture is called DS/CC-CDMA system for description simplicity.

7.1.1 SYSTEM ARCHITECTURE

Assume that the DS/CC-CDMA system we are concerned with here uses a set of orthogonal complementary codes, whose parameters are $N = 2^r$, $M = 2^R$, and $K = 2^R$, where r and R are any positive integers. Therefore, the parameter set for this orthogonal complementary code can be written as $(N, M, K) = (2^r, 2^R, 2^R)$.

An important consideration for any CC-based CDMA is that each user should be assigned a flock of M element codes and different element codes in the same flock should be sent via different channels to a receiver. Therefore, either time division multiplex (TDM) or frequency division multiplex (FDM) can be used to send different element codes. For description simplicity, we will only consider the use of the FDM scheme to send different element codes in this book (unless explained in particular). Therefore, we need to use $M = 2^R$ sub-carriers to send 2^M different element codes in this DS/CC-CDMA system. Due to the fact that the output signal after DS spreading is still binary, we can use the simplest digital modulation scheme, such as BPSK, in the DS/CC-CDMA system. Of course, any other multi-level digital modulation can also be used in DS/CC-CDMA, as long as the link budget is enough for the application of a multi-level digital modulation scheme, such as M-PSK, M-QAM, etc.

A conceptual block diagram for the DS/CC-CDMA system is shown in Figures 7.1 and 7.2, where Figure 7.1 illustrates the multi-user signal formulation process in the DS/CC-CDMA system and Figure 7.2 is the receiver of the DS/CC-CDMA system.

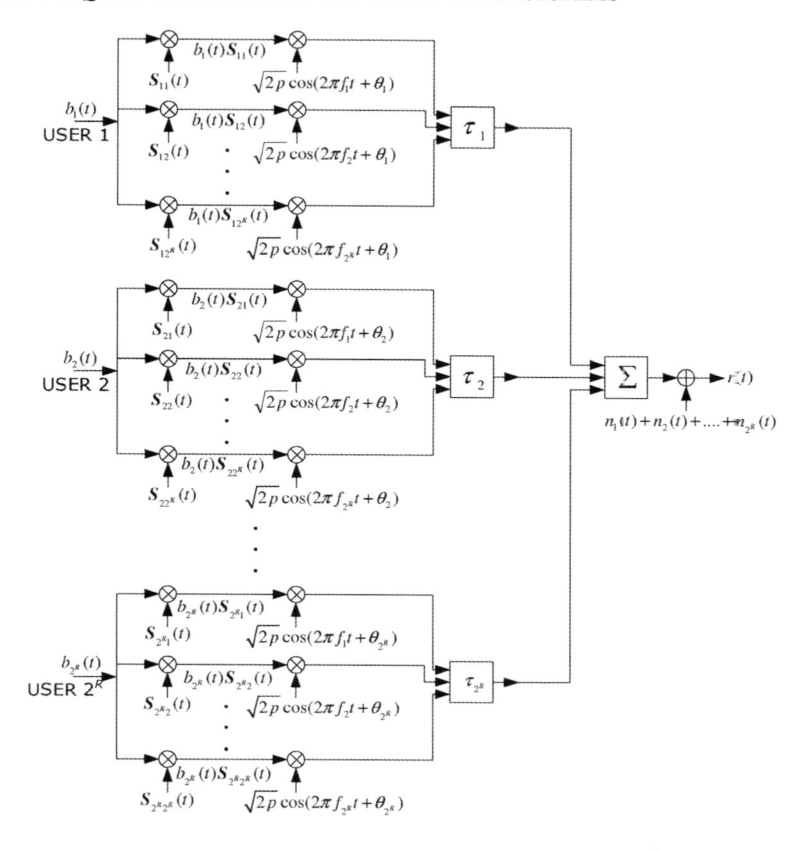

Figure 7.1 DS/CC-CDMA multi-user signal formulation process, where 2^R users are present, each user is assigned a flock of 2^R element codes, and 2^R sub-carriers are needed to send 2^R element codes.

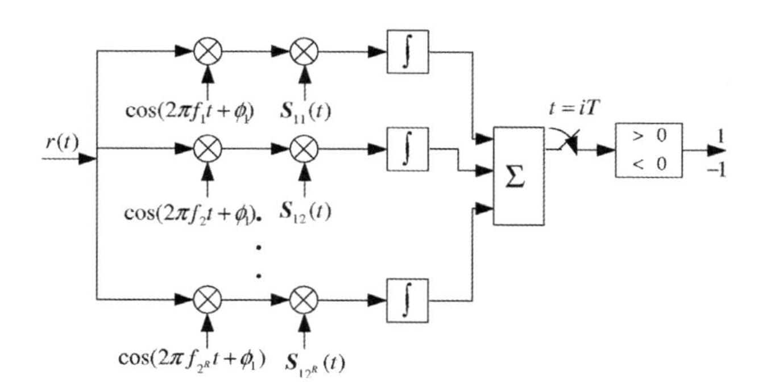

Figure 7.2 DS/CC-CDMA system receiver.

Assume that we will use a set of super complementary codes in this DS/CC-CDMA system, and the parameters for this code set are $N = 4$, $M = 4$, and $K = 4$. Therefore the processing gain for any user will be $4 \times 4 = 16$. Detailed? information about the super complementary code set considered here is given as follows:

$$\begin{array}{lllll} \text{User1:} & \mathbf{C}_{11}: + + + -; & \mathbf{C}_{12}: + + - +; & \mathbf{C}_{13}: + + + -; & \mathbf{C}_{14}: - - + - \\ \text{User 2:} & \mathbf{C}_{21}: + - + +; & \mathbf{C}_{22}: + - - -; & \mathbf{C}_{23}: + - + +; & \mathbf{C}_{24}: - + + + \\ \text{User 3:} & \mathbf{C}_{31}: + + + -; & \mathbf{C}_{32}: + + - +; & \mathbf{C}_{33}: - - - +; & \mathbf{C}_{34}: + + - + \\ \text{User 4:} & \mathbf{C}_{41}: + - + +; & \mathbf{C}_{42}: + - - -; & \mathbf{C}_{43}: - + - -; & \mathbf{C}_{44}: + - - - \end{array} \tag{7.1}$$

In order to illustrate how the DS/CC-CDMA system using the super complementary code set works, we would like to use some figures to demonstrate several important characteristics of the system. To make the illustration easier and clearer, we only consider two users (for instance, User 1 and User 2 with $\mathbf{C}_1 = \{\mathbf{C}_{11}, \mathbf{C}_{12}, \mathbf{C}_{13}, \mathbf{C}_{14}\}$ and $\mathbf{C}_2 = \{\mathbf{C}_{21}, \mathbf{C}_{22}, \mathbf{C}_{23}, \mathbf{C}_{24}\}$, respectively) in the system. In fact, the same results will apply if all four users are taken into account in the illustrations, with all figures becoming more complex. We further assume that the signal from User 1 is useful signal and that from User 2 will be considered as multiple access interference (MAI). Noise is omitted in the illustration for simplicity. In addition, we consider the case with signals being sent in very short packets, which consist of only three bits, $[+1, -1, +1]$, for illustration conciseness.

In the following illustrations, we should first consider the situations where there is no multipath propagation effect, and then we extend the illustrations to the multipath channels. Both synchronous and asynchronous transmissions will be considered here.

7.1.2 ISOTROPIC MAI-FREE OPERATION

In this subsection, we would like to use illustrations to show how a DS/CC-CDMA system (based on a set of super complementary codes as given in Equation (7.1)) could overcome MAI to achieve MAI-free operation. It should be noted that MAI-free operation will be in place for both asynchronous and synchronous transmissions in the DS/CC-CDMA system of concern. Therefore, we would like to name this property 'isotropic MAI-free operation', which is in contrast with a conventional DS-CDMA system based on Walsh-Hadamard codes, which can only offer MAI-free operation in synchronous channels, but not in asynchronous channels, due to the fact that the cross-correlation functions among Walsh-Hadamard codes become very bad in asynchronous transmission mode.

Figure 7.3 shows the process to despread the whole packet sent from Users 1 and 2 at the receiver, which is tuned to User 1. Therefore, the signals from User 2 will contribute as interference in this detection process. In this illustration, we do not consider the noise and multipath and we are only concerned with synchronous transmission, in which all bits from different users are aligned in time.

It is seen from the figure that the detection of the three bits sent from User 1 can be decoded successfully without interference by the signals sent from User 2. Therefore, MAI-free operation is guaranteed in this case.

Figure 7.4 shows the process to despread the whole packet sent from Users 1 and 2 at the receiver, which is tuned to User 1. Therefore, the signals from User 2 will contribute as interference in this detection process. In this illustration, we do not consider the noise and multipath and we are only concerned with asynchronous transmission, in which all bits from different users are not aligned in time. For illustration simplicity, we assume that the relative delay between Users 1 and 2 is only one chip. However, it can be easily shown that the same result will apply if any other relative delays between the transmissions from Users 1 and 2 are assumed.

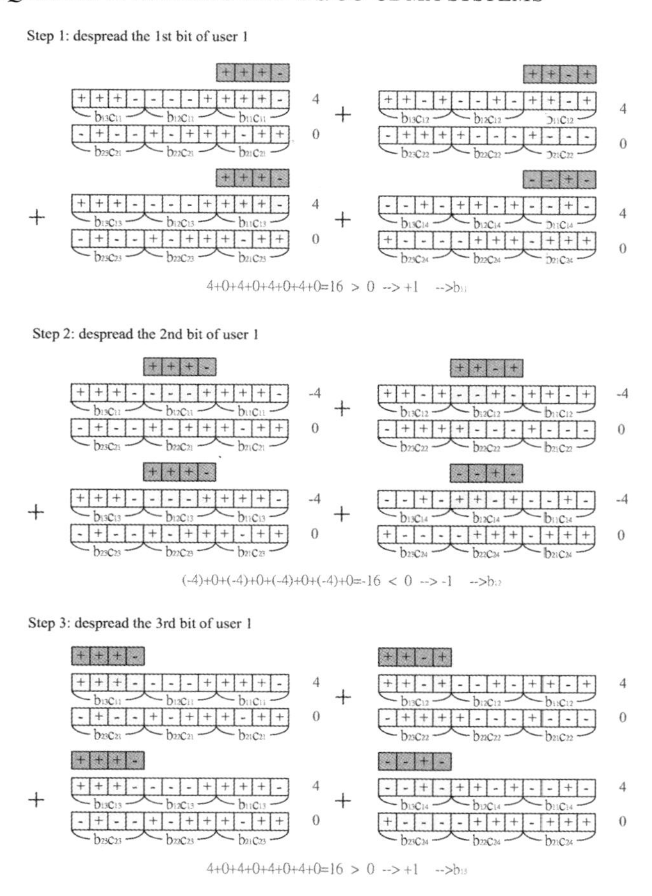

Figure 7.3 Despreading process for the whole packet sent from Users 1 and 2 at the receiver, which is tuned to User 1, where a synchronous channel is considered.

It is seen from the figure that the three bits sent from User 1 can be decoded successfully without interference by the signals sent from User 2. Therefore, MAI-free operation for a DS/CC-CDMA system in asynchronous transmission mode is guaranteed in this case also.

7.1.3 *ISOTROPIC MI-FREE OPERATION*

In the previous subsection, we have used illustrations to show that a DS/CC-CDMA system based on a set of super complementary codes can offer MAI-free operation for both synchronous and asynchronous transmissions, namely isotropic MAI-free operation.

Next, we will illustrate another important operational feature for the DS/CC-CDMA system. More specifically, a DS/CC-CDMA system can also offer an isotropic MI-free operation, where MI stands for multipath interference. Let us consider a simple multipath channel, in which there are only

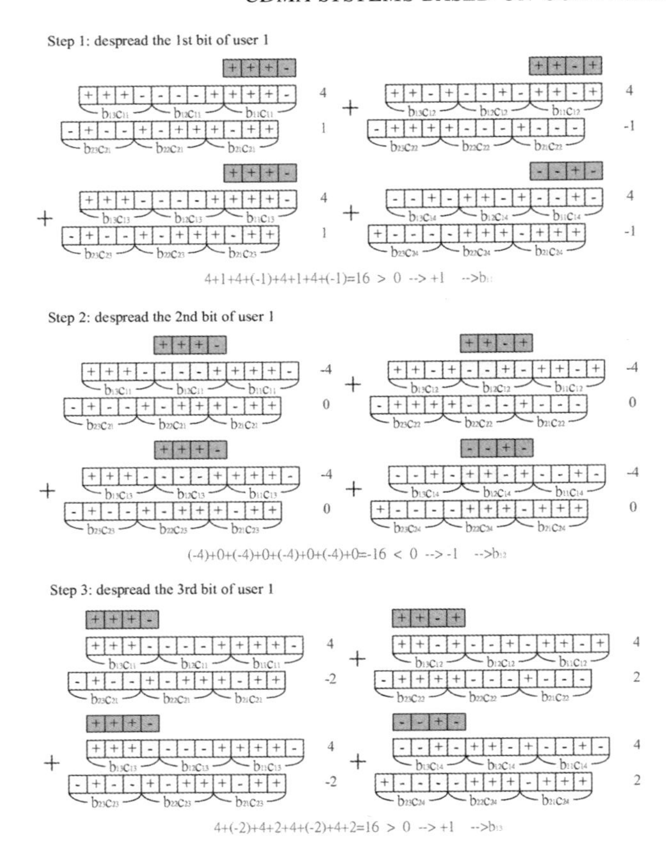

Figure 7.4 Despreading process for the whole packet sent from Users 1 and 2 at the receiver, which is tuned to User 1, where an asynchronous channel is considered.

two multipath returns and their relative delay is only one chip, for illustration simplicity. It can be easily shown that the same results will occur if we use any other relative delays between the two multipath returns. Again, we assume that there are only two users in the system, and each will be assigned a flock of element codes, as given in Equation (7.1). That means the same set of super complementary codes will be used here. In fact, the same results will apply if we use any other type of orthogonal complementary code, and the super complementary code is only a particular type of orthogonal complementary code.

Our discussions will start with a synchronous transmission channel, such as the downlink channels of a mobile cellular system, and then the discussions will be extended to asynchronous transmissions, just as we did in the previous subsection.

Figures 7.5 and 7.6 show the detection of the first, second, and third bits of the packet from User 1, where there are two multipath returns and two users in a downlink transmission channel. It is seen from Figures 7.5 and 7.6 that the DS/CC-CDMA system can offer MI-free operation for this two-user system with synchronous transmissions.

Similarly, Figures 7.7 and 7.8 show the detection of the first, second, and third bits of the packet from User 1, where there are two multipath returns and two users in an uplink (asynchronous)

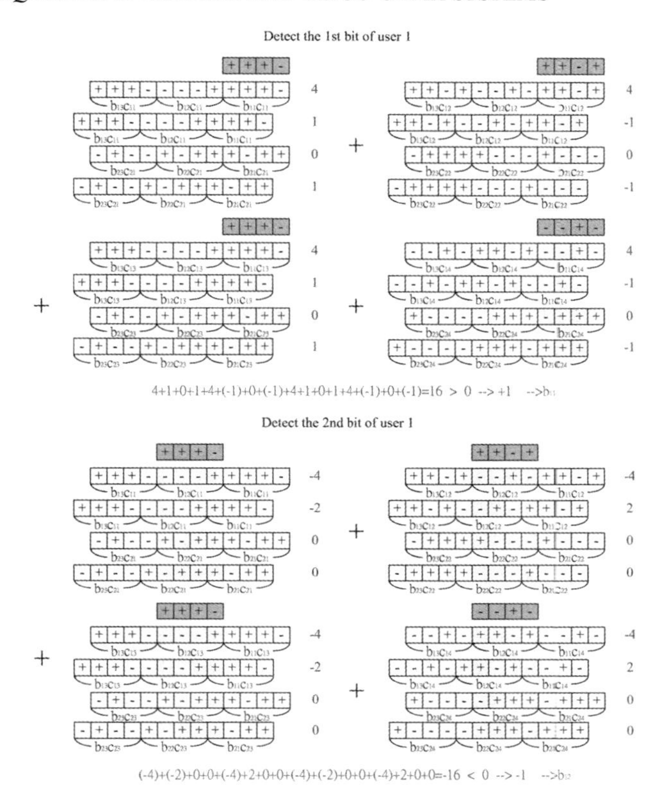

Figure 7.5 Detection of the first and second bits sent from User 1 at the receiver, which is tuned to User 1, where a synchronous two-return multipath channel is considered.

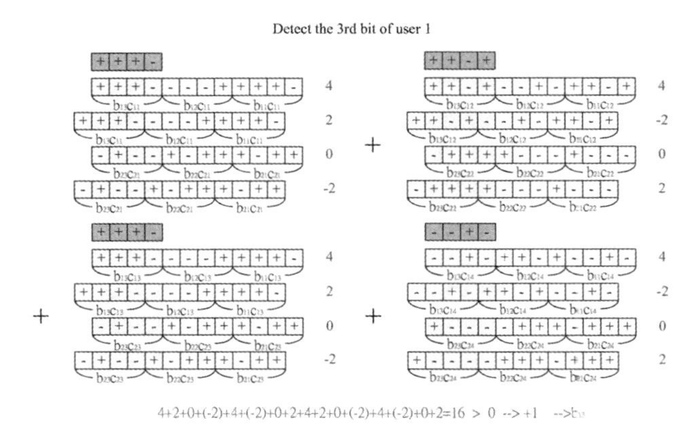

Figure 7.6 Detection of the third bit sent from User 1 at the receiver, which is tuned to User 1, where a synchronous two-return multipath channel is considered.

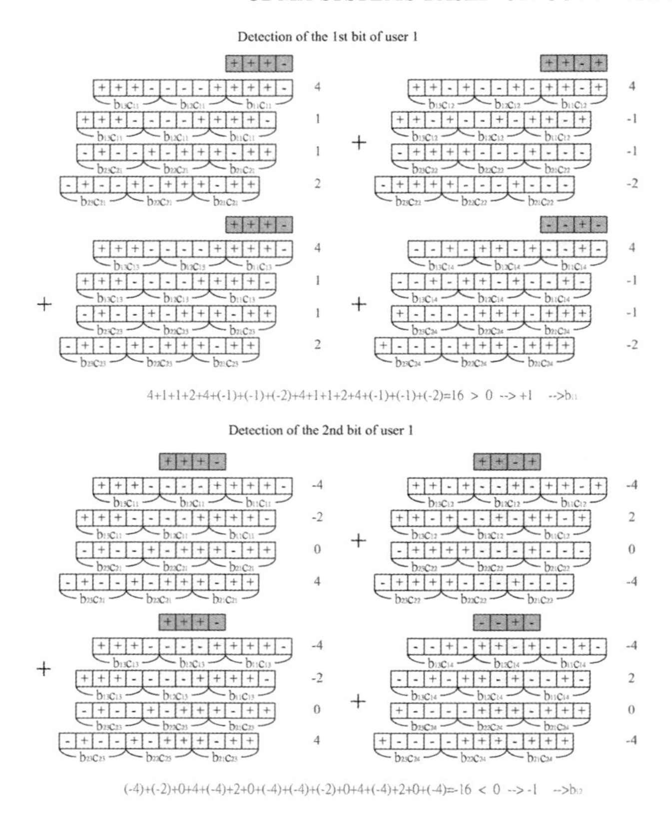

Figure 7.7 Detection of the first and second bits sent from User 1 at the receiver, which is tuned to User 1, where an asynchronous two-way multipath channel is considered.

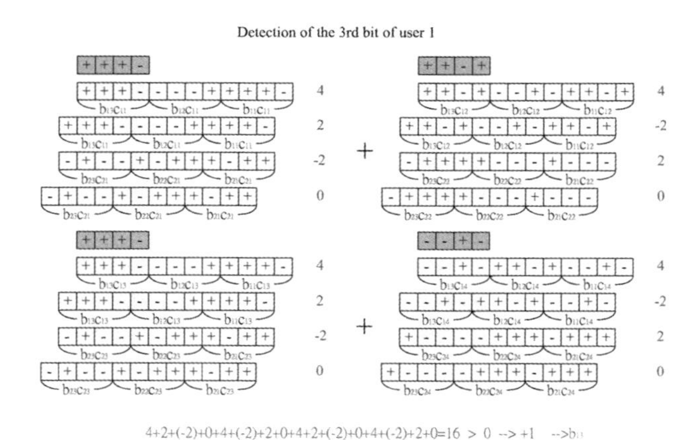

Figure 7.8 Detection of the third bit sent from User 1 at the receiver, which is tuned to User 1, where an asynchronous two-way multipath channel is considered.

transmission channel. It is seen from Figures 7.7 and 7.8 that the DS/CC-CDMA system can offer MI-free operation for this two- user system with asynchronous transmissions as well. Therefore, it is concluded that the DS/CC-CDMA system is capable of providing isotropic MI-free operation.

The information revealed from Figures 7.7 and 7.8 is significant. It is seen from the figures that the detection efficiency at the edges of a short packet can be made exactly the same as that in the middle. In particular, we should note that the detection of the first bit shown in Figures 7.7 and 7.8 actually involves partial correlation functions, instead of periodic correlation functions. This result reflects again the fact that the orthogonal complementary codes can offer unique operational features (which all unitary codes do not have) such that the detection efficiency for a short packet can be made the same as that for a long frame, making them particularly suitable for signal detection in high speed bursty traffic, an important characteristic feature for all wireless communication systems beyond 3G.

7.1.4 ANALYTICAL PERFORMANCE STUDY OF DS/CC-CDMA SYSTEM

After having discussed the issues on isotropic MAI-free and MI-free operation for a DS/CC-CDMA system using illustrations, we would like to go further to study its performance with the help of analysis in this subsection. The analysis can give us a more generic study without being focused only on a few special cases, as we did in the illustrations. Therefore, the results obtained from the analysis will be applicable to all DS/CC-CDMA systems based on any orthogonal complementary codes.

Correlation with Fractional Chip Relative Delay

It is noted that in the illustration study given in the earlier subsection we have assumed that the relative delays between users and multipath returns are always at multiple complete chips. However, it is possible that the relative delay may take any value, which may be a fractional chip, instead of a number of complete chips. Therefore, we should first look at this issue as the first step to extend our study to a more general case.

Figure 7.9 shows arbitrary relative delay between two sequences **P** and **Q**, with their relative delay being $1 + \tau$ chips, where τ is a positive real number less than one. It is noted that Figure 7.9 shows the case with two unitary sequences. However, it can be extended to the cases with complementary codes with multiple element codes.

Now let us look at the cross-correlation function between the two sequences **P** and **Q**, each consisting of four chips, as shown in Figure 7.9. The cross-correlation function between sequences **P**

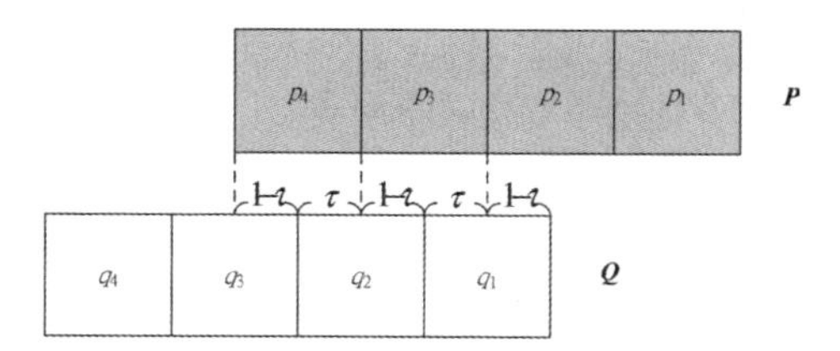

Figure 7.9 Arbitrary relative delay between two sequences **P** and **Q**, each consisting of four chips.

and **Q** with their relative delay being $1 + \tau$ can be written as

$$\begin{aligned}\rho(\mathbf{P}, \mathbf{Q}; 1 + \tau) \\ &= \frac{1}{4}[p_2 q_1^*(1 - \tau) + p_3 q_1^* \tau + p_3 q_2^*(1 - \tau) + p_4 q_2^* \tau + p_4 q_3^*(1 - \tau)] \\ &= \frac{1}{4}[(p_2 q_1^* + p_3 q_2^* + p_4 q_3^*)(1 - \tau) + (p_3 q_1^* + p_4 q_2^*)\tau] \\ &= \rho(\mathbf{P}, \mathbf{Q}; 1)(1 - \tau) + \rho(\mathbf{P}, \mathbf{Q}; 2)\tau\end{aligned} \tag{7.2}$$

It is seen from the above equation that the cross-correlation function with fractional chip delay (which is equal to $1 + \tau$ chips) can be represented by the weighted sum of two decomposed terms, one being the cross-correlation function $\rho(\mathbf{P}, \mathbf{Q}; 1)$ (which stands for the cross-correlation function with one chip relative delay) and the other $\rho(\mathbf{P}, \mathbf{Q}; 2)$ (which stands for the cross-correlation function with two chips relative delay). The weights used in the two decomposed terms are just $(1 - \tau)$ and τ. Therefore, if the cross-correlation functions with relative delays of complete chips are zero, then the cross-correlation functions with any relative delays of fractional chips are also zero, as shown in Equation (7.2).

Based on (7.2), we can have the auto-correlation function for a flock of complementary codes $\mathbf{S}_{nj} = \{\mathbf{S}_{n1}, \mathbf{S}_{n2}, \ldots, \mathbf{S}_{n2^R}\}$ and the cross-correlation functions with relative delay of fractional chips between two flocks of complementary codes $\mathbf{S}_{nj} = \{\mathbf{S}_{n1}, \mathbf{S}_{n2}, \ldots, \mathbf{S}_{n2^R}\}$ and $\mathbf{S}_{mj} = \{\mathbf{S}_{m1}, \mathbf{S}_{m2}, \ldots, \mathbf{S}_{m2^R}\}$ as follows:

$$\begin{aligned}\sum_{j=1}^{2^R} \rho(S_{nj}, S_{nj}; k + \tau) \\ &= \sum_{j=1}^{2^R} \rho(S_{nj}, S_{nj}; k)(1 - \tau) + \sum_{j=1}^{2^R} \rho(S_{nj}, S_{nj}; k + 1)\tau = 0 + 0 = 0\end{aligned} \tag{7.3}$$

where $n = 1, 2, \ldots, 2^R$, $k = 1, 2, \ldots, 2^r - 1$ and $0 < \tau < 1$, and

$$\begin{aligned}\sum_{j=1}^{2^R} \rho(S_{nj}, S_{mj}; k + \tau) \\ &= \sum_{j=1}^{2^R} \rho(S_{nj}, S_{nj}; k)(1 - \tau) + \sum_{j=1}^{2^R} \rho(S_{nj}, S_{nj}; k + 1)\tau = 0 + 0 = 0\end{aligned} \tag{7.4}$$

where $n, m = 1, 2, \ldots, 2^R$, $n \neq m$, $k = 0, 1, \ldots, 2^r - 1$ and $0 < \tau < 1$.

From the discussions given above, we obtain an important conclusion that if the correlation properties for a sequence (any sequence, including unitary codes and complementary codes alike) are perfect with their relative delay being multiple chips, then its correlation properties with fractional relative delays will also be perfect. Therefore, the discussions on the correlation properties of a sequence with a relative delay being multiple chips are general enough. Therefore, also for this reason our discussions given in this book will be limited only to the correlation properties with relative delay being multiple chips, unless for some special purposes.

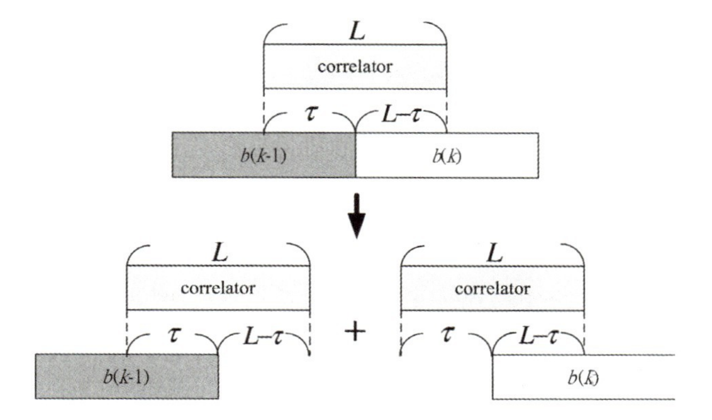

Figure 7.10 Correlation across two consecutive bits, which can be decomposed into two partial correlation functions.

Correlation in Continuous Bit Streams

When discussing correlation functions, we should also consider the real situation where the input data streams can be continuous. Therefore, the correlation functions very often cover two consecutive bits, as shown in Figure 7.10, which shows that the correlation function across two consecutive bits can in fact be decomposed into two partial correlation functions.

It is seen from Figure 7.10 that both even (if two consecutive bits carry the same sign) and odd (if two consecutive bits carry different signs) periodic correlation functions can always be decomposed into two partial correlation functions, followed by a summation of them.

Therefore, if the partial correlation functions for a sequence (no matter what type of code, either unitary or complementary) are ideal, then its even or odd periodic correlation functions should also be ideal. Here, we have used a general term, correlation functions, to include both auto-correlation and cross-correlation functions. We say that a sequence has an ideal correlation function, if and only if its auto-correlation function is always zero except the zero relative shift and its cross-correlation function is also zero for any possible relative shifts. A similar conclusion has also been made in Appendix A.

Analytical Bit Error Rate

Next, we would like to derive the bit error rate (BER) performance of a DS/CC-CDMA system under the influence of both MAI and MI. We will use the results obtained from the discussions given in the previous text to effectively simplify our analysis. The previous discussions have suggested that we need only consider the partial correlation functions in performance analysis, as doing so will not result in a loss of generality. We need only consider the cases with multiple chip delays, as the results will be the same if we include the cases with fractional chip delays.

We still consider the DS/CC-CDMA system model as shown in Figures 7.1 and 7.2. We assume that the receiver wants to receive the signal from User 1 and all other transmissions will be considered as interference. The DS/CC-CDMA system should use a set of super complementary codes with both its flock size and set size being equal to $K = M = 2^R$. The element code length is 2^r. Therefore, we can write the downlink (synchronous) transmission signals via 2^R sub-carriers from a DS/CC-CDMA transmitter as

$$\begin{cases} \mathbf{O}_{f1} &= b_1(t)\mathbf{S}_{11} + b_2(t)\mathbf{S}_{21} + \cdots + b_{2^R}(t)\mathbf{S}_{2^R1} + n_1 \\ \mathbf{O}_{f2} &= b_1(t)\mathbf{S}_{12} + b_2(t)\mathbf{S}_{22} + \cdots + b_{2^R}(t)\mathbf{S}_{2^R2} + n_2 \\ \vdots & \\ \mathbf{O}_{f2^R} &= b_1(t)\mathbf{S}_{12^R} + b_2(t)\mathbf{S}_{22^R} + \cdots + b_{2^R}(t)\mathbf{S}_{2^R2^R} + n_{2^R} \end{cases} \tag{7.5}$$

where $b_i(t)$ is the binary data signal from the ith user at time t, and n_i is the noise term from the ith sub-carrier channel.

Assume that the carrier demodulation process is lossless. Therefore, we can have the output signal from the bank of correlators, each of which is tuned to a particular element code in one sub-carrier frequency, as

$$\begin{cases} \mathbf{O}_{c1} &= b_1(t)\rho(\mathbf{S}_{11}, \mathbf{S}_{11}; 0) + b_2(t)\rho(\mathbf{S}_{21}, \mathbf{S}_{11}; 0) + \cdots + b_{2^R}(t)\rho(\mathbf{S}_{2^R1}, \mathbf{S}_{11}; 0) + \rho(n_1, \mathbf{S}_{11}; 0) \\ \mathbf{O}_{c2} &= b_1(t)\rho(\mathbf{S}_{12}, \mathbf{S}_{12}; 0) + b_2(t)\rho(\mathbf{S}_{22}, \mathbf{S}_{12}; 0) + \cdots + b_{2^R}(t)\rho(\mathbf{S}_{2^R2}, \mathbf{S}_{12}; 0) + \rho(n_2, \mathbf{S}_{12}; 0) \\ \vdots & \\ \mathbf{O}_{c2^R} &= b_1(t)\rho(\mathbf{S}_{12^R}, \mathbf{S}_{12^R}; 0) + b_2(t)\rho(\mathbf{S}_{22^R}, \mathbf{S}_{12^R}; 0) + \cdots + b_{2^R}(t)\rho(\mathbf{S}_{2^R2^R}, \mathbf{S}_{12^R}; 0) \\ & \quad + \rho(n_{2^R}, \mathbf{S}_{12^R}; 0) \end{cases} \tag{7.6}$$

where $\rho(\mathbf{S}_{nj}, \mathbf{S}_{nj}; 0)$ stands for the auto-correlation function of sequence $\mathbf{S}_{nj}$, and $\rho(\mathbf{S}_{nj}, \mathbf{S}_{mj}; 0)$ stands for the cross-correlation function of sequences $\mathbf{S}_{nj}$ and $\mathbf{S}_{mj}$ $(n \neq m)$.

The decision variable is formed after summing over all terms given in (7.6) to yield

$$\begin{aligned} \mathbf{O}_i &= b_1(t)\sum_{j=1}^{2^R} \rho(\mathbf{S}_{1j}, \mathbf{S}_{1j}; 0) + b_2(t)\sum_{j=1}^{2^R} \rho(\mathbf{S}_{2j}, \mathbf{S}_{1j}; 0) + \cdots \\ &\quad + b_{2^R}(t)\sum_{j=1}^{2^R} \rho(\mathbf{S}_{2^Rj}, \mathbf{S}_{1j}; 0) + \sum_{j=1}^{2^R} \rho(n_j, \mathbf{S}_{1j}; 0) \\ &= 2^R b_1(t) + 0 + 0 + \cdots + 0 + \sum_{j=1}^{2^R} \rho(n_j, \mathbf{S}_{1j}; 0) \\ &= 2^R b_1(t) + \sum_{j=1}^{2^R} \rho(n_j, \mathbf{S}_{1j}; 0) \end{aligned} \tag{7.7}$$

which shows that the decision variable is corrupted only by noise, implying that all MAI terms do not affect the overall BER performance of the DS/CC-CDMA system of concern, where multipath propagation is not taken into account.

Next, let us extend our discussion to an asynchronous transmission scenario. We still use the same super complementary code set as considered in the previous text, such that its element code length is 2^r, its set size and flock size are equal to $K = M = 2^R$, $b_i(t)$ is the ith user signal at time t, and the relative delay between the first user and the jth user is $k_{1j}T_b + \tau_{1j}$. Here T_b is bit duration, and $0 < \tau_{1j} < 1$ stands for fractional bit duration. Also, n_i is the noise component carried in the ith sub-carrier frequency.

Therefore, we will have the output signal from different correlators at the receiver as

$$\begin{cases} \mathbf{O}_{c1} = b_1(t)\rho(\mathbf{S}_{11}, \mathbf{S}_{11}; 0) + b_2(t + k_{12} - 1)\rho(\mathbf{S}_{21}, \mathbf{S}_{11}; -(2^r - \tau_{12})) \\ \qquad + b_2(t + k_{12})\rho(\mathbf{S}_{21}, \mathbf{S}_{11}; \tau_{12}) + \cdots + b_{2R}(t + k_{12R} - 1)\rho(\mathbf{S}_{2R1}, \mathbf{S}_{11}; -(2^r - \tau_{12R})) \\ \qquad + b_{2R}(t + k_{12R})\rho(\mathbf{S}_{2R1}, \mathbf{S}_{11}; \tau_{12R}) + \rho(n_1, \mathbf{S}_{11}; 0) \\ \mathbf{O}_{c2} = b_1(t)\rho(\mathbf{S}_{12}, \mathbf{S}_{12}; 0) + b_2(t + k_{12} - 1)\rho(\mathbf{S}_{22}, \mathbf{S}_{12}; -(2^r - \tau_{12})) \\ \qquad + b_2(t + k_{12})\rho(\mathbf{S}_{22}, \mathbf{S}_{12}; \tau_{12}) + \cdots + b_{2R}(t + k_{12R} - 1)\rho(\mathbf{S}_{2R2}, \mathbf{S}_{12}; -(2^r - \tau_{12R})) \\ \qquad + b_{2R}(t + k_{12R})\rho(\mathbf{S}_{2R2}, \mathbf{S}_{12}; \tau_{12R}) + \rho(n_2, \mathbf{S}_{12}; 0) \\ \vdots \\ \mathbf{O}_{c2R} = b_1(t)\rho(\mathbf{S}_{12R}, \mathbf{S}_{12R}; 0) + b_2(t + k_{12} - 1)\rho(\mathbf{S}_{22R}, \mathbf{S}_{12R}; -(2^r - \tau_{12})) \\ \qquad + b_2(t + k_{12})\rho(\mathbf{S}_{22R}, \mathbf{S}_{12R}; \tau_{12}) + \cdots + b_{2R}(t + k_{12R} - 1)\rho(\mathbf{S}_{2R2R}, \mathbf{S}_{12R}; \\ \qquad - (2^r - \tau_{12R})) + b_{2R}(t + k_{12R})\rho(\mathbf{S}_{2R2R}, \mathbf{S}_{12R}; \tau_{12R}) + \rho(n_{2R}, \mathbf{S}_{12R}; 0) \end{cases} \tag{7.8}$$

from which the decision variable can be generated as

$$\begin{aligned} \mathbf{O}_i &= b_1(t)\sum_{j=1}^{2^R} \rho(\mathbf{S}_{1j}, \mathbf{S}_{1j}; 0) + b_2(t + k_{12} - 1)\sum_{j=1}^{2^R} \rho(\mathbf{S}_{2j}, \mathbf{S}_{1j}; -(2^r - \tau_{12})) \\ &\quad + b_2(t + k_{12})\sum_{j=1}^{2^R} \rho(\mathbf{S}_{2j}, \mathbf{S}_{1j}; \tau_{12}) + \cdots + b_{2R}(t + k_{12R} - 1)\sum_{j=1}^{2^R} \rho(\mathbf{S}_{2Rj}, \mathbf{S}_{1j}; -(2^r - \tau_{12R})) \\ &\quad + b_{2R}(t + k_{12R})\sum_{j=1}^{2^R} \rho(\mathbf{S}_{2Rj}, \mathbf{S}_{1j}; \tau_{12R}) + \sum_{j=1}^{2^R} \rho(n_j, \mathbf{S}_{1j}; 0) \\ &= 2^R b_1(t) + 0 + 0 + \cdots + 0 + \sum_{j=1}^{2^R} \rho(n_j, \mathbf{S}_{1j}; 0) \\ &= 2^R b_1(t) + \sum_{j=1}^{2^R} \rho(n_j, \mathbf{S}_{1j}; 0) \end{aligned} \tag{7.9}$$

Again, we can see from the above results that the decision variable contains only useful signal plus noise, meaning that performance of the DS/CC-CDMA system is limited only by noise, but not by the interference.

Finally, we want to look at the most general situation where an asynchronous multipath channel is considered. The multipath channel contains L different propagation paths and we will choose the first path as the useful signal; the rest will be treated as interference (i.e. MI). In the ith propagation path, the relative delay between the first user and the jth user is $k_{ij}T_b + \tau_{ij}$, where T_b is the bit duration and τ_{ij} (which is larger than zero and smaller than one) is the fractional chip delay. Therefore, the output signal from all correlators (corresponding to different carrier frequencies) at the receiver can be written as

$$
\begin{aligned}
\mathbf{O}_{c1} = {} & b_1(t)\rho(\mathbf{S}_{11}, \mathbf{S}_{11}; 0) + b_2(t + k_{12} - 1)\rho(\mathbf{S}_{21}, \mathbf{S}_{11}; -(2^r - \tau_{12})) + b_2(t + k_{12})\rho(\mathbf{S}_{21}, \mathbf{S}_{11}; \tau_{12}) \\
& + \cdots + b_{2^R}(t + k_{12^R} - 1)\rho(\mathbf{S}_{2^R1}, \mathbf{S}_{11}; -(2^r - \tau_{12^R})) + b_{2^R}(t + k_{12^R})\rho(\mathbf{S}_{2^R1}, \mathbf{S}_{11}; \tau_{12^R}) \\
& + b_1(t + k_{21} - 1)\rho(\mathbf{S}_{11}, \mathbf{S}_{11}; -(2^r - \tau_{21})) + b_1(t + k_{21})\rho(\mathbf{S}_{11}, \mathbf{S}_{11}; \tau_{21}) \\
& + b_2(t + k_{22} - 1)\rho(\mathbf{S}_{21}, \mathbf{S}_{11}; -(2^r - \tau_{22})) + b_2(t + k_{22})\rho(\mathbf{S}_{21}, \mathbf{S}_{11}; \tau_{22}) \\
& + \cdots + b_{2^R}(t + k_{22^R} - 1)\rho(\mathbf{S}_{2^R1}, \mathbf{S}_{11}; -(2^r - \tau_{22^R})) + b_{2^R}(t + k_{22^R})\rho(\mathbf{S}_{2^R1}, \mathbf{S}_{11}; \tau_{22^R}) \\
& + \cdots + b_1(t + k_{L1} - 1)\rho(\mathbf{S}_{11}, \mathbf{S}_{11}; -(2^r - \tau_{L1})) + b_1(t + k_{L1})\rho(\mathbf{S}_{11}, \mathbf{S}_{11}; \tau_{L1}) \\
& + b_2(t + k_{L2} - 1)\rho(\mathbf{S}_{21}, \mathbf{S}_{11}; -(2^r - \tau_{L2})) + b_2(t + k_{L2})\rho(\mathbf{S}_{21}, \mathbf{S}_{11}; \tau_{L2}) \\
& + \cdots + b_{2^R}(t + k_{L2^R} - 1)\rho(\mathbf{S}_{2^R1}, \mathbf{S}_{11}; -(2^r - \tau_{L2^R})) + b_{2^R}(t + k_{L2^R})\rho(\mathbf{S}_{2^R1}, \mathbf{S}_{11}; \tau_{L2^R}) \\
& + \rho(n_1, \mathbf{S}_{11}; 0) \\
\\
\mathbf{O}_{c2} = {} & b_1(t)\rho(\mathbf{S}_{12}, \mathbf{S}_{12}; 0) + b_2(t + k_{12} - 1)\rho(\mathbf{S}_{22}, \mathbf{S}_{12}; -(2^r - \tau_{12})) + b_2(t + k_{12})\rho(S_{22}, \mathbf{S}_{12}; \tau_{12}) \\
& + \cdots + b_{2^R}(t + k_{12^R} - 1)\rho(\mathbf{S}_{2^R2}, \mathbf{S}_{12}; -(2^r - \tau_{12^R})) + b_{2^R}(t + k_{12^R})\rho(\mathbf{S}_{2^R2}, \mathbf{S}_{12}; \tau_{12^R}) \\
& + b_1(t + k_{21} - 1)\rho(\mathbf{S}_{12}, \mathbf{S}_{12}; -(2^r - \tau_{21})) + b_1(t + k_{21})\rho(\mathbf{S}_{12}, \mathbf{S}_{12}; \tau_{21}) \\
& + b_2(t + k_{22} - 1)\rho(\mathbf{S}_{22}, \mathbf{S}_{12}; -(2^r - \tau_{22})) + b_2(t + k_{22})\rho(\mathbf{S}_{22}, \mathbf{S}_{12}; \tau_{22}) \\
& + \cdots + b_{2^R}(t + k_{22^R} - 1)\rho(\mathbf{S}_{2^R2}, \mathbf{S}_{12}; -(2^r - \tau_{22^R})) + b_{2^R}(t + k_{22^R})\rho(\mathbf{S}_{2^R2}, \mathbf{S}_{12}; \tau_{22^R}) \\
& + \cdots + b_1(t + k_{L1} - 1)\rho(\mathbf{S}_{12}, \mathbf{S}_{12}; -(2^r - \tau_{L1})) + b_1(t + k_{L1})\rho(\mathbf{S}_{12}, \mathbf{S}_{12}; \tau_{L1}) \\
& + b_2(t + k_{L2} - 1)\rho(\mathbf{S}_{22}, \mathbf{S}_{12}; -(2^r - \tau_{L2})) + b_2(t + k_{L2})\rho(\mathbf{S}_{22}, \mathbf{S}_{12}; \tau_{L2}) \qquad (7.10) \\
& + \cdots + b_{2^R}(t + k_{L2^R} - 1)\rho(\mathbf{S}_{2^R2}, \mathbf{S}_{12}; -(2^r - \tau_{L2^R})) + b_{2^R}(t + k_{L2^R})\rho(\mathbf{S}_{2^R2}, \mathbf{S}_{12}; \tau_{L2^R}) \\
& + \rho(n_2, \mathbf{S}_{12}; 0) \\
\vdots \\
\mathbf{O}_{c2^R} = {} & b_1(t)\rho(\mathbf{S}_{12^R}, \mathbf{S}_{12^R}; 0) + b_2(t + k_{12} - 1)\rho(\mathbf{S}_{22^R}, \mathbf{S}_{12^R}; -(2^r - \tau_{12})) \\
& + b_2(t + k_{12})\rho(\mathbf{S}_{22^R}, \mathbf{S}_{12^R}; \tau_{12}) + \cdots + b_{2^R}(t + k_{12^R} - 1)\rho(\mathbf{S}_{2^R2^R}, \mathbf{S}_{12^R}; -(2^r - \tau_{12^R})) \\
& + b_{2^R}(t + k_{12^R})\rho(\mathbf{S}_{2^R2^R}, \mathbf{S}_{12^R}; \tau_{12^R}) + b_1(t + k_{21} - 1)\rho(\mathbf{S}_{12^R}, \mathbf{S}_{12^R}; -(2^r - \tau_{21})) \\
& + b_1(t + k_{21})\rho(\mathbf{S}_{12^R}, \mathbf{S}_{12^R}; \tau_{21}) + b_2(t + k_{22} - 1)\rho(\mathbf{S}_{22^R}, \mathbf{S}_{12^R}; -(2^r - \tau_{22})) \\
& + b_2(t + k_{22})\rho(\mathbf{S}_{22^R}, \mathbf{S}_{12^R}; \tau_{22}) \\
& + \cdots + + \cdots + b_{2^R}(t + k_{22^R} - 1)\rho(\mathbf{S}_{2^R2^R}, \mathbf{S}_{12^R}; -(2^r - \tau_{22^R})) \\
& + b_{2^R}(t + k_{22^R})\rho(\mathbf{S}_{2^R2^R}, \mathbf{S}_{12^R}; \tau_{22^R}) + \cdots + b_1(t + k_{L1} - 1)\rho(\mathbf{S}_{12^R}, \mathbf{S}_{12^R}; -(2^r - \tau_{L1})) \\
& + b_1(t + k_{L1})\rho(\mathbf{S}_{12^R}, \mathbf{S}_{12^R}; \tau_{L1}) \\
& + b_2(t + k_{L2} - 1)\rho(\mathbf{S}_{22^R}, \mathbf{S}_{12^R}; -(2^r - \tau_{L2})) + b_2(t + k_{L2})\rho(\mathbf{S}_{22^R}, \mathbf{S}_{12^R}; \tau_{L2}) \\
& + \cdots + b_{2^R}(t + k_{L2^R} - 1)\rho(\mathbf{S}_{2^R2^R}, \mathbf{S}_{12^R}; -(2^r - \tau_{L2^R})) \\
& + b_{2^R}(t + k_{L2^R})\rho(\mathbf{S}_{2^R2^R}, \mathbf{S}_{12^R}; \tau_{L2^R}) \\
& + \rho(n_{2^R}, \mathbf{S}_{12^R}; 0)
\end{aligned}
$$

from which we can have the decision variable as

$$
\begin{aligned}
\mathbf{O}_i = {} & b_1(t)\sum_{j=1}^{2^R}\rho(\mathbf{S}_{1j},\mathbf{S}_{1j};0) + b_2(t+k_{12}-1)\sum_{j=1}^{2^R}\rho(\mathbf{S}_{2j},\mathbf{S}_{1j};-(2^r-\tau_{12})) \\
& + b_2(t+k_{12})\sum_{j=1}^{2^R}\rho(\mathbf{S}_{2j},\mathbf{S}_{1j};\tau_{12}) \\
& + \cdots + b_{2^R}(t+k_{12^R}-1)\sum_{j=1}^{2^R}\rho(\mathbf{S}_{2^R j},\mathbf{S}_{1j};-(2^r-\tau_{12^R})) \\
& + b_{2^R}(t+k_{12^R})\sum_{j=1}^{2^R}\rho(\mathbf{S}_{2^R j},\mathbf{S}_{1j};\tau_{12^R}) \\
& + b_1(t+k_{21}-1)\sum_{j=1}^{2^R}\rho(\mathbf{S}_{11},\mathbf{S}_{11};-(2^r-\tau_{21})) + b_1(t+k_{21})\sum_{j=1}^{2^R}\rho(\mathbf{S}_{11},\mathbf{S}_{11};\tau_{21}) \\
& + b_2(t+k_{22}-1)\sum_{j=1}^{2^R}\rho(\mathbf{S}_{2j},\mathbf{S}_{1j};-(2^r-\tau_{22})) + b_2(t+k_{22})\sum_{j=1}^{2^R}\rho(\mathbf{S}_{2j},\mathbf{S}_{1j};\tau_{22}) \\
& + \cdots + b_{2^R}(t+k_{22^R}-1)\sum_{j=1}^{2^R}\rho(\mathbf{S}_{2^R j},\mathbf{S}_{1j};-(2^r-\tau_{22^R})) \\
& + b_{2^R}(t+k_{22^R})\sum_{j=1}^{2^R}\rho(\mathbf{S}_{2^R j},\mathbf{S}_{1j};\tau_{22^R}) \\
& + \cdots + b_1(t+k_{L1}-1)\sum_{j=1}^{2^R}\rho(\mathbf{S}_{11},\mathbf{S}_{11};-(2^r-\tau_{L1})) + b_1(t+k_{L1})\sum_{j=1}^{2^R}\rho(\mathbf{S}_{11},\mathbf{S}_{11};\tau_{L1}) \\
& + b_2(t+k_{L2}-1)\sum_{j=1}^{2^R}\rho(\mathbf{S}_{2j},\mathbf{S}_{1j};-(2^r-\tau_{L2})) + b_2(t+k_{L2})\sum_{j=1}^{2^R}\rho(\mathbf{S}_{2j},\mathbf{S}_{1j};\tau_{L2}) \\
& + \cdots + b_{2^R}(t+k_{L2^R}-1)\sum_{j=1}^{2^R}\rho(\mathbf{S}_{2^R j},\mathbf{S}_{1j};-(2^r-\tau_{L2^R})) \\
& + b_{2^R}(t+k_{L2^R})\sum_{j=1}^{2^R}\rho(\mathbf{S}_{2^R j},\mathbf{S}_{1j};\tau_{L2^R}) \\
& + \sum_{j=1}^{2^R}\rho(n_j,\mathbf{S}_{1j};0) = 2^R b_1(t) + 0 + 0 + \cdots + 0 \\
& + \sum_{j=1}^{2^R}\rho(n_j,\mathbf{S}_{1j};0) = 2^R b_1(t) + \sum_{j=1}^{2^R}\rho(n_j,\mathbf{S}_{1j};0)
\end{aligned}
\tag{7.11}
$$

It is seen from the above results that the decision variable formed from the summation of the signals from the correlator bank of the receiver depends only on the noise but not on the MAI and MI, implying perfect interference-free operation of the DS/CC-CDMA system even under the influence of both MAI and MI. Therefore, the BER performance of the DS/CC-CDMA system can be calculated simply by $P_b = Q(\sqrt{2SNR})$, where BPSK modulation is used.

Simulation Results

After having studied the BER performance of the DS/CC-CDMA system with those of analysis, we would like to counter-check the analytical results with those obtained from computer simulations, in which many practical working environments can be taken into account.

Due to limited space, we can show only two figures here as examples. Figure 7.11 shows the simulation results for a DS/CC-CDMA system, which uses the super complementary codes with $K = M = 8$ and $N = 8$. The performance is compared with a conventional DS-CDMA system using Gold code and OVSF code, whose processing gains are 63 and 64, respectively. All systems carry the same number of users, or eight users, in the simulations. An asynchronous three-ray multipath channel is considered in the simulations, with its inter-path relative delay being two chips and the path gain coefficients vector being [1, 0.8564, 0.107]. The inter-user relative delay is four chips.

Figure 7.12 gives also a comparison among three different CDMA systems, one being DS/CC-CDMA with super complementary code set, and two DS-CDMA systems with Gold and OVSF codes. The same system set-up parameters are used, except for the multipath channel coefficients vector, which is a normalized channel gain of [0.7785, 0.6667, 0.0833] here, as the squared sum of three multipath return elements will be unit. On the other hand, Figure 7.11 uses non-normalized channel coefficients vector. Therefore, the simulation results are different if we compare the two figures. However, the general results are fairly consistent due to the fact that the DS/CC-CDMA system offers a very

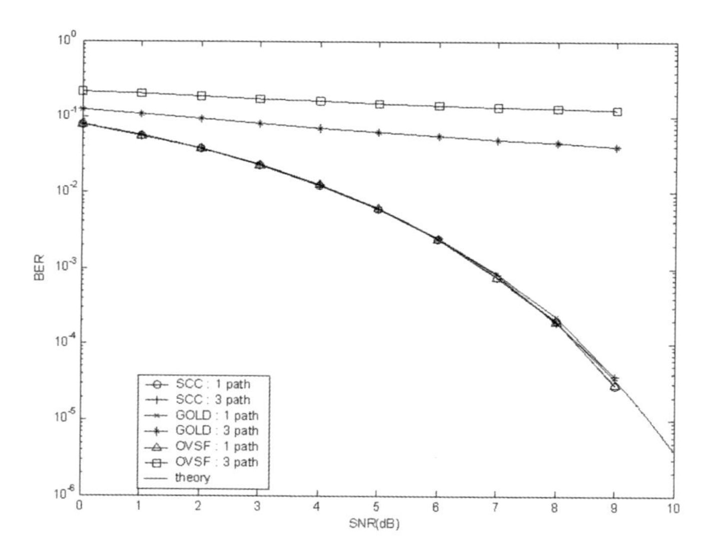

Figure 7.11 BER simulation for a DS/CC-CDMA system with eight users and three-ray multipath channel, where inter-user delay is four chips and inter-path delay is two chips. The performance is compared to that for DS-CDMA systems using Gold and OVSF codes, with their PG values being 63 and 64, respectively and the same channel model considered. The three-ray multipath channel has channel gain coefficients vector [1, 0.8564, 0.107].

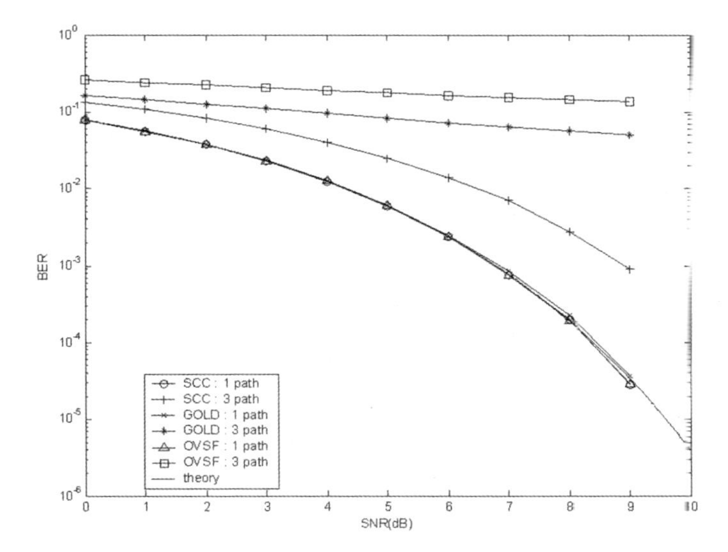

Figure 7.12 BER simulation for a DS/CC-CDMA system with eight users and three-ray multipath channel, where inter-user delay is four chips and inter-path delay is two chips. The performance is compared to that for DS-CDMA systems using Gold and OVSF codes, with their PG values being 63 and 64, respectively and the same channel model considered. The three-ray multipath channel has channel gain coefficients vector [0.7785, 0.6667, 0.0833].

robust BER performance against both MAI and MI, while the performance of the conventional DS-CDMA systems based on either Gold or OVSF codes fails to compete successfully with the DS/CC-CDMA system in terms of their much worse BER, especially when multipath propagation exists.

7.1.5 PROPERTIES OF DS/CC-CDMA SYSTEM

After having taken a look at the performance analysis using both illustrations and analytical derivations, we would like to discuss the major properties of the DS/CC-CDMA system based on orthogonal complementary codes.

It is noted that there are many different types of complementary codes, and not all of them are orthogonal complementary codes. For instance, the generalized pair-wise complementary codes are not orthogonal complementary codes due to their correlation functions not being perfect, as discussed in Subsection 6.2.5. However, many properties we discuss here will also be applicable to these codes in general, subject to some necessary modifications.

We would like also to stress again that, although all results obtained in this section were based on discussions on a particular orthogonal complementary code, namely super complementary code, the same results will be generated if any other orthogonal complementary codes are used, such as complete complementary codes, column-wise complementary codes, etc. It should also be noted that the use of other types of orthogonal complementary codes may affect the properties of a DS/CC-CDMA system in terms of the maximal number of users it can support due to the different set sizes. It may also affect the system implementation complexity due to the different flock sizes and element code lengths, etc. Nevertheless, the correlation properties of a DS/CC-CDMA system based on different types of orthogonal complementary codes will be the same, and they will be discussed as follows.

Suitability for High-Speed Bursty Traffic

As we can see from the results obtained in the previous subsections, a DS/CC-CDMA system based on orthogonal complementary codes can offer many unique desirable characteristic features compared to a conventional DS-CDMA system based on unitary spreading codes, such as Gold codes, m-sequences, Walsh-Hadamard codes, OVSF code, Kasami codes, etc. It is also seen that a great change happens with the introduction of orthogonal complementary codes, and thus we can see how big the impact of the application of different spreading codes can be on the overall performance of a CDMA system.

As an important requirement for a CDMA technology suitable for future gigabit all-IP wireless communications, we should pay sufficient attention to detection efficiency at the edges of a packet or burst. In this sense, a DS/CC-CDMA scheme is in particular well suited for its applications in future gigabit all-IP wireless communications, as shown in Figures 7.7 and 7.8, which illustrate the signal detection process for the first (or the rightmost) bit of a packet. In detecting those bits at the edges of a packet, partial auto-correlation functions (ACFs) and partial cross-correlation functions (CCFs) of the codes will become extremely important. It is seen from the figures that the DS/CC-CDMA system yields zero partial CCFs, which ensures an ideal performance for signal detection even on the edges of a frame or packet. This observation is significant due to the domination of burst traffic in all future wireless systems.

We have to emphasize that the uniquely good detection efficiency for short packets in the DS/CC-CDMA architecture is because of the use of orthogonal complementary codes, which offer a possibility for them to cancel all ACFs side lobes and CCFs in the process of summation of all correlation results generated from different element codes in the same flock. On the other hand, a conventional DS-CDMA system based on unitary codes will not have the opportunity to cancel those ACFs side lobes and CCFs generated from a 'single' correlator structure.

Resilience Against Time-/Frequency-Selective Fading

As demonstrated in the previous discussions, which include those given in Subsections 5.2.2 (about 2D spreading) and 6.2.6 (about column-wise complementary codes), we understand that the spreading process involved in a DS/CC-CDMA system is actually a two-dimensional spreading process, which covers both the time and frequency domains. This gives us a great degree of freedom in choosing how we want to spread the same bit stream. This can never happen if a unitary spreading code is used, as in all conventional DS-CDMA system architecture, where only time-domain spreading is available.

Each user in a DS/CC-CDMA system will be assigned a flock of M element codes, each of which has a length of N chips, thus resulting in a processing gain of $M \times N$.

Due to the use of two-dimensional spreading in the DS/CC-CDMA system, we find an interesting issue on how orthogonality is established based on the two-dimensional spreading. It is well known that for any orthogonal complementary code set the correlation properties for a single element code are never perfect. In fact, the perfect correlation functions in an orthogonal complementary code set are based on the summation of all correlation functions generated from individual element codes. In other words, the non-zero ACF side lobes and CCFs are all canceled in the process of the summation. Then, an interesting question arises: how can these non-zero ACF side lobes and CCFs be canceled? To give an answer to this question, let us look at some simple examples from column-wise complementary codes. It should be noted that although we only take the column-wise complementary codes here as examples, the same results will apply if any other orthogonal complementary codes, such as complete complementary codes, super complementary codes, etc., are considered.

Let us look at two column-wise complementary codes as

$$
\begin{aligned}
\mathbf{C}^{(1)}_{4\times 4} &= \begin{pmatrix} + & + & + & - \\ + & - & + & + \\ + & + & + & - \\ + & - & + & + \end{pmatrix} \\
\mathbf{C}^{(2)}_{4\times 4} &= \begin{pmatrix} + & + & - & + \\ + & - & - & - \\ + & + & - & + \\ + & - & - & - \end{pmatrix}
\end{aligned}
\tag{7.12}
$$

whose parameters are $M = K = 4$ and $N = 4$. We only show the two flocks from the code set (there are in total four flocks in the set) as examples. Figure 7.13 shows the time-domain CCFs cancelation process, based on which the orthogonality of the column-wise code set is established.

On the other hand, we can have another column-wise code set, whose two flocks are given as follows:

$$
\begin{aligned}
\mathbf{C}^{(1)}_{4\times 4} &= \begin{pmatrix} + & + & + & + \\ + & - & + & - \\ + & + & - & - \\ + & - & - & + \end{pmatrix} \\
\mathbf{C}^{(2)}_{4\times 4} &= \begin{pmatrix} + & + & - & - \\ + & - & - & + \\ + & + & + & + \\ + & - & + & - \end{pmatrix}
\end{aligned}
\tag{7.13}
$$

Figure 7.14 shows the CCF cancelation process in the frequency domain for the two column-wise codes given in (7.13). We can see that the cross-correlation functions for these two flocks of element codes are all zero due to the cancelation process carried out purely in the frequency domain.

In most cases, the cancelation can happen in both the time and frequency domains at the same time, although with various proportions, whose percentage varies from code to code. It can be shown

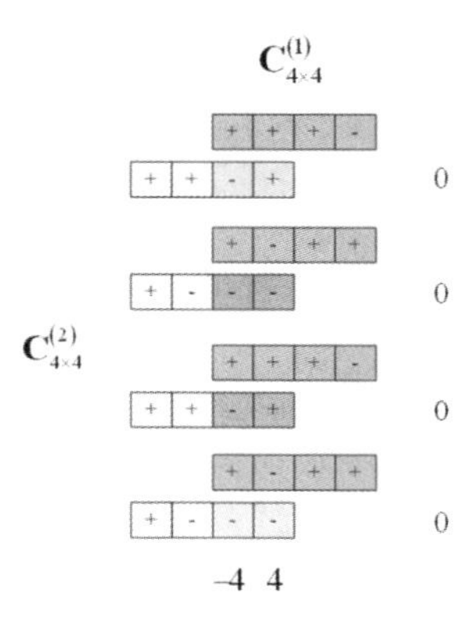

Figure 7.13 Orthogonality of a column-wise code set is based on time-domain CCF cancelation.

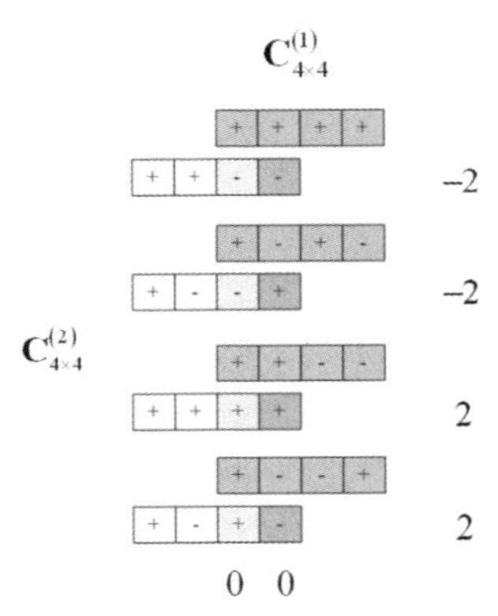

Figure 7.14 Orthogonality of a column-wise code set is based on frequency-domain CCF cancelation.

that the orthogonality of an orthogonal complementary code set can be established based solely on the frequency-domain cancelation (i.e. 100% frequency-domain cancelation), but can be based on at most 50% (which is a maximal value for the time-domain cancelation) on the time-domain cancelation.

The 100% frequency-domain cancelation is an important property for an orthogonal complementary code set, as it can be effectively used in many applications where mobility is an important factor to consider in the system design, such as vehicle-to-vehicle (V2V) communications (the IEEE 802.11p standard is in the process of standardization, in particular for V2V communications). The property of the 100% frequency-domain cancelation in an orthogonal complementary code set can be translated into the fact that the orthogonality of the orthogonal complementary codes will hardly be affected by time-selective fading due to the fact that it is established solely on the frequency-domain CCFs cancelation process, and thus the time-domain changes in channel gains will not affect the detection process at the receiver.

Let us take a look at a particular example to show how time-varying fading will affect the detection process of a DS/CC-CDMA receiver. Assume that time-varying channel coefficients vector at four different chips are $\{h(t_1), h(t_2), h(t_3), h(t_4)\}$, where t_1, t_2, t_3, and t_4 stand for the four chip sampling times. We still use the two codes shown in (7.13) as the examples here. Therefore, we will have the output from the local correlation process at a receiver as

$$\begin{aligned}&-2h^{(1)}(t_3)\,h^{(2)}(t_1) + 2h^{(1)}(t_3)\,h^{(2)}(t_1)\\&-2h^{(1)}(t_4)\,h^{(2)}(t_2) + 2h^{(1)}(t_4)\,h^{(2)}(t_2)\\&=0\end{aligned} \tag{7.14}$$

which tells us that the cross-correlation function is perfect after the frequency-domain cancelation process, which is illustrated in Figure 7.15.

On the other hand, we can also exploit the property of the time-domain CCF cancelation process to design a DS/CC-CDMA system for applications with severe frequency-selective fading. Unfortunately, we can only achieve time-domain CCF cancelation to a maximal percentage of 50%, which is different from the frequency-domain CCF cancelation, which can reach 100%. Therefore, a DS/CC-CDMA system has a stronger time-selective fading resistance than frequency-selective fading resistance. Nevertheless, both time-selective and frequency-selective fading resistance can be exploited to greatly strengthen the capability of a DS/CC-CDMA system against channel impairing factors, which are always formidable to all conventional DS-CDMA systems.

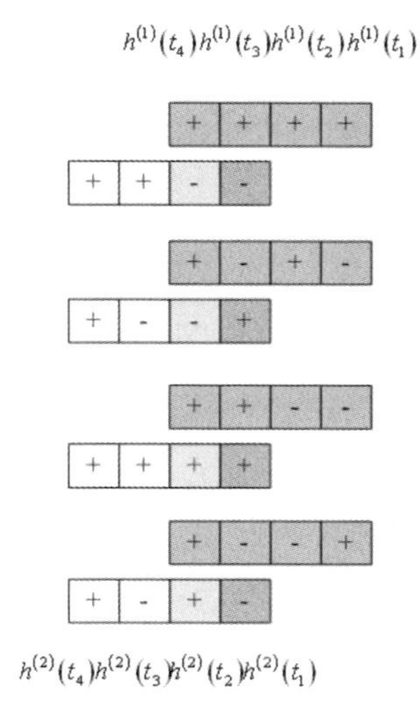

Figure 7.15 Time-varying fading resistance for an orthogonal complementary code set with 100% frequency-domain CCF cancelation property.

Near-Far Effect Resistance

Owing to the isotropic MAI-free and MI-free properties, the near-far effect will virtually cause no harm to the signal detection process at a correlator in a DS/CC-CDMA system, as long as bit synchronization can be achieved prior to the data detection process. In other words, DS/CC-CDMA is a system with excellent near-far resistance. Therefore, complicated open-loop and closed-loop power control is no longer a necessity. More precisely, the power control in a DS/CC-CDMA system is used merely to reduce unnecessary power emission at terminals, whose requirements on response time and accuracy can be made much more relaxed than necessary in a conventional DS-CDMA system. In a DS/CC-CDMA system, a similar conclusion can be drawn with respect to the power control requirement, due to its ideal MAI-free property.

RAKE Versus Matched-Filter

It is well known that all current CDMA systems have to use a RAKE receiver to mitigate otherwise formidable MI, which is due to the imperfect auto-correlation functions of the CDMA codes used. Theoretically speaking, the orthogonal complementary codes virtually do not produce any auto-correlation side lobes if DS spreading modulation is considered, and thus MI-free operation is guaranteed for both uplink and downlink transmissions.

To illustrate clearly how a DS/CC-CDMA system with DS spreading can overcome MI without the help of a RAKE receiver, we used Figures 7.7 and 7.8, where a three-bit data burst is sent into a two-user and two-path uplink asynchronous CDMA channel. The inter-path and inter-user delays are only one chip for illustration simplicity (the same result will be given if any other delays are applied). A simple correlator receiver is used to detect the incoming burst in the presence of both MAI and MI. It is seen from the figures that a simple correlator can successfully recover the original data information $[+1, -1, +1]$ without any impairment caused by either MAI or MI. Similarly, we can show the same result for a downlink multipath channel, which usually causes much less problems than an uplink channel.

Thus, a simple correlator can solve MI and MAI problems in a DS/CC-CDMA system. It should be noted that avoiding the use of a RAKE receiver is significant as it paves the way for a DS/CC-CDMA receiver to work in a truly blind fashion without the need for any prior channel information. On the other hand, a RAKE has to acquire virtually all channel state information, such as delays and amplitudes of all multipath returns, for its maximal-ratio-combining (MRC) operation, whose impact on implementation complexity in a mobile handset should never be underestimated. Obviously, for a DS/CC-CDMA system we can also use a RAKE receiver to further boost the signal-to-interference ratio before decision. It is noted that the use of RAKE here can provide a much higher multipath-diversity gain than achievable in a conventional CDMA system due to the fact that the output signal from each finger now contains only useful auto-correlation peak. On the other hand, the output from a finger of a conventional CDMA RAKE receiver contains both useful and unwanted signals caused by non-trivial auto-correlation side lobes.

Furthermore, if a RAKE has to be used in a DS/CC-CDMA receiver, equal gain combining (EGC) is preferred to yield a satisfactory detection efficiency. The advantage of using EGC rather than MRC in a RAKE is to avoid having to use a complicated multipath amplitude estimation unit.

MUD versus Correlator Receiver

As shown in Figures 7.7 and 7.8, the isotropic MAI-free and MI-free property in a DS/CC-CDMA based on DS spreading makes it unnecessary to use multi-user detection (MUD) to decorrelate transmission signals from different users, due to the fact that the transmissions from different users in the DS/CC-CDMA system have already been pre-decorrelated at the transmitter side because of its unique MAI-free signaling structure.

FDM versus TDM for Element Code Division

To implement a DS/CC-CDMA system, an important requirement is to send a flock of M element codes, which are assigned to a specific user, via separate channels to an intended receiver, where each element code despread separately and their outputs combined to form a decision variable. The most straightforward way to implement element code division in a DS/CC-CDMA system is to use the frequency division multiplex (FDM) method, in which M element codes are sent through M different carriers f_m, where $1 \leq m \leq M$. In this way, a DS/CC-CDMA system looks just like a multi-carrier (MC) CDMA scheme. However, the major difference between a DS/CC-CDMA/FDM system and a traditional MC-CDMA lies in the fact that the former uses different carriers to convey different information without providing any diversity in frequencies, while the latter usually does.

Obviously, another scheme to implement element code division in a DS/CC-CDMA is to send element codes in different time slots or simply via the time division multiplex (TDM) method. The FDM and TDM implementations of element code division offer distinct system operational advantages. One of the benefits of using FDM is to allow a DS/CC-CDMA system to work harmonically with the frequency division duplex (FDD) operation mode used in most mobile cellular standards, such as W-CDMA, cdma2000, etc. On the other hand, the TDM implementation can fit TDD mode naturally, but the TDD operation is only suited to covering a relatively small cell size due to its relatively low average transmission power.

Another salient feature for the FDM option is to reduce overall hardware complexity with the help of OFDM technology. Similar to any MC-CDMA system, a DS/CC-CDMA/FDM system can also be implemented in an OFDM architecture, which can transform a complicated multi-carrier RF transmission system into a baseband signal processing unit.

In this section we have introduced the DS/CC-CDMA system architecture. It is based on a direct sequence spreading modulation scheme, which is a traditional spreading method used by many existing CDMA systems. In the next section we will discuss the issues on a new spreading modulation scheme, namely offset stacking (OS) spreading modulation, which was proposed in our research on next generation CDMA technology.

We have to acknowledge that the CDMA system based on OS spreading and orthogonal complementary codes (we will call it 'OS/CC-CDMA' in the text following) carries many special technical features, which are different from what we have known from DS/CC-CDMA systems. However, there are pros and cons in both DS/CC-CDMA and OS/CC-CDMA schemes and therefore we should select the appropriate one for a particular wireless communication application.

7.2 OFFSET STACKING SPREADING AND OS/CC-CDMA SYSTEMS

The maturing of the 3G mobile communication technologies from their concepts to commercially deliverable systems motivates us to think about possible architectures for future generations of mobile communications. The beyond 3G (B3G) wireless communication systems ought to deliver a much higher data rate than what is achievable in current 3G systems. Some people also expect that the possible data rate for 4G wireless/cellular systems should be roughly at a range of 10 Mbps to 1 Gbps, depending on the channel conditions. Bearing this objective in mind, our question is how to guarantee such a high data rate in a highly unpredictable and hostile mobile/wireless channel, and what types of air-link architecture are qualified to deliver such high data rate services.

Considering the constraints on the available radio spectrum suitable for terrestrial mobile communications (from a few hundred MHz to less than 5 GHz), we would like to argue that probably the most relevant and feasible way to achieve the goal promised by 4G communication systems is to work out some enabling technologies, such as next generation CDMA technology (which is the focus of this book), capable of improving as much as possible the air-link bandwidth efficiency of the systems. In this section, we will tackle this issue comprehensively by proposing a new CDMA architecture based on an innovative spreading technique that will become one of the core techniques of the next generation CDMA technology and has great potential for its applications in future wireless communications.

It is well known that all current CDMA-based 2G and 3G standards (i.e. IS-95, cdma2000, and W-CDMA) use traditional direct sequence (DS) CDMA techniques based on an identical principle that each bit is spread by one single spreading code comprising N contiguous chips to attain certain processing gain or spreading factor (SF). The bandwidth of all these systems is determined by the chip width of the spreading codes used. Thus, it is natural to define a merit parameter called spreading efficiency (SE) in unit of bit(s) per chip to measure the bandwidth efficiency of a CDMA system. Therefore, the SEs of all conventional DS-CDMA based-mobile communication systems, such as IS-95, cdma2000, and W-CDMA, are equal to $1/N$, which is far less than one, leaving us much room for improvement.

In the last few years, we have been working on proposing a possible solution to improve the SE of a CDMA system with the help of a new spreading technique based on orthogonal complementary codes, taking into account various implementation constraints of a practical CDMA system. These implementation constraints will be explained as follows. First, the new CDMA architecture ought to be technically feasible using currently available digital technology. Second, the new system design should not introduce much MAI to ensure a higher capacity potential than that of all conventional CDMA systems. Third, the proposed system should have an inherent ability to mitigate MI problems in mobile channels.

The orthogonal complementary code based CDMA architecture based on offset stacking (OS) spreading is one such proposal, which can meet almost all of the above requirements. For notation simplicity, we call this new CDMA system design OS/CC-CDMA system. In the following text, we will demonstrate its capability to achieve high bandwidth efficiency and satisfactory detection efficiency owing to its innovative spreading technique used in both downlink and uplink transmissions. We will also discuss several peculiarities pertaining to the new spreading modulation scheme in its receiver design. To be more specific, we will show that a traditional RAKE receiver is no longer useful in the proposed OS/CC-CDMA architecture, and a new adaptive recursive filter is introduced particularly to detect signal in the multipath environment. The technical limitations associated with the new scheme will also be addressed at the end of this subsection.

7.2.1 ORTHOGONAL COMPLEMENTARY CODES FOR OS SPREADING

The proposed new spreading modulation scheme should use orthogonal complementary codes as its spreading code, the origin of which can be traced back to the 1960s, when Golay [251] and Turyn [252] first studied pairs of binary complementary codes whose auto-correlation function is zero for all even shifts except the zero shift. Later, the concept of primitive complementary codes was extended to the generation of the complete complementary code families [253, 254] whose auto-correlation function is zero for all even and odd shifts except the zero shift and whose cross-correlation function for any pair is zero for all possible shifts. In this book, along with complete complementary codes, we have also introduced many other complementary codes, such as super complementary codes, generalized pair-wise complementary codes (which is not an ideally orthogonal complementary code), column-wise complementary codes, etc. The introduction of orthogonal complementary codes is a significant step forward in the architecture of next generation CDMA technology. As a matter of fact, the application of various complementary codes has become one of the most important parts of the next generation CDMA technology.

There exist several fundamental distinctions between traditional CDMA codes (such as Gold codes, m-sequences, Walsh-Hadamard codes, etc.) and the orthogonal complementary codes considered in our proposed CDMA system.

- First, orthogonality of the complementary codes is based on a 'flock' of element codes jointly, instead of a single code as the traditional CDMA codes. In other words, every user in the proposed OS/CC-CDMA system will be assigned a flock of element codes as its signature code, which ought to be transmitted, possibly via different carriers, and arrive at different correlators at the same time to produce an auto-correlation peak after the summation. Take a set of complete complementary codes of element code length $L = 4$ as an example, as shown in Table 7.1 (which lists two sets of complete complementary codes: one is for $L = 4$ and the other is for $L = 16$). There are in total four element codes (A_0, A_1, B_0, and B_1) in this case and each user should use two element codes (either A_0, A_1 or B_0, B_1) together, being able to support two users, as shown in Figures 7.16 and 7.17, where both the downlink and uplink channels of the proposed OS/CC-CDMA system are illustrated. In this simple example, both flock size and set size (or family size) are identically two. Table 7.2 shows the flock and set sizes for various complete complementary codes with different element code lengths (L).

- Second, the processing gain of the complete complementary codes is equal to the 'congregated length' of a flock of element codes. For the complete complementary codes of lengths $L = 4$ and $L = 16$, their processing gains are equal to $4 \times 2 = 8$ and $16 \times 4 = 64$, respectively.

- Third, zero cross-correlation and zero out-of-phase auto-correlation are ensured for any relative shifts between two codes. Let us consider $A_0 = (+++-)$, $A_1 = (+-++)$ and $B_0 = (++-+)$, $B_1 = (+---)$, being two flocks of complete complementary codes for an OS/CC-CDMA system of two users A and B. Let $A_0 \otimes A_0$ and $A_1 \otimes A_1$ denote the shift-and-add operations to calculate the auto-correlation function for A_0 and A_1, and $B_0 \otimes B_0$ and $B_1 \otimes B_1$ for B_0 and B_1 likewise. Then we have $A_0 \otimes A_0 + A_1 \otimes A_1 = (0, 0, 0, 8, 0, 0, 0)$, and

Table 7.1 Two examples of complete complementary codes with element code lengths $L = 4$ and $L = 16$.

Element code length L=4		Element code length L=16	
Flock 1	$A_0 : + + + -$	Flock 1	$A_0 : + + + + + - + - + + - - + - - +$
			$A_1 : + - + - + + + + + - - + + + - -$
			$A_2 : + + - - + - - + + + + + + - + -$
			$A_3 : + - - + + + - - + - + - + + + +$
	$A_1 : + - + +$	Flock 2	$B_0 : + + + + - + - + + + - - - + + -$
			$B_1 : + - + - - - - - + - - + - - + +$
			$B_2 : + + - - - + + - + + + + - + - +$
			$B_3 : + - - + - - + + + - + - - - - -$
Flock 2	$B_0 : + + - +$	Flock 3	$C_0 : + + + + + - + - - - + + - + + -$
			$C_1 : + - + - + + + + - + + - - - + +$
			$C_2 : + + - - + - - + - - - - - + - +$
			$C_3 : + - - + + + - - - + - + - - - -$
	$B_1 : + - - -$	Flock 4	$D_0 : + + + + - + - + - - + + + - - +$
			$D_1 : + - + - - - - - - + + - + + - -$
			$D_2 : + + - - - + + - - - - - + - + -$
			$D_3 : + - - + - - + + - + - + + + + +$

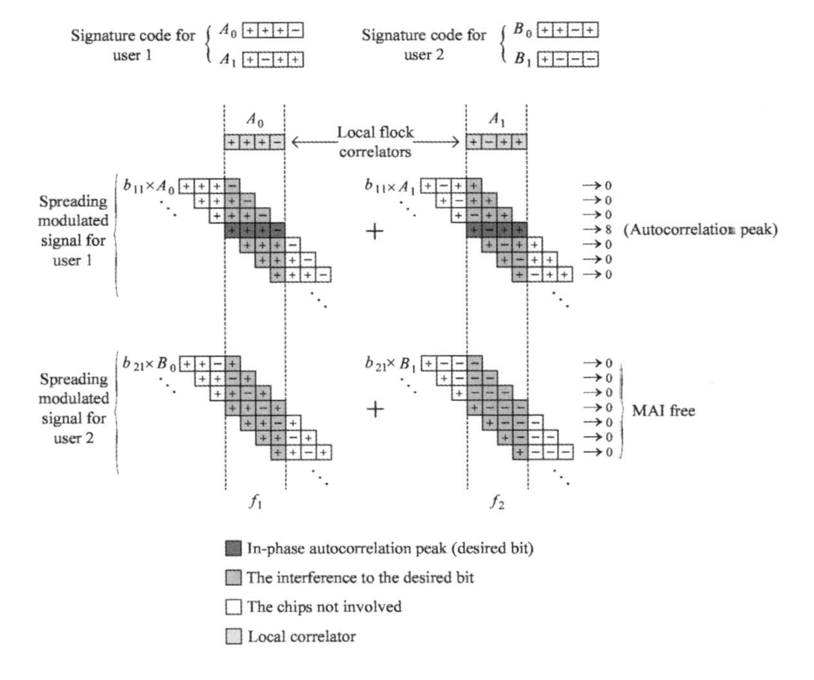

Figure 7.16 Downlink signal reception in a two-user CDMA system in a MAI-AWGN channel using complete complementary codes of length $L = 4$, where User 1 is the intended one.

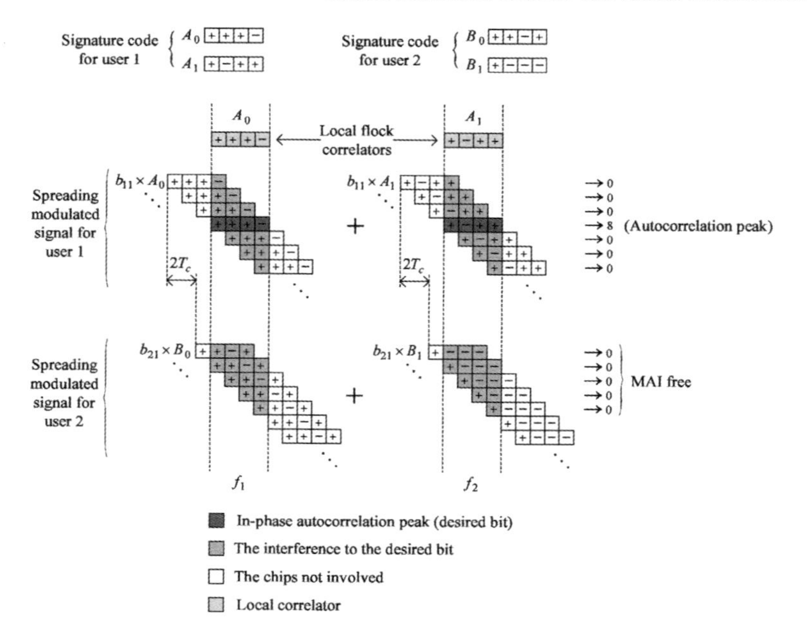

Figure 7.17 Uplink signal reception in a two-user CDMA system in a MAI-AWGN channel using complete complementary codes of length $L = 4$, where User 1 is the intended one.

Table 7.2 Set sizes and flock sizes for complete complementary codes with various element code lengths L.

Element code length ($L = 4^n$)	4	16	64	256	1024	4096
PG ($L\sqrt{L}$)	8	64	512	4096	32768	262144
Family size ($\sqrt{L}$)	2	4	8	16	32	64
Flock size ($\sqrt{L}$)	2	4	8	16	32	64

$B_0 \otimes B_0 + B_1 \otimes B_1 = (0, 0, 0, 8, 0, 0, 0)$. Similarly, we can obtain the cross-correlation function between A and B as $A_0 \otimes B_0 + A_1 \otimes B_1 = (0, 0, 0, 0, 0, 0, 0)$, or $B_0 \otimes A_0 + B_1 \otimes A_1 = (0, 0, 0, 0, 0, 0, 0)$, illustrating the ideal cross-correlation property of the complete complementary codes.

- Fourth, since each user in the proposed OS/CC-CDMA system is assigned a signature code comprising a flock of element codes, those element codes should be sent to a receiver using different carriers. In other words, every signature code is split up into several segments (or element codes) that ought to be transmitted to a receiver via different frequency channels.

For more details on the code generation procedure and other properties of complete complementary codes, readers may refer to Subsection 6.2.2.

7.2.2 OS SPREADING WITH MAI-FREE PROPERTY

The conceptual diagrams for the proposed OS/CC-CDMA system in a multipath-free channel are shown in Figures 7.16 and 7.17, where downlink and uplink spreading modulated signals for a two-user system are illustrated. Each of the two users therein employs two $L = 4$ element codes as its signature code, which is exactly the same as listed in Table 7.1. The information bits $(b_{11}, b_{12}, \ldots)$ and $(b_{21}, b_{22}, \ldots)$, which are assumed to be all +1s for illustration simplicity in the figures, are spreading modulated by element codes that are offset stacked, each being shifted by one chip relative to one another. If compared with the traditional direct sequence (DS) spreading modulation technique used in all conventional CDMA systems, the new system has the following salient features. The most obvious one is that bit streams in the new spreading modulation are no longer aligned in time one bit after another. Instead, a new bit will start right after one chip delay relative to the previous bit, which is spread by an element code of length L. Another important characteristic attribute of the OS/CC-CDMA system is that such an offset stacking spreading modulation method is particularly useful for multi-rate data transmission in multimedia services, whose algorithm is termed as rate-matching in the current 3G mobile communication standards [269–271]. The unique offset stacking spreading method used by the proposed OS/CC-CDMA system can easily slow down the data transmission rate by simply shifting more than one chip (at most L chips) between the neighboring two offset stacked bits. If L chips are shifted between two consecutive bits, the OS spreading reduces to a conventional DS spreading modulation scheme, yielding the lowest data rate. On the other hand, the highest data rate is achieved if only one chip is shifted between the neighboring two offset stacked bits. In doing so, the highest spreading efficiency equal to one can be achieved, implying that every chip is capable of carrying one bit of information. As the bandwidth of a CDMA system is in general determined by the chip width of the spreading codes used, a higher spreading efficiency simply means a higher bandwidth efficiency. Thus, the proposed OS/CC-CDMA architecture is capable of delivering a much higher bandwidth efficiency than a conventional CDMA architecture with the same processing gain.

It should be stressed that the inherent capability of the OS/CC-CDMA system to facilitate multi-rate transmissions is based on its innovative OS spreading technique, which can not be applied to traditional spreading codes. The current 3G W-CDMA architecture has to rely on a complex and sometimes very difficult rate-matching algorithm to adjust the data transmission rate by selecting appropriate orthogonal variable spreading factor (OVSF) codes [269] according to a specific spreading factor and the data rate requirement of the services. On the contrary, the proposed OS/CC-CDMA system is able to change the data transmission rate on the fly, without the need to search for suitable codes with a particular spreading factor. What we should do is just shift more or less chips between two neighboring offset stacked bits to slow down or speed up the data rate. That's it no more complex rate-matching algorithms!

Another important feature of the rate-changing scheme adopted by the OS/CC-CDMA architecture is that the same processing gain will apply to different data transmission rates. However, the rate-matching algorithm in the UMTS W-CDMA standard is processing gain dependent; the slower the transmission rate, the higher the processing gain will be, if transmission bandwidth is kept constant. To maintain an even detection efficiency at a receiver, the transmitter has to adjust the transmitting power for different rate services, which surely complicates both transmitter and receiver hardware.

The OS spreading technique also helps to support asymmetrical transmissions in uplink and downlink channels, pertaining to most Internet services in future all-IP wireless networks. The data rates in a slow uplink and a fast downlink (or vice versa) can be made truly scalable, such that rate-on-demand is achievable by simply adjusting the offset chips between the neighboring two spreading modulated bits.

Figures 7.16 and 7.17 illustrate that the proposed OS/CC-CDMA architecture can offer MAI-free operation in both downlink (synchronous channel) and uplink (asynchronous channel) transmissions because of use of the complete complementary codes. It is assumed that the relative delay between the two users in Figure 7.17 takes multiples of chips. If this assumption does not hold, it can be shown as well that the resultant MAI level is still far less than that of a conventional CDMA system. It should also be pointed out that the rate change through adjusting the number of offset chips between the neighboring two stacked bits does not affect the MAI-free operation of the proposed OS/CC-CDMA system.

The MAI-independent property of the proposed OS/CC-CDMA architecture is significant in terms of its potential to enhance its system capacity in multipath channels. It is well known that a CDMA system is an interference-limited system, whose capacity is dependent on the average co-channel interference contributed from all transmissions using different codes in the same band. The co-channel interference in a conventional CDMA system is caused in principle by non-ideal cross-correlation and out-of-phase auto-correlation functions of the codes considered. In such a system, it is impossible to eliminate the co-channel interference, especially in the uplink channel, where bit streams from different mobiles are asynchronous such that the orthogonality among the codes virtually does not exist. On the contrary, the proposed OS/CC-CDMA system based on orthogonal complementary codes is unique due to the fact that excellent orthogonality among transmitted codes is preserved even in the asynchronous uplink channel, making a truly MAI-independent operation possible for both uplink and downlink channel transmissions. The satisfactory performance in a multipath environment, as shown later in Figures 7.25 and 7.26, is also partly attributable to this property.

It should also be noted that the two element codes for each of the user signature codes concerned in Figures 7.16 and 7.17 have to be sent separately through different carriers, f_1 and f_2. Therefore, the proposed OS/CC-CDMA architecture is basically a multi-carrier CDMA system, bearing many characteristics of a multi-carrier CDMA system. It is also possible to use orthogonal carriers, spaced by $1/T_c$ (where T_c denotes the chip width), to send all those element codes for the same user separately to further enhance the bandwidth efficiency of the system. Therefore, the OS/CC-CDMA system can naturally be implemented by an orthogonal frequency division multiplex (OFDM) structure, saving a lot of cost to implement a complex multi-carrier transceiver.

The bit error rate (BER) of the proposed OS/CC-CDMA system under MAI and AWGN is evaluated using computer simulations. The obtained BER performance of the OS/CC-CDMA system is compared with that of conventional CDMA systems using Gold codes and m-sequences under an identical operation environment. For each of the systems considered here, a matched-filter (single-correlator) is used at a receiver. Both downlink and uplink are simulated considering various numbers of users and processing gains. Figures 7.18 and 7.19 typify the results we have obtained. The former shows the performance for the BER in the downlink (synchronous) channel with a processing gain of 64 for the complete complementary codes, being comparable to that for a Gold code and m-sequence of length 63. The latter gives the BER in the uplink (asynchronous) channel with inter-user delay being equal to three chips. Both figures show the BER only for the First user as the intended one and similar results can be obtained for the BER of the other users. It is observed from Figures 7.18 and 7.19 that gain of at least 3 dB is obtainable from the proposed OS/CC-CDMA system compared to conventional systems using traditional CDMA codes. One of the most interesting observations for the OS/CC-CDMA system is its almost identical BER performance (shown in Figure 7.18), regardless of the number of users, where two curves representing different numbers of users (one and four users) in the system virtually overlap with each other, exemplifying the MAI-independent operation of the OS/CC-CDMA system. On the other hand, the BER for a DS-CDMA using traditional codes is MAI-dependent. The more active users are present, the worse it performs, as shown in Figure 7.18.

7.2.3 OS-SPREADING SIGNAL RECEPTION IN MULTIPATH CHANNELS

Next, let us look at the performance of the proposed OS/CC-CDMA system under multipath channels.

It is well known that a conventional CDMA receiver usually uses a RAKE to collect dispersed energy among different reflection paths to achieve multipath diversity at the receiver. Therefore, the

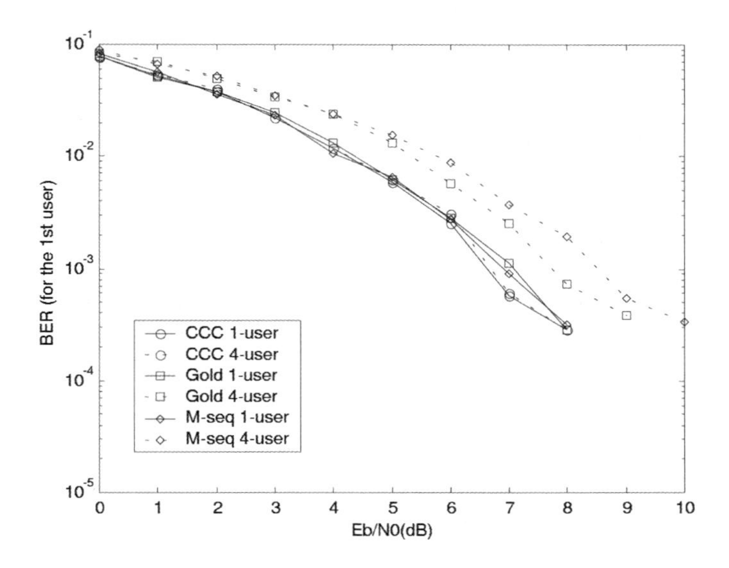

Figure 7.18 Downlink BER comparison for CC code based CDMA and conventional CDMA systems in MAI-AWGN channel using matched filter receiver. The lengths of Gold-code/m-sequence and CC code are 63 and 4×16, respectively.

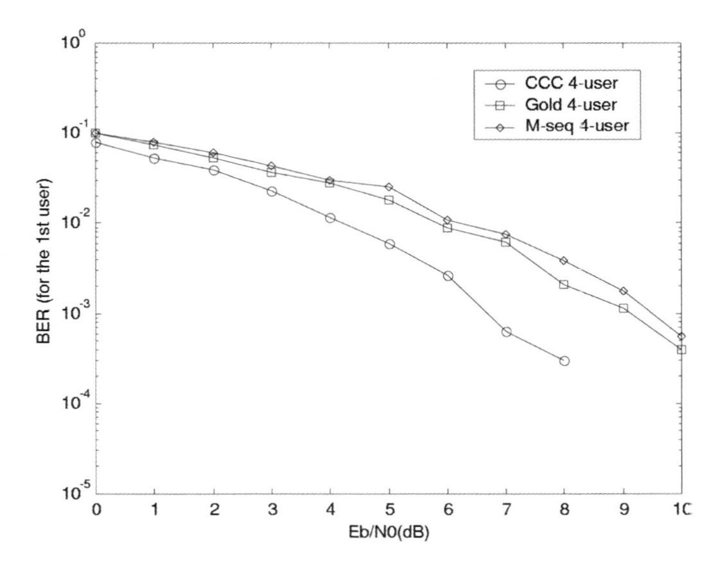

Figure 7.19 Uplink BER comparison for CC code based CDMA and conventional CDMA systems in MAI-AWGN channel using matched filter receiver. The lengths of Gold-code/m-sequence and CC code are 63 and 4×16, respectively.

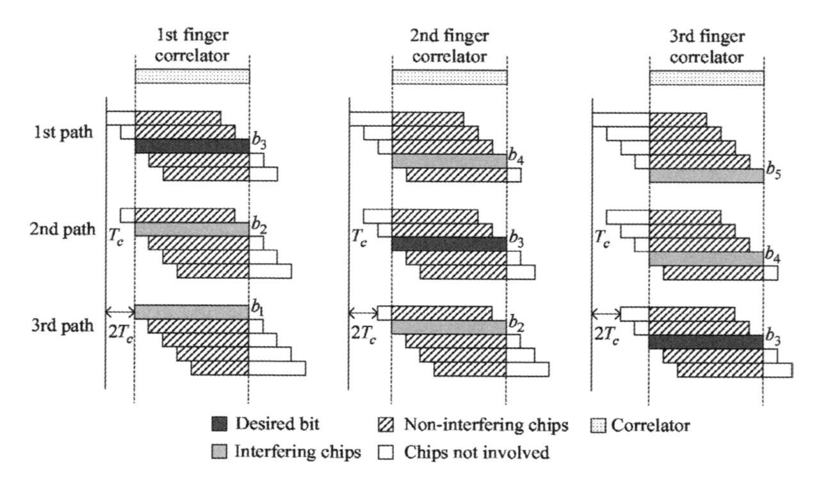

Figure 7.20 A RAKE receiver fails to work satisfactorily in the complete complementary code based CDMA system, where the impulse response of a multipath channel is [1,1,1] with inter-path delay being 1 chip.

RAKE receiver is always a must to all conventional CDMA systems, including currently operational 2G and 3G systems. However, in the OS/CC-CDMA architecture presented in this subsection, the RAKE receiver becomes obsolete due to the nature of the OS spreading modulation technique employed in the system. To illustrate how the proposed OS/CC-CDMA system makes the RAKE receiver fail to work, let us refer to Figure 7.20, where a simple multipath channel consisting of three equally strong reflection rays is considered with its inter-path delay being one chip. The RAKE receiver has three fingers to capture three paths and combine them coherently. The three columns in Figure 7.20 show the output signals from the three fingers and the shaded parts are the chips involved in the RAKE combining algorithm. Due to the use of the offset stacking spreading technique in the proposed OS/CC-CDMA system, there are in total five bits (b_1, b_2, b_3, b_4, and b_5) relevant to the RAKE combining procedure, where it is assumed that $b_3 = +1$ is the desirable bit. Therefore, b_1, b_2, b_4, and b_5 are all interfering terms, whose three possible error-causing patterns are (b_1, $2b_2$, $2b_4$, b_5) = (1,−2,−2,−1), (−1,−2,−2,1), and (−1,−2,−2,−1), respectively. It is noted that among in total 16 possible combinations of binary bits, b_1, b_2, b_4, and b_5, only three of them will cause errors. Therefore, the error probability simply turns out to be $3/16 = 0.1875$ (if each path has the same strength).

It is seen from this simple example that the use of a RAKE receiver in the proposed OS/CC-CDMA system still causes a constant BER = 0.1875 (with three identically strong paths), which is obviously not acceptable. Therefore, an adaptive recursive multipath signal reception filter is designed particularly for the OS/CC-CDMA system based on the orthogonal complementary codes, as shown in Figure 7.21, where the receiver consists of two key modules; the lower part is to estimate the channel impulse response and the upper part is to coherently combine signals in different paths to yield a boosted-up decision variable before the decision device. For this adaptive recursive filter to work, a dedicated pilot signal should be added to the proposed OS/CC-CDMA system, which should be spreading-coded by a signature code different from those used for data channels in the downlink transmission, and should be time-interleaved with user data frames in the uplink channel transmission, as shown in Figure 7.22. The rationale behind the difference in the pilot signals for downlink and uplink channels is explained in the sequel. The downlink transmission is a synchronous channel from

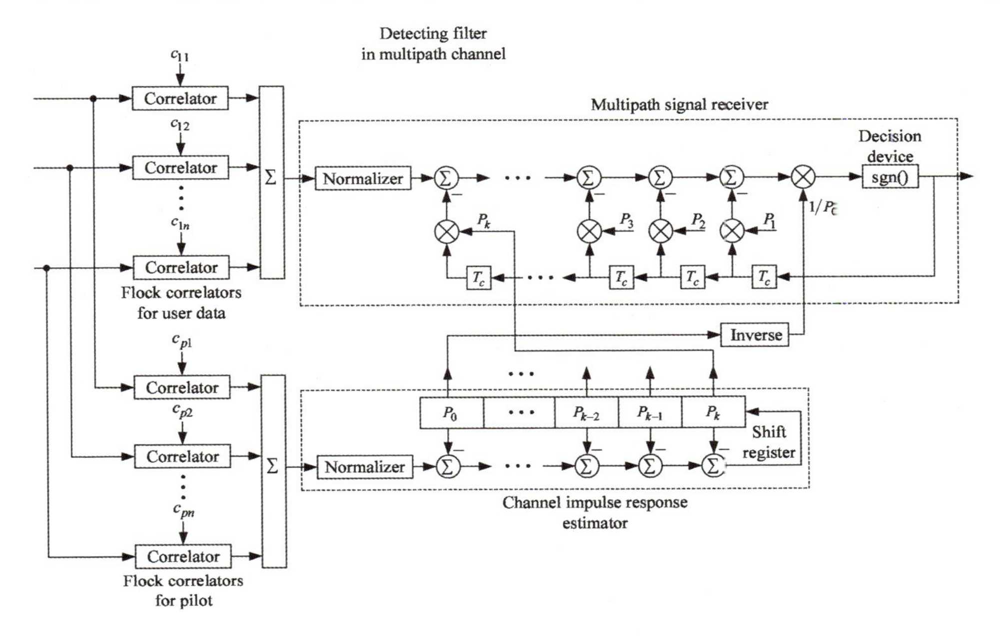

Figure 7.21 A recursive multipath signal reception filter for the complete complementary code based CDMA system, where the upper portion is channel impulse response estimator and the lower portion is signal detection filter.

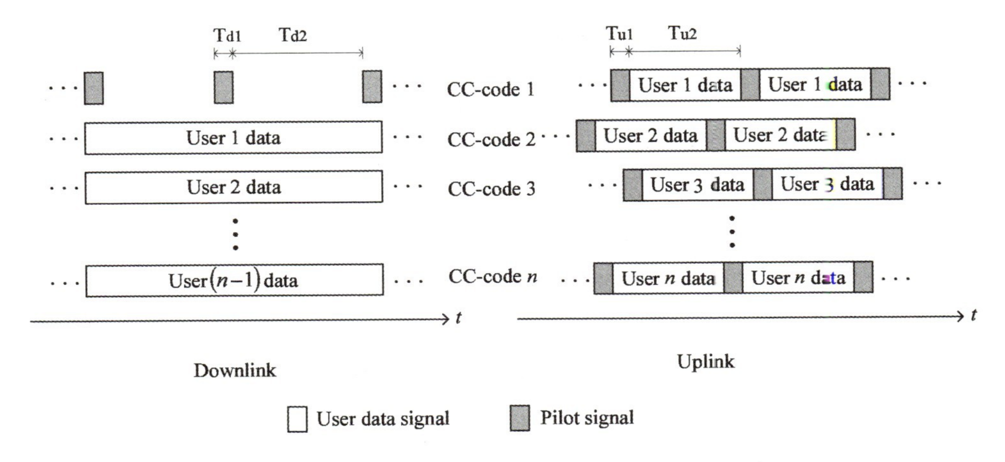

Figure 7.22 Downlink and uplink channel signaling design for the complete complementary code based CDMA system, where the downlink uses a dedicated pilot channel and the pilot signal in the uplink is time-interleaved with data traffic.

the same source (a base station) and thus one dedicated pilot signaling channel is justified, considering that a relatively strong pilot is helpful for all mobiles to lock onto for control information. On the other hand, the uplink transmissions are asynchronous from different mobiles. Therefore, it will consume a lot more signature codes if every mobile is assigned two codes, one for data traffic and the other for pilot signaling. Thus, the pilot signaling has to be time-interleaved with user data traffic in the uplink channels. The time-interleaved pilot signals in the uplink channels can also assist base stations to perform adaptive beamforming, required by a smart antenna system. The signal reception in both downlink and uplink channels can use the same recursive filter (as shown in Figure 7.21) for channel impulse response estimation as long as the receiver achieves frame synchronization with the incoming signal. In fact, the pilot signals in both downlink and uplink channels consist of a series of short pulses, whose durations (T_{d1} and T_{u1}) should be made longer than the delay spread of the channel and whose repetition periods (T_{d2} and T_{u2}) should be made shorter than the coherent time of the channel to adaptively follow the variation of the mobile channel.

The detailed procedure for the recursive multipath signal reception filter to estimate the channel impulse response and to detect signal is illustrated step by step in Figures 7.23 and 7.24, where it has been assumed that a three-ray multipath channel is considered with its mean path strengths being 3, 2, and 1, respectively. It is also assumed that exactly the same orthogonal complementary codes as considered in Figures 7.16 and 7.17 are used, with one signature code ($c_{p1} = A_0$ and $c_{p2} = A_1$)

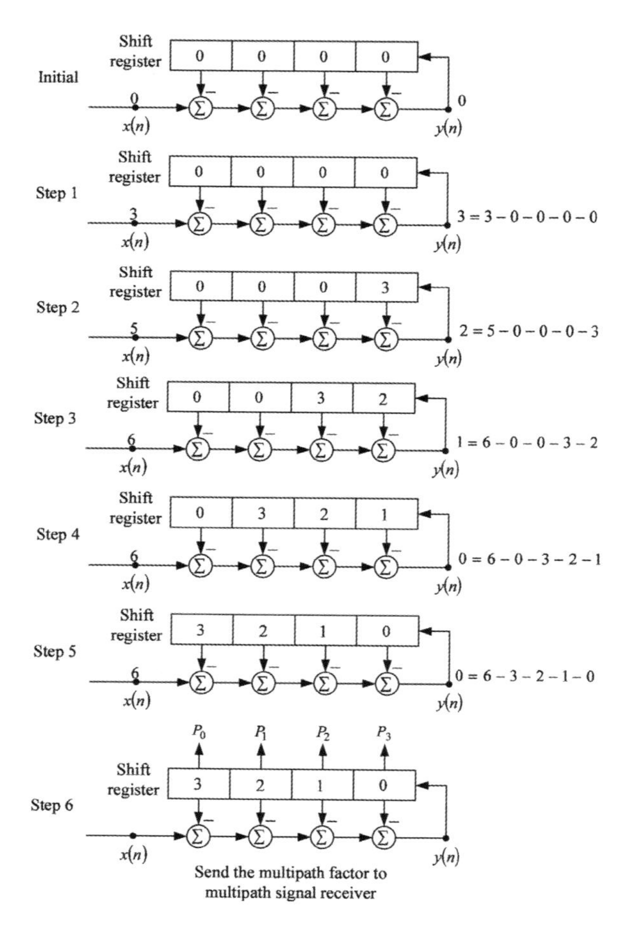

Figure 7.23 Step-by-step illustration for channel impulse response estimation using recursive multipath signal reception filter.

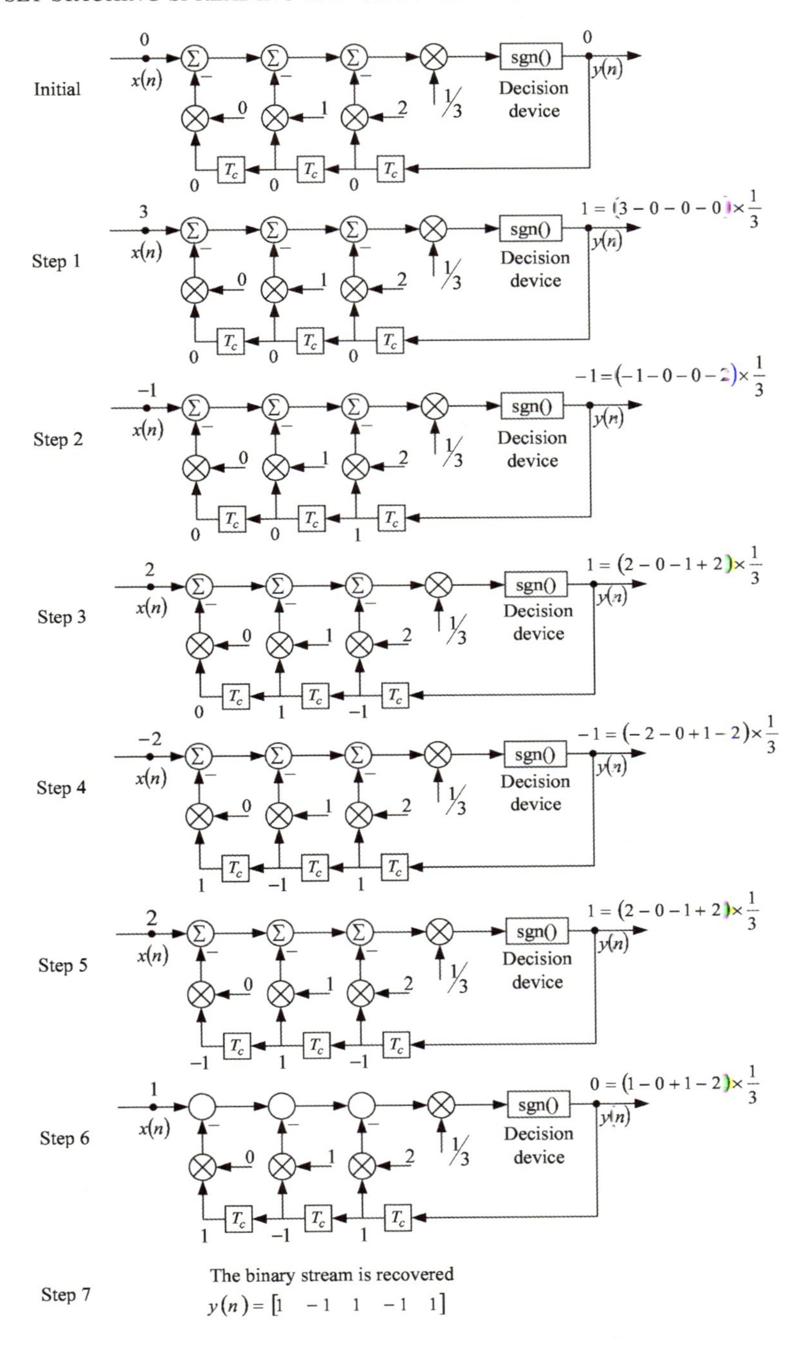

Figure 7.24 Signal detection procedure of the recursive multipath signal reception filter based on channel impulse response estimates with recovered bit stream being $y(n) = (1 - 1 - 11)$.

being used for the pilot channel and the other ($c_{11} = B_0$ and $c_{12} = B_1$) for the user data channel, if only the downlink transmission is considered in this example. In this illustration we presume that each pilot pulse consists of continuously five 1s, which is longer than the channel delay spread (three chips) in this case. The input sequence (3, 5, 6, 6, 6) to the left side of the multipath channel estimator in Figure 7.23 is the received pilot signal after being convoluted with the multipath channel impulse response and local flock correlators. It is seen from Figure 7.23 that the channel impulse response can be estimated accurately and saved in the output register at the end of the algorithm. The obtained channel estimates will then be passed on to the multipath signal receiver in the upper portion of Figure 7.21 to detect the signal contaminated with multipath interference, whose procedure is shown in Figure 7.24, where it is assumed that the originally transmitted binary bit stream is (1,−1,1,−1,1). The input data to the multipath signal receiver, (3, −1, 2, −2, 2, 1), is the received signal of the transmitted bit stream after going through the multipath channel and local flock correlators.

The proposed recursive multipath signal reception filter possesses several advantages. First, it has a very agile structure, the core of which is made up of two transversal filters, one for channel impulse response estimation and the other for data detection. Second, working jointly with the pilot signaling, it performs very well in terms of the accuracy in channel impulse response estimation (as shown in Figure 7.23) and the obtained BER results follow. The multipath channel equalization and signal coherent-combining are actually implemented jointly in the proposed scheme under a relatively simple hardware structure. Third, it operates adaptively to the channel characteristics variation without prior knowledge of the channel state information, such as inter-path delay and relative strength of different paths, etc. On the contrary, a RAKE receiver in a conventional CDMA system requires the path gain coefficients for maximal ratio combining, which themselves are usually unknown and thus have to be estimated by resorting to other complex algorithms.

The performance of the proposed OS/CC-CDMA architecture with the recursive filter for multipath signal reception is shown in Figures 7.25 and 7.26, where two typical scenarios are considered:

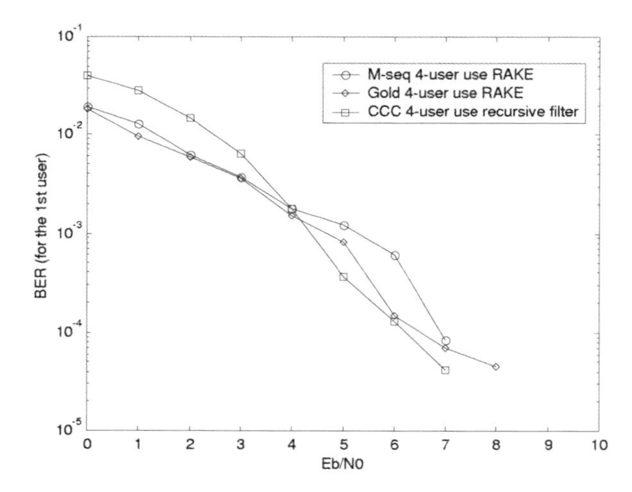

Figure 7.25 Downlink (synchronous) BER for complete complementary code based CDMA and conventional CDMA systems in multipath channel, with normalized multipath power; inter-path delay = 3 chips; multipath channel delay profile = [1.35, 1.08, 0.13]; PG = 63/64; Gold code/m-sequence with MRC-RAKE; complete complementary code based CDMA with the recursive filter.

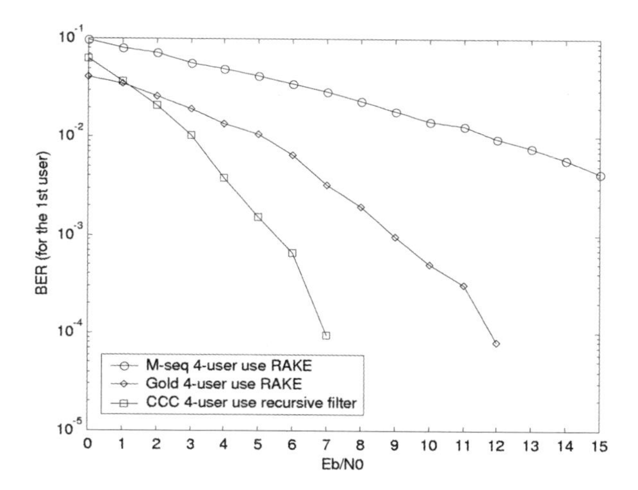

Figure 7.26 Uplink (asynchronous) BER for complete complementary code based and CDMA conventional CDMA systems in multipath channel, with normalized multipath power; inter-path delay=3 chips; inter-user delay=2 chips; multipath channel delay profile=[1.35,1.08,0.13]; PG=63/64; Gold code/m-sequence with MRC-RAKE; complete complementary code based CDMA with recursive filter.

one for downlink performance and the other for uplink performance, similar to the performance comparison made for the MAI-AWGN channel considered in Figures 7.18 and 7.19. It is observed from the figures that, in terms of the BER in the synchronous downlink channel, the Gold code outperforms the m-sequence by a big margin, whereas in the asynchronous uplink channel the Gold code and m-sequence perform almost identically, due to the fact that the orthogonality among either Gold codes or m-sequences has been destroyed by asynchronous bit streams from different mobiles. Nevertheless, the complete complementary code based OS/CC-CDMA system outperforms conventional CDMA systems using either Gold code or m-sequence by a comfortable margin that can be as large as 4–6 dB, because of its superior MAI-independent property.

7.2.4 DISCUSSIONS

In the previous sections, we have demonstrated that the OS/CC-CDMA architecture based on orthogonal complementary codes and adaptive recursive multipath signal reception filter is feasible and well-performing. The OS spreading scheme offers MAI-free operation for both downlink and uplink transmissions in MAI-AWGN channels. Another interesting property of the OS spreading scheme is its agility in changing data transmission rate, which can be finished on the fly without 'stop and search' for a code with a specific spreading factor, as required in the W-CDMA standards [269]. Therefore, the rate-matching algorithm in the proposed system has been greatly simplified.

Yet another important point that has to be mentioned is the bandwidth efficiency of the proposed OS/CC-CDMA architecture. The spreading efficiency (SE) in unit of bit(s) per chip has been used to measure the bandwidth efficiency of a CDMA system due to the fact that the bandwidth of a CDMA system is determined by the chip width of the spreading codes used. Table 7.3 compares the SEs of three systems, the conventional CDMA and the orthogonal complementary code based OS/CC-CDMA with and without using orthogonal carriers. It is clear that the orthogonal complementary code based

Table 7.3 Spreading efficiency (in bit(s) per chip) comparison for conventional CDMA system and complete complementary code based OS/CC-CDMA systems with and without orthogonal carriers

PG	8	64	512	4096	32 768	262 144
Conventional CDMA	1/8	1/64	1/512	1/4096	1/32 768	1/262 144
CC code based CDMA	1/2	1/4	1/8	1/16	1/32	1/64
CC code CDMA (orthogonal carriers)	≈ 1	$\approx 1/2$	$\approx 1/4$	$\approx 1/8$	$\approx 1/16$	$\approx 1/32$

OS/CC-CDMA systems have a much higher SE figure than the conventional CDMA, especially when the processing gain is relatively high.

However, there exist some technical limitations for the proposed OS/CC-CDMA system, which ought to be properly addressed.

Obviously, an orthogonal complementary code based OS/CC-CDMA system needs a multi-level digital modulation scheme to send its baseband information due to the use of the OS spreading modulation technique, as shown in Figures 7.16 and 7.17. If a long orthogonal complementary code is employed in the proposed OS/CC-CDMA system, the number of different levels generated from a baseband spreading modulator can be a problem. For instance, if the orthogonal complementary code of $L = 4$ is used, as shown in Table 7.1, three possible levels will be generated from the OS spreading: 0 and ± 2. However, if the orthogonal complementary code of $L = 16$ in Table 7.1 is considered, the possible levels generated from the OS spreading modulator become $0, \pm 2, \pm 4, \ldots, \pm 14$, comprising 15 different levels. In general, an orthogonal complementary code based OS-spreading scheme using length-L element codes will yield $L - 1$ different levels. Given the element code length (L) of the orthogonal complementary code, it is necessary to choose a digital modem capable of transmitting L different levels in a symbol duration. An L-QAM digital modem can be a suitable choice for its robustness in detection efficiency.

Another concern with the complete complementary code based OS/CC-CDMA system is that a relatively small number of users can be supported with a family of complete complementary codes. Take the $L = 64$ complete complementary code family as an example. It is seen from Table 7.2 that such a family has only eight flocks of codes, each of which can be assigned to one channel (for either pilot or data). If more users must be supported, some long complete complementary codes have to be used. On the other hand, the maximum length of the complete complementary codes is in fact limited by the maximal number of different baseband signal levels manageable in a digital modem, as mentioned earlier in this section. One possible solution to this problem is to introduce frequency divisions on top of the code divisions in each frequency band to create more transmission channels. In fact, the code design issues will be discussed in much more detail in the next few chapters and the problem with the small set size can be solved by using some other effective ways to generate suitable spreading codes for the next generation CDMA systems. By using other orthogonal complementary codes, such as super complementary codes and column-wise complementary codes, the problem can be solved.

7.2.5 SUMMARY

An OS spreading technique based on complete complementary codes is presented and the OS/CC-CDMA system's performance in both MAI-AWGN and multipath channels has been evaluated using simulation. It is to be noted that although the discussions made in this section are based on a particular type of orthogonal complementary code, i.e. complete complementary code, the same results will apply if any other type of orthogonal complementary codes is considered, such as super complementary codes, column-wise complementary codes, etc.

The proposed system possesses several advantages compared with conventional CDMA systems currently available in 2G and 3G standards. First, the system offers a much higher bandwidth efficiency than those achievable in conventional CDMA systems. The system, under the same processing gain, can convey as much as one bit information in each chip width, giving a spreading efficiency equal to one. Second, it offers MAI-free operation in both synchronous and asynchronous MAI-AWGN channels, which contributes to co-channel interference reduction and capacity increase in a mobile cellular system. This excellent property also helps to improve the system performance in multipath channels, as shown by the obtained results. Third, the proposed system is inherently capable of delivering multi-rate/multimedia transmissions due to the use of its OS spreading modulation technique. The rate-matching in the OS spreading modulation becomes very easy, just shifting more or less chips between two consecutive bits to slow down or speed up the data rate, and complex rate-matching algorithms are no longer required.

We have also proposed a novel recursive filter in particular for multipath signal reception of the OS/CC-CDMA system. The recursive filter consists of two modules working jointly, one performing channel impulse response estimation and the other detecting signal contaminated by multipath interferences. The recursive filter has a relatively simple hardware structure compared with a RAKE receiver in conventional CDMA systems, and it performs very well in multipath channels. We have also addressed some technical limitations of the OS/CC-CDMA architecture, such as a relatively small set size of the complete complementary codes and the need for complex multi-level digital modems, etc. Nevertheless, the proposed OS spreading modulation scheme based on the complete complementary codes creates a new option to implement the next generation CDMA technology.

8

Integration of Space-Time Coding with CC-CDMA Technologies

In the previous chapters, we have introduced and discussed orthogonal complementary codes and the basic CDMA architecture based on them. We have also discussed two major orthogonal complementary code based CDMA systems, one being DS/CC-CDMA and the other OS/CC-CDMA.

Based on the discussions given in the earlier chapters, we will discuss in this chapter the issues on space-time (S-T) coding schemes working jointly with complementary code based CDMA systems, namely space-time complementary coding (STCC) CDMA scheme, which takes advantage of both the interference-free property of a DS/CC-CDMA or OS/CC-CDMA system and the full diversity gain of an S-T coding scheme. Each user in a DS/CC-CDMA or OS/CC-CDMA system is assigned a flock of element codes, which should act jointly to ensure an ideal auto-correlation function of any individual flock and an ideal cross-correlation function of any pair of flocks, ensuring desirable multiple access interference (MAI) free operation. In this chapter we will carry out a comprehensive performance analysis of an STCC DS/CC-CDMA system working in an environment where downlink transmission, MAI, Rayleigh fading, and noise are present jointly. The analysis begins with a simple two transmitter antennas scenario and then it will be generalized to the cases with an arbitrary number of transmitter antennas. The performance of an STCC DS/CC-CDMA will also be compared with that of a traditional space-time block coding (STBC) CDMA based on Gold codes using both analysis and simulation, showing its operational advantages in terms of BER performance and capacity.

Directly following the discussions on the STCC DS/CC-CDMA system, we also introduce another joint application of S-T coding and CC-CDMA system, which is called STCC OS/CC-CDMA. With the help of generalized pair-wise complementary codes, a MIMO system based on the proposed STCC OS/CC-CDMA scheme can achieve spatial diversity and parallel transmission at the same time. It will be shown in the discussions that the proposed STCC OS/CC-CDMA-based MIMO system can easily be generalized to achieve an arbitrary combination of diversity and multiplex order. The proposed scheme requires neither bandwidth expansion nor feedback information from the receiver, and it offers a nearly interference-free and multipath-resist performance due mainly to the widely open interference-free window of the generalized pair-wise complementary codes, which are used to implement the STCC OS/CC-CDMA-based MIMO system. The analytical study will be carried out in this chapter to show that it provides a uniquely high diversity-multiplex gain product for its operation in both flat fading and frequency-selective fading channels.

The Next Generation CDMA Technologies Hsiao-Hwa Chen

8.1 MOTIVATIONS

The orthogonal complementary codes can be used as ideal signature codes for a CDMA system, forming DS/CC-CDMA or OS/CC-CDMA, as discussed in the previous chapter. Each of K users in such a DS/CC-CDMA or OS/CC-CDMA system should be assigned a flock of M element codes (each with N chips), which should be sent to a receiver via different channels in either a time-division multiplex (TDM) or frequency-division multiplex (FDM) scheme. After arriving at a particular receiver, each element code should undergo matched filtering separately and their results should be added together to form a decision variable. In this way, an orthogonal complementary code set (with K flocks) can produce a perfect auto-correlation function for every individual flock, which should be assigned to a specific user/channel, and perfect cross-correlation functions for any pair of such flocks. In particular, it should be noted that such ideal correlation properties exist in both synchronous and asynchronous transmission mode. A perfect auto-correlation function means an auto-correlation function with its in-phase value being equal to MN (i.e., its processing gain) and all its side lobes being zero. Perfect cross-correlation functions mean cross-correlation functions with all their in-phase and out-of-phase values being identically zero. Therefore, a CDMA system based on orthogonal complementary codes, either DS/CC-CDMA or OS/CC-CDMA, can offer an attractive isotropic interference-free operation, which can never be made possible in a conventional CDMA system using traditional CDMA codes, such as Gold, Walsh-Hadamard codes, etc.

Recently, many S-T coding schemes [238–244] to enable an open loop transmitter diversity have been reported in the literature. In 1998 Tarokh *et al.* [238] proposed a space-time trellis coding (STTC) scheme, which can maximize both coding gain and diversity gain. The complexity of the scheme, however, increases exponentially with the trellis states, thus prohibiting its feasible hardware realization with currently available digital technologies. Alamouti [239], almost at the same time, suggested a simple but extremely effective STBC method, which can achieve a maximal space diversity gain without the problem with implementation complexity. One year after Alamouti published his STBC scheme, Tarokh *et al.* [240] in 1999 extended Alamouti's STBC algorithm to an even more general form. Later, Hochwald *et al.* [241] successfully applied the STBC scheme to a generic CDMA system to establish a theoretical framework to study an S-T coded CDMA system. Here, we will follow a similar analytical approach first to propose a new S-T coding scheme, namely a space-time complementary coding (STCC) DS/CC-CDMA system. Afterwards, we will also discuss the issues on the implementation of the STCC OS/CC-CDMA system using generalized pair-wise complementary codes.

S-T coding can work jointly with either a DS/CC-CDMA or OS/CC-CDMA system to further enhance its performance due to the available space diversity gain, given that all other inherent merits of a DS/CC-CDMA or OS/CC-CDMA system should not be sacrificed. In this chapter, we will demonstrate that an S-T coding scheme can successfully be integrated into a DS/CC-CDMA or OS/CC-CDMA system, forming an STCC CC-CDMA scheme, in general at an affordable complexity to achieve a full space-diversity gain, in addition to the interference-free property from a DS/CC-CDMA or OS/CC-CDMA system. The focal point of this chapter will be on the analytical treatment of the derivation of bit error probability of an STCC DS/CC-CDMA or STCC OS/CC-CDMA system under MAI and Rayleigh fading. The comparison will also be made between the proposed schemes and traditional STBC-CDMA systems, which use Gold codes as their spreading codes under the same operational conditions.

8.2 STCC DS/CC-CDMA SYSTEM MODEL

Before introducing the STCC DS/CC-CDMA system model, we would like to bring your attention again to the three key parameters for an orthogonal complementary (OC) code set, namely the set size

Table 8.1 The OC code set with $K = 8$, $M = 8$, and $N = 4$ (where k is the flock index and m is the element index).

	m = 1	m = 2	m = 3	m = 4	m = 5	m = 6	m = 7	m = 8
k = 1	+ + + −	+ − + +	+ + − +	+ − − −	+ + + −	+ − + +	− − + −	− + + +
k = 2	+ + + −	− + − −	+ + − +	− + + +	+ + + −	− + − −	− − + −	+ − − −
k = 3	+ − + +	+ + + −	+ − − −	+ + − +	+ − + +	+ + + −	− + + +	− − + −
k = 4	+ − + +	− − − +	+ − − −	− − + −	+ − + +	− − − +	− + + +	+ + − +
k = 5	+ + + −	+ − + +	+ + − +	+ − − −	− − − +	− + − −	+ + − +	+ − − −
k = 6	+ + + −	− + − −	+ + − +	− + + +	− − − +	+ − + +	+ + − +	− + + +
k = 7	+ − + +	+ + + −	+ − − −	+ + − +	− + − −	− − − +	+ − − −	+ + − +
k = 8	+ − + +	− − − +	+ − − −	− − + −	− + − −	+ + + −	+ − − −	− − + −

Table 8.2 The OC code set with $K = 4$, $M = 4$, and $N = 8$ (where k is the flock index and m is the element index).

	m = 1	m = 2	m = 3	m = 4
k = 1	+ + + − + + − +	+ + + − − − + −	+ + + − + + − +	− − − + + + − +
k = 2	+ − + + + − − −	+ − + + − + + +	+ − + + + − − −	− + − − + − − −
k = 3	+ + + − + + − +	+ + + − − − + −	− − − + − − + −	+ + + − − − + −
k = 4	+ − + + + − − −	+ − + + − + + +	− + − − − + + +	+ − + + − + + +

K, the element code length N, and the flock size M. Each user in a DS/CC-CDMA system should be assigned a flock of M element codes, which should be sent via different carrier frequencies. A receiver in a DS/CC-CDMA should use a multi-carrier demodulator to recover M different element codes and use M correlators for matched filtering before summing to form a decision variable. Therefore, the processing gain of a DS/CC-CDMA system is equal to $N \times M$. Tables 8.1 and 8.2 show two example sets of OC codes, whose parameters are $K = M = 8$, $N = 4$ and $K = M = 4$, $N = 8$, respectively. It is noted that the same OC codes may also be given in some other places in this book, but we still list them here for easy reference.

Figure 8.1 shows a generic STCC DS/CC-CDMA system model with K users, each of which is assigned a flock of M element codes, $\{\mathbf{c}_{k,1}, \mathbf{c}_{k,2}, \ldots, \mathbf{c}_{k,M}\}$ for $1 \leq k \leq K$. Each element code, $\mathbf{c}_{k,m}$ where $1 \leq k \leq K$ and $1 \leq m \leq M$, consists of N chips with their chip waveforms being square pulses. In our initial analysis carried out in the text bellow, each transmitter will be assumed to use only two antennas to achieve transmitter diversity. It is to be noted that the underlined orthogonal properties of the OC code sets used in the proposed STCC DS/CC-CDMA system contribute to unique MAI-free operation otherwise impossible in any traditional S-T coded CDMA systems.

The channel coefficients for different transmitters are assumed to be $h_{k,o}$ and $h_{k,e}$, where $1 \leq k \leq K$, for two antennas, respectively. However, in Section 8.5 the analytical results will be generalized to the case with n_t transmitter antennas. In the discussions made in this chapter, we assume that signals from different antennas in a transmitter experience independent Rayleigh fading and additive white Gaussian noise (AWGN). We do not consider the multipath effect here to make the analysis tractable (due to space constraints). A synchronous downlink channel is considered here, which is relevant to most application scenarios with S-T coded systems, where the transmitter diversity is always used in a base station, rather than mobiles.

Figure 8.2 illustrates the architecture of the kth transmitter, where the original data stream will go through STBC encoding, DS spreading modulation, and BPSK carrier modulation before being sent out into the channel via two antennas. It is noted that each transmitter needs M carriers to send

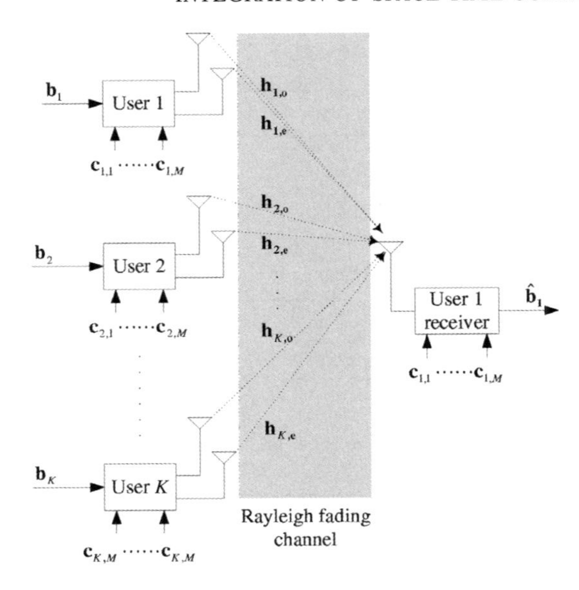

Figure 8.1 A generic STCC DS/CC-CDMA system model with K users, where each user is assigned a flock of M element codes, each element code has N chips, two transmitter antennas (to be extended to n_t transmitter antennas later in this chapter) are used in each user, a Rayleigh fading channel is considered, and the signal of interest to the receiver is that of the first user.

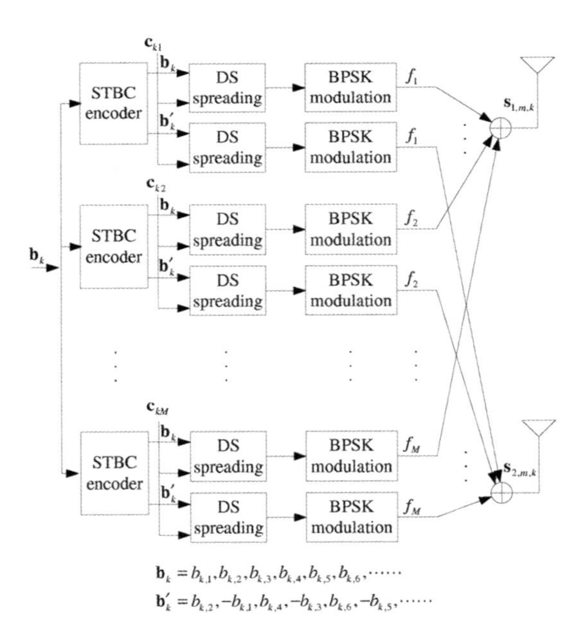

Figure 8.2 Conceptual block diagram for the kth transmitter in an STCC DS/CC-CDMA system, where data stream $\mathbf{b}_k$ is encoded into two sub-streams, $\mathbf{b}_k$ and $\mathbf{b}'_k$, and then DS spreading modulated by a flock of M element codes $\mathbf{c}_{k,1}, \mathbf{c}_{k,2}, \ldots, \mathbf{c}_{k,M}$, followed by BPSK carrier modulation with M different carriers.

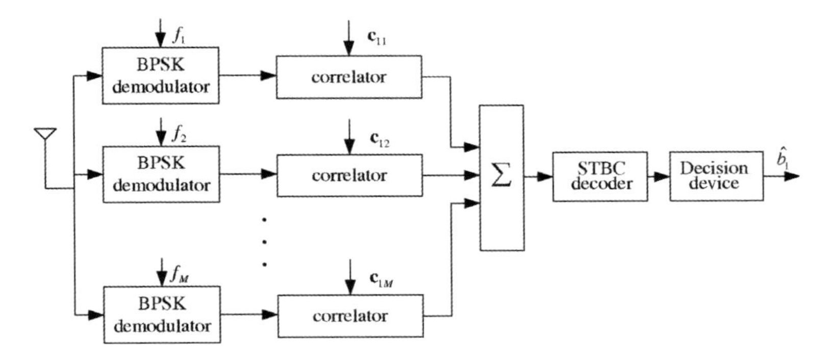

Figure 8.3 Block diagram for an STCC DS/CC-CDMA receiver whose signal of interest is the first user, where M BPSK demodulators are tuned to different carriers, $f_1, f_2, \ldots, f_M$. M element codes should undergo individual matched filtering and then be summed over to form a variable prior to S-T decoding.

M different element codes, which will go through individual matched filtering and then be summed over to reconstruct the orthogonality of the complementary codes at a receiver. Therefore, the DS/CC-CDMA system basically is a multi-carrier CDMA. However, the multiple carriers used herein will not provide any frequency diversity to signal reception at a receiver, as different carriers in fact deliver different signatures, all of which are of equal importance in construction of ideal correlation functions at a receiver, as shown in Figure 8.3.

In this chapter we will focus on an S-T decoder in a receiver intended for the kth user ($k = 1$ is assumed in the analysis followed) in downlink channels, meaning that we need only consider one complete data block consisting of two consecutive symbols, i.e. $b_{k,o}$ and $b_{k,e}$, without losing generality. The subscripts o and e here stand for odd and even symbols, respectively. To facilitate the analysis, two *extended element codes* for the mth element code of user k or $\mathbf{c}_{k,m}$ are formulated as follows:

$$\mathbf{c}_{o,k,m} \triangleq (\mathbf{c}_{k,m}, 0, 0, \ldots, 0) \tag{8.1}$$

$$\mathbf{c}_{e,k,m} \triangleq (0, 0, \ldots, 0, \mathbf{c}_{k,m}) \tag{8.2}$$

each of which covers $2N$ chips. It has to be noted that $\mathbf{c}_{k,m}$ is the mth element code of the kth flock complementary code and its chip level takes either $+1$ or -1.

Two operators, $\otimes$ and $\oplus$, are introduced to perform element-wise product (EWP) and half-length addition (HLA) operations, respectively, with respect to two vectors, $\mathbf{x}$ and $\mathbf{y}$:

$$\begin{cases} \mathbf{x} = (\alpha_1, \alpha_2, \ldots, \alpha_N, \beta_1, \beta_2, \ldots, \beta_N) \\ \mathbf{y} = (\gamma_1, \gamma_2, \ldots, \gamma_N, \delta_1, \delta_2, \ldots, \delta_N) \end{cases} \tag{8.3}$$

as

$$\begin{cases} \mathbf{x} \otimes \mathbf{y} = (\alpha_1\gamma_1, \alpha_2\gamma_2, \ldots, \alpha_N\gamma_N, \beta_1\delta_1, \beta_2\delta_2, \ldots, \beta_N\delta_N) \\ \mathbf{x} \oplus \mathbf{x} = (\alpha_1 + \alpha_2 + \cdots + \alpha_N, \beta_1 + \beta_2 + \cdots + \beta_N) \end{cases} \tag{8.4}$$

where it is noted that the result from the HLA operation is a two-element vector and both sides of the operator $\oplus$ must be the same; otherwise the result should be treated as a null vector, or

$$\mathbf{x} \oplus \mathbf{y} \triangleq (0, 0), \qquad \mathbf{x} \neq \mathbf{y}. \tag{8.5}$$

Thus, the EWP operation never changes the dimension of the vectors involved; while the HLA operation reduces the dimension of the original vectors into two. If each block consists of two symbols, as required by the Alamouti STBC encoder [239], the operation of $(\mathbf{x} \otimes \mathbf{y}) \oplus (\mathbf{x} \otimes \mathbf{y})$ between any two extended element codes $\mathbf{x}$ and $\mathbf{y}$ invokes a block-wise correlation (BWC) operation, which yields a block-wise auto-correlation function if $\mathbf{x} = \mathbf{y}$ and a block-wise cross-correlation function if $\mathbf{x} \neq \mathbf{y}$. Therefore, we have

$$\begin{aligned} f_{bwc}(\mathbf{x}, \mathbf{x}) &= (\mathbf{x} \otimes \mathbf{x}) \oplus (\mathbf{x} \otimes \mathbf{x}) \\ &= \left(\alpha_1^2 + \alpha_2^2 + \cdots + \alpha_N^2, \beta_1^2 + \beta_2^2 + \cdots + \beta_N^2\right) \end{aligned} \tag{8.6}$$

$$\begin{aligned} f_{bwc}(\mathbf{x}, \mathbf{y}) &= (\mathbf{x} \otimes \mathbf{y}) \oplus (\mathbf{x} \otimes \mathbf{y}) \\ &= (\alpha_1\gamma_1 + \alpha_2\gamma_2 + \cdots + \alpha_N\gamma_N, \beta_1\delta_1 + \beta_2\delta_2 + \cdots + \beta_N\delta_N) \end{aligned} \tag{8.7}$$

It can be shown that both the EWP and HLA operators are linear, and so are $f_{bwc}(\mathbf{x}, \mathbf{x})$ and $f_{bwc}(\mathbf{x}, \mathbf{y})$. In particular, we have the following property for the HLA operator:

$$\begin{aligned} (\mathbf{x} + \mathbf{y}) \oplus (\mathbf{x} + \mathbf{y}) &= \mathbf{x} \oplus \mathbf{x} + \mathbf{y} \oplus \mathbf{y} + \mathbf{x} \oplus \mathbf{y} + \mathbf{y} \oplus \mathbf{x} \\ &= (\mathbf{x} \oplus \mathbf{x}) + (\mathbf{y} \oplus \mathbf{y}) \end{aligned} \tag{8.8}$$

where either $\mathbf{x} \oplus \mathbf{y}$ or $\mathbf{y} \oplus \mathbf{x}$ has been defined to be [0, 0]. The above result is reasonable because the following two operations with respect to the extended element codes $\mathbf{x}$ and $\mathbf{y}$ will give exactly the same result, or being equivalent to

1. First add two vectors and then carry out the HLA operation.
2. Then perform the HLA operation for each vector individually, followed by a summation.

8.3 PROPERTIES OF ORTHOGONAL COMPLEMENTARY CODES

As mentioned before, all currently available 2G and 3G CDMA-based systems use unitary signature codes, which always work on a one-code-per-user/channel basis. Those unitary codes can be further classified into two major categories, i.e. quasi-orthogonal codes (such as Gold codes, Kasami codes, m-sequences, etc.), and orthogonal codes (such as Walsh-Hadamard sequences, OVSF codes, etc.). Unfortunately, many technical limitations in traditional CDMA-based 2G and 3G systems stem from the use of these unitary codes, as to be explained below.

Obviously, all these unitary codes were designed based solely on their periodic auto-correlation (PAC) and periodic cross-correlation (PCC) functions. Unfortunately, both PAC and PCC become totally irrelevant when a CDMA system works in an asynchronous transmission mode, let alone in a working environment where inter-symbol interference (ISI) caused by the multipath effect, also called multipath interference (MI), is present. Even for the orthogonal codes, such as Walsh-Hadamard sequences and OVSF codes, multiple access interference (MAI) can be rampant if they work in asynchronous channels, no matter whether multipath propagation exists or not.

It is to be noted that many problem may exist even if a RAKE receiver is applied to an asynchronous CDMA system using traditional spreading codes. First of all, due to their poor PAC and PCC, the output from each finger of a RAKE receiver includes not only useful component (the path signal it is tuned to) but also a lot of unwanted interference generated from the joint effect of asynchronous transmission and multipath propagation. Therefore, the combination of the signals from different fingers will not yield a signal-to-interference ratio sufficiently high to ensure acceptable detection efficiency.

OC codes were proposed as CDMA signature codes due to their following salient features (which have been discussed extensively in Chapter 6):

- The OC codes retain perfect cross-correlation functions for both synchronous and asynchronous transmissions. Here, the word perfect means zero cross-correlation functions for any relative shifts. This property contributes to the desirable MAI-free operation in a CDMA system.
- The OC-code-based CDMA has an inherent immunity against multipath interference due to their perfect auto-correlation functions. This property ensures ISI-free (or MI-free) operation of a CDMA system.
- The OC codes offer perfect partial cross-correlation functions, which make them in particular suitable for applications where traffic is dominated by short bursts, such as in all-IP wireless networks.
- The OC codes work on a flock basis. Every user should use M different carriers to send its M element codes in parallel. Thus, the OFDM technology is a natural solution for implementation of such a multi-carrier CDMA.

Figure 8.4 shows the bit-wise detection process of a two-user DS/CC-CDMA system with the MAI-free property in the downlink channel. Figure 8.5 illustrates the detection process with the MI-free property in the downlink channel. Similarly, we can also show the MAI-free and MI-free operation for such a DS/CC-CDMA system working in asynchronous uplink channels. It is to be noted from Figure 8.5 that the detection of the first bit (or the rightmost bit) involves partial auto-correlation and partial cross-correlation functions, while the detection for the second and third bits concerns complete

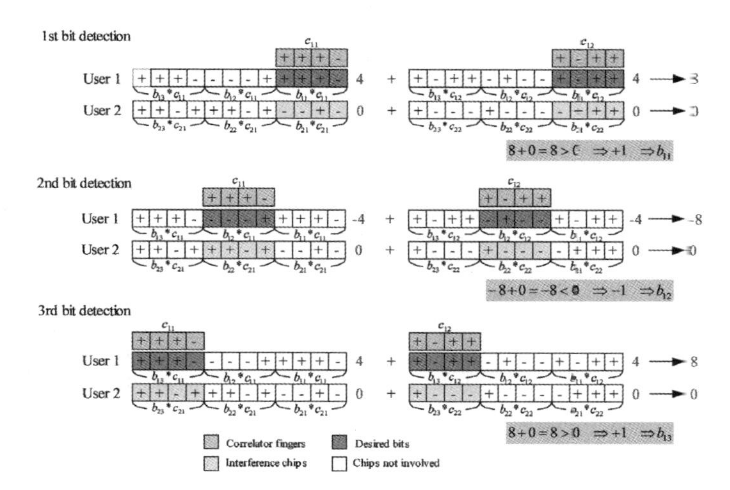

Figure 8.4 Bit-by-bit detection process of a two-user DS/CC-CDMA system with downlink MAI-free operation (based on orthogonal complementary codes and DS spreading modulation), where each user is assigned two 4-chip element codes as its signature code: c_{11} and c_{12} for user 1; c_{21} and c_{22} for user 2. The orthogonal complementary codes of concern are: $c_{11} = + + + -$, $c_{12} = + - + +$, $c_{21} = + + - +$, $c_{22} = + - - -$. Three information bits for users 1 and 2 are $\{b_{11}, b_{12}, b_{13}\} = \{+1, -1, +1\}$ and $\{b_{21}, b_{22}, b_{23}\} = \{+1, +1, -1\}$, respectively. The receiver uses a correlator to detect User 1's signal.

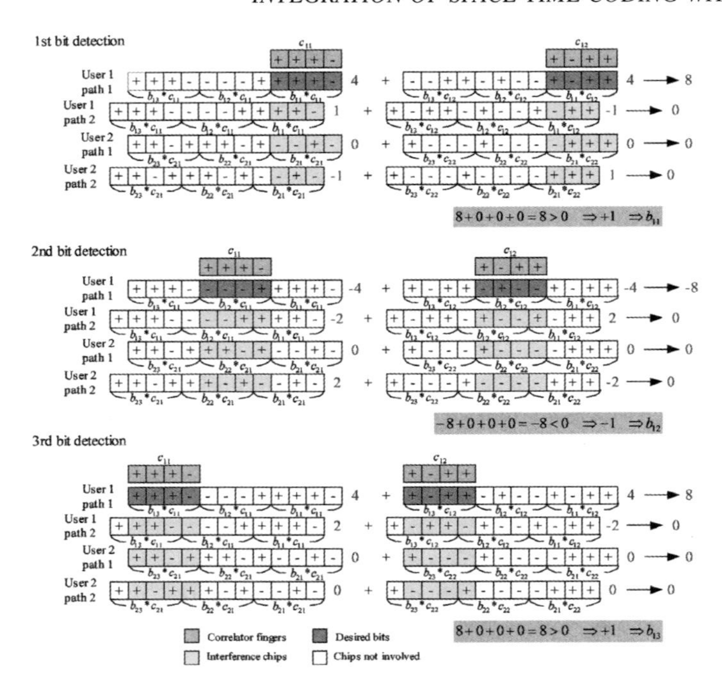

Figure 8.5 Bit-by-bit detection process of a two-user DS/CC-CDMA system with downlink ISI-free operation (based on orthogonal complementary codes and direct sequence spreading modulation), where two multipath returns with identical strength are present with the inter-path delay being one chip. The orthogonal complementary codes of concern are: $c_{11} = + + +-$, $c_{12} = + - ++$, $c_{21} = + + -+$, $c_{22} = + - --$. Each user is assigned two element codes as its signature code: c_{11} and c_{12} for user 1; c_{21} and c_{22} for user 2. Three information bits for users 1 and 2 are $\{b_{11}, b_{12}, b_{13}\} = \{+1, -1, +1\}$ and $\{b_{21}, b_{22}, b_{23}\} = \{+1, +1, -1\}$, respectively. The receiver uses a correlator to detect User 1's signal.

auto-correlation and cross-correlation functions. In both cases, MAI-free and MI-free operation has been demonstrated, as also shown in the previous chapter.

Tables 8.1 and 8.2 give two typical examples of OC code sets, which have the same processing gain ($PG = MN$) with different parameters K, M, and N. The first OC code set is with $K = 8$, $M = 8$, and $N = 4$, while the second has $K = 4$, $M = 4$, and $N = 8$. Using the information given in the tables, we can observe the following orthogonality properties of an OC code set, as discussed in Chapter 6. Take any two flocks of element codes from the tables as examples, for instance

$$\begin{cases} \mathbf{c}_x = (\mathbf{c}_{x,1}(n), \mathbf{c}_{x,2}(n), \ldots, \mathbf{c}_{x,M}(n)) \\ \mathbf{c}_y = (\mathbf{c}_{y,1}(n), \mathbf{c}_{y,2}(n), \ldots, \mathbf{c}_{y,M}(n)) \end{cases} \tag{8.9}$$

where $\mathbf{c}_{x,m}(n)$ and $\mathbf{c}_{y,m}(n)$ are column vectors, and x and y are the flock indexes k in Tables 8.1 and 8.2, and chip index is n, $1 \leq n \leq N$. The in-phase auto-correlation and cross-correlation functions of an OC code set are always perfect (which can be verified by the code sets given in Tables 8.1 and 8.2), such that

$$\sum_{m=1}^{M} \mathbf{c}_{x,m}^T(n)\mathbf{c}_{y,m}(n-i) = 0, \qquad x \neq y \tag{8.10}$$

and

$$\sum_{m=1}^{M} \mathbf{c}_{x,m}^{T}(n)\mathbf{c}_{x,m}(n-i) = \sum_{m=1}^{M} \mathbf{c}_{y,m}^{T}(n)\mathbf{c}_{y,m}(n-i) = \begin{cases} NM, & i = 0 \\ 0, & i \neq 0 \end{cases} \tag{8.11}$$

where $\mathbf{c}_{x,m}^{T}(n)\mathbf{c}_{y,m}(n)$ stands for the inner product between two element codes $\mathbf{c}_{x,m}(n)$ and $\mathbf{c}_{y,m}(n)$. Now let us form two extended element codes as

$$\begin{cases} \mathbf{c}_{o,x,m} = (\mathbf{c}_{x,m}, 0, 0, \ldots, 0) \\ \mathbf{c}_{e,x,m} = (0, 0, \ldots, 0, \mathbf{c}_{x,m}) \end{cases} \tag{8.12}$$

and

$$\begin{cases} \mathbf{c}_{o,y,m} = (\mathbf{c}_{y,m}, 0, 0, \ldots, 0) \\ \mathbf{c}_{e,y,m} = (0, 0, \ldots, 0, \mathbf{c}_{y,m}) \end{cases} \tag{8.13}$$

Thus, we can define

$$\begin{cases} \tilde{\mathbf{c}}_{x,m} = \mathbf{c}_{o,x,m} + \mathbf{c}_{e,x,m} \\ \tilde{\mathbf{c}}_{y,m} = \mathbf{c}_{o,y,m} + \mathbf{c}_{e,y,m} \end{cases} \tag{8.14}$$

Therefore, the block-wise cross-correlation between $\tilde{\mathbf{c}}_{x,m}$ and $\tilde{\mathbf{c}}_{y,m}$ becomes

$$\begin{aligned} f_{bwc}(\tilde{\mathbf{c}}_{x,m}, \tilde{\mathbf{c}}_{y,m}) &= \big((\tilde{\mathbf{c}}_{x,m} \otimes \tilde{\mathbf{c}}_{y,m}\big) \oplus \big(\tilde{\mathbf{c}}_{x,m} \otimes \tilde{\mathbf{c}}_{y,m}\big) \\ &= \big((\mathbf{c}_{o,x,m} + \mathbf{c}_{e,x,m}) \otimes (\mathbf{c}_{o,y,m} + \mathbf{c}_{e,y,m})\big) \\ &\quad \oplus \big((\mathbf{c}_{o,x,m} + \mathbf{c}_{e,x,m}) \otimes (\mathbf{c}_{o,y,m} + \mathbf{c}_{e,y,m})\big) \end{aligned} \tag{8.15}$$

which will give block-wise auto-correlation function if $\tilde{\mathbf{c}}_{x,m} = \tilde{\mathbf{c}}_{y,m}$ and block-wise cross-correlation function otherwise. From the orthogonality property of an OC code set, we have

$$\begin{aligned} \sum_{m=1}^{M} f_{bwc}(\tilde{\mathbf{c}}_{x,m}, \tilde{\mathbf{c}}_{y,m}) &= \sum_{m=1}^{M} \big((\mathbf{c}_{o,x,m} + \mathbf{c}_{e,x,m}) \otimes (\mathbf{c}_{o,y,m} + \mathbf{c}_{e,y,m}\big) \\ &\quad \oplus \big((\mathbf{c}_{o,x,m} + \mathbf{c}_{e,x,m}) \otimes (\mathbf{c}_{o,y,m} + \mathbf{c}_{e,y,m}\big) \\ &= \begin{cases} (NM, NM) & x = y \\ (0, 0) & x \neq y \end{cases} \end{aligned} \tag{8.16}$$

It can be shown that weighted BWC between two different extended element codes $\tilde{\mathbf{c}}_{x,m}$ and $\tilde{\mathbf{c}}_{y,m}$ should also be equal to a two-dimensional null vector or

$$\begin{aligned} &\sum_{m=1}^{M} f_{bwc}(b_{x,o}\mathbf{c}_{o,x,m} + b_{x,e}\mathbf{c}_{e,x,m}, b_{y,o}\mathbf{c}_{o,y,m} + b_{y,e}\mathbf{c}_{e,y,m}) \\ &= \sum_{m=1}^{M} \big\{\big[(b_{x,o}\mathbf{c}_{o,x,m} + b_{x,e}\mathbf{c}_{e,x,m}) \otimes (b_{y,o}\mathbf{c}_{o,y,m} + b_{y,e}\mathbf{c}_{e,y,m})\big] \\ &\quad \oplus \big[(b_{x,o}\mathbf{c}_{o,x,m} + b_{x,e}\mathbf{c}_{e,x,m}) \otimes (b_{y,o}\mathbf{c}_{o,y,m} + b_{y,e}\mathbf{c}_{e,y,m})\big]\big\} \\ &= [0, 0], \qquad x \neq y \end{aligned} \tag{8.17}$$

where $b_{x,o}$, $b_{x,e}$, $b_{y,o}$, and $b_{y,e}$ are binary information data bits, taking either $+1$ or -1 in this chapter.

8.4 DUAL TRANSMITTER ANTENNAS

Based on the Alamouti STBC algorithm [239], an encoded baseband signal block from two transmitter antennas of the kth user in an STCC DS/CC-CDMA system for the mth element code, $\mathbf{s}_{1,k,m}$ and $\mathbf{s}_{2,k,m}$, can be written as

$$\begin{cases} \mathbf{s}_{1,k,m} = (b_{1,o}\mathbf{c}_{o,k,m} + b_{1,e}\mathbf{c}_{e,k,m}) \\ \mathbf{s}_{2,k,m} = (b_{1,e}\mathbf{c}_{o,k,m} - b_{1,o}\mathbf{c}_{e,k,m}) \end{cases} \tag{8.18}$$

where $1 \leq k \leq K$ and $1 \leq m \leq M$. If a perfect demodulation process is assumed, the received signal at a receiver tuned to User 1 in the mth carrier frequency f_m becomes

$$\begin{aligned} \mathbf{r}_{1,m} = \sum_{k=1}^{K} \big[& (b_{k,o}\mathbf{c}_{o,k,m} + b_{k,e}\mathbf{c}_{e,k,m})h_{k,1} \\ & + (b_{k,e}\mathbf{c}_{o,k,m} - b_{k,o}\mathbf{c}_{e,k,m})h_{k,2} + \mathbf{n}_{k,m} \big] \end{aligned} \tag{8.19}$$

where $1 \leq m \leq M$, $h_{k,1}$ and $h_{k,2}$ are independent Rayleigh fading channel coefficients due to two sufficiently spaced antennas at User 1, and $\mathbf{n}_{k,m}$ is the AWGN term with zero mean and variance being $N_o/(2N)$ observed in each chip interval, where N_o is one-sided power spectral density of the noise.

As shown in Figure 8.3, the received signal should first undergo individual matched filtering for different element codes before summation. Taking the first user as the signal of interest or $k = 1$, the received signal $\mathbf{r}_{1,m}$ from different carrier frequencies should undergo matched filtering with respect to different extended element codes or $\mathbf{c}_{o,1,m} + \mathbf{c}_{e,1,m}$, where $1 \leq m \leq M$. For analytical simplicity, we would like to carry out the EWP operation first, followed by the HLA operation, as follows

$$\begin{cases} \mathbf{w}_{1,1} = \mathbf{r}_{1,1} \otimes (\mathbf{c}_{o,1,1} + \mathbf{c}_{e,1,1}) \\ \mathbf{w}_{1,2} = \mathbf{r}_{1,2} \otimes (\mathbf{c}_{o,1,2} + \mathbf{c}_{e,1,2}) \\ \cdots\cdots \\ \mathbf{w}_{1,M} = \mathbf{r}_{1,M} \otimes (\mathbf{c}_{o,1,M} + \mathbf{c}_{e,1,M}) \end{cases} \tag{8.20}$$

or

$$\begin{cases} \mathbf{w}_{1,1} = \sum_{k=1}^{K} \big[(b_{k,o}\mathbf{c}_{o,k,1} + b_{k,e}\mathbf{c}_{e,k,1})h_{k,1} \\ \qquad + (b_{k,e}\mathbf{c}_{o,k,1} - b_{k,o}\mathbf{c}_{e,k,1})h_{k,2} + \mathbf{n}_{k,1} \big] \otimes (\mathbf{c}_{o,1,1} + \mathbf{c}_{e,1,1}) \\ \mathbf{w}_{1,2} = \sum_{k=1}^{K} \big[(b_{k,o}\mathbf{c}_{o,k,2} + b_{k,e}\mathbf{c}_{e,k,2})h_{k,3} \\ \qquad + (b_{k,e}\mathbf{c}_{o,k,2} - b_{k,o}\mathbf{c}_{e,k,2})h_{k,4} + \mathbf{n}_{k,2} \big] \otimes (\mathbf{c}_{o,1,2} + \mathbf{c}_{e,1,2}) \\ \cdots\cdots \\ \mathbf{w}_{1,M} = \sum_{k=1}^{K} \big[(b_{k,o}\mathbf{c}_{o,k,M} + b_{k,e}\mathbf{c}_{e,k,M})h_{k,(2M-1)} \\ \qquad + (b_{k,e}\mathbf{c}_{o,k,M} - b_{k,o}\mathbf{c}_{e,k,M})h_{k,2M} + \mathbf{n}_{k,M} \big] \\ \qquad \otimes (\mathbf{c}_{o,1,M} + \mathbf{c}_{e,1,M}) \end{cases} \tag{8.21}$$

which can be rewritten into

$$\begin{cases} \mathbf{w}_{1,1} = \big[(b_{1,o}\mathbf{c}_{o,1,1} + b_{1,e}\mathbf{c}_{e,1,1})h_{1,1} \\ \qquad + (b_{1,e}\mathbf{c}_{o,1,1} - b_{1,o}\mathbf{c}_{e,1,1})h_{1,2} + \mathbf{I}_{1,1} + \mathbf{n}_{1,1}\big] \\ \qquad \otimes (\mathbf{c}_{o,1,1} + \mathbf{c}_{e,1,1}) \\ \mathbf{w}_{1,2} = \big[(b_{1,o}\mathbf{c}_{o,1,2} + b_{1,e}\mathbf{c}_{e,1,2})h_{1,3} \\ \qquad + (b_{1,e}\mathbf{c}_{o,1,2} - b_{1,o}\mathbf{c}_{e,1,2})h_{1,4} + \mathbf{I}_{1,2} + \mathbf{n}_{1,2}\big] \\ \qquad \otimes (\mathbf{c}_{o,1,2} + \mathbf{c}_{e,1,2}) \\ \cdots\cdots \\ \mathbf{w}_{1,M} = \big[(b_{1,o}\mathbf{c}_{o,1,M} + b_{1,e}\mathbf{c}_{e,1,M})h_{1,(2M-1)} \\ \qquad + (b_{1,e}\mathbf{c}_{o,1,M} - b_{1,o}\mathbf{c}_{e,1,M})h_{1,2M} + \mathbf{I}_{1,M} + \mathbf{n}_{1,M}\big] \\ \qquad \otimes (\mathbf{c}_{o,1,M} + \mathbf{c}_{e,1,M}) \end{cases} \tag{8.22}$$

where $\mathbf{I}_{1,m}$ $(1 \le m \le M)$ is the interference term defined by

$$\mathbf{I}_{1,m} = \sum_{k=2}^{K} \big[(b_{k,o}\mathbf{c}_{o,k,m} + b_{k,e}\mathbf{c}_{e,k,m})h_{k,(2m-1)} \\ + (b_{k,e}\mathbf{c}_{o,k,m} - b_{k,o}\mathbf{c}_{e,k,m})h_{k,2m}\big] \tag{8.23}$$

To proceed with the correlation process, we need to sum over all items given in (8.22) to obtain

$$\begin{aligned} \mathbf{w} &= \sum_{m=1}^{M} \mathbf{w}_{1,m} \\ &= (h_{1,1} + h_{1,3} + \cdots + h_{1,2M-1})\,\{b_{1,o}[1,1,\ldots,1,0,0,\ldots,0] \\ &\quad + b_{1,e}[0,0,\ldots,0,1,1,\ldots,1]\} \\ &\quad + (h_{1,2} + h_{1,4} + \cdots + h_{1,2M})\,\{b_{1,e}[1,1,\ldots,1,0,0,\ldots,0] \\ &\quad - b_{1,o}[0,0,\ldots,0,1,1,\ldots,1]\} \\ &\quad + \sum_{m=1}^{M} (\mathbf{I}_{1,m} + \mathbf{n}_{1,m}) \otimes (\mathbf{c}_{o,1,m} + \mathbf{c}_{e,1,m}) \end{aligned} \tag{8.24}$$

which results in a row vector. To complete the correlation process, we need the HLA operator, which will generate the output from the matched filter as

$$\begin{aligned} (d_{1,1}, d_{1,2}) = \mathbf{w} \oplus \mathbf{w} &= \begin{pmatrix} \xi_1 \\ \xi_2 \end{pmatrix}^T \\ &\quad + \left\{ \sum_{m=1}^{M} (\mathbf{I}_{1,m} + \mathbf{n}_{1,m}) \otimes (\mathbf{c}_{o,1,m} + \mathbf{c}_{e,1,m}) \right\} \\ &\quad \oplus \left\{ \sum_{m=1}^{M} (\mathbf{I}_{1,m} + \mathbf{n}_{1,m}) \otimes (\mathbf{c}_{o,1,m} + \mathbf{c}_{e,1,m}) \right\} \end{aligned} \tag{8.25}$$

where we have

$$\begin{cases} \xi_1 \triangleq (h_{1,1} + h_{1,3} + \cdots + h_{1,2M-1})b_{1,o} + (h_{1,2} + h_{1,4} + \cdots + h_{1,2M})b_{1,e} \\ \xi_2 \triangleq -(h_{1,2} + h_{1,4} + \cdots + h_{1,2M})b_{1,o} + (h_{1,1} + h_{1,3} + \cdots + h_{1,2M-1})b_{1,e} \end{cases} \tag{8.26}$$

In Appendix D we will show that the following equation is valid

$$\left\{\sum_{m=1}^{M} \mathbf{I}_{1,m} \otimes (\mathbf{c}_{o,1,m} + \mathbf{c}_{e,1,m})\right\} \oplus \left\{\sum_{m=1}^{M} \mathbf{I}_{1,m} \otimes (\mathbf{c}_{o,1,m} + \mathbf{c}_{e,1,m})\right\} = (0, 0) \tag{8.27}$$

Let us define

$$[v_{1,1}, v_{1,2}] = \left\{\sum_{m=1}^{M} \mathbf{n}_{1,m} \otimes (\mathbf{c}_{o,1,m} + \mathbf{c}_{e,1,m})\right\} \oplus \left\{\sum_{m=1}^{M} \mathbf{n}_{1,m} \otimes (\mathbf{c}_{o,1,m} + \mathbf{c}_{e,1,m})\right\} \tag{8.28}$$

Equation (8.25) can be rewritten into

$$\begin{cases} d_{1,1} = \xi_1 + v_{1,1} \\ d_{1,2} = \xi_2 + v_{1,2} \end{cases} \tag{8.29}$$

which can be further written into a matrix form as

$$\begin{pmatrix} d_{1,1} \\ d_{1,2} \end{pmatrix} = \begin{pmatrix} h_{1,1,sum} & h_{1,2,sum} \\ -h_{1,2,sum} & h_{1,1,sum} \end{pmatrix} \begin{pmatrix} b_{1,o} \\ b_{1,e} \end{pmatrix} + \begin{pmatrix} v_{1,1} \\ v_{1,2} \end{pmatrix} \tag{8.30}$$

where we have used the following equations

$$\begin{cases} h_{1,1,sum} = h_{1,1} + h_{1,3} + \cdots + h_{1,2M-1} \\ h_{1,2,sum} = h_{1,2} + h_{1,4} + \cdots + h_{1,2M} \end{cases} \tag{8.31}$$

Thus, we obtain

$$\mathbf{d}_{1,sum} = \mathbf{H}_{1,sum}\mathbf{b}_{1,1} + \mathbf{v}_{1,sum} \tag{8.32}$$

where we have used the following definitions:

$$\begin{cases} \mathbf{d}_{1,sum} = \begin{pmatrix} d_{1,1} \\ d_{1,2} \end{pmatrix} \\ \mathbf{H}_{1,sum} = \begin{pmatrix} h_{1,1,sum} & h_{1,2,sum} \\ -h_{1,2,sum} & h_{1,1,sum} \end{pmatrix} \\ \mathbf{b}_{1,1} = \begin{pmatrix} b_{1,o} \\ b_{1,e} \end{pmatrix} \\ \mathbf{v}_{1,sum} = \begin{pmatrix} v_{1,1} \\ v_{1,2} \end{pmatrix} \end{cases} \tag{8.33}$$

Next we can perform S-T decoding by multiplying both sides of (8.32) with $\mathbf{H}_{1,sum}^H$ and retaining only the real part to get the decision variables as

$$\begin{pmatrix} g_{1,o} \\ g_{1,e} \end{pmatrix} = \Re\left(\mathbf{H}_{1,sum}^H \mathbf{d}_{1,sum}\right)$$
$$= \Re\left(\mathbf{H}_{1,sum}^H \mathbf{H}_{1,sum} \mathbf{b}_{1,1}\right) + \Re\left(\mathbf{H}_{1,sum}^H \mathbf{v}_{1,sum}\right)$$
$$= \begin{pmatrix} \eta_1 & 0 \\ 0 & \eta_2 \end{pmatrix} \begin{pmatrix} b_{1,o} \\ b_{1,e} \end{pmatrix} + \Re\left(\mathbf{H}_{1,sum}^H \mathbf{v}_{1,sum}\right) \quad (8.34)$$

where

$$\begin{cases} \eta_1 \triangleq | h_{1,1,sum} |^2 + | h_{1,2,sum} |^2 \\ \eta_2 \triangleq | h_{1,1,sum} |^2 + | h_{1,2,sum} |^2 \end{cases} \quad (8.35)$$

and the operators $\mathbf{x}^H$ and $\Re(\mathbf{x})$ are to calculate the Hermitian form and to retain the real part of a complex vector $\mathbf{x}$, respectively. It should be noted that perfect knowledge of channel information is required to achieve the full diversity gain in the proposed STCC DS/CC-CDMA scheme. The significance of (8.34) is that the S-T decoder in an STCC DS/CC-CDMA system with two transmitter antennas can achieve full diversity gain, in addition to the inherent MAI-free property of the system.

It has to be noted that in the proposed STCC scheme the complementary codes have been used to spread information bit streams which have been encoded by the STBC unit, as shown in Figures 8.1 and 8.2. Therefore, in this sense the proposed STCC system can be viewed as a system which combines STBC and complementary coded CDMA within a unified framework. Therefore, the despreading operation has to be carried out before STBC decoding at a receiver, as shown in Figure 8.3, to eliminate MAI before achieving full diversity gain in the STBC block. It should also be noted that the orthogonality of complementary codes is achieved at chip level for user separation, while the orthogonality of the STBC scheme is achieved at bit level for antenna separation. Thus, the implementation of two-level orthogonality is the main feature that makes STCC excel compared to existing S-T coding schemes in terms of its MAI-free operation and enhanced spatial diversity.

8.5 ARBITRARY NUMBER OF TRANSMITTER ANTENNAS

Next, we would like to extend our previous analysis to the case with n_t transmitter antennas for each user in a K-user STCC DS/CC-CDMA system. Every receiver will still use a single antenna for signal reception.

It can be shown that the generalized form of (8.34), which is the output from an S-T decoder or the decision variable vector, will become

$$\begin{cases} \tilde{g}_{1,o} = \eta_1' b_{1,o} + \sum_{j=1}^{n_t} h_{1,j,sum}^* v_{1,j} \\ \tilde{g}_{1,e} = \eta_2' b_{1,e} + \sum_{j=1}^{n_t} h_{1,j,sum} v_{1,j}^* \end{cases} \quad (8.36)$$

where

$$\begin{cases} \eta_1' = | h_{1,1,sum} |^2 + | h_{1,2,sum} |^2 + \cdots + | h_{1,n_t,sum} |^2 \\ \eta_2' = | h_{1,1,sum} |^2 + | h_{1,2,sum} |^2 + \cdots + | h_{1,n_t,sum} |^2 \end{cases} \quad (8.37)$$

It is noted that now $h_{1,j,sum}$ $(1 \leq j \leq n_t)$ is the summation of M Rayleigh fading channel coefficients, or

$$\begin{cases} h_{1,1,sum} = h_{1,1} + h_{1,1+n_t} + \cdots + h_{1,n_t M-(n_t-1)} \\ h_{1,2,sum} = h_{1,2} + h_{1,2+n_t} + \cdots + h_{1,n_t M-(n_t-2)} \\ \cdots\cdots \\ h_{1,n_t,sum} = h_{1,n_t} + h_{1,2n_t} + \cdots + h_{1,n_t M} \end{cases} \tag{8.38}$$

(8.38) will be reduced to (8.31) if $n_t = 2$. The right-hand side of each equation in (8.38) is the summation of M terms, each of which is an identical and independent distributed ($i.i.d.$) Rayleigh random variable. Let $h_{1,i}$ $(1 \leq i \leq n_t M)$ represent a generic term at the right-hand side of (8.38), whose probability density function (pdf) is

$$f_{h_{1,i}}(r) = \frac{r}{\sigma^2} \exp\left(-\frac{r^2}{2\sigma^2}\right), \quad 0 \leq r < \infty \tag{8.39}$$

with its variance being σ^2. Let $\beta = (h_{1,i})^2$, which obeys exponential distribution as

$$f_\beta(r) = \frac{1}{2\sigma^2} \exp\left(-\frac{r}{2\sigma^2}\right), \quad 0 \leq r < \infty. \tag{8.40}$$

In this STCC DS/CC-CDMA system there are in total K users, each of which is assigned M element codes as its signature codes sent via M different carriers. Therefore, we have

$$Var(h_{1,j,sum}) = M\sigma^2, \quad 1 \leq j \leq n_t. \tag{8.41}$$

The bit error rate (BER) of the system can be derived from (8.36) due to the fact that either $\tilde{g}_{1,o}$ or $\tilde{g}_{1,e}$ is Gaussian under the condition that we first fix all $h_{1,j,sum}$, where $1 \leq j \leq n_t$. As shown in Figure 8.3, BPSK modulations are used in both transmitter and receiver. Therefore, the average BER of an STCC DS/CC-CDMA system can be obtained if we know the signal-to-noise ratio (SNR) at the input side of the decision device shown in Figure 8.3.

Let us define α_M as

$$\alpha_M = \sum_{j=1}^{n_t} \mid h_{1,j,sum} \mid^2 \tag{8.42}$$

From (8.36), fixing $h_{1,j,sum}$ and thus $h^*_{1,j,sum}$ we obtain the variance of the noise terms as

$$\sigma^2_{n-total} = Var\left(\sum_{j=1}^{n_t} h^*_{1,j,sum} v_{1,j}\right) = \sum_{j=1}^{n_t} \mid h_{1,j,sum} \mid^2 Var(v_1) = \alpha_M M N_o, \tag{8.43}$$

where we have used $Var(v_1) = 2M\frac{N_o}{2}$ from (8.28). Thus, the SNR at the output of the STBC decoder becomes

$$\text{SNR} = \frac{\alpha_M^2 E_b}{\sigma^2_{n-total}} = \frac{\alpha_M^2 E_b}{\alpha_M M N_o} = \frac{\alpha_M E_b}{M N_o} \tag{8.44}$$

Therefore, we have the average BER of an STCC DS/CC-CDMA system as

$$P_b = \int_0^\infty Q\left(\sqrt{\frac{2E_b r}{n_t M N_o}}\right) f_{\beta,n_t}(r)\, dr \tag{8.45}$$

where the factor n_t counts for the normalization of transmitting power for n_t antennas and $f_{\beta,n_t}(r)$ is the pdf function for α_M, which takes the form

$$f_{\beta,n_t}(r) = \left(\frac{1}{2M\sigma^2}\right)^{n_t} \frac{r^{n_t-1}}{(n_t-1)!} \exp\left(-\frac{r}{2M\sigma^2}\right), \qquad (0 \le r < \infty) \tag{8.46}$$

Thus, the BER expression can be rewritten into

$$P_b = \int_0^\infty Q\left(\sqrt{\frac{2E_b r}{n_t M N_o}}\right)\left(\frac{1}{2M\sigma^2}\right)^{n_t} \times \frac{r^{n_t-1}}{(n_t-1)!} \exp\left(-\frac{r}{2M\sigma^2}\right) dr \tag{8.47}$$

Letting $z = \frac{E_b r}{n_t M N_o}$, we can simplify (8.47) into

$$\begin{aligned} P_b &= \int_0^\infty Q\left(\sqrt{2z}\right) \frac{(n_t M N_0)^{n_t} z^{n_t-1}}{\left(2E_b M\sigma^2\right)^{n_t} (n_t-1)!} \\ &\quad \times \exp\left(-\frac{n_t M N_0}{2E_b M\sigma^2} z\right) dz \\ &= \int_0^\infty Q\left(\sqrt{2z}\right) \frac{(n_t N_0)^{n_t} z^{n_t-1}}{\left(2E_b\sigma^2\right)^{n_t} (n_t-1)!} \exp\left(-\frac{n_t N_0}{2E_b\sigma^2} z\right) dz \\ &= \left(\frac{1-\mu}{2}\right)^{n_t} \sum_{n=1}^{n_t-1} \binom{n_t-1+n}{n} \left(\frac{1+\mu}{2}\right)^n \end{aligned} \tag{8.48}$$

where μ has been defined as

$$\mu = \sqrt{\frac{\frac{2E_b\sigma^2}{n_t N_o}}{1+\frac{2E_b\sigma^2}{n_t N_o}}} = \sqrt{\frac{\gamma}{1+\gamma}} \tag{8.49}$$

Here, we have introduced the expression

$$\gamma = \frac{2E_b\sigma^2}{n_t N_o} \tag{8.50}$$

as the normalized SNR with respect to the number of transmitter antennas or n_t. As long as $2\sigma^2 = 1$ or $\sigma^2 = 0.5$, we will have $\gamma = \frac{E_b}{n_t N_o}$, which just gives normalized SNR in an STCC DS/CC-CDMA system with n_t transmitter antennas. Therefore, it is seen from (8.48) to (8.50) that the average BER performance of an STCC DS/CC-CDMA system is completely controlled by a single parameter n_t and has nothing to do with other system variables, including K, M, N, etc., implying that it is a noise-limited system with a full diversity gain.

8.6 RESULTS AND DISCUSSIONS ON STCC DS/CC-CDMA

Based on the analysis given in the previous sections, we can evaluate the performance of an STCC DS/CC-CDMA system, which will be compared with an STBC CDMA system using traditional

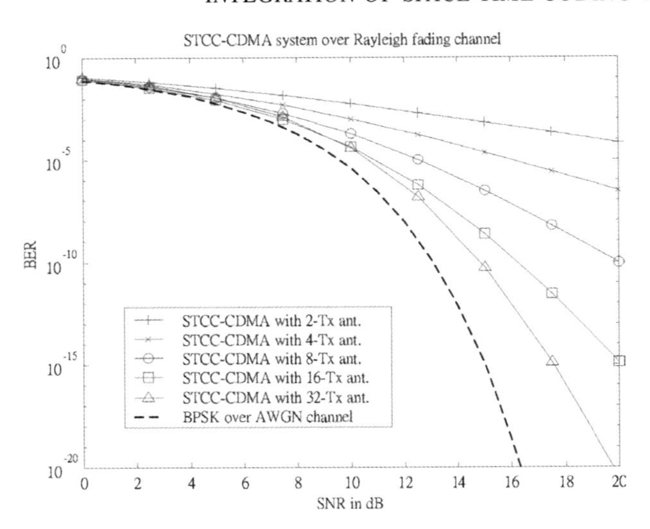

Figure 8.6 BER versus signal-to-noise ratio for STCC DS/CC-CDMA system in Rayleigh fading channels with varying number of transmitter antennas. The performance is independent of number of users and PG values, showing the MAI-free operation of the proposed system.

signature codes, such as Gold codes. To serve as a bench-mark to the theoretical analysis, computer simulations have also been carried out, and the results obtained from both analysis and simulations will be compared.

Figure 8.6 shows BER versus SNR for an STCC DS/CC-CDMA system with variable numbers of transmitter antennas, from 2 to 32 antennas. It shows that a greater advantage can be obtained by using a relatively large number of transmitter antennas. The results illustrates that the BER performance for an STCC DS/CC-CDMA system under Rayleigh fading channels can asymptotically approach that of the single user bound if a sufficiently large number of antennas can be made available. Figure 8.6 gives only theoretical results.

The comparison between theoretical and simulation results is made in Figures 8.7 and 8.8, which show the BER performance versus SNR for two similar system setups, except for the different PG values of the signature codes considered, where OC codes take 32 and 64 and Gold codes take 31 and 63, respectively. Both cases consider an STBC encoder with two transmitter antennas.

Obviously, the proposed STCC DS/CC-CDMA scheme is in particular suitable for its operation in a wireless communication system where a high data throughput is required. As seen from Figures 8.7 and 8.8, the STCC DS/CC-CDMA scheme can offer an almost identical BER performance regardless of the number of users present in the system due to its unique MAI-free operational advantage. In this way, a cell site equipped with an STCC scheme can support many more users than one with conventional S-T coding schemes, such as STBC, under the same BER requirements. A higher capacity per cell can be translated into a higher cell-wise data throughput.

It is seen from Figure 8.7 that the STCC DS/CC-CDMA gives a monotonously decreasing BER regardless of user population in the system, manifesting the superior MAI-free property of the scheme. On the other hand, the BER performance of an STBC DS-CDMA using Gold codes tends to level off at the right-hand side of the figure. The more users are present, the worse the BER of the Gold code STBC DS-CDMA system will be, showing a typical scenario for an interference-limited system. The simulations were carried out for both DS/CC-CDMA and Gold code CDMA systems. Due to the time limitation, simulation results were obtained only for DS/CC-CDMA, 4-user Gold code, and

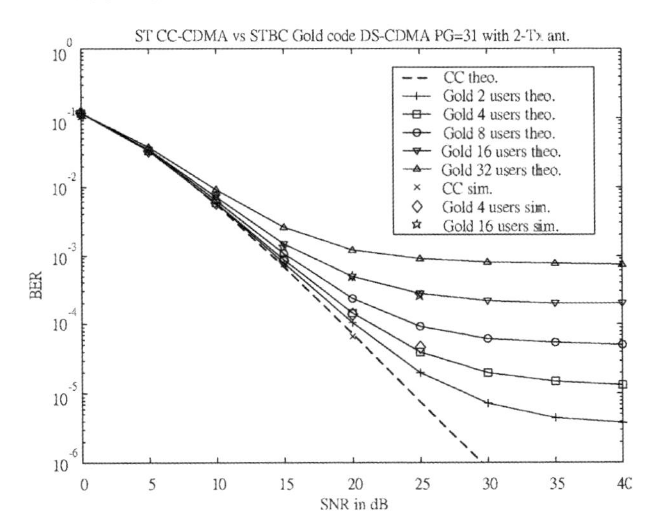

Figure 8.7 Comparison of BER versus signal-to-noise ratio for STCC DS/CC-CDMA and STBC Gold code CDMA systems in Rayleigh fading channels with varying number of users. The PG for OC codes and Gold codes is 32 and 31, respectively. Two transmitter antennas and one receiver antenna are considered. Theoretical results are also compared with simulation results for STCC DS/CC-CDMA, 4-user Gold code CDMA, and 16-user Gold code CDMA systems.

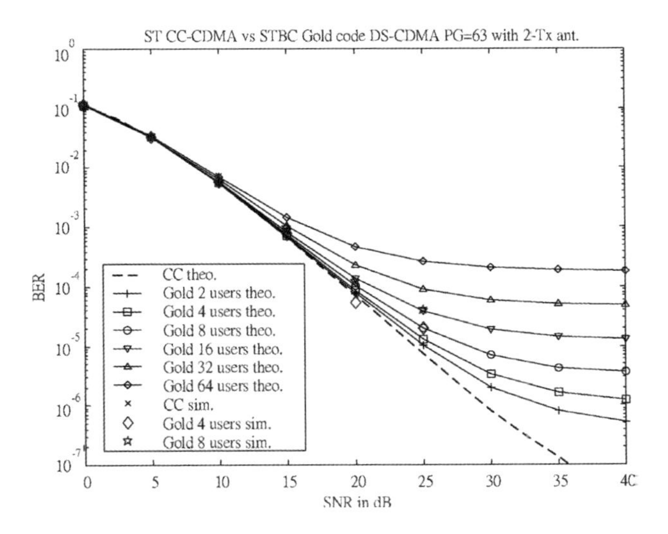

Figure 8.8 Comparison of BER versus signal-to-noise ratio for STCC DS/CC-CDMA and STBC Gold code CDMA systems in Rayleigh fading channels with varying number of users. The PG for OC codes and Gold codes is 64 and 63, respectively. Two transmitter antennas and one receiver antenna are considered. Theoretical results are also compared with simulation results for STCC DS/CC-CDMA, 4-user Gold code CDMA, and 8-user Gold code CDMA systems.

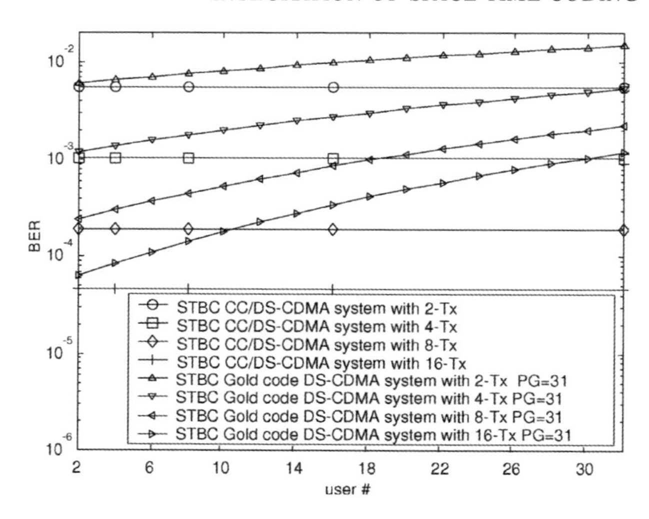

Figure 8.9 Comparison of BER versus number of users for STCC DS/CC-CDMA and STBC Gold code CDMA systems in Rayleigh fading channels with fixed signal-to-noise ratio at 10 dB. The PG for OC codes and Gold codes is 32 and 31, respectively. The number of transmitter antennas is 2, 4, 8, and 16, and a single receiver antenna is considered.

16-user Gold code cases, all of which show a very good match to the corresponding results obtained from analysis. A similar trend can also be observed in Figure 8.8, except for a better BER in general for the curves of the Gold code CDMA systems because of the larger PG value applied therein. It should be noted that a large PG will not affect the BER performance of an STCC DS/CC-CDMA system due to the fact that it works in an interference-free operational mode.

Figures 8.9 and 8.10 illustrate BER versus number of users in both STCC DS/CC-CDMA and Gold code STBC-CDMA systems. The difference between Figures 8.9 and 8.10 lies in their PG values of concern, with the PG used in Figure 8.9 being 32 or 31 and that in Figure 8.10 being 64 or 63 for OC codes or Gold codes, respectively. We can see from the figures that BER performance for the STCC DS/CC-CDMA systems remains constant regardless of varying the number of users in the system. However, it does change with the number of transmitter antennas, as also shown in Figure 8.6 earlier. On the contrary, the STBC DS-CDMA system with Gold codes is in general sensitive to the number of users present in the system, especially if a relatively small PG value is applied to the system, as shown in Figure 8.9.

The capacity gain for an STCC DS/CC-CDMA system over its counterpart Gold code STBC DS-CDMA system can be significant due to its interference-free operation. Assume, for instance, that the required BER is about 10^{-3}. It is observed from Figure 8.9 that an STCC DS/CC-CDMA system with four antennas can support as many as 32 users, which is in fact limited only by the set size of the OC code set. However, a Gold code STBC DS-CDMA with four antennas can only support about two users, differing from that of the STCC DS/CC-CDMA system by as many as 30 users! Alternatively, a Gold code STBC DS-CDMA system has to use up to 16 transmitter antennas to achieve approximately the same BER performance (10^{-3}) at about the same capacity (32 users), thus resulting in much greater complexity.

It has to be admitted that the implementation complexity of the proposed STCC DS/CC-CDMA scheme is higher than that of traditional STBC DS-CDMA with traditional spreading codes, such as Gold codes, etc. The major part of the complexity in the STCC DS/CC-CDMA is due to its use of a multi-carrier modem to send different element codes of the CC code set. However, the

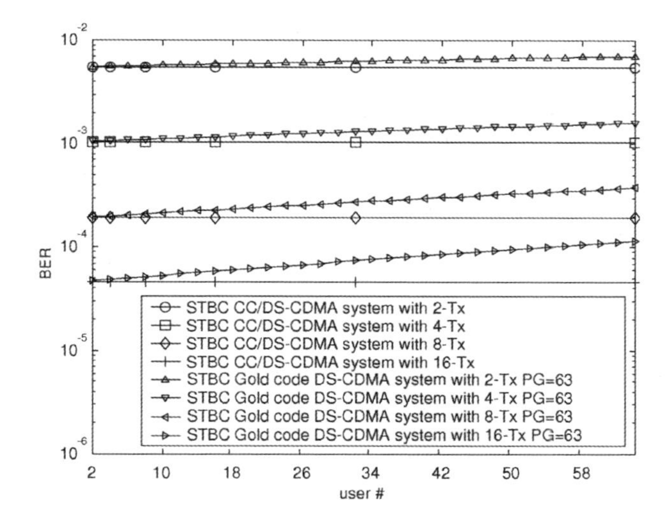

Figure 8.10 Comparison of BER versus number of users for STCC DS/CC-CDMA and STBC Gold code CDMA systems in Rayleigh fading channels with fixed signal-to-noise ratio at 10 dB. The PG for OC codes and Gold codes is 64 and 63, respectively. The number of transmitter antennas is 2, 4, 8, and 16, and one receiver antenna is considered.

implementation complexity of the STCC DS/CC-CDMA scheme can be greatly reduced if we use orthogonal frequency division multiplex (OFDM) to replace the complex multi-carrier module.

8.7 SUMMARY ON STCC DS/CC-CDMA

In this chapter we have proposed and discussed the STCC DS/CC-CDMA system architecture based on orthogonal complementary codes and space-time coding for downlink transmission. A comprehensive analysis has been carried out to evaluate its bit error rate performance under Rayleigh fading channels and multiple access interference (MAI). It has been shown in the analysis that the STCC DS/CC-CDMA system can achieve an ideal MAI-free operation and a full diversity gain jointly under a single system framework. The results obtained from the theoretical analysis have also been compared to those generated from simulations, showing a very good match with each other. It has been concluded that the combinational use of orthogonal complementary code based CDMA and space-time coding is technically feasible, which can be a very attractive option for implementation of beyond 3G wireless systems. The STCC technique will also be an indispensable part of the next generation CDMA technology.

8.8 WHY STCC OS/CC-CDMA?

After having discussed the issues on the STCC DS/CC-CDMA system, we would like also to take a look at its peer, or STCC OS/CC-CDMA, which is the other way to integrate S-T coding with CC-CDMA technology as an effort to exploit joint benefits of the MAI-free operation of an OS/CC-CDMA system and spatial diversity gain and/or spatial multiplex capability. The unique features of this STCC OS/CC-CDMA should be attributed to the use of generalized pair-wise complementary codes in the architecture, such that both spatial diversity gain and spatial multiplexing capability can be achieved jointly.

We will show that, with the help of generalized pair-wise complementary codes, a MIMO system based on the proposed STCC OS/CC-CDMA scheme can achieve spatial diversity and parallel transmission at the same time. It will also be shown that the proposed STCC OS/CC-CDMA MIMO system can easily be generalized to achieve an arbitrary combination of diversity and multiplex order. The scheme requires neither bandwidth expansion nor feedback information from the receiver, and it offers a near interference-free and multipath-resist performance due mainly to the widely open interference-free window of the pair-wise complementary codes, which are used to implement the STCC OS/CC-CDMA MIMO system. The analytical study will be carried out in this chapter to show that it provides a uniquely high diversity-multiplex gain product for its operation in both flat fading and frequency-selective fading channels.

It is a well-known fact that a traditional receiver in a high-speed data-centric wireless system is usually very susceptible to attenuated signals under the influence of frequency-selective fading. To mitigate this effect, the design of the receiver has to make use of some kind of diversity mechanism (which can be achieved in the frequency, time, and space domains) to overcome or compensate the fading. Receiver diversity is a traditional method to generate less-attenuated replicas of transmitted signals by combining the signals with independent statistics at a receiver. However, due to inevitable limitations in hardware implementation complexity, it is usually not feasible for mobile stations (MSs) to use multiple antennas to achieve receiver diversity. Also mainly for this reason, the transmitter diversity, which can be applied to a base station, has emerged as a popular solution to achieve spatial diversity gain at a mobile terminal receiver.

Recently, space-time block coding (STBC) [239, 240] and space-time trellis coding (STTC) [238] schemes have been proposed to implement transmitter diversity. STBC provides full diversity gain and does not sacrifice bandwidth efficiency. Unfortunately, it does not improve bandwidth efficiency directly and it is less tolerant toward frequency-selective fading in the channel. It has been shown in [272] that STBC schemes suffer serious inter-symbol interference (ISI) over multipath channels, and the ISI is proportional to the number of transmitter antennas used in the STBC schemes. In addition, the design of an STBC scheme in particular for time-dispersive multipath channels can be very complex due to the fact that the signals sent from different antennas are mixed not only in space but also in time. If the interference can not be suppressed below a sufficient level, it will destroy all STBC diversity gain achieved. There are numerous reported works [273–282] addressing in particular interference suppression issues in STBC schemes.

The use of multiple transmitter antennas can also provide a multiplexing gain in terms of increased data transmission rate through multiple data pipes formed by different antennas. Parallel transmissions in all transmitter-receiver antenna paths simultaneously can dramatically increase the overall system capacity or data throughput. However, to maximize the multiplexing gain a transmitter usually needs to know the channel state information (CSI) in advance. Hence, the CSI has to be sent via a feedback channel from receiver to transmitter, and doing so will result in throughput degradation and extra implementation complexity in both transmitter and receiver as a whole. Theoretically speaking, the diversity gain and multiplexing capability is a dual for a traditional multiple-input-multiple-output (MIMO) system, which can be traded off for some known numbers of transmitter and receiver antennas. It is still an extremely challenging issue to achieve a diversity and multiplex product as high as possible, where the diversity and multiplex product is defined as the product of spatial diversity order and spatial multiplex order. However, it has to be noted that a particular MIMO system based on traditional S-T coding schemes (such as STBC and STTC) usually can not provide both diversity gain and multiplex capability at the same time under a fixed MIMO configuration.

Here we would like to propose a novel space-time complementary coding (STCC) OS/CC-CDMA scheme, and a MIMO system based on STCC OS/CC-CDMA is capable of providing transmitter diversity and parallel transmission jointly under a unified design. With the help of generalized pair-wise complementary (GPC) codes, the STCC scheme offers a flexible tradeoff mechanism between diversity gain and multiplexing capability to fit varying channel conditions and application requirements. Similar to all other S-T coding based MIMO systems, the STCC OS/CC-CDMA MIMO scheme does not require bandwidth expansion either, and the redundancy is applied in space across multiple antennas, rather than in time or frequency. In addition, it helps also to enhance bandwidth efficiency through parallel transmissions. The STCC scheme works like transmitter-beamforming, but its interference suppression mechanism is uniquely effective due to the use of the GPC codes [263] to encode data streams. Each encoded data stream can be extracted at a receiver by simply correlating its encoding generalized pair-wise complementary code. The unwanted interference caused by frequency-selective fading will be effectively canceled due to the desirable correlation properties of GPC codes, which have a very wide interference-free window (IFW) in their correlation functions. The details of the proposed STCC OS/CC-CDMA MIMO scheme will be discussed in the following section.

8.9 STCC OS/CC-CDMA SYSTEM MODEL

8.9.1 CHANNEL MODEL

In the downlink multiple-input-single-output (MISO) channel model shown in Figure 8.11(a), there are D simultaneous transmission channels, each of which is a multipath fading channel. Different transmission channels suffer from independent multipath fading attenuations. An L-tap transversal filter is used to model the multipath fading channel. The complex low-pass impulse response of the channel for the dth space can be expressed by

$$h_d(t) = \sum_{l=1}^{L} \alpha_{dl} e^{-j\theta_{dl}} \delta(t - \tau_{dl}) = \sum_{l=1}^{L} \beta_{dl} \delta(t - \tau_{dl}) \tag{8.51}$$

where $\beta_{dl} = \alpha_{dl} e^{-j\theta_{dl}}$, and α_{dl}, θ_{dl}, and τ_{dl} are the dth space and lth path's envelope, phase, and delay, respectively. $\delta(t)$ is a Dirac-delta function; α_{dl} is a Rayleigh distributed random variable; β_{dl} obeys complex Gaussian distribution with its variance being $\sigma_{dl}^2/2$ in each dimension; θ_{dl} and τ_{dl} are uniformly distributed over $[0, 2\pi]$ and $[0, T]$, respectively; L is the number of resolvable paths of the channel. The channel delay spread can be written as

$$\tau_{\max} = \max(|\tau_{d'l'} - \tau_{d''l''}|) \tag{8.52}$$

where $\tau_{d'l'}$ and $\tau_{d''l''}$ denote the delays for any two rays observed in the same channel. It is assumed that the fading channel has L resolvable paths, each with a coefficient of α_{dl}. Note that we consider only the normalized multipath channel model, such that $\sum_{l=1}^{L} \sigma_{dl}^2 = 1$ for any d. The multipath channel is modeled by an L-tap transversal filter with its taps being equally spaced by T_c.

8.9.2 GENERALIZED PAIR-WISE COMPLEMENTARY CODES

The GPC codes [268] are used in STCC OS/CC-CDMA MIMO to offer various advantages. Conventional complementary codes, such as complete complementary codes [254], have a serious limitation on code set size K (or supportable channels) that is strictly associated with flock size M. The more element codes each flock has, the more flocks we can generate in a set. In addition, when sending complementary coded signals into a channel, different element codes must be sent separately using either different time slots or different carrier frequencies. If M becomes fairly large, we have to use more time slots or frequencies to encode different element codes for the same flock. To simplify the

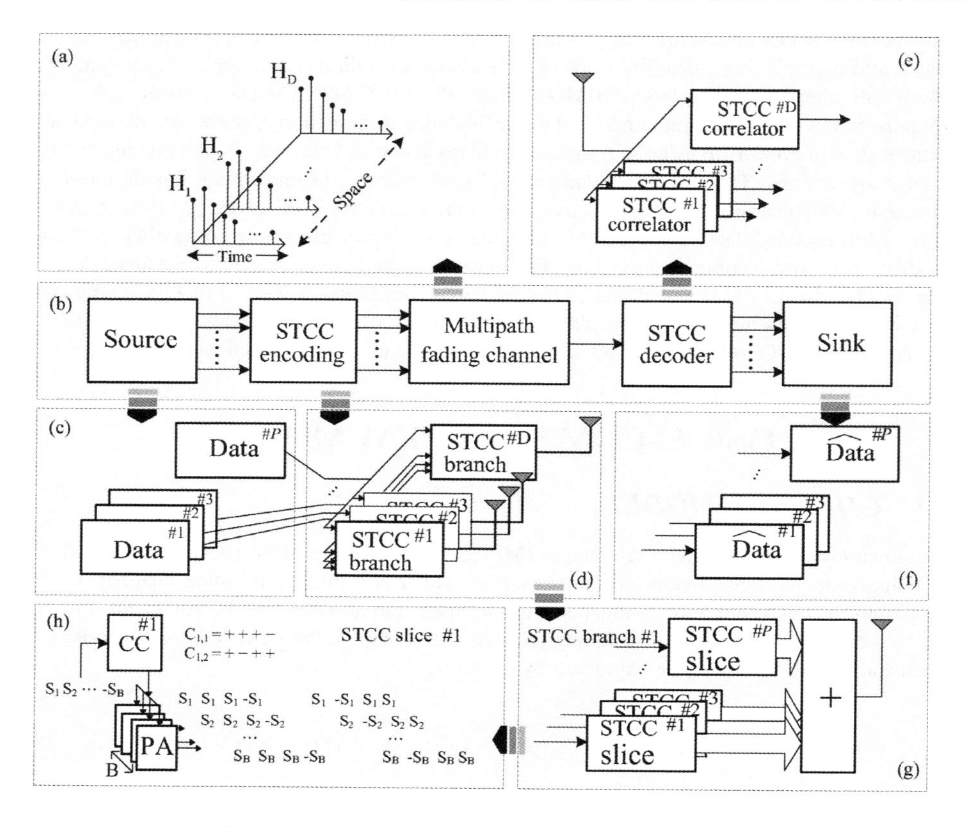

Figure 8.11 Conceptual block diagram for STCC-MIMO scheme. (a) Space-time channel model; (b) system block diagram for a generic STCC-MIMO system; (c) data source block; (d) STCC encoding branches; (e) STCC decoding block; (f) decoded information streams; (g) configuration of STCC branch #1; and (h) STCC encoding algorithm.

system design, we use $M = 2$ GPC codes in this chapter,[1] where only two flocks will be used per user, and $\mathbf{C}_1$ for data transmission and $\mathbf{C}_2$ for pilot or pilot/data multiplexed transmission.

It can be shown that an STCC OS/CC-CDMA scheme using conventional complete complementary codes (with $M = 2$) will never achieve joint diversity and multiplexing at the same time due to its small set size (i.e. $K = 2$). On the other hand, the GPC is particularly well suited for the STCC OS/CC-CDMA MIMO schemes to offer both diversity and multiplexing jointly through orthogonal matrix expansion to generate enough flocks of GPC codes even with a fixed $M = 2$ (more detailed discussions on GPC codes can be found in [268]). Figure 8.12 shows the auto-correlation and cross-correlation properties of a GPC set with its parameters $(K, M, N, E) = (4, 2, 8, 2)$, where N denotes the element code length, $E = N/N'$ is the expansion index, N' denotes the element code length of a seed orthogonal complementary code (OCC) before expansion, and K denotes the set size of the GPC codes and is equal to $2E$ here. The use of different parameter sets (K, M, N, E) will yield different GPC sets. It is seen from Figure 8.12 that the GPC has a fairly large interference-free window (IFW), which is uniformly wide across the entire code set. It is noted that the whole GPC set can be divided

[1]For more information about the GPC codes, refer to Subsection 6.2.5.

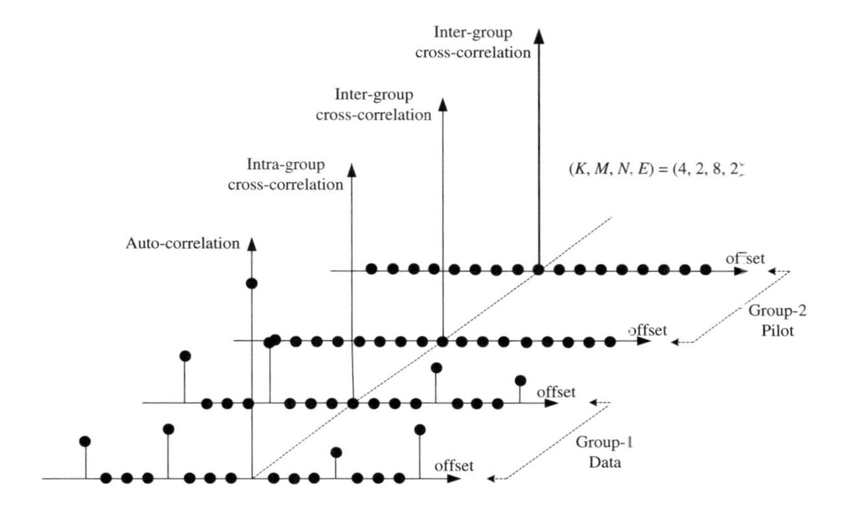

Figure 8.12 Inter-group and intra-group correlation properties of generalized pair-wise complementary codes.

into two groups, G_1 and G_2, of the same size. The codes from the different groups possess ideal cross-correlations, as shown in Figure 8.12.

In the proposed STCC OS/CC-CDMA MIMO scheme, all K GPC codes will be used to achieve maximal diversity gain and parallel transmission capability, i.e. $D \times P \leq K$, where D denotes the diversity order and P is the number of multiplexed symbol streams from the same transmitter. Moreover, due to the correlation properties of the GPC codes, we should avoid some chip offsets or $xN'T_s$, where $x \in Z$ and Z is an integer subset associated with the seed OCC defined in [268], because two GPC codes from the same group have correlation side lobes at these chip offsets. We use null insertion in these chip offsets to avoid the interference, as illustrated in Figure 8.13.

8.9.3 SPACE-TIME COMPLEMENTARY CODING

Figure 8.11 shows a conceptual block diagram for the proposed STCC OS/CC-CDMA MIMO system as a whole. As shown in Figure 8.11(c), the source module contains P parallel independent symbol streams, typically coming from the outputs of digital modulators. The P parallel symbol streams are fed into the STCC encoding module, which consists of D STCC encoding branches, each of which has P STCC slices, as shown in Figure 8.11(g). Each STCC slice has B radio frequency (RF) power amplifiers (PAs), the outputs from which will be sent out through the same antenna. As illustrated in Figure 8.11(h), the STCC slice #1 encodes B symbols with a corresponding GPC code set, which are sent to each PA with one symbol offset relative to its neighboring symbols, as shown in Figure 8.13.

There are in total up to D replicas of P parallel symbol streams encoded by different GPC code sets to implement diversity order of D and parallel transmission order of P. Therefore, the family size of the GPC must be at least $DP \leq K$. Time division multiplex (TDM) is used to separate two GPC element codes from the same code. One STCC encoded block length must be less than the channel coherent time to make the analysis relevant to the assumption of a block-wise time-invariant channel. At the receiver side, the received symbol streams should go through maximum ratio combining from D replicas to extract P parallel symbol streams by correlating the local GPC code. The offset stacking transmission is illustrated in Figure 8.13, which is critical in the STCC OS/CC-CDMA MIMO scheme to improve the overall spectral efficiency. The overlapped signals will not cause any interference due

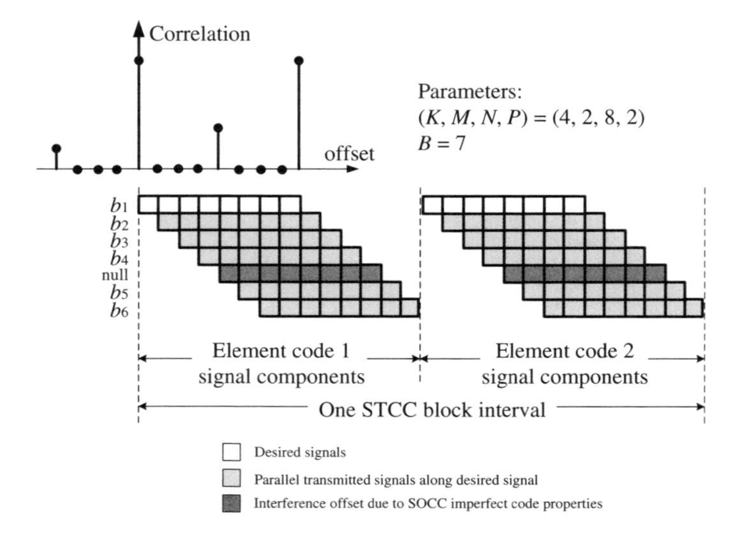

Figure 8.13 Offset stacking transmission mechanism with GPC.

to the unique correlation properties of the GPC codes. However, some particular chip offsets should not be used to carry data information because of the correlation side lobes of the GPC codes. Thus, a NULL signal (no transmission) is inserted at these particular chip offset positions to avoid creating interference. It is shown in Figure 8.13 that $\{b_x\}_{x=1}^{6}$ are encoded by the codes taken from the same group of the GPC code set.

8.10 SLOW FLAT FADING CHANNELS

In this section, we first study the STCC OS/CC-CDMA MIMO scheme working in a flat fading channel. Then the analysis will be extended to a multipath fading channel in Section 8.11. The STCC model of interest is shown in Figure 8.11(b), where 'Source' is the data information prior to STCC encoding, which can be expressed by

$$\mathbf{B}^{(p)} = \left(\begin{array}{cccc} b_{(p-1)B+1}, & b_{(p-1)B+2}, & \ldots, & b_{(p-1)B+B} \end{array} \right)_{1\times B} \tag{8.53}$$

where the subscript (p) represents the pth parallel transmission branch and the right subscript $1 \times B$ denotes the dimension of this vector. The same rule will be applied to all vectors and matrices discussed in this part of text without further explanation. $\mathbf{B}^{(p)}$ denotes the source information of the STCC slice-p, where $p = 1, \ldots, P$. b_x denotes the xth source symbol, and B is the number of source information symbols in one STCC block. The STCC block interval in this case is $2(B + N - 1)T_s$, where T_s denotes source symbol duration. Therefore, the maximum coding rate can be calculated by

$$\frac{BT_s}{2\,(B + N - 1)\,T_s} \times P \log_2 V \quad \text{(bit/sec/Hz)} \tag{8.54}$$

where V denotes the levels of digital modulation of interest, and it increases as B increases. The maximum transmission rate can be approximated by $P/2 \log_2 V$ (bit/sec/Hz). The P transmission branches in the same space are encoded by the STCC encoding module to enhance spectral efficiency through parallel transmissions. The STCC encoding module spreads the input symbol streams with corresponding GPC, and there are in total D replicas of P transmission branches placed in different spaces, which help to provide spatial diversity gain.

Next, we will introduce an element code convolution matrix $\mathbf{C}_{(e),m}^{(B)}$, or

$$\mathbf{C}_{(e),m}^{(B)} = \begin{pmatrix} \mathbf{c}_{e,m}^T & 0 & 0 & 0 & 0 \\ 0 & \mathbf{c}_{e,m}^T & 0 & 0 & 0 \\ 0 & 0 & \mathbf{c}_{e,m}^T & 0 & 0 \\ \vdots & \vdots & \vdots & \ddots & \vdots \\ 0 & 0 & 0 & 0 & \mathbf{c}_{e,m}^T \end{pmatrix}_{(B+N-1)\times B} \tag{8.55}$$

which will be used to transform source symbols into STCC encoded symbol streams. Here, $\mathbf{c}_{e,m}$ is a GPC code with e and m being the GPC code set index and element code index, respectively, and $m \in (1, 2)$. It should be noted that each GPC set here consists of two element codes and the whole GPC family has in total K sets. For description simplicity, we further define a composite element code convolution matrix as

$$\mathbf{C}_{(e)}^{(B)} = \begin{pmatrix} \mathbf{C}_{(e),1}^{(B)} \\ \mathbf{C}_{(e),2}^{(B)} \end{pmatrix}_{2(B+N-1)\times B} \tag{8.56}$$

All composite element code convolution matrixes can be combined to form an STCC encoding matrix $\mathbf{C}^{(B)}$ as

$$\mathbf{C}^{(B)} = \begin{pmatrix} \mathbf{C}_{(1)}^{(B)} & \mathbf{C}_{(2)}^{(B)} & \cdots & \mathbf{C}_{(P)}^{(B)} \\ \mathbf{C}_{(P+1)}^{(B)} & \mathbf{C}_{(P+2)}^{(B)} & \cdots & \mathbf{C}_{(2P)}^{(B)} \\ \vdots & \vdots & \ddots & \vdots \\ \mathbf{C}_{(DP-P+1)}^{(B)} & \mathbf{C}_{(DP-P+2)}^{(B)} & \cdots & \mathbf{C}_{(DP)}^{(B)} \end{pmatrix}_{2D(B+N-1)\times BP} \tag{8.57}$$

whose dimension is $2D\,(B+N-1) \times BP$. Using the above equation, we can encode symbol streams to obtain STCC encoded symbols as follows:

$$\tilde{\mathbf{S}}^{(B)} = \mathbf{C}^{(B)}\mathbf{B} \begin{pmatrix} \tilde{\mathbf{S}}_1^{(B)} \\ \tilde{\mathbf{S}}_2^{(B)} \\ \vdots \\ \tilde{\mathbf{S}}_D^{(B)} \end{pmatrix}_{2D(B+N-1)\times 1} \tag{8.58}$$

where we have

$$\tilde{\mathbf{S}}_d^{(B)} = \sum_{i=1}^{P} \mathbf{C}_{(dP-P+i)}^{(B)} \left(\mathbf{B}^{(i)}\right)^T, \qquad (d = 1, \cdots, D). \tag{8.59}$$

Equation (8.58) can be rearranged to obtain

$$\mathbf{S}^{(B)} = \begin{pmatrix} \left(\tilde{\mathbf{S}}_1^{(B)}\right)^T \\ \left(\tilde{\mathbf{S}}_2^{(B)}\right)^T \\ \vdots \\ \left(\tilde{\mathbf{S}}_D^{(B)}\right)^T \end{pmatrix} = \begin{pmatrix} \mathbf{S}_1^{(B)} \\ \mathbf{S}_2^{(B)} \\ \vdots \\ \mathbf{S}_D^{(B)} \end{pmatrix} \tag{8.60}$$

where we can define

$$\begin{aligned} \mathbf{S}_d^{(B)} &= \sum_{i=1}^{P} \left(\mathbf{C}_{(dP-P+i)}^{(B)} \left(\mathbf{B}^{(i)}\right)^T\right)^T = \sum_{i=1}^{P} \mathbf{B}^{(i)} \left(\mathbf{C}_{(dP-P+i)}^{(B)}\right)^T \\ &= \sum_{i=1}^{P} \mathbf{B}^{(i)} \left(\left(\mathbf{C}_{(dP-P+i),1}^{(B)}\right)^T \quad \left(\mathbf{C}_{(dP-P+i),2}^{(B)}\right)^T \right) \\ &= \sum_{i=1}^{P} \left(\mathbf{B}^{(i)} \otimes \mathbf{c}_{(dP-P+i),1} \quad \mathbf{B}^{(i)} \otimes \mathbf{c}_{(dP-P+i),2} \right), \qquad (d = 1, \cdots, D). \end{aligned} \tag{8.61}$$

Here $\otimes$ denotes element-wise multiplication between two vectors or matrices. The Rayleigh flat fading channel matrix can be written as

$$\mathbf{H}_f = \left(h_1, \quad h_2, \quad \cdots, \quad h_D \right) \tag{8.62}$$

The faded and noisy received signals can be expressed as

$$\begin{aligned} \mathbf{r} = \mathbf{H}_f \mathbf{S} + \mathbf{n} &= \sum_{d=1}^{D} h_d \mathbf{S}_d^{(B)} + \mathbf{n} \\ &= \sum_{d=1}^{D} \sum_{i=1}^{P} h_d \mathbf{B}^{(i)} \left(\mathbf{C}_{(dP-P+i)}^{(B)}\right)^T + \mathbf{n} \\ &= \sum_{d=1}^{D} \sum_{i=1}^{P} h_d \left(\mathbf{B}^{(i)} \otimes \mathbf{c}_{(dP-P+i),1} \; \mathbf{B}^{(i)} \otimes \mathbf{c}_{(dP-P+i),2}\right) + \mathbf{n} \end{aligned} \tag{8.63}$$

where $\mathbf{n}$ is the additive white Gaussian noise (AWGN) vector. Before decoding the received signals, we would like to briefly review the correlation properties of the GPC codes [268]. In particular, the following properties have been shown in [268] such that

1. for intra-group correlations:

$$\sum_{m=1}^{2} f_{corr}\left(\mathbf{c}_{k,m}^{(s)}, \mathbf{c}_{k',m}\right) \triangleq f_{corr}^{BK}\left(\left(\mathbf{c}_{k,1}^{(s)}, \mathbf{c}_{k,2}^{(s)}\right), \mathbf{c}_{k',1}, \mathbf{c}_{k',2}\right)$$

$$= \begin{cases} = 2N, & k = k', s = 0, i \in \mathbf{Z} \\ = 0, & k = k', s \neq iN', i \in \mathbf{Z} \\ \neq 0, & k = k', s = iN', i \in \mathbf{Z} \\ = 0, & k \neq k', s \neq iN', i \in \mathbf{Z} \\ \neq 0, & k \neq k', s = iN', i \in \mathbf{Z} \end{cases} \tag{8.64}$$

2. for inter-group correlations:

$$\sum_{m=1}^{2} f_{corr}\left(\mathbf{c}_{k,m}^{(s)}, \mathbf{c}_{k',m}\right) \triangleq f_{corr}^{BK}\left(\left(\mathbf{c}_{k,1}^{(s)}, \mathbf{c}_{k,2}^{(s)}\right), \mathbf{c}_{k',1}, \mathbf{c}_{k',2}\right) = 0, \qquad k \neq k', s \in \mathbf{Z} \tag{8.65}$$

where N denotes the element code length, $E = N/N'$ is the expansion index, N' denotes the element code length of complete complementary code (CCC) before expansion, and K is the family size of the GPC codes and equal to $2E$ here. Furthermore, $\mathbf{c}_{k,m}^{(s)}$ denotes the cyclic shift version of $\mathbf{c}_{k,m}$ with s chip offset shifts, $\mathbf{Z}$ is the set of all integers, and $f_{corr}(\mathbf{x}, \mathbf{y})$ represents the correlation function between the codes $\mathbf{x}$ and $\mathbf{y}$.

It is seen from (8.64) and (8.65) that the GPC has a large IFW. The whole GPC family can be divided into two groups, G_1 and G_2, of the same size. The codes from the different groups possess ideal cross-correlation functions (CCFs).

It is assumed that all K GPC codes will be used in the STCC OS/CC-CDMA MIMO system to achieve maximal diversity gain and parallel transmission capability, i.e. $DP \leq K$, where D denotes the diversity order and P is the number of different symbol streams transmitted sequentially from the same antenna. Moreover, from the GPC properties, we should avoid those chip offsets $xN'T_s$ (where $x \in \mathbf{Z}$) between any parallel transmitted symbol streams, because two GPC codes from the same group have non-zero correlation side lobes at certain chip offsets. We can now apply the properties of the GPC codes to (8.63) to decode the received signals under flat fading channel as

$$\begin{aligned}
&\sum_{d=1}^{D} h_d^* f_{corr}^{BK}\left(\mathbf{r}, \left(\mathbf{O}_{i-1}^T \mathbf{c}_{(dP-P+i),1} \mathbf{O}_{N-1}^T\right), \mathbf{c}_{(dP-P+i),2}\right) \\
&= \sum_{d=1}^{D} \|h_d\|^2 \mathbf{B}^{(i)} + \mathbf{n}' \\
&= \left(\|h_1\|^2 + \|h_2\|^2 + \cdots + \|h_D\|^2\right) \mathbf{B}^{(i)} + \mathbf{n}', \qquad (i = 1, \cdots, P)
\end{aligned} \tag{8.66}$$

where $*$ denotes the complex conjugation and $\mathbf{O}_{(N-1)}$ is an $(N-1) \times 1$ zero matrix. (8.66) gives a generic decision variable expression with diversity gain [16], and its average error probability can be expressed as

$$\begin{aligned}
P_b &= \int_0^{\infty} Q\left(\sqrt{\frac{2E_b r}{DN_0}}\right) f_{B_D}(r)\, dr \\
&= \int_0^{\infty} Q\left(\sqrt{2x}\right) \frac{(n_t N_0)^D x^{D-1}}{\left(2E_b\sigma^2\right)^{n_t} (D-1)!} \exp\left(-\frac{DN_0}{2E_b\sigma^2} x\right) dx \\
&= \left(\frac{1-\mu}{2}\right)^D \sum_{k=1}^{D-1} \binom{D-1+k}{k} \left(\frac{1+\mu}{2}\right)^k
\end{aligned} \tag{8.67}$$

where $f_{B_D}(r)$ and μ are defined as

$$\begin{cases}
f_{B_D}(r) = \left(\dfrac{1}{2\sigma^2}\right)^D \dfrac{r^{D-1}}{(D-1)!} e^{-\dfrac{r}{2\sigma^2}} u(r) \\
\mu = \sqrt{\dfrac{2E_b\sigma^2/(DN_o)}{1 + 2E_b\sigma^2/(DN_o)}}
\end{cases} \tag{8.68}$$

and $u(r)$ is the unit step function and $\sigma^2 = 0.5$. It is easy to show that each symbol achieves diversity gain of D to mitigate fading channel with P parallel transmission branches carrying different information streams.

Thus, the coding efficiency for the proposed STCC OS/CC-CDMA MIMO scheme using the complete complementary codes (CCC) [254] can be calculated as follows.

$$\frac{T_B}{T_{CC}}K = \frac{B}{M(B+N-1)}K = \frac{B}{B+N-1} \approx 1 \tag{8.69}$$

which holds if B is sufficiently large. The coding efficiency is calculated as follows: total number of symbols before STCC encoding for one STCC block T_B is divided by the length of one STCC coded block T_{CC}, and then multiplied by the family size of the CCC or K. The CCC element code number is constrained as $M = 2$ to lower hardware complexity.

On the other hand, the STCC coding efficiency for $M = 2$ GPC can be calculated as follows.

$$\frac{E-1}{E}\frac{T_B}{T_{CC}}\left\lfloor\frac{K}{D}\right\rfloor \approx \frac{N-N'}{N}\frac{BP}{(B+N-1)} \approx P \tag{8.70}$$

where we have taken into account the fact that the transmitted signals are offset stacked within GPC's IFW, thus causing some spectral efficiency degradation as denoted in the term $(E-1)/E$, compared to the cases with the CCC codes, as shown in (8.69). However, K in (8.70) is much greater than that for CCC codes in (8.69), and this is why GPC can greatly enhance spectral efficiency due to the large number of available GPC codes in a family.

8.11 FREQUENCY-SELECTIVE FADING CHANNELS

This section extends the analysis carried out in the previous section into frequency-selective Rayleigh fading channels. To do so, the aforementioned element code convolution matrix should be modified by zero insertion such that we obtain

$$\mathbf{C}_{(e),m}^{(g,B)} = \begin{pmatrix} \mathbf{c}_{e,m}^T & \mathbf{O}_{(g)} & \mathbf{O}_{(g)} & \mathbf{O}_{(g)} & \mathbf{O}_{(g)} \\ 0 & \mathbf{c}_{e,m}^T & \mathbf{O}_{(g)} & \mathbf{O}_{(g)} & \mathbf{O}_{(g)} \\ 0 & 0 & \mathbf{c}_{e,m}^T & \mathbf{O}_{(g)} & \mathbf{O}_{(g)} \\ \vdots & \vdots & \vdots & \ddots & \vdots \\ 0 & 0 & 0 & 0 & \mathbf{c}_{e,m}^T \end{pmatrix} \tag{8.71}$$

whose dimension becomes $(Bg + N - g) \times B$. Here, $\mathbf{c}_{e,m}$ denotes the $\{e, m\}$th element code of a GPC code, and $\mathbf{O}_{(\mathbf{g})}$ is a $g \times 1$ zero vector. We define $\mathbf{C}_{(e),m}^{(g,B)+}$ as follows:

$$\mathbf{C}_{(e),m}^{(g,B)+} = \begin{pmatrix} \mathbf{C}_{(e),m}^{(g,B)} \\ \mathbf{Q}_{(g-1,B)} \end{pmatrix}_{(Bg+N-1)\times B} \tag{8.72}$$

where $\mathbf{Q}_{(g-1,B)}$ is a zero matrix with its dimension being $(g-1) \times B$. The composite element code convolution matrix after the zero insertion can be written as

$$\mathbf{C}_{(e)}^{(g,B)} = \begin{pmatrix} \mathbf{C}_{(e),1}^{(g,B)+} \\ \mathbf{C}_{(e),2}^{(g,B)+} \end{pmatrix}_{2(Bg+N-1)\times B} \tag{8.73}$$

Finally, all composite element code convolution matrices are combined to form the STCC encoding matrix $\mathbf{C}^{(g,B)}$ as follows

$$\mathbf{C}^{(g,B)} = \begin{pmatrix} \mathbf{C}^{(g,B)}_{(1)} & \mathbf{C}^{(g,B)}_{(2)} & \cdots & \mathbf{C}^{(g,B)}_{(P)} \\ \mathbf{C}^{(g,B)}_{(P+1)} & \mathbf{C}^{(g,B)}_{(P+2)} & \cdots & \mathbf{C}^{(g,B)}_{(2P)} \\ \vdots & \vdots & \ddots & \vdots \\ \mathbf{C}^{(g,B)}_{(DP-P+1)} & \mathbf{C}^{(g,B)}_{(DP-P+2)} & \cdots & \mathbf{C}^{(g,B)}_{(DP)} \end{pmatrix} \tag{8.74}$$

which is a $2D\,(Bg + N - 1) \times BP$ matrix and plays a similar role as the matrix (8.57). Using (8.74), we can encode the input symbol streams to get STCC encoded symbol blocks as

$$\tilde{\mathbf{S}}^{(L,B)} = \mathbf{C}^{(L,B)}\mathbf{B} = \begin{pmatrix} \tilde{\mathbf{S}}^{(L,B)}_1 \\ \tilde{\mathbf{S}}^{(L,B)}_2 \\ \vdots \\ \tilde{\mathbf{S}}^{(L,B)}_D \end{pmatrix}_{2D(BL+N-1)\times 1} \tag{8.75}$$

where we have

$$\tilde{\mathbf{S}}^{(L,B)}_d = \sum_{i=1}^{P} \mathbf{C}^{(L,B)}_{(dP-P+i)} \left(\mathbf{B}^{(i)}\right)^T, \qquad (d = 1, \ldots, D) \tag{8.76}$$

We can rearrange $\tilde{\mathbf{S}}^{(L,B)}$ to form $\mathbf{S}^{(L,B)}$ as

$$\mathbf{S}^{(L,B)} = \begin{pmatrix} \left(\tilde{\mathbf{S}}^{(L,B)}_1\right)^T \\ \left(\tilde{\mathbf{S}}^{(L,B)}_2\right)^T \\ \vdots \\ \left(\tilde{\mathbf{S}}^{(L,B)}_D\right)^T \end{pmatrix} = \begin{pmatrix} \mathbf{S}^{(L,B)}_1 \\ \mathbf{S}^{(L,B)}_2 \\ \vdots \\ \mathbf{S}^{(L,B)}_D \end{pmatrix} \tag{8.77}$$

where

$$\begin{aligned} \mathbf{S}^{(L)}_d &= \sum_{i=1}^{P} \left(\mathbf{C}^{(L,B)}_{(dP-P+i)} \left(\mathbf{B}^{(i)}\right)^T\right)^T \\ &= \sum_{i=1}^{P} \mathbf{B}^{(i)} \left(\mathbf{C}^{(L,B)}_{(dP-P+i)}\right)^T \\ &= \sum_{i=1}^{P} \mathbf{B}^{(i)}\left(\left(\mathbf{C}^{(L,B)}_{(dP-P+i),1}\right)^T \left(\mathbf{C}^{(L,B)}_{(dP-P+i),2}\right)^T\right) \\ &= \sum_{i=1}^{P} \left(\mathbf{B}^{(i)} \otimes \mathbf{c}_{(dP-P+i),1}\ \mathbf{O}^T_{(L-1)}\ \mathbf{B}^{(i)} \otimes \mathbf{c}_{(dP-P+i),2}\right) \\ & (d = 1, \cdots, D) \end{aligned} \tag{8.78}$$

The channel model in the dth space can be defined by a $(2\,(BL+N-1)+L-1)\times 2\,(BL+N-1)$ matrix as

$$\mathbf{H}_d = \begin{pmatrix} \mathbf{h}_d\,(1) & 0 & 0 & 0 \\ \mathbf{h}_d\,(2) & \mathbf{h}_d\,(1) & 0 & \vdots \\ \vdots & \mathbf{h}_d\,(2) & \ddots & 0 \\ \mathbf{h}_d\,(L) & \vdots & \ddots & \mathbf{h}_d\,(1) \\ 0 & \mathbf{h}_d\,(L) & \vdots & \mathbf{h}_d\,(2) \\ \vdots & 0 & \ddots & \vdots \\ 0 & 0 & 0 & \mathbf{h}_d\,(L) \end{pmatrix} \tag{8.79}$$

where $\mathbf{h}_d$ is an L-tap transversal filter coefficient vector in the dth space. Combination of all space channels forms a unified channel matrix $\mathbf{H}$ as

$$\mathbf{H} = \begin{pmatrix} \mathbf{H}_1 & 0 & \ddots & 0 \\ 0 & \mathbf{H}_2 & \ddots & \ddots \\ \ddots & \ddots & \ddots & 0 \\ 0 & \ddots & 0 & \mathbf{H}_D \end{pmatrix} \tag{8.80}$$

which is a $(2\,(BL+N-1)+L-1)\,D\times 2\,(BL+N-1)\,D$ matrix. Over a frequency-selective fading channel, the faded signals can be written as

$$\begin{pmatrix} \mathbf{Z}_1^T \\ \mathbf{Z}_2^T \\ \vdots \\ \mathbf{Z}_D^T \end{pmatrix} = \begin{pmatrix} \mathbf{H}_1 & 0 & \cdots & 0 \\ 0 & \mathbf{H}_2 & \cdots & \vdots \\ \vdots & \vdots & \vdots & 0 \\ 0 & \cdots & 0 & \mathbf{H}_D \end{pmatrix} \begin{pmatrix} \mathbf{S}_1^{(L-1)} \\ \mathbf{S}_2^{(L-1)} \\ \vdots \\ \mathbf{S}_D^{(L-1)} \end{pmatrix} \tag{8.81}$$

where we define

$$\mathbf{Z}_d = \left(\mathbf{H}_d \tilde{\mathbf{S}}_d^{(L-1)}\right)^T = \left(\tilde{\mathbf{S}}_d^{(L-1)}\right)^T \mathbf{H}_d^T = \mathbf{S}_d^{(L-1)} \mathbf{H}_d^T = \mathbf{h}_d \otimes \mathbf{S}_d^{(L-1)} \tag{8.82}$$

The faded and noisy received signals can be expressed by

$$\begin{aligned} \mathbf{r} = \mathbf{Z} + \mathbf{n} &= \sum_{d=1}^{D} \mathbf{Z}_d + \mathbf{n} = \sum_{d=1}^{D} \mathbf{h}_d \otimes \mathbf{S}_d^{(L-1)} + \mathbf{n} \\ &= \sum_{d=1}^{D} \sum_{p=1}^{P} \mathbf{h}_d \otimes \left(\mathbf{B}^{(i)} \otimes \mathbf{c}_{(dP-P+i),1} \quad \mathbf{O}_{(L-1)}^T \quad \mathbf{B}^{(i)} \otimes \mathbf{c}_{(dP-P+i),2}\right) + \mathbf{n} \end{aligned} \tag{8.83}$$

The decoding process is analogous to what we did for flat fading channels in the previous section. Two assumptions should be made here. One is that perfect channel state information (CSI) is available to the receiver, and the other is that the guard interval is greater than or at least equal to the maximum delay spread of the channel, i.e. $gT_s \geq \tau_{\max}$. Under these two assumptions, the decoding result will become the same as (8.66), except for some loss in spectral efficiency due to the use of the guard intervals.

8.12 RESULTS AND DISCUSSIONS ON STCC OS/CC-CDMA

The numerical analysis has been carried out to evaluate the performance of the STCC OS/CC-CDMA MIMO scheme. Four multipath delay profiles have been taken into account for the performance evaluation purpose. In particular, we have used the multipath channel models with two, three, four, and five path delay profiles, which are represented by vectors, $[\sqrt{0.6}, \sqrt{0.4}]$, $[\sqrt{0.45}, \sqrt{0.35}, \sqrt{0.2}]$, $[\sqrt{0.4}, \sqrt{0.25}, \sqrt{0.2}, \sqrt{0.15}]$, and $[\sqrt{0.3}, \sqrt{0.25}, \sqrt{0.2}, \sqrt{0.15}, \sqrt{0.1}]$, respectively, and denoted by the abbreviations, mp-2, mp-3, mp-4, and mp-5, respectively.

In Figure 8.14, STCC and STBC schemes both using BPSK modulation are compared. The figure shows that under the same signal-to-noise ratio (SNR), the STCC OS/CC-CDMA MIMO schemes with different GPC code parameters can achieve a much higher transmission rate than STBC MIMO. The STCC OS/CC-CDMA MIMO scheme provides a unique flexible diversity gain and multiplex capability tradeoff mechanism, which is in particular useful for link adaptation (LA) implementation to meet different operational requirements, such as quality of service (QoS), maximal data throughput, bit error rate (BER), etc. On the other hand, the traditional STBC MIMO system can only offer a fixed transmission rate for whatever spatial diversity order, and is unable to offer joint diversity-multiplex capability.

In Figure 8.15, we have used the cost function defined as the maximum transmission rate times diversity order over a minimum SNR at a given BER $= 0.001$. It is shown that the GPC code set with its parameters $(K, M, N, E) = (16, 2, 64, 8)$ is suitable to work with an STCC OS/CC-CDMA MIMO scheme with $D = 4$ in terms of the cost function. It is shown in Figure 8.16 that even under a relatively severe multipath channel environment (such as the one with five multipath components, the mp-5 model) the STCC OS/CC-CDMA MIMO can still achieve full diversity gain to further improve system performance. This advantage can be attributed to the use of GPC codes in the STCC OS/CC-CDMA MIMO system with a near interference-free operation due to the widely open IFW of the codes.

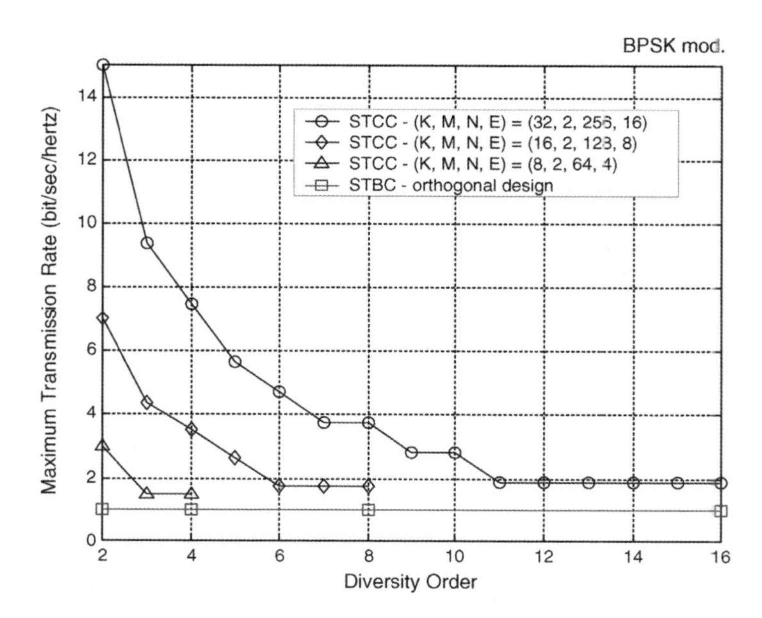

Figure 8.14 Diversity versus parallel transmission tradeoff curves for STCC OS/CC-CDMA MIMO.

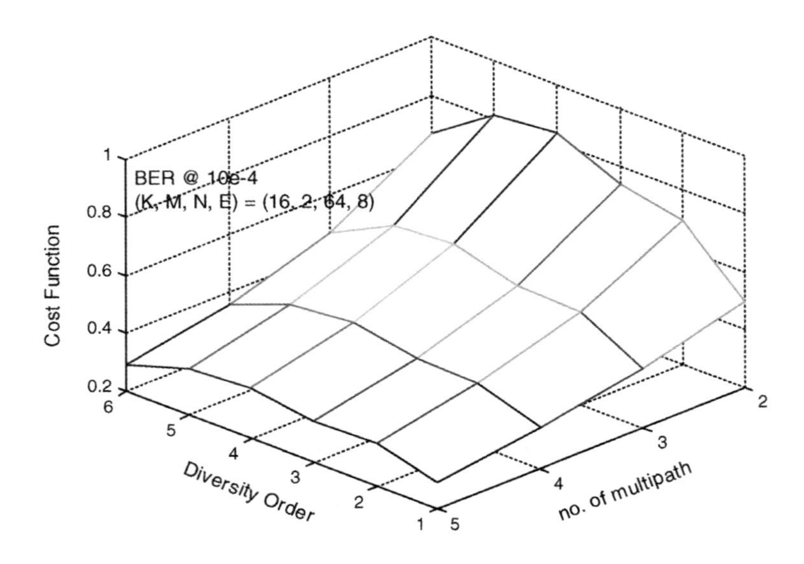

Figure 8.15 Optimized system performance measured by the cost function.

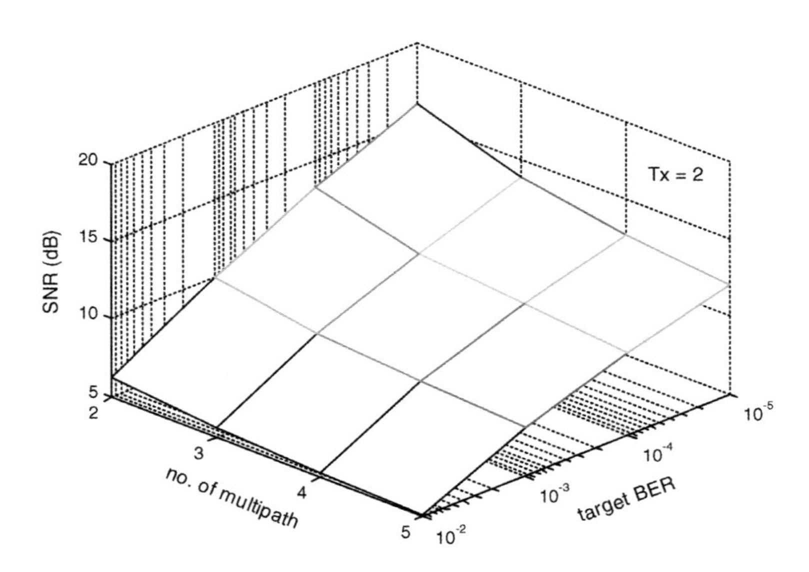

Figure 8.16 BER performance in different multipath channels.

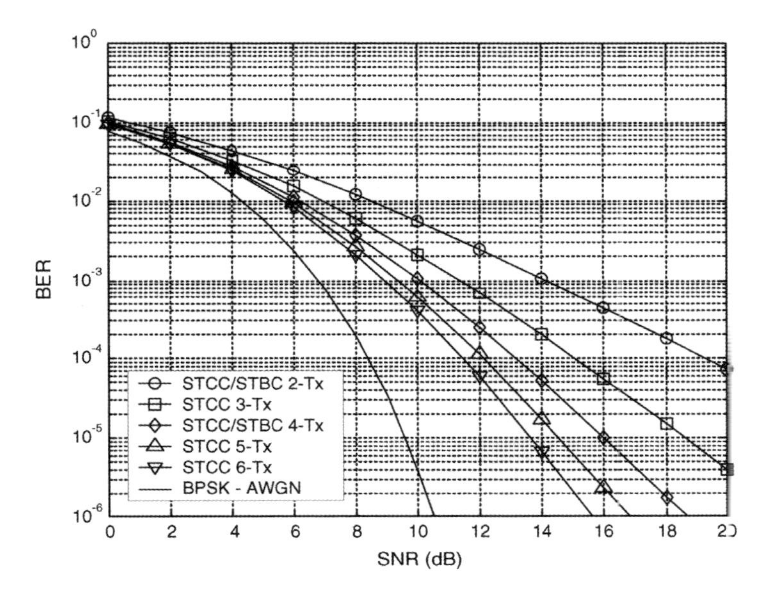

Figure 8.17 Comparison of STCC OS/CC-CDMA and STBC CDMA over flat fading channels.

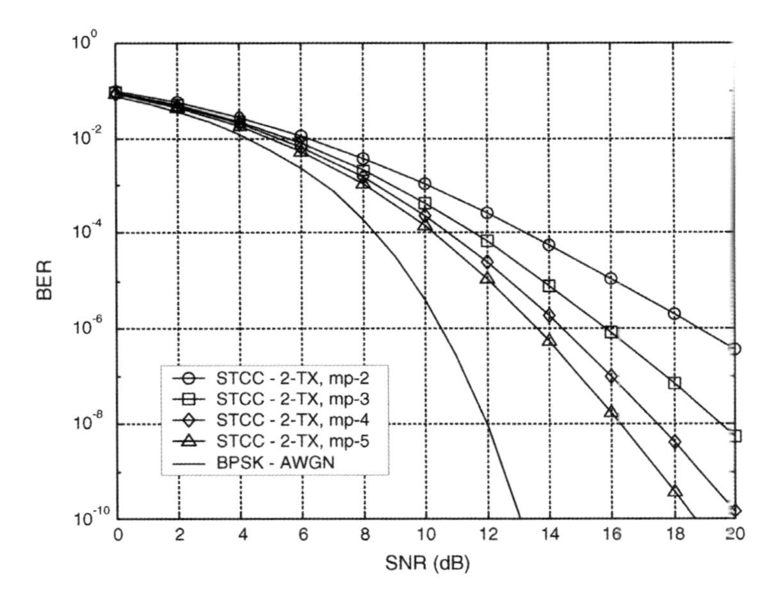

Figure 8.18 BER performance for 2-Tx STCC OS/CC-CDMA MIMO under different multipath channel environments, where $D = 2$.

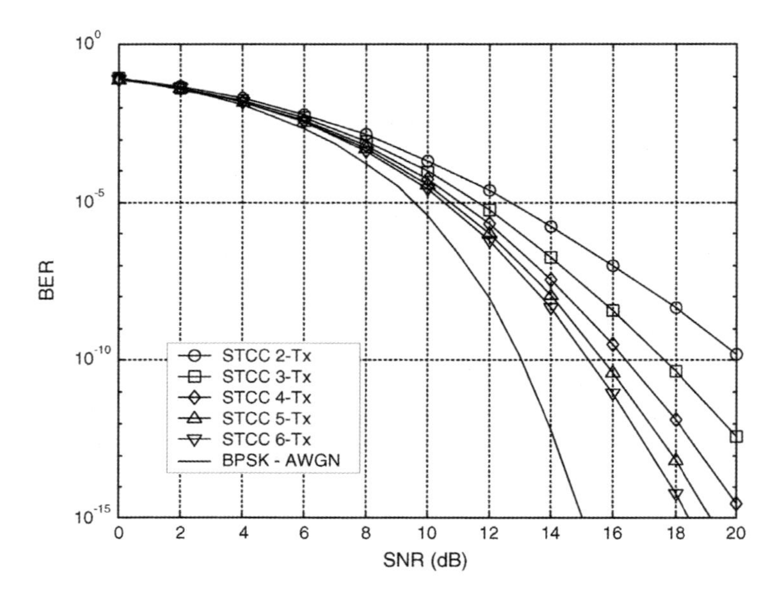

Figure 8.19 BER performance for STCC OS/CC-CDMA MIMO with variable Tx-diversity orders, where mp-4, a four-path multipath channel model, is concerned.

In Figure 8.17, we compare STCC with STBC in terms of spatial diversity gain that affects system performance, in which we consider only flat fading channels. Figure 8.17 shows that the STCC OS/CC-CDMA MIMO can perform nearly as well as STBC MIMO with the same diversity gain. It is shown in Figure 8.18 that under the same diversity order $D = 2$, rich multipath components can be used to further improve the system performance. This again shows that the STCC OS/CC-CDMA MIMO scheme facilitates the achievement of a better multipath diversity as all multipath components in the STCC OS/CC-CDMA MIMO system can be separated and coherently recombined efficiently, without mutual interference generated among the signals from different antennas and different multipath returns. Figure 8.19 shows that under the same multipath channel environment (where the mp-4 channel model is concerned) with the increase in Tx-diversity order (as D becomes very large), the system performance can eventually approach the theoretical bit error rate of BPSK over an AWGN channel.

8.13 SUMMARY ON STCC OS/CC-CDMA

We have introduced another way to integrate S-T coding with the CC-CDMA scheme, namely STCC OS/CC-CDMA MIMO, as an effort to achieve a high diversity-multiplex product gain. The analytical results have shown that the proposed STCC OS/CC-CDMA MIMO scheme is very attractive due to its great flexibility to tradeoff diversity and multiplexing orders. It ensures near interference-free operation in multipath channels owing to the use of GPC codes to implement S-T coding. The STCC OS/CC-CDMA MIMO implementation complexity is slightly higher than that for traditional STBC MIMO due to the use of multiple power amplifiers in its transmitter. It also requires ADC/DAC hardware with a relatively large dynamic range compared with the STBC scheme. Nevertheless, the superb operational performance in terms of BER and capacity even in multipath environments makes it a very competent candidate for applications in future wireless systems.

As a final remark to this chapter, we would like to stress that the S-T coded CC-CDMA schemes can be an extremely important part of the next generation CDMA technologies, as they can help to enhance the overall system performance without consumption of scarce spectral resources. Working jointly with the superb MAI-free property of a CC-CDMA architecture, the S-T coded CC-CDMA (either STCC DS/CC-CDMA or STCC OS/CC-CDMA) can find extensive applications in beyond 3G wireless communication systems.

It is easy to understand that perfect orthogonality of orthogonal complementary codes proposed in this book can be used for both user separation and antenna separation. As a matter of fact, they should work in a much more efficient way to separate the signals sent from different transmitter antennas than any simple block coding schemes, such as the one used in traditional STBC. Therefore, it is not surprising for the STCC CC-CDMA schemes to offer much better spatial diversity gain in any operational environments than other antenna signal separation techniques.

More importantly, the application of the orthogonal complementary codes also paves the way for joint optimization for space-time-frequency three-dimensional signaling design, which can integrate the signal processing for antenna separation and user separation. Under this three-dimensional signaling framework, S-T coding has been combined with CDMA coding. Each user in such a system will be assigned a code volume, which is a three-dimensional data element block, for both user separation and antenna separation purposes. The spreading modulation in such a system will take place in both the time domain and the frequency domain (using an OFDM architecture). The spreading modulation will not be extended to the space domain, but each antenna will use a different matrix for antenna separation, in order to achieve spacial diversity or spacial multiplexing capability.

Then, the problem is how to generate such three-dimensional code volumes for both user separation and antenna separation purposes. One very straightforward approach is to use existing orthogonal complementary codes, which are already capable of implementing time-frequency two-dimensional spreading under a multi-carrier CDMA platform. In this case, each user will use a different code matrix for user separation purposes. Now, we can try to generate a code volume based on each orthogonal complementary code, which forms an $N \times M$ code matrix, where N stands for the element code length and M the flock size of the orthogonal complementary code. One possible way to make a code volume based on one particular two-dimensional orthogonal complementary code is to use a random coding method. For instance, if each user will use two transmitter antennas, then each two-dimensional orthogonal complementary code will be used to generate two code matrices of the same size, being equal to $N \times M$. The elements of the two code matrices will be the randomized versions of the elements of the original two-dimensional orthogonal complementary code. In this way, each $N \times M$ orthogonal complementary code will be converted into one $N \times M \times 2$ code volume, which can be used to separate users as well as two antennas used by each user.

Of course, there should be many other ways to generate the three-dimensional code volumes for this purpose. It is indeed a very interesting topic worthy of much more further study. More discussions on three-dimensional spreading code design can be found in Section 5.2.

9

M-ary CDMA Technologies

In our endeavor to work with the next generation CDMA technology, we have tried to search for some better spreading codes and spreading modulation schemes as an effort to improve the performance of a tradition DS-CDMA system. This approach has been discussed in detail in previous chapters, such as Chapter 6, Chapter 7, and Chapter 8. The content covered in the previous chapters is focused mainly on ways to improve system design based on traditional DS-CDMA systems by using better spreading codes and better spreading modulation schemes, and so forth. In this chapter we will take a different approach to design of a CDMA system. This approach is called M-ary CDMA technology. It is to be noted that M-ary CDMA is not a completely new technology and it can be traced back to the researches conducted in the 1990s [162, 163]. As a matter of fact, a primitive form of the M-ary CDMA technique was successfully used in uplink channel signaling design for the first commercial CDMA system, or the IS-95 standard, where the baseband binary data stream is grouped into symbols, each of which has six bits. Then, each symbol pattern will choose one particular Walsh-Hadamard sequence out of $2^6 = 64$ sequences before carrier modulation. Therefore, at the receiver side a reverse process will take place to decode the original data symbol.

A major difference between conventional DS-CDMA and M-ary CDMA lies in the fact that normal DS-CDMA will use spreading codes to spread the data bit to extend its signal bandwidth and also to act as the signature of a user for CDMA purposes. The spreading codes themselves will not be considered as the information carrier. On the other hand, in M-ary CDMA user data information will be carried directly on the spreading codes, each of which will represent a particular symbol pattern. Therefore, a receiver should immediately know the symbol pattern if it can detect which code is sent by the user. In this sense, we can understand that the spreading code space has been directly used as a signal-representation dimension, which acts very much like the time, frequency, and space domains.

A general introduction to multi-code and M-ary CDMA techniques has been given in Sections 2.7 and 5.7. Therefore, we will not repeat what has been discussed before. In this chapter, we will give some discussions on M-ary CDMA technology in a more in-depth manner. There are four different types of M-ary CDMA techniques: orthogonal code system, multi-code system, parallel combinatorial system, and BPSK M-ary CDMA system. Due to space limitations, we will focus only on BPSK M-ary CDMA system in this chapter.

As to be shown in the discussions covered in this chapter, M-ary CDMA technology can offer some uniquely important features, and thus we strongly believe that it can also contribute as a vital part of next generation CDMA technology for the reasons that will be explained in the following sections.

The Next Generation CDMA Technologies Hsiao-Hwa Chen

9.1 BPSK M-ARY CDMA SYSTEM MODEL

Let us first define the BPSK M-ary CDMA system model, based on which our analysis will be carried out in this chapter. The transmitter diagram is shown in Figure 9.1 and the receiver structure is shown in Figure 9.2. Let us look at the transmitter model first. The original data stream will go through a serial to parallel converter before the symbol mapping operation. Each symbol may contain different numbers of bits, depending on the number of available spreading codes (here we assume that there are H codes available to each user) and the ways of mapping each symbol to the combination of spreading codes to be sent out. Table 9.1 shows the relationship between the number of codes and the code combinatory states, together with the transmission rate in units of bits/symbol. In Table 9.1, the variable H varies only from 2 to 6 to save space. It is seen from the table that an increase in H will result in a fairly large number of states. For instance, if $H = 6$ (or there are six codes available to each user), we have the number of states as large as 728, which can be exploited effectively to pack nine bits of information in each symbol.

It should be noted that with $H = 6$ we in fact can make use of $2^9 = 512$ different states out of the 728 states, thus leaving $728 - 512 = 216$ states unused. We should not consider those 216 states as being wasted as the difference between the actual states in use and those unused will give us a great opportunity to optimize the state constellation. This alone is a very interesting research topic, which is very useful and important in many ways.

Also, it is noted that in an M-ary CDMA system each user will be assigned a group of H_k spreading codes, where $k = 1, 2, \ldots, \tilde{K}$ if there are in total $\tilde{K}$ users in the system. Therefore, if the

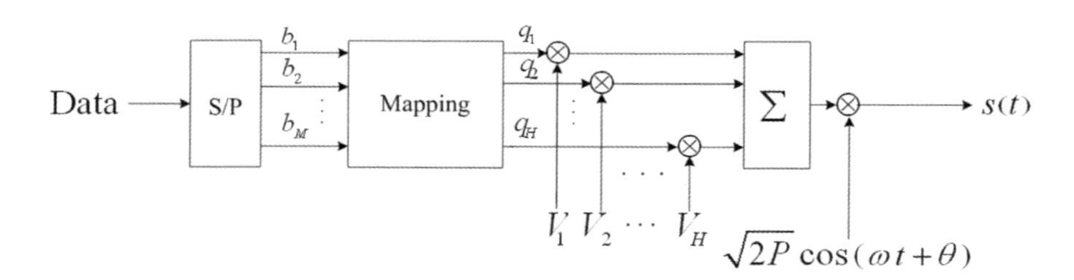

Figure 9.1 Transmitter for the BPSK M-ary CDMA system.

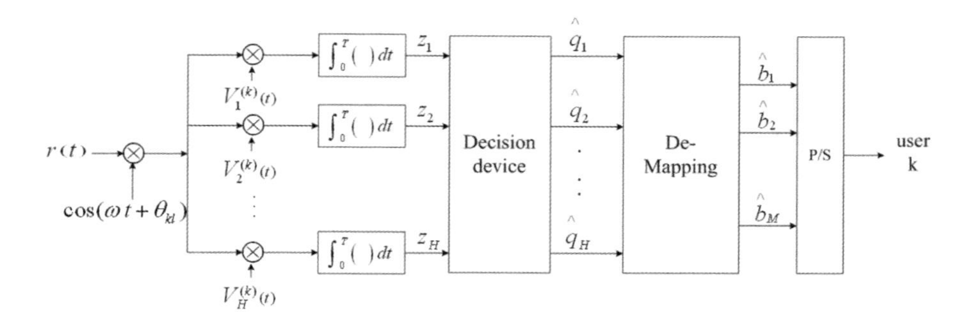

Figure 9.2 Receiver for the BPSK M-ary CDMA system.

Table 9.1 The relation between three parameters in a BPSK M-ary CDMA system.

H	No. of states	Bits/symbol (m)
2	8	3
3	26	4
4	80	6
5	242	7
6	728	9

Table 9.2 Example mapping relation between symbol patterns (in this case each symbol contains four bits) and actually sent codes combination.

Symbol	Sent codes
$d_0\ d_1\ d_2\ d_3$	A, B, C
0000	+1, 0, 0
0001	−1, 0, 0
0010	0, +1, 0
0011	0, −1, 0
0100	0, 0, +1
0101	0, 0, −1
0110	+1, +1, 0
0111	+1,−1, 0
1000	−1, −1, 0
1001	−1, +1, 0
1010	+1, 0, +1
1011	+1, 0, −1
1100	−1, 0, −1
1101	−1, 0, +1
1110	0, +1, +1
1111	0, +1, −1
NA	0, −1, −1
NA	0, −1, +1
NA	+1, +1, +1
NA	−1, +1, +1
NA	+1, −1, +1
NA	+1, +1, −1
NA	−1, −1, +1
NA	−1, +1, −1
NA	+1, −1, −1
NA	−1, −1, −1

set size of the spreading codes is K, then we always have

$$K = \sum_{k=1}^{\tilde{K}} H_k \tag{9.1}$$

which tells us that the requirement on the set size for the spreading codes in an M-ary CDMA system can be much higher than that in a conventional DS-CDMA system, where each user is normally assigned only one code.

It is also seen from the system model (as shown in Figures 9.1 and 9.2) that the transmitter can manipulate the codes in three ways: send it with positive weight, send it with negative weight, or not send it at all. We denote the three weights as $\{+1, -1, 0\}$. Therefore, if we have H codes available, then the total number of states can be $3^H - 1$, as there are in total 3^H states with the H codes, but we must deduct the possibility that the transmitter sends none of the H codes, which is a prohibiting state. In other words, a transmitter must send out something when there is data to convey. Therefore, we always have the number of bits packed in each symbol as

$$m = \lfloor \log_2(3^H - 1) \rfloor \tag{9.2}$$

where the notation $\lfloor x \rfloor$ stands for the largest integer less than x.

For a BPSK M-ary CDMA system with $H = 3$, we have Table 9.2 to show one example of the mapping relation between the symbol patterns (in this case, each symbol contains four bits) and the actually sent codes combination. Of course, there can be many different mapping schemes different from the one given in Table 9.2. Assume that the symbol duration is T. It can be easily shown that the bandwidth efficiency for the BPSK M-ary CDMA system is

$$\eta = \frac{\lfloor \log_2 (3^H - 1) \rfloor}{T} \tag{9.3}$$

9.2 BPSK M-ARY CDMA CONSTELLATION OPTIMIZATION

Having introduced the system model for a BPSK M-ary CDMA system, we can proceed with its performance analysis. To make the analysis manageable in the limited space available, we do not consider the multipath propagation effect here. In fact, the analysis taking into account multipath channels can be done by extending the analysis given here using a similar analytical approach, adding only some more complex derivations.

It is seen from Figure 9.1 that there is a mapping unit in the transmitter model, which decides which code combination should be used for a particular input symbol pattern, which contains m bits.[1]

Assume that there are m bits in each symbol, denoted by a symbol vector as $\{b_1, b_2, \ldots, b_m\}$. This symbol vector will be mapped into a new vector defined as $\{q_1, q_2, \ldots, q_H\}$ corresponding to H codes available at the transmitter. For any possible H, the elements in the vector $\{q_1, q_2, \ldots, q_H\}$ take values based on the following equation

$$\sum_{h=1}^{H} q_h^2 = 1 \tag{9.4}$$

[1]Usually, $3^H - 1$ is always much larger than 2^m, and thus we need to select certain 2^m constellation points from the total $3^H - 1$, leaving us a very nice optimization problem.

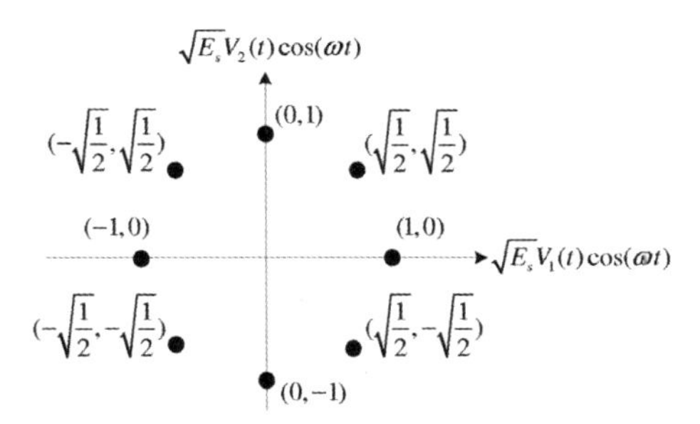

Figure 9.3 Constellation for a BPSK M-ary CDMA system with normalized average power for $H = 2$.

which will ensure that the same average power is used for any different H for fair comparison among different M-ary CDMA systems with different numbers of available spreading codes for each user.

For instance, if $H = 2$ than we will have $\{q_1, q_2\}$, which should satisfy Equation (9.4) such that they will form a constellation as shown in Figure 9.3, in which all constellation points will be distributed around a circle with its radius being unit (due to its normalized power level). It is noted that $H = 2$ is a very special case, in which $3^H - 1 = 8$ is equal exactly to $2^3 = 8$, and thus all eight code combination states will be fully used without leaving unused states. Each symbol consists of three bits ($m = 3$). In this case, transmission in each code duration will convey three bits of information, which is much higher than the case with a conventional DS-CDMA system, where only one bit of information will be sent to a receiver in the duration of a spreading code (here we only consider the short code as the spreading code, such that each spreading code will cover a complete bit duration). Therefore, in general an M-ary CDMA system always provides higher bandwidth efficiency than a conventional DS-CDMA system.

On the other hand, the detection scenario in an M-ary CDMA system is more complex than that in a conventional DS-CDMA system, due to the fact that the M-ary CDMA receiver needs to distinguish different constellation points, as shown in Figure 9.3, within a relatively short Euclidean distance among different constellation points compared to that in a conventional DS-CDMA system. Therefore, the higher bandwidth efficiency of an M-ary CDMA system is obtained in exchange for some degradation in power efficiency.

The constellation for $H = 2$ shown in Figure 9.3 is very special in terms of its possibility to visualize, as it will be very difficult for us to draw the constellation for any other larger H values. For instance, if $H = 3$ we have to use a three-dimensional graph to visualize the constellation, let alone for any other higher H values.

Obviously, for the case with $H = 2$ we do not have a constellation optimization problem, as all possible constellation points will be fully used to convey three bits of information in each sent symbol. However, with the increase in H, there will be a great number of unused constellation points which will be left out, leaving us scope to select those constellation points which can maximize the Euclidian distances among them, thus achieving the lowest possible error rate in the detection process. For example, if $H = 3$ we will have altogether $3^H - 1 = 26$ available constellation points, but we have to choose 16 from them to form a symbol with four bits of information. Therefore, there are in total $C_{26}^{16} = \binom{26}{16} = \frac{26!}{(26-16)!16!}$ choices, which is a huge number! All constellation points for $H = 3$ are given in Figure 9.4, and we need to select 16 from them for an $H = 3$ M-ary CDMA system.

The selection of the best constellation subset from all available ones can be described by a well-known optimization problem. Various algorithms can be used for this purpose, with different

s_1	s_2	s_3	s_4	s_5
$(1,0,0)$	$(-1,0,0)$	$(0,1,0)$	$(0,-1,0)$	$(0,0,1)$
s_6	s_7	s_8	s_9	s_{10}
$(0,0,-1)$	$(\frac{1}{\sqrt{2}},\frac{1}{\sqrt{2}},0)$	$(\frac{1}{\sqrt{2}},\frac{-1}{\sqrt{2}},0)$	$(\frac{-1}{\sqrt{2}},\frac{-1}{\sqrt{2}},0)$	$(\frac{-1}{\sqrt{2}},\frac{1}{\sqrt{2}},0)$
s_{11}	s_{12}	s_{13}	s_{14}	s_{15}
$(\frac{1}{\sqrt{2}},0,\frac{1}{\sqrt{2}})$	$(\frac{1}{\sqrt{2}},0,\frac{-1}{\sqrt{2}})$	$(\frac{-1}{\sqrt{2}},0,\frac{1}{\sqrt{2}})$	$(\frac{-1}{\sqrt{2}},0,\frac{-1}{\sqrt{2}})$	$(0,\frac{1}{\sqrt{2}},\frac{1}{\sqrt{2}})$
s_{16}	s_{17}	s_{18}	s_{19}	s_{20}
$(0,\frac{1}{\sqrt{2}},\frac{-1}{\sqrt{2}})$	$(0,\frac{-1}{\sqrt{2}},\frac{1}{\sqrt{2}})$	$(0,\frac{-1}{\sqrt{2}},\frac{-1}{\sqrt{2}})$	$(\frac{1}{\sqrt{3}},\frac{1}{\sqrt{3}},\frac{1}{\sqrt{3}})$	$(\frac{-1}{\sqrt{3}},\frac{1}{\sqrt{3}},\frac{1}{\sqrt{3}})$
s_{21}	s_{22}	s_{23}	s_{24}	s_{25}
$(\frac{1}{\sqrt{3}},\frac{-1}{\sqrt{3}},\frac{1}{\sqrt{3}})$	$(\frac{1}{\sqrt{3}},\frac{1}{\sqrt{3}},\frac{-1}{\sqrt{3}})$	$(\frac{-1}{\sqrt{3}},\frac{-1}{\sqrt{3}},\frac{1}{\sqrt{3}})$	$(\frac{-1}{\sqrt{3}},\frac{1}{\sqrt{3}},\frac{-1}{\sqrt{3}})$	$(\frac{1}{\sqrt{3}},\frac{-1}{\sqrt{3}},\frac{-1}{\sqrt{3}})$
s_{26}				
$(\frac{-1}{\sqrt{3}},\frac{-1}{\sqrt{3}},\frac{-1}{\sqrt{3}})$				

Figure 9.4 All constellation points for a BPSK M-ary CDMA system with normalized average power for $H = 3$.

constraints and conditions. We would like to use the following criteria to perform this optimization. The best constellation point subset should satisfy the following two conditions:

1. The mean sum of Euclidean distances between any two points should be maximized.
2. If there are multiple subsets which all satisfy the same maximized mean sum of Euclidean distances between any two points, select the one with the smallest variance of the Euclidean distances between any two points.

The first condition is to ensure the lowest detection error and the second condition helps to choose a constellation point subset with the most even distribution in the constellation space, and thus making the most even detection performance among different state points. The mean sum of Euclidean distances between any two points and the variance of the Euclidean distances between any two points can be defined as follows:

$$\mu = E[d_i] = \frac{1}{n}\sum_{i=1}^{n} d_i \tag{9.5}$$

and

$$\begin{aligned} Var(d_i) &= E[(d_i - \mu)^2] = E[d_i^2] - (E[\mu])^2 \\ &= \frac{1}{n}\sum_{i=1}^{n} d_i^2 - \left(\frac{1}{n}\sum_{i=1}^{n} d_i\right)^2 \end{aligned} \tag{9.6}$$

respectively, where $n = 2^m$ is the number of states in each symbol, and d_i is the Euclidean distance between a particular pair of two points. Based on the above algorithm we have obtained the two best subsets (which have the same performance under the above criteria) of constellation points for $H = 3$ as follows:

$$(s_1, s_2, s_3, s_4, s_5, s_6, s_7, s_8, s_9, s_{10}, s_{11}, s_{12}, s_{13}, s_{14}, s_{15}, s_{16}) \tag{9.7}$$

$$(s_1, s_2, s_3, s_4, s_5, s_6, s_7, s_{17}, s_{19}, s_{20}, s_{21}, s_{22}, s_{23}, s_{24}, s_{25}, s_{26}) \tag{9.8}$$

The worst subset of constellation points for $H = 3$ is

$$(s_7, s_8, s_9, s_{10}, s_{12}, s_{13}, s_{14}, s_{16}, s_{17}, s_{18}, s_{21}, s_{22}, s_{23}, s_{24}, s_{25}, s_{26}) \tag{9.9}$$

9.3 PRELIMINARIES FOR PERFORMANCE ANALYSIS

Next, we will go further to study the bit error rate performance of a BPSK M-ary CDMA system under a multiple user environment, where K users will be present and each user will use multiple codes for M-ary CDMA communications.

For easy reference, we present the BPSK M-ary CDMA system model with K users in Figures 9.5 and 9.6, in which Figure 9.5 shows the K-user signal formulation process and Figure 9.6 illustrates the receiver model tuned to the kth user.

Let us consider a BPSK M-ary CDMA system working under an asynchronous AWGN channel with K users, each with multiple codes for M-ary CDMA transmission. In particular, we further assume that User 1 uses A codes, and the rest use H codes for M-ary CDMA transmissions. We let User 1 use A codes, unlike all other users, in an effort to make the analysis more general compared

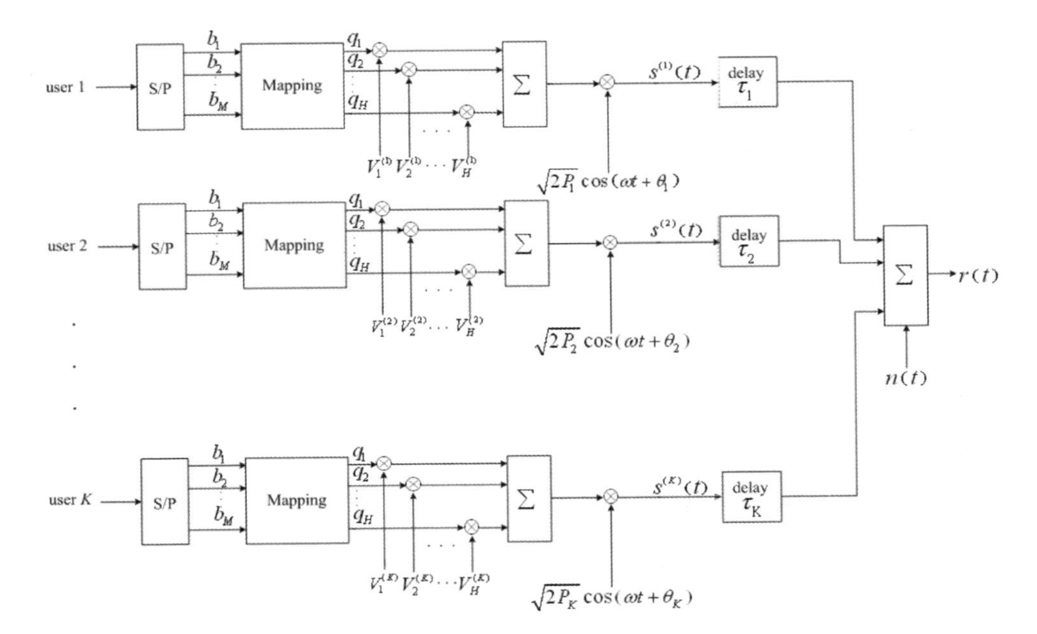

Figure 9.5 K-user signal formulation process for a BPSK M-ary CDMA system with each user assigned H codes.

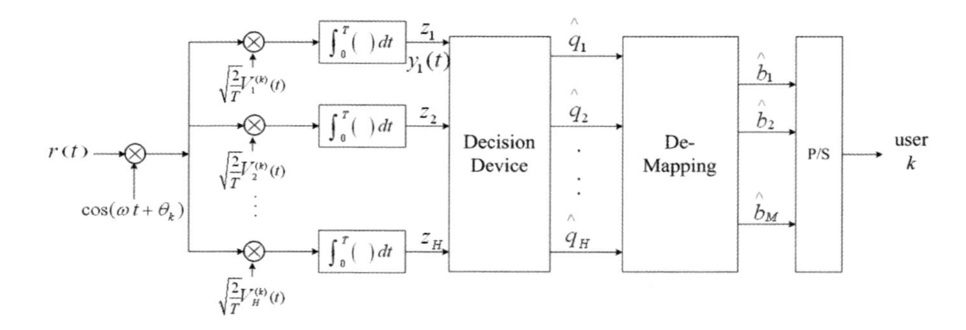

Figure 9.6 Receiver of a BPSK M-ary CDMA system with its target being the kth user.

to the case where all users use the same number of codes. The receiver will be tuned to the signal from User 1.

The transmitting signal from User 1 can be written as

$$s^{(1)}(t) = \sqrt{2P_1}\left[\sum_{h=1}^{A} V_h^{(1)}(t-\tau_1)q_h^{(1)}\right]\cos(\omega t+\theta_1) \tag{9.10}$$

and all other transmissions become

$$s^{(k)}(t) = \sqrt{2P_k}\left[\sum_{h=1}^{H} V_h^{(k)}(t-\tau_k)q_h^{(k)}\right]\cos(\omega t+\theta_k) \tag{9.11}$$

where $k = 1, 2, \ldots, K$, $s^{(k)}(t)$ is the signal sent from the kth user, V_h is the hth code available to the transmitter, q_h is the weight coefficient for the hth code after signal mapping, τ_k is the propagation delay for the kth user, θ_k is the initial carrier phase for the kth user, and $n(t)$ is the AWGN with its power spectral density being $N_0/2$.

Assume that the receiver wants to decode the signal from User 1. The received signal can be written as

$$r(t) = \sqrt{2P_1}\sum_{h=1}^{A} V_h^{(1)}(t-\tau_1)q_h\cos(\omega t+\theta_1) + r_I(t) + n(t) \tag{9.12}$$

where $r_I(t)$ is the interference contributed from all other $K-1$ unwanted transmissions. If User 1 uses A ($A = 1, 2, \ldots$) codes and the others use H codes, we have

$$r_I(t) = \sum_{k=2}^{K-1}\sqrt{2P_k}\left[\sum_{h=1}^{H} V_h^{(k)}(t-\tau_k)q_h^{(k)}\right]\cos(\omega t+\theta_k) \tag{9.13}$$

9.4 MAI ANALYSIS

Assume that there are a relatively large number of unwanted transmissions in the system. Therefore, we can approximate the interference component as a Gaussian random variable to simplify the analysis carried out in this chapter. As a matter of fact, as the number of interferers increases, the mismatch between the real distribution and the Gaussian distribution to model the interfering signals will be within an acceptable range. Also, we assume that the receiver has achieved perfect synchronization with the incoming signal from User 1, and thus we let $\tau_1 = 0$ and $\theta_1 = 0$ without losing generality.

Now we consider the interference caused by the kth unwanted transmission which uses the hth code, which can be expressed by

$$\begin{aligned} y_1(t) &= \int_0^T \sqrt{2P_k} q_h V_h^{(k)}(t-\tau_k)\cos(\omega t+\theta_k)\sqrt{\frac{2}{T}}V_1^{(1)}(t)\cos(\omega t)dt \\ &= \sqrt{\frac{P_k}{2}}\cos(\theta_k)\sqrt{\frac{2}{T}}\left(q_h^{(-1)}\int_0^{\tau_k} V_h^{(k)}(t-\tau_k)V_1^{(1)}(t)dt \right. \\ &\left. + q_h^{(0)}\int_{\tau_k}^{T} V_h^{(k)}(t-\tau_k)V_1^{(1)}(t)dt\right) = I_{k,h} \end{aligned} \tag{9.14}$$

Therefore, if taking into account all H codes used in the kth user, we will have

$$\begin{aligned} I_k &= \sum_{h=1}^{H} I_{h,k} \\ &= \sum_{h=1}^{H}\left[\sqrt{\frac{P_k}{2}}\cos(\theta_k)\sqrt{\frac{2}{T}}\left(q_h^{(-1)}\int_0^{\tau_k} V_h^{(k)}(t-\tau_k)V_1^{(1)}(t)dt \right.\right. \\ &\left.\left. + q_h^{(0)}\int_{\tau_k}^{T} V_h^{(k)}(t-\tau_k)V_1^{(1)}(t)dt\right)\right] \end{aligned} \tag{9.15}$$

Obviously, every code can take any value from $\{0, +1, -1\}$ (except for the all-zero state or no code sent, which is not allowed here), and there is no interference to the receiver if 0 is sent. Let us define a new variable $D_k^{(-1)}$, which denotes the event that the kth user indeed sent at least one non-zero code value (either $+1$ or -1, but not 0) among all H available codes in the previous symbol duration. Thus, we have

$$D_k^{(-1)} = \left\{h|\ |q_{k,h}^{(-1)}| \neq 0,\quad 1 \le h \le H\right\} \tag{9.16}$$

where $q_{k,h}^{(-1)}$ is the amplitude coefficient for the kth user and the hth code in the previous symbol. Similarly, we use $D_k^{(0)}$ to denote the event that the kth user sends at least one non-zero code value (either $+1$ or -1, but not 0) among all H available codes in the current symbol duration, being defined as

$$D_k^{(0)} = \left\{h|\ |q_{k,h}^{(0)}| \neq 0,\quad 1 \le h \le H\right\} \tag{9.17}$$

where $q_{k,h}^{(0)}$ is the amplitude coefficient for the kth user and the hth code in the current symbol duration. Therefore, we have

$$\begin{aligned} I_k = \sqrt{\frac{P_k}{2}}\cos(\theta_k)\sqrt{\frac{2}{T}}&\left[\sum_{D_k^{(-1)}\in h} q_h^{(-1)}\int_0^{\tau_k} V_h^{(k)}(t-\tau_k)V_1^{(1)}(t)dt \right. \\ &\left. + \sum_{D_k^{(0)}\in h} q_h^{(0)}\int_{\tau_k}^{T} V_h^{(k)}(t-\tau_k)V_1^{(1)}(t)dt\right] \end{aligned} \tag{9.18}$$

Therefore, we can use $D^{(-1)}$ to denote the event that all $K-1$ users send at least one non-zero code value (either $+1$ or -1, but not 0) among H available codes in the previous symbol duration, and thus we have

$$D^{(-1)} = \left\{(k,h)|\ |q_{k,h}^{(-1)}| \neq 0,\quad 1 < k \le K, 1 \le h \le H\right\} \tag{9.19}$$

Similarly, we have $D^{(0)}$ to denote the event that all $K-1$ users send at least one non-zero code value (either $+1$ or -1, but not 0) among H available codes in the current symbol duration as

$$D^{(0)} = \left\{(k,h)|\ |q_{k,h}^{(0)}| \neq 0,\ 1 \le k \le K, 1 \le h \le H\right\} \tag{9.20}$$

Taking into account all $K-1$ interferences, we obtain

$$\begin{aligned}
y_1(t) &= \int_0^T \sum_{k=2}^{K-1} \sum_{h=1}^{H} \sqrt{2P_k} V_h^{(k)}(t-\tau_k) q_h^{(k)} \cos(\omega t + \theta_k) V_1^{(1)}(t) \cos(\omega t) dt \\
&= \sqrt{\frac{P_k}{2}} \cos(\theta_k) \sqrt{\frac{2}{T}} \left[\sum_{D^{(-1)} \in (k,h)} q_{k,h}^{(-1)} \int_0^{\tau_k} V_h^{(k)}(t-\tau_k) V_1^{(1)}(t) dt \right. \\
&\quad \left. + \sum_{D^{(0)} \in (k,h)} q_{k,h}^{(0)} \int_0^{\tau_k} V_h^{(k)}(t-\tau_k) V_1^{(1)}(t) dt \right] = I
\end{aligned} \tag{9.21}$$

where I is all $K-1$ interference to the detection of the first user signal at the receiver. Under the Gaussian approximation assumption, we can derive the conditional variance (assume that $D_{(-1)}$ and $D_{(0)}$ are fixed at the moment) of I based on Equation (9.21) as follows.

$$\begin{aligned}
Var(I|D^{(-1)}, D^{(0)}) &= E\left[I^2\right] - (E[I])^2 \\
&= E\left[\left(\sqrt{\frac{P_k}{2}} \cos(\theta_k) \sqrt{\frac{2}{T}} \left(\sum_{D^{(-1)} \in (k,h)} q_{k,h}^{(-1)} \int_0^{\tau_k} V_h^{(k)}(t-\tau_k) V_1^{(1)}(t) dt \right.\right.\right. \\
&\quad \left.\left.\left. + \sum_{D^{(0)} \in (k,h)} q_{k,h}^{(0)} \int_0^{\tau_k} V_h^{(k)}(t-\tau_k) V_1^{(1)}(t) dt \right)\right)^2\right] \\
&= \sum_{D^{(-1)} \in (k,h)} \frac{E[P_k](q_{k,h}^{(-1)})^2}{2T} E\left[\left(\int_0^{\tau_k} V_h^{(k)}(t-\tau_k) V_1^{(1)}(t) dt\right)^2\right] \\
&\quad + \frac{E[P_k]}{T} \left(\sum_{D^{(-1)} \in (k,h)} E[q_{k,h}^{(-1)}] \int_0^{\tau_k} V_h^{(k)}(t-\tau_k) V_1^{(1)}(t) dt \right] \\
&\quad \times \left. \sum_{D^{(0)} \in (k,h)} E[q_{k,h}^{(0)}] \int_0^{\tau_k} V_h^{(k)}(t-\tau_k) V_1^{(1)}(t) dt \right) \\
&\quad + \sum_{D^{(0)} \in (k,h)} \frac{E[P_k](q_{k,h}^{(0)})^2}{2T} E\left[\left(\int_{\tau_k}^{T} V_h^{(k)}(t-\tau_k) V_1^{(1)}(t) dt\right)^2\right] \\
&= \sum_{D^{(-1)} \in (k,h)} \frac{E[P_k](q_{k,h}^{(-1)})^2}{2T} E\left[\left(\int_0^{\tau_k} V_h^{(k)}(t-\tau_k) V_1^{(1)}(t) dt\right)^2\right] \\
&\quad + \sum_{D^{(0)} \in (k,h)} \frac{E[P_k](q_{k,h}^{(0)})^2}{2T} E\left[\left(\int_{\tau_k}^{T} V_h^{(k)}(t-\tau_k) V_1^{(1)}(t) dt\right)^2\right]
\end{aligned} \tag{9.22}$$

The discrete aperiodic cross-correlation function between any two codes $V^{(m)}$ and $V^{(n)}$ can be written as

$$V_{m,n}(i) = \begin{cases} \sum_{j=0}^{N-1-i} a_j^{(m)} a_{j+1}^{(n)}, & 0 \le i \le N-1 \\ \sum_{j=0}^{N-1+i} a_{j-i}^{(m)} a_j^{(n)}, & -(N-1) \le i \le 0 \\ 0, & \text{otherwise} \end{cases} \tag{9.23}$$

where the two sequences of interest are

$$\begin{cases} V^{(m)} = \left(a_0^{(m)}, a_1^{(m)}, \ldots a_{N-1}^{(m)}\right) \\ V^{(n)} = \left(a_0^{(n)}, a_1^{(n)}, \ldots a_{N-1}^{(n)}\right) \end{cases} \tag{9.24}$$

The discrete aperiodic cross-correlation function between two codes $V^{(m)}$ and $V^{(n)}$ is shown in Figure 9.7.

With the help of discrete cross-correlation functions, we can decompose the calculation of either even or odd (depending on the signs of two consecutive bits) periodic cross-correlation functions into two parts, expressed in terms of the discrete cross-correlation functions, as illustrated in Figure 9.8.

Using the following relations,

$$\begin{cases} E(\cos\theta_k) = 0 \\ E[q_{k,h}^{(-1)} q_{k,h}^{(0)}] = E[q_{k,h}^{(-1)}]E[q_{k,h}^{(0)}] = 0 \\ \int_0^{\tau_k} V_m(t-\tau_k)V_n(t)dt \\ = T_c\left[V_{m,n}(-(N-i-1))\gamma_k + V_{m,n}(-(N-i))(1-\gamma_k)\right] \\ = T_c\left[V_{m,n}(i-N) + (V_{m,n}(1+i-N) - V_{m,n}(i-N))\gamma_k\right] \\ \int_{\tau_k}^{T} V_m(t-\tau_k)V_n(t)dt \\ = T_c\left[V_{m,n}(i)(1-\gamma_k) + V_{m,n}(i+1)\gamma_k\right] \\ = T_c\left[V_{m,n}(i) + (V_{m,n}(1+i) - V_{m,n}(i))\gamma_k\right] \end{cases} \tag{9.25}$$

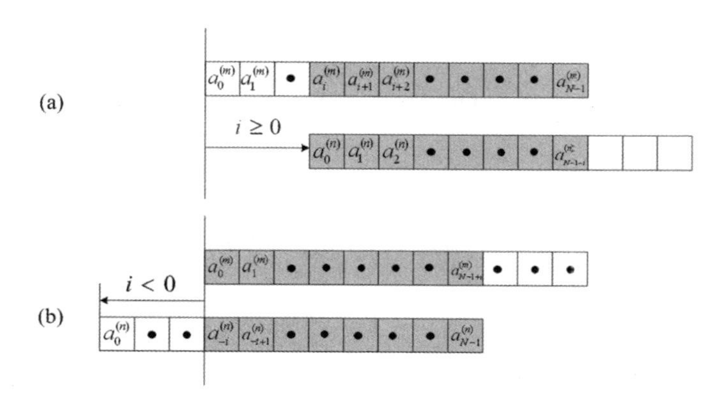

Figure 9.7 Discrete aperiodic cross-correlation function between two codes $V^{(m)}$ and $V^{(n)}$.

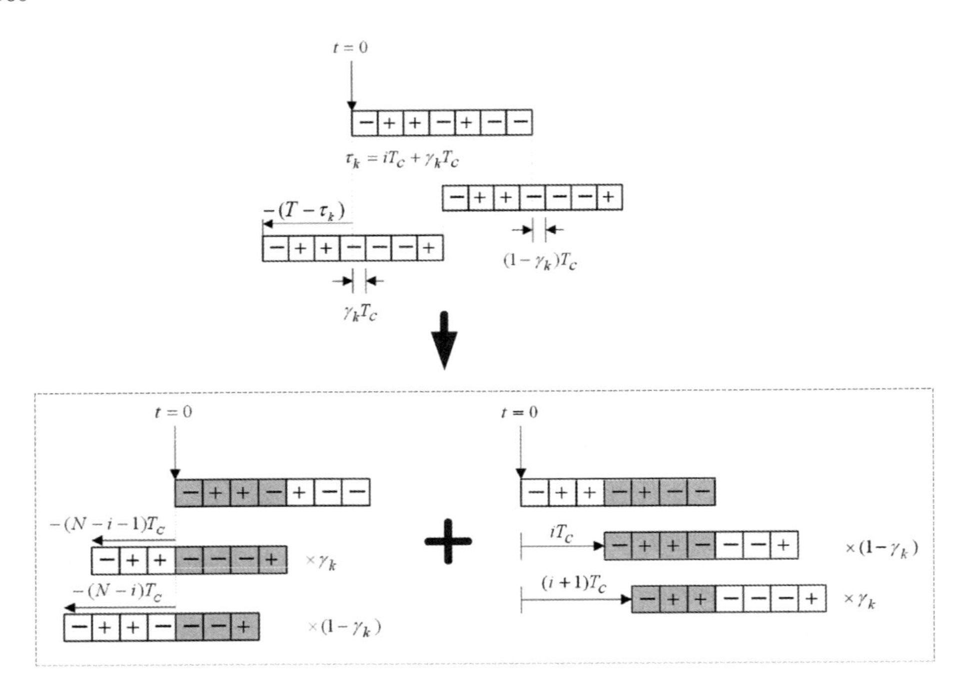

Figure 9.8 Decomposition of periodic cross-correlation function into two partial cross-correlation functions with the help of discrete cross-correlation functions between the two sequences.

we can rewrite Equation (9.22) into

$$
\begin{aligned}
&Var\left(I|D^{(-1)}, D^{(0)}\right)\\
&= \sum_{D^{(-1)}} \frac{E[P_k](q_{k,h}^{(-1)})^2}{2T} E\left[\left(T_c\left[V_{h,1}^{(k,1)}(i-N) + \left(V_{h,1}^{(k,1)}(1+i-N) - V_{h,1}^{(k,1)}(i-N)\right)\gamma_k\right]\right)^2\right]\\
&\quad + \sum_{D^{(0)}} \frac{E[P_k](q_{k,h}^{(0)})^2}{2T} E\left[\left(T_c\left[V_{h,1}^{(k,1)}(i) + \left(V_{h,1}^{(k,1)}(i+1) - V_{h,1}^{(k,1)}(i)\right)\gamma_k\right]\right)^2\right]\\
&= \sum_{D^{(-1)}} \frac{E[P_k](q_{k,h}^{(-1)})^2}{2T} \Bigg\{\frac{T_c^2}{N}\sum_{i=0}^{N-1}\Bigg(\left(V_{h,1}^{(k,1)}(i-N)\right)^2 + 2E[\gamma_k]\Big(V_{h,1}^{(k,1)}(i-N)\\
&\quad \times V_{h,1}^{(k,1)}(1+i-N) - \left(V_{h,1}^{(k,1)}(i-N)\right)^2\Big) + E[\gamma_k^2]\Big\{\left(V_{h,1}^{(k,1)}(1+i-N)^2\right)\\
&\quad - 2V_{h,1}^{(k,1)}(1+i-N)\,V_{h,1}^{(k,1)}(i-N) + \left(V_{h,1}^{(k,1)}(i-N)\right)^2\Big\}\Bigg)\Bigg\}\\
&\quad + \sum_{D^{(0)}} \frac{E[P_k](q_{k,h}^{(0)})^2}{2T} \Bigg\{\frac{T_c^2}{N}\sum_{i=0}^{N-1}\Bigg(\left(V_{h,1}^{(k,1)}(i)\right)^2 + 2E[\gamma_k]\Big(V_{h,1}^{(k,1)}(i)\,V_{h,1}^{(k,1)}(i+1)\\
&\quad - \left(V_{h,1}^{(k,1)}(i)\right)^2\Big) + E[\gamma_k^2]\Big\{\Big(\left(V_{h,1}^{(k,1)}(i+1)\right)^2 - 2V_{h,1}^{(k,1)}(i+1)\,V_{h,1}^{(k,1)}(i) + \left(V_{h,1}^{(k,1)}(i)\right)^2\Big)\Big\}\Bigg)\Bigg\}
\end{aligned}
\tag{9.26}
$$

After inserting $E[\gamma_k] = \frac{1}{2}$ and $E[\gamma_k^2] = \frac{1}{3}$ into the above equation, we obtain

$$\begin{aligned} V(I|D^{(-1)}, D^{(0)}) &= \sum_{D^{(-1)}\in(k,h)} \frac{E[P_k](q_{k,h}^{(-1)})^2 T_c^2}{12N} \left\{ \sum_{i=0}^{N-1} \left(\left(V_{h,1}^{(k,1)}(i-N) \right)^2 \right. \right. \\ &\left. \left. + V_{h,1}^{(k,1)}(i-N)\, V_{h,1}^{(k,1)}(1+i-N) + \left(V_{h,1}^{(k,1)}(1+i-N) \right)^2 \right) \right\} \\ &+ \sum_{D^{(0)}\in(k,h)} \frac{E[P_k](q_{k,h}^{(0)})^2 T_c^2}{12N} \sum_{i=0}^{N-1} \left[\left(V_{h,1}^{(k,1)}(i) \right)^2 + V_{h,1}^{(k,1)}(i)\, V_{h,1}^{(k,1)}(i-1) \right] \end{aligned} \tag{9.27}$$

9.5 BER ANALYSIS FOR BPSK M-ARY CDMA

Assume that we will use purely random sequences for their application in the BPSK M-ary CDMA system of concern. Then, we will have the following relations

$$\begin{aligned} &\sum_{i=0}^{N-1} E\left[V_{m,n}^2(1+i-N)\right] \\ &= \sum_{i=0}^{N-1} \sum_{j=0}^{N-1+(1+i-N)} \sum_{l=0}^{N-1+(1+i-N)} E\left[a_{j-(1+i-N)}^{(m)} a_j^{(n)} a_{l-(1+i-N)}^{(m)} a_l^{(n)} \right] \\ &= \sum_{i=0}^{N-1} \sum_{l=0}^{i} E(1) = \sum_{i=0}^{N-1} (i+1) = \frac{(N+1)N}{2} \end{aligned} \tag{9.28}$$

as we always have

$$\begin{aligned} E\left[a_{j-(1+i-N)}^{(m)} a_j^{(n)} a_{l-(1+i-N)}^{(m)} a_l^{(n)} \right] &= E\left[a_{j-(1+i-N)}^{(m)} a_{l-(1+i-N)}^{(m)} \right] E\left[a_j^{(n)} c_l^{(n)} \right] \\ &= E\left[a_{j-(1+i-N)}^{(m)} \right] E\left[a_{l-(1+i-N)}^{(m)} \right] E\left[a_j^{(n)} \right] E\left[a_l^{(n)} \right] \end{aligned} \tag{9.29}$$

where $l \neq j$. In a similar way we can show that the following equations are always true

$$\begin{cases} \sum_{i=0}^{N-1} E\left[V_{m,n}^2(i)\right] = \frac{N(N+1)}{2} \\ \sum_{i=0}^{N-1} E\left[V_{m,n}^2(i-N)\right] = \frac{N(N-1)}{2} \\ \sum_{i=0}^{N-1} E\left[V_{m,n}^2(i+1)\right] = \frac{N(N-1)}{2} \end{cases} \tag{9.30}$$

The co-variance term can also be calculated as

$$\begin{aligned} &\sum_{i=0}^{N-1} E(V_{m,n}(1+i-N) V_{m,n}(i-N)) \\ &= \sum_{i=1}^{N-1} E(V_{m,n}(1+i-N) V_{m,n}(i-N)) \\ &= \sum_{i=1}^{N-1} \sum_{j=0}^{N-1+(1+i-N)} \sum_{l=0}^{N-1+(i-N)} E(a_{j-(1+i-N)}^{(m)} a_j^{(n)} a_{l-(i-N)}^{(m)} a_l^{(n)}) \\ &= \sum_{i=1}^{N-1} \sum_{l=0}^{i-1} E(a_{l-(1+i-N)}^{(m)} a_l^{(n)} a_{l-(i-N)}^{(m)} a_l^{(n)}) = 0 \end{aligned} \tag{9.31}$$

where we have used $V_{m,n}(-N) = 0$, $E\left[a_j^{(n)} a_l^{(n)}\right] = 0$ and $E\left[a_{l-(1+i-N)}^{(m)} a_{l-(i-N)}^{(m)}\right] = 0$. We can also show that we have $E\left[V_{m,n}(i) V_{m,n}(i+1)\right] = 0$. Therefore, we get

$$Var\left(I|D^{(-1)}, D^{(0)}\right) = \frac{E[P_k]T}{6N}\left(\sum_{D^{(-1)} \in (k,h)} (q_{k,h}^{(-1)})^2 + \sum_{D^{(0)} \in (k,h)} (q_{k,h}^{(0)})^2\right) \tag{9.32}$$

To unconditional the two random variables $D^{(-1)}$ and $D^{(0)}$, we define

$$\begin{cases} x = \sum_{D^{(-1)} \in (k,h)} (q_{k,h}^{(-1)})^2 \\ y = \sum_{D^{(0)} \in (k,h)} (q_{k,h}^{(0)})^2 \end{cases} \tag{9.33}$$

It is noted that x and y obey the same distribution, and thus we have

$$Var(I) = \frac{E[P_k]T}{6N} E\left[x + y\right] = \frac{E[P_k]T}{3N} E[x] \tag{9.34}$$

As we have $E\left[\sum_{h=1}^{H} q_{k,h}^2\right] = 1$, we obtain

$$E[x] = E\left[\sum_{D^{(-1)} \in (k,h)} (q_{k,h}^{(-1)})^2\right] = E\left[\sum_{k=1}^{K-1}\sum_{h=1}^{H} (q_{k,h}^{(-1)})^2\right] = E\left[\sum_{k=1}^{K-1} 1\right] = K - 1 \tag{9.35}$$

After some simplifications, we have $Var(I) = \frac{E[P_k]T(K-1)}{3N}$. Using $E_k = TE[P_k]$ we can have the equivalent variance that includes both noise and interference as

$$\sigma^2 = Var[I] + \frac{N_0}{2} = \frac{E_k(K-1)}{3N} + \frac{N_0}{2} \tag{9.36}$$

The bit error rate for a BPSK M-ary CDMA system can be calculated by using the maximal likelihood (ML) detection algorithm [17] as

$$\begin{aligned} P_e &= \frac{1}{F}\sum_{i=1}^{F} P_e(m_i) \le \frac{1}{2F}\sum_{i=1}^{F}\sum_{k=1,k\neq i}^{F} \operatorname{erfc}\left(\frac{d_{ik}}{2\sqrt{2\sigma^2}}\right) \\ &= \frac{1}{2F}\sum_{i=1}^{F}\sum_{k=1,k\neq i}^{F} \operatorname{erfc}\left(\frac{d_{ik}}{2\sqrt{\frac{2E_k(K-1)}{3N} + N_0}}\right) \\ &= \frac{1}{2F}\sum_{i=1}^{F}\sum_{k=1,k\neq i}^{F} \operatorname{erfc}\left(\frac{d_{ik}}{2\sqrt{E_1\left(\frac{2E_k(K-1)}{3E_1 N} + \frac{N_0}{E_1}\right)}}\right) \\ &= \frac{1}{2F}\sum_{i=1}^{F}\sum_{k=1,k\neq i}^{F} \operatorname{erfc}\left(\frac{d_{ik}}{2\sqrt{E_1\left(\frac{2m_k\psi(K-1)}{3m_1 N} + \frac{1}{SNR}\right)}}\right), \quad d_{ik} = ||s_i - s_k|| \end{aligned} \tag{9.37}$$

where $d_{ik} = ||s_i - s_k||$ is the Euclidean distance between the ith and kth constellation points, N is the length of spreading code, m_k is the number of bits per symbol for the kth user, K is the number of users supported by the M-ary CDMA system, and we have $\psi = E_{bk}/E_{b1}$, indicating that the kth user has the same energy per bit as that of the first user.

9.6 RESULTS AND DISCUSSION

It is to be noted that in the analysis carried out above for the BPSK M-ary CDMA system we do not consider the orthogonal complementary codes. Instead, we choose a set of much simpler purely random spreading sequences here for analysis simplicity. Of course, we can also make use of the orthogonal complementary codes as the spreading codes in M-ary CDMA systems, which may result in analytical results much more complicated than those using purely random spreading sequences. Another shortcoming for using purely random spreading sequences for the BER performance analysis is that it will give a BER result that is independent of the type of spreading codes used. Therefore, all results shown in this section are also spreading code independent.

Figure 9.9 shows the bit error rate performance for BPSK M-ary CDMA systems, in which there are 1, 5, 10, and 15 users. The channel model is AWGN without multipath propagation effect and the spreading code length is 128. User 1 (the signal of interest to the receiver) uses two spreading codes and the rest use only one code. $\psi = E_{bk}/E_{b1} = 1$ tells us that the kth user has the same energy per bit as that of the first user.

The bit error rate performance for this case is quite normal (without any surprise), with the BER curves varying smoothly as the SNR changes from 0 dB to 20 dB.

Figure 9.10 shows the bit error rate performance for a BPSK M-ary CDMA system, in particular with different lengths of spreading codes, being 128, 256, and 512, while the number of users is fixed at 15. Again, User 1 uses two codes, while the rest use only one code.

Figure 9.11 shows the bit error rate performance for a BPSK M-ary CDMA system, in particular with different lengths of spreading codes, being 128, 256, and 512, while the number of users is treated as an x-axis variable. Again, User 1 uses two codes, while the rest use only one code. SNR is fixed at 15 dB and once again we have $\psi = E_{bk}/E_{b1} = 1$. The channel is also AWGN without multipath effect, as in the previous two figures.

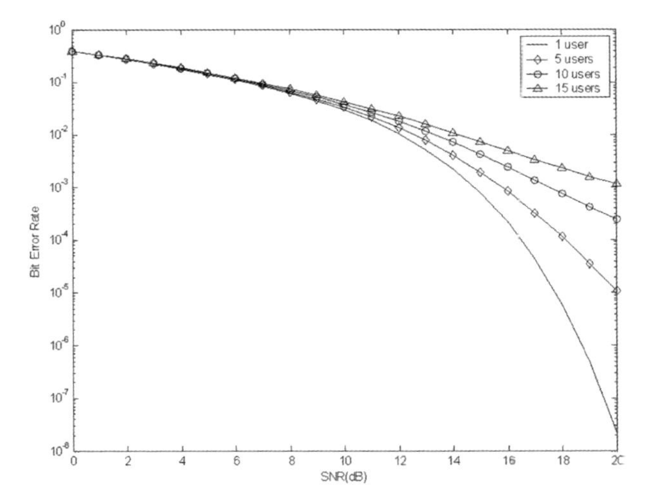

Figure 9.9 Bit error rate performance for a BPSK M-ary CDMA system with spreading code length being 128. The channel model is AWGN. The number of users is 1, 5, 10, and 15. User 1 (the signal of interest to the receiver) uses two spreading codes and the rest use only one code. $\psi = E_{bk}/E_{b1} = 1$.

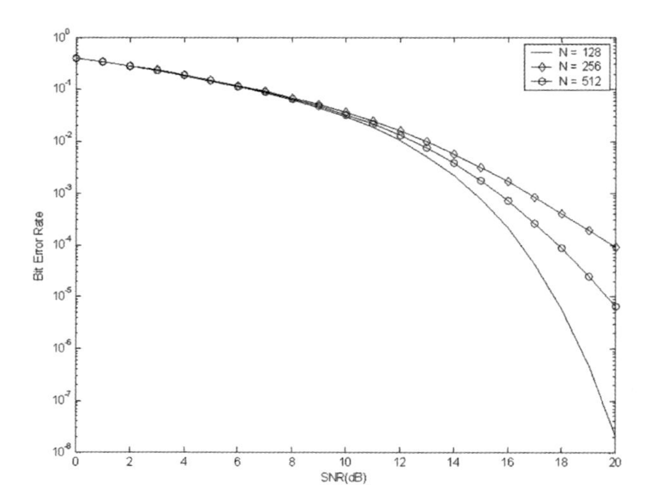

Figure 9.10 Bit error rate performance for a BPSK M-ary CDMA system with spreading code length being 128, 256, and 512. The channel model is AWGN. The number of users is fixed at 15. User 1 (the signal of interest to the receiver) uses two spreading codes and the rest use only one code. $\psi = E_{bk}/E_{b1} = 1$.

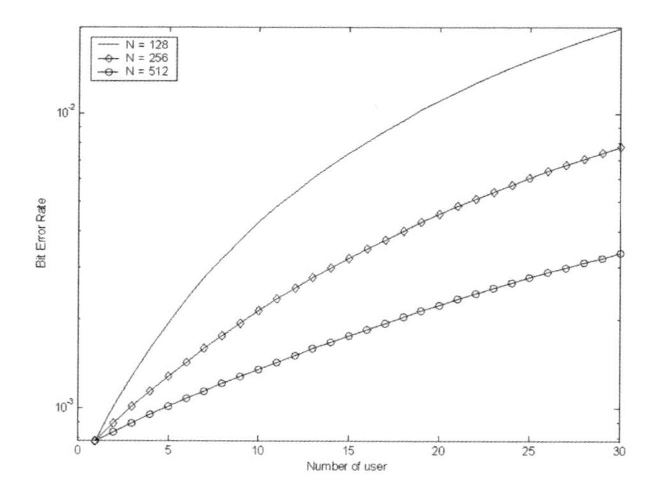

Figure 9.11 Bit error rate performance for a BPSK M-ary CDMA system versus the number of users. The spreading code length takes 128, 256, and 512. The channel model is AWGN. The SNR is fixed at 15 dB. User 1 (the signal of interest to the receiver) uses two spreading codes and the rest use only one code. $\psi = E_{bk}/E_{b1} = 1$.

In Figure 9.12, we would like to see how the BER performance changes with the value of H, which is the number of codes used by the other users. User 1 still uses two codes. The length of the spreading code is 256 and the number of users is fixed at 15, and again $\psi = E_{bk}/E_{b1} = 1$.

Now we would like to take a look at the issues on how constellation optimization can affect the bit error rate performance of a BPSK M-ary CDMA system. In this case, we assume that User 1

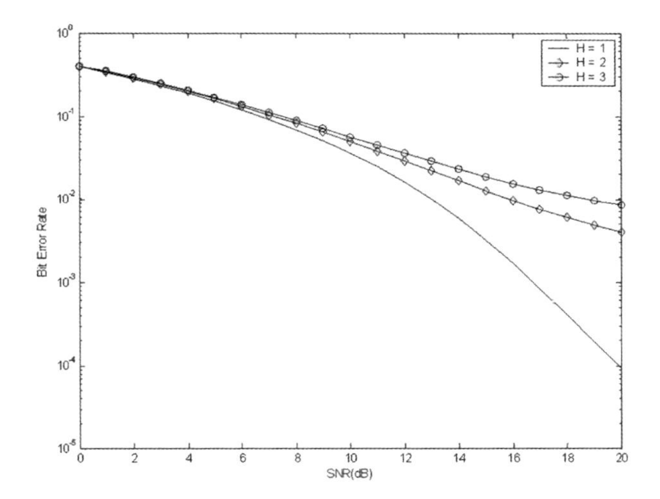

Figure 9.12 Bit error rate performance for a BPSK M-ary CDMA system versus the number of codes used by other users. The spreading code length is 256. The channel model is AWGN. The SNR is fixed at 15 dB. User 1 (the signal of interest to the receiver) uses two spreading codes. $\psi = E_{bk}/E_{b1} = 1$.

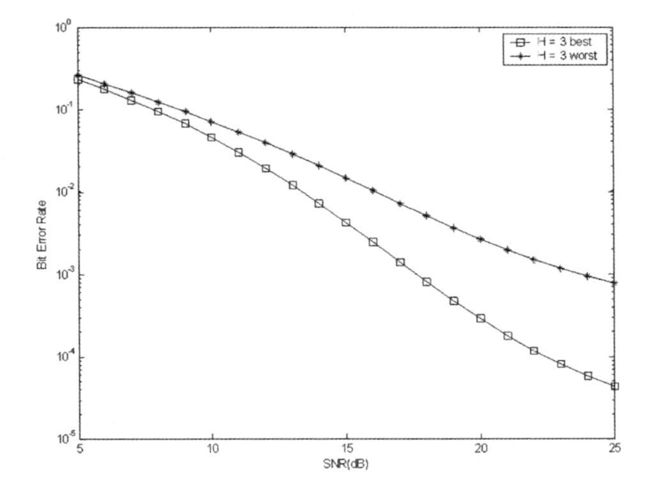

Figure 9.13 Bit error rate performance for a BPSK M-ary CDMA system using the best and worst constellations. The spreading code length is 256. The channel model is AWGN. User 1 (the signal of interest to the receiver) uses two spreading codes, while the rest use three codes. $\psi = E_{bk}/E_{b1} = 1$.

uses three codes and the rest use two codes only. The code length is the same, being 256 chips. We still assume $\psi = E_{bk}/E_{b1} = 1$, and the number of users is fixed at 10. We will use the best (as given in Equation (9.7) or (9.8)) and worst (as given in Equation (9.9)) constellations to illustrate the performance at two extremes. It is seen from Figure 9.13 that the BER performance of a BPSK M-ary CDMA system can vary substantially for different constellations used, and its SNR gain can

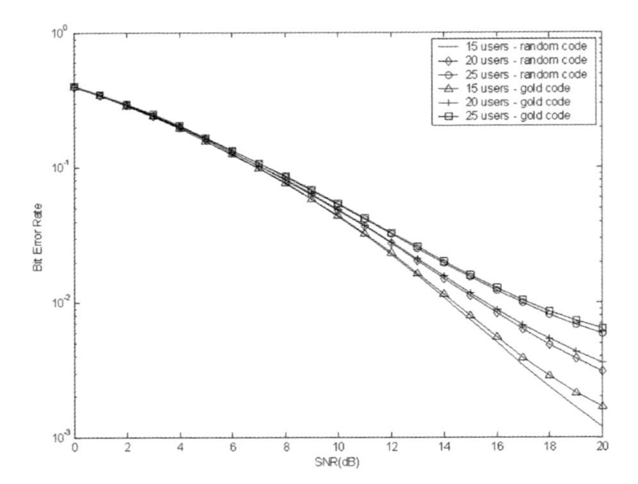

Figure 9.14 Bit error rate performance for a BPSK M-ary CDMA system with Gold code and purely random code. The spreading code length is 127. The channel model is AWGN. User 1 (the signal of interest to the receiver) uses two spreading codes, while the rest use one code. $\psi = E_{bk}/E_{b1} = 1$.

be as large as 5-6 dB! Therefore, it is shown that the constellation optimization is significant and should be carried out very carefully in the design process of an M-ary CDMA system.

Finally, Figure 9.14 compares the BER difference between Gold code and purely random code, with their code lengths being 127 and 128, respectively. User 1 uses two codes, and the rest use one code only. Again, we have $\psi = E_{bk}/E_{b1} = 1$ and the number of users in the system is 15, 20, and 25. It is seen from the figure that the BER results for both Gold code and purely random code are very close, verifying that the use of purely random codes in the analysis is meaningful and acceptable. It can help us to greatly simplify the overall analytical process.

9.7 SUMMARY

Due to space limitations, the study on the M-ary CDMA system is limited to a simple operational environment, a channel without multipath propagation effect. However, the study can be easily extended to the multipath environment using the analytical framework presented in this chapter.

M-ary CDMA technology offers another dimension to implement a highly bandwidth efficient CDMA system. Different from a conventional DS-CDMA system, an M-ary CDMA system makes use of the spreading code as a direct medium to encode data information. Therefore, the transmission time within a code length can carry a lot more information than that of a conventional DS-CDMA system, where each code can carry strictly only one bit of information, and no more than that is possible. On the other hand, in an M-ary CDMA system the duration of one spreading code in principle can carry m bits of information, if there are H codes available to a user and the relation $2^m \leq (3^H - 1)$ is satisfied.

As mentioned earlier, the M-ary CDMA technology has its own limitations. Due to the use of multiple codes at each terminal, the detection of signal at a receiver will become the detection of a symbol based on the type of combinational codes sent from the transmitter. Therefore, the Euclidean distances among the constellation points formed by the sent-code patterns will play a critical role in the detection efficiency. At the same time, it also raises the requirements on the power efficiency of the system. In other words, the bandwidth efficiency of an M-ary CDMA system is obtained in exchange for power efficiency. Therefore, to strike a good balance will be an important issue in the design of an M-ary CDMA system.

Basically, an M-ary CDMA system requires more spreading codes than a conventional DS-CDMA system, in which each user is assigned only one code for data spreading and signature division purposes. However, an M-ary CDMA system needs at least two codes for each user, and thus the number of codes required in an M-ary CDMA system can be several times more than what is needed in a conventional DS-CDMA system. For this reason, great pressure exists to search for some good spreading code sets with a relatively large set size. This raises another interesting issue, which is relevant to next generation CDMA technology. Some complementary codes, such as super complementary codes, column-wise complementary codes, and pair-wise complementary codes, can be very suitable for their applications in M-ary CDMA systems due to their relatively large set sizes.

On the other hand, the use of multiple codes in each terminal also creates some concern on the increase in multiple access interference in the whole M-ary CDMA system, especially if the codes used have relatively poor cross-correlation functions. Therefore, the choice of spreading codes for an M-ary CDMA system should be exercised in a very careful way, and we have to select those codes with minimal cross-correlation levels. Therefore, again under this context, the orthogonal complementary codes are suitable candidates for application in M-ary CDMA systems, because they are able to offer a unique interference-free operation, which is impossible if any other conventional spreading codes are used instead. Unfortunately, due to the limited space in this book we could not do detailed investigations on the performance if orthogonal complementary codes are applied to an M-ary CDMA system.

It should be noted that study of M-ary CDMA systems [163–169,246,247,249,250,283–285] has not been well emphasized compared to research on other CDMA-related topics, such as DS-CDMA, FH-CDMA, multi-carrier CDMA, etc. We have seen very few research reports on M-ary CDMA systems. We feel that much more research should be done to thoroughly investigate the great potential of M-ary CDMA and its related technologies due to the simple fact that an M-ary CDMA system can offer much better bandwidth efficiency than all other conventional CDMA technologies, such as DS-CDMA, etc. Radio spectrum is a scarce resource which can never be replaced. The increasing demand for wireless communications anywhere and any time has pushed us hard to find some more competent wireless technology which can offer better use of limited bandwidth resource. The beyond 3G wireless communication systems ought to provide users with up to 1 Gbps data transmission rate, and thus more spectrum should be allocated for future wireless services. Unfortunately, the suitable radio frequency bandwidth is very limited, as it is unrealistic to expect that the beyond 3G wireless applications should be operating in millimeter wavelength bands, as very short wavelengths will become extremely susceptible to environmental conditions, such that dust, vapor, and raindrops will effectively impair the propagation in the channels. Therefore, it seems to us that the only way to enable future high-speed wireless applications is to find some air-link technologies which can provide greatly improved bandwidth efficiency. Therefore, in this sense, M-ary CDMA technology definitely will play an important role.

10

Next Generation Optical CDMA Communications

As this book addresses the issues on the next generation CDMA technology, it will be appropriate if it can also cover the part of our work done recently on optical CDMA systems. While writing this book, it took me a while to think about the issues on whether the subject of optical CDMA is suitable for this book or not. The supporting idea is that what we have done on optical CDMA is a completely new methodology to design CDMA codes for optical communications, which is relevant to the scope of this book: the next generation CDMA technology, and the results are very exciting, as the method can be used to effectively improve the overall performance of an optical CDMA system. However, we have to admit that the discussions covered in this chapter will shift our thought flow away from radio or wireless communications to optical communications all of a sudden. Hopefully, this shift will not cause too many problems to understand the contents covered in this chapter. Of course, the good thing about including optical CDMA systems in this book is that it can be made more complete under the context of the next generation CDMA technologies, which should address the issues not only on wireless communications, but also on optical communications, which have been witnessed extremely fast advances in recent years. As a matter of fact, optical fibers have become the most important backbone trunks for the telecommunication infrastructure in the world. We simply should not underestimate the great importance of optical communication and leave it unaddressed in this book.

10.1 PECULIARITIES IN OPTICAL COMMUNICATIONS

We have to admit that topics on optical communications are very much different from what we can perceive in wireless communications. There are many peculiarities in optical communication systems compared to radio or wireless, as summarized below.

1. The mechanism and properties of noise generated in optical communication systems are very different from those in wireless or radio communication systems. Those noise components include shot noise, dark current, thermal noise, etc. Therefore, we need to use different approaches to model and characterize them.

The Next Generation CDMA Technologies Hsiao-Hwa Chen

2. The mechanism and properties of interference generated in optical communication systems are also very much different from those in wireless or radio communication systems. In general, the propagation environment in an optical communication system is much simpler and easier to predict than what we have in a wireless communication system, where there are many channel impairing factors to deal with, such as multipath propagation, Doppler effect, external interferences, etc., making it very hard to predict its performance accurately.

3. An optical communication system is not able to send binary data streams using -1 and $+1$ signal levels. Instead, it will send binary information using directly 0 and 1 states. This is because it is extremely difficult for an optical system to distinguish the phases of the optical or light signals. Thus, only amplitude will be the medium used to carry information data. In a more precise term, an optical system usually detects signal via detecting the energy or power of the light signals.

4. There is usually relatively abundant bandwidth in optical systems compared with radio or wireless systems, and thus the issues of bandwidth efficiency improvement in an optical system become less critical than in a wireless or radio communication system, in which the spectral resource has become so scarce that a great effort has been made recently in order to improve bandwidth efficiency.

5. On the other hand, an optical communication system cares much more about its power efficiency (which is more important than its bandwidth efficiency) because optical communications always involve relatively long transmission distances, in particular for some applications like under-ocean cables, etc. Therefore, the distance-related attenuation in an optical system can be substantial and ought to be compensated properly using many repeaters on the optical trunk systems.

6. Signal modeling in an optical system is always scalar-based, instead of vector-based or complex-valued due to the fact that signal detection always proceeds based on energy or power level (which is a scaler) detection. Therefore, system modeling in an optical system is usually much simpler than in a wireless communication system.

7. There is no multipath propagation on Doppler effect in any optical communication system, but there are synchronous/asynchronous transmission problems. Therefore, propagation delays should also be considered in optical system modeling.

8. It is noted that the attenuation loss for different wavelengths in an optical fiber cable is almost the same. The optical power transmission in an optical fiber cable can be well contained inside the fiber with almost no energy emission to the outside world. Therefore, optical fiber is a very good medium for communications with high security requirements.

9. The signal transmissions in an optical fiber cable are usually much more stable than those in a wireless medium. In addition, the signals in optical fiber cables will not be easily affected or interfered with by external radio frequency transmissions, and thus optical fiber is in particular suitable for very high quality trunk communications.

10. Finally, most optical systems use different wavelengths to divide different signal channels (namely wavelength division multiple access or WDMA), while wireless systems often use frequencies to divide different signal channels. It should also be noted that the application of optical time division multiple access (OTDMA) is not as common as that of wavelength division multiple access (WDMA) in optical communications. However, code division multiple access (CDMA) has been considered for applications in optical communication systems. Therefore, multiple access interference (MAI) suppression is also an issue of interest in optical CDMA communications.

Due to those peculiarities, the design of optical communication systems is usually very different from that of wireless or radio communication systems. An optical communication system based on CDMA will also be very different from a CDMA wireless communication system. As a matter of fact, optical CDMA communication systems have not received enough attention due largely to the fact that there is enough bandwidth available to optical communication systems, which use optical fiber cables with great capacity to support many more users than wireless or radio communication systems. Only very recently, research activities on optical CDMA communications have picked up quickly, as shown in the references given at the end of this book [286–306].

10.2 PREVIOUS RESEARCH ON OCDMA COMMUNICATIONS

Research on optical CDMA communications is not new and it has been around for quite some time, as shown in the references [286–306], although its intensity could never be comparable to that of research activities on wireless CDMA communications. Therefore, it will be useful for us to take a look at what has been done in the optical CDMA communications literature before introducing our own work on this subject.

Traditionally, there are two major issues in optical CDMA communication system design. One is the design of spreading codes with good correlation properties, and the other is the number of users/channels which can be supported by the optical CDMA system. The correlation properties of the spreading codes used in an optical CDMA system can directly influence the system bit error rate (BER), and the number of users supportable in an optical CDMA system will affect the system capacity. In an optical CDMA system (which is very different from a wireless CDMA system), it is usually impossible for optical spreading codes to offer ideal correlation properties, especially for the cross-correlation functions, due to the fact that the chip levels for optical CDMA sequences have to take either 0 or 1 (instead of $+1$ and -1, as in wireless CDMA systems). Therefore, it is obviously impossible to completely eliminate auto-correlation function side lobes and cross-correlation functions of the spreading codes used in any optical CDMA system. This is a serious issue we have to deal with when we want to design an optical CDMA system with satisfactory performance.

There have been some spreading codes available for possible applications in optical CDMA communication systems. Based on those codes, different optical CDMA system architectures have also been proposed in the literature [286–306] . However, it is noted that all existing optical CDMA codes were proposed for their applications in optical CDMA systems without being optimized in terms of their performance under a practical optical CDMA system framework. As a matter of fact, the design approaches for all those existing optical CDMA codes were proposed in an ad hoc fashion. In other words, the design of optical CDMA codes was not carried out in conjunction with the design of an optical CDMA system as a whole, such that the code design and system design could be integrated to obtain an ultimate solution, which is optimized in terms of the whole system performance.

In this chapter, we are motivated to propose an innovative approach to integrate optical CDMA system and code design approaches, based on a thorough review of all existing optical code design approaches. In a more precise term, we would like to say that we have successfully transplanted the REAL approach, which is based on complementary codes and was proposed for wireless CDMA communications as discussed in Subsection 6.3.5, into an optical CDMA system, such that the optical CDMA codes will be generated based on the requirements of an optical CDMA system with many real operational requirements as a whole, ensuring that the system that uses these codes will offer optimal performance. Therefore, it is a breakthrough to apply complementary codes to optical CDMA communications. In this sense, the work presented in this chapter should be considered as an important part of next generation CDMA technology.

Before introducing our own work on optical CDMA communication systems, we would like to take a look at the existing research carried out so far in the literature.

There are two major techniques widely used in optical CDMA communication systems: direct sequence (DS) and spectral encoding (SE). The DS optical CDMA works in a very similar way to a DS-CDMA system in wireless communications: that is, by spreading data bits directly using some time-domain sequences. On the other hand, the SE technique performs spreading in the frequency domain, similar to the method used in a multi-carrier CDMA system, where the spreading sequence will be modulated into data in terms of the sub-carrier index. Of course, OCDMA can also be used jointly with other techniques, such as OCDMA/WDMA and OCDMA/OTDMA. Due to its relatively wide applications, this chapter will focus only on the DS spreading optical CDMA technique.

DS-OCDMA Technique

Figure 10.1 shows a simple diagram to illustrate a conceptual block diagram for a DS-OCDMA system using a delay line for time-domain spreading encoding. It is seen that the DS spreading modulation is implemented through a delay line configuration, which is very different to the DS spreading modulator used in a wireless DS-CDMA system. This delay line configuration was first proposed by Prucnal *et al.* in 1986 [288]. Therefore, the spread modulated signal is mono-polar signal, consisting of only 0 and 1, while the length of the delay lines should be adjusted to fit a particular chip delay requirement.

Obviously, the spreading modulated signal in a DS-OCDMA system always contains a lot of zeros compared to the number of 1s in order to control the multiple access interference (MAI), due to the fact that in an optical system the -1 level does not exist and thus it is impossible to cancel the interferences based on the fact that their levels have the same amplitude but different signs. This property of the spreading codes used in DS-OCDMA systems prevails, such as the optical orthogonal code (OOC) proposed by Salehi and colleagues [287,290], and the prime code suggested by Fuji-Hara and miao in [293].

Therefore, a DS-OCDMA optical communication system based on time-domain spreading modulation can not eliminate all MAI, which is governed by the property of the optical spreading codes, which are made up of zeros and ones. Therefore, the spreading code optimization for a DS-OCDMA system can never achieve what we can for wireless DS-CDMA systems, such as DS/CC-CDMA as discussed in the previous chapters.

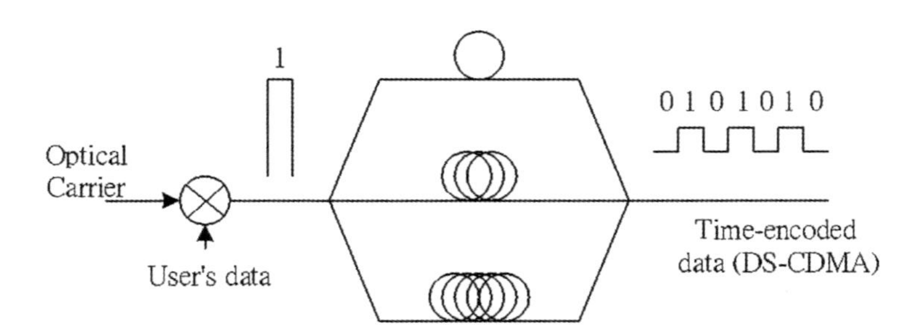

Figure 10.1 Conceptual block diagram for DS-OCDMA system using delay line for time-domain spreading encoding.

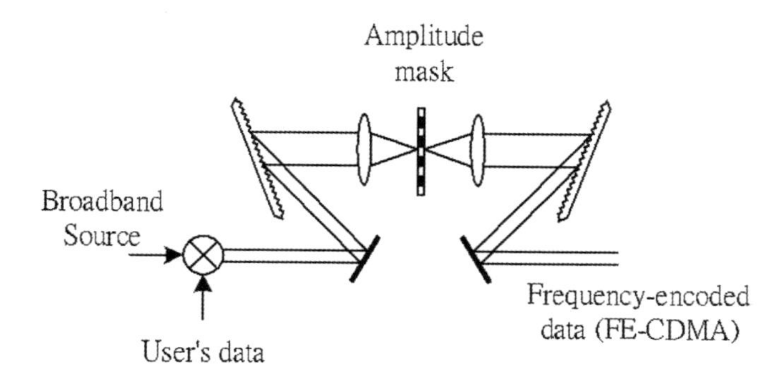

Figure 10.2 Implementation of a spectral encoded DS-OCDMA system.

Spectral Encoded OCDMA Technique

Another type of OCDMA system is based on spectral encoding techniques, which were proposed by Kavehrad and Zaccarin in [298]. The spectral encoded techniques include spectral-phase encoding and spectral-amplitude encoding schemes. Figure 10.2 shows the basic ideas for the spectral encoded OCDMA technique.

Hybrid OCDMA Technique

The DS-OCDMA technique can also work jointly with other techniques to form hybrid OCDMA systems. For instance, time-domain spreading can work with a frequency hopping scheme to form an FH-DS-OCDMA system, such as the one proposed by Tancevski in [299]. In this proposed scheme, prime codes are used for time-domain spreading modulation. To overcome the shortcomings of prime codes, which can support relatively few users, the scheme uses frequency hopping to increase the number of users supportable in the system. The use of an FH unit will also complicate the overall system hardware implementation and thus this scheme has not been widely used due to its implementation difficulties.

10.3 EXISTING SEQUENCES FOR OPTICAL CDMA

In this chapter, our focal point is on the design of next generation OCDMA systems with the help of innovative spreading codes. Therefore, the discussions on sequence design for DS-OCDMA systems will be very important. The comparison will also be made among different DS-OCDMA systems based on several major spreading codes, which are briefly introduced in this section.

There are three major OCDMA sequences reported in the literature so far: optical orthogonal codes (OOCs), prime codes, and multi-length codes. We will introduce them one by one as follows.

10.3.1 OPTICAL ORTHOGONAL CODES

An optical orthogonal code (OOC) set can be denoted by $\Phi(N, w, \lambda_a, \lambda_c)$, which constitutes a series of zeros and ones. The code length of the OOC is N, the weight of the code is w, and the auto-correlation and cross-correlation levels are λ_a and λ_c, respectively. An OOC sequence has to satisfy the following two conditions:

- It should possess a good auto-correlation function with any arbitrary relative chip shifts, such that the detection efficiency of a DS-OCDMA system is acceptable.
- It should also have a satisfactory cross-correlation function between any two sequences in the same set to minimize the MAI.

Assume that there are two OCDMA sequences $x(t)$ and $y(t)$, which can be represented as

$$\begin{aligned} x(t) &= \frac{1}{T_c}\sum_{n=-\infty}^{\infty} x_n P_{T_c}(t - nT_c) \\ y(t) &= \frac{1}{T_c}\sum_{n=-\infty}^{\infty} y_n P_{T_c}(t - nT_c) \end{aligned} \tag{10.1}$$

where x_n and y_n are two periodic sequences with their period being $N = T/T_c$, and P_{T_c} is the square waveform with its duration being T_c. The sequences x_n and y_n should also satisfy their auto-correlation and cross-correlation functions given as follows

- Auto-correlation requirement:

$$\rho(\mathbf{x}; l) = \sum_{i=0}^{N-1} x_i x_{i+l} = \begin{cases} w, & l = 0 \\ \leq \lambda_a, & 1 \leq l \leq N-1 \end{cases} \tag{10.2}$$

where $\mathbf{x} \in \mathcal{C}$, and $\mathbf{x} = \{x_0, x_1, \ldots, x_{N-1}\}$. $\mathcal{C}$ denotes the code set we concern here.

- Cross-correlation requirement:

$$\rho(\mathbf{x}, \mathbf{y}; l) = \sum_{i=0}^{N-1} x_i y_{i+l} \leq \lambda_c, \quad \leq l \leq N-1, \tag{10.3}$$

where $(\mathbf{x}, \mathbf{y}) \in \mathcal{C}$, $\mathbf{x} \neq \mathbf{y}$, and we have

$$\begin{cases} \mathbf{x} = (x_0, x_1, \ldots, x_{N-1}) \\ \mathbf{y} = (y_0, y_1, \ldots, y_{N-1}) \end{cases} \tag{10.4}$$

and $\mathcal{C}$ denotes the code set.

λ_a and λ_c are the maximally allowable auto-correlation and cross-correlation levels. In the design process of the OOC code, $\lambda_a = \lambda_c = 1$ is required. Therefore, we can simplify the notation for an OOC set as $\Phi(N, w, \lambda)$ with $\lambda_a = \lambda_c = \lambda = 1$. According to the Johnson Bond, we can derive the relation between the code length N, the weight w (the number of ones in an OOC code), and set size K as

$$\Phi(N, w, \lambda) \leq \left\lfloor \frac{1}{w} \left\lfloor \frac{n-1}{w-1} \left\lfloor \frac{n-2}{w-2} \cdots \left\lfloor \frac{n-\lambda}{w-\lambda} \cdots \right\rfloor \right\rfloor \right\rfloor \right\rfloor \tag{10.5}$$

where $\lfloor x \rfloor$ denotes the largest integer less than x. We can further obtain the relation between the set size K, code length N, and the weight w as

$$K \leq \left\lfloor \frac{N-1}{w(w-1)} \right\rfloor \tag{10.6}$$

Now let us take a look at an example of $\Phi(32, 4, 1)$, in which $N = 32$, $w = 4$, and $\lambda = 1$, as shown in Figure 10.3. Codes A and B have their maximal auto-correlation side lobes, being one, and

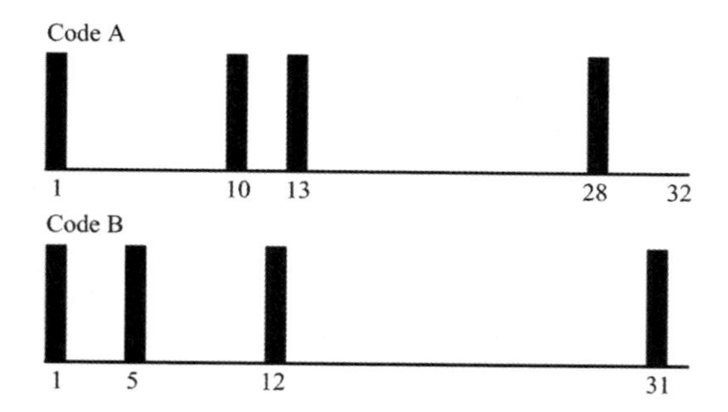

Figure 10.3 Two example optical orthogonal codes from set $\Phi(32, 4, 1)$, with code A having ones in 1, 10, 13, and 28 chip positions and code B having ones in 1, 5, 12, and 31 chip positions. The other chip positions are zeros.

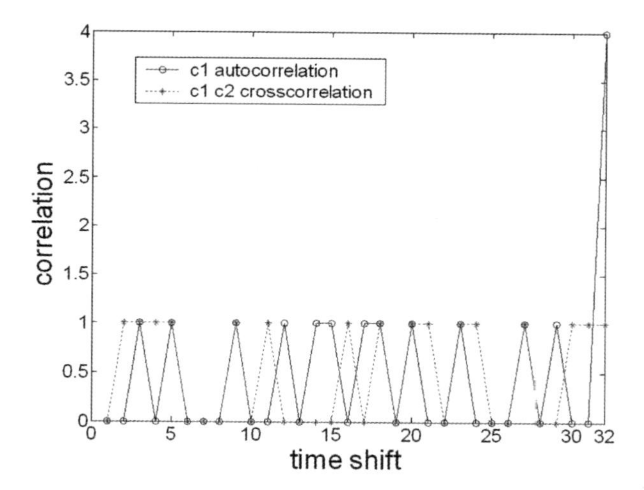

Figure 10.4 The auto-correlation of either code A or B, and cross-correlation functions between codes A and B, where c1 denotes code A and c2 denotes code B.

their maximal cross-correlation function, also one. It is noted that the inter-pulse distance for code A is $\{9, 3, 15, 5\}$, and that for code B is $\{4, 7, 19, 2\}$. Therefore, none of the inter-pulse distances are the same for the two codes. Thus, no matter what the relative chip shift between them is, no more than one overlapped pulse can happen. In this way, the maximal cross-correlation level will be equal to one, or $\lambda_c = 1$.

On the other hand, we can make a similar observation for their auto-correlation functions. Let us take code A as an example. It is easy to see that the inter-pulse distance for code A is $\{9, 3, 15, 5\}$, in which the four values are all different. Therefore, no matter what the relative chip shift of the same code (code A in this case) can be, the maximal number of overlapped pulses is never more than one, meaning that $\lambda_a = 1$. The auto-correlation of either code A or B, and cross-correlation functions between code A and code B are shown in Figure 10.4, where it is shown that the maximal auto-correlation and cross-correlation functions are equal to one only, or $\lambda_a = \lambda_c = 1$.

10.3.2 PRIME CODES

The prime code is another important type of spreading sequence widely used in optical CDMA systems. The prime codes are derived from Galois field:

$$GF(p) = (0, 1, 2, \ldots, p-1) \tag{10.7}$$

where p is a prime number. In any prime code, p will act as the weight of the code, its code length will be p^2, and its set size is always p. The maximal auto-correlation side lobe and maximal cross-correlation level are the same as the OOC codes, being one, or $\lambda_a = \lambda_c = 1$.

The procedure to generate a prime code set can be explained as follows. First, we can generate a set of sequences (denoted by $S_i^P : i = 0 \sim P-1$) based on one particular prime number P as

$$S_i^P = [ij]_P, \quad \text{for } i = 0 \sim P-1, \; j = 0 \sim P-1 \tag{10.8}$$

where $[x]_P$ stands for the modulo-P calculation for x. For every different i and j we can generate a prime code C_i^P which has a length of P^2, and can be defined as

$$C_i^P(n) = \begin{cases} 1, & n = jP + S_i^P(j) \\ 0, & \text{otherwise} \end{cases} \tag{10.9}$$

where S_i^P has been defined in Equation (10.8). Therefore, we understand that the format of the prime codes looks very much like the optical orthogonal codes (OOCs). It is also seen from the above discussions that the major difference between them lies in the fact that the prime codes can be generated using a very systematic approach, as given in Equations (10.8) and (10.9). On the other hand, the generation of the OOC codes can be done based on a set of rules and there is no systematic approach to generate all OOC codes. Strictly speaking, the prime code is also a type of optical orthogonal code.

10.3.3 MULTI-LENGTH CODES

The third type of sequence suitable for optical CDMA systems is a multi-length code. It is noted that normally the spreading codes used by different users in a CDMA system have the same length or spreading factors, unless for some very special applications such as variable rate transmission systems, etc. The same code length among all CDMA users will also make the system design much simpler. The same is true of an optical CDMA system, as we have seen from the above discussions on OOC codes and prime codes, which are all equal length codes for OCDMA systems. However, in this subsection we will introduce a different type of sequence for OCDMA applications. The name of the sequence is multi-length codes or unequal-length OOCs, which were first proposed by Jian-Guo Zhang in [306]

The encoding procedure for multi-length codes or unequal-length OOC codes can be explained as follows. First, we should decide a suitable length for a code based on the application requirements. During the generation procedure, we still want to have some constraints on the maximal auto-correlation and cross-correlation functions as $\lambda_a = \lambda_c = 1$. The basic idea in generation of unequal-length OOC codes is very similar to that of normal OOC codes, except for their lengths. Denote an unequal-length OOC set as $\Phi(n_i, w, 1)$, where n_i is the code length for the ith user ($i = 1, 2, \ldots, K$), w is the weight for the codes, and the maximal auto-correlation and cross-correlation functions are $\lambda_a = \lambda_c = 1$. According to the Johnson Bond, we can also calculate the set size K or the maximal number of users that this OCDMA can support.

The ith code in the set can be denoted as

$$C_i = \left(a_1^{(i)}, a_2^{(i)}, a_3^{(i)}, \ldots, a_w^{(i)}\right) \tag{10.10}$$

where $i = 1, 2, \ldots, K$, and $a_j^{(i)}$ denotes the position of the jth pulse ($j = 1, 2, \ldots, w$) for the ith user. Thus we have

$$1 \leq a_1^{(i)} < a_2^{(i)} < \cdots < a_j^{(i)} < \cdots < a_w^{(i)} \leq n_i \tag{10.11}$$

Let us define the inter-pulse duration as

$$d_{jk}^{(i)} = a_k^{(i)} - a_j^{(i)} - 1 \geq 0 \tag{10.12}$$

where $j = 1, 2, \ldots, k-1$ and $k = 2, 3, \ldots, w$. The biggest inter-pulse duration for C_i is

$$D^{(i)} = d_{1w}^{(i)} = a_w^{(i)} - 2 \tag{10.13}$$

The biggest inter-pulse duration in this code set can be written as

$$D = \max\left\{D^{(i)} | i = 1, 2, \ldots, K\right\} \tag{10.14}$$

The following two fundamental rules have to be observed in the design process of unequal-length OOC codes.

- Rule 1: The maximal auto-correlation and cross-correlation levels should be one, or $\lambda_a = \lambda_c = 1$.

  Under this rule, similar to the OOC codes, the inter-pulse distances for different codes should never be identical, or

$$d_{kl}^{(i)} \neq d_{k'l'}^{(i')} \tag{10.15}$$

  where

$$\begin{cases} i = 1, 2, \ldots, K \\ i' = 1, 2, \ldots, K \\ d_{kl}^{(i)} \in C_i \\ d_{k'l'}^{(i')} \in C_{i'} \\ k = 1, 2, \ldots, l-1 \\ l = 2, 3, \ldots, w \\ k' = 1, 2, \ldots, l'-1 \\ l' = 2, 3, \ldots, w \end{cases} \tag{10.16}$$

  The code length of each user should be

$$n_i \geq D^{(i)} + D + 3, \qquad i = 1, 2, \ldots, K \tag{10.17}$$

- Rule 2: The maximal inter-pulse distance in a set of unequal-length OOC codes can be defined as

$$D \geq \frac{w(w-1)}{2} K - 1 \tag{10.18}$$

  where each code C_i has $\frac{w(w-1)}{2}$ different inter-pulse distances.

After having given the rules for generation of unequal-length OOC codes, we are ready to introduce the generation procedure of the codes. To simplify the discussions, we would like to define the following two parameters, one being 'select pulse distance' or $d_{k,k+1}^{(i)}$, where $k = 1, 2, \ldots, w-1$, and the other being 'associated pulse distance' or

$$d_{kl}^{(i)} = \sum_{j=k}^{l-2} \left(d_{j,j+1}^{(i)} + 1\right) + d_{l-1,l}^{(i)} \tag{10.19}$$

where $k = 1, 2, \ldots, l-2$ and $l = 3, 4, \ldots, w$.

Next, we show the generation process for an example unequal-length OOC code set $\Phi(n_i, 4, 1)$, whose weight is $w = 4$ and set size is $K = 4$.

1. First, we should choose the pulse distance as $d_{1,2}^{(i)} = i - 1$, where $i = 1, 2, 3, 4$. We then obtain a set of codes with two ones, or $C_{i,2} = \{1, i + 1\}$, where $i = 1, 2, 3, 4$ and

$$\begin{cases} n_{1,2} \geq 6 \\ n_{2,2} \geq 7 \\ n_{3,2} \geq 8 \\ n_{4,2} \geq 9 \end{cases} \tag{10.20}$$

2. Choose four different distances as $d_{2,3}^{(i)}$, which denotes the distance between the second and third pulses in each code.

3. After having decided the position for the third pulse, we need to calculate the associated pulse distance of $d_{1,3}^{(i)}$, to check if there is any violation against the two rules given above. Also, it is not allowed to use the same inter-pulse distance among the codes. If yes, then we have to go back to Step 2 to re-decide $d_{2,3}^{(i)}$, where $i = 1, 2, 3, 4$. In our example, we choose

$$\begin{cases} d_{2,3}^{(1)} = 10 \\ d_{2,3}^{(2)} = 5 \\ d_{2,3}^{(3)} = 6 \\ d_{2,3}^{(4)} = 4 \end{cases} \tag{10.21}$$

and we can verify that

$$\begin{cases} d_{1,3}^{(1)} = 11 \\ d_{1,3}^{(2)} = 7 \\ d_{1,3}^{(3)} = 9 \\ d_{1,3}^{(4)} = 8 \end{cases} \tag{10.22}$$

in which there is no violation of the above two rules. Therefore, we obtain a set of codes as

$$\begin{cases} C_{1,3} = \{1, 2, 13\} \\ C_{2,3} = \{1, 3, 9\} \\ C_{3,3} = \{1, 4, 11\} \\ C_{4,3} = \{1, 5, 10\} \end{cases} \tag{10.23}$$

4. After having decided the position for the third pulse and verified that there is no violation of either the two rules, we can proceed to decide the position for the fourth pulse. In this case, we can select

$$\begin{cases} d_{3,4}^{(1)} = 14 \\ d_{3,4}^{(2)} = 16 \\ d_{3,4}^{(3)} = 13 \\ d_{3,4}^{(4)} = 12 \end{cases} \tag{10.24}$$

Table 10.1 An example set of unequal-length OOC codes $\Phi(n_i, 4, 1)$ with ten codes, where $D(i)$, n_i, and C_i denote the inter-pulse distance, the code length, and the code representation for the ith code, respectively

$D(i)$	n_i	C_i
59	121	$C_1 = (1, 5, 34, 61)$
58	120	$C_2 = (1, 11, 45, 60)$
57	119	$C_3 = (1, 9, 48, 59)$
56	118	$C_4 = (1, 18, 32, 58)$
54	116	$C_5 = (1, 8, 36, 56)$
53	115	$C_6 = (1, 13, 37, 55)$
52	114	$C_7 = (1, 2, 24, 54)$
50	112	$C_8 = (1, 7, 20, 52)$
45	107	$C_9 = (1, 10, 26, 27)$
42	104	$C_{10} = (1, 3, 6, 44)$

and then calculate

$$\begin{cases} d_{1,4}^{(1)} = 26 \\ d_{1,4}^{(2)} = 24 \\ d_{1,4}^{(3)} = 23 \\ d_{1,4}^{(4)} = 21 \\ d_{2,4}^{(1)} = 25 \\ d_{2,4}^{(2)} = 22 \\ d_{2,4}^{(3)} = 20 \\ d_{2,4}^{(4)} = 17 \end{cases} \tag{10.25}$$

all of which are different and thus they are just the codes we want. Therefore, we finally obtain a set of codes $\Phi(n_i, 4, 1)$, in which each code contains four ones and altogether four users can be supported. The code set is listed as follows:

$$\begin{cases} C_1 = (1, 2, 13, 28) \\ C_2 = (1, 3, 9, 26) \\ C_3 = (1, 4, 11, 25) \\ C_4 = (1, 5, 10, 23) \end{cases} \tag{10.26}$$

in which the lengths for different codes are $n_1 \geq 55, n_2 \geq 53, n_3 \geq 52, n_4 \geq 50$. Table 10.1 shows a set of unequal-length OOC codes $\Phi(n_i, 4, 1)$ with ten codes.

10.4 COMPLEMENTARY CODES FOR OCDMA

It can be seen from the discussions made in the previous sections that all existing sequences suitable for optical CDMA systems are unitary codes, which work on a one-code-per-user basis. Therefore, it is amazing to note that the idea to assign one user a code has very deep roots in all traditional

CDMA system designs, including both wireless and optical CDMA systems. Obviously, the design of all existing CDMA sequences for optical CDMA systems was strongly influenced by the sequence design methodologies for wireless communication systems. Therefore, it is totally expected that there is a strong similarity between the approaches used to design sequences for both optical and wireless systems.

As we have already seen from the earlier discussions, there are some shortcomings in the traditional methodologies used to generate sequences for optical CDMA systems. Therefore, we were motivated to find some better ways to design the sequences for optical CDMA applications, as explained below.

- First, it can be noted that the methods used to generate traditional optical CDMA sequences, such as OOC codes, prime codes, etc., were proposed based only on partial correlation functions, just as given in Equations (10.2) and (10.3) in the previous sections. Therefore, the code design process never takes into account odd and even periodic correlation functions, which may depend on the patterns of continuous bit streams or even the transmission modes (high-speed bursty or continuous-time traffic) in an optical communication system.

- Second, the sequence design based on unitary codes lacks degrees-of-freedom in terms of optimization of auto-correlation and cross-correlation functions. In this respect, complementary codes are much better. It is natural that we had a strong curiosity before we started to use complementary codes for optical CDMA applications. It certainly will be an interesting research topic to investigate the possibility of applying complementary codes in an optical CDMA system.

- Third, the code structure of complementary codes is much more general than that of unitary codes, as the unitary codes are only special cases of complementary codes, with their flock sizes been reduced to one. Therefore, study of an optical CDMA system based on complementary codes offers broad scope to view the whole vista in order to achieve some results which can be more general and relevant to the applications for optical CDMA communications. Therefore, choosing complementary codes as a new starting point for an optical CDMA system is logically correct and theoretically significant.

The generic block diagrams for an OCDMA system based on optical complementary codes are shown in Figures 10.5 and 10.6, where Figure 10.5 illustrates the signal generation process and Figure 10.6 shows the receiver structure tuned to a particular signal.

In this section, we would like to investigate ways to design an optimal sequence for its applications in optical CDMA systems. We choose complementary codes as the code of interest, because they offer a perfect code structure, which is general enough to include all existing sequences (which are always unitary codes and thus become special cases of complementary codes with their flock sizes being one). It should be noted that the complementary codes discussed in this section should not be considered as orthogonal complementary codes. As a matter of fact, all sequences used for optical communications are never orthogonal to one another due to the fact that it is impossible to cancel or completely eliminate all auto-correlation side lobes and cross-correlation levels because the sequences consist of zeros and ones, instead of -1 and $+1$, as in wireless systems. However, what we can do here is to minimize the auto-correlation side lobes and cross-correlation levels in the code design process.

10.4.1 PARAMETERS OF OPTICAL COMPLEMENTARY CODES

Let us define a few important variables before the discussions on the design of OCCs. Each user in an optical CDMA system is assigned a flock of M element codes for CDMA purposes. Each element

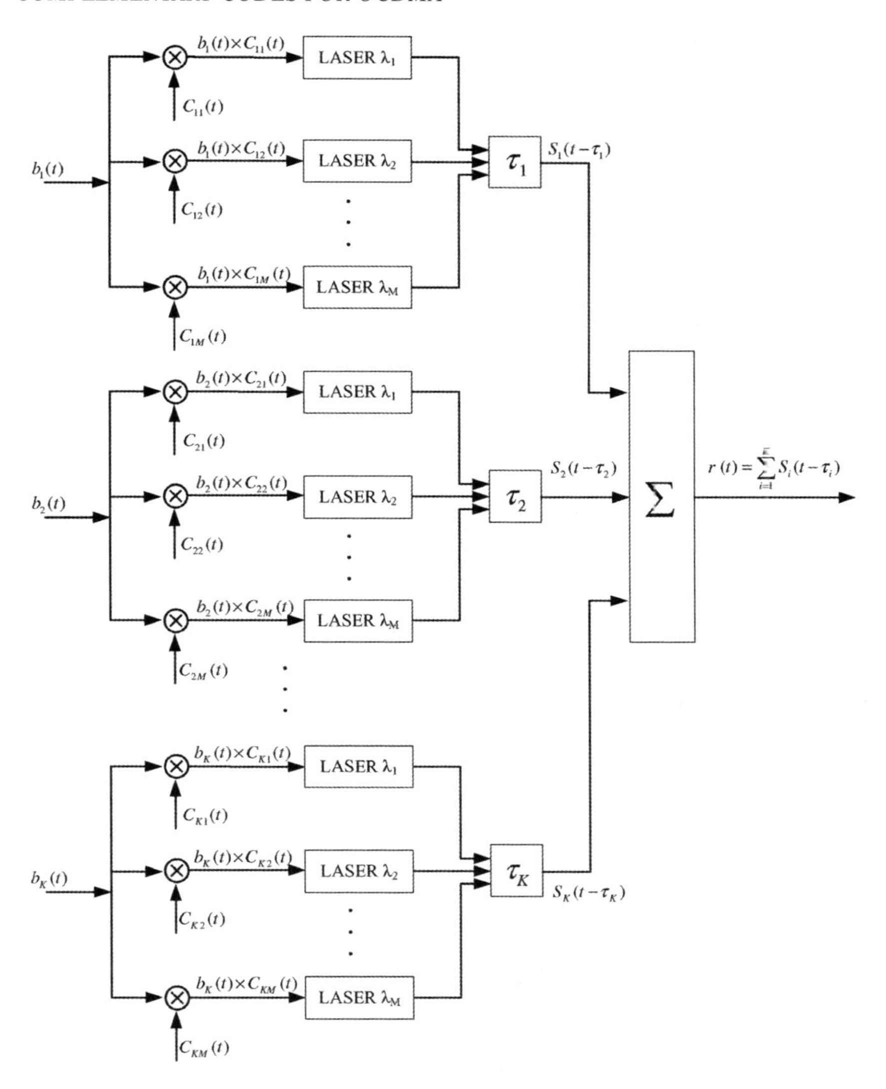

Figure 10.5 Generic block diagram for an OCDMA system based on optical complementary codes, where M flocks of OCCs are present.

code has a length of N chips. The set size of the occs is K, which is also the maximal number of users that the system can support.

A set of occs can be denoted by $\Phi(N, W, M, \lambda_a, \lambda_c)$, where N is the element code length, w is the number of ones in an element code and is also called the weight of the code (in the case we will discuss, we set $w = 1$), W is the number of ones in one flock of M element codes (obviously, if $w = 1$ then we have $W = M$), and M and K are the flock size and set size, respectively. λ_a and λ_c are the maximal auto-correlation side lobe and maximal cross-correlation level for the OCC set, respectively.

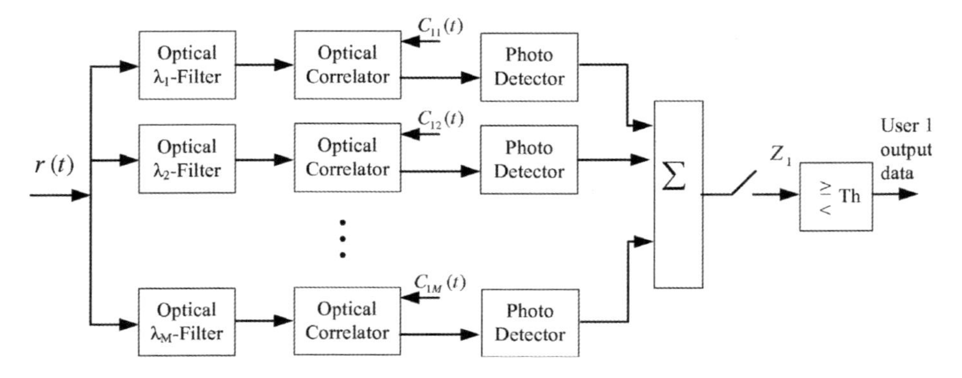

Figure 10.6 Generic block diagram for an OCDMA receiver based on optical complementary codes, where the first flock of the OCC is the signal of interest.

Let **X** and **Y** be the two flocks of the optical complementary codes, with their element code length and flock size being N and M, respectively. We can denote them as

$$\begin{aligned} \mathbf{X} &= (X_{1,0}X_{1,1}\ldots X_{1,N-1};\ X_{2,0}X_{2,1}\ldots X_{2,N-1};\ \ldots;\ X_{M,0}X_{M,1}\ldots X_{M,N-1}) \\ \mathbf{Y} &= (Y_{1,0}Y_{1,1}\ldots Y_{1,N-1};\ Y_{2,0}Y_{2,1}\ldots Y_{2,N-1};\ \ldots;\ Y_{M,0}Y_{M,1}\ldots Y_{M,N-1}) \end{aligned} \tag{10.27}$$

where $X_{i,j}$ and $Y_{i,j}$ (where $i, j = 0, 1, \ldots, N-1$) stand for the jth chip in the ith element code of **X** and **Y**, respectively. Then, we can set up the following requirements on their auto-correlation and cross-correlation functions as

- Requirements on the auto-correlation function ($\lambda_a = 0$)

$$\sum_{\alpha=1}^{M}\sum_{\beta=0}^{N-1} X_{\alpha,\beta}X_{\alpha,\beta+l} = \begin{cases} W, & l = 0 \\ \leq \lambda_a, & 1 \leq l \leq N-1 \end{cases} \tag{10.28}$$

  where $\beta + l$ should be calculated based on modulo-N.

- Requirements on the cross-correlation function ($\lambda_c = 1$)

$$\sum_{\alpha=1}^{M}\sum_{\beta=0}^{N-1} X_{\alpha,\beta}Y_{\alpha,\beta+l} \leq \lambda_c, \quad 0 \leq l \leq N-1 \tag{10.29}$$

  where $\beta + l$ should be calculated based on modulo-N.

It is not difficult to prove that the requirements $\lambda_a = 0$ and $\lambda_c = 1$ on the auto-correlation and cross-correlation functions used here have been optimal in terms of their performance in an optical CDMA system. The issue will be to see if the OCC set $\Phi(N, W, M, \lambda_a, \lambda_c)$ exists with the requirements of $\lambda_a = 0$ and $\lambda_c = 1$ satisfied.

10.4.2 CORRELATION PROPERTIES OF OPTICAL COMPLEMENTARY CODES

Let us first look at the auto-correlation functions of a flock of optical complementary codes, or

$$\mathbf{X} = (X_{1,0}X_{1,1}\ldots X_{1,N-1};\ X_{2,0}X_{2,1}\ldots X_{2,N-1};\ \ldots;\ X_{M,0}X_{M,1}\ldots X_{M,N-1}) \tag{10.30}$$

whose flock size is M and element code length is N. It can be shown that the auto-correlation and cross-correlation conditions, or $\lambda_a = 0$ and $\lambda_c = 1$, can only be met if we let $w = 1$. Therefore, each element code contains only one pulse, with the remaining chips being zeros.

Now we would like to formulate a matrix, whose rows are formed by cyclic shifts of each element code of $\mathbf{X}$, as

$$\begin{pmatrix} X_{10}X_{11}\cdots X_{1N-2}X_{1N-1}; & X_{20}X_{21}\cdots X_{2N-2}X_{2N-1}; & \cdots; & X_{M0}X_{M1}\cdots X_{MN-2}X_{MN-1} \\ X_{1N-1}X_{10}\cdots X_{1N-3}X_{1N-2}; & X_{2N-1}X_{20}\cdots X_{2N-3}X_{2N-2}; & \cdots; & X_{MN-1}X_{M0}\cdots X_{MN-3}X_{MN-2} \\ X_{1N-2}X_{1N-1}\cdots X_{1N-4}X_{1N-3}; & X_{2N-2}X_{2N-1}\cdots X_{2N-4}X_{2N-3}; & \cdots; & X_{MN-2}X_{MN-1}\cdots X_{MN-4}X_{MN-3} \\ \vdots & \vdots & \vdots & \vdots \\ X_{11}X_{12}\cdots X_{1N-1}X_{10}; & X_{21}X_{22}\cdots X_{2N-1}X_{20}; & \cdots; & X_{M1}X_{M2}\cdots X_{MN-1}X_{M0} \end{pmatrix} \tag{10.31}$$

where it is noted that the semi-colon is used to separate different element codes and thus the above matrix has a dimension of $N \times (NM)$. Please also note that we have omitted the commas used to separate two subscript indexes in each chip variable to reduce the space occupied by the equation. For example, we have used X_{2N-1} to denote $X_{2,N-1}$, which has the same meaning. Therefore, please be careful with that, and the same notation will be applied in the equations given in this chapter, unless explained in particular. With this matrix we can easily calculate the auto-correlation functions with all possible relation shifts as follows

$$\begin{pmatrix} X_{10}X_{12}\ldots X_{1N-2}X_{1N-1};\ldots;X_{M0}X_{M1}\ldots X_{MN-2}X_{MN-1} \\ X_{1N-1}X_{10}\ldots X_{1N-3}X_{1N-2};\ldots;X_{MN-1}X_{M0}\ldots X_{MN-3}X_{MN-2} \\ \vdots \\ X_{11}X_{12}\ldots X_{1N-1}X_{10};\ldots;X_{M1}X_{M2}\ldots X_{MN-1}X_{M0} \end{pmatrix} \begin{pmatrix} X_{10} \\ \vdots \\ X_{1N-1} \\ \vdots \\ X_{M0} \\ \vdots \\ X_{MN-1} \end{pmatrix} =$$

$$\begin{pmatrix} (X_{10}^2 + X_{11}^2 + \cdots + X_{1N-1}^2) + \cdots + (X_{M0}^2 + X_{M1}^2 + \cdots + X_{MN-1}^2) \\ (X_{1N-1}X_{10} + X_{10}X_{11} + \cdots + X_{1N-2}X_{1N-1}) + \cdots + (X_{MN-1}X_{M0} + X_{M0}X_{M1} + \cdots + X_{MN-2}X_{MN-1}) \\ (X_{1N-2}X_{10} + X_{1N-1}X_{11} + \cdots + X_{1N-3}X_{1N-1}) + \cdots + (X_{MN-2}X_{M0} + X_{MN-1}X_{M1} + \cdots - X_{MN-3}X_{MN-1}) \\ \vdots \\ (X_{12}X_{10} + X_{13}X_{11} + \cdots + X_{11}X_{1N-1}) + \cdots + (X_{M2}X_{M0} + X_{M3}X_{M1} + \cdots + X_{M1}X_{MN-1}) \\ (X_{11}X_{10} + X_{12}X_{11} + \cdots + X_{10}X_{1N-1}) + \cdots + (X_{M1}X_{M0} + X_{M2}X_{M1} + \cdots + X_{M0}X_{MN-1}) \end{pmatrix} \tag{10.32}$$

In order to satisfy the conditions $\lambda_a = 0$ and $w = 1$, we have to let all rows of the left-hand side of Equation (10.32) be zero except the first row, which represents the zero-shift auto-correlation peak. Thus, we can let

$$\begin{aligned} &(X_{10}^2 + X_{11}^2 + \cdots + X_{1N-1}^2) + \cdots + (X_{M0}^2 + X_{M1}^2 + \cdots + X_{MN-1}^2) \\ &= M = W \end{aligned} \tag{10.33}$$

The results from all other rows should be zero in order to satisfy the requirement specified in

$$\sum_{\alpha=1}^{M}\sum_{\beta=0}^{N-1} X_{\alpha,\beta}X_{\alpha,\beta+l} = \begin{cases} W, & l = 0 \\ \leq \lambda_a, & 1 \leq l \leq N-1 \end{cases} \tag{10.34}$$

The auto-correlation functions for all M element codes should be added as required by signal detection on all complementary codes, and thus we have

$$\begin{aligned}&\left(X_{10}\sum_{i=0}^{N-1}X_{1i}+X_{11}\sum_{i=0}^{N-1}X_{1i}+\cdots+X_{1N-1}\sum_{i=0}^{N-1}X_{1i}\right)+\\&\left(X_{20}\sum_{i=0}^{N-1}X_{2i}+X_{21}\sum_{i=0}^{N-1}X_{2i}+\cdots+X_{2N-1}\sum_{i=0}^{N-1}X_{2i}\right)+\cdots\\&\cdots+\left(X_{M0}\sum_{i=0}^{N-1}X_{Mi}+X_{M1}\sum_{i=0}^{N-1}X_{Mi}+\cdots+X_{MN-1}\sum_{i=0}^{N-1}X_{Mi}\right)\end{aligned} \tag{10.35}$$

where we have

$$\begin{aligned}&X_{10}\sum_{i=0}^{N-1}X_{1i}+X_{11}\sum_{i=0}^{N-1}X_{1i}+\cdots+X_{1N-1}\sum_{i=0}^{N-1}X_{1i}\\&=\sum_{i=0}^{N-1}X_{1i}(X_{10}+X_{11}+\cdots+X_{1N-1})\\&=\sum_{i=0}^{N-1}X_{1i}\sum_{j=0}^{N-1}X_{1j}\end{aligned} \tag{10.36}$$

As there is only one pulse in each element code (or $w=1$), thus we have

$$\begin{cases}\displaystyle\sum_{i=0}^{N-1}X_{1i}=1\\\displaystyle\sum_{j=0}^{N-1}X_{1j}=1\end{cases} \tag{10.37}$$

Therefore, Equation (10.35) can be simplified into

$$\begin{aligned}&\sum_{i=0}^{N-1}X_{1i}\sum_{j=0}^{N-1}X_{1j}+\sum_{i=0}^{N-1}X_{2i}\sum_{j=0}^{N-1}X_{2j}+\cdots+\sum_{i=0}^{N-1}X_{Mi}\sum_{j=0}^{N-1}X_{Mj}\\&=\overbrace{1+1+\cdots 1}^{M1's}=M=W\end{aligned} \tag{10.38}$$

which is just what we expected. Based on the discussions given above, we can rewrite the auto-correlation function with all possible relative chip shifts as

$$\begin{pmatrix}X_{10}X_{12}\cdots X_{1N-2}X_{1N-1};\ldots;X_{M0}X_{M1}\ldots X_{MN-2}X_{MN-1}\\X_{1N-1}X_{10}\ldots X_{1N-3}X_{1N-2};\ldots;X_{MN-1}X_{M0}\ldots X_{MN-3}X_{MN-2}\\\vdots\\X_{11}X_{12}\ldots X_{1N-1}X_{10};\ldots;X_{M1}X_{M2}\ldots X_{MN-1}X_{M0}\end{pmatrix}\begin{pmatrix}X_{10}\\\vdots\\X_{1N-1}\\\vdots\\X_{M0}\\\vdots\\X_{MN-1}\end{pmatrix}=\begin{pmatrix}W\\0\\\vdots\\0\end{pmatrix} \tag{10.39}$$

which is the auto-correlation function expression to be used in the optical complementary codes later.

Then, we would like also to take a look at the cross-correlation functions for optical complementary codes. Assume that there are two optical complementary codes, **X** and **Y**, whose element code length is N and flock size is M, as defined by

$$\begin{aligned}\mathbf{X} &= (X_{1,0}X_{1,1}\ldots X_{1,N-1};\ X_{2,0}X_{2,1}\ldots X_{2,N-1};\ \ldots;\ X_{M,0}X_{M,1}\ldots X_{M,N-1})\\ \mathbf{Y} &= (Y_{1,0}Y_{1,1}\ldots Y_{1,N-1};\ Y_{2,0}Y_{2,1}\ldots Y_{2,N-1};\ \ldots;\ Y_{M,0}Y_{M,1}\ldots Y_{M,N-1})\end{aligned} \tag{10.40}$$

in which each element code contains only one pulse due to the constraint of $w = 1$. Therefore, we can write down the cross-correlation functions between **X** and **Y** for any possible relative chip shifts as

$$\begin{pmatrix} X_{10}X_{12}\ldots X_{1N-2}X_{1N-1};\ldots;X_{M0}X_{M1}\ldots X_{MN-2}X_{MN-1} \\ X_{1N-1}X_{10}\ldots X_{1N-3}X_{1N-2};\ldots;X_{MN-1}X_{M0}\ldots X_{MN-3}X_{MN-2} \\ \vdots \\ X_{11}X_{12}\ldots X_{1N-1}X_{10};\ldots;X_{M1}X_{M2}\ldots X_{MN-1}X_{M0} \end{pmatrix} \begin{pmatrix} Y_{10} \\ \vdots \\ Y_{1N-1} \\ \vdots \\ Y_{M0} \\ \vdots \\ Y_{MN-1} \end{pmatrix} =$$

$$\begin{pmatrix} (X_{10}Y_{10}+X_{11}Y_{11}+\cdots+X_{1N-1}Y_{1N-1})+\cdots+(X_{M0}Y_{M0}+X_{M1}Y_{M1}+\cdots+X_{MN-1}Y_{MN-1}) \\ (X_{1N-1}Y_{10}+X_{10}Y_{11}+\cdots+X_{1N-2}Y_{1N-1})+\cdots+(X_{MN-1}Y_{M0}+X_{M0}Y_{M1}+\cdots+X_{MN-2}Y_{MN-1}) \\ (X_{1N-2}Y_{10}+X_{1N-1}Y_{11}+\cdots+X_{1N-3}Y_{1N-1})+\cdots+(X_{MN-2}Y_{M0}+X_{MN-1}Y_{M1}+\cdots+X_{MN-3}Y_{MN-1}) \\ \vdots \\ (X_{12}Y_{10}+X_{13}Y_{11}+\cdots+X_{11}Y_{1N-1})+\cdots+(X_{M2}Y_{M0}+X_{M3}Y_{M1}+\cdots+X_{M1}Y_{MN-1}) \\ (X_{11}Y_{10}+X_{12}Y_{11}+\cdots+X_{10}Y_{1N-1})+\cdots+(X_{M1}Y_{M0}+X_{M2}Y_{M1}+\cdots+X_{M0}Y_{MN-1}) \end{pmatrix} \tag{10.41}$$

According to the requirements for ideal cross-correlation functions, we have

$$\sum_{\alpha=1}^{M}\sum_{\beta=0}^{N-1} X_{\alpha,\beta}Y_{\alpha,\beta+l} \le \lambda_c \tag{10.42}$$

where $0 \le l \le N-1$ and $\lambda_c = 1$. Therefore, every equation on the right-hand side of (10.41) should be no larger than one, or

$$\begin{cases} [X_{11}Y_{11}+X_{12}Y_{12}+\cdots+X_{1N}Y_{1N}]+\cdots+[X_{M1}Y_{M1}+X_{M2}Y_{M2}+\cdots+X_{MN}Y_{MN}] \le 1 \\ \left[X_{1N}Y_{11}+X_{11}Y_{12}+\cdots+X_{1N-1}Y_{1N}\right]+\cdots+\left[X_{MN}Y_{M1}+X_{M1}Y_{M2}+\cdots+X_{MN-1}Y_{MN}\right] \le 1 \\ \left[X_{1N-1}Y_{11}+X_{1N}Y_{12}+\cdots+X_{1N-2}Y_{1N}\right]+\cdots+\left[X_{MN-1}Y_{M1}+X_{MN}Y_{M2}+\cdots+X_{MN-2}Y_{MN}\right] \le 1 \\ \vdots \\ [X_{13}Y_{11}+X_{14}Y_{12}+\cdots+X_{12}Y_{1N}]+\cdots+[X_{M3}Y_{M1}+X_{M4}Y_{M2}+\cdots+X_{M2}Y_{MN}] \le 1 \\ [X_{12}Y_{11}+X_{13}Y_{12}+\cdots+X_{11}Y_{1N}]+\cdots+[X_{M2}Y_{M1}+X_{M3}Y_{M2}+\cdots+X_{M1}Y_{MN}] \le 1 \end{cases} \tag{10.43}$$

In an OCDMA system based on optical complementary codes, different element codes should be encoded by the same data and sent at the same time in parallel via different wavelengths in the same fibre. Each element code contains at most one pulse (due to $w = 1$), and thus the summation of all outputs from the M element code correlators becomes

$$\begin{aligned}&\left(Y_{10}\sum_{i=0}^{N-1}X_{1i}+Y_{11}\sum_{i=0}^{N-1}X_{1i}+\cdots+Y_{1N-1}\sum_{i=0}^{N-1}X_{1i}\right)+\\&\left(Y_{20}\sum_{i=0}^{N-1}X_{2i}+Y_{21}\sum_{i=0}^{N-1}X_{2i}+\cdots+Y_{2N-1}\sum_{i=0}^{N-1}X_{2i}\right)+\cdots\\&\cdots+\left(Y_{M0}\sum_{i=0}^{N-1}X_{Mi}+Y_{M1}\sum_{i=0}^{N-1}X_{Mi}+\cdots+Y_{MN-1}\sum_{i=0}^{N-1}X_{Mi}\right)\end{aligned}\tag{10.44}$$

where we have

$$\begin{aligned}&Y_{10}\sum_{i=0}^{N-1}X_{1i}+Y_{11}\sum_{i=0}^{N-1}X_{1i}+\cdots+Y_{1N-1}\sum_{i=0}^{N-1}X_{1i}\\&=\sum_{i=0}^{N-1}X_{1i}(Y_{10}+Y_{11}+\cdots+Y_{1-1})\\&=\sum_{i=0}^{N-1}X_{1i}\sum_{i=0}^{N-1}Y_{1i}\end{aligned}\tag{10.45}$$

As each element code has only one pulse, we have

$$\begin{cases}\sum_{i=0}^{N-1}X_{1i}=1\\\sum_{j=0}^{N-1}Y_{1j}=1\end{cases}\tag{10.46}$$

Therefore, the summation of all element-wise cross-correlation functions becomes

$$\begin{aligned}&\sum_{i=0}^{N-1}X_{1i}\sum_{i=0}^{N-1}Y_{1i}+\sum_{i=0}^{N-1}X_{2i}\sum_{i=0}^{N-1}Y_{2i}+\cdots+\sum_{i=0}^{N-1}X_{Mi}\sum_{i=0}^{N-1}Y_{Mi}\\&=\overbrace{1+1+\cdots 1}^{M1'\text{s}}=M\end{aligned}\tag{10.47}$$

which shows that in an optical complementary code set the maximal cross-correlation level of a flock of M element codes is M.

10.4.3 GENERATION OF OPTICAL COMPLEMENTARY CODES

After having introduced the properties of optical complementary codes in the previous subsection, we are ready to discuss the in generation.

An optical complementary code can be denoted by

$$\mathbf{C}_{occ}(i)=(a_i^{(1)},a_i^{(2)},\ldots,a_i^{(M)}),\tag{10.48}$$

where $a_i^{(j)}$ denotes the pulse position of the jth element code in the ith code, and we also have $i=1,2,\ldots,K$, $j=1,2,\ldots,M$, and $a_i^{(j)}=1,2,\ldots,N$.

There are four basic principles we have to observe in generation of an optical complementary code set, as explained below.

$$A = \{\boxed{1}, 1, \boxed{1}, 1\}$$
$$B = \{\boxed{1}, 2, \boxed{1}, 4\}$$

Figure 10.7 Two optical complementary codes **A** and **B**. The first and third element codes have their pulses overlapped in the first chip position.

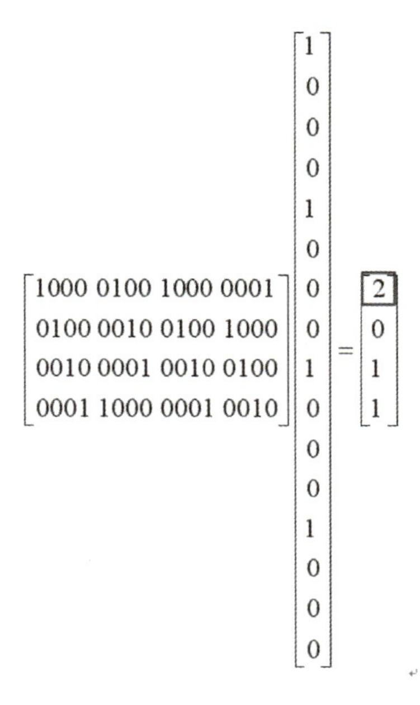

Figure 10.8 The cross-correlation functions for the two optical complementary codes **A** and **B**. The first and third element codes have their pulses overlapped in the first chip position.

1. For any pair of optical complementary codes, their pulses can be overlapped only in one element code. Take two optical complementary codes with $N = 4$ and $M = 4$, **A** and **B**, as examples. If their pulses overlap in two element codes, as shown in Figure 10.7, where the first and third element codes have their pulses overlapped in the first chip position, then the cross-correlation function between **A** and **B** will be larger than one, violating the requirement for $\lambda_c = 1$, as shown by the 2 within the square block in Figure 10.8. Therefore, to make it easier for us to encode optical complementary codes, we always put a pulse in the first chip position in the first element code, or $a_i^{(1)}$, where $i = 1, 2, \ldots, K$. Then, the pulses that appear in all other element codes (from the second to the Mth element codes) should never be overlapped.

2. The relative distances between the pulses that appear in two different element codes for two optical complementary codes should never be the same. To illustrate this, Figure 10.9 shows that the pulses between the second and third element codes of **A** have the same relative distance (which is $3 - 2 = 1$) as that $(4 - 3 = 1)$ between the second and third element codes of **B**.

$$A = \{\, 1\,, \boxed{2}\,, \boxed{3}\,, 4 \,\}$$
$$B = \{\, 1\,, \boxed{3}\,, \boxed{4}\,, 2 \,\}$$

Figure 10.9 The pulses between the second and third element codes of **A** have the same relative distance (which is $3 - 2 = 1$) as that $(4 - 3 = 1)$ between the second and third element codes of **B**.

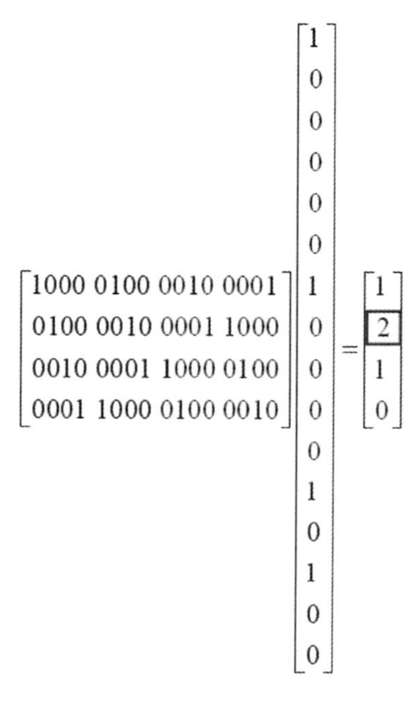

Figure 10.10 The cross-correlation function between **A** and **B** can be larger than one, as shown by the 2 within the square.

Therefore, the cross-correlation function between **A** and **B** may be larger than one, as shown in Figure 10.10, where the second row will yield 2, which is larger than one.

3. Each element code contains only one pulse, or $w = 1$. Thus, the summation of the auto-correlation peaks of all element codes is equal to the flock size or $W = M$.

4. The flock size of a set of optical complementary codes should be $2 \leq M \leq N$.

In order to observe the first two principles as described earlier (to ensure that (1) the pulses should never appear at the same chip position in the same element codes of two different flocks, and (2) the relative pulse distances between two element codes for two different flocks should not be the same), we can let the code length be a prime number and the pulse positions in different element codes can be determined by the Galois field approach, or

$$GF(p) = (0, 1, 2, \ldots, p - 1) \tag{10.49}$$

$\odot$	$j=0$	$j=1$	$j=2$	$\cdots$	$j=N-2$	$j=N-1$
$i=0$	0	0	0	$\cdots$	0	0
$i=1$	0	1	2	$\cdots$	$N-2$	$N-1$
$i=2$	0	2	4	$\cdots$	$N-4$	$N-2$
$\vdots$				$\vdots$		
$i=N-2$	0	$N-2$	$N-4$	$\cdots$	4	2
$i=N-1$	0	$N-1$	$N-2$	$\cdots$	2	1

Figure 10.11 The results from the modulo-N calculation defined as $[ij]_N$ (where $i = \{0, 1, 2, \ldots, N-1\}$ and $j = \{0, 1, 2, \ldots, N-1\}$) over the Galois field $GF(N) = \{0, 1, 2, \ldots, N-1\}$, where N is a prime number.

$a_i^{(j)}$	$j=0$	$j=1$	$j=2$	$\cdots$	$j=N-2$	$j=N-1$
$i=0$	1	1	1	$\cdots$	1	1
$i=1$	1	2	3	$\cdots$	$N-1$	N
$i=2$	1	3	5	$\cdots$	$N-3$	$N-1$
$\vdots$				$\vdots$		
$i=N-2$	1	$N-1$	$N-3$	$\cdots$	5	3
$i=N-1$	1	N	$N-1$	$\cdots$	3	2

Figure 10.12 The results from the modulo-N calculation defined as $[ij]_N + 1$ (where $i = \{0, 1, 2, \ldots, N-1\}$ and $j = \{0, 1, 2, \ldots, N-1\}$) over the Galois field $GF(N) = \{0, 1, 2, \ldots, N-1\}$, where N is a prime number. The row index can be viewed as flock or user index, the column index can be viewed as element code index, and the contents in the table given the pulse positions for all different flocks and all different element codes. The optical complementary codes encoded using this approach will always satisfy the first two principles.

where p is a prime number. A prime number is defined as a number which can not be divided by any other numbers except one and itself. For instance, 1, 2, 3, 5, 7, 11, 13, 17, 19, 23, ... are prime numbers. The approach is introduced as follows.

Let us first examine the modulo-N calculation defined as $[ij]_N$ (where $i = \{0, 1, 2, \ldots, N-1\}$ and $j = \{0, 1, 2, \ldots, N-1\}$) over the Galois field $GF(N) = \{0, 1, 2, \ldots, N-1\}$, where N is a prime number. The result of $[ij]_N$ is shown in Figure 10.11, based on which we can easily obtain another table as shown in Figure 10.12, where the contents are generated using $[ij]_N + 1$ The row index in

$a_i^{(j)}$	$j=0$	$j=1$	$j=2$	$j=3$	$j=4$	$j=5$	$j=6$
$i=0$	1	1	1	1	1	1	1
$i=1$	1	2	3	4	5	6	7
$i=2$	1	3	5	7	2	4	6
$i=3$	1	4	7	3	6	2	5
$i=4$	1	5	2	6	3	7	4
$i=5$	1	6	4	2	7	5	3
$i=6$	1	7	6	5	4	3	2

Figure 10.13 Example set of optical complementary codes generated using prime number $N = 7$, where $GF(7) = \{0, 1, 2, \ldots, 6\}$, and $i, j = 0, 1, \ldots, 6$. The variable i can be viewed as user index and j as element code index.

Figure 10.12 can be viewed as flock or user index, the column index can be viewed as element code index, and the contents in the table given the pulse positions for all different flocks and all different element codes. The optical complementary codes encoded using this approach will always satisfy the first two principles: (1) the pulses should never appear at the same chip position in the same element codes of two different flocks, and (2) the relative pulse distances between two element codes for two different flocks should not be the same.

Now let us look at an example. Let $N = 7$, $GF(7) = \{0, 1, 2, \ldots, 6\}$, where $i, j = 0, 1, \ldots, 6$. The variable i can be viewed as user index and j as element code index. Therefore, it is very easy to generate all pulse positions for this particular set of optical complementary codes, as shown in Figure 10.13.

From the above discussions, we can conclude that, when the element code length is a prime number N, we always can formulate a set of optical complementary codes using $GF(N)$. The flock size will be $2 \leq M \leq N$ (if $M < N$, we can simply delete some columns to make the flock size smaller), and the set size will always be N, or $K = N$ if N is a prime number.

On the other hand, if the element code length is not a prime number, we should not be able to use the Galois field approach to decide the pulse positions in all element codes. It can be proved that, if and only if $M = 2$, we will have $K = N$. With the increase in the flock size (M), the set size of the optical complementary codes will decrease. In this case, we have to resort to some exhaustive search approaches to find all valid codes.

Now we can summarize the generation procedure for a set of optical complementary codes as follows. To make the procedure as general as possible, we should not limit the element code length to a prime number. Instead, it can be any positive number larger than two. Still let $w = 1$, which is the number of ones in one element code. Assume that the element code length is N and flock size is M. As we can only put one pulse in each element code, and we have M element codes in each flock, then we have altogether N^M ways to generate a flock of M element codes which ensures $w = 1$.

The generation procedure for a set of optical complementary codes is given as follows:

1. Generate an auto-correlation function matrix as shown in Equation (10.31), which is formed from all N^M possible candidate flocks and has N^M rows and NM columns.

2. Based on all N^M possible candidate flocks, select one as the flock for the first user. Then, calculate the cross-correlation function between the first one and any one from the remaining $N^M - 1$ flocks, to check if their cross-correlation function satisfies our pre-set requirements. If yes, then the second flock is determined. Next, we check the cross-correlations between the first and all other $N^M - 1$ flocks, as well as checking that between the second and all other $N^M - 1$ flocks, to see if we can find the third flock. The procedure continues until we find the number of flocks we need. It is noted that different choices of the first flock may lead to different sets of optical complementary codes, but with the same set size.

We should show the generation procedure by using an example. Assume that we have $N = 4$ and $M = 2$. The step by step procedure is explained as follows.

1. We generate all flocks whose auto-correlation functions satisfy our pre-set requirements ($\lambda_a = 0$) as follows:

$$\begin{cases} \{1000; 1000\} \\ \{0100; 1000\} \\ \{0010; 1000\} \\ \{0001; 1000\} \\ \{1000; 0100\} \\ \{0100; 0100\} \\ \{0010; 0100\} \\ \{0001; 0100\} \\ \{1000; 0010\} \\ \{0100; 0010\} \\ \{0010; 0010\} \\ \{0001; 0010\} \\ \{1000; 0001\} \\ \{0100; 0001\} \\ \{0010; 0001\} \\ \{0001; 0001\} \end{cases} \tag{10.50}$$

in which there are in total 16 flocks.

2. Then, select {1000; 1000} as the first flock, and let the other 15 flocks counter-check with it in terms of their cross-correlation functions. The results are that there are four flocks satisfying the condition of $\lambda_c = 1$. Thus, we obtain the four flocks of optical complementary codes with $N = 4$ and $M = 2$. They are

$$\begin{cases} A = \{1000; 1000\} \\ B = \{0100; 1000\} \\ C = \{0010; 1000\} \\ D = \{0001; 1000\} \end{cases} \tag{10.51}$$

which forms a set of codes we want.

We can check their auto-correlation and cross-correlation functions to see if they indeed satisfy our pre-set requirements as follows. Figure 10.14 shows their auto-correlation functions and Figure 10.15 shows their cross-correlation functions. It is seen from the figures that the generated code flocks indeed satisfy all our conditions on their auto-correlation functions and cross-correlation functions, or $\lambda_a = 0$ and $\lambda_c = 1$. It can also be proved that there are only four available flocks from this example. If we add a fifth flock, the pre-set conditions will not be satisfied.

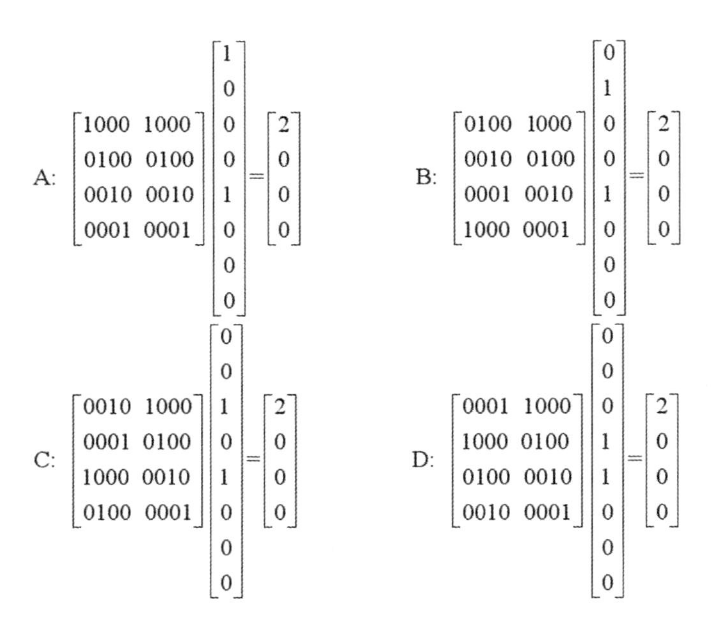

Figure 10.14 The auto-correlation functions for the set of optical complementary codes with $N = 4$ and $M = 2$.

10.4.4 PERFORMANCE COMPARISON

Since we have introduced the OCCs and their generation method, it is time to compare them with other existing sequences for optical CDMA applications. We will use optical orthogonal codes (OOC) and prime codes as the benchmarks for the comparisons made here.

It can be seen from the discussions on the OOCs and prime codes (both of which are unitary codes) in the previous sections that both OOCs and prime codes have achieved that their maximal auto-correlation and cross-correlation functions are not greater than one, but their auto-correlation functions are not ideal in the sense that the auto-correlation side lobes are not zero. On the other hand, the OCCs have made great progress to achieve perfect auto-correlation functions: the values of auto-correlation functions are zero, except at the zero shift. Therefore, we can gladly say that the occs have achieved what we wanted to achieve to develop next generation optical CDMA technology.

It is noted that the OOCs and prime codes are unitary codes, while the OCCs are complementary codes. To make an objective comparison, let us define the following four comparing performance parameters, r_a, r'_a, r_c, and r'_c, defined by

$$r_a = \frac{\text{No. of interfering pulses in ACF}}{\text{Processing gain}} \tag{10.52}$$

$$r'_a = \frac{\text{No. of interfering pulses in ACF}}{\text{Weight of the code}} \tag{10.53}$$

$$r_c = \frac{\text{No. of interfering pulses in CCF}}{\text{Processing gain}} \tag{10.54}$$

$$r'_c = \frac{\text{No. of interfering pulses in CCF}}{\text{Weight of the code}} \tag{10.55}$$

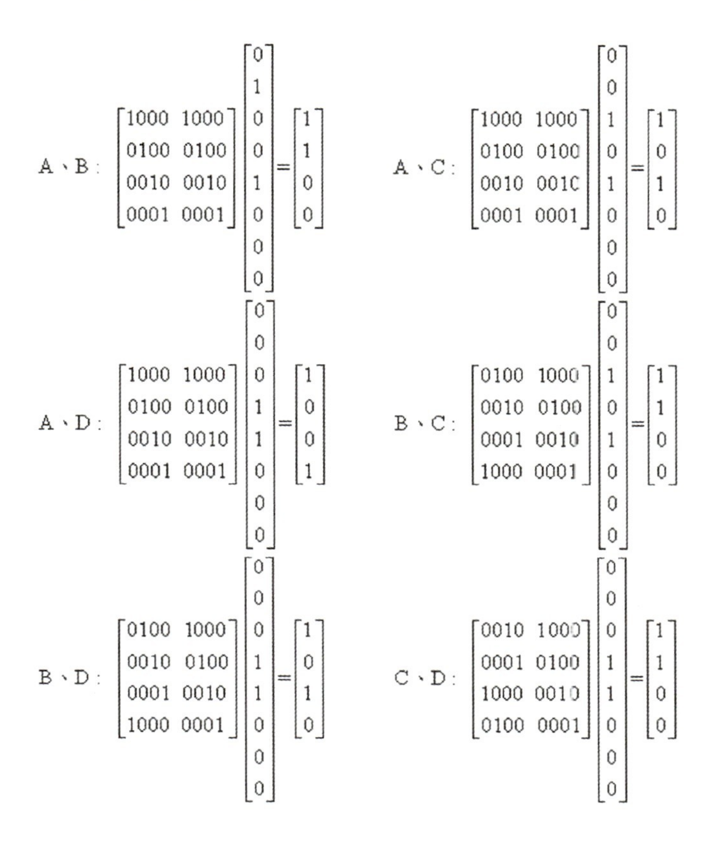

Figure 10.15 The cross-correlation functions for the set of optical complementary codes with $N = 4$ and $M = 2$.

where 'No. of interfering pulses in ACF' and 'No. of interfering pulses in CCF' for both OOCs and prime codes are $w(w-1)$ and w^2, respectively. On the other hand, 'No. of interfering pulses in ACF' and 'No. of interfering pulses in CCF' for optical complementary codes are zero and M, respectively.

Figure 10.16 compares the parameter r_a for OOCs and OCCs, with the processing gain being a parameter, which changes from zero to 250. It is seen from the figure that the OCCs clearly excel their counterparts with a very big margin.

Figure 10.17 shows the performance parameter r'_a also for the OOCs, prime codes, and OCCs, with their weight being a parameter, which changes from two to seven. Again, this figure shows that the OCCs will not generate any interference for any weight value. This is obvious as the code design procedure of the OCCs has ensured that the zero auto-correlation side lobe condition or $\lambda_a = 0$ is satisfied. Figures 10.18 and 10.19 give the results for performance parameters r_c and r'_c again with their PG and weights being the parameters. The superior performance of the OCCs can be easily seen.

The number of users supportable in an optical CDMA system is also compared in Figure 10.20, where the number of users versus processing gains of the codes is shown. It is seen from the figure

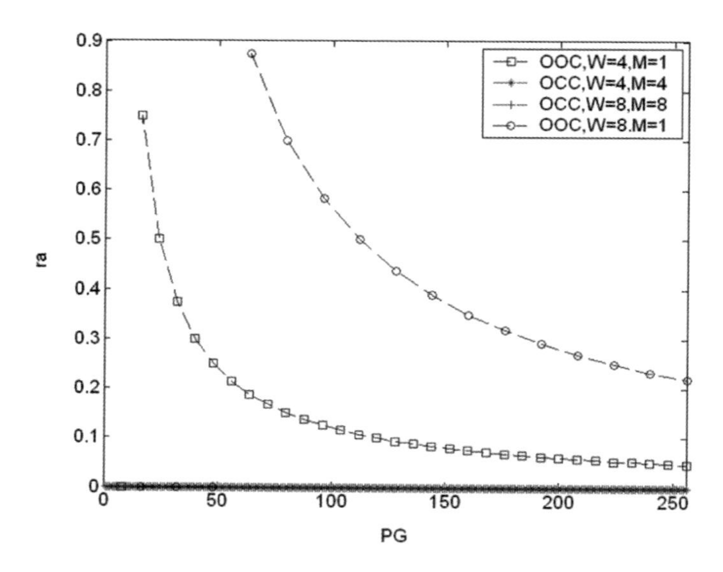

Figure 10.16 Comparison of the parameter r_a for OOCs and OCCs, with the processing gain being a parameter.

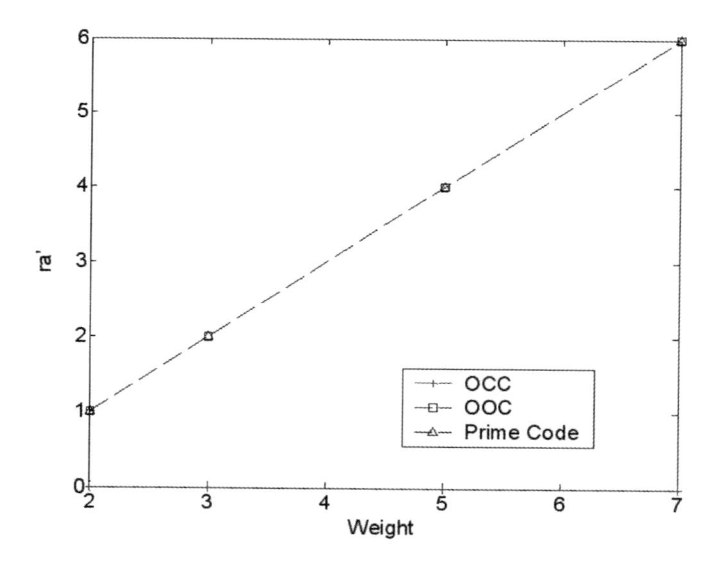

Figure 10.17 Comparison of the parameter r_a' for OOCs,prime codes, and OCCs, with their weights being a parameter, which change from two to seven.

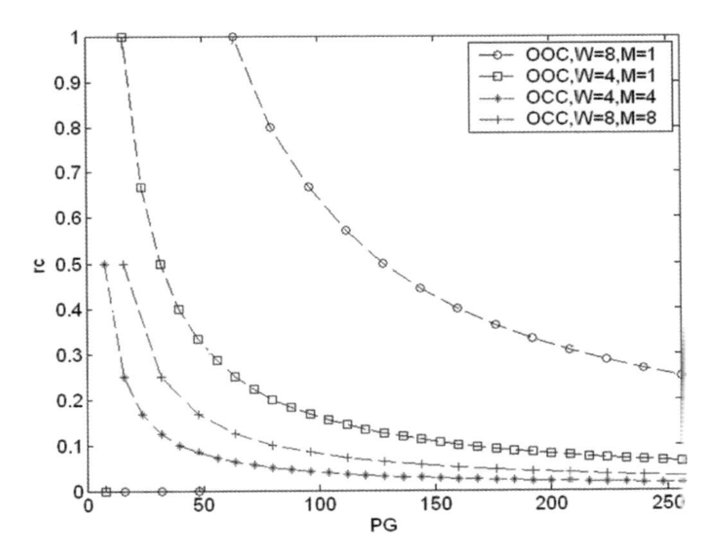

Figure 10.18 Comparison of the parameter r_c for OOCs and OCCs, with the processing gain being a parameter.

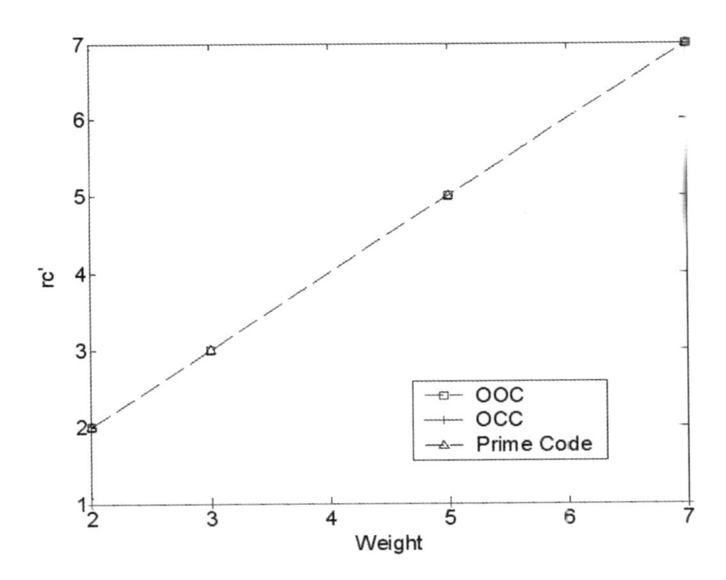

Figure 10.19 Comparison of the parameter r_c' for OOCs, prime codes, and OCCs, with their weights being a parameter, which change from two to seven.

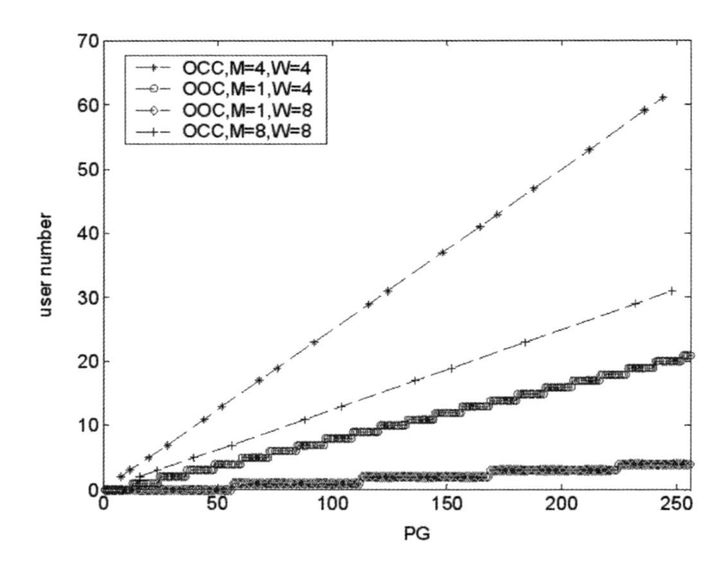

Figure 10.20 The number of users supportable in an optical CDMA system using OCCs and OOCs.

that under the same processing gain the OCDMA system using OCCs can provide a much larger capacity than that using OOCs.

To give an example with more detailed information used in the comparison, we take two typical codes, one being OOC and the other OCC, to show their correlation properties as follows. The two optical orthogonal codes of interest here are $C_1 = (1, 10, 13, 28)$ and $C_2 = (1, 5, 12, 31)$, with their code length being $N = 32$, weight $W = 4$, flock size $M = 1$ (a unitary code), and set size $K = 2$. For these two OOCs, we can calculate their numbers of interfering pulses in ACF being 12, and thus $r_a = 0.375$. The number of interfering pulses in CCF is 16, and thus $r_c = 0.5$.

On the other hand, we have two OCCs,which are $C_1 = (1; 1; 1; 1)$ and $C_2 = (8; 2; 3; 4)$, with each element code separated by semi-colon. Their element code length is $N = 8$, the weight is $W = 4$, the flock size is $M = 4$, and the set size is $K = 6$. For these two OCCs, we can calculate their numbers of interfering pulses in ACF being 0, and thus $r_a = 0$. The number of interfering pulses in CCF is 4, and thus $r_c = 0.125$.

Based on the above data, we can plot their ACF and CCF in Figure 10.21. It is clearly seen from the figure that the ACF for the OCCs is substantially better than that for the OOCs, while the CCFs of both are very close to each other.

The bit error rate performance for both OOC and OCC is compared in Figure 10.22, where Figure 10.22(a) shows the BER for the two codes versus decision threshold, which changes from 1 to 7, with the number of users being fixed at $K = 20$, weight being $W = 5$, and PG being 1000 and 2000, respectively.

Figure 10.22(b) illustrates the BER performance for both OOC and OCC codes versus decision threshold, which varies from 1 to 11, with the number of users being fixed at $K = 10$, weight changing $W = 1, 4$, and 8, and PG being 1000.

Figure 10.22(c) gives BER comparison between OOCs and OCCs versus decision threshold, which varies from 1 to 8. The PG is fixed at PG $= 1000$, weight is $W = 5$, and K takes 10, 30, and 50.

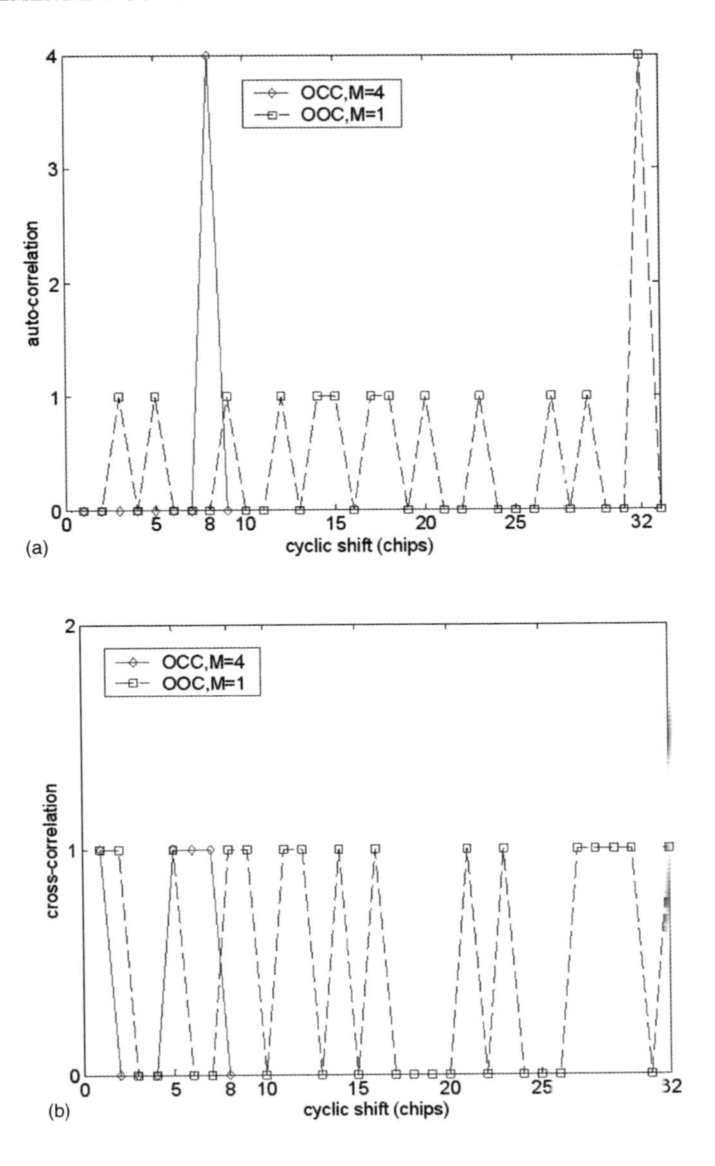

Figure 10.21 (a) Auto-correlation functions for OOCs ({1, 10, 13, 28}) and OCCs ({1; 1; 1; 1}), where PG = 32 for both codes. (b) Cross-correlation functions for OOCs ({1, 10, 13, 28} and {1, 5, 12, 31}) and OCCs ({1; 1; 1; 1} and {8; 2; 3; 4}), where PG = 32 for both codes.

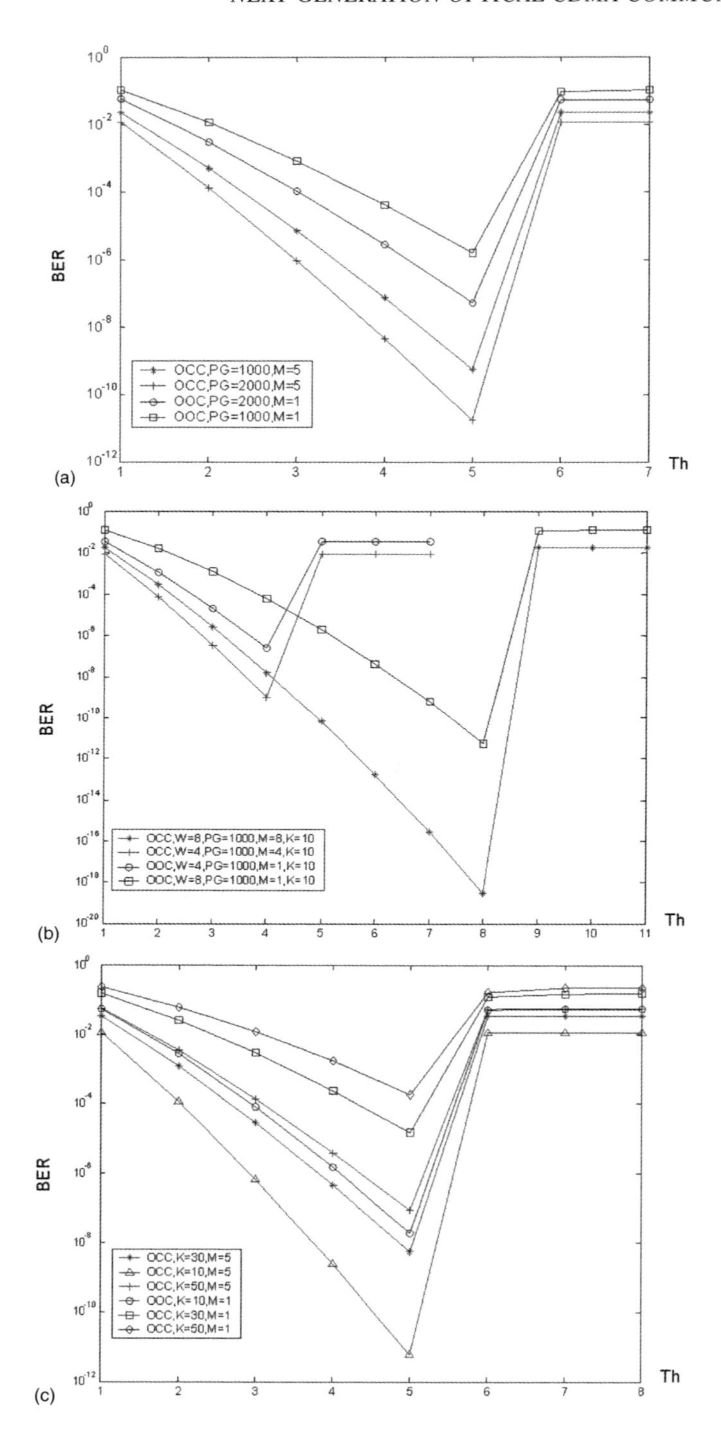

Figure 10.22 BER comparison between OOCs and OCCs versus decision threshold. (a) Decision threshold varies from 1 to 7. The number of users is fixed at $K = 20$, weight is $W = 5$, PG is 1000 and 2000. (b) Decision threshold varies from 1 to 11. The number of users is fixed at $K = 10$, weight changes $W = 1$, 4, and 8, PG is 1000. (c) Decision threshold varies from 1 to 8. The PG is fixed at PG = 1000, weight is $W = 5$, K takes 10, 30, and 50.

It is noted that the threshold selection is very important in an optical CDMA receiver, as it will affect its performance if some wrong thresholds are chosen. In most cases, we need to do a lot of simulation experiments to decide which threshold can offer the best performance for a particular OCDMA system setup. All figures have shown that the best performance of an OCDMA system using OCCs can be achieved when the threshold is selected to be the same as the weight W. Also, all figures show that the OCDMA system using the OCCs outperforms that using OOC by a very comfortable margin.

A

Relation between Periodic and Aperiodic Correlation Functions

In this book, much of the content has been focused on the spreading codes and their correlation properties, which have a great impact on the overall performance of a CDMA system. There are two different types of correlation functions which are often used in this book: periodic and aperiodic correlation functions. It is noted that what we mean here in terms of correlation functions obviously includes auto-correlation and cross-correlation functions.

In this appendix, first we would like to give the definitions for both periodic and aperiodic correlation functions. Afterwards, we will continue to show the relation between them. The main objective of this appendix is to show the close relationship between the periodic and the aperiodic correlation function for the same spreading code set of concern. We will prove that if an ideally orthogonal code set has perfect aperiodic correlation properties the code set will also have perfect periodic correlation properties. More specifically, it is to be proved that perfect aperiodic correlation properties of a code set is a sufficient and necessary condition to achieve perfect periodic correlation properties of the same code set. The perfect correlation properties for a code set means that the code set possesses both ideal auto-correlation function of any code and ideal cross-correlation function for any two codes in the same set.

The proof shown in this appendix is significant as the aperiodic correlation functions (also called partial correlation functions in this book) can always be expressed in a much simpler form than the periodic correlation functions. Thus, if we can use aperiodic correlation functions to study a spreading code set, we can greatly reduce the overall complexity of the analysis.

A.1 APERIODIC CORRELATION FUNCTIONS

It is noted that throughout this book we assume that the aperiodic correlation functions are exactly equal to the partial correlation functions of the same code set to simplify the discussions. The definition of the partial correlation functions is derived based mainly on a single bit instead of a bit stream (as the case in a real application scenario). Therefore, the discussions on the partial correlation can be substantially simplified compared to those based on the periodic correlation functions of the same code set.

Let us take a look at a simple example with two codes, their flock size and element code length being M and N, respectively, which is written as follows:

$$\mathbf{X} = \begin{pmatrix} x_{1,1} & x_{1,2} & \cdots & x_{1,N} \\ x_{2,1} & x_{2,2} & \cdots & x_{2,N} \\ & & \vdots & \\ x_{M,1} & x_{M,2} & \cdots & x_{M,N} \end{pmatrix} \tag{A.1}$$

$$\mathbf{Y} = \begin{pmatrix} y_{1,1} & y_{1,2} & \cdots & y_{1,N} \\ y_{2,1} & y_{2,2} & \cdots & y_{2,N} \\ & & \vdots & \\ y_{M,1} & y_{M,2} & \cdots & y_{M,N} \end{pmatrix} \tag{A.2}$$

The aperiodic auto-correlation function or the partial auto-correlation function for code $\mathbf{X}$ is shown in Figure A.1.

Therefore, the partial auto-correlation function of $\mathbf{X}$ can be written as

$$\rho_{aperiodic}(\mathbf{X}; i) = \sum_{m=1}^{M} \sum_{n=1}^{N-i} x_{m,n} x_{m,n+i} \tag{A.3}$$

Assume that there is a relative chip shift between codes $\mathbf{X}$ and $\mathbf{Y}$ (in which code $\mathbf{X}$ is shifted to the right-hand side of code $\mathbf{Y}$). We can obtain their partial cross-correlation function with their offset chip shift being i chips, as shown in Figure A.2, or

$$\rho_{aperiodic}(\mathbf{X}, \mathbf{Y}; i) = \sum_{m=1}^{M} \sum_{n=1}^{N-i} x_{m,n} y_{m,n+i} \tag{A.4}$$

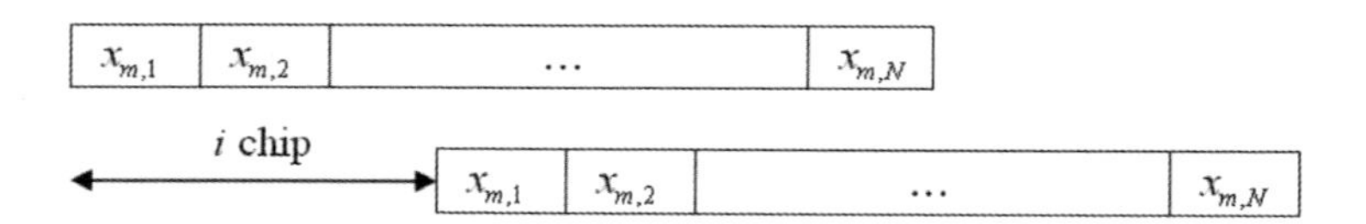

Figure A.1 The aperiodic auto-correlation function or partial auto-correlation function for the m element code (where $m \in (1, M)$) in the code flock $\mathbf{X}$, where there is a relative chip shift of i chips.

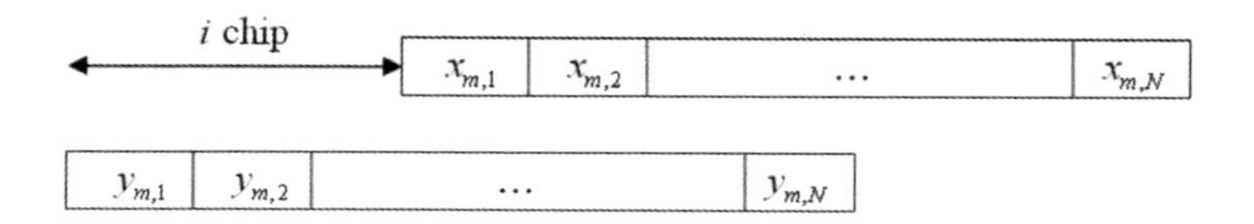

Figure A.2 The aperiodic cross-correlation function or partial cross-correlation function of the m element codes (where $m \in (1, M)$) between codes $\mathbf{X}$ and $\mathbf{Y}$, in which code $\mathbf{X}$ is shifted i chips to the right side of code $\mathbf{Y}$.

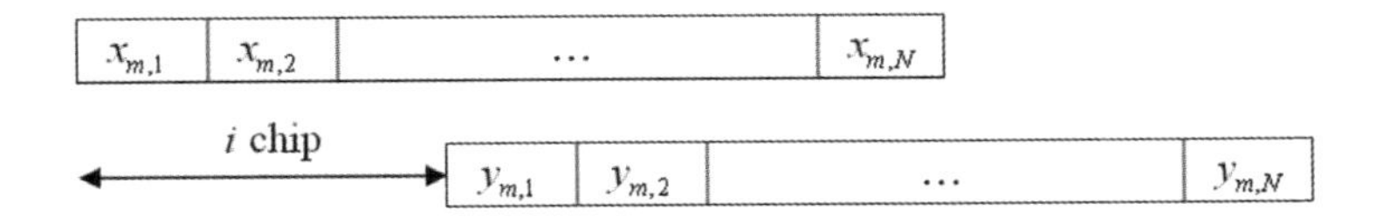

Figure A.3 The aperiodic cross-correlation function or partial cross-correlation function of the m element codes (where $m \in (1, M)$) between codes **X** and **Y**, in which code **X** is shifted i chips to the left side of code **Y**.

In the same way, we can have the partial cross-correlation between codes **X** and **Y** (in which code **X** is shifted to the left-hand side of code **Y**, as shown in Figure A.3) as follows

$$\rho_{aperiodic}(\mathbf{Y}, \mathbf{X}; i) = \sum_{m=1}^{M} \sum_{n=1}^{N-i} y_{m,n} x_{m,n+i} \tag{A.5}$$

A.2 PERIODIC CORRELATION FUNCTIONS

Different from the aperiodic correlation functions which cover only one element code length, periodic correlation functions should usually consider the real data stream, in which consecutive bits may take different signs, and thus will generate different correlation patterns.

Therefore, we have to take into account at least two neighboring bits when calculating the periodic correlation functions. If the two consecutive bits carry the same sign, we have even periodic correlation functions; otherwise odd periodic correlation functions will be generated.

Let us again consider the two complementary codes given in (A.1) and (A.2), with its parameters being $(N, M, K) = (N, M, M)$.

The even auto-correlation function with i chips relative shift of code **X** (as shown in Figure A.4) can be written as

$$\rho_{even\ periodic}(\mathbf{X}; i) = \sum_{m=1}^{M} \sum_{n=1}^{N-i} x_{m,n} x_{m,n+i} + \sum_{m=1}^{M} \sum_{n=1}^{i} x_{m,n} x_{m,N-i+n} \tag{A.6}$$

The odd auto-correlation function with i chips relative shift of the code **X** (as shown in Figure A.5) can be written as

$$\rho_{odd\ periodic}(\mathbf{X}; i) = \sum_{m=1}^{M} \sum_{n=1}^{N-i} x_{m,n} x_{m,n+i} + \sum_{m=1}^{M} \sum_{n=1}^{i} x_{m,n} \overline{x_{m,N-i+n}} \tag{A.7}$$

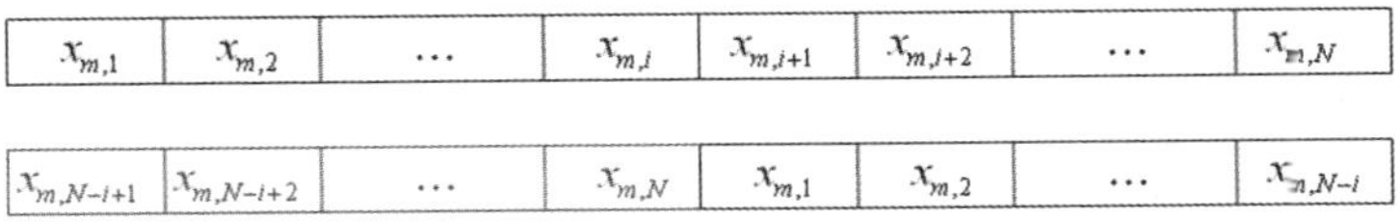

$x_{m,1}$	$x_{m,2}$	...	$x_{m,i}$	$x_{m,i+1}$	$x_{m,i+2}$	...	$x_{m,N}$
$x_{m,N-i+1}$	$x_{m,N-i+2}$	...	$x_{m,N}$	$x_{m,1}$	$x_{m,2}$	...	$x_{m,N-i}$

Figure A.4 The even periodic auto-correlation function of the mth element code of code **X**, where $m \in (1, M)$.

$x_{m,1}$	$x_{m,2}$	$\cdots$	$x_{m,i}$	$x_{m,i+1}$	$x_{m,i+2}$	$\cdots$	$x_{m,N}$
$\overline{x_{m,N-i+1}}$	$\overline{x_{m,N-i+2}}$	$\cdots$	$\overline{x_{m,N}}$	$x_{m,1}$	$x_{m,2}$	$\cdots$	$x_{m,N-i}$

Figure A.5 The odd periodic auto-correlation function of the mth element code of code **X**, where $m \in (1, M)$.

$x_{m,1}$	$x_{m,2}$	$\cdots$	$x_{m,i}$	$x_{m,i+1}$	$x_{m,i+2}$	$\cdots$	$x_{m,N}$
$y_{m,N-i+1}$	$y_{m,N-i+2}$	$\cdots$	$y_{m,N}$	$y_{m,1}$	$y_{m,2}$	$\cdots$	$y_{m,N-i}$

Figure A.6 The even periodic cross-correlation function of the mth element codes of codes **X** and **Y**, where $m \in (1, M)$.

$x_{m,1}$	$x_{m,2}$	$\cdots$	$x_{m,i}$	$x_{m,i+1}$	$x_{m,i+2}$	$\cdots$	$x_{m,N}$
$\overline{y_{m,N-i+1}}$	$\overline{y_{m,N-i+2}}$	$\cdots$	$\overline{y_{m,N}}$	$y_{m,1}$	$y_{m,2}$	$\cdots$	$y_{m,N-i}$

Figure A.7 The odd periodic cross-correlation function of the mth element codes of codes **X** and **Y**, where $m \in (1, M)$.

In a similar way, we can obtain the even and odd periodic cross-correlation functions between codes **X** and **Y** with i relative chips shift between them (as shown in Figures A.6 and A.7), as

$$\rho_{even\ periodic}(\mathbf{X}, \mathbf{Y}; i) = \sum_{m=1}^{M} \sum_{n=1}^{N-i} y_{m,n} x_{m,n+i} + \sum_{m=1}^{M} \sum_{n=1}^{i} x_{m,n} y_{m,N-i+n} \tag{A.8}$$

and

$$\rho_{odd\ periodic}(\mathbf{X}, \mathbf{Y}; i) = \sum_{m=1}^{M} \sum_{n=1}^{N-i} y_{m,n} x_{m,n+i} + \sum_{m=1}^{M} \sum_{n=1}^{i} x_{m,n} \overline{y_{m,N-i+n}} \tag{A.9}$$

respectively, where the notation $\bar{x}$ represents the negative value of x, or $-x$, for expression simplicity.

A.3 PROOF

With all the above definitions, now we can proceed to show that the following two claims

- the periodic correlation functions of a CDMA code set are perfect;
- the aperiodic correlation functions of a CDMA code set are perfect;

in fact express mutually sufficient and necessary conditions for each other. That means we can prove the first claim based on the second claim, or vice versa. We can simply write the above relationship as

$$\begin{array}{c}\textbf{Perfect periodic correlation functions}\\ \Updownarrow \\ \textbf{Perfect aperiodic correlation functions}\end{array} \tag{A.10}$$

Here, the periodic correlation functions of a CDMA code set include both periodic auto-correlation functions and periodic cross-correlation functions. Also, the aperiodic correlation functions of a CDMA code set include both aperiodic auto-correlation functions and aperiodic cross-correlation functions.

Next, the proof of relationship (A.10) will proceed in two steps. First, we should prove that if a CDMA code set has perfect periodic correlation functions, then we can derive that the code set has also perfect aperiodic correlation functions. Second, we will show that if a CDMA code set has perfect aperiodic correlation functions, then we can derive that the code set has also perfect periodic correlation functions. This will conclude our proof.

Now, we would like to show that if a CDMA code set has perfect periodic correlation functions, then we can derive that the code set has also perfect aperiodic correlation functions, or the following relation exists

$$\begin{array}{c}\textbf{Perfect periodic correlation functions}\\ \Downarrow \\ \textbf{Perfect aperiodic correlation functions}\end{array} \tag{A.11}$$

It is known that if a code set has perfect periodic auto-correlation functions, then its even and odd periodic auto-correlation functions must be perfect at the same time, or

$$\begin{cases}\rho_{even\ periodic}(\mathbf{X}; i=0) = \sum_{m=1}^{M}\sum_{n=1}^{N} x_{m,n}x_{m,n} = MN\\ \rho_{even\ periodic}(\mathbf{X}; i\neq 0) = \sum_{m=1}^{M}\sum_{n=1}^{N-i} x_{m,n}x_{m,n+i} + \sum_{m=1}^{M}\sum_{n=1}^{i} x_{m,n}x_{m,N-i+n} = 0\\ \rho_{odd\ periodic}(\mathbf{X}; i=0) = \sum_{m=1}^{M}\sum_{n=1}^{N} x_{m,n}x_{m,n} = MN\\ \rho_{odd\ periodic}(\mathbf{X}; i\neq 0) = \sum_{m=1}^{M}\sum_{n=1}^{N-i} x_{m,n}x_{m,n+i} + \sum_{m=1}^{M}\sum_{n=1}^{i} x_{m,n}\overline{x_{m,N-i+n}} = 0\end{cases} \tag{A.12}$$

On the other hand, the aperiodic auto-correlation function for code **X** is

$$\rho_{aperiodic}(\mathbf{X}; i=0) = \sum_{m=1}^{M}\sum_{n=1}^{N} x_{m,n}x_{m,n} = MN \tag{A.13}$$

and

$$\begin{aligned}\rho_{aperiodic}(\mathbf{X}; i\neq 0) &= \sum_{m=1}^{M}\sum_{n=1}^{N-i} x_{m,n}x_{m,n+i}\\ &= \frac{1}{2}\left(\sum_{m=1}^{M}\sum_{n=1}^{N-i} x_{m,n}x_{m,n+i} + \sum_{m=1}^{M}\sum_{n=1}^{i} x_{m,n}x_{m,N-i-n}\right.\end{aligned}$$

$$
+\sum_{m=1}^{M}\sum_{n=1}^{N-i}x_{m,n}x_{m,n+i}+\sum_{m=1}^{M}\sum_{n=1}^{i}x_{m,n}\overline{x_{m,N-i+n}}\Bigg)
$$

$$
=\frac{1}{2}\left(\rho_{even\ periodic}\left(\mathbf{X};i\neq 0\right)+\rho_{odd\ periodic}\left(\mathbf{X};i\neq 0\right)\right)
$$

$$
=0 \tag{A.14}
$$

where we have used the condition of perfect periodic auto-correlation functions of code $\mathbf{X}$ in the derivation process. Therefore, we have proved that the perfect aperiodic auto-correlation function can be derived directly from the perfect periodic auto-correlation function.

Next, we will show that the condition of perfect periodic cross-correlation functions of a code set can be used to directly derive its perfect aperiodic cross-correlation functions.

If a code set has perfect periodic cross-correlation functions, its even and odd cross-correlation functions should also be perfect, or we always have

$$
\begin{cases}
\rho_{even\ periodic}\left(\mathbf{X},\mathbf{Y};i\right)=\sum_{m=1}^{M}\sum_{n=1}^{N-i}y_{m,n}x_{m,n+i}+\sum_{m=1}^{M}\sum_{n=1}^{i}x_{m,n}y_{m,N-i+n}=0\\
\rho_{odd\ periodic}\left(\mathbf{X},\mathbf{Y};i\right)=\sum_{m=1}^{M}\sum_{n=1}^{N-i}y_{m,n}x_{m,n+i}+\sum_{m=1}^{M}\sum_{n=1}^{i}x_{m,n}\overline{y_{m,N-i+n}}=0
\end{cases} \tag{A.15}
$$

Then, its aperiodic cross-correlation functions can be written as

$$
\begin{aligned}
\rho_{aperiodic}\left(\mathbf{X},\mathbf{Y};i\neq 0\right)&=\sum_{m=1}^{M}\sum_{n=1}^{N-i}x_{m,n}x_{m,n+i}\\
&=\frac{1}{2}\left(\sum_{m=1}^{M}\sum_{n=1}^{N-i}x_{m,n}y_{m,n+i}+\sum_{m=1}^{M}\sum_{n=1}^{i}x_{m,n}y_{m,N-i+n}\right.\\
&\quad\left.+\sum_{m=1}^{M}\sum_{n=1}^{N-i}x_{m,n}y_{m,n+i}+\sum_{m=1}^{M}\sum_{n=1}^{i}x_{m,n}\overline{y_{m,N-i+n}}\right)\\
&=\frac{1}{2}\left(\rho_{even\ periodic}\left(\mathbf{X},\mathbf{Y};i\neq 0\right)+\rho_{odd\ periodic}\left(\mathbf{X},\mathbf{Y};i\neq 0\right)\right)=0
\end{aligned} \tag{A.16}
$$

where we have used the condition of perfect periodic cross-correlation functions between codes $\mathbf{X}$ and $\mathbf{Y}$. Therefore, we have proved that the relation given in (A.10) is true.

Similarly, we can also prove the relation

$$
\begin{gathered}
\textbf{Perfect periodic correlation functions}\\
\Uparrow\\
\textbf{Perfect aperiodic correlation functions}
\end{gathered} \tag{A.17}
$$

as follows. If a code set has perfect aperiodic auto-correlation functions, we always have

$$
\begin{cases}
\rho_{aperiodic}\left(\mathbf{X};i=0\right)=\sum_{m=1}^{M}\sum_{n=1}^{N}x_{m,n}x_{m,n}=MN\\
\rho_{aperiodic}\left(\mathbf{X};i\neq 0\right)=\sum_{m=1}^{M}\sum_{n=1}^{N-i}x_{m,n}x_{m,n+i}=0
\end{cases} \tag{A.18}
$$

Then, we can obtain the even periodic auto-correlation function as

$$
\rho_{even\ periodic}\left(\mathbf{X};i=0\right)=\sum_{m=1}^{M}\sum_{n=1}^{N-i}x_{m,n}x_{m,n+i}=MN \tag{A.19}
$$

and

$$\rho_{even\ periodic}(\mathbf{X}; i \neq 0) = \sum_{m=1}^{M}\sum_{n=1}^{N-i} x_{m,n}x_{m,n+i} + \sum_{m=1}^{M}\sum_{n=1}^{i} x_{m,n}x_{m,N-i+n}$$
$$= \rho_{aperiodic}(\mathbf{X}; i) + \rho_{aperiodic}(\mathbf{X}; N-i) = 0 \quad (A.20)$$

Similarly, we can obtain the odd periodic auto-correlation function as

$$\rho_{odd\ periodic}(\mathbf{X}; i = 0) = \sum_{m=1}^{M}\sum_{n=1}^{N} x_{m,n}x_{m,n} = MN \quad (A.21)$$

and

$$\rho_{odd\ periodic}(\mathbf{X}; i) = \sum_{m=1}^{M}\sum_{n=1}^{N-i} x_{m,n}x_{m,n+i} + \sum_{m=1}^{M}\sum_{n=1}^{i} x_{m,n}\overline{x_{m,N-i+n}}$$
$$= \rho_{aperiodic}(\mathbf{X}; i) - \rho_{aperiodic}(\mathbf{X}; N-i) = 0 \quad (A.22)$$

Therefore, the perfect periodic auto-correlation function for code **X** is there.

Now, we look at the cross-correlation functions. If a code set has perfect aperiodic cross-correlation functions between codes **X** and **Y**, which are defined in (A.1) and (A.2), we always have

$$\begin{cases} \rho_{aperiodic}(\mathbf{X}, \mathbf{Y}; i) = \sum_{m=1}^{M}\sum_{n=1}^{N-i} x_{m,n}y_{m,n+i} \\ \rho_{aperiodic}(\mathbf{Y}, \mathbf{X}; i) = \sum_{m=1}^{M}\sum_{n=1}^{N-i} y_{m,n}x_{m,n+i} \end{cases} \quad (A.23)$$

Then, the even and odd periodic cross-correlation functions can be written as

$$\rho_{even\ periodic}(\mathbf{X}, \mathbf{Y}; i) = \sum_{m=1}^{M}\sum_{n=1}^{N-i} y_{m,n}x_{m,n+i} + \sum_{m=1}^{M}\sum_{n=1}^{i} x_{m,n}y_{m,N-i+n}$$
$$= \rho_{aperiodic}(\mathbf{Y}, \mathbf{X}; i) + \rho_{aperiodic}(\mathbf{X}, \mathbf{Y}; N-i) = 0 \quad (A.24)$$

and

$$\rho_{odd\ periodic}(\mathbf{X}, \mathbf{Y}; i) = \sum_{m=1}^{M}\sum_{n=1}^{N-i} y_{m,n}x_{m,n+i} + \sum_{m=1}^{M}\sum_{n=1}^{i} x_{m,n}\overline{y_{m,N-i+n}}$$
$$= \rho_{aperiodic}(\mathbf{Y}, \mathbf{X}; i) - \rho_{aperiodic}(\mathbf{X}, \mathbf{Y}; N-i) = 0 \quad (A.25)$$

respectively. Thus, we have proved that the periodic cross-correlation function between codes **X** and **Y** is also perfect. Therefore, we have proved also

Perfect periodic correlation functions

$\Uparrow$ (A.26)

Perfect aperiodic correlation functions

Finally, to combine the results shown in (A.11) and (A.26), we obtain (A.10).

The conclusion made in this appendix is significant as based on this we are justified to only consider the aperiodic or partial correlation functions for a CDMA code set, instead of their periodic correlation functions. The use of aperiodic or partial correlation functions for a CDMA code set in our analysis can significantly reduce our analysis complexity, in particular in the process of orthogonality analysis of a spreading code set.

B

Proof of Flock-Wise Orthogonality of CC Codes

Complete complementary (CC) codes and their generation method have been discussed in Subsections 6.2.2 and 6.3.1. However, no rigorous proof on their flock-wise orthogonality has ever been given in the literature. In this appendix, we would like to give the proof of their orthogonality.

To generate a CC code set, we need first introduce two $N \times N$ orthogonal matrices, **A** and **B**, whose elements take unitary values or $| a_{ij} |= 1$ and $| b_{ij} |= 1$, as follows:

$$\mathbf{A} = \begin{pmatrix} a_{11} & a_{12} & \cdots & a_{1N} \\ a_{21} & a_{22} & \cdots & a_{2N} \\ \vdots & \vdots & \vdots & \vdots \\ a_{N1} & a_{N2} & \cdots & a_{NN} \end{pmatrix} = \begin{pmatrix} \mathbf{a}_1 \\ \mathbf{a}_2 \\ \vdots \\ \mathbf{a}_N \end{pmatrix} \tag{B.1}$$

$$\mathbf{B} = \begin{pmatrix} b_{11} & b_{12} & \cdots & b_{1N} \\ b_{21} & b_{22} & \cdots & b_{2N} \\ \vdots & \vdots & \vdots & \vdots \\ b_{N1} & b_{N2} & \cdots & b_{NN} \end{pmatrix} = \begin{pmatrix} \mathbf{b}_1 \\ \mathbf{b}_2 \\ \vdots \\ \mathbf{b}_N \end{pmatrix} \tag{B.2}$$

from which we can form a matrix **C** as

$$\mathbf{C} = \begin{pmatrix} b_{11}\mathbf{a}_1 & b_{12}\mathbf{a}_2 & \cdots & b_{1N}\mathbf{a}_N \\ b_{21}\mathbf{a}_1 & b_{22}\mathbf{a}_2 & \cdots & b_{2N}\mathbf{a}_N \\ \vdots & \vdots & \vdots & \vdots \\ b_{N1}\mathbf{a}_1 & b_{N2}\mathbf{a}_2 & \cdots & b_{NN}\mathbf{a}_N \end{pmatrix} \tag{B.3}$$

It is to be noted that both periodic and aperiodic ACF of any row or column and CCF of any two rows or two columns should be zero. Now, let us introduce another $N \times N$ orthogonal matrix **D** with

The Next Generation CDMA Technologies Hsiao-Hwa Chen

the absolute value of all its elements being unit or $| d_{ij} |= 1$, as follows:

$$\mathbf{D} = \begin{pmatrix} d_{11} & d_{12} & \cdots & d_{1N} \\ d_{21} & d_{22} & \cdots & d_{2N} \\ \vdots & \vdots & \vdots & \vdots \\ d_{N1} & d_{N2} & \cdots & d_{NN} \end{pmatrix} = \begin{pmatrix} \mathbf{d}_1 \\ \mathbf{d}_2 \\ \vdots \\ \mathbf{d}_N \end{pmatrix} \tag{B.4}$$

Based on **A** and **D**, we can have a vector $\mathbf{Z}_{xy}$ or

$$\mathbf{Z}_{xy} = \begin{pmatrix} a_{x1}d_{y1} & a_{x2}d_{y2} & \cdots & a_{xN}d_{yN} \end{pmatrix}, \qquad (x, y = 1, \ldots, N) \tag{B.5}$$

such that the kth code in a CC code set can be expressed as

$$\mathbf{E}_k = \begin{pmatrix} b_{k1}\mathbf{Z}_{11} & b_{k2}\mathbf{Z}_{21} & \cdots & b_{kN}\mathbf{Z}_{N1} \\ b_{k1}\mathbf{Z}_{12} & b_{k2}\mathbf{Z}_{22} & \cdots & b_{kN}\mathbf{Z}_{N2} \\ \vdots & \vdots & \vdots & \vdots \\ b_{k1}\mathbf{Z}_{1N} & b_{k2}\mathbf{Z}_{2N} & \cdots & b_{kN}\mathbf{Z}_{NN} \end{pmatrix} = \begin{pmatrix} \mathbf{e}_{k1} \\ \mathbf{e}_{k2} \\ \vdots \\ \mathbf{e}_{kN} \end{pmatrix}, \qquad (k = 1, \ldots, N) \tag{B.6}$$

In general, the partial ACF of $\mathbf{c}_{k,m}$ and partial CCF between $\mathbf{c}_{k,m}$ and $\mathbf{c}_{k',m}$ can be defined as

$$\rho\left(\mathbf{c}_{k,m}; i\right) = \begin{cases} \frac{1}{N}\sum_{j=1}^{N-i} \mathbf{c}_{k,m}(j+i)\,\mathbf{c}_{k,m}^{*}(j), & i = 0, 1, \ldots, N-1 \\ \frac{1}{N}\sum_{j=1-i}^{N} \mathbf{c}_{k,m}(j+i)\,\mathbf{c}_{k,m}^{*}(j), & i = -1, -2, \ldots, -N+1 \end{cases} \tag{B.7}$$

$$\rho\left(\mathbf{c}_{k,m}, \mathbf{c}_{k',m}; i\right) = \begin{cases} \frac{1}{N}\sum_{j=1}^{N-i} \mathbf{c}_{k,m}(j+i)\,\mathbf{c}_{k',m}^{*}(j), & i = 0, 1, \ldots, N-1 \\ \frac{1}{N}\sum_{j=1-i}^{N} \mathbf{c}_{k,m}(j+i)\,\mathbf{c}_{k',m}^{*}(j), & i = -1, -2, \ldots, -N+1 \end{cases} \tag{B.8}$$

respectively. It can be shown easily that either periodic or aperiodic ACFs and CCFs can be expressed as some linear combination of partial ACF or (B.7) and partial CCF or (B.8). Therefore, zero partial ACF and zero partial CCF will always give zero periodic or aperiodic ACFs and CCFs. For this reason, all proofs following will be carried out based on partial ACF and partial CCF only.

To facilitate the proof, let us define

$$\begin{cases} g_1(x) = \mathrm{mod}(x, N) + N\delta(\mathrm{mod}(x, N)) \\ g_2(x) = (x - \mathrm{mod}(x, N))/N + (1 - \delta(\mathrm{mod}(x, N))) \end{cases} \tag{B.9}$$

where $\mathrm{mod}(x, N)$ stands for the modulo-N operation for x, and $\delta(x)$ will yield one at $x = 0$ and zero else where. Also, for notation simplicity we define

$$\begin{cases} p = g_1(x) \\ p' = g_1(x') \\ q = g_2(x) \\ q' = g_2(x') \end{cases} \tag{B.10}$$

Thus, from (B.5) and (B.6) we have

$$\begin{aligned} \mathbf{E}_k(m,x)\,\mathbf{E}_{k'}(m,x') &= \left(b_{kq}\mathbf{Z}_{qm}(1,p)\right)\left(b_{kq'}\mathbf{Z}_{q'm}(1,p')\right) \\ &= \left(b_{kq}a_{qp}d_{mp}\right)\left(b_{k'q'}a_{q'p'}d_{mp'}\right). \end{aligned} \tag{B.11}$$

When the relative time shift or i is not equal to multiples of N, the ACF of a CC code $\mathbf{E}_k$ becomes:

$$\begin{aligned} \rho(\mathbf{E}_k;i) &= \frac{1}{N}\sum_{m=1}^{N}\frac{1}{N^2}\sum_{j=1}^{N^2-i}\mathbf{E}_k(m,j+i)\,\mathbf{E}_k(m,j) \\ &= \frac{1}{N^3}\sum_{j=1}^{N^2-i}\sum_{m=1}^{N}\mathbf{E}_k(m,j+i)\,\mathbf{E}_k(m,j) \\ &= \frac{1}{N^3}\sum_{j=1}^{N^2-i}\sum_{m=1}^{N}\left(b_{kg_2(j+i)}a_{g_2(j+i)g_1(j+i)}d_{mg_1(j+i)}\right)\left(b_{kg_2(j)}a_{g_2(j)g_1(j)}d_{mg_1(j)}\right) \\ &= \frac{1}{N^3}\sum_{j=1}^{N^2-i}\left(b_{kg_2(j+i)}a_{g_2(j+i)g_1(j+i)}\right)\left(b_{kg_2(j)}a_{g_2(j)g_1(j)}\right)\sum_{m=1}^{N}\left(d_{mg_1(j+i)}\right)\left(d_{mg_1(j)}\right) \\ &= \frac{1}{N^3}\sum_{j=1}^{N^2-i}\left(b_{k(g_2(j+i))}a_{(g_2(j+i))(g_1(j+i))}\right)\left(b_{k(g_2(j))}a_{(g_2(j))(g_1(j))}\right)\mathbf{d}_{(g_1(j+i))}\mathbf{d}^*_{(g_1(j))} \\ &= \begin{cases} 1, & \text{if } i=0 \\ 0, & \text{if } (i\neq 0)\cap(\mathrm{mod}(i,N)\neq 0)\cap\left(\mathrm{mod}\left(i,N^2\right)\neq 0\right) \end{cases} \end{aligned} \tag{B.12}$$

where N is the dimension of orthogonal matrix $\mathbf{D}$ defined in (B.4), whose orthogonality has been used to obtain the result here.

Similarly, the CCF of two CC codes, $\mathbf{E}_k$ and $\mathbf{E}'_k$, with their relative time shift being not equal to multiples of N can be written as

$$\begin{aligned} \rho(\mathbf{E}_k,\mathbf{E}_{k'};i) &= \frac{1}{N}\sum_{m=1}^{N}\frac{1}{N^2}\sum_{j=1}^{N^2-i}\mathbf{E}_k(m,j+i)\,\mathbf{E}_{k'}(m,j) \\ &= \frac{1}{N^3}\sum_{j=1}^{N^2-i}\sum_{m=1}^{N}\mathbf{E}_k(m,j+i)\,\mathbf{E}_{k'}(m,j) \\ &= \frac{1}{N^3}\sum_{j=1}^{N^2-i}\sum_{m=1}^{N}\left(b_{kg_2(j+i)}a_{g_2(j+i)g_1(j+i)}d_{mg_1(j+i)}\right)\left(b_{k'g_2(j)}a_{g_2(j)g_1(j)}d_{mg_1(j)}\right) \\ &= \frac{1}{N^3}\sum_{j=1}^{N^2-i}\left(b_{kg_2(j+i)}a_{g_2(j+i)g_1(j+i)}\right)\left(b_{k'g_2(j)}a_{g_2(j)g_1(j)}\right)\sum_{m=1}^{N}\left(d_{mg_1(j+i)}\right)\left(d_{mg_1(j)}\right) \\ &= \frac{1}{N^3}\sum_{j=1}^{N^2-i}\left(b_{k(g_2(j+i))}a_{(g_2(j+i))(g_1(j+i))}\right)\left(b_{k'(g_2(j))}a_{(g_2(j))(g_1(j))}\right)\mathbf{d}_{(g_1(j+i))}\mathbf{d}^*_{(g_1(j))} \\ &= \begin{cases} 1, & \text{if } (i=0)\cap\left(k=k'\right) \\ 0, & \text{if } (i\neq 0)\cap(\mathrm{mod}(i,N)\neq 0)\cap\left(\mathrm{mod}\left(i,N^2\right)\neq 0\right) \end{cases} \end{aligned} \tag{B.13}$$

If relative time shift i is equal to multiples of N but not equal to multiples of N^2, we have $g_1(j+i) = g_1(j)$ for any j different i. In this case, the ACF of a code $\mathbf{E}_k$ and CCF between codes $\mathbf{E}_k$ and $\mathbf{E}'_k$ become

$$\rho(\mathbf{E}_k; i) = \frac{1}{N}\sum_{m=1}^{N}\frac{1}{N^2}\sum_{j=1}^{N^2-i}\mathbf{E}_k(m, j+i)\,\mathbf{E}_k(m, j)$$

$$= \frac{1}{N^3}\sum_{j=1}^{N^2-i}\sum_{m=1}^{N}\mathbf{E}_k(m, j+i)\,\mathbf{E}_k(m, j)$$

$$= \frac{1}{N^3}\sum_{j=1}^{N^2-i}\sum_{m=1}^{N}\left(b_{k(g_2(j+i))}a_{(g_2(j+i))(g_1(j+i))}d_{m(g_1(j+i))}\right)\left(b_{k(g_2(j))}a_{(g_2(j))(g_1(j))}d_{m(g_1(j))}\right)$$

$$= \frac{1}{N^3}\sum_{j=1}^{N^2-i}\sum_{m=1}^{N}\left(b_{k(g_2(j+i))}a_{(g_2(j+i))(g_1(j))}d_{m(g_1(j))}\right)\left(b_{k(g_2(j))}a_{(g_2(j))(g_1(j))}d_{m(g_1(j))}\right)$$

$$= \frac{1}{N^3}\sum_{j=1}^{N^2-i}\left(b_{k(g_2(j+i))}a_{(g_2(j+i))(g_1(j))}\right)\left(b_{k(g_2(j))}a_{(g_2(j))(g_1(j))}\right)\sum_{m=1}^{N}\left(d_{m(g_1(j))}\right)\left(d_{m(g_1(j))}\right)$$

$$= \frac{1}{N^3}\sum_{j'=1}^{N-i/N}\sum_{j=1}^{N}\left(b_{kg_2(j'+(j-1)N+i)}b_{kg_2(j'+(j-1)N)}\right)\left(a_{g_2(j'+(j-1)N+i)g_1(j)}a_{g_2(j'+(j-1)N)g_1(j)}\right)$$

$$= \frac{1}{N^2}\sum_{j'=1}^{N-i/N}\sum_{j=1}^{N}\left(b_{k(g_2(j'+i))}b_{k(g_2(j'))}\right)\left(a_{(g_2(j'))(g_1(j))}a_{(g_2(j'))(g_1(j))}\right)$$

$$= \frac{1}{N^2}\sum_{j'=1}^{N-i/N}\left(b_{k(g_2(j'+i))}b_{k(g_2(j'))}\right)\sum_{j=1}^{N}\left(a_{(g_2(j'+i))(g_1(j))}a_{(g_2(j'))(g_1(j))}\right)$$

$$= \frac{1}{N^2}\sum_{j'=1}^{N-i/N}\left(b_{k(g_2(j'+i))}b_{k(g_2(j'))}\right)\mathbf{a}_{(g_2(j'+i))}\mathbf{a}^*_{(g_2(j'))}$$

$$= \begin{cases} 1, & \text{if } (i=0)\cap(\mathrm{mod}(i,N)=0)\cap\left(\mathrm{mod}\left(i,N^2\right)\neq 0\right) \\ 0, & \text{if } (i\neq 0)\cap(\mathrm{mod}(i,N)=0)\cap\left(\mathrm{mod}\left(i,N^2\right)\neq 0\right)\end{cases} \tag{B.14}$$

and

$$\rho(\mathbf{E}_k, \mathbf{E}_{k'}; i) = \frac{1}{N}\sum_{m=1}^{N}\frac{1}{N^2}\sum_{j=1}^{N^2-i}\mathbf{E}_k(m, j+i)\,\mathbf{E}_{k'}(m, j)$$

$$= \frac{1}{N^3}\sum_{j=1}^{N^2-i}\sum_{m=1}^{N}\mathbf{E}_k(m, j+i)\,\mathbf{E}_{k'}(m, j)$$

$$= \frac{1}{N^3}\sum_{j=1}^{N^2-i}\sum_{m=1}^{N}\left(b_{k(g_2(j+i))}a_{(g_2(j+i))(g_1(j+i))}d_{m(g_1(j+i))}\right)\left(b_{k'(g_2(j))}a_{(g_2(j))(g_1(j))}d_{m(g_1(j))}\right)$$

$$= \frac{1}{N^3} \sum_{j=1}^{N^2-i} \sum_{m=1}^{N} \left(b_{k(g_2(j+i))} a_{(g_2(j+i))(g_1(j))} d_{m(g_1(j))} \right) \left(b_{k'(g_2(j))} a_{(g_2(j))(g_1(j))} d_{m(g_1(j))} \right)$$

$$= \frac{1}{N^3} \sum_{j=1}^{N^2-i} \left(b_{k(g_2(j+i))} a_{(g_2(j+i))(g_1(j))} \right) \left(b_{k'(g_2(j))} a_{(g_2(j))(g_1(j))} \right) \sum_{m=1}^{N} \left(d_{m(g_1(j))} \right) \left(d_{m(g_1(j))} \right)$$

$$= \frac{1}{N^3} \sum_{j'=1}^{N-i/N} \sum_{j=1}^{N} \left(b_{kg_2(j'+(j-1)N+i)} b_{k'g_2(j'+(j-1)N)} \right) \left(a_{g_2(j'+(j-1)N+i)g_1(j)} a_{g_2(j'-(j-1)N)g_1(j)} \right)$$

$$= \frac{1}{N^2} \sum_{j'=1}^{N-i/N} \sum_{j=1}^{N} \left(b_{k(g_2(j'+i))} b_{k'(g_2(j'))} \right) \left(a_{(g_2(j'+i))(g_1(j))} a_{(g_2(j'))(g_1(j))} \right)$$

$$= \frac{1}{N^2} \sum_{j'=1}^{N-i/N} \left(b_{k(g_2(j'+i))} b_{k'(g_2(j'))} \right) \sum_{j=1}^{N} \left(a_{(g_2(j'+i))(g_1(j))} a_{(g_2(j'))(g_1(j))} \right)$$

$$= \frac{1}{N^2} \sum_{j'=1}^{N-i/N} \left(b_{k(g_2(j'+i))} b_{k'(g_2(j'))} \right) \mathbf{a}_{(g_2(j'+i))} \mathbf{a}^*_{(g_2(j'))}$$

$$= \begin{cases} 1, & \text{if } (i=0) \cap \left(\text{mod}\left(i, N^2\right)=0\right) \cap \left(k=k'\right) \\ 0, & \text{if } (i \neq 0) \cap (\text{mod}(i, N) \neq 0) \cap \left(\text{mod}\left(i, N^2\right)=0\right) \end{cases} \tag{B.15}$$

respectively. In the above proof we have used the orthogonality of matrix **A** defined in (B.1).

Finally, we want to calculate $\rho(\mathbf{E}_k, \mathbf{E}_{k'}; i)$ with i being equal to multiples of N^2 as

$$\rho(\mathbf{E}_k, \mathbf{E}_{k'}; i) = \frac{1}{N} \sum_{m=1}^{N} \frac{1}{N^2} \sum_{j=1}^{N^2-i} \mathbf{E}_k(m, j+i) \mathbf{E}_{k'}(m, j)$$

$$= \frac{1}{N^3} \sum_{j=1}^{N^2-i} \sum_{m=1}^{N} \mathbf{E}_k(m, j+i) \mathbf{E}_{k'}(m, j)$$

$$= \frac{1}{N^3} \sum_{j=1}^{N^2-i} \sum_{m=1}^{N} \left(b_{k(g_2(j+i))} a_{(g_2(j+i))(g_1(j+i))} d_{m(g_1(j+i))} \right) \left(b_{k'(g_2(j))} a_{(g_2(j))(g_1(j))} d_{m(g_1(j))} \right)$$

$$= \frac{1}{N^3} \sum_{j=1}^{N^2-i} \sum_{m=1}^{N} \left(b_{k(g_2(j+i))} a_{(g_2(j+i))(g_1(j))} d_{m(g_1(j))} \right) \left(b_{k'(g_2(j))} a_{(g_2(j))(g_1(j))} d_{m(g_1(j))} \right)$$

$$= \frac{1}{N^3} \sum_{j=1}^{N^2-i} \left(b_{kg_2(j+i)} a_{g_2(j+i)g_1(j)} \right) \left(b_{k'(g_2(j))} a_{(g_2(j))(g_1(j))} \right) \sum_{m=1}^{N} \left(d_{m(g_1(j))} \right) \left(d_{m(g_1(j))} \right)$$

$$= \frac{1}{N^3} \sum_{j'=1}^{N-i/N} \sum_{j=1}^{N} \left(b_{kg_2(j'+(j-1)N+i)} b_{k'g_2(j'+(j-1)N)} \right) \left(a_{g_2(j'+(j-1)N+i)g_1(j)} a_{g_2(j'+(j-1)N)g_1(j)} \right)$$

$$= \frac{1}{N^2} \sum_{j'=1}^{N-i/N} \sum_{j=1}^{N} \left(b_{kg_2(j')} b_{k'g_2(j')} \right) \left(a_{g_2(j')g_1(j)} a_{g_2(j')g_1(j)} \right)$$

$$= \frac{1}{N^2} \sum_{j'=1}^{N-i/N} \left(b_{k(g_2(j'))} b_{k'(g_2(j'))} \right) \sum_{m=1}^{N} 1^2 = \frac{1}{N} \sum_{j'=1}^{N} \left(b_{k(g_2(j'))} b_{k'(g_2(j'))} \right)$$

$$= \frac{1}{N} \mathbf{b}_k \mathbf{b}_{k'}^* = \begin{cases} 1, & \text{if } k = k' \\ 0, & \text{if } k \neq k' \end{cases} \tag{B.16}$$

where the orthogonality of matrix **B**, defined in (B.2), has been used here to obtain the result.

To summarize, we have proved that the CC codes have ideal ACF and CCF for whatever relative time shifts i, or

$$\rho\left(\mathbf{E}_k, \mathbf{E}_k; i\right) = \frac{1}{N} \sum_{m=1}^{N} \frac{1}{N^2} \sum_{j=1}^{N^2 - i} \mathbf{E}_k\left(m, j + i\right) \mathbf{E}_k^*\left(m, j\right)$$

$$= \begin{cases} 1, & \text{if } (k = k') \cap (i = 0) \\ 0, & \text{if } (k \neq k') \cap (i \neq 0) \end{cases} \tag{B.17}$$

C

Proof of n-Chip Orthogonality of CC Codes

It will be of interest to show that a complete complementary (CC) code set has a property that it has zero chip-wise auto-correlation function (ACF) and cross-correlation function (CCF). Figure C.1 shows the scenario where n chips are involved in the correlation process between code k and k'. In this appendix we are in particular interested in two cases, one being $n = 1$ and the other $n = N$.

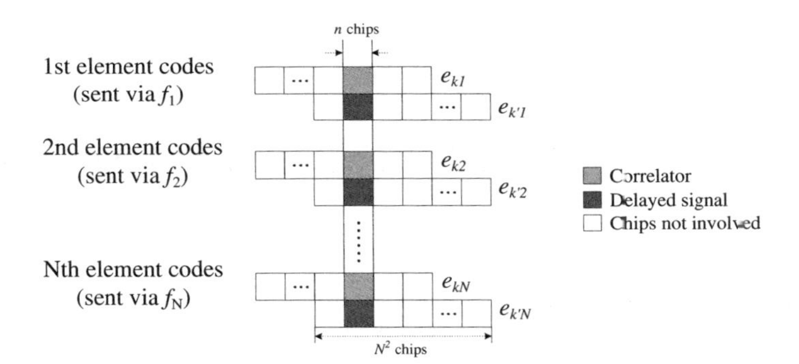

Figure C.1 Illustration of n-chip CCF for two CC codes $\mathbf{E}_k = (\mathbf{e}_{k1}, \mathbf{e}_{k2}, \ldots, \mathbf{e}_{kN})$ and $\mathbf{E}_{k'} = (\mathbf{e}_{k'1}, \mathbf{e}_{k'2}, \ldots, \mathbf{e}_{k'N})$. It is to be noted that if $k = k'$, the chip-wise CCF becomes chip-wise ACF. Assume that different element codes are sent via different carriers or f_1, f_2, and f_N.

The Next Generation CDMA Technologies Hsiao-Hwa Chen

C.1 SINGLE CHIP ($n = 1$) ORTHOGONALITY OF A CC CODE SET

Let us first consider the case with $n = 1$. Thus, only one chip will be involved in the orthogonality proof here. Based on the definitions given in (B.9) and (B.10), the one-chip CCF between codes k and k' can be written as

$$\sum_{m=1}^{N} \mathbf{E}_k(m,x)\,\mathbf{E}_{k'}(m,x') = \sum_{m=1}^{N} \left(b_{kq}\mathbf{Z}_{qm}(1,p)\right)\left(b_{k'q'}\mathbf{Z}_{q'm}(1,p')\right)$$
$$= \sum_{m=1}^{N} \left(b_{kq}a_{qp}d_{mp}\right)\left(b_{k'q'}a_{q'p'}d_{mp'}\right) = \sum_{m=1}^{N} \left(b_{kq}a_{qp}b_{k'q'}a_{q'p'}\right)\left(d_{mp}d_{mp'}\right)$$
$$= \left(b_{kq}b_{k'q'}a_{qp}a_{q'p'}\right)\sum_{m=1}^{N}\left(d_{mp}d_{mp'}\right) = \begin{cases} \pm N, & \text{if } p' = p \\ 0, & \text{if } p' \neq p \end{cases} \tag{C.1}$$

in which we have used the orthogonality property of matrix $\mathbf{D}$, defined in (B.4).

It is to be noted that the same result will be yielded no matter whether $k = k'$ or $k \neq k'$, which means that the same result will apply for either ACF or CCF of a CC code set. The result tells us that, as long as the relative time shifts, x and x', between two CC codes, $\mathbf{E}_k(m,x)$ and $\mathbf{E}_k(m,x')$, satisfy the following condition

$$\text{mod}(x,N) + N\delta(\text{mod}(x,N)) = \text{mod}(x',N) + N\delta(\text{mod}(x',N)), \tag{C.2}$$

both chip-wise ACF and chip-wise CCF of a CC code set are always zero.

C.2 N-CHIP ($n = N$) ORTHOGONALITY OF A CC CODE SET

Now let us look at the case with $n = N$, which stands for the scenario of $n = N$ in Figure C.1. Now, we should consider the orthogonality of a CC code set that involves N consecutive chips. It is noted that each element code in a CC code set contains N^2 chips, and thus N chips span only $\frac{1}{N}$ of the whole element code length. Also, it is assumed that the relative time shift between any two codes of concern takes multiples of N chips. For two CC codes given by

$$\mathbf{E}_k = \begin{pmatrix} b_{k1}\mathbf{Z}_{11} & b_{k2}\mathbf{Z}_{21} & \cdots & b_{kN}\mathbf{Z}_{N1} \\ b_{k1}\mathbf{Z}_{12} & b_{k2}\mathbf{Z}_{22} & \cdots & b_{kN}\mathbf{Z}_{N2} \\ \vdots & \vdots & \vdots & \vdots \\ b_{k1}\mathbf{Z}_{1N} & b_{k2}\mathbf{Z}_{2N} & \cdots & b_{kN}\mathbf{Z}_{NN} \end{pmatrix} = \begin{pmatrix} \mathbf{e}_{k1} \\ \mathbf{e}_{k2} \\ \vdots \\ \mathbf{e}_{kN} \end{pmatrix} \tag{C.3}$$

$$\mathbf{E}_{k'} = \begin{pmatrix} b_{k'1}\mathbf{Z}_{11} & b_{k'2}\mathbf{Z}_{21} & \cdots & b_{k'N}\mathbf{Z}_{N1} \\ b_{k'1}\mathbf{Z}_{12} & b_{k'2}\mathbf{Z}_{22} & \cdots & b_{k'N}\mathbf{Z}_{N2} \\ \vdots & \vdots & \vdots & \vdots \\ b_{k'1}\mathbf{Z}_{1N} & b_{k'2}\mathbf{Z}_{2N} & \cdots & b_{k'N}\mathbf{Z}_{NN} \end{pmatrix} = \begin{pmatrix} \mathbf{e}_{k'1} \\ \mathbf{e}_{k'2} \\ \vdots \\ \mathbf{e}_{k'N} \end{pmatrix} \tag{C.4}$$

we can readily calculate their N-chip CCF as

$$\sum_{y=1}^{N} b_{kx}\mathbf{Z}_{xy}\left(b_{k'x'}\mathbf{Z}_{x'y}\right)^H = \sum_{y=1}^{N} b_{kx}b_{k'x'}\sum_{z=1}^{N}\left(a_{xz}a_{x'z}\right)\left(d_{yz}\right)^2$$

$$= \sum_{y=1}^{N} b_{kx}b_{k'x'}\sum_{z=1}^{N}\left(a_{xz}a_{x'z}\right)\times 1 = \sum_{y=1}^{N} b_{kx}b_{k'x'}\times 0 = 0 \qquad \text{(C.5)}$$

where $\mathbf{x}^H$ stands for the Hermitia transform of $\mathbf{x}$. The above result shows that N-chip shifted ACF and CCF of a CC code set are always zero. It is noted that in the above proof we have used the orthogonality property of matrix $\mathbf{A}$, defined in (B.1).

D

Proof of Equation (8.27)

In this appendix we will show the validity of Equation (8.27), rewritten as follows:

$$\left\{\sum_{m=1}^{M} \mathbf{I}_{1,m} \otimes (\mathbf{c}_{o,1,m} + \mathbf{c}_{e,1,m})\right\} \oplus \left\{\sum_{m=1}^{M} \mathbf{I}_{1,m} \otimes (\mathbf{c}_{o,1,m} + \mathbf{c}_{e,1,m})\right\} = (0, 0) \tag{D.1}$$

where, as also given in (8.2) and (8.23), we have

$$\left\{ \begin{array}{l} \mathbf{c}_{o,1,m} = (\mathbf{c}_{1,m}, 0, 0, \ldots, 0) \\ \mathbf{c}_{e,1,m} = (0, 0, \ldots, 0, \mathbf{c}_{1,m}) \end{array} \right. \tag{D.2}$$

and

$$\begin{aligned} \mathbf{I}_{1,m} = \sum_{k=2}^{K} \big[& (b_{k,o}\mathbf{c}_{o,k,m} + b_{k,e}\mathbf{c}_{e,k,m}) h_{k,(2m-1)} \\ & + (b_{k,e}\mathbf{c}_{o,k,m} - b_{k,o}\mathbf{c}_{e,k,m}) h_{k,2m} \big] \end{aligned} \tag{D.3}$$

Thus, we get

$$\begin{aligned} & \sum_{m=1}^{M} \mathbf{I}_{1,m} \otimes (\mathbf{c}_{o,1,m} + \mathbf{c}_{e,1,m}) \\ & \quad = \sum_{m=1}^{M} \sum_{k=2}^{K} \big[(b_{k,o}\mathbf{c}_{o,k,m} + b_{k,e}\mathbf{c}_{e,k,m}) h_{k,(2m-1)} \\ & \quad\quad + (b_{k,e}\mathbf{c}_{o,k,m} - b_{k,o}\mathbf{c}_{e,k,m}) h_{k,2m} \big] \otimes (\mathbf{c}_{o,1,m} + \mathbf{c}_{e,1,m}) \\ & \quad = \sum_{k=2}^{K} \sum_{m=1}^{M} h_{k,(2m-1)} (b_{k,o}\mathbf{c}_{o,k,m} + b_{k,e}\mathbf{c}_{e,k,m}) \otimes (\mathbf{c}_{o,1,m} + \mathbf{c}_{e,1,m}) \\ & \quad\quad + \sum_{k=2}^{K} \sum_{m=1}^{M} h_{k,2m} (b_{k,e}\mathbf{c}_{o,k,m} - b_{k,o}\mathbf{c}_{e,k,m}) \otimes (\mathbf{c}_{o,1,m} + \mathbf{c}_{e,1,m}) \\ & \quad = \mathbf{A} + \mathbf{B} \end{aligned} \tag{D.4}$$

The Next Generation CDMA Technologies Hsiao-Hwa Chen

where

$$\begin{cases} \mathbf{A} = \sum_{k=2}^{K}\sum_{m=1}^{M} h_{k,(2m-1)}(b_{k,o}\mathbf{c}_{o,k,m} + b_{k,e}\mathbf{c}_{e,k,m}) \otimes (\mathbf{c}_{o,1,m} + \mathbf{c}_{e,1,m}) \\ \mathbf{B} = \sum_{k=2}^{K}\sum_{m=1}^{M} h_{k,2m}(b_{k,e}\mathbf{c}_{o,k,m} - b_{k,o}\mathbf{c}_{e,k,m}) \otimes (\mathbf{c}_{o,1,m} + \mathbf{c}_{e,1,m}) \end{cases} \tag{D.5}$$

and $(b_{k,o}\mathbf{c}_{o,k,m} + b_{k,e}\mathbf{c}_{e,k,m}) \otimes (\mathbf{c}_{o,1,m} + \mathbf{c}_{e,1,m})$ represents the EWP operation between two extended element codes that belong to two different flocks. The same result applies to $(b_{k,e}\mathbf{c}_{o,k,m} - b_{k,o}\mathbf{c}_{e,k,m}) \otimes (\mathbf{c}_{o,1,m} + \mathbf{c}_{e,1,m})$, where $b_{k,o}$ and $b_{k,e}$ take either $+1$ or -1, and will not change the orthogonality property of the OC codes sets involved. Therefore, applying HLA operation to $\sum_{m=1}^{M} \mathbf{I}_{1,m} \otimes (\mathbf{c}_{o,1,m} + \mathbf{c}_{e,1,m})$ is just equal to applying HLA operation to **A** and **B** individually, followed by a summation, or

$$\begin{aligned} &\left\{\sum_{m=1}^{M} \mathbf{I}_{1,m} \otimes (\mathbf{c}_{o,1,m} + \mathbf{c}_{e,1,m})\right\} \oplus \left\{\sum_{m=1}^{M} \mathbf{I}_{1,m} \otimes (\mathbf{c}_{o,1,m} + \mathbf{c}_{e,1,m})\right\} \\ &= (\mathbf{A} + \mathbf{B}) \oplus (\mathbf{A} + \mathbf{B}) = (\mathbf{A} \oplus \mathbf{A}) + (\mathbf{B} \oplus \mathbf{B}) \\ &= (0, 0) + (0, 0) = (0, 0) \end{aligned} \tag{D.6}$$

where $\mathbf{A} \oplus \mathbf{A}$ and $\mathbf{B} \oplus \mathbf{B}$ calculate weighted in-phase cross-correlation functions between two different flocks of extended element codes, both of which should be equal to a zero vector or [0,0], as shown in (8.8) and (8.17).

E

List of Complete Complementary Codes (PG = 8 ∼ 512)

This appendix gives a list of complete complementary codes with PG values from 8 to 512. Commas are used to separate different element codes within a specific flock, and brackets are used to denote flocks in a complete complementary code set. It is noted that the processing gain for a complete complementary code set should be the product of its element code length (N) and its flock size (M), or $N \times M$.

1. PG = 8

   Length of element codes = 4, flock size = 2

   (+++−, +−++)

   (++−+, +−−−)

2. PG = 16

   Length of element codes = 8, flock size = 2

   (+++−++−+, +−+++−−−)

   (+++−−−+−, +−++−+++)

3. PG = 32

   Length of element codes = 16, flock size = 2

   (+++−++−++++−−−+−, +−+++−−−+−++−+++)

   (+++−++−+−−−+++−+, +−+++−−−−+−−+−−−)

4. PG = 64

   (1) Length of element codes = 32, flock size = 2

   (+++−++−++++−−−+−+++−++−+−−−+++−+, +−+++−−−+−++−++++−+++−−−−+−−+−−−)

   (+++−++−++++−−−+−−−−+−−+−+++−−−+−, +−+++−−−+−++−+++−+−−−++++−++−+++)

The Next Generation CDMA Technologies Hsiao-Hwa Chen

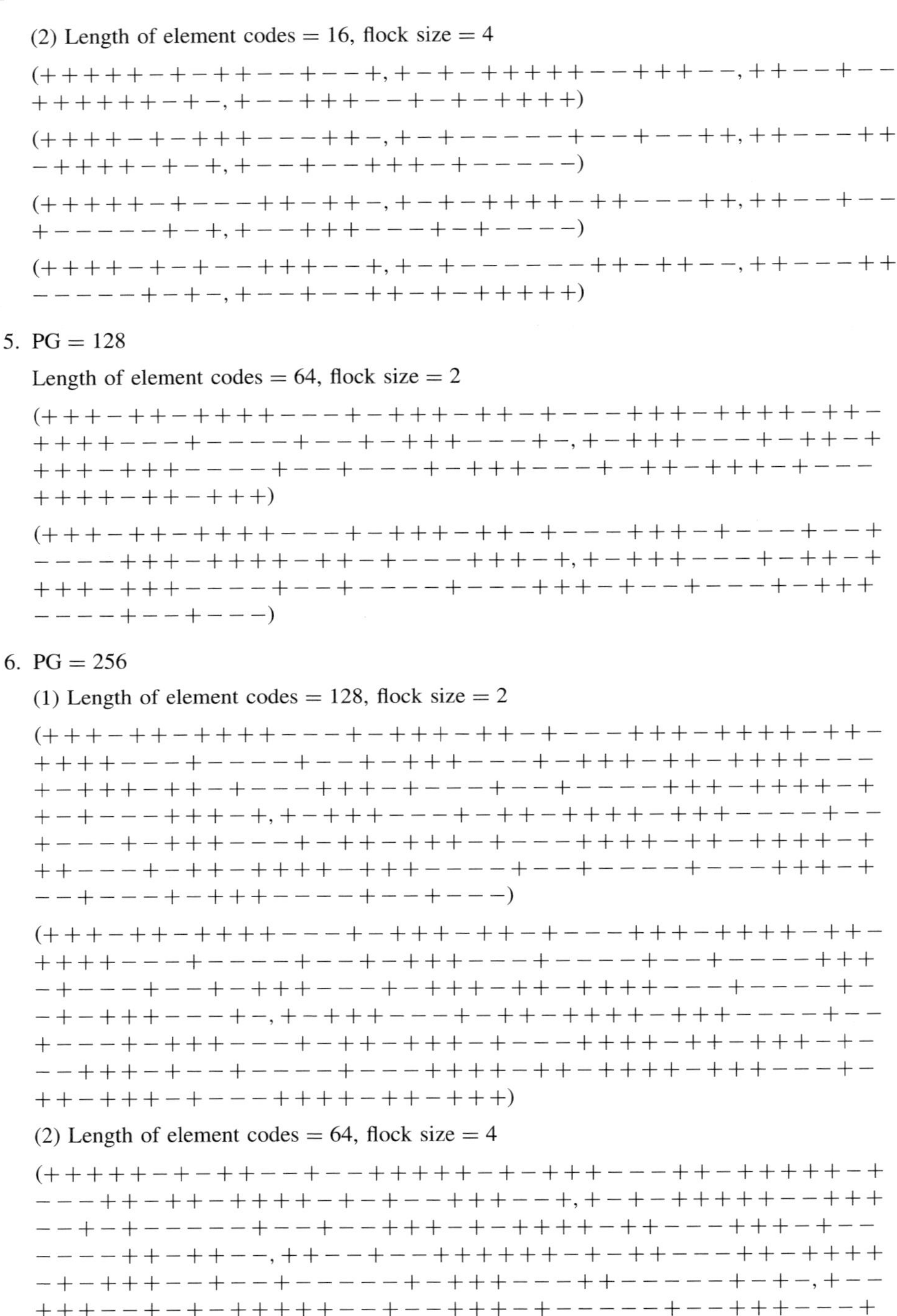

(2) Length of element codes = 16, flock size = 4

(+++++−+−++−−+−−+,+−+−+++++−−+++−−,++−−+−−
++++++−+−,+−−+++−−+−+−++++)

(++++−+−+++−−−++−,+−+−−−−−+−−+−−++,++−−−++
−++++−+−+,+−−+−−+++−+−−−−−)

(+++++−+−−−++−++−,+−+−++++−++−−−++,++−−+−−
+−−−−−+−+,+−−+++−−−+−+−−−−)

(++++−+−+−−+++−−+,+−+−−−−−−++−++−−,++−−−++
−−−−−+−+−,+−−+−−++−+−+++++)

5. PG = 128

Length of element codes = 64, flock size = 2

(+++−++−++++−−−+−+++−++−+−−−+++−++++−++−
++++−−−+−−−−+−−+−+++−−−+−,+−+++−−−+−++−+
+++−+++−−−−+−−+−−−+−+++−−−+−++−+++−+−−−
++++−++−+++)

(+++−++−++++−−−+−+++−++−+−−−+++−+−−−+−−+
−−−−+++−++++−++−+−−−+++−+,+−+++−−−+−++−+
+++−+++−−−−+−−+−−−−+−−−+++−+−−+−−−+−+++
−−−−+−−+−−−)

6. PG = 256

(1) Length of element codes = 128, flock size = 2

(+++−++−++++−−−+−+++−++−+−−−+++−++++−++−
++++−−−+−−−−+−−+−+++−−−+−+++−++−++++−−−
+−+++−++−+−−−+++−+−−−+−−+−−−−+++−++++−+
+−+−−−+++−+,+−+++−−−+−++−++++−+++−−−−+−−
+−−−+−+++−−−+−++−+++−+−−−++++−++−++++−+
++−−−+−++−++++−+++−−−−+−−+−−−−+−−−+++−+
−−+−−−+−+++−−−−+−−+−−−)

(+++−++−++++−−−+−+++−++−+−−−+++−++++−++−
++++−−−+−−−−+−−+−+++−−−+−−−−+−−+−−−−+++
−+−−−+−−+−+++−−−+−+++−++−++++−−−+−−−−+−
−+−+++−−−+−,+−+++−−−+−++−++++−+++−−−−+−−
+−−−+−+++−−−+−++−+++−+−−−++++−++−+++−+−
−−+++−+−−+−−−−+−−−++++−++−++++−+++−−−+−
++−+++−+−−−++++−++−+++)

(2) Length of element codes = 64, flock size = 4

(+++++−+−++−−+−−+++++−+−+++−−−++−+++++−+
−−−++−++−++++−+−+−−+++−−+,+−+−+++++−−+++
−−+−+−−−−−+−−+−−+++−+−++++−++−−−+++−+−−
−−−−++−++−−,++−−+−−++++++−+−++−−−++−++++
−+−+++−−+−−+−−−−−+−+++−−−++−−−−−+−+−,+−−
+++−−+−+−+++++−−+−−+++−+−−−−−+−−+++−−−+
−+−−−−+−−+−−++−+−+++++)

(+++++−+−++−−+−−+−−−−+−+−−−+++−−++++++−+
−−−++−++−−−−−+−+−++−−−++−,+−+−+++++−−+++
−−−+−+++++−++−++−−+−+−++++−++−−−++−+−++
++++−−+−−++,++−−+−−++++++−+−−−+++−−+−−−−
+−+−++−−+−−+−−−−−+−+−−+++−−+++++−+−+,+−−
+++−−+−+−++++−++−++−−−+−++++++−−+++−−−+
−+−−−−−++−++−−+−+−−−−−)

(+++++−+−++−−+−−+++++−+−+++−−−++−−−−−−+−
+++−−+−−+−−−−+−+−++−−−++−,+−+−+++++−−+++
−−+−+−−−−−+−−+−−++−+−+−−−−+−−+++−−−+−++
++++−−+−−++,++−−+−−++++++−+−++−−−+−−++++
−+−+−−++−++−+++++−+−−−+++−−+++++−+−+,+−−
+++−−+−+−+++++−−+−−+++−+−−−−−−++−−−+++−
+−++++−++−++−−+−+−−−−−)

(+++++−+−++−−+−−+−−−−+−+−−−+++−−+−−−−−+−,
+++−−+−−++++++−+−+−−+++−−+,+−+−++++−−−+++
−−−+−+++++−++−++−−−+−+−−−−+−−+++−−+−+−−
−−−−++−++−−,++−−+−−++++++−+−−−+++−−+−−−−
+−+−−−++−++−+++++−+−++−−−++−−−−−+−+−,+−−
+++−−+−+−++++−++−++−−−+−+++++−++−−−+++−
+−+++++−−+−−++−+−+++++)

7. PG = 512

(1) Length of element codes = 256, flock size = 2

(+++−++−++++−−−+−+++−++−+−−−+++−++++−++−
++++−−−+−−−−+−−+−+++−−−+−+++−++−+−++−−−
+−+++−++−+−−−+++−+−−−+−−+−−−−+++−−+++−+
+−+−−−++++−++++−++−++++−−−+−+−+−++−+−−−+
++−++++−++−++++−−−+−−−−+−−+−+++−−−+−−−−
+−−+−−−−+++−+−−−+−−+−+++−−−+−+++−−+−+++
+−−−+−−−−+−−+−+++−−−+−,+−+++−−−+−++−++++
−+++−−−−+−−+−−−+−+++−−−+−++−+++−+−−−+++
+−++−++++−+++−−−+−++−++++−+++−−−−+−−+−−
−−+−−−+++−+−−+−−−+−+++−−−−+−−+−−−+−+++−
−−+−++−++++−+++−−−−+−−+−−−+−+++−−−+−++−
+++−+−−−++++−++−+++−+−−−+++−+−−+−−−−+−−
−++++−++−++++−+++−−−+−++−+++−+−−−++++−+
+−+++)

(+++−++−++++−−−+−+++−++−+−−−+++−++++−++−
++++−−−+−−−−+−−+−+++−−−+−+++−++−++++−−−
+−+++−++−+−−−+++−+−−−+−−+−−−−+++−++++−+
+−+−−−+++−+−−−+−−+−−−−+++−+−−−+−−+−+++−
−−+−−−−+−−+−−−−+++−++++−++−+−−−+++−++++
−++−++++−−−+−+++−++−+−−−+++−+−−−+−−+−−−
−+++−+++++−++−+−−−+++−+,+−+++−−−+−++−++++
−+++−−−−+−−+−−−+−+++−−−+−++−+++−+−−−+++
+−++−++++−+++−−−+−++−++++−+++−−−−+−−+−−

−−+−−−+++−+−−+−−−+−+++−−−−+−−+−−−−+−−−+
++−+−−+−−−−+−−−++++−++−+++−+−−−+++−+−−+
−−−+−+++−−−−+−−+−−−+−+++−−−+−++−++++−++
+−−−−+−−+−−−−+−−−+++−+−−+−−−+−+++−−−−+−
−+−−−)

(2) Length of element codes = 64, flock size = 8

(+++++++++−+−+−+−++−−++−−+−−++−−+++++−−−
−+−+−−+−+++−−−−+++−−+−++−,+−+−+−+−++++++
+++−−++−−+++−−++−−+−+−−+−+++++−−−−+−−+−
++−++−−−−++,++−−++−−+−−++−−++++++++++−+−
+−+−++−−−−+++−−+−++−++++−−−−+−+−−+−+,+−−
++−−+++−−++−−+−+−+−+−++++++++++−−+−++−++
−−−−+++−+−−+−+++++−−−−,++++−−−−+−+−−+−++
+−−−−+++−−+−++−++++++++++−+−+−+−++−−++−−
+−−++−−+,+−+−−+−+++++−−−−+−−+−++−++−−−−+
++−+−+−+−++++++++++−−++−−+++−−++−−,++−−−−
+++−−+−++−++++−−−−+−+−−+−+++−−++−−+−−++
−−++++++++++−+−+−+−,+−−+−++−++−−−−+++−+−
−+−++++++−−−−+−−++−−+++−−++−−+−+−+−+−+++
+++++)

(+++++++++−+−+−+−+++−−++−−−++−−++−++++−−−
−−+−++−+−++−−−−++−++−+−−+,+−+−+−+−−−−−−−−
−−+−−++−−+−−++−−+++−+−−+−+−−−−++++++−−+−
++−−−++++−−,++−−++−−−++−−++−++++++++−+−+
−+−+++−−−−++−++−+−−+++++−−−−−+−++−+−,+−−
++−−+−−++−−+++−+−+−+−−−−−−−−−−+−−+−++−−−
++++−−+−+−−+−+−−−−++++,++++−−−−−+−++−+−+
+−−−−++−++−+−−++++++++++−+−+−+−+++−−++−−
−++−−++−,+−+−−+−+−−−−+++++−−+−++−−−++++−
−+−+−+−+−−−−−−−−−−+−−++−−+−−++−−++,++−−−−
++−++−+−−+++++−−−−−+−++−+−++−−++−−−++−−
++−+++++++++−+−+−+−+,+−−+−++−−−++++−−+−+−
−+−+−−−−+++++−−++−−+−−++−−+++−+−+−+−−−−
−−−−−)

(+++++++++−+−+−+−−−++−−++−++−−++−++++−−−
−+−+−−+−+−−++++−−−++−+−−+,+−+−+−+−++++++
++−++−−++−−−++−−+++−+−−+−++++++−−−−−++−+
−−+−−++++−−,++−−++−−+−−++−−+−−−−−−−−−−+−+
−+−+++−−−−+++−−+−++−−−−−++++−+−++−+−,+−−
++−−+++−−++−−−+−+−+−+−−−−−−−−+−−+−++−++
−−−−++−+−++−+−−−−−−++++,++++−−−−+−+−−+−+−
−++++−−−++−+−−++++++++++−+−+−+−−−++−−++
−++−−++−,+−+−−+−+++++−−−−−++−+−−+−−++++−
−+−+−+−+−++++++++−++−−++−−−++−−++,++−−−−
+++−−+−++−−−−−++++−+−++−+−++−−++−−+−−++
−−+−−−−−−−−−+−+−+−+,+−−+−++−++−−−−++−+−+
+−+−−−−−+++++−−++−−+++−−++−−−+−+−+−+−−−
−−−−−)

(++++++++−+−+−+−+−−++−−+++−−++−−+++++−−−
−−+−++−+−−−++++−−+−−+−++−,+−+−+−+−−−−−−−
−−−++−−++−++−−++−−+−+−−+−+−−−−++++−++−+
−−+++−−−−++,++−−++−−−++−−++−−−−−−−−−−+−+−
+−+−++−−−−++−++−+−−+−−−−+++++−+−−+−+,+−−
++−−+−−++−−++−+−+−+−++++++++++−−+−++−−−
++++−−−+−++−+−++++−−−−,++++−−−−−+−−+−+−−
−++++−−+−−+−++−++++++++−+−+−+−+−−++−−++
+−−++−−+,+−+−−+−+−−−−++++−++−+−−+++−−−−−+
++−+−+−+−−−−−−−−−−++−−++−++−−++−−,++−−−−−
++−++−+−−+−−−−+++++−+−−+−+++−−++−−−++−−
++−−−−−−−−−+−+−+−+−,+−−+−++−−−++++−−−+−+
+−+−++++−−−−+−−++−−+−−++−−++−+−+−−−++++
+++++)

(+++++++++−+−+−+−++−−++−−+−−++−−+−−−−+++
+−+−++−+−−−++++−−−++−+−−+,+−+−+−+−++++++
+++−−++−−+++−−++−−−+−++−+−−−−−+++−−++−+
−−+−−++++−−,++−−++−−+−−++−−++++++++++−+−
+−+−−−++++−−−++−+−−+−−−−++++−+−++−+−,+−−
++−−+++−−++−−+−+−+−+−++++++++−++−+−−+−−
++++−−−+−++−+−−−−−++++,++++−−−−+−+−−+−++
+−−−−+++−−+−++−−−−−−−−−−+−+−+−+−−++−−++
−++−−++−,+−+−−+−+++++−−−+−−+−++−+−−−−−+
+−+−+−+−+−−−−−−−−−++−−++−−−++−−++++−−−−
+++−−+−++−++++−−−−+−+−−+−+−−++−−++−++−−
++−−−−−−−−−−+−+−+−+,+−−+−++−++−−−−+++−+−
−+−+++++−−−−−++−−++−−−++−−++−+−+−+−+−−−
−−−−−)

(++++++++−+−+−+++−−++−−−++−−++−−−−−+++
++−+−−+−+−−++++−−+−−+−++−,+−+−+−+−−−−−−−
−−+−−++−−+−−++−−++−+−++−+−++++−−−−−++−+
−−+++−−−−++,++−−++−−−++−−++−−++++−++−+−+
−+−+−−++++−−+−−+−++−−−−−+++++−+−−+−+,+−−
++−−+−−++−−+++−+−+−+−−−−−−−−−−++−+−−+++
−−−−++−+−++−+−++++−−−−,++++−−−−−+−++−+−+
+−−−−++−++−+−−+−−−−−−−−+−+−+−+−−−++−−++
+−−++−−+,+−+−−+−+−−−−+++++−−+−++−−−++++−
−−+−+−+−+++++++++−++−−++−++−−++−−,++−−−−
++−++−+−−+++++−−−−−+−++−+−−−++−−−++−−++
−−+−−−−−−−−+−+−+−+−,+−−+−++−−−++++−−+−+−
−+−+−−−−++++−++−−++−++−−++−−−+−+−+−++++
+++++)

(+++++++++−+−+−−−++−−++−++−−++−−−−−+++
+−+−++−+−++−−−−+++−−+−++−,+−+−+−+−++++++
++−++−−++−−−++−−++−+−++−+−−−−−+++++−−+−
++−++−−−−++,++−−++−−+−−++−−+−−−−−−−−−+−+
−+−+−−++++−−−++−+−−+++++−−−−+−+−−+−+,+−−
++−−+++−−++−−−+−+−+−+−−−−−−−−−++−+−−+−−
++++−−+−+−−+−+++++−−−−,++++−−−−+−+−−+−+−

−++++−−−++−+−−+−−−−−−−−−+−+−+−+++−−++−−
+−−++−−+,+−+−−+−+++++−−−−−++−+−−+−−++++−
−−+−+−+−+−−−−−−−−+−−++−−+++−−++−−,++−−−−
+++−−+−++−−−−−++++−+−++−+−−−++−−++−++−−
++−+++++++++−+−+−+−,+−−+−++−++−−−−++−+−+
+−+−−−−−++++−++−−++−−−++−−+++−+−+−+−+++
+++++)

(++++++++−+−+−+−+−−++−−+++−−++−−+−−−−+++
++−+−−+−+++−−−−++−++−+−−+,+−+−+−+−−−−−−−
−−−++−−++−++−−++−−−+−++−+−++++−−−−+−−+−
++−−−++++−−,++−−++−−−++−−++−−−−−−−−−+−+−
+−+−−−++++−−+−−+−++−++++−−−−−+−++−+−,+−−
++−−+−−++−−++−+−+−+−+++++++++−++−+−−+++
−−−−+++−+−−+−+−−−−++++,++++−−−−−+−++−+−−
−++++−−+−−+−++−−−−−−−−−+−+−+−+−++−−++−−
−++−−++−,+−+−−+−+−−−−++++−++−+−−+++−−−−+
+−+−+−+−++++++++++−−++−−+−−++−−++,++−−−−
++−++−+−−+−−−−+++++−+−−+−+−−++−−+++−−++
−−+++++++++−+−+−+−+,+−−+−++−−−++++−−−+−+
+−+−++++−−−−−++−−++−++−−++−−+−+−+−+−−−−
−−−−−)

F

List of Super Complementary Codes (PG $= 4 \sim 64$)

This appendix gives a list of super complementary codes with their PG values being from 4 to 64. Commas are used to separate different element codes within a flock and brackets to denote flocks in a super complementary code set. It is noted that the processing gain for a super complementary code set should be calculated as the product of its element code length (N) and its flock size (M), or $N \times M$.

1. PG = 4

   (1) Length of element codes = 1, flock size = 4

   (+, +, +, −)

   (+, −, +, +)

   (+, +, −, +)

   (+, −, −, −)

2. PG = 8

   (1) Length of element codes = 2, flock size = 4

   (+ +, + −, + +, − +)

   (+ −, + +, + −, − −)

   (+ +, + −, − −, + −)

   (+ −, + +, − +, + +)

   (2) Length of element codes = 1, flock size = 8

   (+, +, +, −, +, +, −, +)

   (+, −, +, +, +, −, −, −)

   (+, +, −, +, +, +, +, −)

   (+, −, −, −, +, −, +, +)

   (+, +, +, −, −, −, +, −)

The Next Generation CDMA Technologies Hsiao-Hwa Chen

(+, −, +, +, −, +, +, +)

(+, +, −, +, −, −, −, +)

(+, −, −, −, −, +, −, −)

3. PG = 16

(1) Length of element codes = 4, flock size = 4

(+ + + −, + + − +, + + + −, − − + −)

(+ − + +, + − − −, + − + +, − + + +)

(+ + + −, + + − +, − − − +, + + − +)

(+ − + +, + − − −, − + − −, + − − −)

(2) Length of element codes = 2, flock size = 8

(+ +, + −, + −, + +, + +, + −, − +, − −)

(+ +, − +, + −, − −, + +, − +, − +, + +)

(+ −, + +, + +, + −, + −, + +, − −, − +)

(+ −, − −, + +, − +, + −, − −, − −, + −)

(+ +, + −, + −, + +, − −, − +, + −, + +)

(+ +, − +, + −, − −, − −, + −, + −, − −)

(+ −, + +, + +, + −, − +, − −, + +, + −)

(+ −, − −, + +, − +, − +, + +, + +, − +)

(3) Length of element codes = 1, flock size = 16

(+, +, +, −, +, +, −, +, +, +, +, −, −, −, +, −)

(+, −, +, +, +, −, −, −, +, −, +, +, −, +, +, +)

(+, +, −, +, +, +, +, −, +, +, −, +, −, −, −, +)

(+, −, −, −, +, −, +, +, +, −, −, −, −, +, −, −)

(+, +, +, −, −, −, +, −, +, +, +, −, +, +, −, +)

(+, −, +, +, −, +, +, +, +, −, +, +, +, −, −, −)

(+, +, −, +, −, −, −, +, +, +, −, +, +, +, +, −)

(+, −, −, −, −, +, −, −, +, −, −, −, +, −, +, +)

(+, +, +, −, +, +, −, +, −, −, −, +, +, +, −, +)

(+, −, +, +, +, −, −, −, −, +, −, −, +, −, −, −)

(+, +, −, +, +, +, +, −, −, −, +, −, +, +, +, −)

(+, −, −, −, +, −, +, +, −, +, +, +, +, −, +, +)

(+, +, +, −, −, −, +, −, −, −, −, +, −, −, +, −)

(+, −, +, +, −, +, +, +, −, +, −, −, −, +, +, +)

(+, +, −, +, −, −, −, +, −, −, +, −, −, −, −, +)

(+, −, −, −, −, +, −, −, −, +, +, +, −, +, −, −)

4. PG = 32

(1) Length of element codes = 8, flock size = 4

(+ + + − + + − +, + + + − − − + −, + + + − + + − +, − − − + + + − +)

(+ − + + + − − −, + − + + − + + +, + − + + + − − −, − + − − + − − −)

(+ + + − + + − +, + + + − − − + −, − − − + − − + −, + + + − − − + −)

(+ − + + + − − −, + − + + − + + +, − + − − − + + +, + − + + − + + +)

(2) Length of element codes = 4, flock size = 8

(+ + + −, + − + +, + + − +, + − − −, + + + −, + − + +, − − + −, − + + +)

(+ + + −, − + − −, + + − +, − + + +, + + + −, − + − −, − − + −, + − − −)

(+ − + +, + + + −, + − − −, + + − +, + − + +, + + + −, − + − +, − − + −)

(+ − + +, − − − +, + − − −, − − + −, + − + +, − − − +, − + + +, + + − +)

(+ + + −, + − + +, + + − +, + − − −, − − − +, − + − −, + + − +, + − − −)

(+ + + −, − + − −, + + − +, − + + +, − − − +, + − + +, + + − +, − + + +)

(+ − + +, + + + −, + − − −, + + − +, − + − −, − − − +, + − − −, + + − +)

(+ − + +, − − − +, + − − −, − − + −, − + − −, + + + −, + − − −, − − + −)

(3) Length of element codes = 2, flock size = 16

(+ +, + +, + −, − +, + −, + −, + +, − −, + +, + +, + −, − +, − +, − +, − −, + +)

(+ +, − −, + −, + −, + −, − +, + +, + +, + +, − −, + −, + −, − +, + −, − −, − −)

(+ +, + +, − +, + −, + −, + −, − −, + +, + +, + +, − +, + −, − +, − +, + +, − −)

(+ +, − −, − +, − +, + −, − +, − −, − −, + +, − −, − +, − +, − +, + −, + +, + +)

(+ −, + −, + +, − −, + +, + +, + −, − +, + −, + −, + +, − −, − −, − −, − +, + −)

(+ −, − +, + +, + +, + +, − −, + −, + −, + −, − +, + +, + +, − −, + +, − +, − +)

(+ −, + −, − −, + +, + +, + +, − +, + −, + −, + −, − −, + +, − −, − −, + −, − +)

(+ −, − +, − −, − −, + +, − −, − +, − +, + −, − +, − −, − −, − −, + +, + −, + −)

(+ +, + +, + −, − +, + −, + −, + +, − −, − −, − −, − +, + −, + −, + −, + +, − −)

(+ +, − −, + −, + −, + −, − +, + +, + +, − −, + +, − +, − +, + −, − +, + +, + +)

(+ +, + +, − +, + −, + −, + −, − −, + +, − −, − −, + −, − +, + −, + −, − −, + +)

(+ +, − −, − +, − +, + −, − +, − −, − −, − −, + +, + −, + −, + −, − +, − −, − −)

(+ −, + −, + +, − −, + +, + +, + −, − +, − +, − +, − −, + +, + +, + +, + −, − +)

(+ −, − +, + +, + +, + +, − −, + −, + −, − +, + −, − −, − −, + +, − −, + −, + −)

(+ −, + −, − −, + +, + +, + +, − +, + −, − +, − +, + +, − −, + +, + +, − +, + −)

(+ −, − +, − −, − −, + +, − −, − +, − +, − +, + −, + +, + +, + +, − −, − +, − +)

(4) Length of element codes = 1, flock size = 32

(+, +, +, −, +, +, −, +, +, +, +, −, −, −, +, −, +, +, +, −, +, +, −, +, −, −, −, +, +, +, −, +)

(+, −, +, +, +, −, −, −, +, −, +, +, −, +, +, +, +, −, +, +, +, −, −, −, −, +, −, −, +, −, −, −)

(+, +, −, +, +, +, +, −, +, +, −, +, −, −, −, +, +, +, −, +, +, +, +, −, −, −, +, −, +, +, +, −)

(+, −, −, −, +, −, +, +, +, −, −, −, −, +, −, −, +, −, −, −, +, −, +, +, −, +, +, +, +, −, +, +)

(+, +, +, −, −, −, +, −, +, +, +, −, +, +, −, +, +, +, +, −, −, −, +, −, −, −, −, +, −, −, +, −)

(+, −, +, +, −, +, +, +, +, −, +, +, +, −, −, −, +, −, +, +, −, +, +, +, −, +, −, −, −, +, +, +)

(+, +, −, +, −, −, −, +, +, +, −, +, +, +, +, −, +, +, −, +, −, −, −, +, −, −, +, −, −, −, −, +)

(+, −, −, −, −, +, −, −, +, −, −, −, +, −, +, +, +, −, −, −, −, +, −, −, −, +, +, +, −, +, −, −)

(+, +, +, −, +, +, −, +, −, −, −, +, +, +, −, +, +, +, +, −, +, +, −, +, +, +, +, −, −, −, +, −)

(+, −, +, +, +, −, −, −, −, +, −, −, +, −, −, −, +, −, +, +, +, −, −, −, +, −, +, +, −, +, +, +)

(+, +, −, +, +, +, +, −, −, −, +, −, +, +, +, −, +, +, −, +, +, +, +, −, +, +, −, +, −, −, −, +)

(+, −, −, −, +, −, +, +, −, +, +, +, +, −, +, +, +, −, −, −, +, −, +, +, +, −, −, −, −, +, −, −)

(+, +, +, −, −, −, +, −, −, −, −, +, −, −, +, −, +, +, +, −, −, −, +, −, +, +, +, −, +, +, −, +)

(+, −, +, +, −, +, +, +, −, +, −, −, −, +, +, +, +, −, +, +, −, +, +, +, +, −, +, +, +, −, −, −)

(+, +, −, +, −, −, −, +, −, −, +, −, −, −, −, +, +, +, −, +, −, −, −, +, +, +, −, +, +, +, +, −)

(+, −, −, −, −, +, −, −, −, +, +, +, −, +, −, −, +, −, −, −, −, +, −, −, +, −, −, −, +, −, +, +)

(+, +, +, −, +, +, −, +, +, +, +, −, −, −, +, −, −, −, −, +, −, −, +, −, +, +, +, −, −, −, +, −)

(+, −, +, +, +, −, −, −, +, −, +, +, −, +, +, +, −, +, −, −, −, +, +, +, +, −, +, +, −, +, +, +)

(+, +, −, +, +, +, +, −, +, +, −, +, −, −, −, +, −, −, +, −, −, −, −, +, +, +, −, +, −, −, −, +)

(+, −, −, −, +, −, +, +, +, −, −, −, −, +, −, −, −, +, +, +, −, +, −, −, +, −, −, −, −, +, −, −)

(+, +, +, −, −, −, +, −, +, +, +, −, +, +, −, +, −, −, −, +, +, +, −, +, +, +, +, −, +, +, −, +)

(+, −, +, +, −, +, +, +, +, −, +, +, +, −, −, −, −, +, −, −, +, −, −, −, +, −, +, +, +, −, −, −)

(+, +, −, +, −, −, −, +, +, +, −, +, +, +, +, −, −, −, +, −, +, +, +, −, +, +, −, +, +, +, +, −)

(+, −, −, −, −, +, −, −, +, −, −, −, +, −, +, +, −, +, +, +, +, −, +, +, +, −, −, −,
+, −, +, +)

(+, +, +, −, +, +, −, +, −, −, −, +, +, +, −, +, −, −, −, +, −, −, +, −, −, −, −, +,
+, +, −, +)

(+, −, +, +, +, −, −, −, −, +, −, −, +, −, −, −, −, +, −, −, −, +, +, +, −, +, −, −,
+, −, −, −)

(+, +, −, +, +, +, +, −, −, −, +, −, +, +, +, −, −, −, +, −, −, −, −, +, −, −, +, −,
+, +, +, −)

(+, −, −, −, +, −, +, +, −, +, +, +, +, −, +, +, −, +, +, +, −, +, −, −, −. +, +, +,
+, −, +, +)

(+, +, +, −, −, −, +, −, −, −, −, +, −, −, +, −, −, −, −, +, +, +, −, +, − −, −, +,
−, −, +, −)

(+, −, +, +, −, +, +, +, −, +, −, −, −, +, +, +, −, +, −, −, +, −, −, −, −, +, −, −,
−, +, +, +)

(+, +, −, +, −, −, −, +, −, −, +, −, −, −, −, +, −, −, +, −, +, +, +, −, −, −, +, −,
−, −, −, +)

(+, −, −, −, −, +, −, −, −, +, +, +, −, +, −, −, −, +, +, +, +, −, +, +, −, +, +, +,
−, +, −, −)

5. PG = 64

(1) Length of element codes = 16, flock size = 4

(+ + + − + + − + + + + − − − + −, + + + − + + − + − − − + + + − +, + + + − + + −
+ + + + − − − + −, − − − + − − + − + + + − − − + −)

(+ − + + + − − − + − + + − + + +, + − + + + − − − − + − − + − − −, + − + + + − −
− + − + + − + + +, − + − − − + + + + − + + − + + +)

(+ + + − + + − + + + + − − − + −, + + + − + + − + − − − − + + − +, − − − + − − +
− − − − + + + − +, + + + − + + − + − − − + + + − +)

(+ − + + + − − − + − + + − + + +, + − + + + − − − − + − − + − − −, − + − − − + +
+ − + − − + − − −, + − + + + − − − − + − − + − − −)

(2) Length of element codes = 8, flock size = 8

(+ + + − + + − +, + − + + + − − −, + + + − − − + −, + − + + − + + −, + + + − + + −
+ + − + + + − − −, − − − + + + − +, − + − − + − − −)

(+ + + − + + − +, − + − − − + + +, + + + − − − + −, − + − − + − − −, + + + − + + −
+, − + − − − + + +, − − − + + + − +, + − + + − + + +)

(+ − + + + − − −, + + + − + + − +, + − + + − + + +, + + + − − − + −, + − + + + − −
−, + + + − + + − +, − + − − + − − −, − − − + + + − +)

(+ − + + + − − −, − − − + − − + −, + − + + − + + +, − − − + + + − +, + − + + + − −
−, − − − + − − + −, − + − − + − − −, + + + − − − + −)

(+ + + − + + − +, + − + + + − − −, + + + − − − + −, + − + + − + + +, − − − + − − +
−, − + − − − + + +, + + + − − − + −, + − + + − + + +)

(+ + + − + + − +, − + − − − + + +, + + + − − − + −, − + − − + − − −, − − − + − − +
−, + − + + + − − −, + + + − − − + −, − + − − + − − −)

(+ − + + + − − −, + + + − + + − +, + − + + − + + +, + + + − − − + −, − + − − − + +
+, − − − + − − + −, + − + + − + + +, + + + − − − + −)

(+ − + + + − − −, − − − + − − + −, + − + + − + + +, − − − + + + − +, − + − − − + +
+, + + + − + + − +, + − + + − + + +, − − − + + + − +)

(3) Length of element codes = 4, flock size = 16

(+ + + −, + + + −, + − + +, − + − −, + + − +, + + − +, + − − −, − + + +, + + + −, +
+ + −, + − + +, − + − −, − − + −, − − + −, − + + +, + − − −)

(+ + + −, − − − +, + − + +, + − + +, + + − +, − − + −, + − − −, + − − −, + + + −, −
− − +, + − + +, + − + +, − − + −, + + − +, − + + +, − + + +)

(+ + + −, + + + −, − + − −, + − + +, + + − +, + + − +, − + + +, + − − −, + + + −, +
+ + −, − + − −, + − + +, − − + −, − − + −, + − − −, − + + +)

(+ + + −, − − − +, − + − −, − + − −, + + − +, − − + −, − + + +, − + + +, + + + −, −
− − +, − + − −, − + − −, − − + −, + + − +, + − − −, + − − −)

(+ − + +, + − + +, + + + −, − − − +, + − − −, + − − −, + + − +, − − + −, + − + +, +
− + +, + + + −, − − − +, − + + +, − + + +, − − + −, + + − +)

(+ − + +, − + − −, + + + −, + + + −, + − − −, − + + +, + + − +, + + − +, + − + +, −
+ − −, + + + −, + + + −, − + + +, + − − −, − − + −, − − + −)

(+ − + +, + − + +, − − − +, + + + −, + − − −, + − − −, − − + −, + + − +, + − + +, +
− + +, − − − +, + + + −, − + + +, − + + +, + + − +, − − + −)

(+ − + +, − + − −, − − − +, − − − +, + − − −, − + + +, − − + −, − − + −, + − + +, −
+ − −, − − − +, − − − +, − + + +, + − − −, + + − +, + + − +)

(+ + + −, + + + −, + − + +, − + − −, + + − +, + + − +, + − − −, − + + +, − − − +, −
− − +, − + − −, + − + +, + + − +, + + − +, + − − −, − + + +)

(+ + + −, − − − +, + − + +, + − + +, + + − +, − − + −, + − − −, + − − −, − − − +, +
+ + −, − + − −, − + − −, + + − +, − − + −, + − − −, + − − −)

(+ + + −, + + + −, − + − −, + − + +, + + − +, + + − +, − + + +, + − − −, − − − +, −
− − +, + − + +, − + − −, + + − +, + + − +, − + + +, + − − −)

(+ + + −, − − − +, − + − −, − + − −, + + − +, − − + −, − + + +, − + + +, − − − +, +
+ + −, + − + +, + − + +, + + − +, − − + −, − + + +, − + + +)

(+ − + +, + − + +, + + + −, − − − +, + − − −, + − − −, + + − +, − − + −, − + − −, −
+ − −, − − − +, + + + −, + − − −, + − − −, + + − +, − − + −)

(+ − + +, − + − −, + + + −, + + + −, + − − −, − + + +, + + − +, + + − +, − + − −, +
− + +, − − − +, − − − +, + − − −, − + + +, + + − +, + + − +)

(+ − + +, + − + +, − − − +, + + + −, + − − −, + − − −, − − + −, + + − +, − + − −, −
+ − −, + + + −, − − − +, + − − −, + − − −, − − + −, + + − +)

(+ − + +, − + − −, − − − +, − − − +, + − − −, − + + +, − − + −, − − + −, − + − −, +
− + +, + + + −, + + + −, + − − −, − + + +, − − + −, − − + −)

(4) Length of element codes = 2, flock size = 32

(+ +, + +, + +, − −, + −, + −, − +, + −, + −, + −, + −, − +, + +, + +, − −, + +, + +,
+ +, + +, − −, + −, + −, − +, + −, − +, − +, − +, + −, − −, − −, + +, − −)

(+ +, − −, + +, + +, + −, − +, − +, − +, + −, − +, + −, + −, + +, − −, − −, − −, + +,
− −, + +, + +, + −, − +, − +, − +, − +, + −, − +, − +, − −, + +, + +, + +)

(++, ++, −−, ++, +−, +−, +−, −+, +−, +−, −+, +−, ++, ++, ++, −−, ++,
++, −−, ++, +−, +−, +−, −+, −+, −+, +−, −+, −−, −−, −−, +−)

(++, −−, −−, −−, +−, −+, +−, +−, +−, −+, −+, −+, ++, −−, ++, ++, ++,
−−, −−, −−, +−, −+, +−, +−, −+, +−, +−, +−, −−, ++, −−, −−)

(++, ++, ++, −−, −+, −+, +−, −+, +−, +−, +−, −+, −−, −−, ++, −−, ++,
++, ++, −−, −+, −+, +−, −+, −+, −+, −+, +−, ++, ++, −−, ++)

(++, −−, ++, ++, −+, +−, +−, +−, +−, −+, +−, +−, −−, ++, ++, ++, ++,
−−, ++, ++, −+, +−, +−, +−, −+, +−, −+, −+, ++, −−, −−, −−)

(++, ++, −−, ++, −+, −+, −+, +−, +−, +−, −+, +−, −−, −−, −−, ++, ++,
++, −−, ++, −+, −+, −+, +−, −+, −+, +−, −+, ++, ++, ++, −−)

(++, −−, −−, −−, −+, +−, −+, −+, +−, −+, −+, −+, −−, ++, −−, −−, ++,
−−, −−, −−, −+, +−, −+, −+, −+, +−, +−, +−, ++, −−, ++, ++)

(+−, +−, +−, −+, ++, ++, −−, ++, ++, ++, ++, −−, +−, +−, −+, +−, +−,
+−, +−, −+, ++, ++, −−, ++, −−, −−, −−, ++, −+, −+, +−, −+)

(+−, −+, +−, +−, ++, −−, −−, −−, ++, −−, ++, ++, +−, −+, −+, −+, +−,
−+, +−, +−, ++, −−, −−, −−, −−, ++, −−, −−, −+, +−, +−, +−)

(+−, +−, −+, +−, ++, ++, ++, −−, ++, ++, −−, ++, +−, +−, −−, −+, +−,
+−, −+, +−, ++, ++, ++, −−, −−, −−, ++, −−, −+, −+, −+, +−)

(+−, −+, −+, −+, ++, −−, ++, ++, ++, −−, −−, −−, +−, −+, +−, +−, +−,
−+, −+, −+, ++, −−, ++, ++, −−, ++, ++, ++, −+, +−, −+, −+)

(+−, +−, +−, −+, −−, −−, ++, −−, ++, ++, ++, −−, −+, −+, +−, −+, +−,
+−, +−, −+, −−, −−, ++, −−, −−, −−, −−, ++, +−, +−, −+, +−)

(+−, −+, +−, +−, −−, ++, ++, ++, ++, −−, ++, ++, −+, +−, +−, +−, +−,
−+, +−, +−, −−, ++, ++, ++, −−, ++, −−, −−, +−, −+, −+, −+)

(+−, +−, −+, +−, −−, −−, −−, ++, ++, ++, −−, ++, −+, −+, −+, +−, +−,
+−, −+, +−, −−, −−, −−, ++, −−, −−, ++, −−, +−, +−, +−, −+)

(+−, −+, −+, −+, −−, ++, −−, −−, ++, −−, −−, −−, −+, +−, −+, −+, +−,
−+, −+, −+, −−, ++, −−, −−, −−, ++, ++, ++, +−, −+, +−, +−)

(++, ++, ++, −−, +−, +−, −+, +−, +−, +−, +−, −+, ++, ++, −−, ++, −−,
−−, −−, ++, −+, −+, +−, −+, +−, +−, +−, −+, ++, ++, −−, ++)

(++, −−, ++, ++, +−, −+, −+, −+, +−, −+, +−, +−, ++, −− −−, −−, −−,
++, −−, −−, −+, +−, +−, +−, +−, −+, +−, +−, ++, −−, −−, −−)

(++, ++, −−, ++, +−, +−, +−, −+, +−, +−, −+, +−, ++, ++, ++, −−, −−,
−−, ++, −−, −+, −+, −+, +−, +−, +−, −+, +−, ++, ++, ++, −−)

(++, −−, −−, −−, +−, −+, +−, +−, +−, −+, −+, −+, ++, −−, ++, ++, −−,
++, ++, ++, −+, +−, −+, −+, +−, −+, −+, −+, ++, −−, ++, ++)

(++, ++, ++, −−, −+, −+, +−, −+, +−, +−, +−, −+, −−, −−, ++, −−, −−,
−−, −−, ++, +−, +−, −+, +−, +−, +−, +−, −+, −−, −−, ++, −−)

(++, −−, ++, ++, −+, +−, +−, +−, +−, −+, +−, +−, −−, ++, ++, ++, −−,
++, −−, −−, +−, −+, −+, −+, +−, −+, +−, +−, −−, ++, ++ ++)

(++, ++, −−, ++, −+, −+, −+, +−, +−, +−, −+, +−, −−, −−, −−, ++, −−,
−−, ++, −−, +−, +−, +−, −+, +−, +−, −+, +−, −−, −−, −−, ++)

(+ +, − −, − −, − −, − +, + −, − +, − +, + −, − +, − +, − +, − −, + +, − −, − −, − −,
+ +, + +, + +, + −, − +, + −, + −, + −, − +, − +, − +, − −, + +, − −, − −)

(+ −, + −, + −, − +, + +, + +, − −, + +, + +, + +, + +, − −, + −, + −, − +, + −, − +,
− +, − +, + −, − −, − −, + +, − −, + +, + +, + +, − −, + −, + −, − +, + −)

(+ −, − +, + −, + −, + +, − −, − −, − −, + +, − −, + +, + +, + −, − +, − +, − +, − +,
+ −, − +, − +, − −, + +, + +, + +, + +, − −, + +, + +, + −, − +, − +, − +)

(+ −, + −, − +, + −, + +, + +, + +, − −, + +, + +, − −, + +, + −, + −, + −, − +, − +,
− +, + −, − +, − −, − −, − −, + +, + +, + +, − −, + +, + −, + −, + −, − +)

(+ −, − +, − +, − +, + +, − −, + +, + +, + +, − −, − −, − −, + −, − +, + −, + −, − +,
+ −, + −, + −, − −, + +, − −, − −, + +, − −, − −, − −, + −, − +, + −, + −)

(+ −, + −, + −, − +, − −, − −, + +, − −, + +, + +, + +, − −, − +, − +, + −, − +, − +,
− +, − +, + −, + +, + +, − −, + +, + +, + +, + +, − −, − +, − +, + −, − +)

(+ −, − +, + −, + −, − −, + +, + +, + +, + +, − −, + +, + +, − +, + −, + −, + −, − +,
+ −, − +, − +, + +, − −, − −, − −, + +, − −, + +, + +, − +, + −, + −, + −)

(+ −, + −, − +, + −, − −, − −, − −, + +, + +, + +, − −, + +, − +, − +, − +, + −, − +,
− +, + −, − +, + +, + +, + +, − −, + +, + +, − −, + +, − +, − +, − +, + −)

(+ −, − +, − +, − +, − −, + +, − −, − −, + +, − −, − −, − −, − +, + −, − +, − +, − +,
+ −, + −, + −, + +, − −, + +, + +, + +, − −, − −, − −, − +, + −, − +, − +)

(5) Length of element codes = 1, flock size = 64

(+, +, +, −, +, +, −, +, +, +, +, −, −, −, +, −, +, +, +, −, +, +, −, +, −, −, −, +, +, +,
−, +, +, +, +, −, +, +, −, +, +, +, +, −, −, −, +, −, −, −, −, +, −, −, +, −, +, +, +, −,
−, −, +, −)

(+, −, +, +, +, −, −, −, +, −, +, +, −, +, +, +, +, −, +, +, +, −, −, −, −, +, −, −, +, −,
−, −, +, −, +, +, +, −, −, −, +, −, +, +, −, +, +, +, −, +, −, −, −, +, +, +, +, −, +, +,
−, +, +, +)

(+, +, −, +, +, +, +, −, +, +, −, +, −, −, −, +, +, +, −, +, +, +, +, −, −, −, +, −, +, +,
+, −, +, +, −, +, +, +, +, −, +, +, −, +, −, −, −, +, −, −, +, −, −, −, −, +, +, +, −, +,
−, −, −, +)

(+, −, −, −, +, −, +, +, +, −, −, −, −, +, −, −, +, −, −, −, +, −, +, +, −, +, +, +, +, −,
+, +, +, −, −, −, +, −, +, +, +, −, −, −, −, +, −, −, −, +, +, +, −, +, −, −, +, −, −, −,
−, +, −, −)

(+, +, +, −, −, −, +, −, +, +, +, −, +, +, −, +, +, +, +, −, −, −, +, −, −, −, −, +, −, −,
+, −, +, +, +, −, −, −, +, −, +, +, +, −, +, +, −, +, −, −, −, +, +, +, −, +, +, +, +, −,
+, +, −, +)

(+, −, +, +, −, +, +, +, +, −, +, +, +, −, −, −, +, −, +, +, −, +, +, +, −, +, −, −, −, +,
+, +, +, −, +, +, −, +, +, +, +, −, +, +, +, −, −, −, −, +, −, −, +, −, −, −, +, −, +, +,
+, −, −, −)

(+, +, −, +, −, −, −, +, +, +, −, +, +, +, +, −, +, +, −, +, −, −, −, +, −, −, +, −, −, −,
−, +, +, +, −, +, −, −, −, +, +, +, −, +, +, +, +, −, −, −, +, −, +, +, +, −, +, +, −, +,
+, +, +, −)

(+, −, −, −, −, +, −, −, +, −, −, −, +, −, +, +, +, −, −, −, −, +, −, −, −, +, +, +, −, +,
−, −, +, −, −, −, −, +, −, −, +, −, −, −, +, −, +, +, −, +, +, +, +, −, +, +, +, −, −, −,
+, −, +, +)

(+, +, +, −, +, +, −, +, −, −, −, +, +, +, −, +, +, +, +, −, +, +, −, +, +, +, +, −, −, −,
+, −, +, +, +, −, +, +, −, +, −, −, −, +, +, +, −, +, −, −, −, +, −, −, +, −, −, −, −, +,
+, +, −, +)

(+, −, +, +, +, −, −, −, −, +, −, −, +, −, −, −, +, −, +, +, +, −, −, −, +, −, +, +, −, +,
+, +, +, −, +, +, +, −, −, −, −, +, −, −, +, −, −, −, −, +, −, −, −, +, +, −, −, +, −, −,
+, −, −, −)

(+, +, −, +, +, +, +, −, −, −, +, −, +, +, +, −, +, +, −, +, +, −, +, −, +, +, −, +, −, −,
−, +, +, +, −, +, +, +, +, −, −, −, +, −, +, +, +, −, −, −, +, −, −, −, −, +, −, −, +, −,
+, +, +, −)

(+, −, −, −, +, −, +, +, −, +, +, +, +, −, +, +, +, −, −, −, +, −, +, +, +, −, −, −, −, +,
−, −, +, −, −, −, +, −, +, +, −, +, +, +, +, −, +, +, −, +, +, +, −, +, −, −, −, +, +, +,
+, −, +, +)

(+, +, +, −, −, −, +, −, −, −, −, +, −, −, +, −, +, +, +, −, −, −, +, −, +, +, +, −, +, +,
−, +, +, +, +, −, −, −, +, −, −, −, −, +, −, −, +, −, −, −, −, +, +, +, −, +, −, −, −, +,
−, −, +, −)

(+, −, +, +, −, +, +, +, −, +, −, −, −, +, +, +, +, −, +, +, −, +, +, +, +, −, +, +, +, −,
−, −, +, −, +, +, −, +, +, +, −, +, −, −, −, +, +, +, −, +, −, −, +, −, −, −, −, +, −, −,
−, +, +, +)

(+, +, −, +, −, −, −, +, −, −, +, −, −, −, −, +, +, +, −, +, −, −, −, +, +, +, −, +, +, +,
+, −, +, +, −, +, −, −, −, +, −, −, +, −, −, −, −, +, −, −, +, −, +, +, +, −, −, −, +, −,
−, −, −, +)

(+, −, −, −, −, +, −, −, −, +, +, +, −, +, −, −, +, −, −, −, −, +, −, −, +, −, −, −, +, −,
+, +, +, −, −, −, −, +, −, −, −, +, +, +, −, +, −, −, −, +, +, +, +, −, + +, −, +, +, +,
−, +, −, −)

(+, +, +, −, +, +, −, +, +, +, +, −, −, −, +, −, −, −, −, +, −, −, +, −, +, +, +, −, −, −,
+, −, +, +, +, −, +, +, −, +, +, +, +, −, −, −, +, −, +, +, +, −, +, +, −, +, −, −, −, +,
+, +, −, +)

(+, −, +, +, +, −, −, −, +, −, +, +, −, +, +, +, −, +, −, −, −, +, +, +, −, −, +, +, −, +,
+, +, +, −, +, +, +, −, −, −, +, −, +, +, −, +, +, +, +, −, +, +, +, −, −, −, −, +, −, −,
+, −, −, −)

(+, +, −, +, +, +, +, −, +, +, −, +, −, −, −, +, −, −, +, −, −, −, −, +, +, +, −, +, −, −,
−, +, +, +, −, +, +, +, +, −, +, +, −, +, −, −, −, +, +, +, −, +, +, +, +, −, −, −, +, −,
+, +, +, −)

(+, −, −, −, +, −, +, +, +, −, −, −, −, +, −, −, −, +, +, +, −, +, −, −, +, −, −, −, −, +,
−, −, +, −, −, −, +, −, +, +, +, −, −, −, −, +, −, −, +, −, −, −, +, −, −, +, −, +, +, +,
+, −, +, +)

(+, +, +, −, −, −, +, −, +, +, +, −, +, +, −, +, −, −, −, +, −, +, −, +, +, +, +, −, +, +,
−, +, +, +, +, −, −, −, +, −, +, +, +, −, +, +, −, +, +, +, +, −, −, −, +, −, −, −, −, +,
−, −, +, −)

(+, −, +, +, −, +, +, +, +, −, +, +, +, −, −, −, −, +, −, −, +, −, −, −, +, −, +, +, +, −,
−, −, +, −, +, +, −, +, +, +, +, −, +, +, +, −, −, −, +, −, +, +, −, +, +, +, −, +, −, −,
−, +, +, +)

(+, +, −, +, −, −, −, +, +, +, −, +, +, +, +, −, −, −, +, −, +, +, +, −, +, +, −, +, +, +,
+, −, +, +, −, +, −, −, −, +, +, +, −, +, +, +, +, −, +, +, −, +, −, −, −, +, −, −, +, −,
−, −, −, +)

(+, −, −, −, −, +, −, −, +, −, −, −, +, −, +, +, −, +, +, +, +, −, +, +, +, −, −, −, +, −,
+, +, +, −, −, −, −, +, −, −, +, −, −, −, +, −, +, +, +, −, −, −, −, +, −, −, −, +, +, +,
−, +, −, −)

(+, +, +, −, +, +, −, +, −, −, −, +, +, +, −, +, −, −, −, +, −, −, +, −, −, −, −, +, +, +,
−, +, +, +, +, −, +, +, −, +, −, −, −, +, +, +, −, +, +, +, +, −, +, +, −, +, +, +, +, −,
−, −, +, −)

(+, −, +, +, +, −, −, −, −, +, −, −, +, −, −, −, −, +, −, −, −, +, +, +, −, +, −, −, +, −,
−, −, +, −, +, +, +, −, −, −, −, +, −, −, +, −, −, −, +, −, +, +, +, −, −, −, +, −, +, +,
−, +, +, +)

(+, +, −, +, +, +, +, −, −, −, +, −, +, +, +, −, −, −, +, −, −, −, −, +, −, −, +, −, +, +,
+, −, +, +, −, +, +, +, +, −, −, −, +, −, +, +, +, −, +, +, −, +, +, +, +, −, +, +, −, +,
−, −, −, +)

(+, −, −, −, +, −, +, +, −, +, +, +, +, −, +, +, −, +, +, +, −, +, −, −, −, +, +, +, +, −,
+, +, +, −, −, −, +, −, +, +, −, +, +, +, +, −, +, +, +, −, −, −, +, −, +, +, +, −, −, −,
−, +, −, −)

(+, +, +, −, −, −, +, −, −, −, −, +, −, −, +, −, −, −, −, +, +, +, −, +, −, −, −, +, −, −,
+, −, +, +, +, −, −, −, +, −, −, −, −, +, −, −, +, −, +, +, +, −, −, −, +, −, +, +, +, −,
+, +, −, +)

(+, −, +, +, −, +, +, +, −, +, −, −, −, +, +, +, −, +, −, −, +, −, −, −, −, +, −, −, −, +,
+, +, +, −, +, +, −, +, +, +, −, +, −, −, −, +, +, +, +, −, +, +, −, +, +, +, +, −, +, +,
+, −, −, −)

(+, +, −, +, −, −, −, +, −, −, +, −, −, −, −, +, −, −, +, −, +, +, +, −, −, −, +, −, −, −,
−, +, +, +, −, +, −, −, −, +, −, −, +, −, −, −, −, +, +, +, −, +, −, −, −, +, +, +, −, +,
+, +, +, −)

(+, −, −, −, −, +, −, −, −, +, +, +, −, +, −, −, −, +, +, +, +, −, +, +, −, +, +, +, −, +,
−, −, +, −, −, −, −, +, −, −, −, +, +, +, −, +, −, −, +, −, −, −, −, +, −, −, +, −, −, −,
+, −, +, +)

(+, +, +, −, +, +, −, +, +, +, +, −, −, −, +, −, +, +, +, −, +, +, −, +, −, −, −, +, +, +,
−, +, −, −, −, +, −, −, +, −, −, −, −, +, +, +, −, +, +, +, +, −, +, +, −, +, −, −, −, +,
+, +, −, +)

(+, −, +, +, +, −, −, −, +, −, +, +, −, +, +, +, +, −, +, +, +, −, −, −, −, +, −, −, +, −,
−, −, −, +, −, −, −, +, +, +, −, +, −, −, +, −, −, −, +, −, +, +, +, −, −, −, −, +, −, −,
+, −, −, −)

(+, +, −, +, +, +, +, −, +, +, −, +, −, −, −, +, +, +, −, +, +, +, +, −, −, −, +, −, +, +,
+, −, −, −, +, −, −, −, −, +, −, −, +, −, +, +, +, −, +, +, −, +, +, +, +, −, −, −, +, −,
+, +, +, −)

(+, −, −, −, +, −, +, +, +, −, −, −, −, +, −, −, +, −, −, −, +, −, +, +, −, +, +, +, +, −,
+, +, −, +, +, +, −, +, −, −, −, +, +, +, +, −, +, +, +, −, −, −, +, −, +, +, −, +, +, +,
+, −, +, +)

(+, +, +, −, −, −, +, −, +, +, +, −, +, +, −, +, +, +, +, −, −, −, +, −, −, −, −, +, −, −,
+, −, −, −, −, +, +, +, −, +, −, −, −, +, −, −, +, −, +, +, +, −, −, −, +, −, −, −, −, +,
−, −, +, −)

(+, −, +, +, −, +, +, +, +, −, +, +, +, −, −, −, +, −, +, +, −, +, +, +, −, +, −, −, −, +,
+, +, −, +, −, −, +, −, −, −, −, +, −, −, −, +, +, +, +, −, +, +, −, +, +, +, −, +, −, −,
−, +, +, +)

(+, +, −, +, −, −, −, +, +, +, −, +, +, +, +, −, +, +, −, +, −, −, −, +, −, −, +, −, −, −, −, +, −, −, +, −, +, +, +, −, −, −, +, −, −, −, −, +, +, +, −, +, −, −, −, +, −, −, +, −, −, −, −, +)

(+, −, −, −, −, +, −, −, +, −, −, −, +, −, +, +, +, −, −, −, −, +, −, −, −, +, +, +, −, +, −, −, −, +, +, +, +, −, +, +, −, +, +, +, −, +, −, −, +, −, −, −, −, +, −, −, −, +, +, +, −, +, −, −)

(+, +, +, −, +, +, −, +, −, −, −, +, +, +, −, +, +, +, +, −, +, +, −, +, −, +, +, −, −, −, +, −, −, −, −, +, −, −, +, −, +, +, +, −, −, −, +, −, +, +, +, −, +, +, −, +, +, +, +, −, −, −, +, −)

(+, −, +, +, +, −, −, −, −, +, −, −, +, −, −, −, +, −, +, +, +, −, −, −, +, −, +, +, −, +, +, +, −, +, −, −, −, +, +, +, +, −, +, +, −, +, +, +, +, −, +, +, +, −, −, −, +, −, +, +, −, +, +, +)

(+, +, −, +, +, +, +, −, −, −, +, −, +, +, +, −, +, +, −, +, +, +, +, −, +, +, −, +, −, −, −, +, −, −, +, −, −, −, −, +, +, +, −, +, −, −, −, +, +, +, −, +, +, +, +, −, +, +, −, +, −, −, −, +)

(+, −, −, −, +, −, +, +, −, +, +, +, +, −, +, +, +, −, −, −, +, −, +, +, +, −, −, −, −, +, −, −, −, +, +, +, −, +, −, −, +, −, −, −, −, +, −, −, +, −, −, −, +, −, +, +, +, −, −, −, −, +, −, −)

(+, +, +, −, −, −, +, −, −, −, −, +, −, −, +, −, +, +, +, −, −, −, +, −, +, +, +, −, +, +, −, +, −, −, −, +, +, +, −, +, +, +, +, −, +, +, −, +, +, +, +, −, −, −, +, −, +, +, +, −, +, +, −, +)

(+, −, +, +, −, +, +, +, −, +, −, −, −, +, +, +, +, −, +, +, −, +, +, +, +, −, +, +, +, −, −, −, −, +, −, −, +, −, −, −, +, −, +, +, +, −, −, −, +, −, +, +, −, +, −, +, +, −, +, +, +, −, −, −)

(+, +, −, +, −, −, −, +, −, −, +, −, −, −, −, +, +, +, −, +, −, −, −, +, +, +, −, +, +, +, +, −, −, −, +, −, +, +, +, −, +, +, −, +, +, +, +, −, +, +, −, +, −, −, −, +, +, +, −, +, +, +, +, −)

(+, −, −, −, −, +, −, −, −, +, +, +, −, +, −, −, +, −, −, −, −, +, −, −, +, −, −, −, +, −, +, +, −, +, +, +, +, −, +, +, +, −, −, −, +, −, +, +, +, −, −, −, −, +, −, −, +, −, −, −, +, −, +, +)

(+, +, +, −, +, +, −, +, +, +, +, −, −, −, +, −, −, −, −, +, −, −, +, −, +, +, +, −, −, −, +, −, −, −, −, +, −, −, +, −, −, −, −, +, +, +, −, +, −, −, −, +, −, −, +, −, +, +, +, −, −, −, +, −)

(+, −, +, +, +, −, −, −, +, −, +, +, −, +, +, +, −, +, −, −, −, +, +, +, +, −, +, +, −, +, +, +, −, +, −, −, −, +, +, +, −, +, −, −, +, −, −, −, −, +, −, −, −, +, +, +, +, −, +, +, −, +, +, +)

(+, +, −, +, +, +, +, −, +, +, −, +, −, −, −, +, −, −, +, −, −, −, −, +, +, +, −, +, −, −, −, +, −, −, +, −, −, −, −, +, −, −, +, −, +, +, +, −, −, −, +, −, −, −, −, +, +, +, −, +, −, −, −, +)

(+, −, −, −, +, −, +, +, +, −, −, −, −, +, −, −, −, +, +, +, −, +, −, −, +, −, −, −, −, +, −, −, −, +, +, +, −, +, −, −, −, +, +, +, +, −, +, +, −, +, +, +, −, +, −, −, +, −, −, −, −, +, −, −)

(+, +, +, −, −, −, +, −, +, +, +, −, +, +, −, +, −, −, −, +, +, +, −, +, +, +, +, −, +, +, −, +, −, −, −, +, +, +, −, +, −, −, −, +, −, −, +, −, −, −, −, +, +, +, −, +, +, +, +, −, +, +, −, +)

(+, −, +, +, −, +, +, +, +, −, +, +, +, −, −, −, −, +, −, −, +, −, −, −, +, −, +, +, +,
−, −, −, −, +, −, −, +, −, −, −, −, +, −, −, −, +, +, +, −, +, −, −, +, −, −, −, +, −, +,
+, +, −, −, −)

(+, +, −, +, −, −, −, +, +, +, −, +, +, +, +, −, −, −, +, −, +, +, +, −, +, +, −, +, +,
+, +, −, −, −, +, −, +, +, +, −, −, −, +, −, −, −, −, +, −, −, +, −, +, +, +, −, +, +, −,
+, +, +, +, −)

(+, −, −, −, −, +, −, −, +, −, −, −, +, −, +, +, −, +, +, +, +, −, +, +, +, −, −, −, +,
−, +, +, −, +, +, +, +, −, +, +, −, +, +, +, −, +, −, −, −, +, +, +, +, −, +, +, +, −, −,
−, +, −, +, +)

(+, +, +, −, +, +, −, +, −, −, −, +, +, +, −, +, −, −, −, +, −, −, +, −, −, −, −, +, +,
+, −, +, −, −, −, +, −, −, +, −, +, +, +, −, −, −, +, −, −, −, −, +, −, −, +, −, −, −, −,
+, +, +, −, +)

(+, −, +, +, +, −, −, −, −, +, −, −, +, −, −, −, −, +, −, −, −, +, +, +, −, +, −, −, +,
−, −, −, −, +, −, −, −, +, +, +, +, −, +, +, −, +, +, +, −, +, −, −, −, +, +, +, −, +, −,
−, +, −, −, −)

(+, +, −, +, +, +, +, −, −, −, +, −, +, +, +, −, −, −, +, −, −, −, −, +, −, −, +, −, +,
+, +, −, −, −, +, −, −, −, −, +, +, +, −, +, −, −, −, +, −, −, +, −, −, −, −, +, −, −, +,
−, +, +, +, −)

(+, −, −, −, +, −, +, +, −, +, +, +, +, −, +, +, −, +, +, +, −, +, −, −, −, +, +, +, +,
−, +, +, −, +, +, +, −, +, −, −, +, −, −, −, −, +, −, −, −, +, +, +, −, +, −, −, −, +, +,
+, +, −, +, +)

(+, +, +, −, −, −, +, −, −, −, −, +, −, −, +, −, −, −, −, +, +, +, −, +, −, −, −, +, −,
−, +, −, −, −, −, +, +, +, −, +, +, +, +, −, +, +, −, +, −, −, −, +, +, +, −, +, −, −, −,
+, −, −, +, −)

(+, −, +, +, −, +, +, +, −, +, −, −, −, +, +, +, −, +, −, −, +, −, −, −, −, +, −, −, −,
+, +, +, −, +, −, −, +, −, −, −, +, −, +, +, +, −, −, −, −, +, −, −, +, −, −, −, −, +, −,
−, −, +, +, +)

(+, +, −, +, −, −, −, +, −, −, +, −, −, −, −, +, −, −, +, −, +, +, +, −, −, −, +, −, −,
−, −, +, −, −, +, −, +, +, +, −, +, +, −, +, +, +, +, −, −, −, +, −, +, +, +, −, −, −, +,
−, −, −, −, +)

(+, −, −, −, −, +, −, −, −, +, +, +, −, +, −, −, −, +, +, +, +, −, +, +, −, +, +, +, −,
+, −, −, −, +, +, +, +, −, +, +, +, −, −, −, +, −, +, +, −, +, +, +, +, −, +, +, −, +, +,
+, −, +, −, −)

References

[1] H.-H. Chen and M. Guizani (2006) *Next Generation Wireless Systems and Networks*, Chichester: John Wiley & Sons.
[2] H.-H. Chen, M. Guizani and Josef F. Huber (2005) Guest Editorial of Feature Topic on Multiple Access Technologies for B3G Wireless Communications, *IEEE Communications Magazine*, 43(2), 65–67.
[3] H.-H. Chen, A. J. Han Vinck, Qi Bi and F. Adachi (2006) Guest Editorial of Special Issue on The Next Generation of CDMA Technologies. *IEEE Journal of Selected Areas in Communications*, 24(1), 1–3.
[4] H.-H. Chen, D. Wong and P. Mueller (2006) Guest Editorial of Special Issue on Evolution of Air-Interface Technologies for 4G Wireless Communications. *IEEE Vehicular Technology Magazine*, September.
[5] H.-H. Chen, M. Guizani and W. Mohr (2007) Guest Editorial for Special Issue on Evolution toward 4G Wireless Networking. *IEEE Network Magazine*, January.
[6] H.-H. Chen, Xi Zhang and Wen Xu (2007) Special Issue on Next Generation CDMA vs. OFDMA for 4G Wireless Applications. *IEEE Wireless Communications Magazine*, in preparation.
[7] R. L. Peterson, R. E. Ziemer and D. E. Borth (1995) *Introduction to Spread Spectrum Communications*. Englewood Cliffs, NJ: Prentice Hall.
[8] K. S. Zigangirov (2004) *Theory of Code Division Multiple Access Communications*. Chichester: John Wiley & Sons.
[9] A. J. Viterbi (1995) *CDMA–Principles of Spread Spectrum Communications*. Harlow: Addison-Wesley.
[10] Robert C. Dixon (1994) *Spread spectrum systems with commercial applications*, 3rd edn. Chichester: John Wiley & Sons.
[11] V. K. Garg (2000) *IS-95 CDMA and cdma2000: Cellular/PCS Systems Implementation*. Englewood Cliffs, NJ: Prentice Hall.
[12] R. Steele, Chin-Chun Lee and P. Gould (2001) *GSM, cdmaOne and 3G Systems*. Chichester: John Wiley & Sons.
[13] L. Hanzo, L.-L. Yang, E.-L. Kuan and K. Yen (2003) *Single and Multi-Carrier DS-CDMA*. Chichester: John Wiley & Sons.
[14] L. Hanzo, M. Munster, B. J. Choi and T. Keller (2003) *OFDM and MC-CDMA for Broadband Multi-User Communications, WLAN and Broadcasting*. Chichester: John Wiley & Sons.
[15] W. W. Lu (2002) Broadband Wireless Mobile: 3G and Beyond. Chichester: John Wiley & Sons.
[16] J. G. Proakis (1995) *Digital Communications*, 3rd edn. New York: McGraw-Hill.
[17] J. G. Proakis (2000) *Digital Communications*, 4th edn. Harlow: Addison Wesley.
[18] K. Pahlavan and A. H. Levesque (1995) *Wireless Information Networks*. Chichester John Wiley & Sons.
[19] R. E. Ziemer and R. L. Peterson (1990) *Digital Communications*. Englewood Cliffs, NJ: Prentice Hall.
[20] R. E. Ziemer and R. L. Peterson (1992) *Introduction to Digital Communications*. London: Macmillan.
[21] M. D. Yacoub (1993) *Foundations of Mobile Radio Engineering*. Boca Raton, FL: CRC Press.
[22] G. Calhuon (1988) *Digital Cellular Radio*. Norwood, MA: Artech House.
[23] R. Steele ed. (1994) *Mobile Radio Communications*. New York: IEEE Press.
[24] G. L. Stuber (1996) *Principle of Mobile Communications*. Dordrecht: Kluwer.
[25] TIA/EIA IS-19-B, Recommended Minimum Standards for 800-MHZ Cellular Subscriber Units, May 1988.
[26] TIA/EIA IS-20-A, Recommended Minimum Standards for 800-MHZ Cellular Land Stations, May 1988.
[27] EIA/TIA IS-41-B, Cellular Radio-Telecommunications Intersystem Operations, December 1992.
[28] TIA/EIA IS-41-C, Cellular Radiotelecommunications Intersystem Operations, January 1996.

The Next Generation CDMA Technologies Hsiao-Hwa Chen

[29] EIA/TIA IS-52, Uniform Dialing Procedures and Call Processing Treatment for Use in Cellular Radio Telecommunications, November 1989.

[30] EIA/TIA IS-53, Cellular Features Description, August 1991.

[31] EIA/TIA IS-54-B, Cellular System Dual-Mode Mobile Station—Base Station Compatibility Standard, April 1992.

[32] TIA/EIA IS-91, Mobile Station-Base Station Compatibility Standard for 800 MHz Analog Cellular, October 1994.

[33] TIA/E1A IS-95, Mobile Station—Base Station Compatibility Standard for Dual-Mode Wideband Spread Spectrum Cellular System, July 1993.

[34] TIA/EIA IS-95-B 1997 Mobile Station-Base Station Compatibility Standard for Dual-Mode Wideband Spread Spectrum Cellular Systems, Baseline Version, July 31, 1997.

[35] TIA/EIA/IS-96-A, Speech Service Option Standard for Wideband Spread Spectrum Digital Cellular System.

[36] TIA/EIA IS-99, Data Services Option Standard for Wideband Spread Spectrum Digital Cellular System.

[37] TIA/EIA IS-124, Cellular Radio Telecommunications Intersystem Non-Signaling Data Message Handlers (DMH), 1994.

[38] TIA/EIA IS-125, Recommended Minimum Performance Standard for Digital Cellular Wideband Spread Spectrum Speech Service Option

[39] TIA/EIA IS-126, Service Option 2: Mobile Station Loopback Service Option Standard, December 1994. Page G-2 TIA/EIA/IS-95-A.

[40] TIA/EIA IS-634, MSC-BS Interface for Public 800 MHz, Revision 0, 1995.

[41] TIA/EIA IS-637, Short Message Services for Wideband Spread Spectrum Cellular Systems.

[42] TIA/EIA IS-657, Packet Data Services Option for Wideband Spread Spectrum Cellular System.

[43] TIA/EIA IS-687, Data Services Inter-Working Function Interface Standard for Wideband Spread Spectrum Digital Cellular System.

[44] TIA/EIA IS-707-A, Data Services Options for Spread Spectrum Digital Cellular 24 Systems.

[45] TIA 232E, Interface between DTE and DCE Employing Serial Binary Data Interchange.

[46] TIA/EIA SP-2977, Cellular Features Description, Prepublication Version, March 14, 1995.

[47] TIA/EIA SP-3693, Mobile Station-Base Station Compatibility Standard for Dual-Mode Wideband Spread Spectrum Cellular Systems, November 18, 1997.

[48] TIA TR-45 Reference Model, 1990.

[49] TIA TR-46 Reference Model, 1991.

[50] JTC(AIR)/94.08.01-022R2, PN-3384, Personal Station—Base Station Compatibility Requirements for 1.8 to 2.0 GHz Code Division Multiple Access (CDMA) Personal Communications Systems, August 1, 1994.

[51] Qualcomm Inc., An Overview of the Application of Code Division Multiple Access (CDMA) to Digital Cellular Systems and Personal Cellular Networks, Qualcomm Inc. Doc. No. EX60-10010, May 21, 1992.

[52] R. Padovani (1994) Reverse link performance of IS-95 based cellular systems. *IEEE Personal Communications*, **1**(3), 28–34.

[53] TIA TR 45.5, The cdma2000 ITU-RTT Candidate Submission, TR 45-ISD/98.06.02.03, May 15, 1998.

[54] 1xEV-DO Inter-Operability Specification (IOS) for CDMA 2000 Access Network Interfaces, Release 0, 3GPP2 A.S0007, Ballot Version, June 14, 2001.

[55] 1xEV-DV Evaluation methodology—Addendum (V6), 3GPP2 WG5 Evaluation Ad Hoc, July 25, 2001.

[56] Introduction to cdma2000 Standards for Spread Spectrum Systems, 3GPP2, C.S0001-0, Version 1.0, Version Date: July 1999.

[57] Physical Layer Standard for cdma2000 Spread Spectrum Systems, 3GPP2 C.S0002-0, Version 1.0, Version Date: July 1999.

[58] Medium Access Control (MAC) Standard for cdma2000 Spread Spectrum Systems, 3GPP2 C.S0003-0, Version 1.0, Version Date: October 1999.

[59] Signaling Link Access Control (LAC) Standard for cdma2000 Spread Spectrum Systems, 3GPP2, C.S0004-0, Version 1.0, Version Date: July 1999.

[60] Upper Layer (Layer 3) Signaling Standard for cdma2000 Spread Spectrum Systems, 3GPP2 C.S0005-0, Version 1.0, Version Date: July 1999.

[61] Analog Signaling Standard for cdma2000 Spread Spectrum Systems, 3GPP2 C.S0006-0, Version 1.0, Version Date: July 1999.

[62] Introduction to cdma2000 Spread Spectrum Systems Revision D, 3GPP2 C.S0001-D, Version 1.0, Date: February 2004.

[63] Physical Layer Standard for cdma2000 Spread Spectrum Systems Revision D, 3GPP2 C.S0002-D, Version 1.0, Date: February 13, 2004.

[64] Medium Access Control (MAC) Standard for cdma2000 Spread Spectrum Systems Release D, 3GPP2 C.S0003-D, Version 1.0, Date: February 13, 2004.

[65] Signaling Link Access Control (LAC) Standard for cdma2000 Spread Spectrum Systems Release D, 3GPP2 C.S0004-D Version 1.0 Date: February 13, 2004.

[66] Upper Layer (Layer 3) Signaling Standard for cdma2000 Spread Spectrum Systems Release D, 3GPP2 C.S0005-D, Version 1.0, Date: February 2004.

[67] Analog Signaling Standard for cdma2000 Spread Spectrum Systems Release D, 3GPP2 C.S0006-D, Version 1.0, Date: February 2004.

[68] cdma2000 High Rate Packet Data Air Interface Specification, 3GPP2 C.S20024 v2.0. October 2000.

[69] cdma2000 High Rate Packet Data Air Interface Specification, 3GPP2 C.S0024-A, Version 1.0, Date: March 2004.

[70] Rec. ITU-R M.1225 Guidelines for Evaluation of Radio Transmission Technologies for IMT-2000.

[71] P. Bender, P. Black, M. Grob, R. Padovani, N. Sindhushayana and A. Viterbi (2000) CDMA/HDR: a bandwidth efficient high-speed data service for nomadic users. *IEEE Communications Magazine*, **38**, 70–77.

[72] E. Esteves (2000) The high data rate evolution of the cdma2000 cellular system In *Multiaccess, Mobility and Teletraffic for Wireless Communications*, vol. 5 (eds) G. Stuber and B. Jabbari). Dordrecht: Kluwer, pp. 61–72.

[73] A. Jalali, R. Padovani and R. Pankaj (2000) Data throughput of CDMA/HDR a high efficiency high data rate personal communication wireless system. In *Proceedings of IEEE 51st Vehicular Technology Conference*, Tokyo, Japan, May.

[74] P. J. Black and M. I. Gurelli (2001) Capacity simulation of cdma2000 1xEV wireless Internet access system. In *3rd IEEE International Conference on Mobile and Wireless Communications Networks*, Recife, Brazil, August.

[75] E. Esteves, P. J. Black and M. I. Gurelli (2002) Link adaptation techniques for high-speed packet data in third generation cellular systems. In European Wireless Conference.

[76] Youn-Kwan Kim and Byung K. Yi (2005) 3G wireless and cdma2000 1x evolution in Korea, *IEEE Communications Magazine*, **43**(4), 36–40.

[77] R. T. Derryberry and Zhouyue Pi (2005) Reverse high-speed packet data physical layer enhancements in cdma2000 1xEV-DV. *IEEE Communications Magazine*, **43**(4), 41–47.

[78] D. Comstock, R. Vannithamby, S. Balasubbamanin, L. A. Hsu and M. W. Cheng (2005) Reverse high-speed packet data support in cdma2000 1xEV-DV: upper layer protocols, *IEEE Communications Magazine*, 43(4), 48–56.

[79] Young Kim, Jungsoo Jung, Beomsik Bae, Daegyun Kim, P. R. Rajkotia and Young Kyun Kim (2005) Upper layer enhancements for fast call setup in cdma2000 Revision D. *IEEE Communications Magazine*, **43**(4), 57–66.

[80] Hwanjoon Kwon, Younsun Kim, Jin-Kyu Han, Donghee Kim, Hyeon Woo Lee and Young Kyun Kim (2005) Performance evalution of high-speed packet enhancement of cdma2000 1xEV-DV. *IEEE Communications Magazine*, **43**(4), 67–76.

[81] Soonyil Kwon, Kijun Kim, Youngwoo Yun, Sanh G. Kim and Byung K. Yi (2005) Power controlled H-ARQ in cdma2000 1xEV-DV. *IEEE Communications Magazine*, **43**(4), 77–81.

[82] Young-Hoon Choi, Laesung Park, Beomjoon Kim and M. A. Shayman (2005) A framework for elastic QoS provisioning in the cdma2000 1xEV-DV packet core network. *IEEE Communications Magazine*, **43**(4), 82–88.

[83] ETSI (1998) *The ETSI UMTS Terrestrial Radio Access (UTRA) ITU-R RTT Candidate Submission*, January 29.

[84] H. Holma and A. Toskala (eds) (2000) *WCDMA for UMTS: Radio Access for Third Generation Mobile Communications*, New York, Wiley, 2000.

[85] H. Kaarianen, A. Ahtiainen, L. Laitinen, S. Naghian and V. Niemi (2001) *UMTS Networks: Architecture, Mobility and Services*. Chichester: John Wiley & Sons.

[86] J. Laiho, A. Wacker and T. Novosad (2002) *Radio Network Planning and Optimisation for UMTS*. Chichester: John Wiley & Sons.

[87] J. P. Castro (2001) *The UMTS Network and Radio Access Technology*. Chichester: John Wiley & Sons.

[88] J. Korhonen (2001) *Introduction to 3G Mobile Communications*, 2nd edn. Norwood, MA: Artech House.

[89] ARIB (1998) *Japan's Proposal for Candidate Radio Transmission Technology on IMT-2000: W-CDMA*, June 26.
[90] CATT (1998) *TD-SCDMA Radio Transmission Technology for IMT-2000 Candidate submission*, Draft V.0.4, September.
[91] H.-H. Chen, C. X. Fan and W. W. Lu (2002) China's perspectives on 3G mobile communications and beyond: TD-SCDMA technology. *IEEE Wireless Communications*, **9**(2), 48–59.
[92] CWTS-WG1, *Pyhysical layer—General description*, TS C101, V3.1.1, September 2000.
[93] CWTS-WG1, *Physical channels and mapping of transport channels onto physical channels*, TS C102, V3.3.0, September 2000.
[94] CWTS-WG1, *Multiplexing and channel coding*, TS C103, V3.1.0, September 2000.
[95] CWTS-WG1, *Spreading and modulation*, TS C104, V3.3.0, September 2000.
[96] CWTS-WG1, *Physical layer procedures*, TS C105, V3.2.0, September 2000.
[97] CWTS-WG1, *Physical layer—Measurements (TD-SCDMA)*, TS C106, V3.0.0, May 2000.
[98] CWTS WG1 LAS-CDMA, 2001 *Physical channels and mapping of transport channels onto physical channels*, LAS TS 25.221, V1.0.0, July 17, 2001.
[99] CWTS-SWG2, LAS-CDMA, *Physical layer aspects of TD-LAS high speed packet technology*, LAS-TR 25.951, V1.0.0, July 2001.
[100] CWTS-SWG2, LAS-CDMA, *Multiplexing and channel coding*, LAS-TS 25.222, V1.0.0, July 2001.
[101] CWTS-SWG2, LAS-CDMA, *Spreading and modulation*, LAS-TS 25.223, V1.0.0, July 2001.
[102] CWTS-SWG2, LAS-CDMA, *Physical layer procedures*, LAS-TS 25.224, V1.0.0, July 2001.
[103] CWTS-SWG2, LAS-CDMA, *Physical layer—Measurements*, LAS-TS 25.225, V1.0.0, July 2001.
[104] CWTS-SWG2, LAS-CDMA, *TD-LAS high level system design document*, LAS-TR 25.960, V1.0.0, July 2001.
[105] J. A. Audestad (1988) Network aspects of the GSM system. In *EUROCON 88*, June.
[106] D. M. Balston (1993) The pan-European system: GSM. In D. M. Balston and R.C.V. Macario (eds) *Cellular Radio Systems*. Boston, MA: Artech House.
[107] D. M. Balston (1991) The pan-European cellular technology. In R.C.V. Macario (ed.) *Personal and Mobile Radio Systems*. London: Peter Peregrinus.
[108] M. Bezler *et al.* (1993) GSM base station system. *Electrical Communication*, 2nd Quarter.
[109] D. Cheeseman (1991) The pan-European cellular mobile radio system. In R.C.V. Macario (ed.) *Personal and Mobile Radio Systems*. London: Peter Peregrinus.
[110] C. Dechaux and R. Scheller (1993) What are GSM and DCS? *Electrical Communication*, 2nd Quarter.
[111] M. Feldmann and J. P. Rissen (1993) GSM network systems and overall system integration. *Electrical Communication*, 2nd Quarter.
[112] J. M. Griffiths (1992) *ISDN Explained: Worldwide Network and Applications Technology*, 2nd edn. Chichester: John Wiley & Sons.
[113] I. Harris (1993) Data in the GSM cellular network. In D. M. Balston and R.C.V. Macario (eds) *Cellular Radio Systems*. Boston, MA: Artech House.
[114] I. Harris (1993) Facsimile over cellular radio. In D. M. Balston and R.C.V. Macario (eds) *Cellular Radio Systems*. Boston, MA: Artech House.
[115] T. Haug (1988) Overview of the GSM project. In *EUROCON 88*, June.
[116] J.-F. Huber (1992) Advanced equipment for an advanced network. *Telcom Report International*, **15**(3/4).
[117] H. Lobensommer and H. Mahner (1992) GSM—a European mobile radio standard for the world market. *Telcom Report International*, **15**(3/4).
[118] B. J. T. Mallinder (1988) Specification methodology applied to the GSM system. In *EUROCON 88*, June.
[119] S. Mohan and R. Jain (1994) Two user location strategies for personal communication services. *IEEE Personal Communications*, **1**(1), 42–50.
[120] M. Mouly and M.-B. Pautet (1992) *The GSM System for Mobile Communications*. Palaiseav: Telecom Publishing.
[121] J. E. Natvig, S. Hansen and J. de Brito (1989) Speech processing in the pan-European digital mobile radio system (GSM)—system overview. In *IEEE GLOBECOM 1989*, November.
[122] T. Nilsson (n.d.) Toward a new era in mobile communications. http://193.78.100.33/ (Ericsson WWW server).
[123] M. Rahnema (1993) Overview of the GSM system and protocol architecture. *IEEE Communications Magazine*, **31**(4), 92–100.

[124] E. H. Schmid and M. Kahler (1993) GSM operation and maintenance. *Electrical Communication*, 2nd Quarter, 164–171.

[125] M. Silventoinen (1995) Personal email, quoted from European Mobile Communications Business and Technology Report, March and December.

[126] C. B. Southcott, D. Freeman, G. Cosier *et al.* (1989) Voice control of the pan-European digital mobile radio system. In *IEEE GLOBECOM 1989*, November, vol. 2, pp. 1070–1074.

[127] K. Hellwig, P. Vary, D. Massaloux, *et al.* (1989) Speech codec for the European mobile radio system. In *IEEE GLOBECOM 1989*, November, vol. 2, pp. 1065–1069.

[128] C. Watson (1993) Radio equipment for GSM. In D. M. Balston and R. C. V. Macario (eds) *Cellular Radio Systems*. Boston, MA: Artech House.

[129] R. G. Winch (1993) *Telecommunication Transmission Systems*. New York: McGraw-Hill.

[130] H.-H. Chen, Cheng-Hsien Cai, Chien-Yao Chao and Yu-Hsin Lin (2006) Synthesization of pulse shaping waveforms for spectral efficient digital modulations—some practical approaches. *European Transactions on Telecommunications*, **17**, 99–110.

[131] F. Hill and Won Lee (1974) PAM pulse generation using binary transversal filters. *IEEE Transactions on Communications*, **22**(7), 904–913.

[132] M. R. Civanlar and R. A. Nobakht (1988) Optimal pulse shape design using projections onto convex sets. In *1988 International Conference on Acoustics, Speech, and Signal Processing*, 11–14 April, New York, vol. 3, pp. 1874–1877.

[133] N. R. Sollenberger (1990) Pulse design and efficient generation circuits for linear TDMA modulation. In *1990 IEEE 40th Vehicular Technology Conference*, 6–9 May, Orlando, FL, pp. 616–621.

[134] D. S. Dias and K. Feher (1991) Baseband pulse shaping techniques for $\pi/4$-DQPSK in non-linearly amplified land mobile channels. In *Proceedings of INFOCOM 1991*, pp. 759–764.

[135] G. Madhusudhana Rao, K. V. V. S. Reddy and E. M. D. E. A. Reddy (1993) Waveform design of FH-BFSK communication system useful for mobile telephone subscribers. In *Proceedings of IEEE 1993 National Aerospace and Electronics Conference*, 24–28 May, Dayton, OH, vol. 1, pp. 447–453.

[136] S. G. Schock, L. R. Leblanc and S. Panda (1994) Spatial and temporal pulse design considerations for a marine sediment classification sonar. *IEEE Journal of Oceanic Engineering*, **19**(3), 406–415.

[137] Z. Cvetkovic (1999) Modulating waveforms for OFDM. In *Proceedings of 1999 IEEE International Conference on Acoustics, Speech, and Signal Processing*, 15–19 March, Phoenix, AZ, vol. 5, pp. 2463–2466.

[138] Xinjun Zhang, Xin Jiang, Wentao Song and Hanwen Luo (2002) A novel direct waveform synthesis technique with carrier frequency programmable. In *2002 IEEE Wireless Communications and Networking Conference*, 17–21 March, vol. 1, pp. 150–154.

[139] L. Milstein, R. Pickholtz and D. Schilling (1982) Comparison of performance of digital modulation techniques in the presence of adjacent channel interference. *IEEE Transactions on Communications*, **30**(8), 1984–1993.

[140] M. C. Austin and Ming U. Chang (1981) Quadrature overlapped raised-cosine modulation. *IEEE Transactions on Communications*, **29**(3), 237–249.

[141] E. R. Berlekamp (1968) *Algebraic Coding Theory*. New York: McGraw-Hill, p. 84.

[142] T. L. Booth (1963) An analytical representation of signals in sequential networks. In *Proceedings of the Symposium on Mathematical Theory of Automata*. New York, NY, April 24–26, 1962. Brooklyn, NY: Polytechnic Press of Polytechnic Institute of Brooklyn, pp. 301–324.

[143] R. Church (1935) Tables of irreducible polynomials for the first four prime moduli. *Annals of Mathematics*, **36**, 198–209.

[144] P. Fan and M. Darnell (1996) *Sequence Design for Communications Applications*. New York: John Wiley & Sons, p. 118, Table 5.1.

[145] W. W. Peterson and E. J. Weldon, Jr. (1972) *Error-Correcting Codes*, 2nd ed. Cambridge, MA: MIT Press, pp. 476, 560.

[146] M. P. Ristenblatt (1965) Pseudo-random binary coded waveforms In *Modern radar* (ed.) R. S. Berkowitz). New York: John Wiley & Sons, pp. 274–314.

[147] N. J. A. Sloane (n.d.) Sequence A011260/M0107. In *The On-Line Encyclopedia of Integer Sequences*. www.research.att.com/~njas/sequences.

[148] N. Zierler and J. Brillhart (1968) On primitive trinomials. *Information and Control*, **13**, 541–544.

[149] N. Zierler and J. Brillhart (1969) On primitive trinomials (II). *Information and Control*, **14**, 566–569.

[150] H.-H. Chen, Daoben Li and Qi Bi (2003) Editorial, Special Issue, Ultra-Broadband Wireless Communications for the Future. *Journal of Wireless Communications and Mobile Computing (WCMC)*, **3**(6), 659–662.

[151] http://www.swedetrack.com/usblue1.htm
[152] http://www.embedded.com/internet/0007/0007ia1.htm
[153] http://www.comsoc.org/pubs/surveys/4q98issue/prasad.html
[154] http://www.ee.iitb.ernet.in/uma/ aman/bluetooth/
[155] http://www.cambridgesiliconradio.com/nf/bluecore.htm
[156] http://www.parthus.com/products/parthus_bluestream/parthus_bluestream_platform.html
[157] http://www.oki.co.jp/semi/english/t-blue.htm
[158] http://www.infineon.com/cgi/ecrm.dll/ecrm/scripts/prod_ov.jsp?oid=13722
[159] http://www.nikkeibp.asiabiztech.com/nea/200006/srep_103397.html
[160] http://www.motorola.com/SPS/HPESD/docs/pr/hpesd/111896_SEC000.html
[161] http://www.siliconwave.com/pdf/73_0005_R00A_Bluetooth_802_11.pdf
[162] H. Monson, Bluetooth Technology and Implications. http://www.sysopt.com/articles/bluetooth/index1.html
[163] P. K. Enge and D. V. Sarwate (1987) Spread-spectrum multiple-access performance of orthogonal codes: linear receivers. *IEEE Transactions Communications*, **35**, 1309–1319.
[164] P. K. Enge and D. V. Sarwate (1988) Spread-spectrum multiple-access performance of orthogonal codes: impulsive noise. *IEEE Transactions on Communications*, **36**(1), 98–106.
[165] K. Pahlavan, M. Chase and M. Sawahashi (1990) Spread-spectrum multiple-access performance of orthogonal codes for indoor radio communications. *IEEE Transactions on Communications*, **38**(5), 574–577.
[166] C.-L. I and R. D. Gitlin (1995) Multi-code CDMA wireless personal communications networks. In *Proceedings of the IEEE International Conference on Communications* (ICC 95), Seattle, WA, vol. 2, pp. 1060–1064.
[167] Seung Joon Lee, Tai Suk Kim and Dan Keun Sung (2001) Bit-error probabilities of multicode direct-sequence spread-spectrum multiple-access systems. *IEEE Transactions on Communications*, **49**(1), 31–34.
[168] R. Zhang and T. T. Tjhung (2002) BER performance comparison of single code and multicode DS/CDMA channelization schemes for high rate data transmission. *Communications Letters*, **5**(2), 67–69.
[169] N. Guo and L. B. Milstein (1999) On rate-variable multidimensional DS/SSMA with sequence sharing. *IEEE Journal on Selected Areas in Communications*, **17**(5), 902–917.
[170] E. A. Sourour and M. Nakagawa (1996) Performance of orthogonal multicarrier CDMA in a multipath fading channel. *IEEE Transactions on Communications*, **44**(3), 356–367.
[171] S. Kondo and L. B. Milstein (1996) Performance of multicarrier DS CDMA systems. *IEEE Transactions on Communications*, **44**(2), 238–246.
[172] B. M. Popovic (1999) Spreading sequences for multicarrier CDMA systems. *IEEE Transactions on Communications*, **47**(6), 918–926.
[173] A. M. Tulino, Linbo Li and S. Verdu (2005) Spectral efficiency of multicarrier CDMA. *IEEE Transactions on Information Theory*, **51**(2), 479–505.
[174] S. Hara and R. Prasad (1997) Overview of multicarrier CDMA. *IEEE Communications Magazine*, **35**(12), 126–133.
[175] . N. Yee, J. P. M. G. Linnartz and G. Fettweis (1993) Multi-carrier CDMA in indoor wireless radio networks. *IEEE Personal Indoor and Mobile Radio Communications (PIMRC) International Conference*, September, Yokohama, Japan, pp. 109–113.
[176] L. Yun, M. Couture, J. R. Camagna and J. P. M. G. Linnartz (1993) BER for QPSK DS-CDMA downlink in an indoor Riciean dispersive pico-cellular channel. In *Signals, Systems and Computers, 1993. Conference Record of the 27th Asilomar Conference*, November 1–3, Pacific Grove, CA, vol. 2, pp. 1417–1421.
[177] N. Yee and J. P. M. G. Linnartz (1993) BER for multi-carrier CDMA in indoor Ricean-fading channel. In *Signals, Systems and Computers, 1993. Conference Record of the 27th Asilomar Conference*, November 1–3, Pacific Grove, CA, vol. 2, pp. 426–430.
[178] N. Yee and J. P. M. G. Linnartz (1994) Controlled equalization for MC-CDMA in Rician fading channels. In *44th IEEE Vehicular Technology Conference*, June, Stockholm, pp. 1665–1669.
[179] N. Yee and J. P. M. G. Linnartz (1994) Wiener filtering for multi-carrier CDMA. In *IEEE/ICCC Conference on Personal Indoor Mobile Radio Communications (PIMRC) and Wireless Computer Networks (WCN)*, The Hague, September 19–23, vol. 4, pp. 1344–1347.
[180] N. Yee, J. P. M. G. Linnartz and G. Fettweis (1994) Multi-carrier-CDMA in indoor wireless networks. *IEICE Transactions on Communications, Japan*, E77-B(7), 900–904.
[181] J. P. M. G. Linnartz (2001) Performance analysis of synchronous MC-CDMA in mobile Rayleigh channels with both delay and Doppler spreads. *IEEE Transactions on Vehicular Technology*, **50**(6), 1375–1387.

[182] A. Gorokhov and J. P. M. G. Linnartz (2004) Robust OFDM receivers for dispersive time varying channels: equalization and channel acquisition. *IEEE Transactions on Communications*, **52**(4), 572–583.

[183] S. Tomasin, A. Gorokhov, H. Yang and J. P. M. G. Linnartz (2005) Iterative Interference cancellation and channel estimation for mobile OFDM. *IEEE Transactions on Wireless Communications*, **4**(1), 238–245.

[184] S. Tomasin, A. Gorokhov, H. Yang and J.-P. Linnartz (2002) Reduced complexity doppler compensation for mobile DVB-T. In *Proceedings of the 13th IEEE International Symposium on Personal, Indoor and Mobil Radio Communications*, September 15–18, Lisbon, vol. 5, pp. 2077–2081.

[185] J.-P. Linnartz, A. Gorokhov, S. Tomasin and H. Yang (2002) Achieving mobility for DVB-T by signal processing for Doppler compensation. In *Proceedings of the IBC Conference*, September 14, Amsterdam, pp. 412–420.

[186] Dongwook Lee, Hun Lee and K. B. Milstein (1998) Direct sequence spread spectrum Walsh-QPSK modulation. *IEEE Transactions on Communications*, **46**(9), 1227–1232.

[187] Joonyoung Cho, Youhan Kim and Kyungwhoon Cheun (2000) A novel FHSS multiple-access network using M-ary orthogonal Walsh modulation. In *Vehicular Technology Conference, IEEE VTS-Fall VTC 2000, 52nd*, vol. 3, pp. 1134–1141.

[188] S. Tsai, F. Khaleghi, Seong-Jun Oh and V. Vanghi (2001) Allocation of Walsh codes and quasi-orthogonal functions in cdma2000 forward link. In *Vehicular Technology Conference, Fall, IEEE VTS 54th*, vol. 2, pp. 747–751.

[189] J. Granlund, A. R. Thompson, and B. G. Clark (1978) An application of Walsh functions in radio astronomy instrumentation. *IEEE Transactions on Electromagnetic Compatibility*, **20**(3), 451–453.

[190] R. Gold (1968) Maximal recursive sequences with 3-valued recursive cross-correlation functions. *IEEE Transactions on Information Theory*, **14**, 154–156.

[191] R. Gold (1967) Optimal binary sequences for spread spectrum multiplexing. *IEEE Transactions on Information Theory*, **13**(5), 619–621.

[192] T. Kasami (1966) Weight distribution formula for some class of cyclic codes. Tech. Rep. R-285 (AD632574), Coordinated Science Laboratory, University of illionos, Urbana, IL.

[193] T. Kasami (1969) Weight distribution of Bose-Chaudhuri-Hocquenghem codes. In *Combinatorial Mathematics and its Applications*. Chapel Hill, NC: University of North Carolina Press.

[194] D. V. Sarwate and M. B. Pursley (1980) Crosscorrelation properties of pseudorandom and related sequences. *Proceedings of the IEEE*, **68**(5), 593–620.

[195] J. Lahtonen (1995) On the odd and the aperiodic correlation properties of the Kasami sequences. *IEEE Transactions on Information Theory*, **41**(5), 1506–1508.

[196] S. W. Golomb (1967) *Shift Register Sequences*. San Francisco, CA: Holden-Day.

[197] X. H. Chen, T. Lang and J. Oksman (1996) Searching for quasi-optimal subfamilies of m-sequences for CDMA systems. In *Seventh IEEE International Symposium on Personal, Indoor and Mobile Radio Communications (PIMRC'96)*, October 15–18, vol. 1, pp. 113–117.

[198] A. Z. Tirkel (1996) Cross correlation of m-sequences—some unusual coincidences. In *Proceedings of IEEE 4th International Symposium on Spread Spectrum Techniques and Applications*. September 22–25, vol. 3, pp. 969–973.

[199] T. Ito, Sampei, S. and N. Morinaga (2000) M-sequence based M-ary/SS scheme for high bit rate transmission in DS/CDMA systems. *Electronics Letters*, **36**(6), 574–576.

[200] K. Imamura and Guo-Zhen Xiao (1992) On periodic sequences of the maximum linear complexity and M-sequences. In *Singapore ICCS/ISITA '92, Communications on the Move*, November 16–20, vol. 3, pp. 1219–1221.

[201] S. Uehara and K. Imamura (1992) Some properties of the partial correlation of M-sequences. In *Singapore ICCS/ISITA '92, Communications on the Move*. November 16–20, vol. 3, p. 1223.

[202] P. A. N. Briggs and K. R. Godfrey (1963) Autocorrelation function of a 4-level m-sequence. *Electronics Letters*, **4**, 232–233.

[203] N. Zierler (1959) Linear recurring sequences. *Journal of the society for Industrial and Applied Mathematics*, **7**, 31–48.

[204] S. W. Golomb (1982) *Shift Register Sequences*. 2nd edn. Laguna Hills, CA: Aegean Park Press.

[205] K. H. A. Karkkainen (2001) Linear complexity of Kronecker sequences. *IEICE Transactions on Fundamentals*, E84-A(5), 1348–1351.

[206] W. E. Stark and D. V. Sarwate (1981) Kronecker sequences for spread spectrum communication. *IEE Proceedings*, Part F, **128**(2), 104–109.

[207] M. Beale and T. C. Tozer (1979) A class of composite sequences for spread-spectrum communications. *IEE Journal of Computers and Digital Technology*, **2**(2), 87–92.

[208] S. A. Faulkner and J. S. Wight (1991) Structure of composite codes for rapid acquisition of DS-SS signals. In *Proceedings of Spread Spectrum Workshop: Potential Commercial Applications—Myth or Reality?*, Montebello, Quebec, Canada, May, pp. 7.3.1–7.3.3.

[209] S. Uehara and K. Imamura (1997) Characteristic polynomials of binary complementary sequences. *IEICE Transactions on Fundamentals*, E80-A(1), 193–196.

[210] R. A Scholtz and L. R. Welch (1984) GMW sequences. *IEEE Transactions on Information Theory*, **30**(3), 548–553.

[211] H.-H. Chen, T. Lang and J. Oksman (1997) Constructing quasi-optimal subfamilies of GMW sequences suitable for CDMA applications. *IEE Proceedings-Communications*, **144**(2), 99–106.

[212] H.-H. Chen, T. Lang and J. Oksman (1996) Constructing quasi-optimal GMW and M-sequence subfamilies with minimized bit error rate. *IEICE Transactions on Communications*, E79-B(7), 963–973.

[213] X. H. Chen and J. Oksman (1992) BER performance analysis of 4-CCL and 5-CCL codes in slotted indoor CDMA systems. *Communications, Speech and Vision, IEE Proceedings*, **139**, 79–84.

[214] P. V. Kumar and R. A. Scholtz, (1983) Bounds on the linear span of Bent sequences. *IEEE Transactions on Information Theory*, **29**(6), 854–862.

[215] J.-S. No and P. V. Kumar (1989) A new family of binary pseudorandom sequences having optimal periodic correlation properties and large linear span. *IEEE Transactions on Information Theory*, **35**(2), 371–379.

[216] H.-H. Chen (2001) Multi-band wavelet packet spreading codes with intra-code subband diversity for communications in multipath fading channels. *IEICE Transactions on Communications*, E84-B(7), 1876–1884.

[217] W. T. V. William, H. Press, S. A. Teukolsky and B. P. Flannery (1992) *Numerical Recipes in C.* Cambridge: Cambridge University Press.

[218] L. Welch (1974) Lower bounds on the maximum cross correlation of signals. *IEEE Transactions on Information Theory*, **20**(3), 397–399.

[219] J. Mitola, III and G. Q. Maguire, Jr. (1999) Cognitive radio: making software radios more personal, *Personal Communications, IEEE* **6**(4), 13—18 [see also *IEEE Wireless Communications*].

[220] J. Mitola III (2000) Cognitive radio: an integrated agent architecture for software defined radio, Dissertation of Doctor of Technology, Royal Institute of Technology (KTH), SE-164 40 Kista, Sweden, 8 May.

[221] D. Cabric, S. M. Mishra and R. W. Brodersen, (2004) Implementation issues in spectrum sensing for cognitive radios. *Conference Record of the Thirty-Eighth Asilomar Conference on Signals, Systems and Computers, 2004*, November 7–10, vol. 1, pp. 772–776.

[222] T. A. Weiss and F. K. Jondral (2004) Spectrum pooling: an innovative strategy for the enhancement of spectrum efficiency. *IEEE Communications Magazine*, **42**(3), 8–14.

[223] IEEE 802.15.2-2003 IEEE Recommended Practice for Telecommunications and Information exchange between systems–Local and metropolitan area networks. Specific Requirements Part 15.2: Coexistence of Wireless Personal Area Networks with Other Wireless Devices Operating in Unlicensed Frequency Band, Monday, 6 November 2006. http://ieee802.org/15/pub/TG2.html.

[224] IEEE 802.19 Coexistence Technical Advisory Group (TAG). http://grouper.ieee.org/groups/802/19.

[225] J. Hillenbrand, T. A. Weiss and F. K. Jondral (2005) Calculation of detection and false alarm probabilities in spectrum pooling systems. *IEEE Communications Letters*, **9**(4), 349–351.

[226] B. Fette (ed.) (2006) *Cognitive Radio Technology*. Oxford: Elsevier.

[227] E. C. Van Der Meulen (1971) Three-terminal communication channels. *Advances Applied probability* **3**(1), 120–154.

[228] T. Cover and A. E. Gamal (1979) Capacity theorems for relay channel. *IEEE Transactions on Information Theory*, **25**(5), 572–584.

[229] A. Nosratinia, T. E. Hunter, and A. Hedayat (2004) Cooperative communication in wireless networks. *IEEE Communications Magazine*, **42**(10), 74–80.

[230] http://cmc.rice.edu/docs/docs/Ahm2005Mar5Cooperativ.pdf.

[231] N. Ahmed, M. A. Khojastepour and B. Aazhang (2004) Outage minimization and optimal power control for the fading relay channel. In *IEEE Information Theory Workshop*, October 24–29, pp. 458–462.

[232] B. Zhao and M. C. Valenti (2003) Distributed turbo coded diversity for relay channel. *Electronic Letters*, **39**(10), 786–787.

[233] M. C. Valenti and B. Zhao (2003) Distributed turbo codes: toward the capacity of the relay channel, In *IEEE 58th Vehicular Technology Conference, VTC-Fall*, vol. 1, October 6–9, pp. 322–326.

[234] J. N. Laneman, D. N. C. Tse and G. W. Wornell (2004) Cooperative diversity in wireless networks: efficient protocols and outage behaviour. *IEEE Transactions on Information Theory*, **50**(12), 3062–3080.

[235] Xiaohua (Edward) Li (2005) Cooperative Communications for Wireless Information Assurance, AFRL-IF-RS-TR-2005-279, Final Technical Report, Air Force Research Laboratory, July.

[236] J. N. Laneman and G. W. Wornell (2003) Distributed space-time-coded protocols for exploiting cooperative diversity in wireless networks. *IEEE Transactions on Information Theory*, **49**(10), 2415–2425.

[237] H.-H. Chen, J. F. Yeh and N. Seuhiro (2001) A multi-carrier CDMA architecture based on orthogonal complementary codes for new generations of wideband wireless communications. *IEEE Communications Magazine*, **39**(10), 126–135.

[238] V. Tarokh, A. Naguib, N. Seshadri and A. R. Calderbank (1998) Space-time codes for high data rate wireless communication: performance criterion and code construction. *IEEE Transactions on Information Theory*, **44**(2), 744–765.

[239] S. M. Alamouti (1998) A simple transmit diversity technique for wireless communications. *IEEE Journal on Selected Areas in Communications*, **16**(8), 1451–1458.

[240] V. Tarokh, H. Jafarkhani and A. R. Calderbank (1999) Space-time block codes from orthogonal designs. *IEEE Transactions on Information Theory*, **45**(5), 1456–1467.

[241] B. M. Hochwald, T. L. Marzetta and C. B. Papadias (2001) A transmitter diversity scheme for wideband CDMA systems based on space-time spreading. *IEEE Journal on Selected Areas in Communications*, **19**(1), 48–60.

[242] J. Geng, U. Mitra and M. P. Fitz (2001) Space-time block codes in multipath CDMA systems. In *Proceedings of the 2001 IEEE International Symposium on Information Theory*, Washington, DC, June 24–29, p. 151.

[243] M. Nagatsuka and R. Kohno (1995) A spatially and temporally optimal multiuser receiver using an antenna array for DS/CDMA. *IEICE Transactions on Communications*, E78-B, 1489–1497.

[244] F. Petre, G. Leus, L. Deneire, M. Engels, M. Moonen and H. De Man (2003) Space-time block coding for single-carrier block transmission DS-CDMA downlink. *IEEE Journal on Selected Areas in Communications*, **21**(3), 350–361.

[245] CATT/China, TD-SCDMA Radio Transmission Technology for IMT-2000, June 1998.

[246] M.-S. Alouini, Sang Wu Kim and A. Goldsmith (1997) Rake reception with maximal-ratio and equal-gain combining for DS-CDMA systems in nakagami fading. In *Record, 1997. Conference Record, 1997 IEEE 6th International Conference on Universal Personal Communications*, October 12–16, vol. 2, pp. 708–712.

[247] T. F. Wong and T. M. T. M. Lok (2001) Transmitter adaptation in multicode DS-CDMA systems. *IEEE Journal on Selected Areas in Communications*, **19**(1), 69–82.

[248] T. Eng and L. B. Milstein, (1995) Coherent DS-CDMA performance in Nakagami multipath fading. *IEEE Transactions on Communications*, **43**(2–4234), 1134–1143.

[249] S. Sasaki, H. Kikuchi, H. Watanabe and J. Zhu (1994) Performance evaluation of parallel combinatory SSMA systems in Rayleigh fading channel. In *Proceedings of the IEEE 3rd International Symposium on Spread Spectrum Techniques and Applications (ISSSTA'94)*, Oulu, Finland, vol. 1, pp. 198–202.

[250] Dong In Kim (1995) Analysis of a direct-sequence CDMA mobile radio system with reduced set of code sequences. *IEEE Transactions on Vehicular Technology*, 44(3) 525–534.

[251] M. J. E. Golay (1961) Complementary series. *IRE Transactions on Information Theory*, IT-7, 82–87.

[252] R. Turyn (1963) Ambiguity function of complementary sequences. *IEEE Transactions on Information Theory*, IT-9, 46–47.

[253] N. Suehiro (1982) Complete complementary code composed of N-multiple-shift orthogonal sequences. *Transactions of IECE of Japan* [in Japanese], vol. J65-A, 1247–1253.

[254] N. Suehiro and M. Hatori (1988) N-Shift cross-orthogonal sequences. *IEEE Transactions on Information Theory*, **34**(1), 143–146.

[255] Y. Taki, H. Miyakawa, M. Hatori and S. Namba (1969) Even-shift orthogonal sequences. *IEEE Transactions on Information Theory*, **15**, 295–300.

[256] C. C. Tseng and C. L. Liu (1972) Complementary sets of sequences. *IEEE Transactions on Information Theory*, 18, 644–652.

[257] R. Sivaswamy (1978) Multiphase complementary codes. *IEEE Transactions on Information Theory*, 24, 546–552.

[258] R. L. Frank (1980) Polyphase complementary codes. *IEEE Transactions on Information Theory*, 26, 641–647.

[259] H.-H. Chen and Yu-Ching Yeh (2005) Capacity of space-time block-coded CDMA systems: comparison of unitary and complementary codes. *IEE Proceedings—Communications*, **152**(2), 203–214.

[260] H.-H. Chen, Yu-Ching Yeh, Chien-Yao Chao and Jun-Feng Yeh. (2004). A pilot-added signal detection algorithm and its application in OCC-CDMA systems under multipath interference. *IEE Electronics Letters*, **40**(8), 488–489.

[261] H.-H. Chen, (2004) On next generation CDMA technology for future wireless networking (invited paper). In *Wireless Ad Hoc and Sensor Networks Workshop, IEEE Globecom 2004*, Dallas, TX, November 29–December 3.

[262] H.-H. Chen and Hsin-Wei Chiu (2004) Generation of super-set of perfect complementary codes for next generation CDMA systems. In *IEEE Military Communications Conference (IEEE MILCOM) 2004*, Monterey, CA, October 31—November 3.

[263] H.-H. Chen and Hsin-Wei Chiu (2004) Design of perfect complementary codes to implement an interference-free CDMA system. In *IEEE Globecom 2004*, Dallas, TX, November 29–December 3.

[264] H.-H. Chen and Hsin-Wei Chiu (2004) Generation of perfect orthogonal complementary codes for their applications in interference-free CDMA systems. In *Record of PIMRC 04, 15th IEEE International Symposium on Personal, Indoor and Mobile Radio Communications*, Barcelona, Spain, September 5–8.

[265] H.-H. Chen, Yu-Ching Yeh, Chien-Yao Chao and Kuo-Sheng Chen (2003) Interference-free CDMA air-link technology promising noise-limited performance. In *Proceedings of IEEE VTC 2003-Fall*, Orlando, FL, October 4–9.

[266] H.-H. Chen, Jin-Xiao Lin, Shin-Wei Chu, Chi-Feng Wu and Guo-Sheng Chen (2003) Isotropic air-interface technologies for fourth generation wireless communications. *Wireless Communications and Mobile Computing (WCMC) Journal*, **3**(6), 687–704.

[267] H.-H. Chen and Jun-Feng Yeh (2003) A complementary codes based CDMA architecture for wideband mobile Internet with high spectral efficiency and exact rate-matching. *International Journal of Communication Systems*, **16**, 497–512.

[268] H.-H. Chen, Yu-Ching Yeh, Xi Zhang, Aiping Huang, Yang Yang, Jie Li, Yang Xiao, H. R. Sharif and A. J. Han Vinck (2006) Generalized pairwise complementary codes with set-wise uniform interference-free windows. *IEEE Journal of Selected Areas in Communications*, **24**(1), 65–74.

[269] E. H. Dinan and B. Jabbari (1998) Spreading codes for direct sequence CDMA and wideband CDMA cellular networks. *IEEE Communications Magazine*, **36**(9), 48–54.

[270] E. Dahlman, B. Gudmundson, M. Nilsson *et al.* (1998) UMTS/IMT-2000 based on wideband CDMA. *IEEE Communications Magazine*, **36**(9), 70–80.

[271] P. Chaudhury, W. Mohr and S. Onoe (1999) The 3gpp proposal for IMT-2000. *IEEE Communications Magazine*, **37**(12), 72–81.

[272] S. N. Diggavi, N. Al-Dhahir and A. R. Calderbank (2003) Algebraic properties of space-time block codes in intersymbol interference multiple-access channels. *IEEE Transactions on Information Theory*, **49**(10), 2403–2414.

[273] W. Choi and J. M. Cioffi (1999) Multiple input/multiple output (MIMO) equalization for space-time block coding. In *Proceedings of IEEE Pacific Rim Conference on Communications, Computers and Signal Processing*, pp. 341–344.

[274] W. Choi and J. M. Cioffi (1999) Space-time block codes over frequency selective Rayleigh fading channels. In *Proceedings of Vehicular Technology Conference*, Amsterdam, vol. **5**, pp. 2541–2545.

[275] Y. Li, J. C. Chung and N. R. Sollenberger (1999) Transmitter diversity for OFDM systems and its impact on high-rate data wireless networks. *IEEE Journal of Selected Areas in Communications*, **17**, 1233–1243.

[276] Y. Li, N. Seshadri and S. Ariyavisitakul (1999) Channel estimation for OFDM systems with transmitter diversity in mobile wireless channels. *IEEE Journal of Selected Areas in Communications*, **17**, 461–471.

[277] A. F. Naguib, N. Seshadri and A. R. Calderbank (1998) Applications of space-time block codes and interference suppression for high capacity and high data rate wireless systems. In *Proceedings of 32nd Asilomar Conference on Signals, Systems and Computers*, pp. 1803–1810.

[278] S. Zhou and G. B. Giannakis (2003) Single-carrier space-time block-coded transmissions over frequency-selective fading channels. *IEEE Transactions Information Theory*, **49**(1), 164–179.

[279] Y. Liu, M. P. Fitz and O. Y. Takeshita (2000) Space-time codes for frequency selective channel: outage probability, performance criteria, and code design. In *Proceedings of 38th Annual Allerton Conference on Communication, Control, and Computing*, Monticello, IL, October.

[280] Y. Liu, M. P. Fitz and O. Y. Takeshita (2001) Space-time codes performance criteria and design for frequency selective fading channels. In *Proceedings of International Conference on Communications*, Helsinki, Finland, June, vol. 1, pp. 2800–2804.

[281] Z. Liu and G. B. Giannakis (2000) Space-time coding with transmit antennas for multiple access regardless of frequency-selective multipath. In *Proceedings of Sensor Array and Multichannel Signal Processing Workshop*, Boston, MA, March, pp. 178–182.

[282] Z. Liu and G. B. Giannakis (2001) Space-time block coded multiple access through frequency selective fading channels. *IEEE Transactions on Communications*, **49**, 1033–1044.

[283] S. Verdu (1998) *Multiuser Detection*. Cambridge: Cambridge University Press, p. 10[illegible].

[284] S. Haykin (2001) *Communication Systems*, 4th edn. New York: John Wiley & Sons.

[285] M. K. Simon, M.-S. Alouini (2000) *Digital Communication over Fading Channels*. New York: John Wiley & Sons.

[286] J. A. Salehi (1989) Code division multiple-access techniques in optical fiber networks. I. Fundamental principles. *IEEE Transactions on, Communications*, **37**(8), 824–833.

[287] J. A. Salehi, C. A. Brackett (1989) Code division multiple-access techniques in optical fiber networks. II. Systems performance analysis. *IEEE Transactions on Communications*, **37**(8), 834–842.

[288] P. Prucnal, M. Santoro, and Ting Fan (1986) Spread spectrum fiber-optic local area network using optical processing. *Journal of Lightwave Technology*, **4**(5), 547–554.

[289] Wei Huang, M. H. M. Nizam, I. Andonovic and M. Tur (2000) Coherent optical CDMA (OCDMA) systems used for high-capacity optical fiber networks-system description, OTDMA comparison, and CDMA/WDMA networking. *Journal of Lightwave Technology*, **18**(6), 765–778.

[290] F. R. K. Chung, J. A. Salehi and V. K. Wei (1989) Optical orthogonal codes: design, analysis and applications. *IEEE Transactions on Information Theory*, **35**(3), 595–604.

[291] J.A. Salehi (1989) Emerging optical code-division multiple access communication systems. *IEEE Network*, **3**(2), 31–39.

[292] H. Chung and P. V. Kumar (1990) Optical orthogonal codes—new bounds and an optimal construction. *IEEE Transactions on Information Theory*, **36**(4), 866–873.

[293] R. Fuji-Hara and Y. Miao (2000) Optical orthogonal codes: their bounds and new optimal constructions. *IEEE Transactions on Information Theory*, **46**(7), 2396–2406.

[294] G. Ge and Yin, J (2001) Constructions for optimal (v, 4, 1) optical orthogonal codes. *IEEE Transactions on Information Theory*, **47**(7), 2998–3004.

[295] Yanxun Chang, R. Fuji-Hara and Ying Miao (2003) Combinatorial constructions of optimal optical orthogonal codes with weight 4. *IEEE Transactions on Information Theory*, **49**(5), 1283–1292.

[296] A. S. Holmes and R. R. A. Syms (1992) All-optical CDMA using quasi-prime codes. *Journal of Lightwave Technology*, **10**(2), 279–286.

[297] A. A. Shaar, C. E. Woodcock and P. A. Davies (1999) Prime sequences for asynchronous pulse repetition interval agile radar. *IEEE Transactions on Aerospace and Electronic Systems*, **35**(2), 543–548.

[298] M. Kavehrad and D. Zaccarin (1995) Optical code-division-multiplexed systems based on spectral encoding of noncoherent sources. *Journal of Lightwave Technology*, **13**(3), 534–545.

[299] L. Tancevski, I. Andonovic, M. Tur and J. Budin (1996) Hybrid wavelength hopping/time spreading code division multiple access systems. *IEE Proceedings–Optoelectronics*, **143**(3), 161–166.

[300] L. Tancevski, I. Andonovic, M. Tur and J. Budin (1996) Massive optical LANs using wavelength hopping/time spreading with increased security. *IEEE Photonics Technology Letters*, **8**(7), 935–937.

[301] Guu-Chang Lang and T. E. Fuja (1995) Optical orthogonal codes with unequal auto- and cross-correlation constraints. *IEEE Transactions on Information Theory*, **41**(1), 96–106

[302] G.-C. Yang (1995) Some new families of optical orthogonal codes for code-division multiple access fibre-optic networks. *IEE Proceedings—Communications*, **142**(6), 363–368.

[303] Xhi-Shun Weng and Jingshown Wu (2001) Perfect difference codes for synchronous fiber-optic CDMA communication systems. *Journal of Lightwave Technology*, **19**(2), 186–194.

[304] Sangin Kim, Kyungsik Yu and N. Park (2000) A new family of space/wavelength/time spread three-dimensional optical code for OCDMA networks. *Journal of Lightwave Technology*, **18**(4), 502–511.

[305] W. C. Kwong and Guu-Chang Yang (2002) Design of multilength optical orthogonal codes for optical CDMA multimedia networks. *IEEE Transactions on Communications*, **50**(8), 1258–1265.

[306] Jian-Guo Zhang (2002) Optical CDMA codes for use in a lightwave communication network with multiple data rates. *European Transactions on Telecommunications*, **13**(3), 257–267.

Index

The Next Generation CDMA Technologies Hsiao-Hwa Chen